MRI Atlas of the Musculoskeletal System

MRI Atlas of the Musculoskeletal System

Lawrence W Bassett, MD
Professor of Radiological Sciences
UCLA School of Medicine

Richard H Gold, MD
Professor of Radiological Sciences
UCLA School of Medicine

Leanne L Seeger, MD
Assistant Professor of Radiological Sciences
UCLA School of Medicine

MARTIN DUNITZ

© **Martin Dunitz Ltd 1989**

First published in the United Kingdom in 1989
by Martin Dunitz Ltd, 154 Camden High Street, London NW1 0NE

Distributed in the United States of America and Canada by CRC Press, Inc.
2000 Corporate Blvd., N.W., Boca Raton, Florida 33431

All rights reserved. No part of this publication may be reproduced, stored in a
retrieval system, or transmitted, in any form or by any means, without the prior
permission of the publisher.

Editorial Board

William G Bradley MD PhD
Director, NMR Imaging Laboratory, Huntington Research Institute,
Pasadena, California, USA

Masahiro Iio MD
Professor and Chairman of Radiology, Faculty of Medicine, University of Tokyo,
Japan

Herbert Y Kressel MD
Director of David W Devon Medical Imaging Center, University of
Pennsylvania Hospitals, Philadelphia, USA

Francis W Smith MB ChB FFR
Director of Clinical NMR Research Unit, Aberdeen Royal Infirmary, UK

Library of Congress Cataloging-in-Publication Data

MRI atlas of the musculoskeletal system.
  (MRI atlas series)
  Bibliography: p.
  1. Musculoskeletal system–Diseases–Diagnosis.
2. Musculoskeletal system–Magnetic resonance imaging.
I. Bassett, Lawrence W. (Lawrence Wayne), 1942–
II. Gold, Richard H. III. Seeger, Leanne L.
IV. Series. [DNLM: 1. Magnetic Resonance Imaging
–atlases. 2. Musculoskeletal system–anatomy &
histology–atlases. 3. Musculoskeletal System
–pathology–atlases. WE 17 M9395]
RC925.7.M75   1988   616.7′0757   88–26198
ISBN 0–8493–2751–2

Laserset by Scribe Design, Gillingham, Kent
Origination by Adroit Photolitho Ltd, Birmingham
Printed and bound in Great Britain, at the University Press, Cambridge

Contributors

Lawrence W Bassett MD
Professor of Radiological Sciences
UCLA School of Medicine
Los Angeles, California

Akbar Bonakdarpour MD
Professor of Radiology and Orthopedic Surgery
Temple University, Pennsylvania

Daniel H Bunnell MD
Fellow, Department of Radiological Sciences
UCLA School of Medicine
Los Angeles, California

Julia R Crim MD
Resident Physician, Department of Radiological Sciences
UCLA School of Medicine
Los Angeles, California

Richard H Gold MD
Professor of Radiological Sciences
UCLA School of Medicine
Los Angeles, California

Theodore R Hall MD
Assistant Professor of Radiological Sciences
UCLA School of Medicine
Los Angeles, California

Steven E Harms MD
Director of Magnetic Resonance Imaging
Department of Radiology
Baylor University Medical Center
Dallas, Texas

Steven Hartzman MD
Assistant Clinical Professor of Radiological Sciences
UCLA School of Medicine
Los Angeles, California

Hooshang Kangarloo MD
Professor of Pediatrics and Radiology
Chairman of Radiological Sciences
UCLA School of Medicine
Los Angeles, California

Robert B Lufkin MD
Assistant Professor of Radiological Sciences
UCLA School of Medicine
Los Angeles, California

Charles T McGlade MD
Fellow, Department of Radiological Sciences
UCLA School of Medicine
Los Angeles, California

Mamed Mesgarzadeh MD
Assistant Professor of Diagnostic Imaging
Temple University Medical School
Philadelphia, Pennsylvania

Carson D Schneck MD PhD
Professor of Anatomy and Diagnostic Imaging
Temple University Medical School
Philadelphia, Pennsylvania

Leanne L Seeger MD
Assistant Professor of Radiological Sciences
UCLA School of Medicine
Los Angeles, California

Cynthia S Sherry MD
Fellow, Magnetic Resonance Imaging
Department of Radiology
Baylor University Medical Center
Dallas, Texas

John S H Tang MD
Fellow in Musculoskeletal Radiology
UCLA School of Medicine
Los Angeles, California

To Sandy, Gittelle, Rudi, Rawls, Lara, David, Blanche, Blazer, Rupert and Pixel

We are very grateful to Mary Frazier, our administrative assistant, for her invaluable assistance and dedication to the completion of this book.

LWB
RHG
LLS

Contents

Introduction
Lawrence W Bassett, Richard H Gold and Leanne L Seeger   1

1 **Physical principles of MRI**
Leanne L Seeger and Robert B Lufkin   11

2 **The spine**
Robert B Lufkin   25

3 **The temporomandibular joint**
Cynthia S Sherry and Steven E Harms   59

4 **The shoulder**
Leanne L Seeger   95

5 **The elbow**
Daniel H Bunnell and Lawrence W Bassett   129

6 **The wrist and hand**
Mamed Mesgarzadeh, Carson D Schneck and Akbar Bonakdarpour   139

7 The hip
Charles T McGlade and Lawrence W Bassett   175

8 The knee
Steven Hartzman and Richard H Gold   215

9 The ankle and foot
Julia R Crim and Lawrence W Bassett   266

10 Pediatrics
Theodore R Hall and Hooshang Kangarloo   279

11 Tumors and tumor-like conditions
Leanne L Seeger and Lawrence W Bassett   317

12 Musculoskeletal infection
John S H Tang and Richard H Gold   379

Index   413

Introduction

Lawrence W Bassett
Richard H Gold and
Leanne L Seeger

The potential of magnetic resonance imaging (MRI) in the evaluation of musculoskeletal disorders was recognized in early clinical trials.[1-3] Although still relatively new and not yet developed to anywhere near its full potential, MRI has already made a profound impact on the evaluation and management of many musculoskeletal disorders.[4] MRI has been shown to be useful in the evaluation of traumatic, inflammatory, neoplastic and degenerative conditions involving the spine and appendicular musculoskeletal system.

The features of MRI that have proved useful include: (1) excellent depiction of bone marrow, (2) high contrast discrimination of soft tissues, and (3) ability to produce thin-section images in any plane. In some cases, MRI is less susceptible to metal artifacts than CT when examining patients with internal fixation devices or prostheses. However, MRI examinations like those of CT are expensive, and MRI scanners are limited in number. Therefore, MRI should be used only when it can be expected to provide information that is unavailable from less expensive noninvasive diagnostic methods, and when the results of the examination may significantly affect patient management. MRI examinations yield the most information when they are tailored to a specific clinical problem.

Because the findings in MR images are relatively non-specific (Figures 1–2), it cannot be overstressed that MRI examinations should only be performed and interpreted with the full knowledge of the findings of other imaging examinations, especially plain radiography.

Tissue parameters ultimately affect the appearance of the MR image, and these are discussed in the chapter on physics. There are advantages and disadvantages associated with different field strengths, and the controversy as to what is the optimal magnetic field strength is still unresolved.[5] Although high-field-strength systems provide a stronger signal compared to low-field-strength systems, motion artifacts and chemical shift artifacts are more common in the T_1-weighted images of high-field-strength systems.[6] Intermediate- and low-field-strength systems produce a lower signal but less noise. High-field-strength units also require greater shielding and are more expensive to install. Image quality also depends on other equipment parameters, such as the design of radiofrequency coils, and the homogeneity of the magnetic field. Although most of the images in this atlas were obtained with a 0.3 Tesla imaging system, they should prove instructive to radiologists using either low- or high-field-strength systems.

Bone marrow

Bone marrow T_1- and T_2-relaxation times vary with age.[7] In adults, tumor infiltrates in the bone marrow are recognized in T_1-weighted images by abnormally decreased signal intensity within the normally high-signal-intensity bone marrow. While MRI is extremely sensitive to abnormalities in the marrow, and especially in depicting their extent, the MRI findings are usually non-specific; traumatic, inflammatory and neoplastic disorders may have similar appearances (Figures 1–2). Hence the importance of plain radiographs which, although not as accurate as MRI in revealing lesional extent, are often far more helpful to the formation of the differential diagnosis.

Muscles and tendons

T_1-weighted images provide excellent contrast between muscles and surrounding tissues. However, because normal muscle has a short T_2-relaxation time, and lesions usually have a long T_2-relaxation time, T_2-weighted images are more suitable for depicting intramuscular lesions. In T_2-weighted images, increased signal intensity has been shown to occur in skeletal muscle adjacent to neoplasms and inflammatory processes,[8] most likely secondary to edema.

Injuries of ligaments and tendons can be identified on MR images. The high-contrast resolution of MRI and its capability for multiplanar imaging are important factors favoring the identification of these lesions.[9-11] T_2-weighted images reveal the injuries by their high-signal intensity within the normally low-signal-intensity tendinous and ligamentous structures (Figure 3). Occasionally, it is helpful to include the contralateral extremity for comparison.

Tumors

MRI is proving to be a sensitive method for detecting and staging primary bone tumors,[12] and shows their intramedullary and extramedullary extent as well as or better than CT. The capability of imaging in the sagittal, coronal, axial, or any oblique plane, allows optimal assessment of the relationship of the tumor to such adjacent normal structures as physes, joints, and neurovascular bundles. Thus, MRI is an excellent modality for the preoperative or preradiotherapy evaluation of musculoskeletal tumors.[2,12-15] However, because the specificity of MRI is generally poor, plain radiography continues to be the most important imaging tool for the differential diagnosis of bone tumors and for the determination of their aggressiveness. Plain radiography and CT are preferable to MRI for demonstrating calcific or osseous deposits in tumor matrix, and for disclosing subtle cortical involvement and pathologic fractures.

MRI is superior to non-contrast-enhanced CT and equal to contrast-enhanced CT in delineating the characteristics of spinal tumors, including the degree of bone involvement, spinal canal invasion, and paraspinal soft-tissue extension.[16]

MRI is more effective than CT in demonstrating the size and extent of most soft-tissue masses, as well as their relationship to neighboring structures.[17-18] Lesions within fat are usually shown better in T_1-weighted images, while lesions within muscle are better depicted in T_2-weighted images. Subtle cortical destruction, however, may be difficult to appreciate with MRI, and, with the exception of its characteristic depiction of fatty tumors, MRI is not tissue specific.

Osteomyelitis

Osteomyelitis causes changes in the bone marrow that are reflected in MR images.[19-20] A reduction of signal intensity in T_1-weighted images corresponds to infiltration of the marrow by inflammatory exudate. In T_2-weighted images, foci of active infection are characterized by their high-signal intensity. A normal T_1-weighted MR examination excludes the diagnosis of osteomyelitis.

Spine

Between 1985 and 1987, the region of the body undergoing the greatest increase in MRI utilization was the spine.[21] High-quality images, negligible risk, and multiplanar imaging in any orthogonal or non-orthogonal plane make MRI ideal for spine imaging. MRI provides a noninvasive method for depiction of the spinal cord and nerve roots, and is the most sensitive and accurate modality for screening patients with symptoms of a spinal cord lesion, frequently obviating the need for myelography and CT.[22] MRI reveals information about degenerated disks comparable to that derived from CT and myelography.[23-24] However, MRI is currently not as useful as these two modalities for the evaluation of the bony changes of spondylosis.[25] MRI may also aid in the evaluation of vertebral osteomyelitis.[26]

Temporomandibular joints

Investigators have shown MRI to be superior to other imaging methods in depicting the anatomy and pathology of the temporomandibular articular meniscus.[27-28]

Shoulder

Surface-coil imaging allows excellent depiction of the anatomy of the shoulder,[29-30] and is useful in the evaluation of impingement syndrome, rotator cuff disease, and shoulder stability.[31]

Elbow

Although MRI depicts elbow anatomy in exquisite detail,[32-33] clinically useful applications for the elbow have yet to be demonstrated.

Hand and wrist

MRI provides superb depiction of the hand and wrist, allowing discrimination of soft-tissue structures including nerves, ligaments, tendons and blood vessels.[34] MRI reliably depicts the structures within the carpal tunnel, and can disclose some of the findings associated with the carpal tunnel syndrome, including swelling of the median nerve and thickening of the tendon sheaths.[35] Marrow and articular abnormalities have also been identified.[36-37]

Hip

MRI is the most accurate method for the early detection of ischemic necrosis of the femoral head.[38-41] Although the MR appearance of ischemic necrosis is subject to some variability, in all cases the abnormality is characterized in T_1-weighted images by a decrease in the normally high-intensity signal of the bone marrow within the femoral head. The changes feature a ring- or band-like area of diminished signal, or a homogeneous or inhomogeneous focal region of diminished signal in the subarticular region of the femoral head. A pattern of diffusely decreased signal intensity may also be seen. The reasons for the diminished signal have not yet been satisfactorily explained, nor have the pathophysiological mechanisms behind the various MRI changes been proved, but replacement of marrow fat by a histiocytic, fibrous connective tissue response in combination with new bone proliferation undoubtedly plays a contributing role.

Knee

High-resolution, thin-section, multiplanar MR images are capable of depicting the important soft-tissue structures of the knee joint, including the cruciate and collateral ligaments, and the menisci.[42-43] Injuries of the cruciate and collateral ligaments are recognized by their distortion or disruption in T_1-weighted images, and by an abnormally high-signal intensity within the ligaments in T_2-weighted images.[44] MRI is also an effective technique for the depiction of meniscal tears. The meniscus is normally devoid of signal, and a tear is characterized by a linear or complex region of signal within the meniscus, extending to an articular surface.[45-46] The predictive value of an MR examination that is negative for meniscal tears is close to 100 per cent. MRI can provide useful information about other disorders that affect the knee, including osteochondritis dissecans,[47] synovial and meniscal cysts,[48] intraarticular tumors, fluid collections and osteomyelitis.

Ankle

MRI provides excellent depiction of the anatomy of the ankle,[49] and has proven useful in the evaluation of injured ligaments and ischemic necrosis of the talus. Further clinical trials are needed to determine the usefulness of MRI for the evaluation of other ankle abnormalities.

References

1. MOON KL, GENANT HK, HELMS CA et al, Musculoskeletal applications of nuclear magnetic resonance, *Radiology* (1983) **147**:161–71.
2. BRADY TJ, ROSEN BR, PYKETT IL et al, NMR imaging of leg tumors, *Radiology* (1983) **149**:181–7.
3. BERQUIST TH, Preliminary experience in orthopedic radiology, *Magn Reson Imaging* (1984) **2**:41–52.
4. EHMAN RL, BERQUIST TH, MCLEOD RA, MR imaging of the musculoskeletal system: a 5-year appraisal, *Radiology* (1988) **166**:313–20.
5. MARGULIS AR, CROOKS LE, Present and future status of MR imaging, *AJR* (1988) **150**:487–92.
6. HOULT DI, Field, contrast and sensitivity in imaging. In: James TL, Margulis AR, eds. *Biomedical magnetic resonance*. (Radiology Research and Education Foundation: San Francisco 1984) 35–45.
7. DOOMS GC, FISHER MR, HRICAK H et al, Bone marrow imaging: magnetic resonance studies related to age and sex, *Radiology* (1985) **155**:429–32.

8 BELTRAN J, SIMON DC, KATZ W et al, Increased MR signal intensity in skeletal muscle adjacent to malignant tumors: pathologic correlation and clinical relevance, *Radiology* (1987) **162**:251–5.

9 FULLERTON GD, CAMERON IL, ORD VA, Orientation of tendons in the magnetic field and its effect on T_2 relaxation times, *Radiology* (1985) **155**:433–5.

10 BELTRAN J, NOTO AM, HERMAN LJ et al, Tendons: high-field-strength, surface coil MR imaging, *Radiology* (1987) **162**:735–40.

11 EHMAN RL, BERQUIST TH, Magnetic resonance imaging of musculoskeletal trauma, *Radiol Clin North Am* (1986) **24**:291–319.

12 RICHARDSON ML, KILCOYNE RF, GILLESPY T III et al, Magnetic resonance imaging of musculoskeletal neoplasms, *Radiol Clin North Am* (1986) **24**:259–67.

13 AISEN AM, MARTEL W, BRAUNSTEIN EM et al, MRI and CT evaluation of primary bone and soft-tissue tumors, *AJR* (1986) **146**:749–56.

14 ZIMMER WD, BERQUIST TH, MCLEOD RA et al, Bone tumors: magnetic resonance imaging versus computed tomography, *Radiology* (1985) **155**:709–18.

15 BLOEM JL, BLUEMM RG, TAMINIAU AHM et al, Magnetic resonance imaging of primary malignant bone tumors, *RadioGraphics* (1987) **7**:425–45.

16 BELTRAN J, NOTO AM, CHAKERES DW et al, Tumors of the osseous spine: staging with MR imaging versus CT, *Radiology* (1987) **162**:565–9.

17 TOTTY WG, MURPHY WA, LEE JK, Soft-tissue tumors: MR imaging, *Radiology* (1986) **160**:135–41.

18 PETASNICK JP, TURNER DA, CHARTERS JR et al, Soft-tissue masses of the locomotor system: comparison of MR imaging with CT, *Radiology* (1986) **160**:125–33.

19 BELTRAN J, NOTO AM, MCGHEE RB et al, Infections of the musculoskeletal system: high-field-strength MR imaging, *Radiology* (1987) **164**:449–54.

20 TANG JS, GOLD RH, BASSETT LW et al, Musculoskeletal infection of the extremities: evaluation with MR Imaging, *Radiology* (1988) **166**:205–9.

21 EVENS RG, EVENS RG JR, Economic and utilization analysis of MR imaging units in the United States in 1987, *Radiology* (1988) **166**:27–30.

22 HAUGHTON VM, MR imaging of the spine, *Radiology* (1988) **166**:297–301.

23 MODIC MT, PAVLICEK W, WEINSTEIN MA et al, Magnetic resonance imaging of intervertebral disk disease. Clinical and pulse sequence considerations, *Radiology* (1984) **152**:103–11.

24 EDELMAN RR, SHOULIMAS GM, STARK DD et al, High-resolution surface-coil imaging of lumbar disk disease, *AJR* (1985) **144**:1123–9.

25 MODIC MT, MASARYK T, BOUMPHREY F et al, Lumbar herniated disk disease and canal stenosis: prospective evaluation by surface coil MR, CT, and myelography, *AJR* (1986) **147**:757–65.

26 MODIC MT, FEIGLIN DH, PIRAINO DW et al, Vertebral osteomyelitis: assessment using MR, *Radiology* (1985) **157**:157–66.

27 HARMS SE, WILK RM, WOLFORD LM et al, The temporomandibular joint: magnetic resonance imaging using surface coils, *Radiology* (1985) **157**:133–6.

28 KATZBERG RW, BESSETTE RW, TALLENTS RH et al, Normal and abnormal temporomandibular joint: MR imaging with surface coil, *Radiology* (1986) **158**:183–9.

29 HUBER DJ, SAUTER R, MUELLER E et al, MR imaging of the normal shoulder, *Radiology* (1986) **158**:405–8.

30 SEEGER LL, RUSZKOWSKI JT, BASSETT LW et al, MR imaging of the normal shoulder: anatomic correlation, *AJR* (1987) **148**:83–91.

31 SEEGER LL, GOLD RH, BASSETT LW et al, Shoulder impingement syndrome: MR findings in 53 shoulders, *AJR* (1988) **150**:343–7.

32 MIDDLETON WD, MACRANDER S, KNEELAND JB et al, MR imaging of the normal elbow: anatomic correlation, *AJR* (1987) **149**:543–7.

33 BUNNELL DH, FISHER DA, BASSETT LW et al, Elbow joint: normal anatomy on MR images, *Radiology* (1987) **165**:527–31.

34 WEISS KL, BELTRAN J, SHAMAM OM et al, High-field MR surface-coil imaging of the hand and wrist. Part I. Normal anatomy, *Radiology* (1986) **160**:143–6.

35 MIDDLETON WD, KNEELAND JB, KELLMAN GM et al, MR imaging of the carpal tunnel: normal anatomy and preliminary findings in the carpal tunnel syndrome, *AJR* (1987) **148**:307–16.

36 KOENIG H, LUCAS D, MEISSNER R, The wrist: a preliminary report on high-resolution MR imaging, *Radiology* (1986) **160**:463–7.

37 WEISS KL, BELTRAN J, LUBBERS LM, High-field surface-coil imaging of the hand and wrist. Part II. Pathologic correlations and clinical relevance, *Radiology* (1986) **160**:147–52.

38 TOTTY WG, MURPHY WA, GANZ WI et al, Magnetic resonance imaging of the normal and ischemic femoral head, *AJR* (1984) **143**:1273–80.

39 MARKISZ JA, KNOWLES RJR, ALTCHEK DW et al, Segmental patterns of avascular necrosis of the femoral heads: early detection with MR imaging, *Radiology* (1987) **162**:717–20.

40 MITCHELL MD, KUNDEL HL, STEINBERG ME et al, Avascular necrosis of the hip: comparison of MR, CT, and scintigraphy, *AJR* (1986) **147**:67–71.

41 MITCHELL DG, RAO VM, DALINKA MK et al, Femoral head avascular necrosis: correlation of MR imaging, radiographic staging, radionuclide imaging, and clinical findings, *Radiology* (1987) **162**:709–15.

42 LI DK, ADAMS ME, MCCONKEY JP, Magnetic resonance imaging of the ligaments and menisci of the knee, *Radiol Clin North Am* (1986) **24**:209–27.

43 REICHER MA, RAUSCHNING W, GOLD RH et al, High-resolution MR imaging of the knee joint: normal anatomy, *AJR* (1985) **145**:895–902.

44 REICHER MA, BASSETT LW, GOLD RH, High-resolution magnetic resonance imaging of the knee joint: pathologic correlations, *AJR* (1985) **145**:903–9.

45 REICHER MA, HARTZMAN S, BASSETT LW et al, MR imaging of the knee. Part I. Traumatic disorders, *Radiology* (1987) **162**:547–51.

46 BURK DL JR, KANAL E, BRUNBERG JA et al, 1.5 T surface-coil MRI of the knee, *AJR* (1986) **147**:293–300.

47 HARTZMAN S, REICHER MA, BASSETT LW et al, MR imaging of the knee. Part II. Chronic disorders, *Radiology* (1987) **162**:553–7.

48 BURK DL JR, DALINKA MK, KANAL E et al, Meniscal and ganglion cysts of the knee: MR evaluation, *AJR* (1988) **150**:331–6.

49 HAJEK PC, BAKER LL, BJORKENGREN A et al, High-resolution magnetic resonance imaging of the ankle: normal anatomy, *Skeletal Radiol* (1986) **15**:536–40.

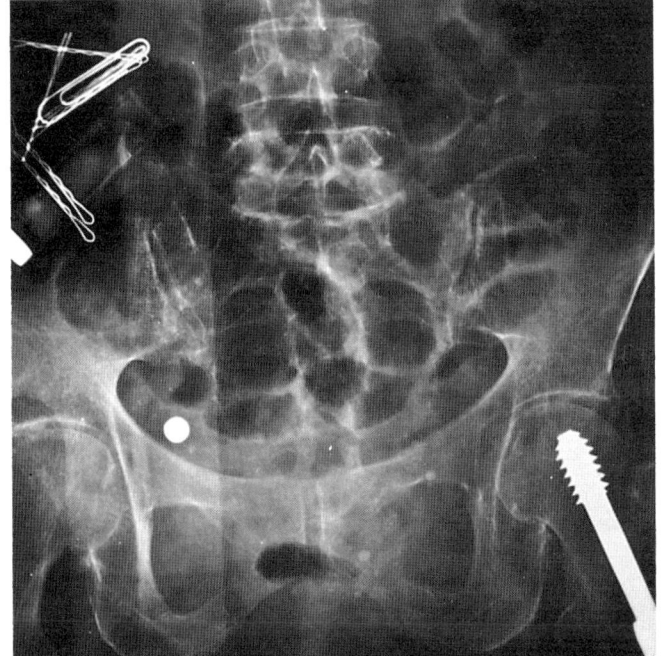

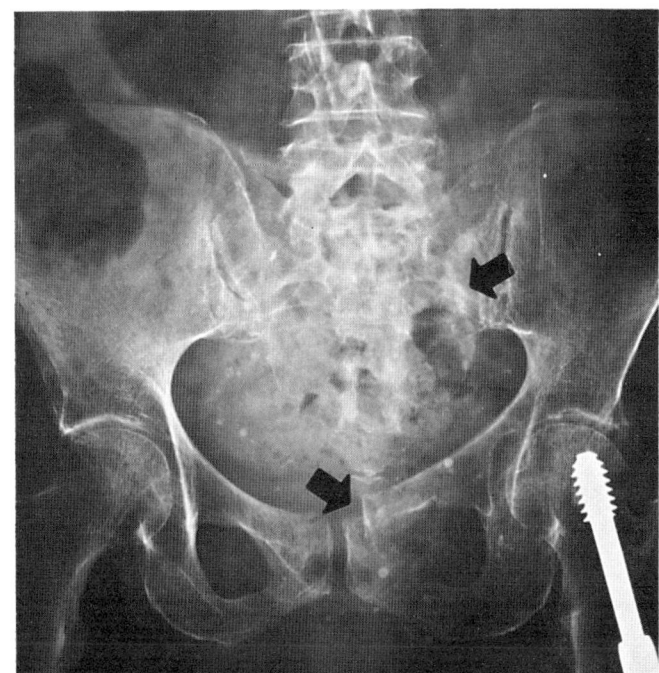

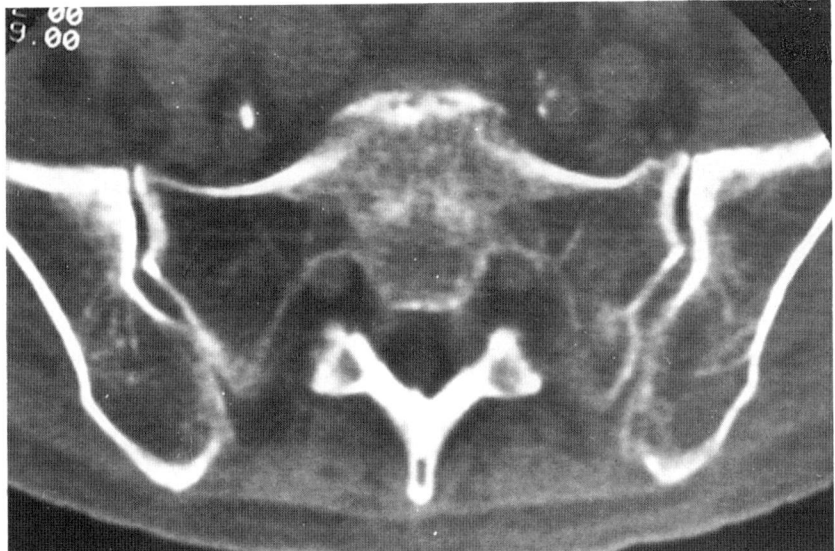

Figure 1

Non-specificity of MRI. A 78-year-old woman with palpable breast mass and sacral pain after a fall. (**a**) Anteroposterior emergency room radiograph of pelvis. Aside from metal fixation of previous left hip fracture and paraphernalia in her right pocket, the radiograph was interpreted as showing only osteopenia. (**b**) CT scan was interpreted as normal. (**c**) Anteroposterior radiograph 3 weeks after CT scan. Previously occult fractures of left sacrum and pubic rami have become evident due to displacement and callus (arrows).

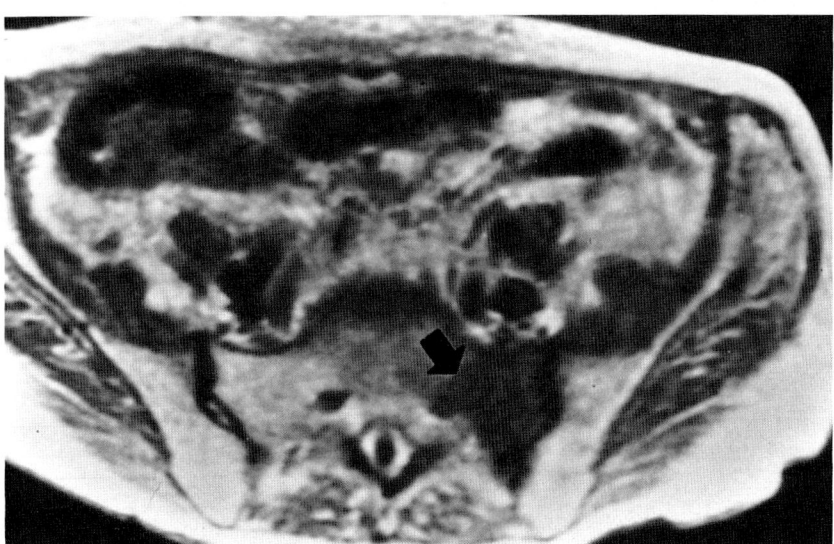

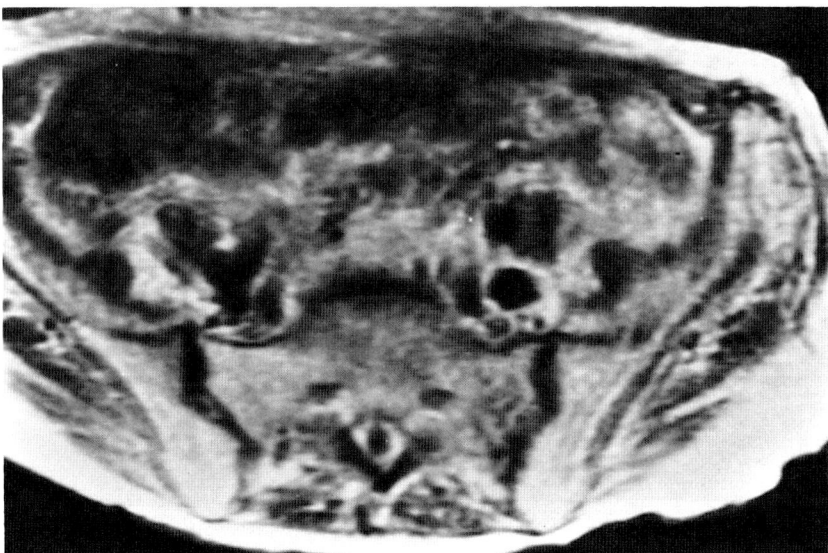

Figure 1 *continued*

(**d,e**) Axial T_1-weighted image (SE 600/28) (**d**) discloses region of decreased signal intensity in left sacrum (arrow). On axial T_2-weighted image (SE 1500/56) (**e**), same area has signal intensity similar to that of surrounding marrow. MR findings in sacrum, evaluated without benefit of concurrent plain radiographic correlation, were misinterpreted as representing metastatic disease. (**f**) Radionuclide bone scan discloses increased isotope uptake only at fracture sites. Biopsy of breast mass revealed benign disease.

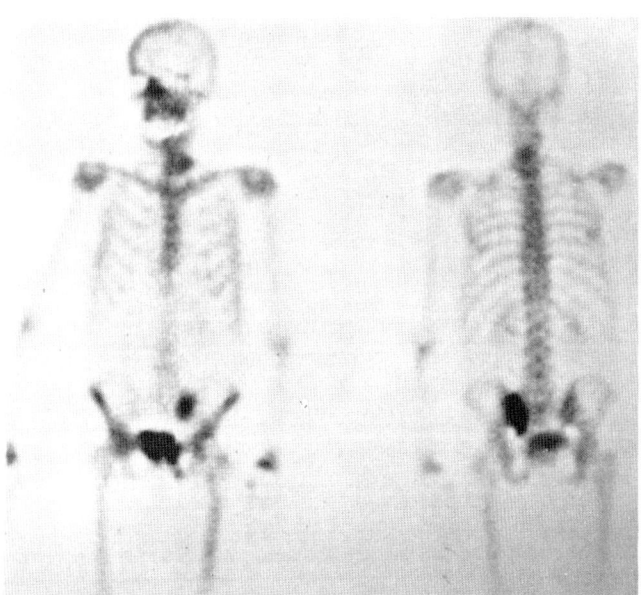

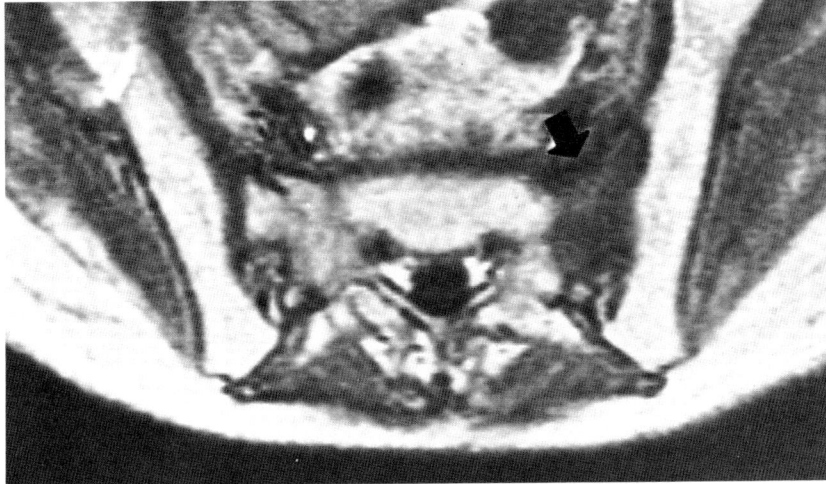

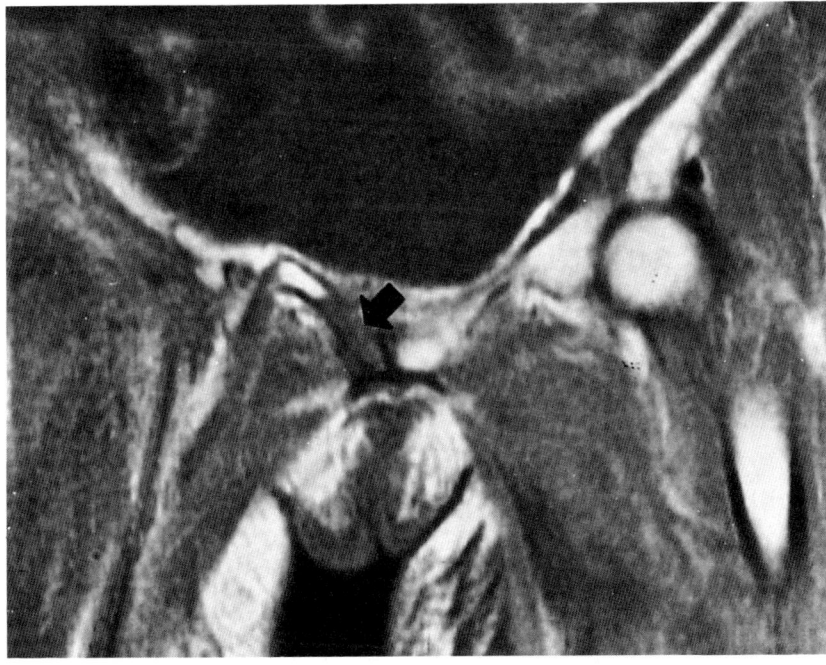

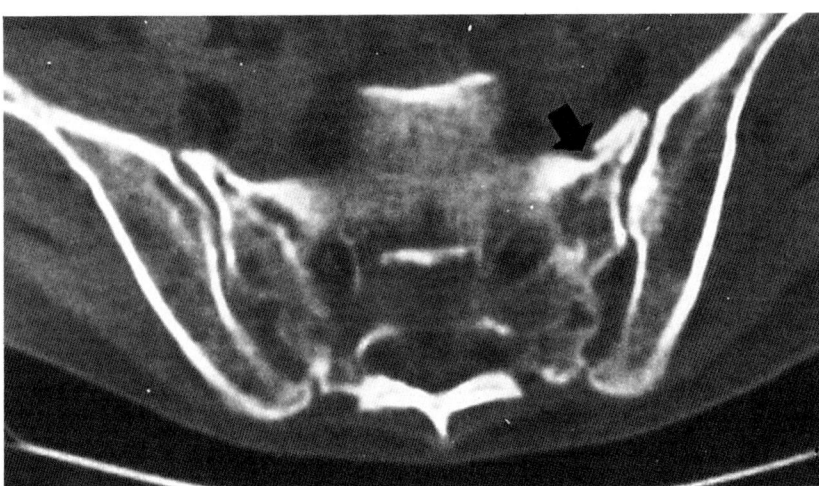

Figure 2

Non-specificity of MRI. A 72-year-old woman with colon carcinoma and recent onset of pain in lower back and right hip. Radiographs showed osteoporosis. (**a**) T_1-weighted axial image (SE 400/28) discloses region of low-signal intensity in left sacral ala (arrow). (**b**) T_1-weighted coronal image (SE 500/28) reveals region of low-signal intensity in right pubic bone (arrow). (**c,d**) CT scan reveals that these regions of low-signal intensity represent fractures, not metastases (arrow).

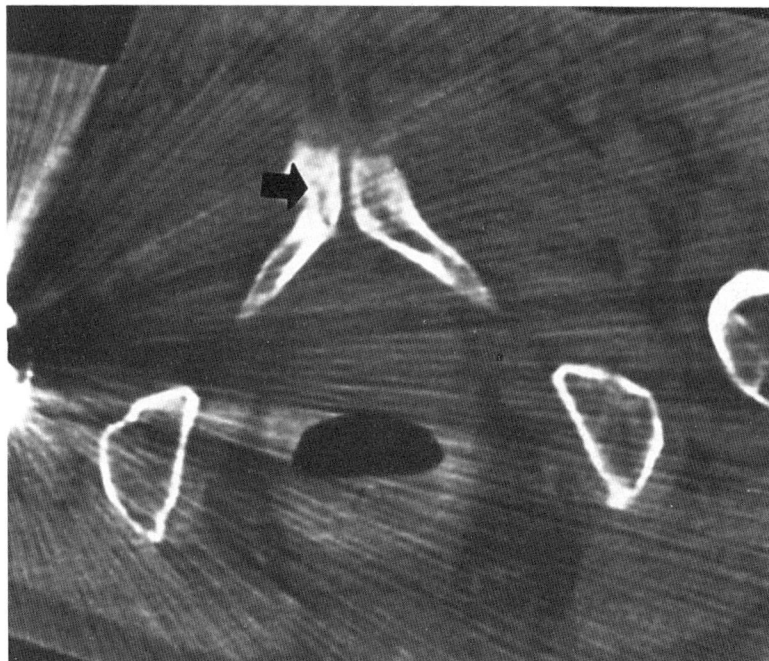

Figure 2 continued

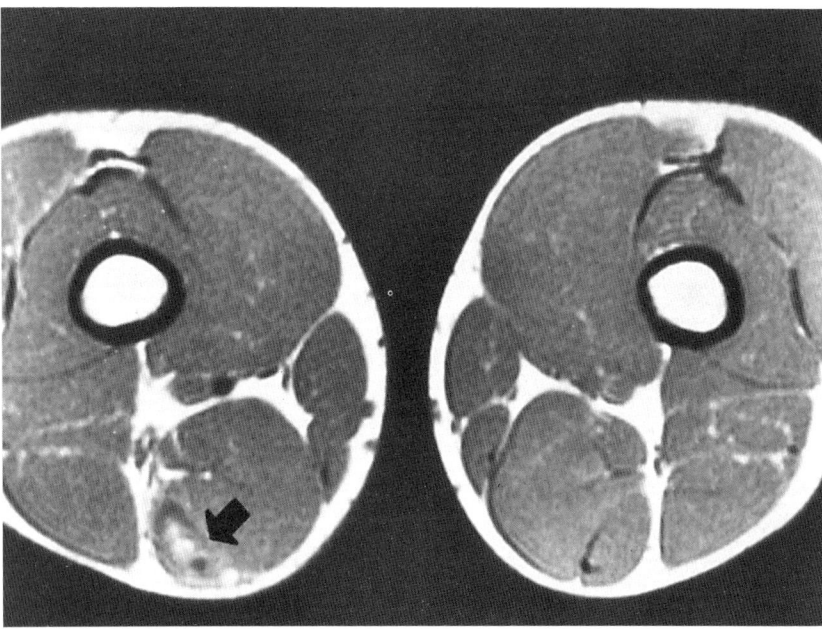

Figure 3

Tendon rupture. A 22-year-old athlete 'pulled his hamstring' on 2 occasions and was unable to run. (**a,b**) 7 cm above patella. Axial T_1-weighted (SE 600/20) (**a**) and T_2-weighted (SE 2000/85) (**b**) images show abnormally high-signal intensity at site of semitendinosus muscle (arrow). (**c,d**) 3.5 cm above patella. Axial T_1-weighted image (SE 600/20) (**c**) discloses region of homogeneous low-signal intensity in semitendinosus musculotendinous junction. T_2-weighted image (SE 2000/85) (**d**) reveals high-signal intensity in this region. The combined changes in the T_1- and T_2-weighted images reflect replacement of the torn tendon by fluid. (**e**) Sagittal T_2-weighted image (SE 2000/85) discloses that the retracted semitendinosis muscle has abnormally high-signal intensity. Arrow indicates musculotendinous junction.

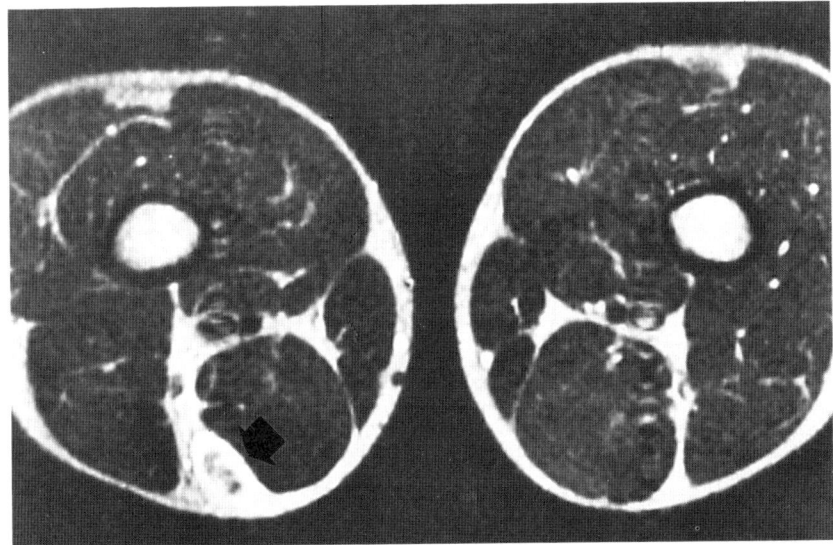

b

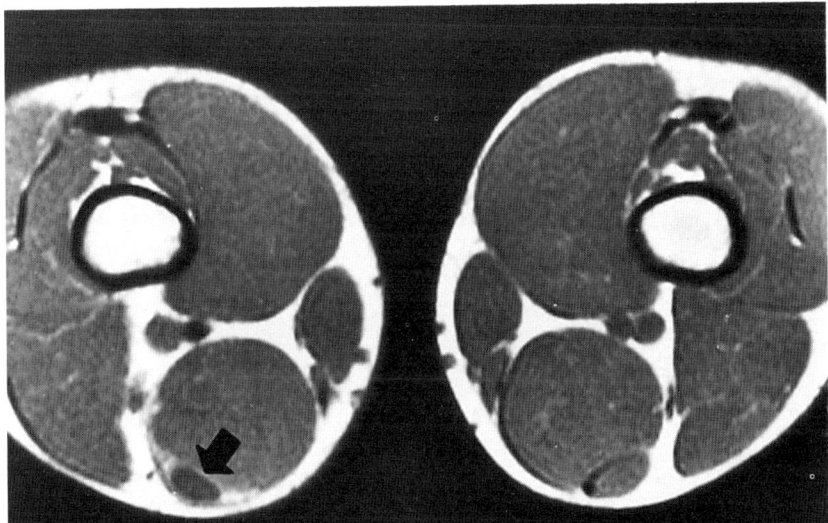

c

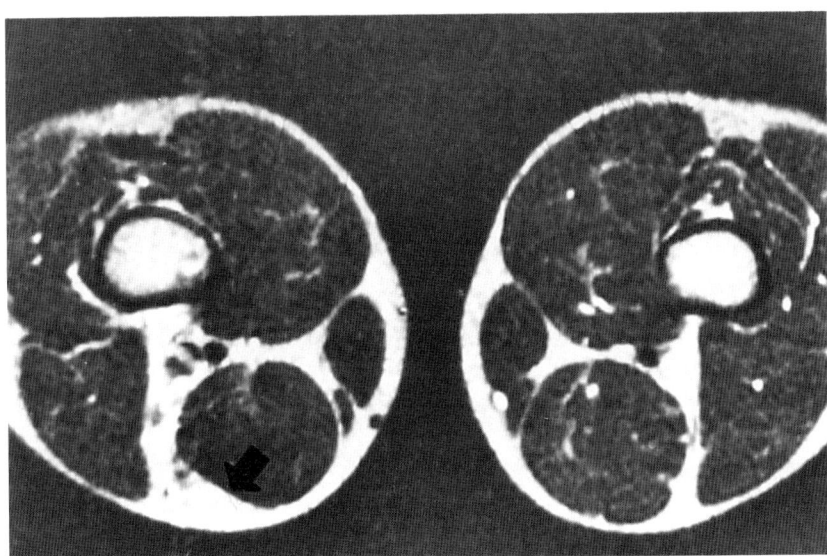

d

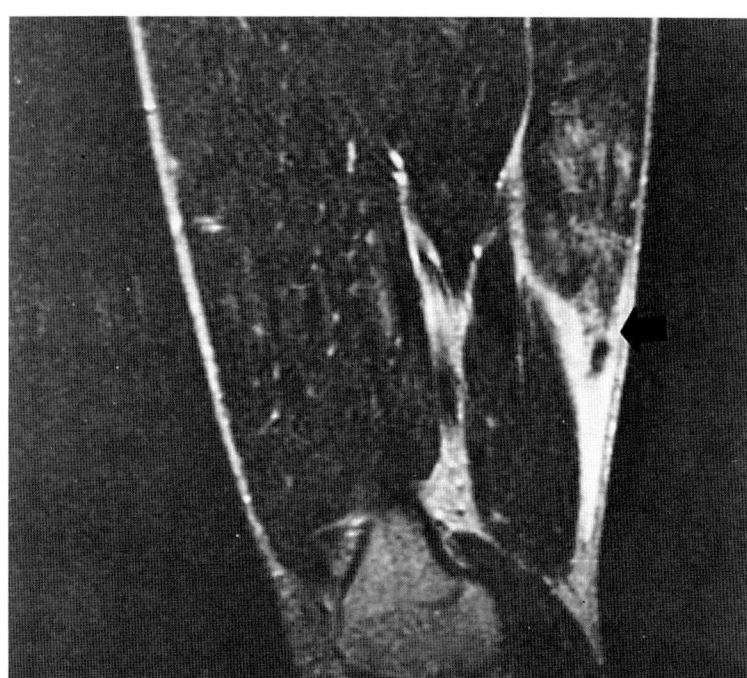

e

Figure 3 *continued*

1
Physical principles of MRI

Leanne L Seeger and
Robert B Lufkin

General principles

Rather than providing a detailed survey of the physics of magnetic resonance imaging (MRI), this chapter is intended to review briefly only those basic principles which are required to understand the MR appearance of the musculoskeletal images in this atlas. So as not to discourage those with a limited background in physics, we have excluded all equations from this discussion. Many excellent references are available for those who desire a more complete understanding of the physics of MRI.[1-8]

Although several different scanning techniques may be used in clinical MRI, we will discuss only spin-echo (SE) MRI.

It is useful to note the position of magnetic resonance along the electromagnetic spectrum, and appreciate the concept of energy windows for biological imaging (Figure 1.1). X-rays are at one end of the spectrum, and radiowaves are at the other end. It is within the energy windows of these two extremes that medical imaging is possible, as the body is 'transparent' to very high (X-ray) and very low (radiowave) energy levels, but not to those levels in the middle of the spectrum, i.e. visible, infrared, and ultraviolet light.

The original name given to this imaging modality was 'nuclear magnetic resonance' (NMR), but because of confusion and public concern with regard to the term 'nuclear', the name was changed to 'magnetic resonance imaging' (MRI) for medical applications of NMR.

The term 'nuclear' in NMR refers to the nucleus of the hydrogen atom. Although the chemical properties of an atom depend on its electron structure, its physical properties are mainly dependent on the nucleus, which accounts for almost all of the mass of the atom. With the exception of hydrogen, the nucleus of all atoms contains both protons and neutrons. These particles within the nucleus are collectively called nucleons. The number of positively charged nuclear protons and negatively charged orbital electrons is usually the same in order to maintain electrical neutrality. The number of protons and neutrons, however, may not necessarily be the same. If a nucleus contains either unpaired protons, neutrons, or both, it has a net spin and angular momentum. Angular momentum describes the rotational motion of a body, and must be non-zero in order for the MR phenomenon to occur. Although roughly one-third of the almost 300 stable atoms have unpaired nuclei (and therefore have angular momentum), only a select few of these are of interest for medical imaging. Hydrogen is the ideal atom for medical imaging, as it is not only the simplest (only one nucleon: a proton), but it makes up two-thirds of all the atoms in the human body. For the remainder of this discussion, we will refer only to MRI using hydrogen, although other atoms which are less abundant may also be used (^{23}sodium, ^{31}phosphorus, ^{13}carbon, and others). The terms 'proton' and 'hydrogen' will be used interchangably.

The term 'magnetic' in NMR refers to the magnetic (dipole) moment, or tendency to produce motion which is present in the proton as a result of its angular momentum. The spin of the atom represents a current loop, with north and south poles. In the absence of an externally applied magnetic field, the vectors of these magnetic dipole moments are randomly oriented. Once exposed to a magnetic field such as is present in a magnetic resonance scanner, these dipoles tend to align with the field like tiny bar magnets.

Protons, because of their small size, do not follow the rules of classical Newtonian physics, but rather those of quantum mechanics. In Newtonian physical theory, a bar magnet must align precisely with the applied magnetic field. In quantum theory, the protons may be in one of two positions which differ from each other by 180 degrees; that is, either parallel or antiparallel to the applied field. The parallel orientation is the low-energy condition, and is called the ground state. The antiparallel orientation is the high-energy state, and is referred to as the excited state. Slightly more protons align in the ground state parallel to the magnetic field. Thus, a population of protons placed in a static magnetic field will have a net magnetic field vector parallel to the applied field due to the slightly greater number of spins in the parallel (low-energy) orientation. While this difference in the direction of alignment depends on the strength of the applied magnetic field, it is small in clinical MRI. In a population of one million atoms, only one more atom will be in the parallel than in the antiparallel position, a difference that, while minute, is sufficient to allow clinical MR imaging.

The term 'resonance' in NMR incorporates the two preceding concepts of an atom's nucleus (the proton of hydrogen) and its magnetic moment. In addition to the spinning action, protons precess, or wobble a few degrees off the axis of the applied magnetic field, similar to the spinning of a gyroscope under the influence of the Earth's gravitational field (Figure 1.2). The frequency of this precession is known as the resonant frequency, and is proportional to the strength of the applied magnetic field.

In order to detect an MR signal, the orientation of the net magnetic vector of the protons must be rotated 90 degrees, and the spins of the protons must be in phase, or spinning together. Both of these conditions are accomplished by the application of a second magnetic field, which is oscillated with a frequency synchronized to the resonant frequency of the precessing spins. This oscillating magnetic field is a radiofrequency (RF) pulse. The RF pulse will push the net magnetization vector from the parallel (longitudinal) position into the transverse plane, allowing detection of the signal by the receiver coils of the scanner. The application of the RF pulse also brings the spins of the protons into phase so that they are coherent, and the spins are moved into a higher energy level such that there are an equal number of spins in the high- and low-energy states (Figure 1.3). In clinical practice, the 90-degree pulse is followed by a 180-degree pulse in order to compensate for inherent inhomogeneities in the static magnetic field of the scanning system. This pulse, however, is not necessary to generate an MR signal.

After the second magnetic field is turned off, the nuclei move back to the lower energy state and emit RF energy of their own. This detected RF output is the MR signal, and explains all MR contrast effects, including those due to T_1 and T_2 (see below). The amplitude of the MR signal is proportional to the number of spins in the sample, or its proton density, as well as several other factors. When more protons are present, the intensity of the magnetization is greater and the signal detected by the RF receiver coils is greater. Multiple signals can be obtained by reapplication of the RF pulse, giving so-called 'multi-echo' images. This in a sense provides a 'free' second look at the same tissue without the additional time required for a second pulse sequence. As will be explained below, however, this type of pulse sequencing may be misleading when evaluating certain types of musculoskeletal pathology, especially tumors and infection.

In order to obtain an anatomical image, the spins are subjected to three magnetic field gradients within the magnet (Figure 1.4). The slice select gradient locates the position and specifies the thickness of an individual image within a scan sequence. The remaining two gradients are used to construct the two-dimensional image. The frequency, or read gradient, locates the proton by the frequency of its spin. The phase-encoding gradient specifies the location in space of a given proton within the slice according to the phase of its spin. Depending on the selection of imaging plane and type of imaging sequence chosen, any of these three gradients may be along the X, Y, or Z axis. An appreciation for these different gradients, or axes, is important in understanding MR artifacts, which are discussed below.

Pulse sequencing

For any sample, the recovered signal intensity is less than its original amplitude as determined by the proton density. This is due to irreversible signal losses within the sample itself, known as T_1- and T_2-relaxation. The rate of these relaxations depends on the local environment of the sample and thus reflects its chemical structure, and is responsible for the anatomical image.

T_1-relaxation, also called longitudinal or spin-lattice relaxation, characterizes the interaction of a nucleus with its environment. After the RF stimulation, the spins are moved into the higher transverse energy state. As they fall back into the lower energy longitudinal state, energy is dissipated into the surrounding environment. T_1 in milliseconds is the time required for 63 per cent of the longitudinal magnetization to recover following the RF pulse. T_1-relaxation times vary with the main magnetic field strength of the imaging system, and increase slightly with stronger magnets. In clinical practice, however, the effect on image contrast is minimal and of little consequence when compared to that which is possible with pulse sequence manipulation.

T_2-relaxation, also known as transverse or spin-spin relaxation, is due to randomly varying inhomogeneities in the magnetic field created by adjacent nuclei within the sample. It characterizes the interaction of a nucleus with surrounding nuclei of the same kind. T_2 in milliseconds is the time necessary to reduce the transverse magnetization to 37 per cent of its original value following the RF pulse. Because it reflects the chemical environment of a proton, T_2-relaxation is independent of magnetic field strength.

Image contrast

The MR signal intensity, as represented by pixel brightness on an image, is a reflection of T_1- and T_2-relaxation values, as well as several other parameters which are beyond the scope of this discussion. The relative contributions of some of these factors may be manipulated by controlling the timing of the RF pulses. Variations in this timing are determined by the pulse sequence chosen for imaging. The effect of manipulating the pulse sequence of the examination is illustrated in Figure 1.5. The timing values themselves are the TR (repetition time) and TE (echo time). TR refers to the time in milliseconds (msec) between the 90-degree RF pulses. TE is the time in milliseconds between application of the 90-degree RF pulse and recording the signal (echo) produced by the sample. A short TR and TE will result in an image which emphasizes the T_1 characteristics and minimizes the T_2 characteristics of the tissue, and is said to be 'T_1-weighted', for example, TR = 200 and TE = 15. A longer TR and TE will result in a more 'T_2-weighted' image, for example TR = 2000 and TE = 85. Figure 1.6 shows how varying the TR and TE affect the tissue contrast of an image. Generally, an image is said to be T_1-weighted if the TR is less than approximately 1000 msec, and the TE is less than 50 msec. T_2-weighted imaging is accomplished with a TR greater than approximately 1000 msec, and a TE greater than 50 msec. This convention was used for the images in this atlas, although the reader should keep in mind that there are 'degrees' of T_1- and T_2-weighting. For example, most of the images in the knee chapter were acquired with a TR of 800 msec, thus imparting a small amount of T_2-weighting to the image. 'Proton density' images are obtained by using a pulse sequence with a short TE and a long TR. This type of pulse sequencing may be the first echo of a double-echo sequence (a long TR and two different TEs: the first short, and the second long), or it may be used in situations in which true T_1- or T_2-weighting is not needed.

As shown in Table 1.1, different tissues behave differently according to the pulse sequence chosen for imaging. This property assists in characterization of normal as compared to abnormal tissue when both T_1- and T_2-weighted imaging are utilized.

Table 1.1 Signal intensity of different tissues according to pulse sequence chosen for imaging.

Signal intensity	Tissue
(a) T_1-weighting	
High (white)	Fat
	Normal adult marrow
Intermediate	Muscle
	Tumor
	Abnormal fluid
	Infection
Low (black)	Cortical bone
	Fibrous tissue
	Most normal fluid
(b) T_2-weighting	
High (white)	Fat
	Normal adult marrow
	Tumor
	Infection
	All fluid (including edema)
Low to intermediate	Muscle
Low (black)	Cortical bone
	Fibrous tissue

T_1-weighted imaging is useful for identifying abnormal tissue within structures of high fat content, such as marrow and subcutaneous fat. T_1-weighted imaging also provides the highest signal-to-noise ratio, and is therefore optimal for evaluating anatomic subtleties. Substances with a short T_1 value (high-signal intensity) include fat and lipid-containing materials, and proteinaceous fluid. Tissues with a long T_1 value (low-signal intensity) include normal body fluids (CSF, urine), cortical bone, and most ligaments and tendons. Most other soft tissues, including muscle, tumor, and infection, have an intermediate signal intensity in T_1-weighted images.

T_2-weighted images assist in distinguishing normal from abnormal soft tissues. Substances with a short T_2 value (low-signal intensity) include cortical bone, and most ligaments and tendons. Note that in reference to normal tissues of the musculoskeletal system, these include the same tissues which had a low-signal intensity in T_1-weighted images. Tissues with a long T_2 (high-signal intensity) include neoplasms, inflammation, and most fluids. The signal intensity of normal muscle on T_2-weighting is low to intermediate, allowing the identification of boundaries of tumor and infection within the soft tissues.

There is little published data on the MR features of extracranial hematomas, but the signal intensity patterns seem to follow that for intracranial bleeds.[9] The appearance of blood varies with respect to the timing of image acquisition relative to bleeding episodes. In the acute stage (less than one week), blood has an intermediate or slightly decreased signal intensity in T_1-weighted images, and markedly decreased signal intensity in T_2-weighted images, due to the T_2-shortening effects of deoxyhemoglobin. Between 90 hours and seven weeks, the blood will have intermediate- to high-signal intensity on T_1-weighted images and high-signal intensity on T_2-weighted images, due to the T_1-shortening effects of paramagnetic methemoglobin. A low-signal intensity rim may be present around these subacute hematomas. Chronic hematomas appear hypointense with both T_1- and T_2-weighted imaging, due to T_2-shortening by hemosiderin-laden macrophages.

Although it is common in clinical practice routinely to employ multi-echo imaging instead of both T_1- and T_2-weighted images, we have found that this practice may provide misleading information in certain situations. The effect of the long TR, even though combined with a short TE, will increase the overall T_2-weighting of the image. Thus, lesions which have a high-signal intensity in T_2-weighted images (e.g. tumor) and lie within tissue which also has a high-signal intensity in T_2-weighted images (e.g. bone marrow) are easily obscured. This pitfall is of no concern when imaging for pathology such as meniscal tears. In this situation, the longer TR will increase the signal intensity of fluid so that it is easily identified within the normally low-signal-intensity meniscus, and imaging time is kept relatively short. It is, however, extremely important to consider the ramifications of proton density imaging in the evaluation of musculoskeletal tumors or infection (Figures 1.7–1.8). We strongly urge that in these situations, true T_1- and T_2-weighted images be obtained rather than relying on double-echo imaging for a 'complete examination'.

Artifacts

Because magnetic resonance uses an approach to imaging totally different from that of ionizing radiation, several unfamiliar artifacts may be encountered. Recognizing these artifacts and understanding their source will prevent non-diagnostic images and misdiagnoses. Several excellent articles have been written on the subject of MR artifacts,[10-13] and the interested reader is encouraged to consult them. The discussion below will concentrate on those artifacts which are either common in musculoskeletal imaging, or can be a source of critical misdiagnoses.

Porter et al have classified the majority of MR artifacts into one of four general categories, according to that aspect of image acquisition which is primarily responsible: (1) static magnetic field artifacts, (2) radiofrequency magnetic field artifacts, (3) gradient field artifacts, and (4) motion artifacts.[12]

Static magnetic field artifacts are due either to intrinsic inhomogeneities within the static magnetic field itself, or external environmental influences on this field. Intrinsic field inhomogeneities most frequently result from improper coil design or shimming, or thermal changes in the case of resistive magnets. Environmental interference may be secondary to large metallic objects far removed from the scanner (e.g. passing automobiles), or small metallic objects introduced into the scanner with the patient. Even when metallic objects are not attracted to the magnet and therefore present no potential danger to the patient during scanning, severe image degradation may occur secondary to minute amounts of impurities. Careful screening of patients prior to scanning should be routine so as to avoid image degradation due to such things as snaps on clothing or jewelery. Metal artifact characteristically appears as a region of signal void with adjacent very high-signal intensity.

One should be aware of internally implanted metallic materials such as prostheses and fixation devices prior to image acquisition, as in some cases the artifacts they produce may be minimized or altered so as to allow diagnostic imaging in specific regions of concern. The geometric image distortion from the magnetic field non-uniformity caused by ferromagnetic materials is worse along the frequency encoding axis (Figure 1.9). It is therefore essential for the radiologist to have access to plain radiographs at the time of image acquisition.

It is not unusual to encounter significant artifacts in images of patients who have undergone prior orthopedic procedures during which no metallic foreign bodies were implanted and whose radiographs display no evidence of retained metal. Heindel et al have shown that these artifacts may be caused by minute metallic particles from surgical instruments.[14] We have also encountered artifacts, at times quite severe, in patients who have undergone removal of internal fixation devices. In these cases, the metallic fragments may have been left at the time of operation, or may have been left within pin, nail, rod or screw holes after removal of the appliance (Figure 1.10).

Heindel et al have shown that methylmethacrylate cement does not cause an MR artifact, and is imaged as a region of signal void (Figure 1.11).[14]

The second type of artifact, radiofrequency field artifact, is produced because MRI uses RF pulses which are within overlapping frequencies from many common sources, including commercial radios, computers, and a variety of motors. While great efforts are made to shield scanning systems from these extraneous sources, interference is not uncommon. The appearance of RF interference will depend on the specific frequency and the bandwidth of the outside source. In cases of wide bandwidth interference, the entire image will be distorted. A narrow bandwidth source will be evident in the image as a line perpendicular to the frequency-encoding axis. The exact position of this line will be determined by the frequency of the noise, the resonant frequency of the scanning system, and the strength of the frequency-encoding gradient.

Gradient field artifacts arise within the three gradients used to construct the two-dimensional image. In general, the phase-encoding axis is the most resistant to artifacts which are due to inherent inhomogeneities within the static magnetic field. Image quality is therefore usually optimized if the chosen pulse sequence aligns the phase-encoding axis along the main magnetic field of the system.

Two common gradient-induced artifacts include chemical shift misregistration artifact and aliasing. Chemical shift misregistration artifact is found at the interface between tissues with a high fat content and those with a high water content. Because the protons in fat are more tightly bound than those in water, their resonance frequencies differ. The effect is a localized misregistration in the formation of the image at the fat/water junction. Dark and light bands are found at this interface, which are offset along the frequency-encoding axis. In musculoskeletal imaging, chemical shift misregistration artifact is often seen at the interface between facial planes containing loose areolar tissue and adjacent muscle. This effect is more pronounced at high field strengths.

Aliasing, or wrap-around artifact, is found in situations where the body part being imaged is substantially larger than the field of view chosen for image acquisition (Figure 1.12). Those parts of the subject which are outside the imaging field will be 'folded back' in the image, and superimposed on the area of interest. Filters are often used to eliminate this artifact from the frequency-encoding axis. Therefore, aliasing in clinical MR images is usually found only along the phase-encoding axis.

Motion artifact is probably the most common cause of MR image degradation. Unlike CT scanning, where one image is acquired at a time, MRI data are incorporated into all images simultaneously. For this reason, any type of motion degrades all images in a given imaging sequence. Because of differing sensitivities of the MR image axes to motion, no matter what direction the motion takes, motion artifacts occur along the phase-encoding axis. In situations where it is either impractical or impossible to suppress movement, scans may be planned so as to avoid obscuring specific regions of interest.

There are several unavoidable types of physiologic movement, such as cardiac, respiratory, and vascular motion. Cardiac and respiratory motion may degrade image quality when scanning the spine, but are generally of no consequence in imaging extremities. Vascular flow may at times be evident in imaging of extremities. Its pulsatile nature may cause linear bands of noise along the phase-encoding axis. Flow may be evident as foci of very high-signal intensity within the vessel, an effect that is seen when unsaturated blood enters the imaging region. Because this volume of blood has not previously been subjected to RF pulses, it has a higher signal than the blood in the central part of the imaging volume. As the center of the scan is approached, the high-signal intensity representing flow will no longer be evident. This type of artifact may be eliminated by techniques which use RF pulses to saturate the spins on all sides of the imaging volume.

Patient motion, even though seemingly minimal, can serve to render the examination useless. Every effort should be made to explain to the patient the importance of being as still as possible, and to assure his or her comfort in order to avoid all random motion during image acquisition.

Conclusion

As experience with MRI increases, the applications of this modality to the musculoskeletal system continue to soar. It is not unreasonable to consider that MRI will eventually replace most invasive diagnostic procedures, including imaging-guided biopsy in selected cases. However, because of the possibility of obscuring diagnostic information by the wrong choice of pulsing sequences or imaging planes, MRI provides an additional challenge. It is essential for the radiologist to be familiar with the general principles governing the appearance of MR images in order to provide the maximum amount of information to the referring clinician. An appropriate clinical history, preliminary plain radiographs, and knowledge of the suspected pathology are essential for appropriate image acquisition and interpretation. It should always be kept in mind that the trade-off for the high sensitivity of MRI in the detection of tissue abnormalities is a lower specificity.

In most instances, a full evaluation of suspected musculoskeletal pathology (especially tumor and infection) requires both T_1- and T_2-weighted imaging. In these situations, marrow abnormalities demand T_1-weighting for detection, and the full extent of soft-tissue pathology can only be delineated with T_2-weighted imaging.

References

1 BALTER S, An introduction to the physics of magnetic resonance imaging, *Radiographics* (1987) **7**:371–83.
2 FULLERTON GD, Magnetic resonance imaging signal concepts, *Radiographics* (1987) **7**:579–96.
3 PAVLICEK W, MR instrumentation and image formation, *Radiographics* (1987) **7**:809–14.
4 MERRITT CRB, Magnetic resonance imaging – a clinical perspective: image quality, safety and risk management, *Radiographics* (1987) **7**:1001–16.
5 FULLERTON GD, Basic concepts for nuclear magnetic resonance imaging, *Magn Reson Imaging* (1982) **1**:39–55.
6 PYKETT IL, NEWHOUSE JH, BUONANNO FS et al, Principles of nuclear magnetic resonance imaging, *Radiology* (1982) **143**:157–68.
7 YOUNG SW, *Nuclear magnetic resonance imaging: basic principles.* (Raven Press: New York 1984.)
8 CURREY TS, DOWDEY JE, MURRY RC (eds), Nuclear magnetic resonance. In *Christensen's introduction to the physics of diagnostic radiology*, 3rd edn. (Lea and Febiger: Philadelphia 1984.)
9 RUBIN JI, GOMORI JM, GROSSMAN RI et al, High-field MR imaging of extracranial hematomas, *AJR* (1987) **148**:813–17.
10 PUSEY E, LUFKIN RB, BROWN RKJ et al, Magnetic resonance imaging artifacts: mechanism and clinical significance, *Radiographics* (1986) **6**:891–911.
11 BELLON EM, HAACKE EM, COLEMAN PE et al, MR artifacts: a review, *AJR* (1986) **147**:1271–81.
12 PORTER BA, HASTRUP W, RICHARDSON ML et al, Classification and investigation of artifacts in magnetic resonance imaging, *Radiographics* (1987) **7**:271–87.
13 PATTON JA, KULKARNI MV, CRAIG JK et al, Techniques, pitfalls and artifacts in magnetic resonance imaging, *Radiographics* (1987) **7**:505–19.
14 HEINDEL W, FRIEDMANN G, BUNKE J et al, Artifacts in MR imaging after surgical intervention, *J Comput Assist Tomogr* (1986) **10**:596–9.

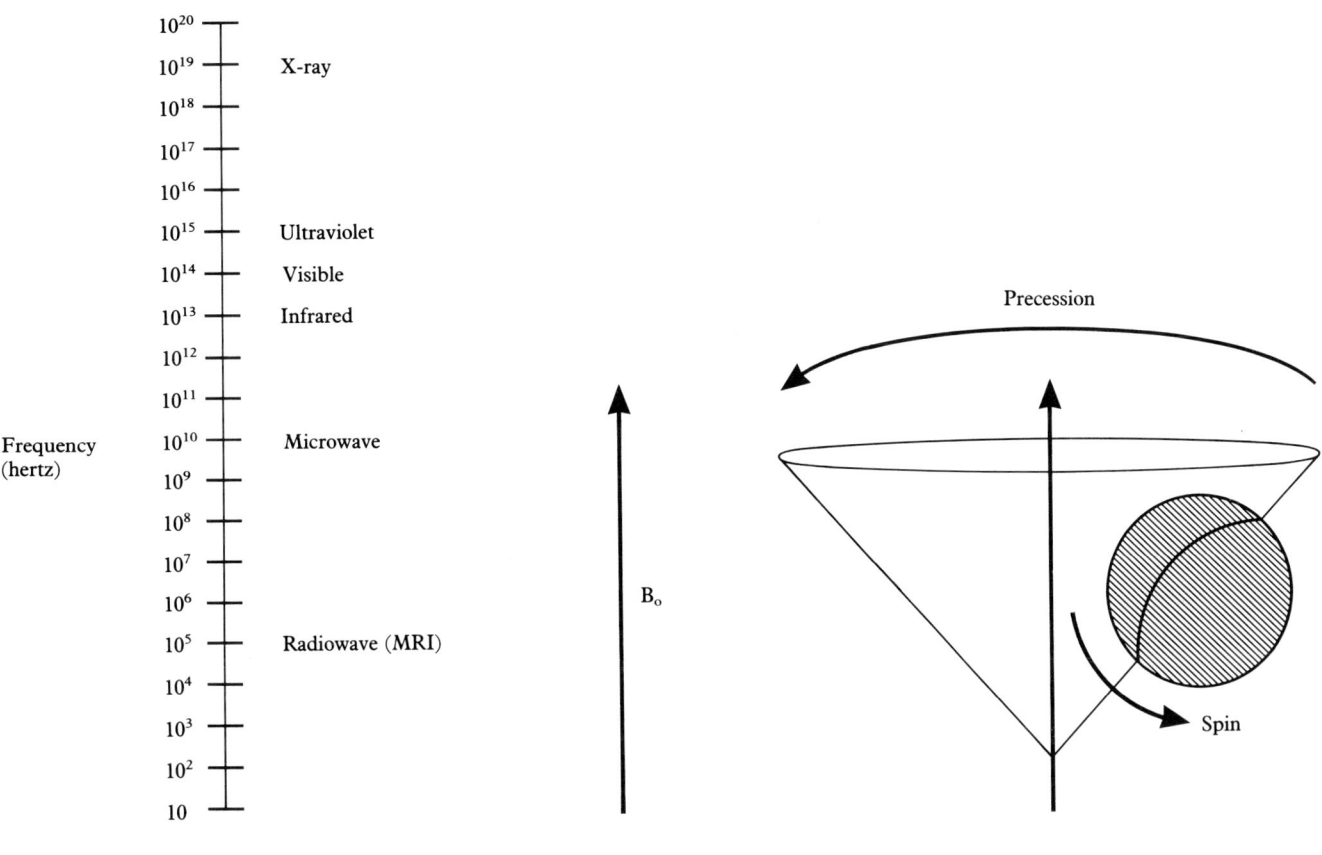

Figure 1.1

The electromagnetic spectrum. The 'windows' for imaging the human body are at the 2 extremes of the spectrum, with high energy X-rays at one extreme, and low energy radiowaves at the other.

Figure 1.2

Precession of spinning proton about static magnetic field (B_o).

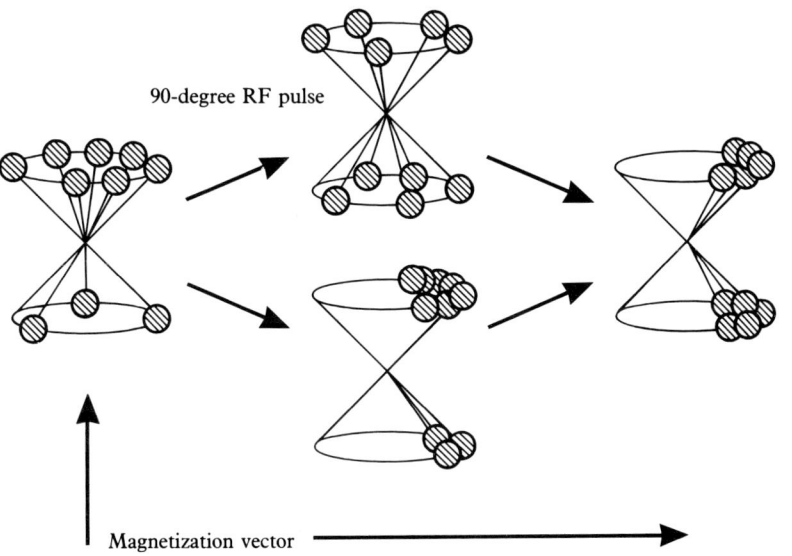

Figure 1.3

Effect of 90-degree RF pulse is to add energy to the system and bring the spins into phase.

18 MRI atlas of the musculoskeletal system

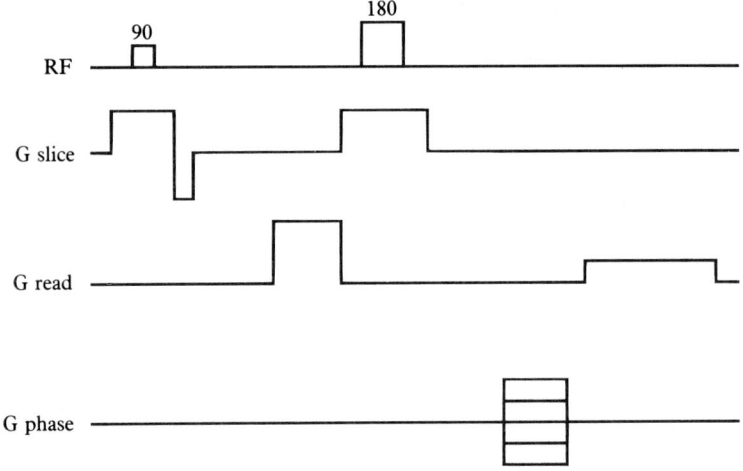

Figure 1.4

Standard spin-echo pulse sequence diagram showing timing for application of slice select, frequency (read), and phase-encoding gradients.

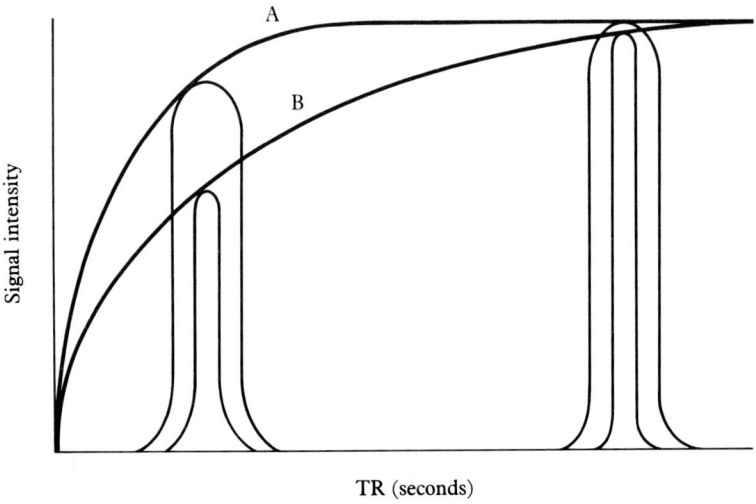

a

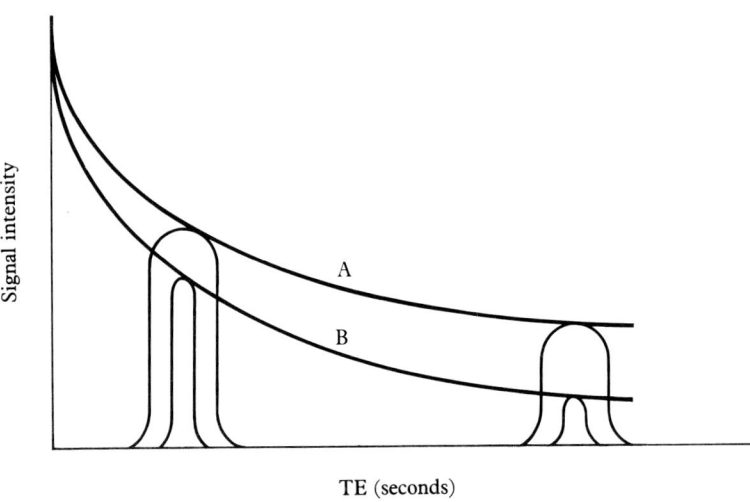

b

Figure 1.5

(**a**) Effect of TR and tissue T_1 values on image contrast. Upper curve (A) represents a substance with a short T_1 as compared to the lower curve (B). For pulse sequences with a relatively short TR, substance A will have a shorter T_1 (higher signal intensity) than material B. (**b**) Effect of TE and tissue T_2 values on image contrast. Substance A (upper curve) has a longer T_2 than substance B. For pulse sequences with a relatively long TE, substance A will have a higher signal intensity than substance B.

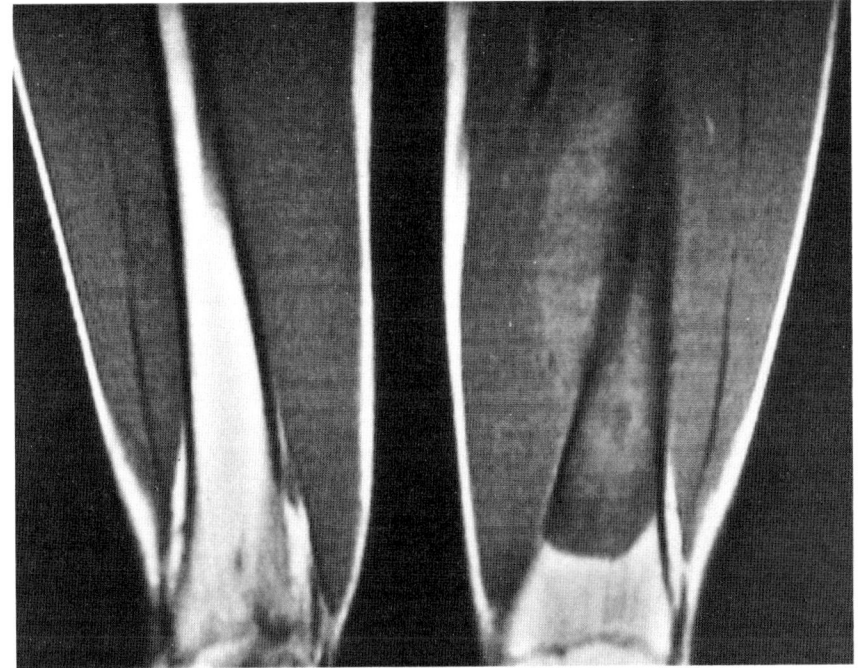

a

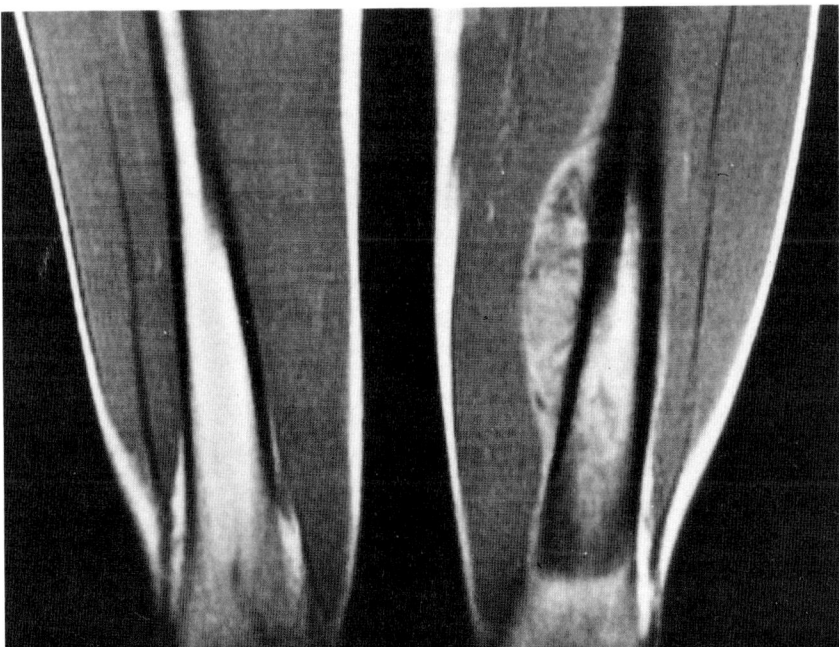

b

Figure 1.6

Effect of varying TE and TR on tissue contrast. A 17-year-old man with osteogenic sarcoma of left distal femur. (**a**) Coronal SE 300/28 image. Intra- and extraosseous tumor manifests intermediate-signal intensity. T_1-weighting is ideal for defining marrow extent of the tumor. (**b**) Coronal SE 500/28 image. Although TR was only increased from 300 to 500, tumor signal intensity has markedly increased. (**c**) Axial SE 2000/56 image. Tumor now has a very high-signal intensity. Soft-tissue extent is readily apparent due to marked differences in tissue contrast between normal muscle and tumor. Because tumor signal intensity is similar to marrow fat, infiltration of marrow by tumor cannot be appreciated.

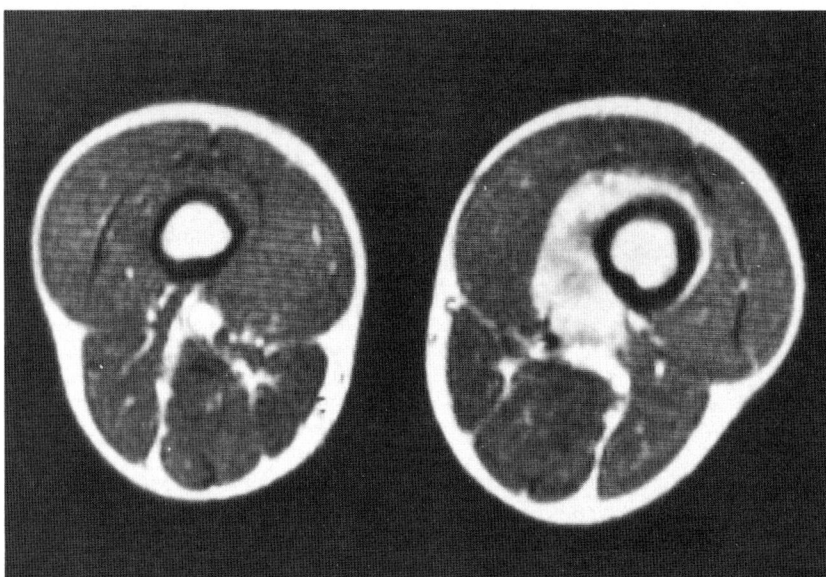

Figure 1.6 *continued*

c

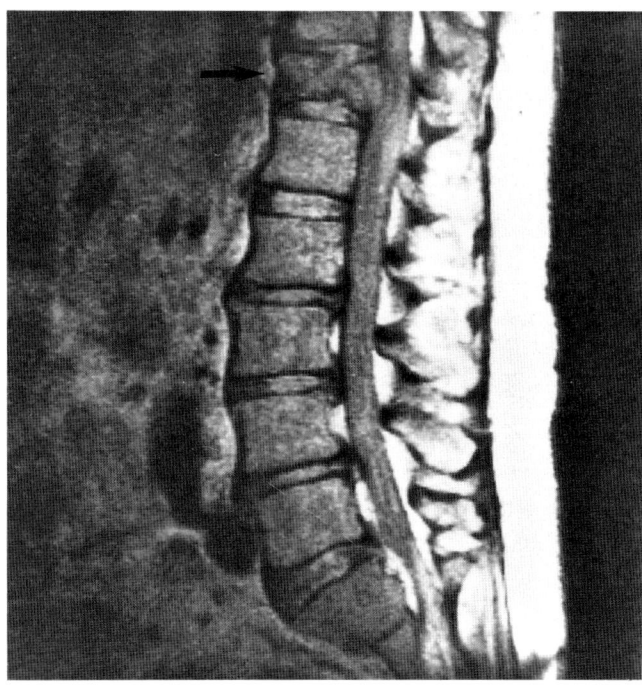

a

Figure 1.7

A 54-four-year-old male truck driver with sudden onset of back pain. Radiographs revealed collapse of T12 vertebra. (**a,b**) Sagittal double-echo images (**a** = SE 1500/35, **b** = SE 1500/70). Collapsed T12 vertebra (arrow) is appreciated, as is retropulsed fragment and compression on thecal sac. Bone marrow signal intensity is homogeneous. (**c**) Sagittal T_1-weighted image (SE 400/30). Multiple foci of abnormally low-signal intensity are now evident within bone marrow. This was later proven to represent metastatic disease from previously undiagnosed lung carcinoma. (*Courtesy of Livia Bohman, Los Angeles, CA*)

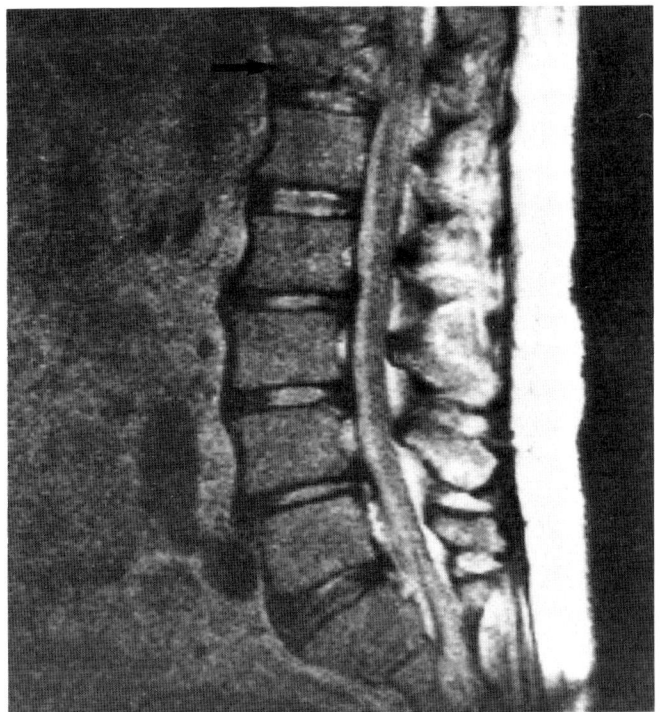

b

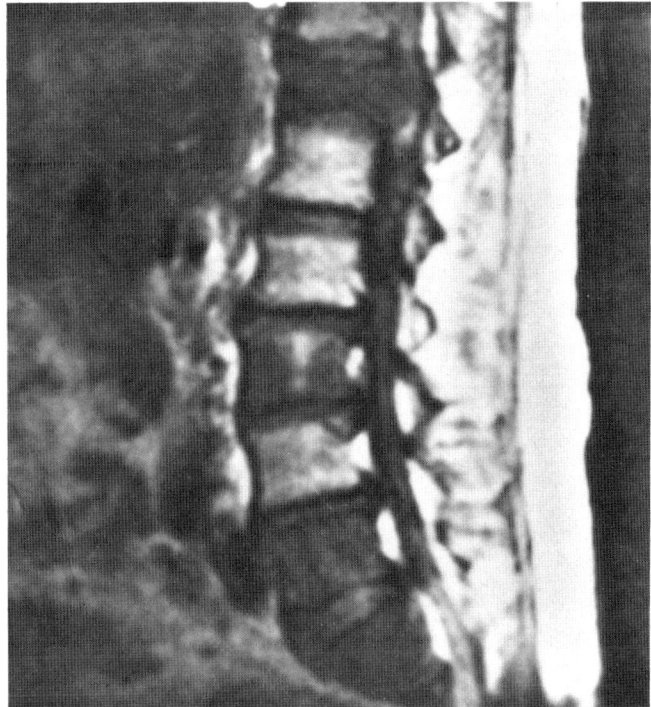

c

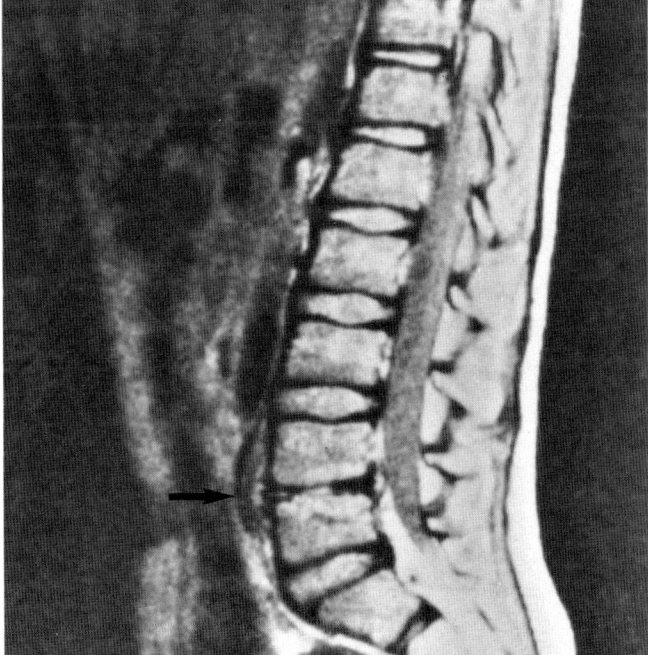

a

Figure 1.8

An 11-year-old boy with back pain. (**a,b**) Sagittal double-echo images (**a** = SE 1500/35, **b** = SE 1500/70). Superior endplate of L5 vertebra is indistinct, and L4–5 disk displays irregular contour. Anterior longitudinal ligament is displaced anteriorly (arrow). No abnormalities are appreciated in the adjacent bone marrow. (**c**) Sagittal T_1-weighted image (SE 400/30). Abnormal signal intensity is now appreciated throughout entire L4 and L5 vertebra. This finding is non-specific, and could represent either infection or edema. Subsequent biopsy following antibiotic therapy revealed only chronic inflammatory tissue, and cultures were negative. (*Courtesy of Livia Bohman, Los Angeles, CA*)

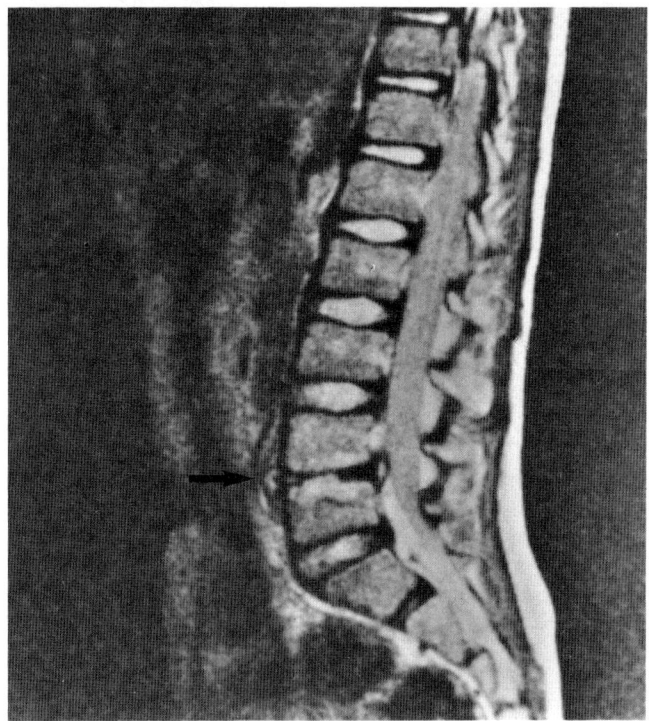
b

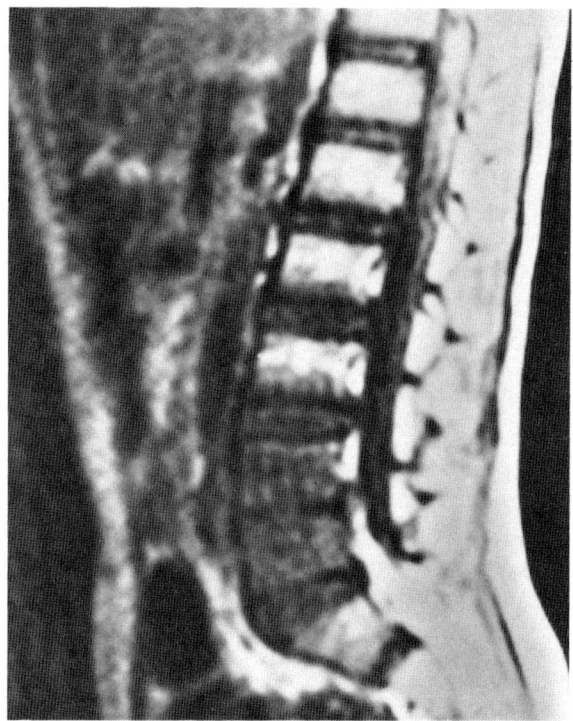
c

Figure 1.8 *continued*

Figure 1.9

A 45-year-old man with healed supracondyler femoral fracture. Blade-plate had been removed, but 1 screw was left in place. (**a**) Anteroposterior radiograph showing screw traversing femoral condyles. (**b**) Sagittal image (SE 800/30) for which the sagittal plane was used for the frequency-encoding axis. Artifact from screw primarily runs in an anterior-to-posterior direction. (**c**) Sagittal image (SE 800/30) for which the coronal plane was used for the frequency-encoding axis. Artifact from screw primarily runs in a superior-to-inferior direction.

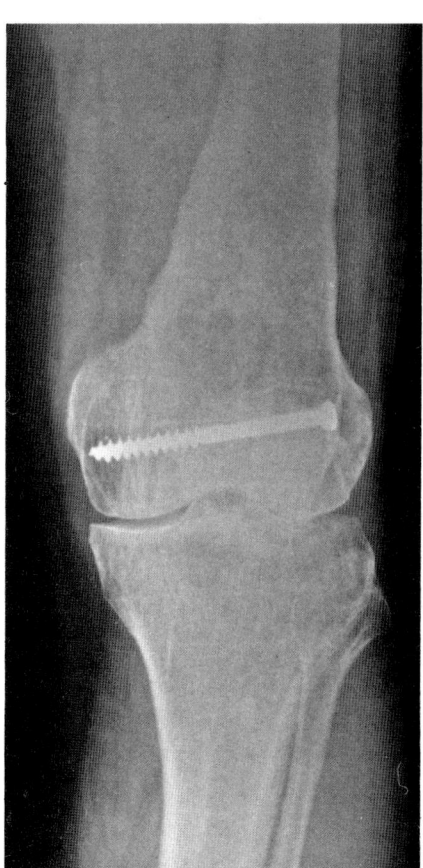
a

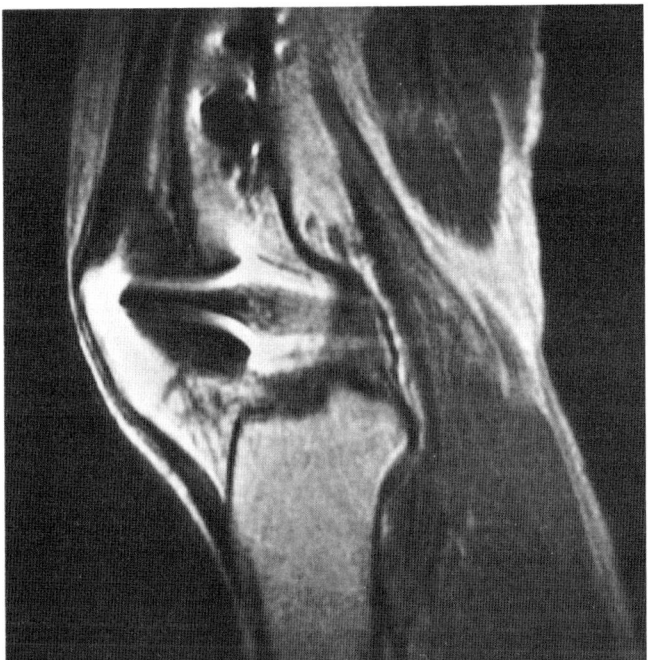

b

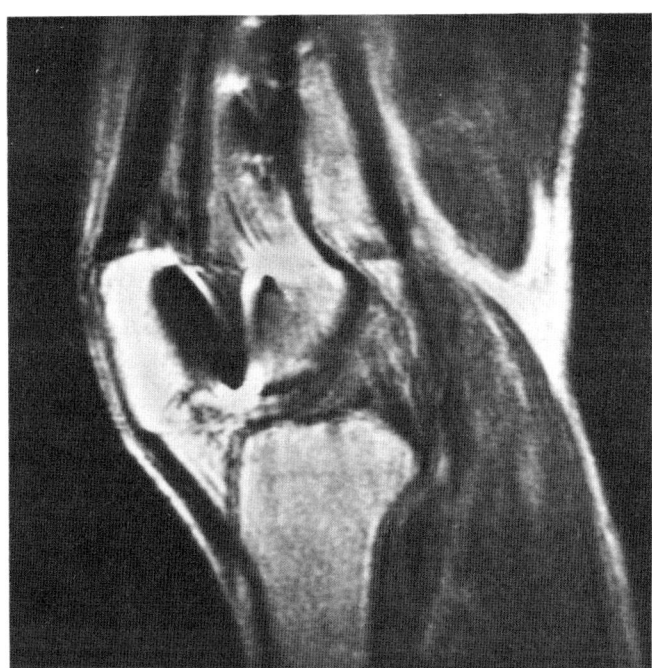

c

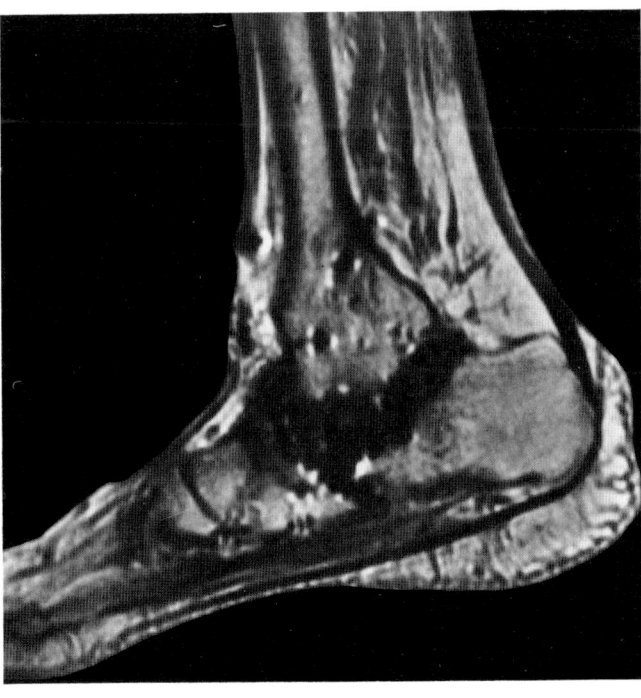

Figure 1.10

A 69-year-old woman following triple arthrodesis and removal of hardware (SE 700/30). No metal was evident radiographically. Multiple areas of signal void and adjacent high-signal intensity are present in areas corresponding to previous screws as well as in anterior ankle soft tissues.

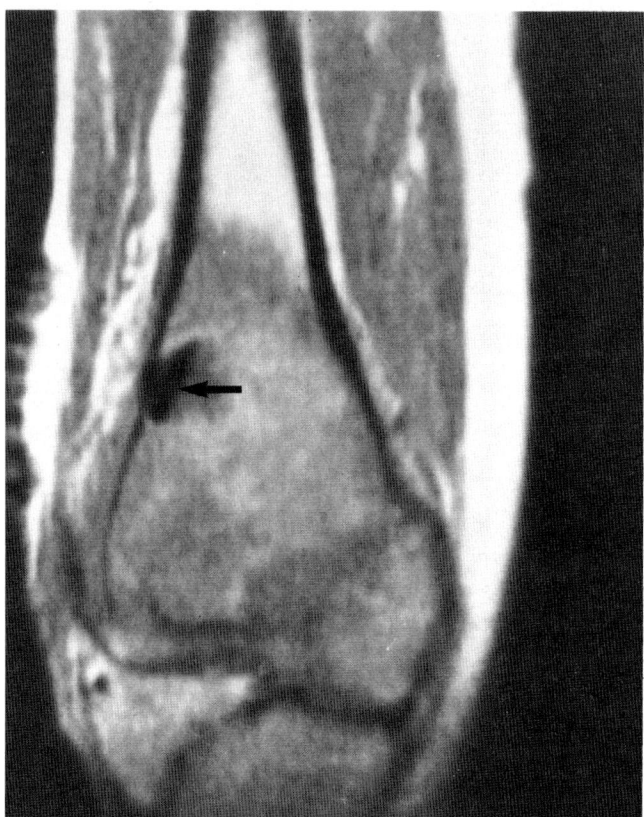

Figure 1.11

A 30-year-old man with giant cell tumor of distal femur. This image (SE 500/28) was acquired after open biopsy. Plug of methylmethacrylate (arrow) was used to occlude cortical defect made at the time of surgery. Bone cement is imaged as a focus of signal void.

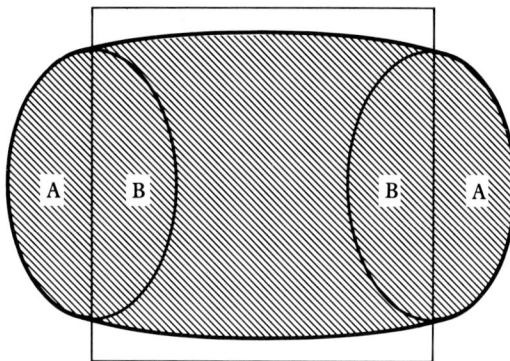

Display matrix

Figure 1.12

Aliasing artifact. The areas represented as A are outside the imaging field of view, and will be superimposed on the image in the area marked B.

2
The spine

Robert B Lufkin

A variety of techniques have been developed for the evaluation of disorders of the spine. Many of these techniques are associated with the hazards of ionizing radiation (plain radiography, CT), and some also require the use of potentially noxious radio-opaque contrast material (myelography, CT myelography, diskography). Although magnetic resonance imaging (MRI) is still young, because it provides superior soft-tissue contrast resolution compared to CT and is noninvasive it has already come into extensive use for the evaluation of spinal abnormalities. While other imaging modalities may be necessary to complement MRI in complex or confusing situations, MRI deserves to be the primary imaging modality for the majority of cases, and the sole imaging technique for many of them.

Whereas older imaging techniques for the evaluation of the spine are in a state of relative technological maturity, MRI is still evolving. Ancillary techniques, such as the use of intravenous paramagnetic contrast agents to enhance differences in signal intensity, and fast scanning strategies, are currently under investigation and may increase the effectiveness of MRI.[1] This chapter addresses the current role and future application of MRI in the evaluation of the spine.

Technique and normal anatomy of the spine (*Figures 2.1–2.3*)

The technique for imaging the spine with MRI requires optimization of signal-to-noise ratio, spatial resolution, contrast resolution, and scan plane. Because these various factors are interrelated, alteration of one will usually result in changes to the others.

A number of techniques have been used to improve the signal-to-noise ratio in an image, including increasing the number of excitations to allow signal averaging, and reducing the width of the image band. However, the most significant technique in this respect has been the use of surface coils (Figure 2.1). Because of the smallness of the neck and cervical spine relative to the size of body coils, and the relative proximity of the thoracic and lumbar spine to the surface of the body, the use of either planar or circumferential surface radio receiver coils has been an invaluable aid to improved image quality.[2-4]

While surface coils themselves do not directly affect spatial or contrast resolution, the increase in the signal-to-noise ratio that they provide may be translated into higher spatial or contrast resolution as well as a shorter scan time. Surface coils can be used in all pulse sequences, including field echo, and may also be used during oblique image formation. They may be employed with both high- and low-field-strength systems. With low-field-strength systems, surface coils allow significant improvement in the signal-to-noise ratio without the production of the image artifacts so common to high-field-strength systems.

High spatial resolution is a priority in MRI studies of the spine because of the fine anatomic detail that must be depicted. In images with a sufficient signal-to-noise ratio, spatial resolution may be improved through the use of high-gradient amplitudes to decrease the field of view. An increase in the number of phase-encoding steps as well as finer sampling in the frequency axis result in a larger matrix, which implies higher spatial resolution for a given field of view. In general, slice thicknesses ranging from 3 to 5 mm are optimal for imaging the spine. When the echo time

(TE) is long enough, sinc (sin x/x) RF pulse profile wave forms are used to allow contiguous slices. Because high in-plane spatial resolution is also important, a pixel size of 1 mm or less is desirable. Lower in-plane spatial resolution and thicker slices are tolerated with more T_2-weighted sequences because of the inherent lower signal-to-noise ratio in these acquisitions.

Image contrast may be manipulated through alterations in repetition time (TR) and TE. With the latest field-echo imaging techniques, image contrast may also be manipulated with an additional parameter, the flip angle (theta). The final combination of contrast parameters in the pulse sequence and the choice of imaging planes complete the protocols for spine imaging.

In general, T_1-weighted images (TR 300 to 500 msec, TE 20 to 30 msec) in the sagittal plane are used as the first survey sequence for studying the cervical, thoracic and lumbar spine (Figures 2.2–2.3). Some investigators advocate the use of an asymmetric double echo (TR 1000 msec, TE 20 to 70 msec) for T_1- and T_2-weighted images of the spine. However, the time-saving advantage of this simultaneous acquisition sequence is offset by the relatively small variations in contrast between T_1- and T_2-weighted images.[5]

Sagittal T_1-weighted images provide the best anatomic detail, and are excellent survey images with which to begin the examination of patients with suspected spinal pathology.[6] Midline sections show high-signal intensity from the fat-containing marrow in the vertebral bodies, surrounded by low-signal, proton-poor cortical bone. Normal vertebral marrow is usually homogeneous in signal intensity, although occasional islands of acellular marrow may show higher signal intensity.[7] In infants, a transverse line of intermediate signal intensity crossing each vertebral body represents the horizontal cartilaginous plate, the plane of junction between the two mesenchymal sclerotomes which have fused to form the vertebral body.[8] With maturation, this intermediate signal zone decreases in size and undergoes ossification. In adults, the intersegmental veins persist as small areas of flow void in the basivertebral plexus.

The intervertebral disks appear as homogeneous structures, with a signal intensity slightly less than the adjacent vertebral bodies. The annulus, ligaments, posterior cortex, and dura are all of low-signal intensity on T_1-weighted images and are usually indistinguishable. A line of low-signal intensity through the base of the dens represents the subdental synchrondrosis and is present in many normal individuals. Because this normal variant does not extend to the adjacent cortical bone, it may be distinguished from a fracture.

Cerebrospinal fluid (CSF) is of low-signal intensity on T_1-weighted images and provides excellent contrast with the adjacent spinal cord and nerve roots. Because the nerve roots are surrounded by contrasting high-intensity fat as they pass laterally through the neural foramina and bony canals, they are easily seen. The epidural veins may appear as linear areas of signal void anterosuperior to the nerves. The bony canals are well seen *en face* in the lumbar region in standard sagittal images. The use of oblique sagittal views through the canals of the cervical spine to demonstrate the course of the cervical nerves is under investigation.

A high-intensity region posterolateral to the vertebral bodies represents a combination of epidural fat and blood flow through the venous plexus. In the cervical region, where this area is primarily lateral, elevation of the plexus in midline sagittal images is a sensitive indicator of disk protrusion.[9]

Images are next obtained in the sagittal plane with T_2-weighting (TR 2000 to 3000 msec, TE 60 to 120 msec) (Figure 2.2). As mentioned previously, an alternative sequence is occasionally obtained as the second echo of an asymmetric double-echo sequence (TR 1000, TE 20, 70 msec). With T_2-weighted sequences the cortical bone of the vertebral bodies continues to be of low-signal intensity while the marrow remains fairly high in intensity, although occasionally the basivertebral veins may be of even higher intensity due to flow phenomena. With this pulse sequence, metastases with long T_2-relaxation times may occasionally be masked by the high-signal intensity of the fatty marrow.

The normally hydrated intervertebral disk shows increased signal intensity centrally in the vicinity of the nucleus pulposus, surrounded by lower signal in the region occupied by the annulus fibrosus. However, because of the changes in their appearance that accompany changes in pulse sequence, it is believed that these signals do not correspond exactly to the nucleus pulposus and annulus fibrosus.[10-11]

The cervical spinal fluid assumes a high-signal intensity in T_2-weighted images due to its long T_2-relaxation time. This allows sensitive identification of extradural structures such as protruding disks and/or osteophytes extending into the canal. Because of transmitted pulsations the CSF may have areas of low signal due to turbulence and/or other flow effects, creating the appearance of filling defects. These may be particularly troublesome in images with longer echo delays and in those obtained with high magnetic field strength systems. Various strategies to reduce or eliminate the effects of such motion will be discussed below. While the increased CSF signal in T_2-weighted images may obscure CSF/parenchymal interfaces, these images are valuable for identifying intramedullary abnormalities such as edema, demyelination, gliosis, and neoplasia.

Relatively T_1-weighted images (TR 1000, TE 20 to 30 msec) are next obtained in the axial plane (Figures

2.2–2.3) at levels through the disk spaces and/or through areas of abnormality identified from the sagittal images. Thus the standard evaluation of the spine includes sagittal T_1- and T_2-weighted images through the area of interest, followed by angled axial oblique sections through the disks and/or through areas of pathology. The use of multiple-slice, multiple-angle, variable interval, non-orthogonal (MAVIN) sections for the axial placement greatly shortens the scan time required for the evaluation of intervertebral disk disease.[12] This technique allows oblique scan planes to be set individually for each section, so that paraxial planes orthogonal to the disk may be obtained in a single multislice sequence (Figure 2.3). The relatively long TR of this sequence allows an adequate number of slices to be obtained to cover the areas of interest. The axial images show the spinal canal to be a low-intensity region lined by high-intensity epidural fat. The ligamentum flavum in axial views is intermediate in signal intensity and easily distinguishable from the high-intensity epidural fat and low-intensity adjacent lamina. The apophyseal joints appear as linear structures of intermediate intensity due to their combination of hyaline cartilage and synovial fluid. The spinal cord and nerve roots can be identified within the spinal canal as structures of relatively high signal. Nerve roots surrounded by high-intensity fat may be followed into the neural foramina and neural canals. The axial plane is particularly useful for identifying lateral disk disease and assessing the lateral extent of pathology. The low-intensity signals of the nerves may be traced laterally until they form ganglia. Epidural veins are low-intensity structures adjacent to the nerves. Despite the superb definition of the neural canals on sagittal and oblique images, significant disk disease may be overlooked if axial images through the area of abnormality are omitted.

Magnetic field gradient refocusing techniques are presently under investigation for spine imaging (Figure 2.2). These pulse sequences allow fast scanning and provide increased sensitivity to flow and T_2^* magnetic susceptibility phenomena. They are presently used as an adjunct to spin-echo pulse sequences and in situations involving flow and/or the presence of acute or chronic hemorrhage. They also represent a potential tool to distinguish between disk and osteophyte protrusion into the spinal canal. The effects of phase cancellation with gradient echoes of fat and water signals are also of interest in defining some states of marrow pathology.[13] The role of these new pulse sequences in spine imaging still remains to be defined.

Artifacts (*Figures 2.1, 2.2, 2.4*)

There are a number of image artifacts which are of clinical significance.[14] The first general class of these relates to chemical shift, and arises from the fact that fat and water protons resonate at slightly different frequencies due to the effects of their local magnetic environment. The most common type of chemical shift artifact results in a spatial misregistration along the frequency-encoding axis. Since the chemical shift frequency separation is proportional to the magnetic field strength, this effect increases with higher field strength systems.[15] This artifact is manifested on spine images as artifactual black and white lines at fat/water interfaces oriented along the frequency-encoding axis. Although the artifact may occur at any fat/water interface (Figure 2.4), the lines are most evident in sagittal T_1-weighted images where they produce asymmetric thicknesses of the vertebral end plates.

A second type of chemical shift artifact occurs primarily with field-echo images. This is because magnetic field gradients refocus the fat and water protons differentially, based on their Larmor frequencies, as compared with the more uniform refocusing that takes place with the 180-degree RF pulse used in spin-echo sequences. This artifact results in a potential cancellation between the fat and water signals.[13] Depending on the TE selection, variation in fat/water signal cancellation occurs in voxels which contain both fat and water protons. In spine imaging this is most evident in marrow, which may be completely void of signal with full phase cancellation effects. These chemical shift errors also result in low-signal black lines at fat/water interfaces in surrounding structures. Unlike the first type of chemical shift misregistration errors, which occur primarily in the frequency axis, chemical shift phase cancellation errors are found in both the phase-encoding and frequency axes on gradient-echo imaging.

Because of the propensity for motion and flow artifacts to be propagated in the form of image harmonics along the phase-encoding axis, it is important to select the appropriate direction for phase encoding in imaging the spine[3] (Figures 2.1–2.2). In the sagittal plane, for example, vertical phase encoding allows a considerable number of motion artifacts from anterior structures such as the large blood vessels, gut and larynx to be projected over the area of interest in the spinal region. Although this projection occasionally induces aliasing artifacts with extremely large fields of view, the improvement in the definition of the spinal contents justifies this approach. In the axial plane, horizontal phase encoding moves the flow and motion artifacts off the spine by projecting them laterally.

Although judicious selection of the phase-encoding axis will effectively project motion and flow artifacts of surrounding structures off of the spine, significant flow and motion artifacts are still produced from the transmitted pulsations of the cerebrospinal fluid within the canal itself. Because of the central location of the spinal canal, simple phase axis alterations do not adequately serve to minimize these artifacts. Instead, more complex suppression strategies have been developed.

Synchronization of the imaging sequence with the cardiac cycle is possible with EKG electrodes or peripheral plethysmographic detectors. This allows acquisition of spine images during the point in the cardiac cycle when the transmitted pulsations are minimal.[16] This strategy effectively minimizes artifacts due to within-view motion or motion occurring between each phase acquisition. The drawbacks to this approach are of course the necessity for additional instrumentation in the form of EEG electrodes and/or peripheral plethysmography as well as the constrained repetition time based on the patient's heart rate.

Because of the difficulty of performing cardiac-gated spine work, many users have adopted a different approach. By using specially designed compensatory or nulling gradients it is possible to remove artifacts caused by constant velocity, acceleration, or even change in acceleration times in the phase expansion of the imaging sequence.[17] Although these strategies limit the minimum echo time possible due to the additional time required for the rephasing gradients, no additional hardware is required and no constraint on TR time occurs. The effects can be dramatic, even though this approach corrects only within-view motion artifacts or artifacts that occur between the 90-degree pulse and the echo. In some situations with high-field scanners and/or long TE sequences where motion artifacts are a particular problem, a combination of cardiac gating and gradient nulling strategies can be used to reduce further motion artifacts.

Developmental abnormalities (*Figures 2.5–2.8*)

The most common developmental anomaly of the spine is the Chiari malformation. The most common type of this anomaly is the Chiari I malformation, in which downward displacement of the cerebellar tonsils is associated with a normal position of the fourth ventricle and remainder of the posterior fossa[18] (Figure 2.5). A considerable variation exists in the normal position of cerebellar tonsils, and displacement of up to 2 mm below the level of the foramen magnum may be seen in asymptomatic individuals.[19] Because of the superior definition of this anatomic region in T_1-weighted sagittal images, MRI is the examination of choice for this problem. The less common Chiari II and Chiari III malformations are also best evaluated with MRI.

Associated conditions such as hydrocephalus and cystic cavities of the spinal cord can be demonstrated in the same examination.[20-22] Cystic changes of the spinal cord may be divided into syringomyelia and hydromyelia. Syringomyelia is a cavitation, without an ependymal lining, developing eccentric to the central canal, and which occasionally communicates with it. Hydromyelia is a dilatation of the central canal, and has an ependymal lining. Because these two, as well as other forms of 'cystic myelomalacia' are difficult to distinguish in MR images, they will be considered together in this discussion.

Syringohydromyelia is associated with Chiari I malformation in up to 50 per cent of cases (Figure 2.6). Its full extent is best evaluated in sagittal T_1-weighted images which maximize CSF/parenchymal contrast. Occasionally, axial T_1-weighted images may be needed to confirm a suspected abnormality seen on the sagittal images (Figure 2.7).

Dysraphic states are also eminently suitable for evaluation with MRI.[8] The soft-tissue abnormalities of meningocele, myelomeningocele, and tethered cord are well depicted in T_1-weighted images (Figure 2.8). The high-intensity signal of the fat in lipomas is also well seen.

MRI is also excellent for the evaluation of diastematomyelia and associated abnormalities. A major drawback, however, is the inability of MRI to differentiate the various components of the spur, such as cortical bone, fibrous tissue and other structures. When marrow is present within the bony septum, it is seen as a high-intensity signal; however, marrow is present only in a minority of cases.[23]

Degenerative changes (*Figures 2.9–2.11*)

The normal nucleus pulposus in young individuals contains 85 to 95 per cent water, resulting in a high-intensity signal in T_2-weighted images. With normal aging and degenerative disease, the fluid of the nucleus pulposus may diminish to 70 per cent or less. This is associated with a shortening of T_2-relaxation time and a lessening of signal intensity in T_2-weighted images. The degeneration is associated with a loss of disk height. In the lumbar spine, not all degenerated disks herniate, but most herniated disks are degenerated.[24] This is not true in the cervical area, and acute cervical disk herniations of non-degenerated disks are frequently seen. Focal elevation of the prevertebral venous plexus seen in midline sagittal images of the cervical spine may provide a sensitive indicator of disk protrusion into the spinal canal (Figure 2.9).

The differentiation between disk herniation and bulge is made through the same morphologic criteria as for CT scanning. Herniation is a focal extension of the disk due to a disruption of the annulus fibrosus, and is best recognized in T_1-weighted images. Disk bulge occurs with an intact annulus, and results in a gradually rounded contour extending a short distance into the spinal canal.[25]

When portions of disk material herniate through the posterior longitudinal ligament and extend above or below the disk space within the spinal canal they are referred to as extruded fragments (Figures 2.10–2.11). Axial and sagittal images are both valuable in identifying disk bulge, herniation and extruded fragments. A drawback of MRI is its low sensitivity to calcification in herniated disks; CT scanning is more sensitive to disk calcification. Nevertheless, the calcification may have little or no clinical significance.

Osteophytes may also produce extradural defects as they extend into the spinal canal. They are often recognized on MRI by the presence of high-intensity fatty marrow within the protruding structures. In this situation, it is a simple matter to distinguish between the protrusion of an osteophyte and that of a herniated disk by their differing signal intensities in T_1-weighted images. However, if the osteophyte is composed entirely of compact, cortical bone the distinction is difficult. It is hoped that gradient-echo imaging may aid in this differentiation by depicting the disk as a high-intensity structure, as opposed to the low-intensity cortical bone of an osteophyte.

Stenosis of the spinal canal is easily evaluated in sagittal and axial images, and is characterized by replacement of epidural fat and a decrease in the overall size of the spinal canal. CT is superior to MRI in defining the anatomy of the bony facets. Although hypertrophic facet disease is usually obvious in MR images, a clear distinction between ligamentum flavum hypertrophy and osteophytosis may be difficult.

In a prospective evaluation of 60 patients with lumbar disk herniation and/or stenosis of the spinal canal, MRI provided information equivalent to that of CT and myelography.[24] In a similar study in patients with disease in the cervical region, the accuracy of the various imaging modalities in depicting the type and location of disease was: CT-myelography 84 per cent, MRI 74 per cent, and conventional myelography 66 per cent.[26] Because it is noninvasive and able to survey the entire cervical region in a single sagittal image study, MRI is the optimal tool for the initial evaluation of these patients. Moreover, it is the superior technique for the evaluation of myelopathy. Although experience is more limited in the thoracic region, MRI is at least as accurate as CT and myelography in the evaluation of thoracic disk herniation.

It is important to remember that significant degenerative disk disease and spondylosis may be present coincidentally in individuals whose spines are undergoing MRI evaluation for unrelated problems. In one MRI study of the cervical spine involving 100 asymptomatic patients, a variety of degenerative changes were found at multiple levels, and the incidence of these changes increased with increasing age.[27] It is noteworthy that 7 per cent of these asymptomatic patients showed frank compression of the spinal cord.

In cases of spondylolisthesis, MRI easily shows the alignment abnormalities in sagittal T_1-weighted images. Although MRI can depict the pars defect in spondylosis, CT is currently more sensitive in detecting the associated bony abnormalities.

Trauma (*Figures 2.12–2.15*)

Experience with MRI of spinal trauma has been limited largely by the restricted access of acutely traumatized patients to MRI facilities. Many older superconducting and air core resistant magnets are relatively intolerant of ferromagnetic-containing life support systems, IV poles and crash carts. With the increasing use of permanent and iron core resistive magnets with iron flux return paths the problems of fringe fields are practically eliminated, and increased MRI of spinal trauma should be possible. The great potential of MRI in the evaluation of spinal trauma lies in its ability to demonstrate the spinal cord and its relationship to fracture fragments, and to accomplish this in any projection, quickly and noninvasively, without moving the patient from the horizontal position.[28]

Early experience has shown that MRI is sensitive to cord contusion and/or hemorrhage. Contusions are visible as high-intensity changes in relatively T_2-weighted images. The blood breakdown products that arise secondary to hemorrhage are visible in T_2-weighted images as the preferential T_2-shortening effects of deoxyhemoglobin and/or hemosiderin, and on T_1-weighted images as the T_1-shortening effects of the proton–electron dipole–dipole interactions of paramagnetic methemoglobin. Due to its increased $T_2^\star$ sensitivity, field-echo imaging will undoubtedly play a role in the detection of acute hemorrhage of the spinal cord.

The chronic sequelae of cord injury, including myelomalacia and frank cystic changes, can be easily monitored with MRI (Figure 2.12). T_1-weighted images are particularly useful for maximizing fluid/cord contrast in the evaluation of patients with post-traumatic syringomyelia.

Vertebral body fractures are recognized in T_1-weighted images by their morphological alterations. Early experience suggests that MRI is useful in distinguishing osteoporotic compression fractures from pathologic fractures resulting from infiltrative processes of the marrow (Figure 2.13). With fractures secondary to osteoporosis, the normal signal intensity of the vertebral body marrow is usually maintained in T_1-weighted images, even though the height of the vertebrae is reduced.

The marrow in the vertebral body is exquisitely sensitive to radiation. As early as one month following a course of radiotherapy there commences a gradual decrease in the normal cellular elements of the marrow, with replacement by lipid. In T_1-weighted images this results in a characteristic increase in marrow signal corresponding in its dimensions to the radiation port. These changes have been detected as long as ten years following therapy.[29-30] In T_1-weighted images, the high-signal intensity of the fatty marrow that follows radiation actually increases the sensitivity of MRI by serving as a contrasting background for recurrent tumor, which usually is manifested as low-intensity foci of marrow replacement.

Postoperative changes

Evaluation of the postoperative spine is a challenge to any imaging technique (Figure 2.14). In this regard, experience with MRI is limited and surgical confirmation often has not been possible. T_1-weighted images in the sagittal and axial planes are useful for defining the extent of postoperative pseudomeningoceles. Arachnoiditis may be identified in T_1-weighted images by an increase in signal intensity from the normally low intensity of the CSF, and, in axial images by a clumping together of the nerve roots (Figure 2.15). The increased signal intensity is presumably due to an increased protein content in association with water binding in the CSF, resulting in T_1 shortening. In general, arachnoiditis is still depicted better with CT or conventional myelography than with MRI. In patients who have received intraspinal pantopaque, characteristic high-intensity foci may be seen in T_1-weighted images.[31] The distinction on MRI between a fibrous scar and a recurrent extruded disk is difficult. Early work suggests that gadolinium-DTPA may enhance preferentially the signal intensity of fibrous scar tissue compared to disk material.

Neoplasia (*Figures 2.16–2.23*)

The majority of neoplastic conditions of the spine, benign and malignant, exhibit similar MR signal characteristics, with prolongation of both T_1- and T_2-relaxation times. Exceptions to this rule are tumors with unusual tissue components such as fat, hemorrhage, or calcification, and those exhibiting flow effects. All tumors are generally best evaluated with a combination of T_1- and T_2-weighted images in both axial and sagittal planes.[32]

Of all the epidural neoplasias, metastasis is the most common. In T_1-weighted images, metastases of the vertebral bodies show focal replacement of normally high-intensity marrow with low-intensity tumor (Figures 2.16–2.17). In T_2-weighted images, metastases have a high-signal intensity, approximating or exceeding that of the marrow. This is true of the majority of cell types, although blastic metastases occasionally are of low-signal intensity in both T_1- and T_2-weighted images. MRI is valuable for assessing the site and extent of a complete block of the spinal canal by metastatic disease, and has virtually replaced myelography for emergent evaluation of this situation. T_1-weighted images clearly show the extension of osseous metastases into the spinal canal by the contrast of the metastases with the adjacent lower intensity CSF.

Hemangiomas of the vertebral body have a somewhat characteristic MR appearance due to their flow effects. While in T_1-weighted images their high-intensity appearance may be confused with focal islands of yellow acellular marrow, in T_2-weighted images they have a striking high-signal intensity due to flow enhancement (Figure 2.18). Primary vertebral tumors with a high calcium content may yield low-intensity signal intensity on both T_1- and T_2-weighted images[33] (Figure 2.19).

Intradural extramedullary tumors represent 55 per cent of all spinal neoplasms. The two most common are meningioma and neurofibroma (Figures 2.20–2.21). The majority have similar MR signal characteristics and are best depicted using T_1-weighted images in which the tumors are well contrasted with the adjacent lower intensity CSF. Both axial and sagittal views are useful to define the full extent of the tumors.[34]

MRI is useful for evaluating patients with intramedullary neoplasia.[35] T_1-weighted images allow the distinction between solid and cystic portions of a tumor, and disclose secondary cystic changes in the spinal cord. T_2-weighted images are valuable for identifying areas of T_2 prolongation due to tumor and/or edema and to differentiate simple syringohydromyelia from syringohydromyelia secondary to tumor. Although ependymomas and astrocytomas represent 90 per cent of intramedullary tumors, the possibility of hemangioblastoma should be considered when a large cyst is found to be associated with one or more draining vessels and a small tumor nodule (Figure 2.22).

Although experience with 'drop' metastases is limited, they appear to be difficult to detect with conventional MRI. However, contrast-enhancing gadolinium-DTPA shows promise of overcoming this limitation, and should simplify the recognition of intraspinal metastases.

Inflammatory disease (*Figure 2.23*)

Small epidural and subdural fluid collections are easily detected with MRI. Although their signal intensities vary according to protein concentration, their relaxation times are generally shorter than CSF. The pannus of rheumatoid arthritis at the C1–C2 articulation is clearly visualized in sagittal images.

In the detection of vertebral osteomyelitis and diskitis, MRI has been shown to be more specific than radioisotope studies and more sensitive than plain film radiography or CT scanning.[36] Vertebral osteomyelitis results in a decrease in the normal high-intensity signal of the marrow in T_1-weighted images and a corresponding increase in T_2-weighted images, and produces irregularity of the adjacent vertebral endplates (Figure 2.24). There may be associated soft-tissue paravertebral masses. Involvement of the intervening disk is characteristic and leads to a loss of disk height, a clue that the process is inflammatory and not neoplastic.

Involvement of the disk results in a prolonged T_2-relaxation time. The resultant increased intensity of the disk in T_2-weighted images allows differentiation from degenerative disk disease, in which the disk has a characteristically low-signal intensity. Similar findings may be seen following chymopapain diskectomy (chemonucleosis). Another advantage of MRI in the evaluation of vertebral osteomyelitis and disk space infection is that the administration of antibiotics, which can lead to false-negative gallium citrate or tagged leukocyte radioisotope studies, has no effect on MRI.

Demyelination due to multiple sclerosis (MS) may be recognized in the cervical region as foci of high-signal intensity on T_2-weighted images.[37] Diagnostic mistakes may occur when the image sections are so thick that the high-intensity CSF overlying the cord is misdiagnosed as a cord lesion. The lesions of MS may also be associated with a focal mass effect which may simulate a primary neoplasm of the cord. In this situation, T_2-weighted images of the brain that demonstrate focal areas of demyelination will reinforce the diagnosis of MS. However, it must not be forgotten that occasional patients may exhibit lesions of MS in the spinal cord but not the brain.

Vascular malformations

Primary vascular malformations of the cord are well depicted with MRI just as they are well depicted in other areas of the central nervous system. Due to high blood flow, the vascular components appear as focal lesions of signal void.[38] The smaller cryptic malformations of the cord appear similar to those in the brain, with a low-intensity peripheral region representing hemosiderin-laden macrophages surrounding a high-intensity central core representing methemoglobin.

Dural arteriovenous malformations are difficult to evaluate with MRI due to their lack of signal contrast with the adjacent CSF. In T_1-weighted images, the low-intensity CSF provides little contrast with the low-intensity signal of flowing blood. In T_2-weighted images, the varied appearance of CSF flow phenomena may mimic the appearance of blood flow in extramedullary vessels. The use of gradient-echo pulse sequences to cause paradoxical enhancement of blood in T_1-weighted images may someday play a significant role in the detection of dural arteriovenous malformations of the spine.

Conclusion

Despite the short time that has elapsed since its inception, MRI has been accepted as the examination of choice for the evaluation of the craniocervical junction and for abnormalities of the spinal cord. The technology has also gained a premier position in the evaluation of spinal degenerative disease and infection.

Although there is less experience with MRI in the evaluation of arteriovenous malformations, postoperative conditions and trauma, it appears to have great potential in the investigation of these disorders. The current major limitations of MRI are its high cost, relatively long scan time and low sensitivity to calcification. Fortunately, these limitations will decrease with the further development of new pulse sequences and faster scanning strategies, and with the diminishing cost of MR scanners and contrast agents. In the near future, three-dimensional Fourier transformation techniques combined with gradient-echo pulse sequences also promise to advance the role of MRI in the study of the spine.

References

1 BYDDER GM, BROWN J, NIENDORF HP et al, Enhancement of cervical intraspinal tumors in MR imaging with intravenous gadolinium-DPTA, *J Comput Assist Tomogr* (1985) **9**:847–51.

2 EDELMAN RR, SHOUKIMAS GM, STARK DD et al, High-resolution surface-coil imaging of lumbar disk disease, *AJR* (1985) **144**(6):1123–9.

3 REICHER MA, GOLD RH, HALBACH VV et al, MR imaging of the lumbar spine: anatomic correlations and the effects of technical variations, *AJR* (1986) **147**:891–8.

4 LUFKIN R, VOTRUBA J, REICHER M et al, Solenoid surface coils in magnetic resonance imaging, *AJR* (1986) **146**:409–12.

5 HYMAN RA, EDWARDS JH, VACIRCA JJ et al, 0.6 T MR imaging of the cervical spine: multislice and multiecho techniques, *AJNR* (1985) **6**:229–36.

6 NORMAN D, MILLS CM, GRANT-ZAWADZKI M et al, Magnetic resonance imaging of the spinal cord and canal: potentials and limitations, *AJR* (1983) **141**:1147–52.

7 HAJEK PC, BAKER LL, GOOBAR JE et al, Focal fat deposition in axial bone marrow: MR characteristics, *Radiology* (1987) **162**:245–9.

8 WALKER HS, DIETRICH RB, FLANNIGAN BD et al, Magnetic resonance imaging of the pediatric spine, *Radiographics* (1987) **10**(7): 1129–52.

9 FLANNIGAN B, LUFKIN R, RAUSCHNING W et al, MRI of the cervical spine, *AJR* (1987) **148** (4): 785–90.

10 AGUILA LA, PIRAINO DW, MODIC MT et al, The intranuclear cleft of the intervertebral disk: magnetic resonance imaging, *Radiology* (1985) **155**:155–8.

11 PECH P, HAUGHTON VM, Lumbar intervertebral disk: correlative MR and anatomic study, *Radiology* (1985) **156**:699–701.

12 REICHER M, LUFKIN R, SMITH S et al, Multiple-angle variable-interval non-orthogonal MRI, *AJR* (1986) **147**:363–6.

13 WEHRLI FW, PERKINS TG, SHIMAKAWA A et al, Chemical shift-induced amplitude modulations in images obtained with gradient refocusing, *Magn Reson Imaging* (1987) **5**:157–8.

14 PUSEY E, LUFKIN R, BROWN R et al, Magnetic resonance imaging artifacts: mechanisms and clinical significance, *Radiographics* (1986) **6**:891–911.

15 LUFKIN R, ANSELMO M, CRUES J et al, Magnetic field strength dependence of chemical shift artifacts, *Comput Radiol* (in press).

16 RUBIN JB, ENZMANN DR, WRIGHT A, CSF-gated MR imaging of the spine: theory and clinical implementation, *Radiology* (1987) **163**:784–92.

17 PATTANY PM, PHILLIPS JJ, CHIU LC et al, Motion artifact suppression technique (MAST) for MR imaging, *J Comput Assist Tomogr* (1987) **11**(3):369–77.

18 DELAPAZ RL, BRADY TJ, BUONANNO FS et al, Nuclear magnetic resonance (NMR) imaging of Arnold-Chiari type I malformation with hydromyelia, *J Comput Assist Tomogr* (1983) **7**(1):126–9.

19 BARKOVICH AJ, WIPOLD FJ, SHERMAN JL et al, Significance of cerebellar tonsillar position on MR, *AJNR* (1986) **7**:795–9.

20 YATES A, BRANT-ZAWADZKI M, NORMAN D et al, Nuclear magnetic resonance imaging of syringomyelia, *AJNR* (1983) **4**:234–7.

21 LEE BC, ZIMMERMAN RD, MANNING JJ et al, MR imaging of syringomyelia and hydromyelia, *AJR* (1985) **144**(6):1149–56.

22 SHERMAN JL, BARKOVICH AJ, CITRIN CM, The MR appearance of syringomyelia: new observations, *AJNR* (1986) **7**:985–95.

23 HAN JS, BENSON JE, KAUFMAN B et al, Demonstration of diastematamaelia and associated abnormalities with MR imaging, *AJNR* (1985) **6**:215–9.

24 MODIC MT, MASARYK T, BOUMPHREY F et al, Lumbar herniated disk disease and canal stenosis: prospective evaluation by surface coil MR, CT, and myelography, *AJR* (1986) **147**: 757–65.

25 MARAVILLA KR, LESH P, WEINREB JC et al, Magnetic resonance imaging of the lumbar spine with CT correlation, *AJNR* (1985) **6**:237–45.

26 MODIC MT, MASARYK TJ, MULOPULOS GP et al, Cervical radiculopathy: prospective evaluation with surface coil MR imaging, CT with metrizamide, and metrizamide myelography, *Radiology* (1986) **161**:753–9.

27 TERESI LM, LUFKIN RB, REICHER MA et al, Asymptomatic degenerative disk disease and spondylosis of the cervical spine: MR imaging, *Radiology* (1987) **164**:83–8.

28 GEBARSKI SS, MAYNARD FW, GABRIELSEN TO et al, Posttraumatic progressive myelopathy. Clinical and radiologic correlation employing MR imaging delayed, CT metrizamide myelography, and intraoperative sonography, *Radiology* (1985) **157**: 379–85.

29 RAMSEY RG, ZACHARIAS CE, MR imaging of the spine after radiation therapy: easily recognizable effects, *AJR* (1985) **144**:1131–5.

30 DOOMS GC, FISHER MR, HRICACK H et al, Bone marrow imaging: magnetic resonance studies related to age and sex, *Radiology* (1985) **155**:429–32.

31 MAMOURIAN AC, BRIGGS RW, The appearance of pantopaque on MR images, *Radiology* (1986) **158**:457–60.

32 MASARYK TJ, MODIC MT, GEISINGER MA et al, Cervical myelopathy: a comparison of magnetic resonance and myelography, *J Comput Assist Tomogr* (1986) **10**:184–94.

33 BELTRAN J, NOTO AM, CHAKERES DW et al, Tumors of the osseous spine: staging with MR imaging versus CT, *Radiology* (1987) **162**:565–9.

34 BURK DL, BRUNBERG JA, KANAL E et al, Spinal and paraspinal neurofibromatosis: surface coil MR imaging at 1.5 T, *Radiology* (1987) **162**:797–801.

35 GOY AMC, PINTO RS, RAGHAVENDRA BN et al, Intramedullary spinal cord tumors: MR imaging, with emphasis on associated cysts, *Radiology* (1986) **161**:381–6.

36 MODIC MT, FEIGLAN DH, PIRAINO BW et al, Vertebral osteomyelitis: assessment using MR, *Radiology* (1985) **157**:157-66.

37 MARAVILLA KR, WEINRAB JC, SUSS R et al, Magnetic resonance demonstration of multiple sclerosis plaques in the cervical cord, *AJNR* (1984) **5**:685–9.

38 DI CHIRO G, DOPPMAN JL, DWYER AJ et al, Tumors and arteriovenous malformations of the spinal cord: assessment using MR, *Radiology* (1985) **156**:6890–97.

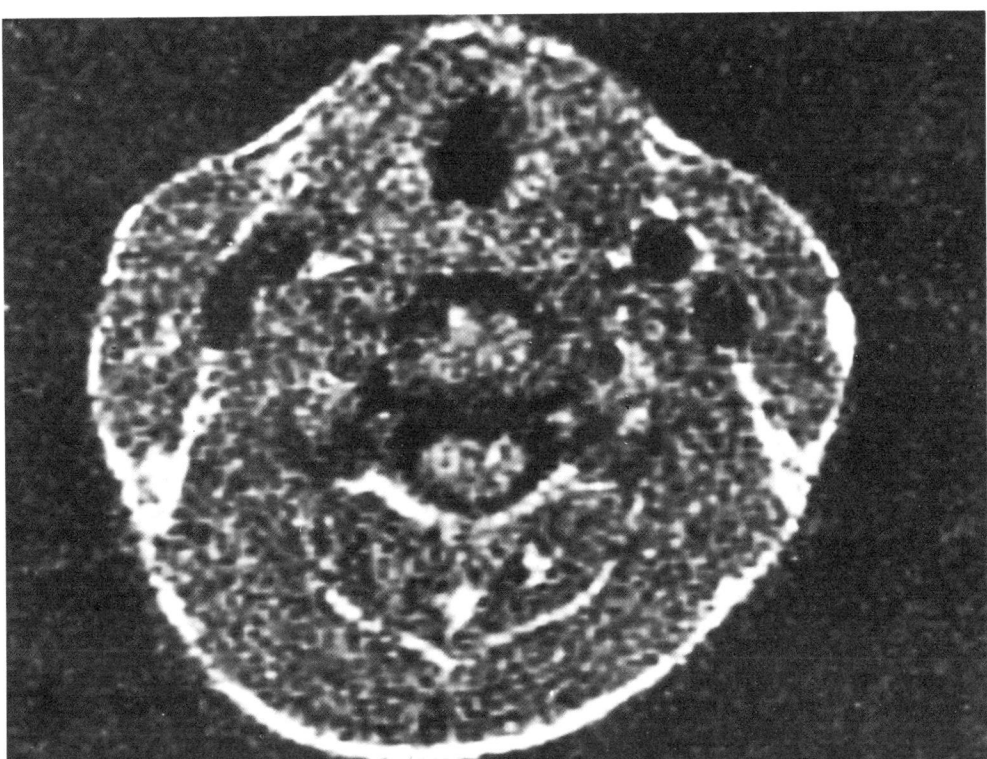

a

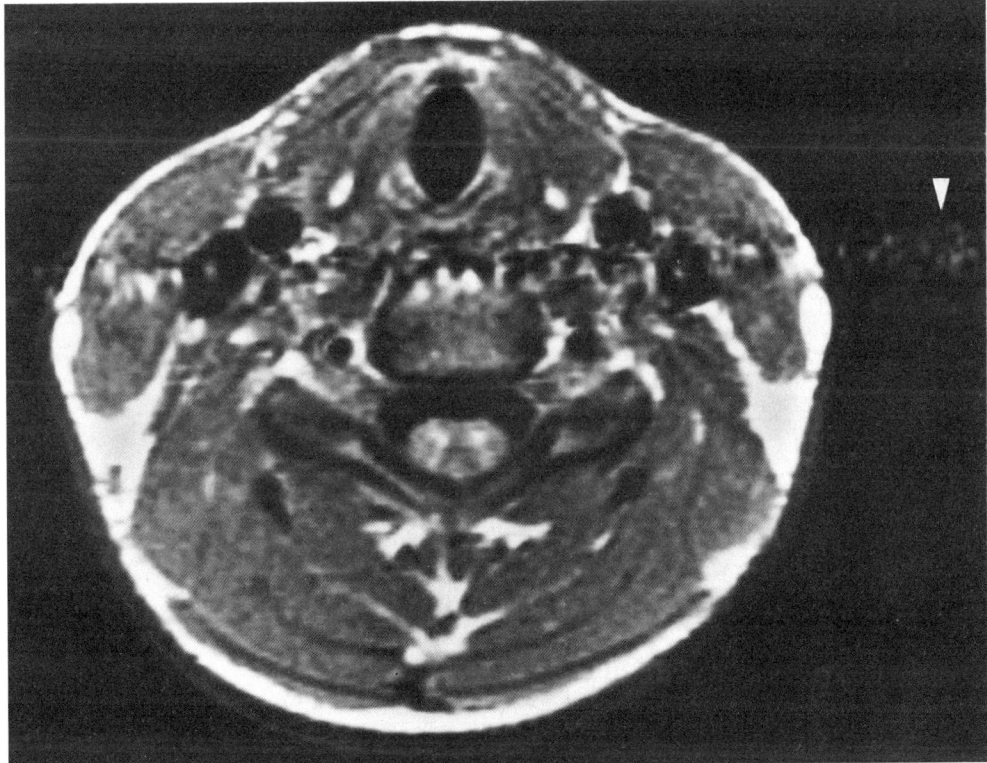

b

Figure 2.1

Comparison of images obtained with standard body radiofrequency (RF) receiver coil and circumferential solenoid surface coil. (**a**) Axial spin-echo image (SE 500/30) NEX = 1 of normal cervical spine, obtained with standard body coil. (**b**) Axial spin-echo image (SE 500/30) NEX = 1 at same level, using solenoid surface coil. Note dramatic improvement in signal-to-noise ratio. Horizontal phase encoding is used to project image harmonics due to flow (arrowhead) off spinal canal. (**c**) Solenoidal circumferential surface coil in use.

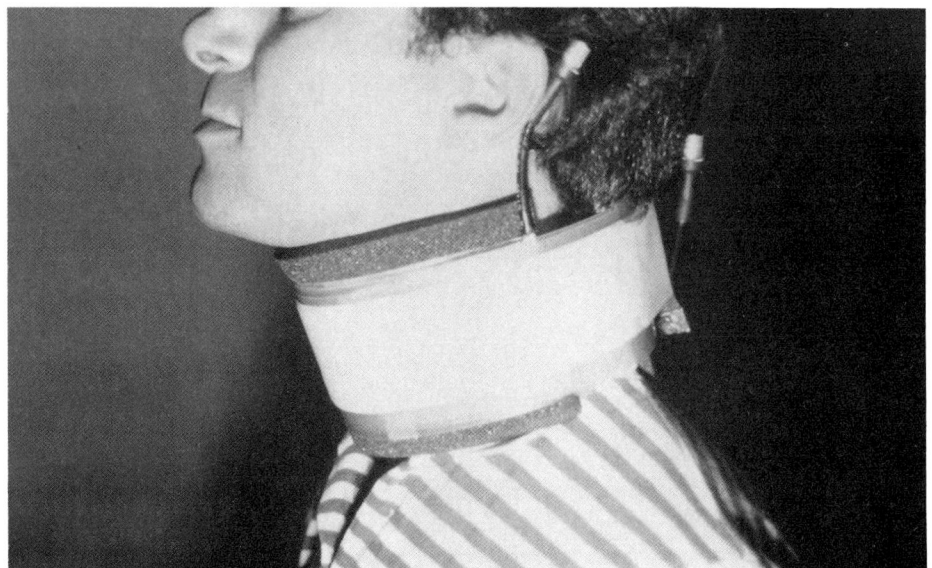

Figure 2.1 *continued*

Figure 2.2

Normal cervical spine.
(**a**) Sagittal T$_1$-weighted (SE 500/30) spin/echo scan through midline. (**b**) Sagittal T$_2$-weighted (SE 2000/85) spin-echo image in the same plane. Areas of low-signal intensity anterior to the cord result from flow/motion artifacts from transmitted cardiac and respiratory pulsations (arrowhead). (**c**) Sagittal T$_2$*-weighted field-echo image (FE 280/20/10) for comparison. Flow artifacts are greatly reduced. (**d**) Axial T$_1$-weighted spin-echo image (SE 800/28) through C3–4 neural canal. (**e**) Axial T$_1$-weighted spin-echo image (SE 100/28) obtained through level of C5 body. (**f**) Scout view showing placement of oblique scan planes to demonstrate neural canals *en face*. (**g**) Resulting oblique T$_1$-weighted spin-echo image (SE 500/28) shows low-signal intensity nerve roots surrounded by high-intensity fat within neural canal.

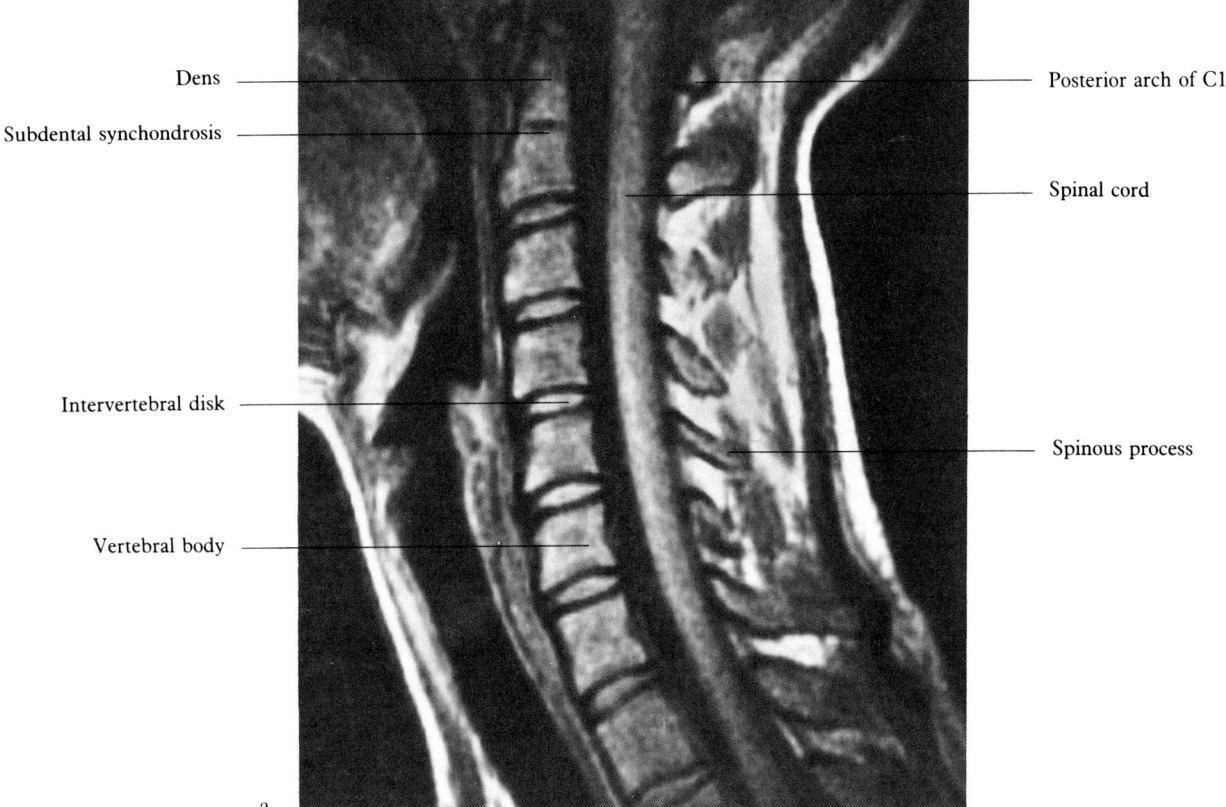

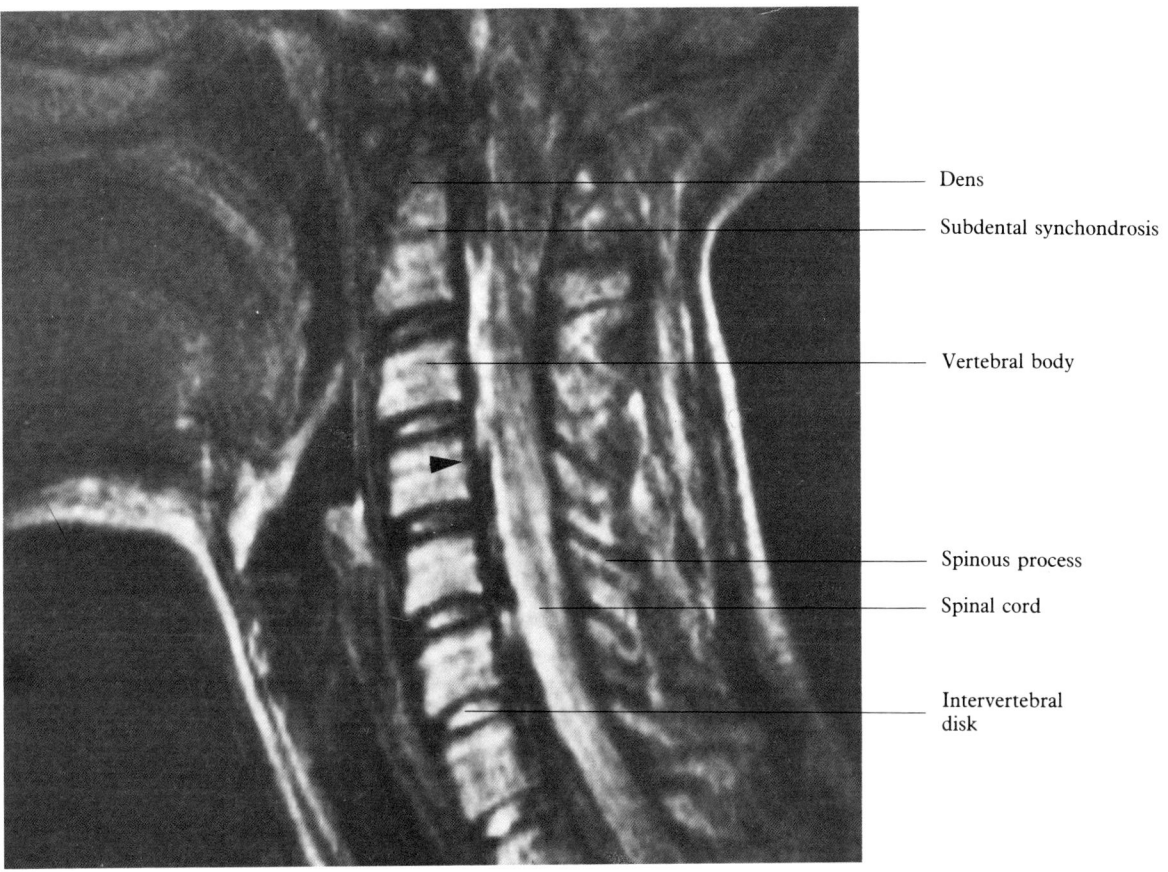

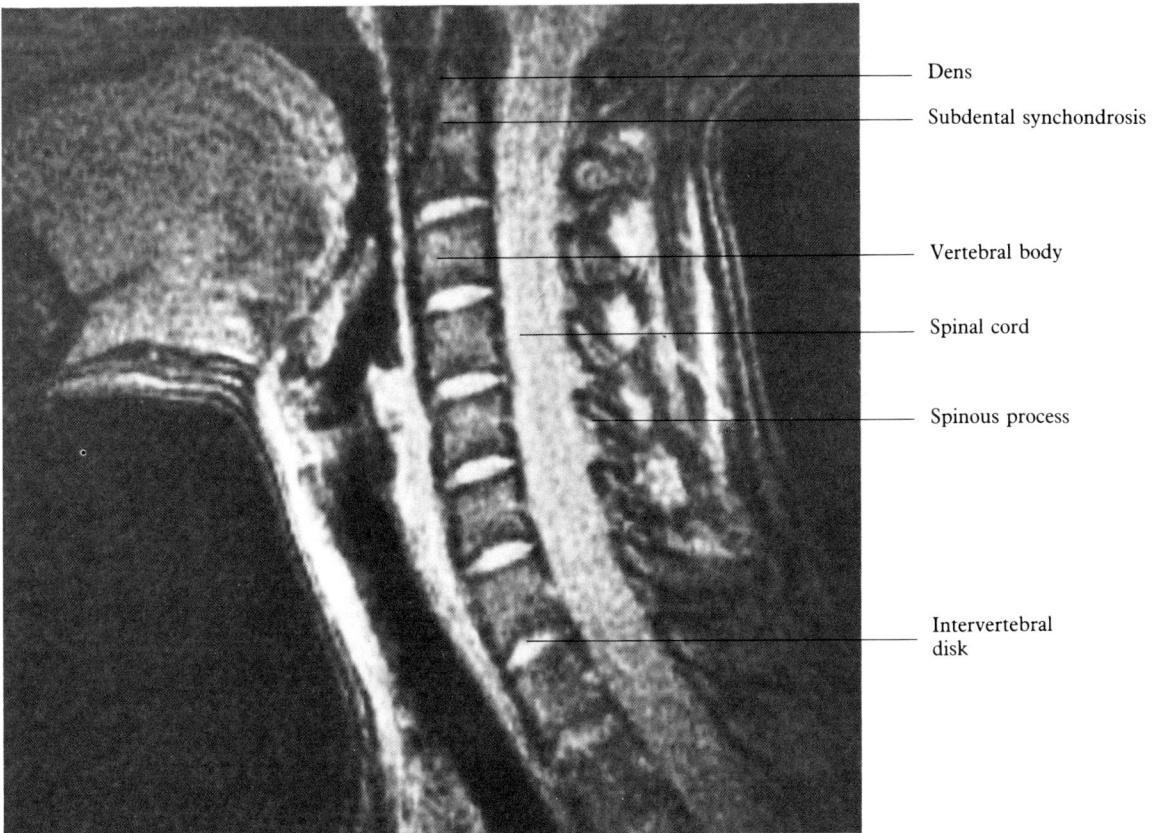

Figure 2.2 *continued*

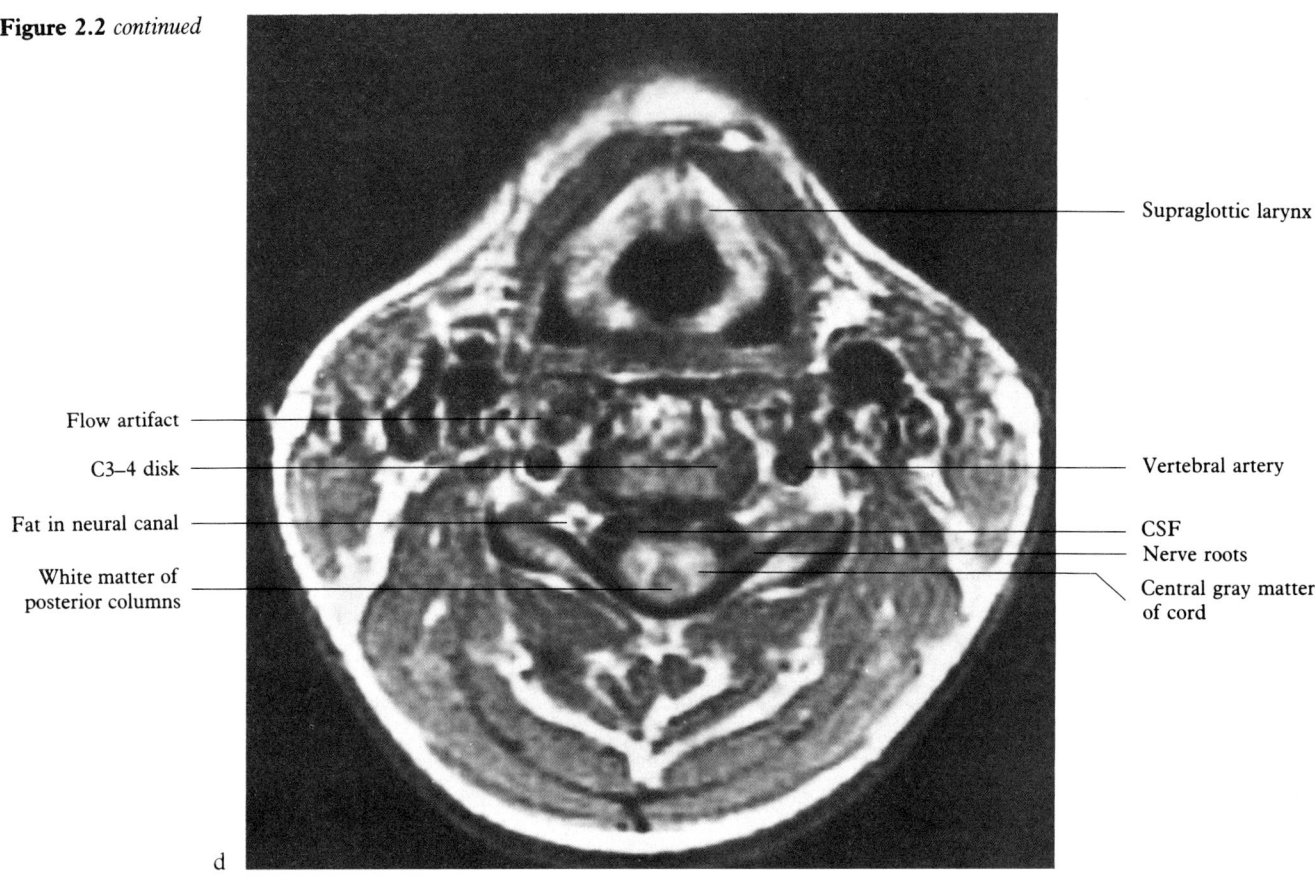

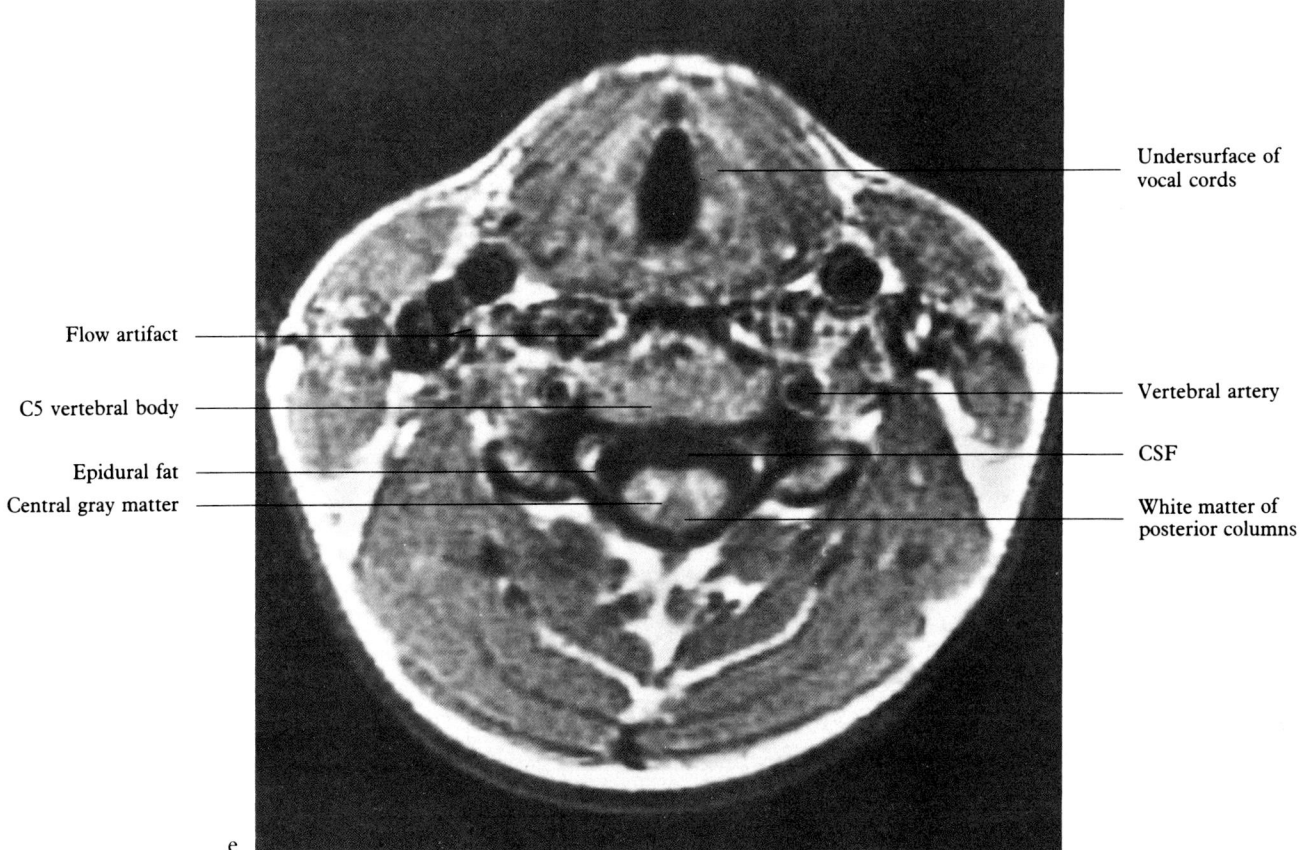

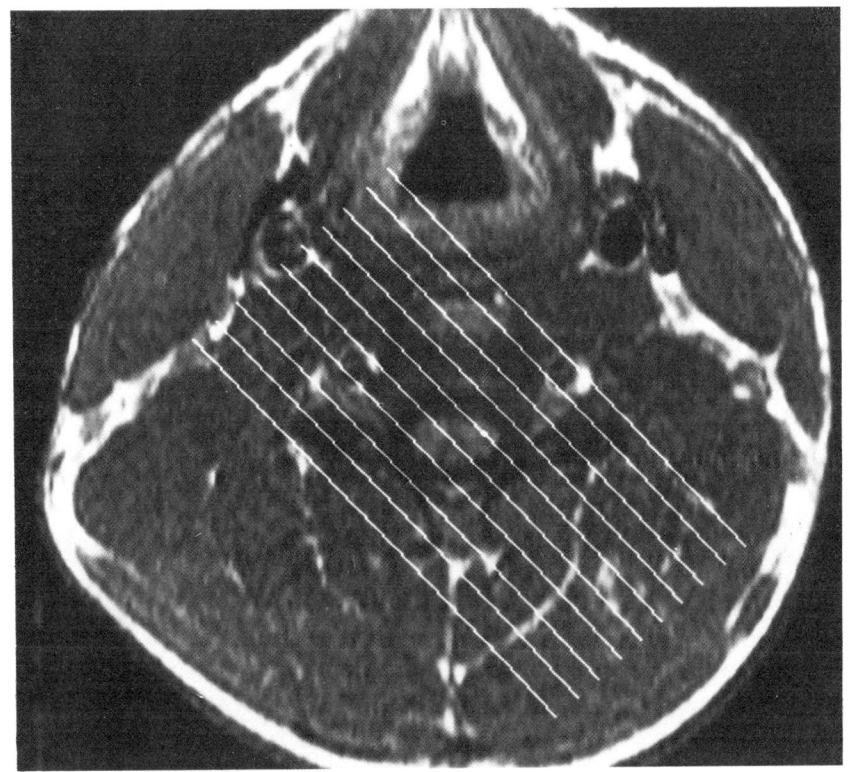

f

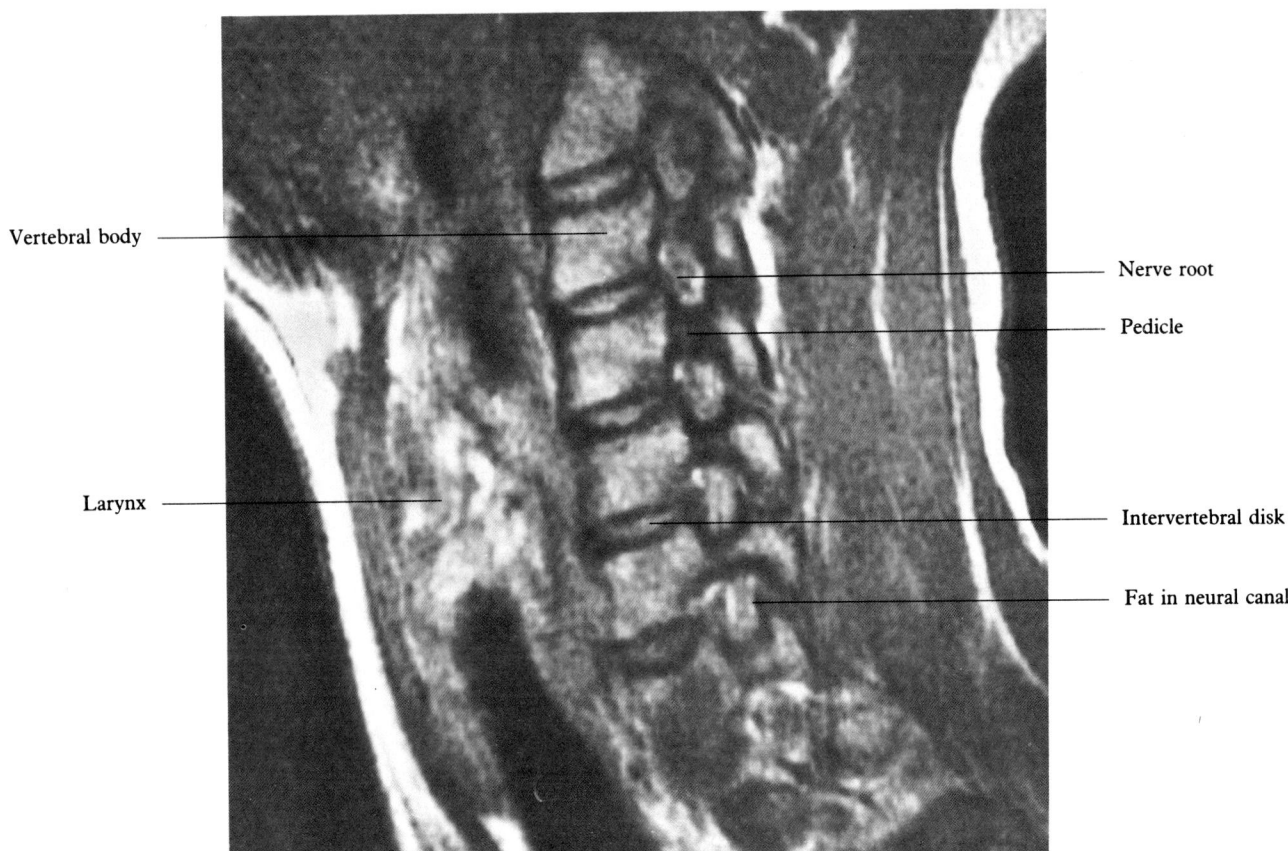

g

Figure 2.3

Normal lumbar spine. (**a**) Midline sagittal T_1-weighted image (SE 500/30). (**b**) Sagittal T_1-weighted image (SE 500/30) slightly lateral to (**a**). (**c**) Parasagittal T_1-weighted image (SE 500/30) through level of neural canals. (**d**) Scout view for multiple angle variable interval non-orthogonal (MAVIN) oblique views to be obtained in single multislice sequence. (**e**) Angled oblique T_1-weighted spin-echo image (SE 800/30) through level of neural canal. (**f**) Angled oblique T_1-weighted spin-echo image (SE 800/30) through level of pedicle.

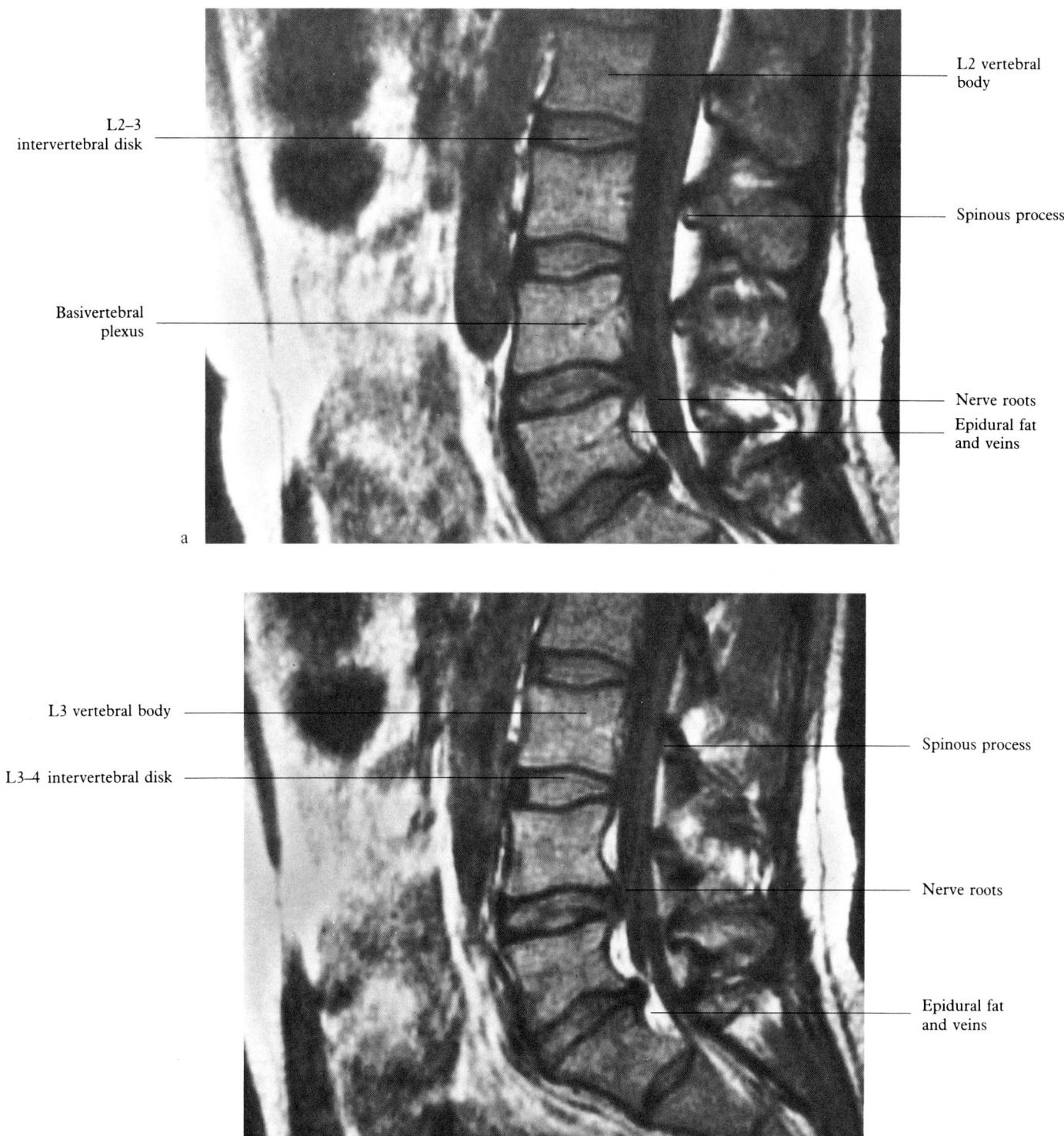

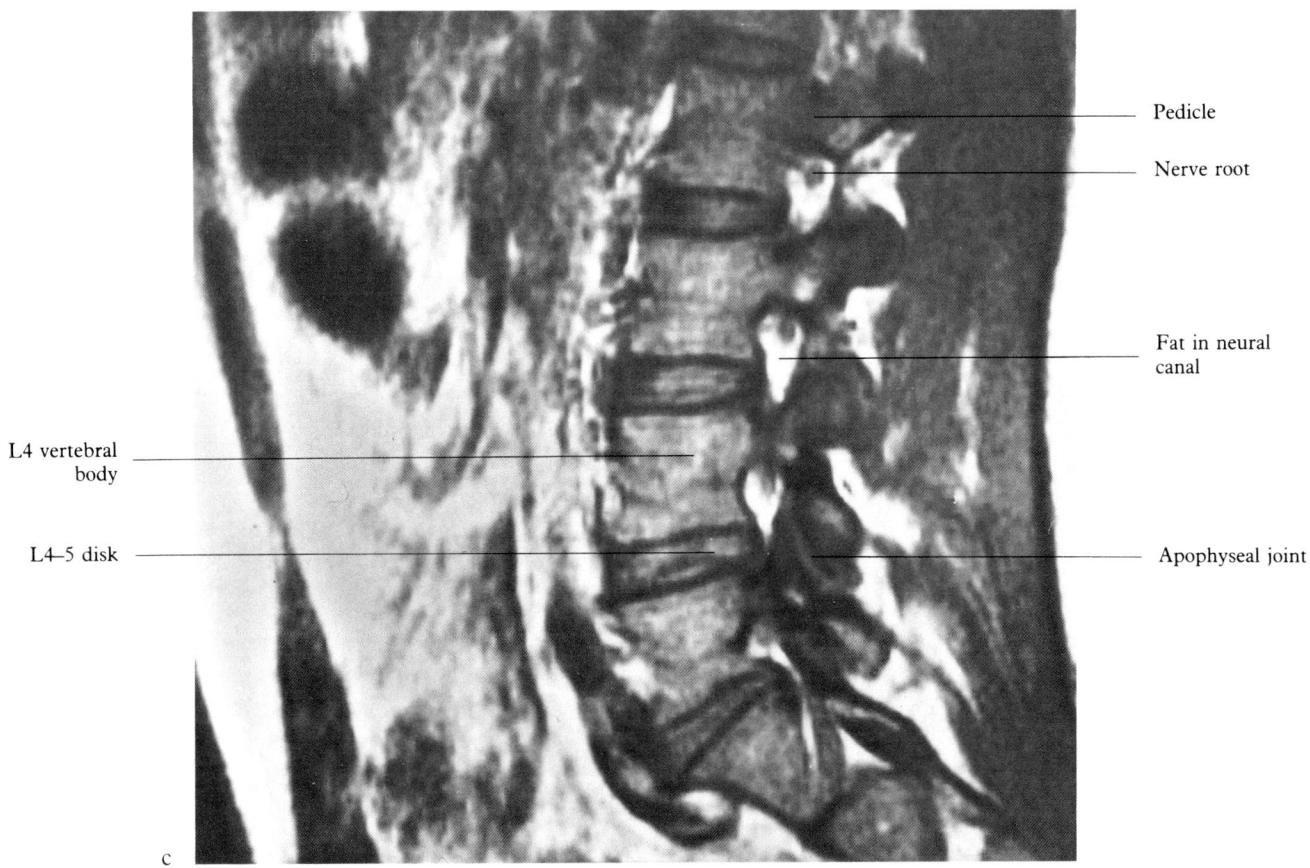

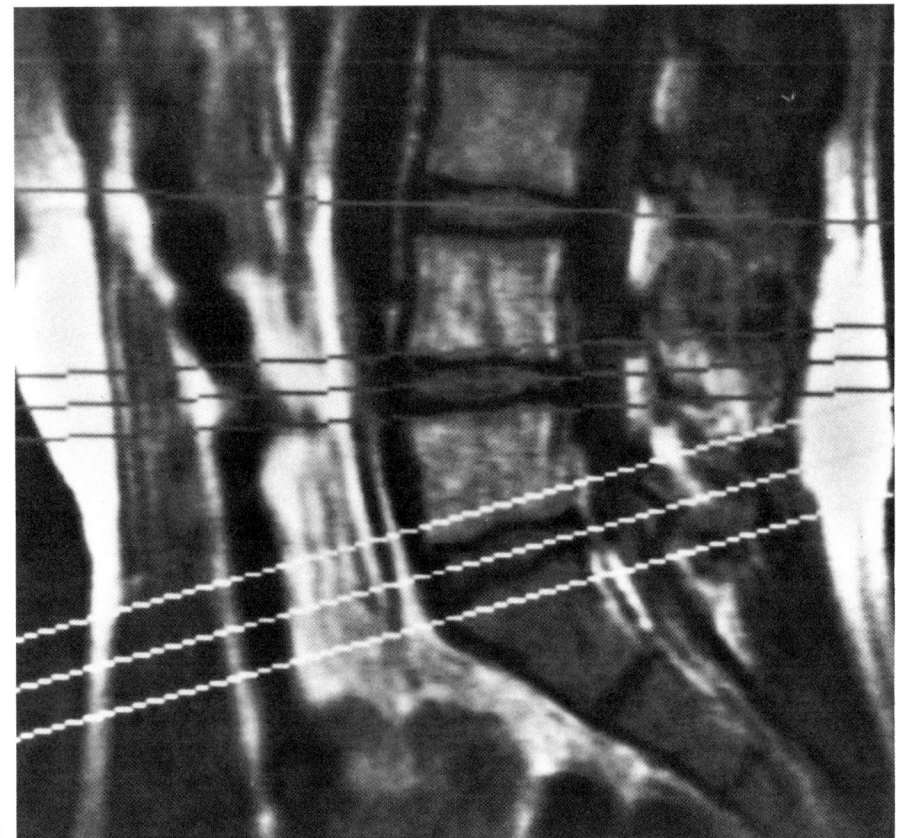

40 MRI atlas of the musculoskeletal system

Figure 2.3 continued

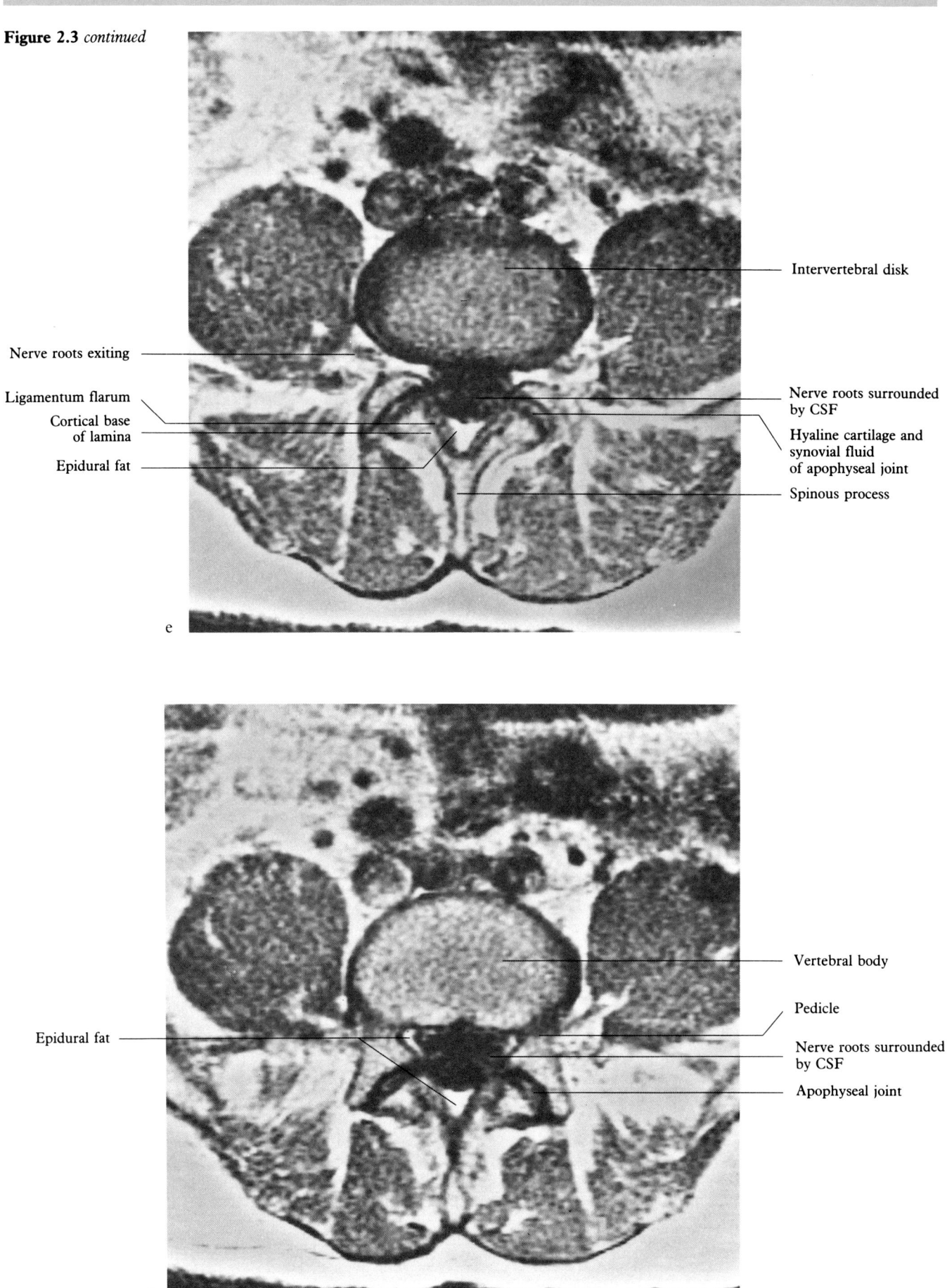

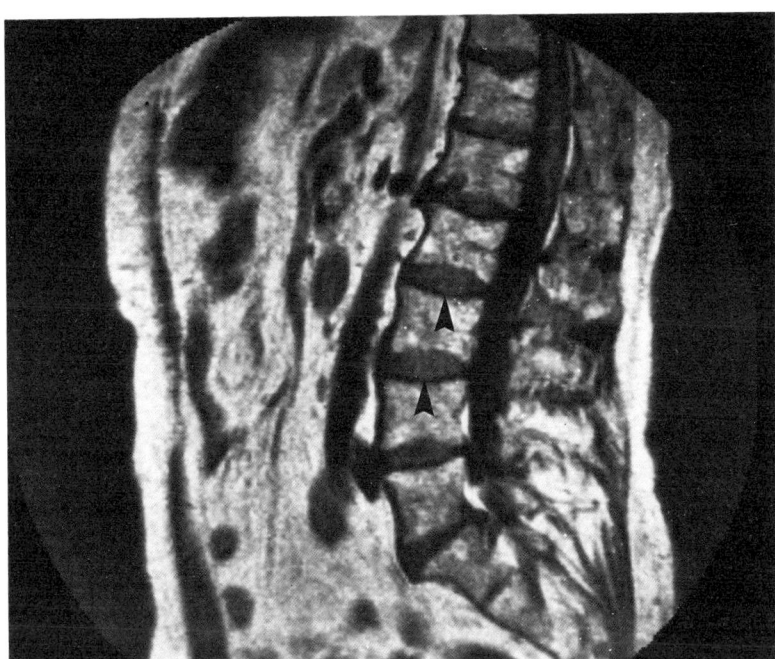

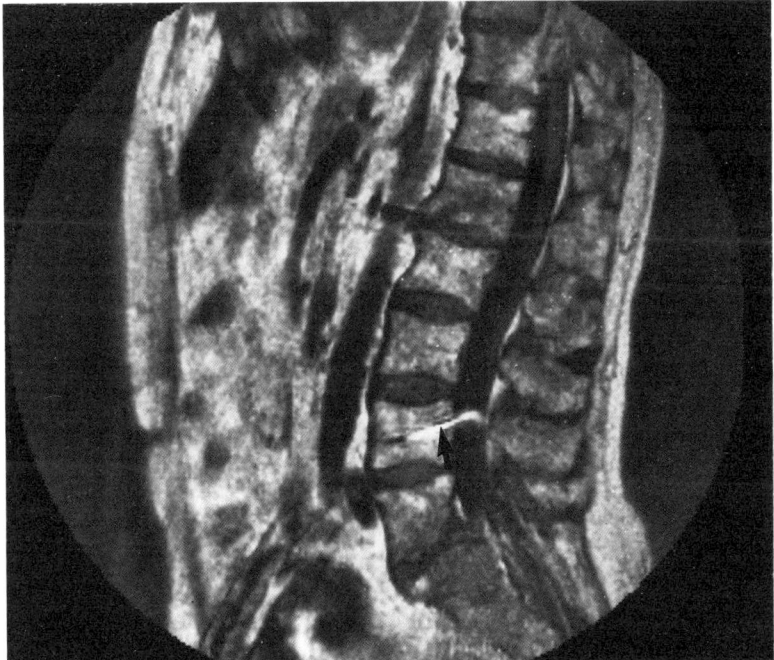

Figure 2.4

Effect of reversing phase- and frequency-encoding axes on chemical shift artifacts. (**a**) Sagittal T_1-weighted image (SE 600/26) of lumbar spine with vertical frequency-encoding axis reveals artifactural asymmetric vertebral end plates due to chemical shift misregistration errors (arrowheads). (**b**) Use of horizontal frequency encoding now minimizes this artifact. A small aliasing artifact is present (arrow). (*Courtesy of Michael Anselmo*)

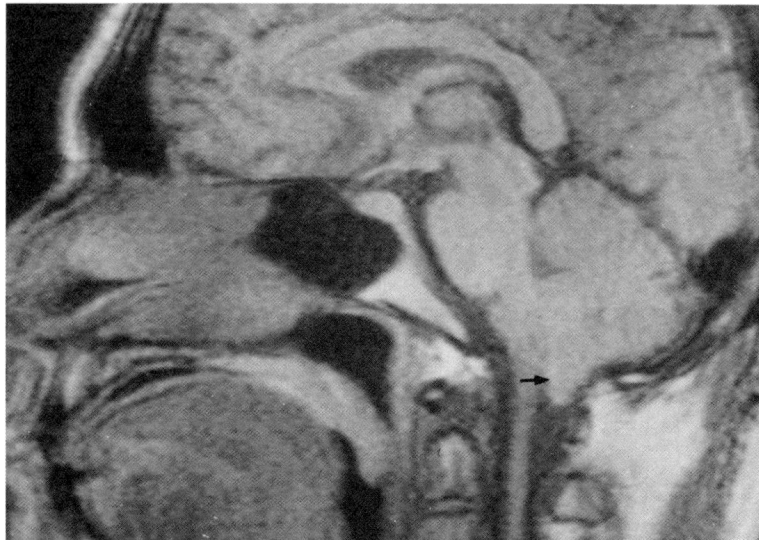

Figure 2.5

Chiari I malformation. Sagittal T_1-weighted image (SE 500/30) shows downward displacement of the cerebellar tonsils (arrow) below level of foramen magnum. The 4th ventricle and remainder of posterior fossa are unremarkable.

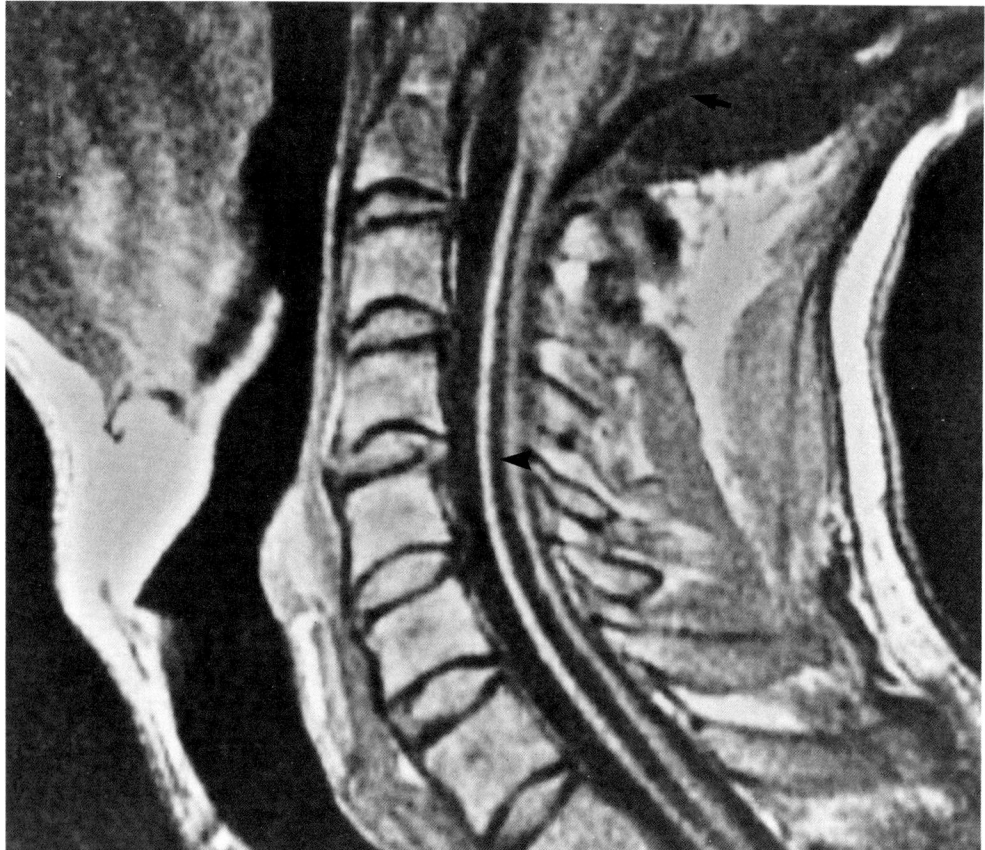

Figure 2.6

Syringohydromyelia associated with Chiari malformation. (**a**) Sagittal T_1-weighted image (SE 500/30) shows central low-signal intensity cystic region (arrowhead) in cervical cord. Postsurgical changes following repair of Chiari malformation are noted at foramen magnum (arrow). (**b**) Sagittal T_1-weighted field-echo image (SE/200/10/60) provides similar information with reduced scanning time. Note low-signal-intensity bone marrow due to phase cancellation chemical shift effects.

a

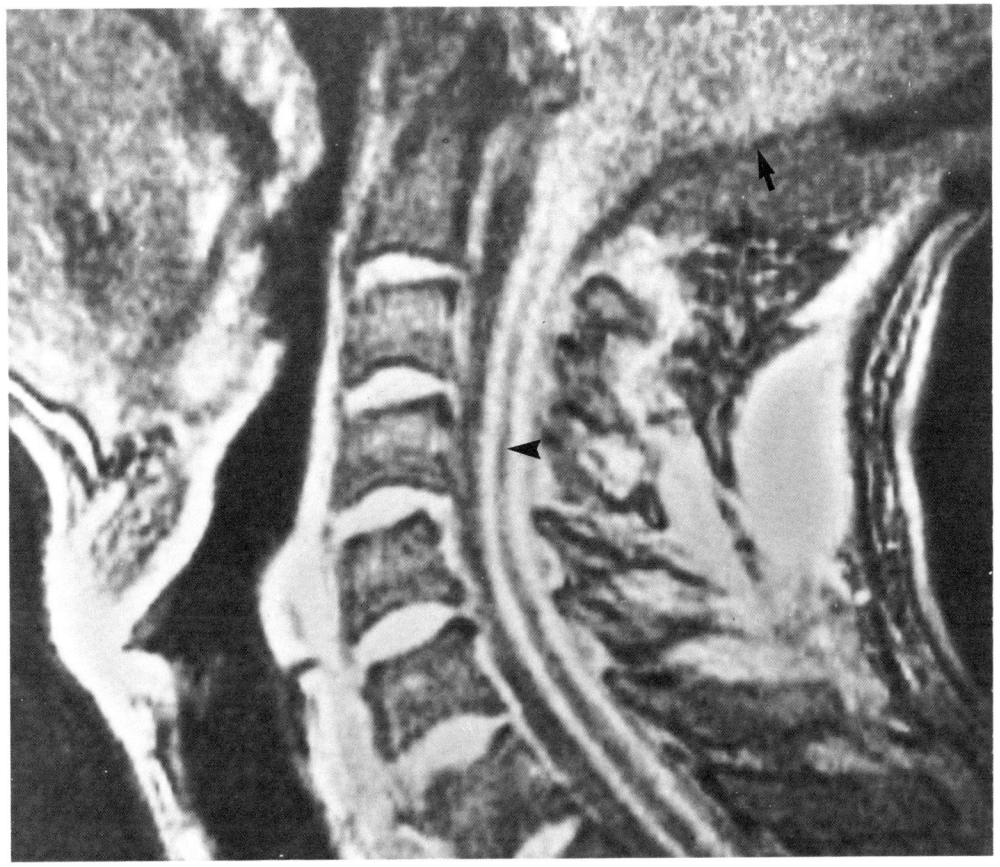

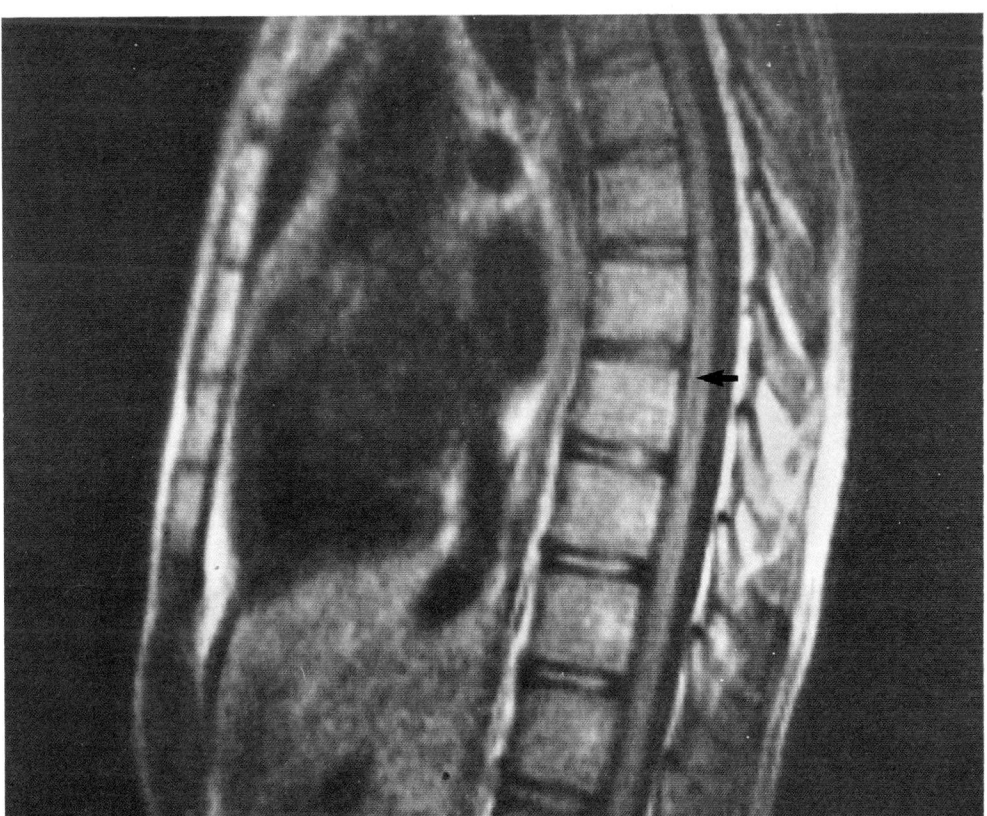

Figure 2.7
Syringohydromyelia.
(**a**) Sagittal T_1-weighted image (SE 500/30) of patient with suspected syringohydromyelia of thoracic spine. Small area of linear low-signal intensity is noted (arrow). (**b**) Axial T_1-weighted image (SE 500/30) through suspected abnormality confirms presence of low-signal-intensity fluid collection in center of thoracic cord (arrowhead).

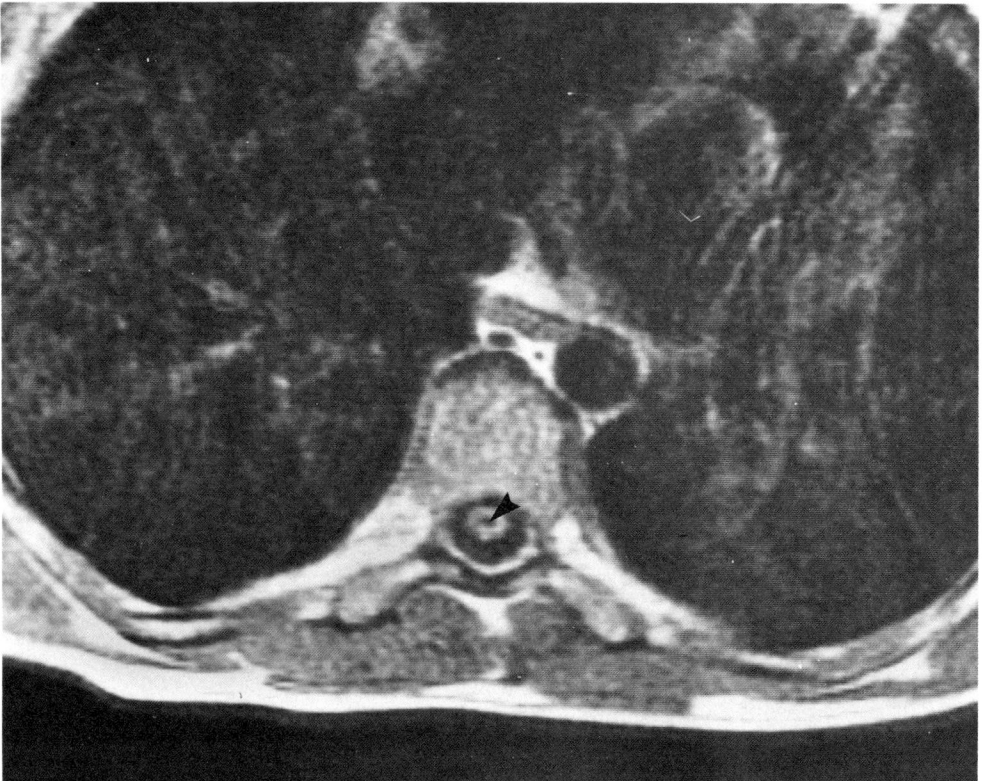

Figure 2.7 continued

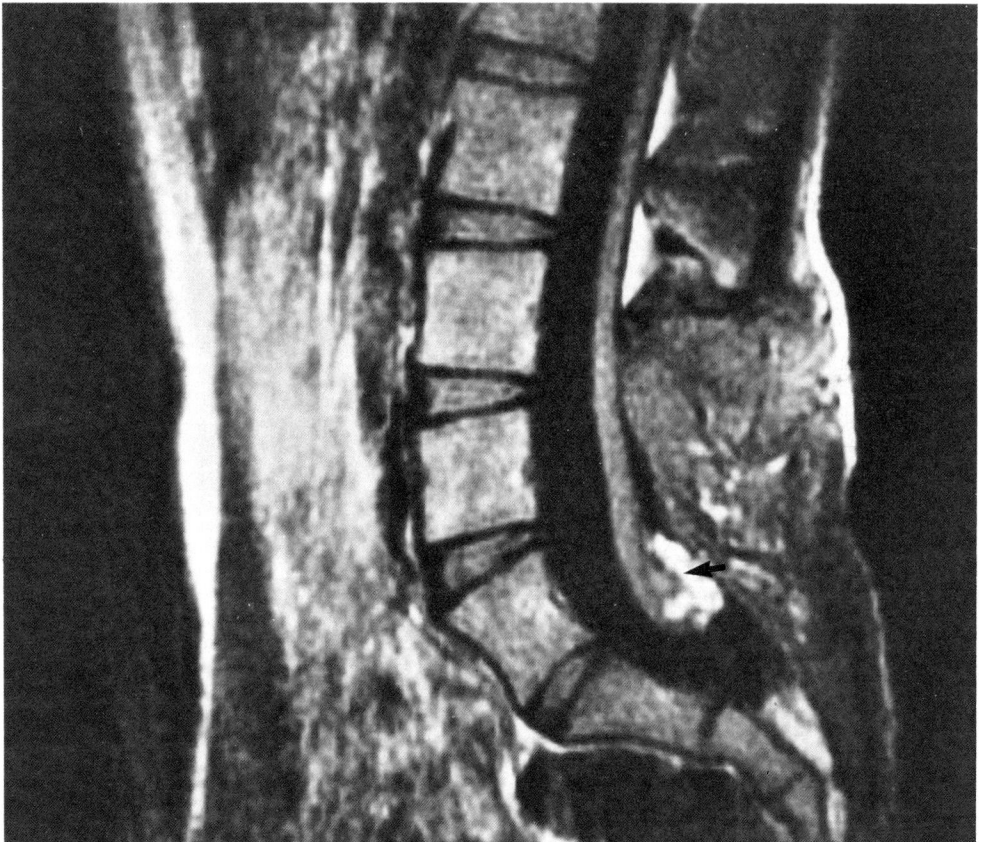

Figure 2.8

Lipomeningomyelocele. (a) Sagittal T_1-weighted spin-echo image (SE 500/30) reveals low lying spinal cord, dilated thecal sac, and dysraphism. The lipoma is easily recognized by its high-intensity signal (arrow). (b) Sagittal T_1-weighted field-echo image (FE 200/10/60) in the same plane shows similar anatomy. Note prominent effects of low-signal chemical shift phase cancellation at lipoma/CSF interface (arrowhead) as well as in vertebral marrow. (c) Axial T_1-weighted spin-echo image (SE 800/30) confirms findings (arrowhead). (d) T_1-weighted field-echo image (FE 200/10/60) in axial plane discloses similar phase-cancellation errors (arrowhead).

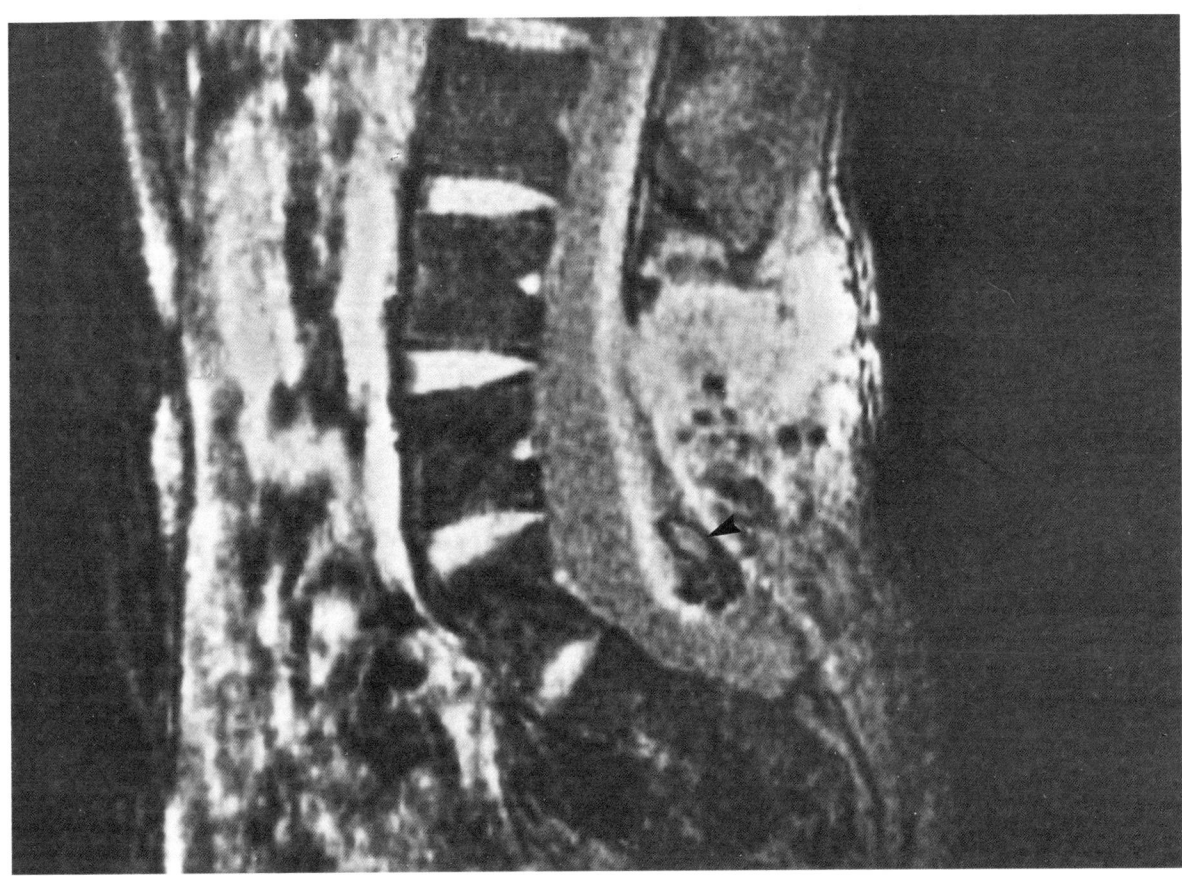

b

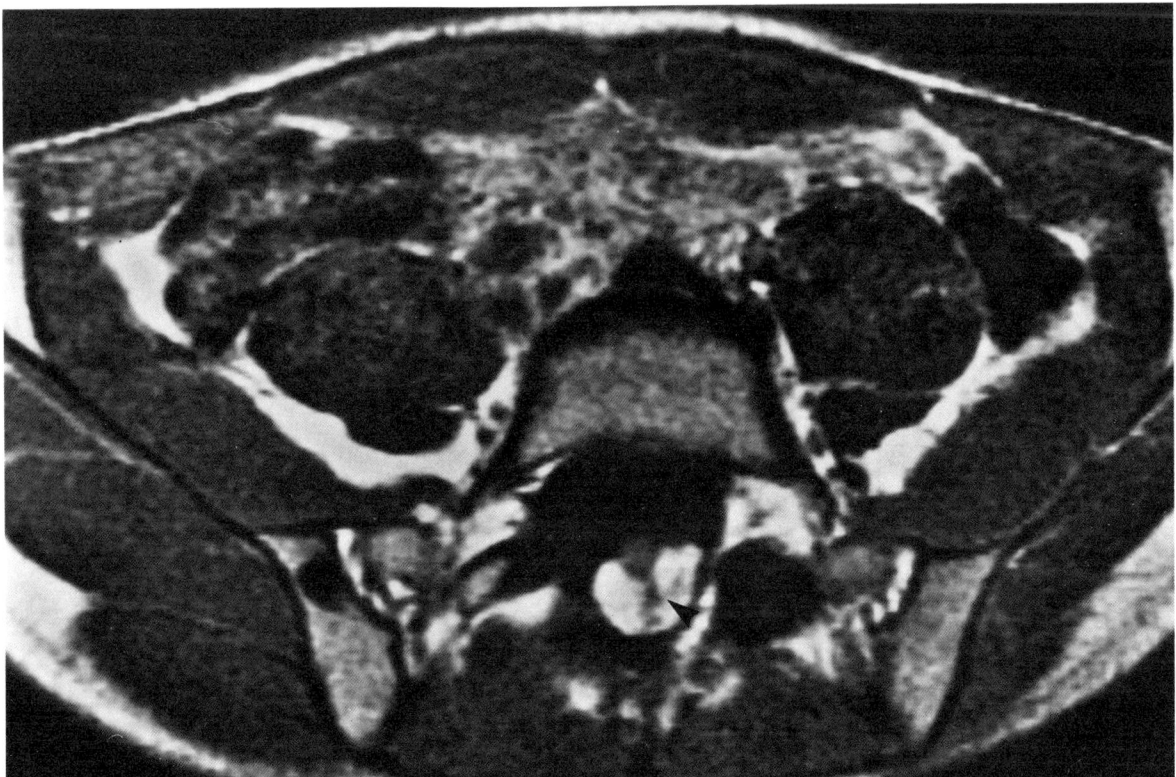

c

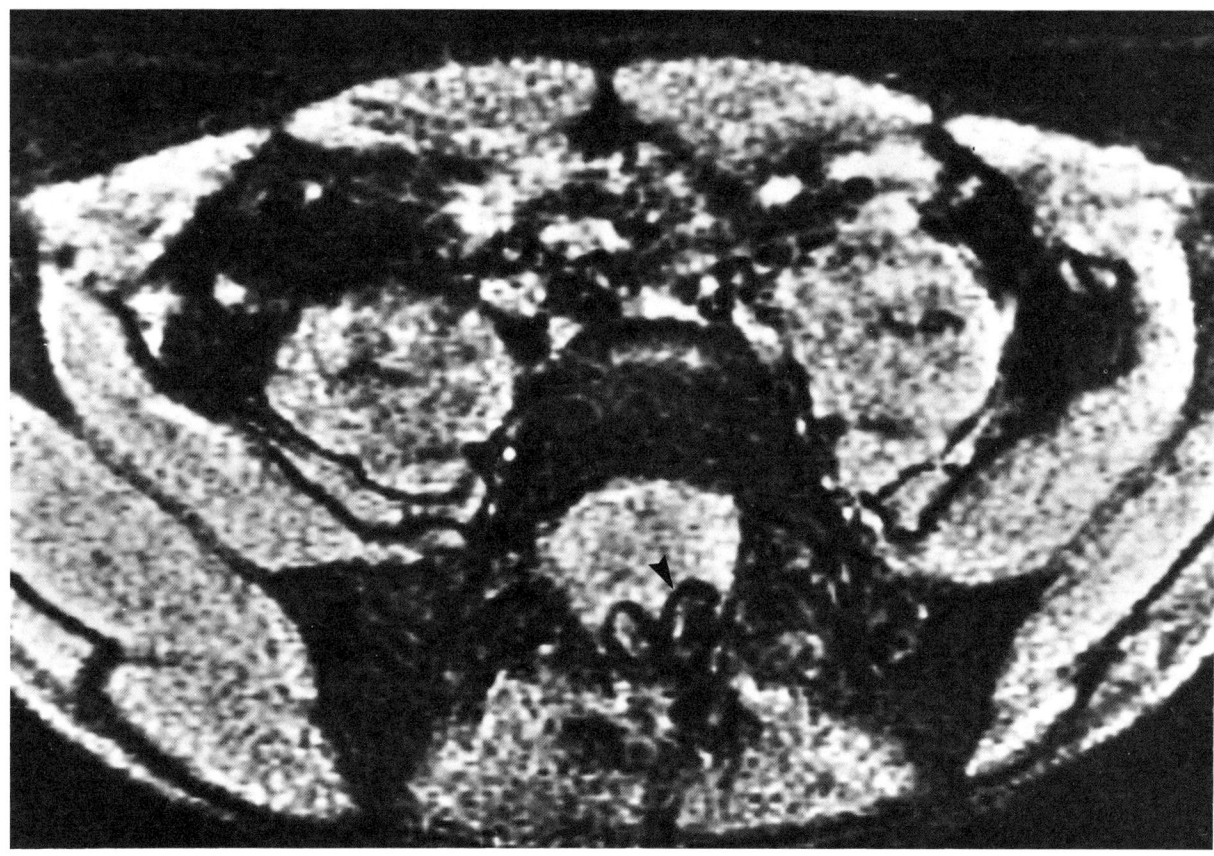

d

Figure 2.8 continued

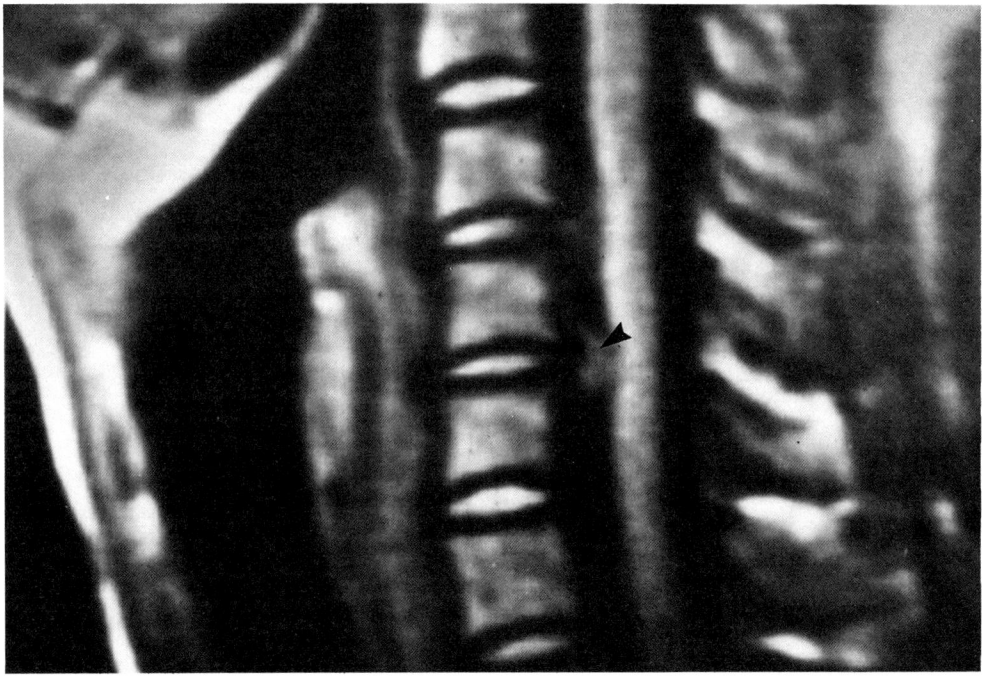

Figure 2.9

Herniated cervical disk. Sagittal T_1-weighted spin-echo image (arrowhead) adjacent to herniated cervical disk. Adjacent cord is unremarkable.

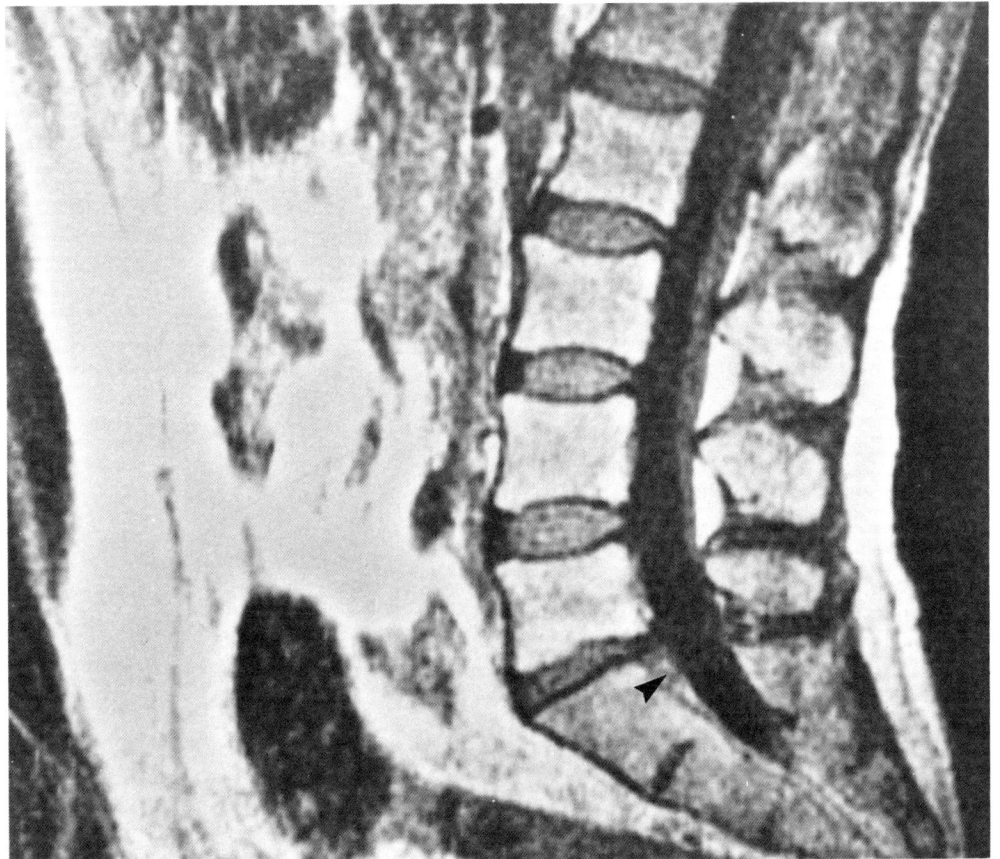

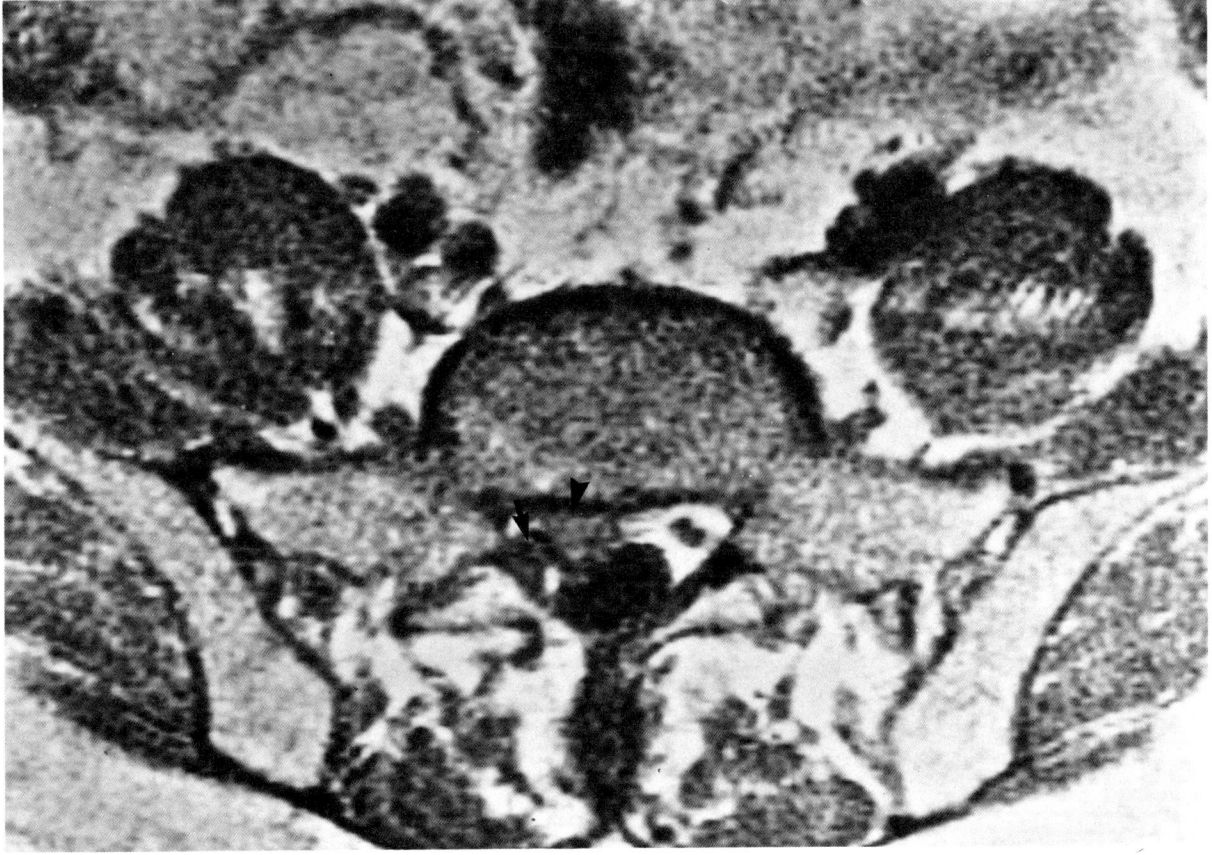

Figure 2.10
Herniated lumbar nucleus pulposus with extruded fragment. (**a**) Sagittal image (SE 500/30) reveals abnormal low-signal-intensity material (arrowhead) in L5–S1 interspace, replacing high-intensity epidural fat and high-intensity venous plexus, and extending inferiorly. (**b**) Axial image (SE 800/30) at same level confirms presence of herniated nucleus pulposus, with downwardly projecting disk fragment (arrowhead) displacing right lateral nerve roots (arrow).

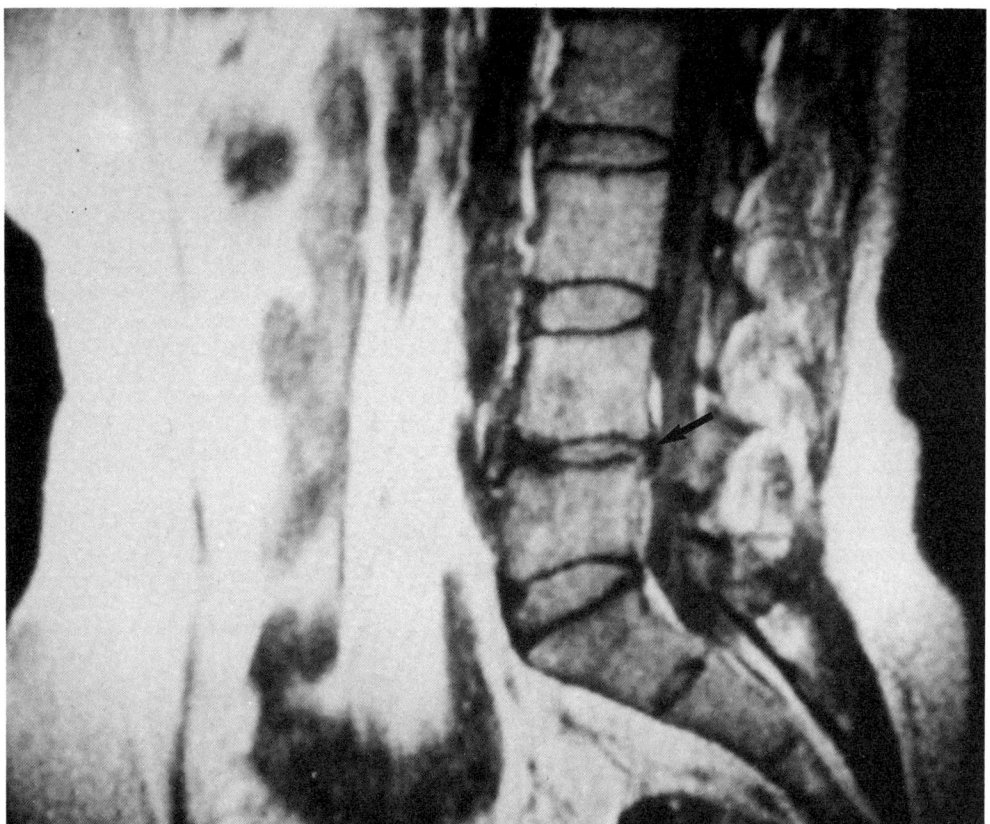

Figure 2.11
Herniated lumbar disk with extruded fragment. (**a**) T_1-weighted sagittal spin-echo image (SE 500/28) shows the herniated disk and downward migration of disk fragment (arrow). (**b**) Gradient-echo image (FE 200/10/60) allows clear distinction between herniated disk material (arrow), which has high-signal intensity, adjacent low-intensity osteophyte and marrow.

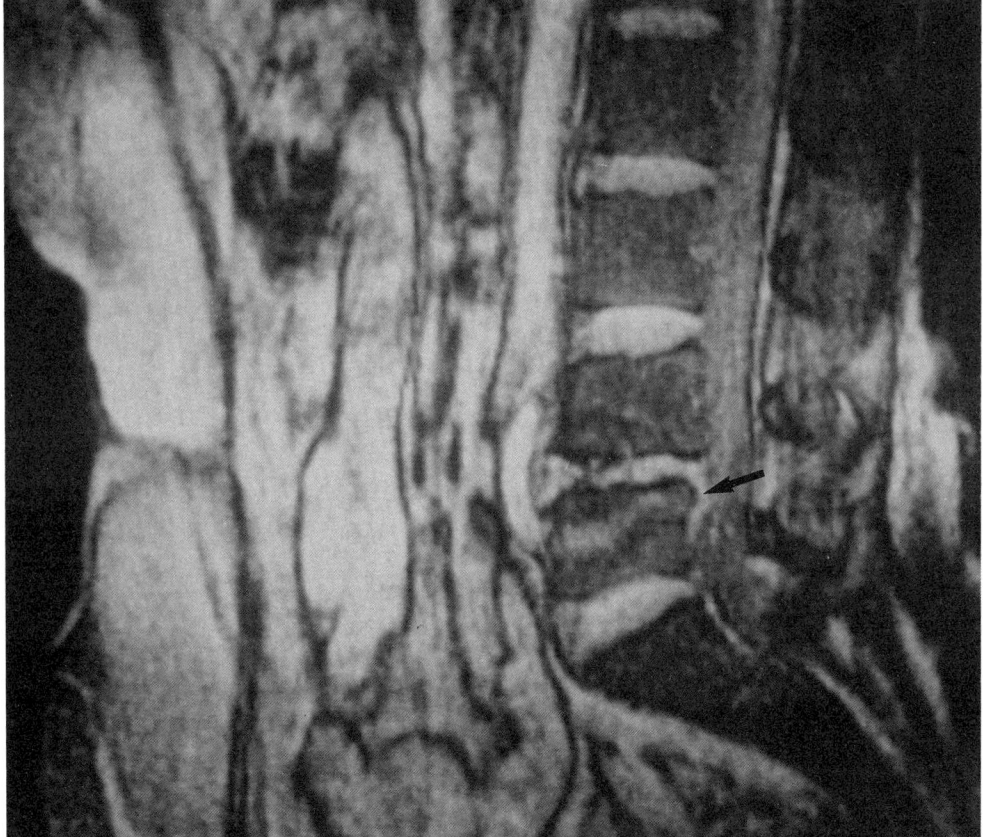

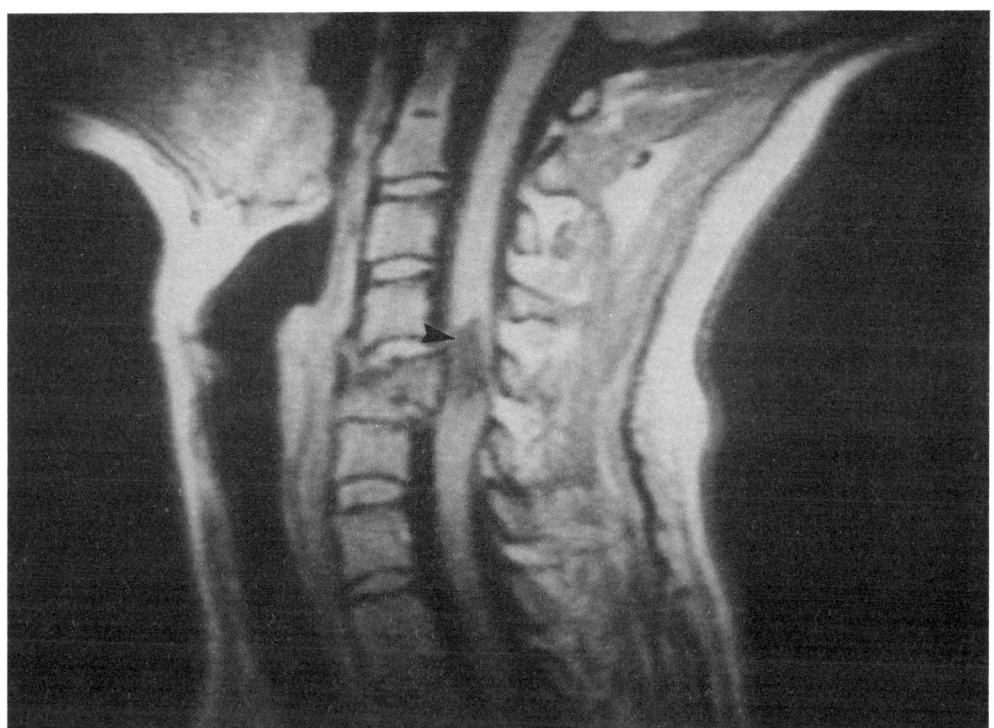

Figure 2.12

Fracture of cervical spine with associated syringohydromyelia. Sagittal T_1-weighted image (SE 500/28) reveals fracture of C5 body with posterior migration of fragment into cord, and associated cystic changes (arrowhead). The patient is quadriplegic, having sustained trauma 3 months previously.

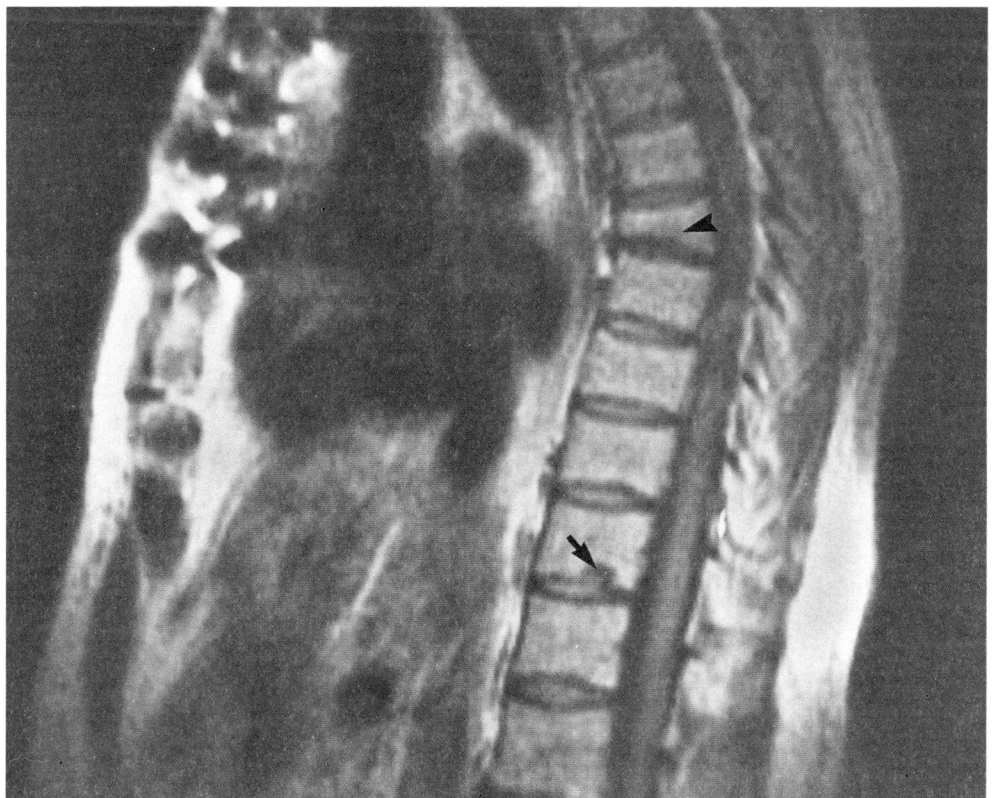

Figure 2.13

Osteoporotic compression fracture of thoracic spine. Midline sagittal T_1-weighted image (SE 500/30) shows compression fracture of upper thoracic vertebral body (arrowhead), indicated by anterior wedging. Marrow signal intensity is maintained (arrowhead). Schmorl's node is incidentally noted at a lower level (arrow).

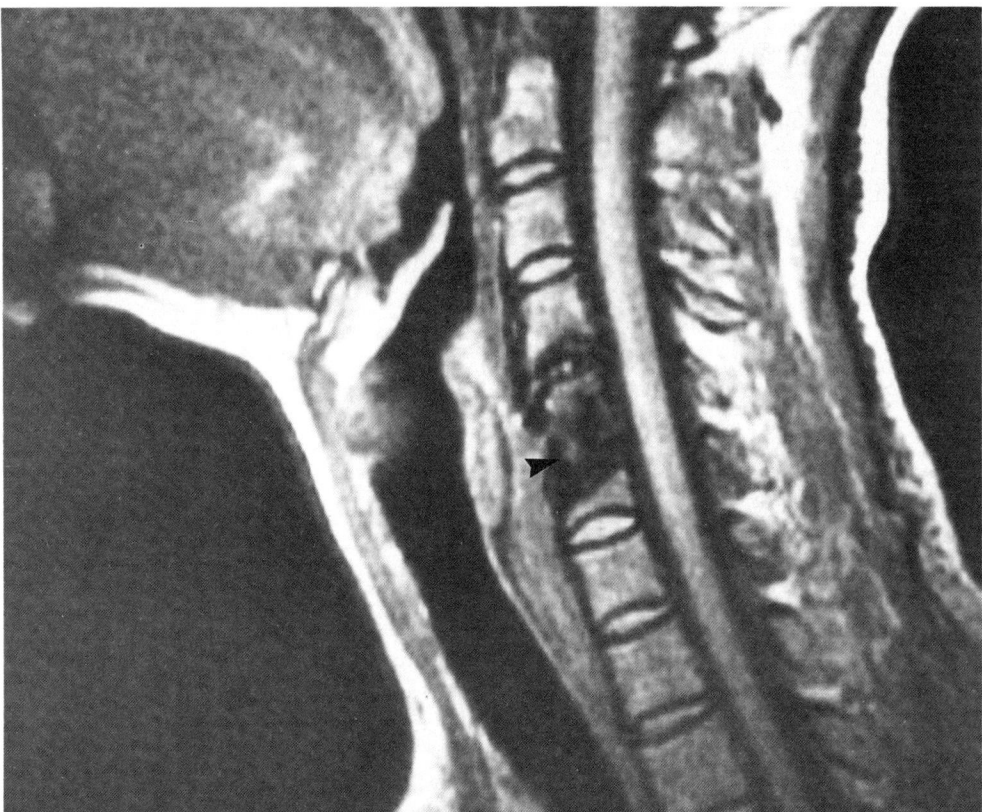

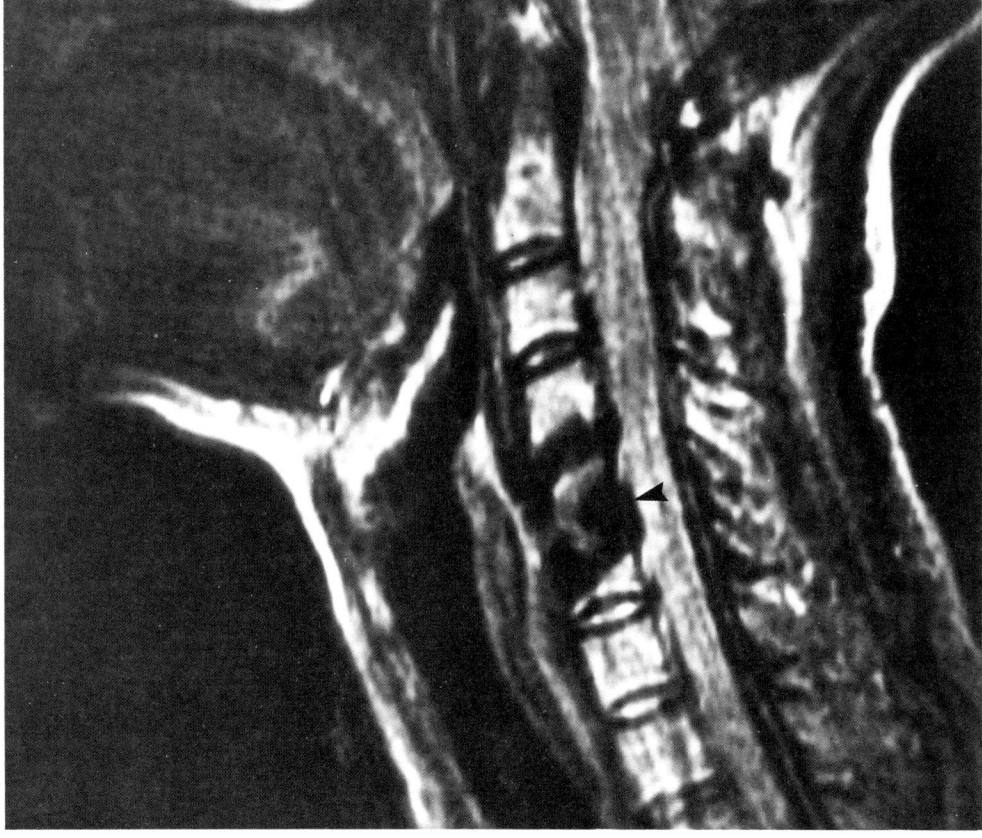

Figure 2.14

Postoperative cervical spine. Patient with anterior fusion had MRI to evaluate possible residual cord compression. (a) T_1-weighted spin-echo image (SE 500/28) discloses postsurgical changes (arrowhead), with no evidence of cord deformity. Evaluation of CSF/bone interface is difficult because both are of low-signal intensity on this pulse sequence. (b) T_2-weighted spin-echo image (SE 2000/28) shows better definition of CSF/bone interface (arrowhead); however, low-signal-intensity CSF pulsation artifacts can be misleading. (c) T_2-weighted field-echo image (FE 223/20/10) provides information similar to that in (a) but does so in shorter time. In addition, CSF/bone interface (arrowhead) is seen more clearly due to fewer pulsation artifacts.

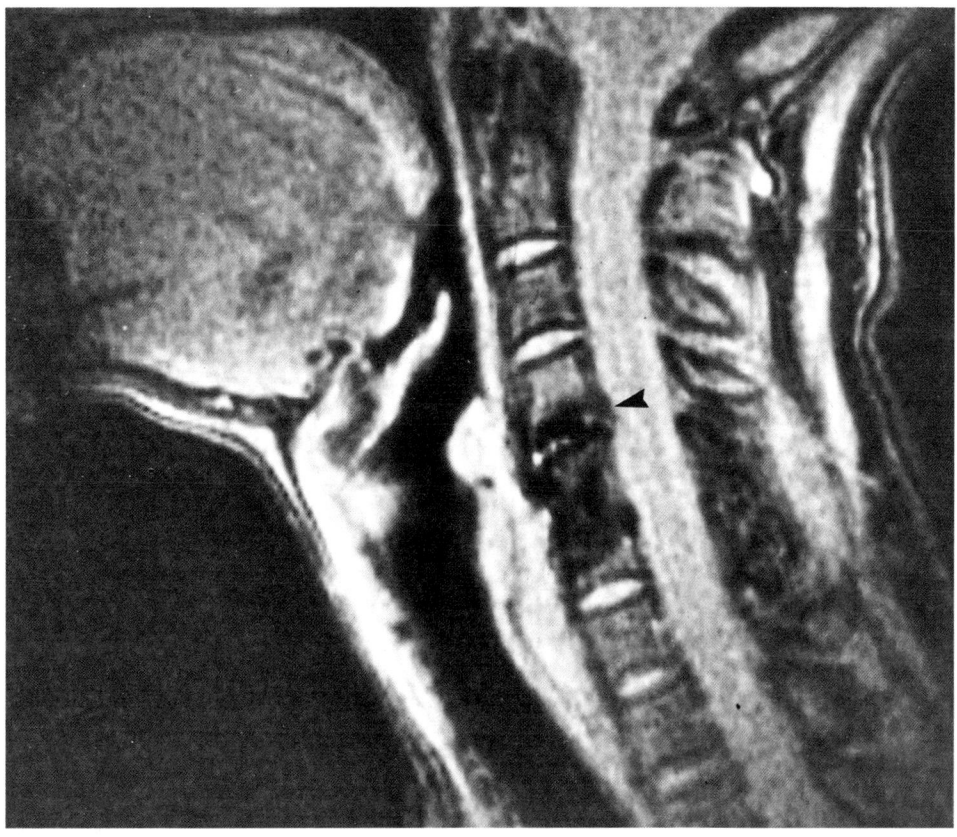

c

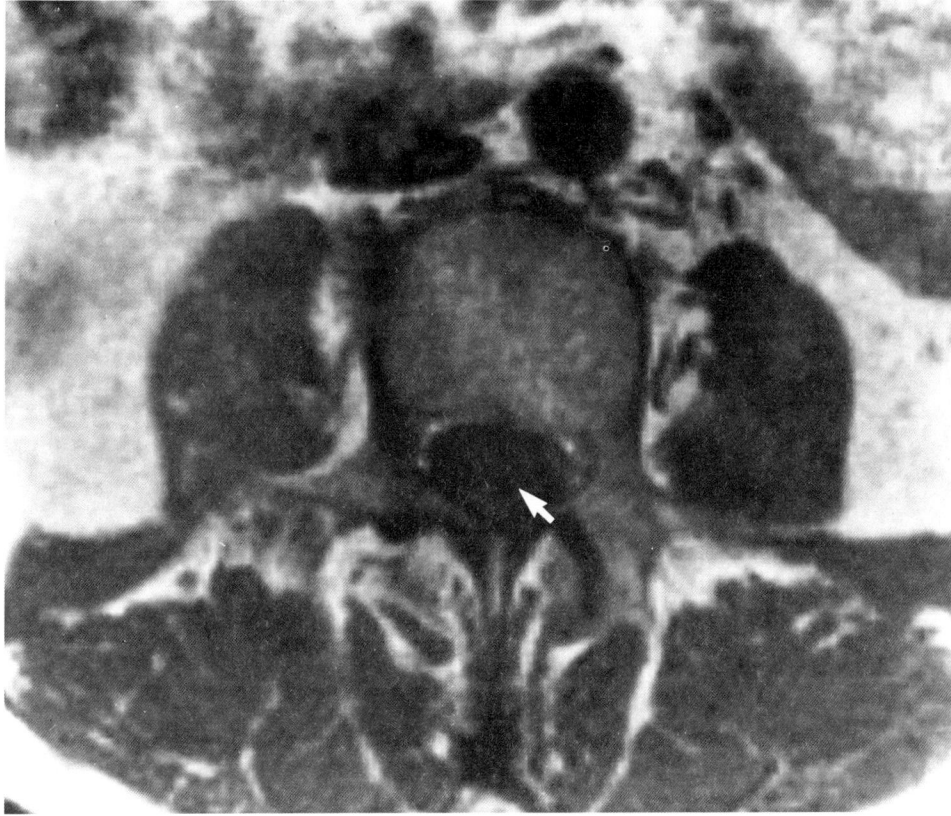

Figure 2.15

Arachnoiditis. Axial T_1-weighted image (SE 500/30) shows characteristic posterior clumping of nerve roots (arrow).

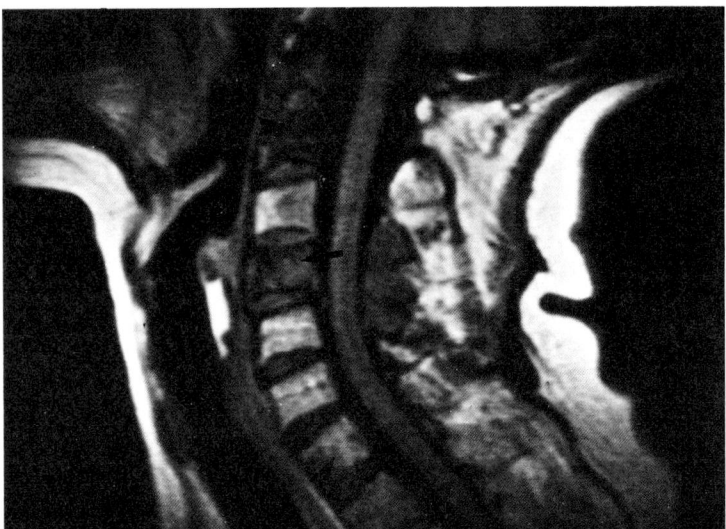

Figure 2.16

Metastasis. Sagittal T_1-weighted image (SE 500/28) of cervical spine in patient with metastatic adenocarcinoma. C4 vertebra is expanded due to metastasis (arrow). In other regions normal high-signal intensity marrow has been replaced by low-intensity metastases (arrowhead).

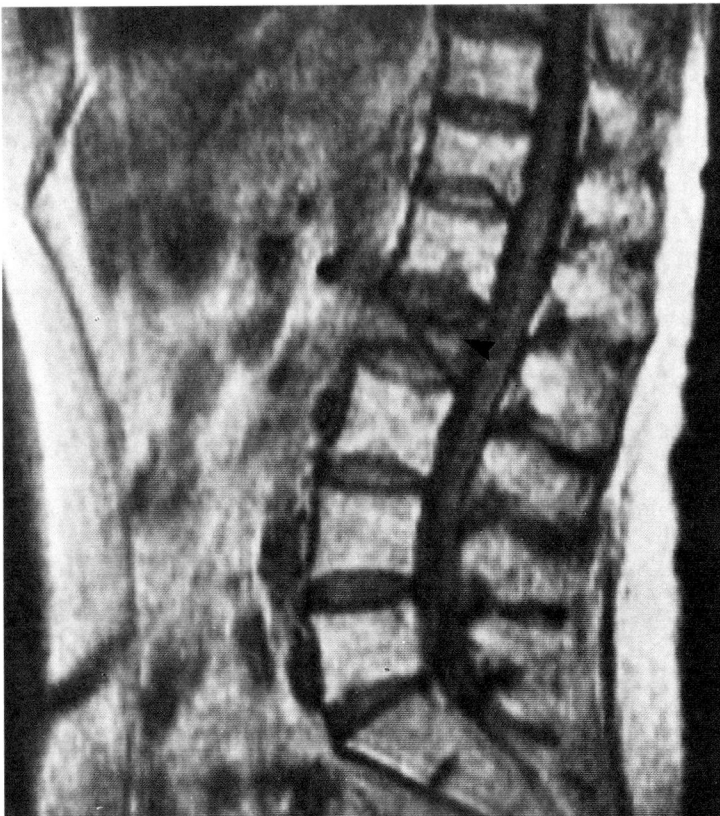

Figure 2.17

Pathologic compression fracture of L2 due to metastatic adenocarcinoma. Sagittal T_1-weighted image (SE 500/28) discloses loss of high-signal-intensity marrow (arrowhead). (Compare with Figure 2.13.)

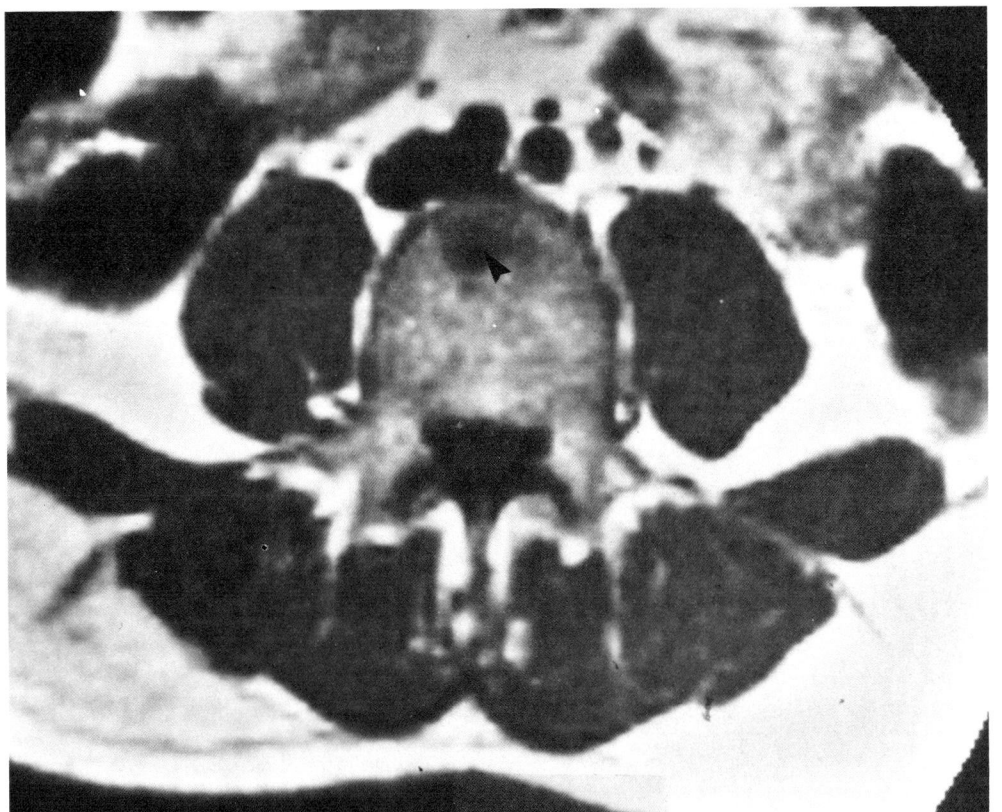

Figure 2.18
Vertebral hemangioma.
(**a**) Axial T_1-weighted image (SE 600/30) reveals focal low-intensity region in anterior part of vertebral body (arrowhead). (**b**) Sagittal scan with the same pulse sequence shows lesion (arrowhead) located anteriorly in vertebra. (**c**) Axial T_2-weighted image (SE 2000/90) shows high-signal intensity consistent with hemangioma. (**d**) Sagittal T_2-weighted image (SE 2000/90) confirms findings. (*Courtesy of David Schale*)

a

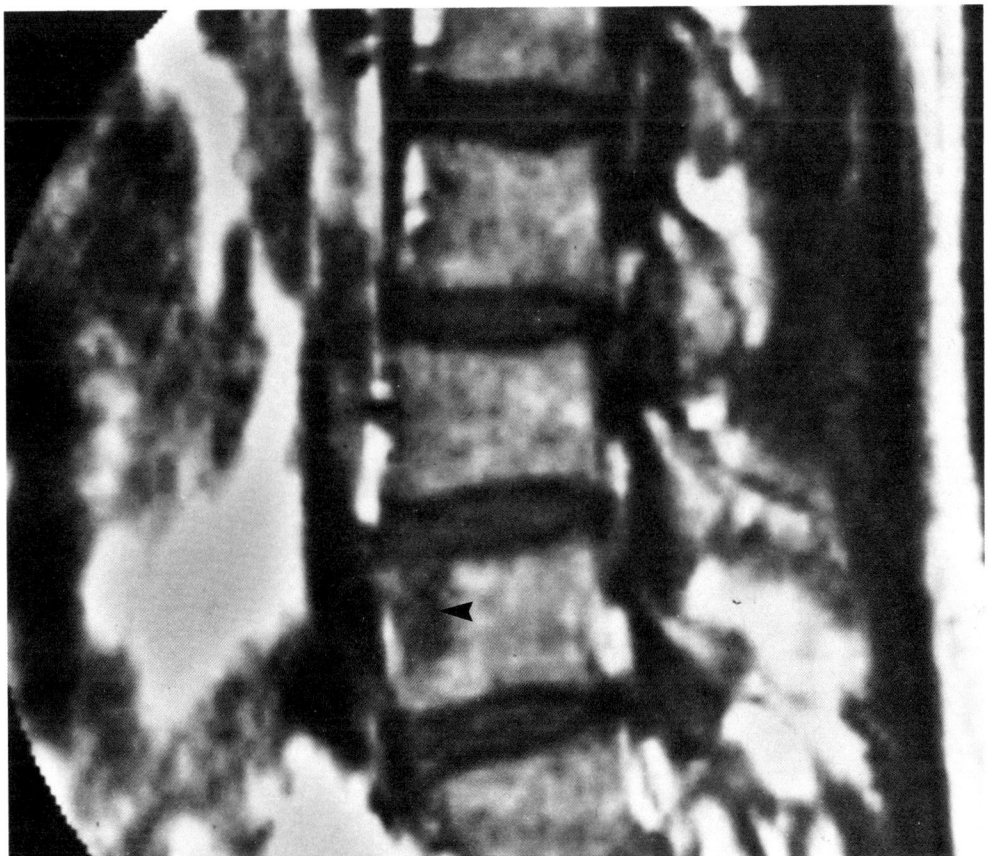

b

54 MRI atlas of the musculoskeletal system

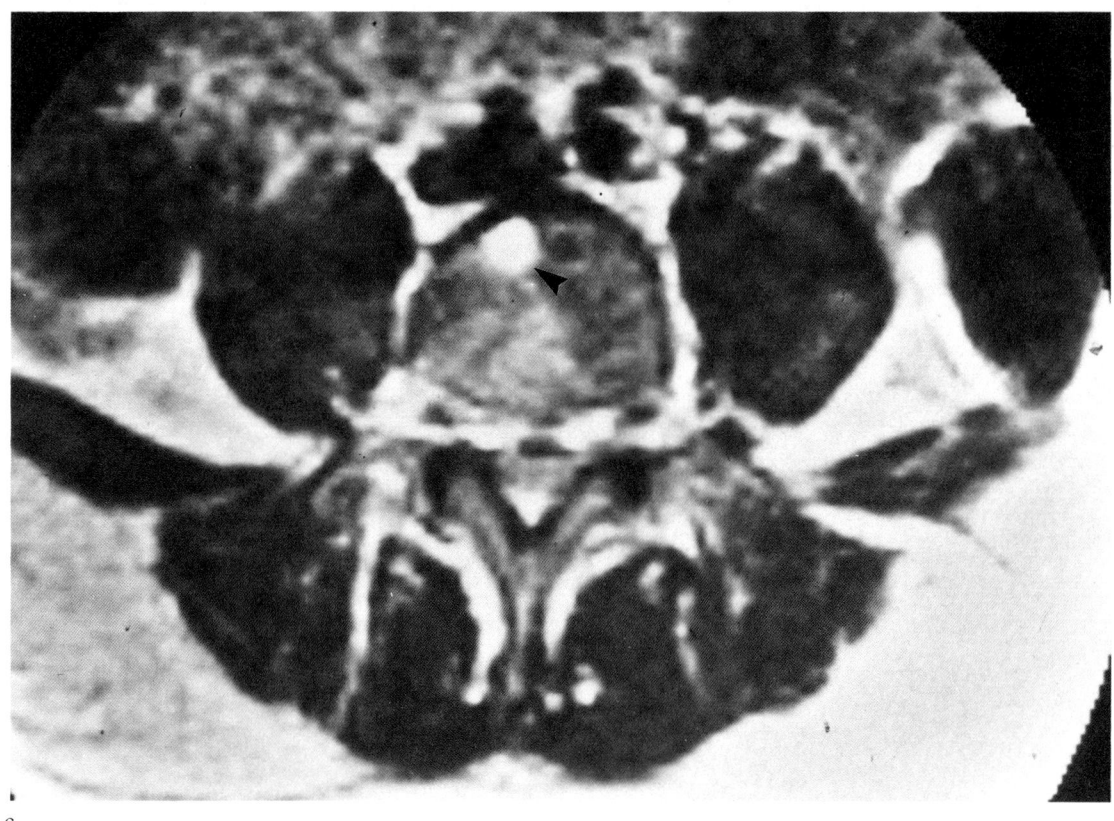

c

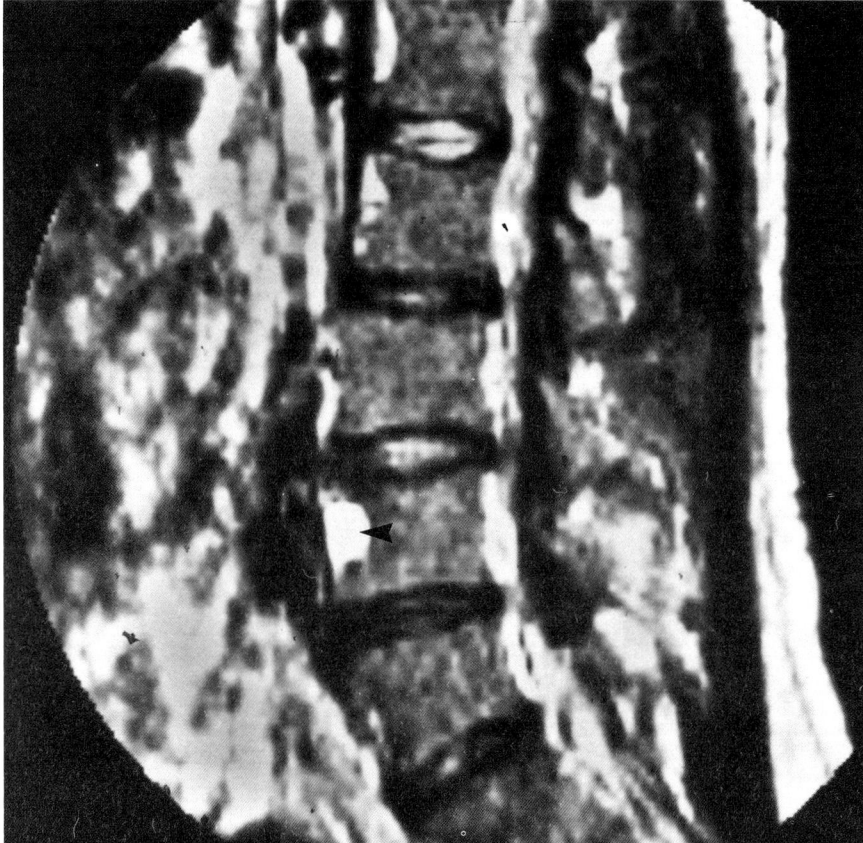

d

Figure 2.18 continued

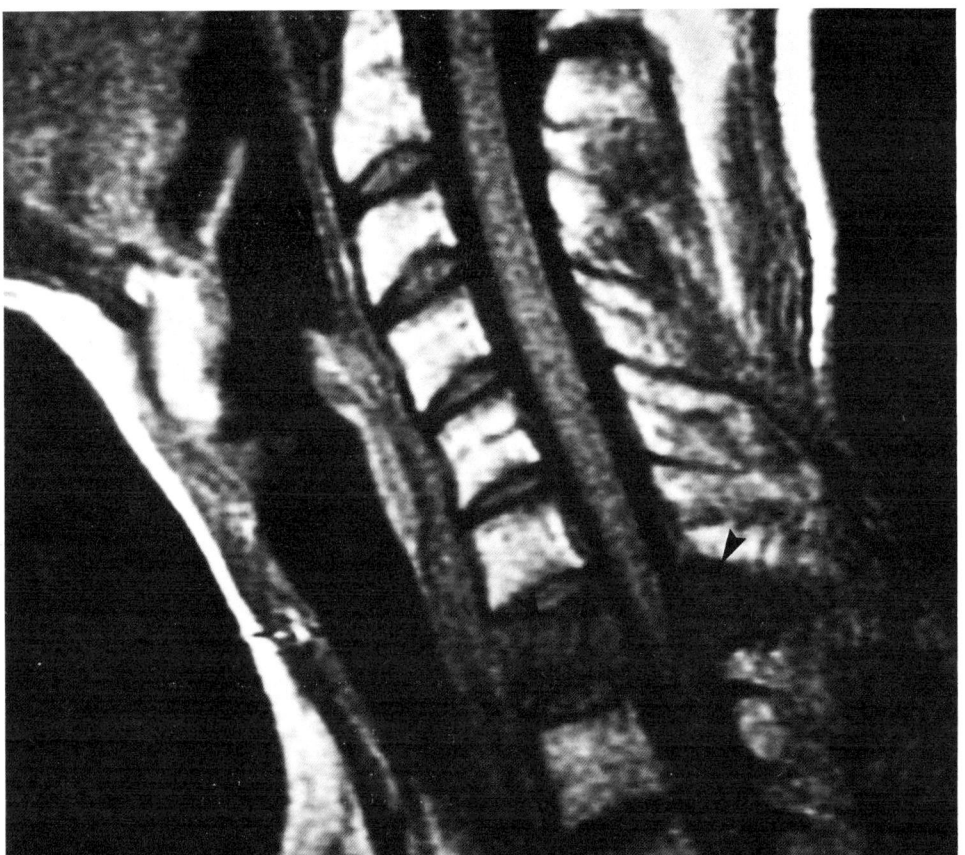
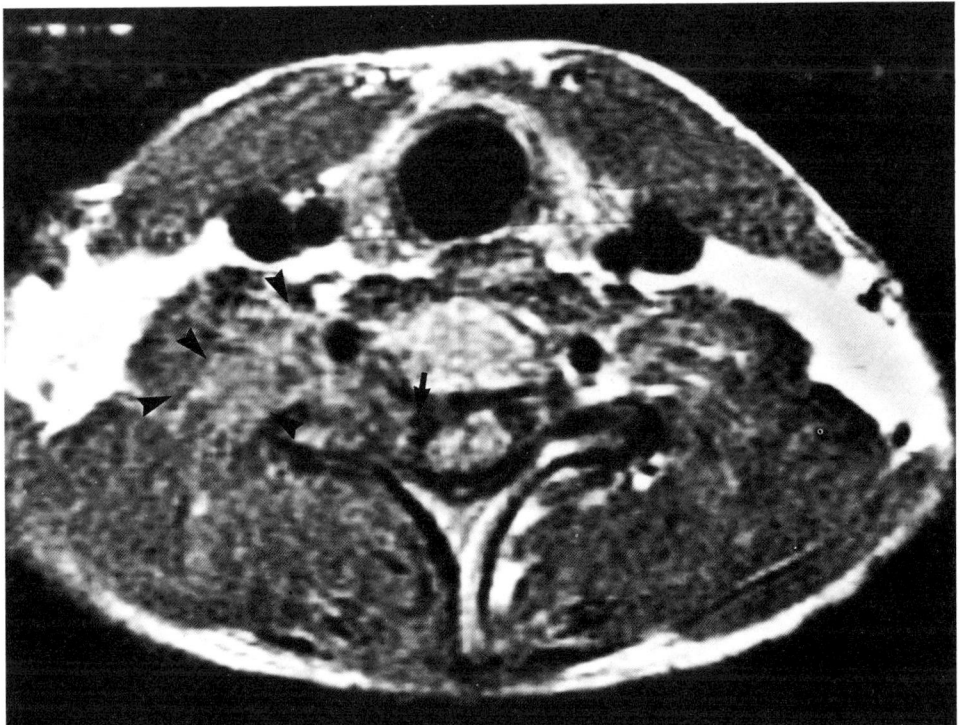

Figure 2.19

Ewing's sarcoma of C7. (**a**) Sagittal T_1-weighted image (SE 500/30) shows focal replacement of high-signal intensity marrow of C7 body and spinous process (arrowhead). (**b**) Axial T_1-weighted section (SE 800/30) discloses significant soft-tissue extension (arrowhead), with displacement of ipsilateral nerve roots (arrow).

a

b

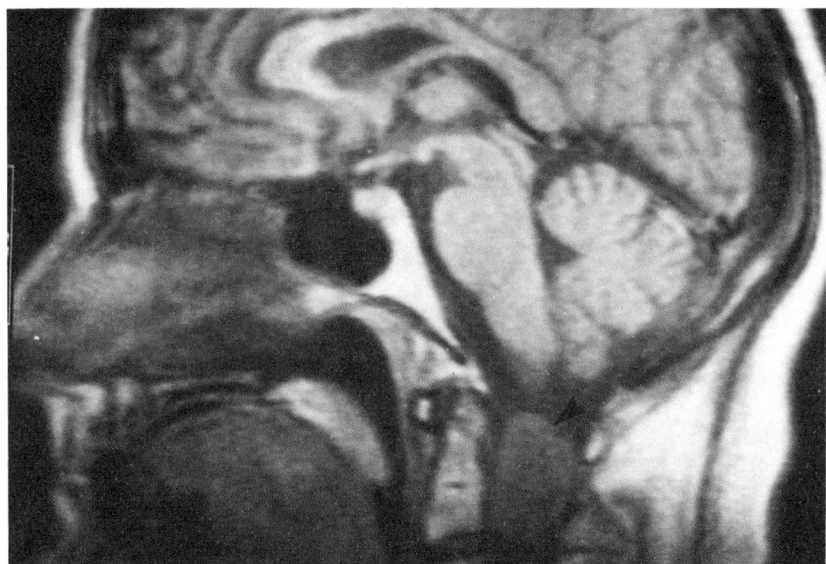

Figure 2.20

Cervical meningioma. Sagittal T_1-weighted image (SE 500/28) through craniocervical junction shows homogeneous mass (arrowhead) at C2 level. Axial images suggested that mass was intradural and extramedullary. At surgery, meningioma was found.

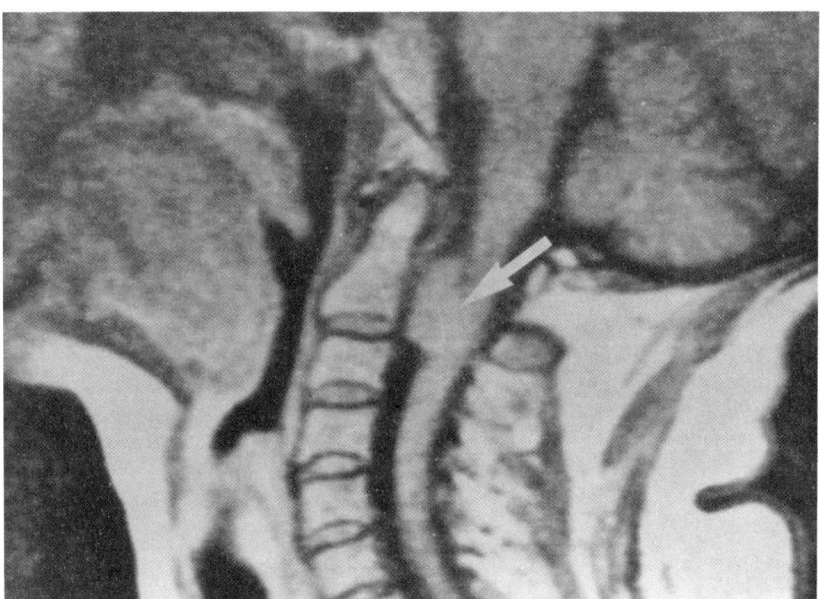

Figure 2.21

Neurofibroma. Sagittal T_1-weighted image (SE 500/28) shows focal homogeneous intradural extramedullary soft-tissue mass (arrow) similar in appearance to previous case. At surgery, however, this represented neurofibroma.

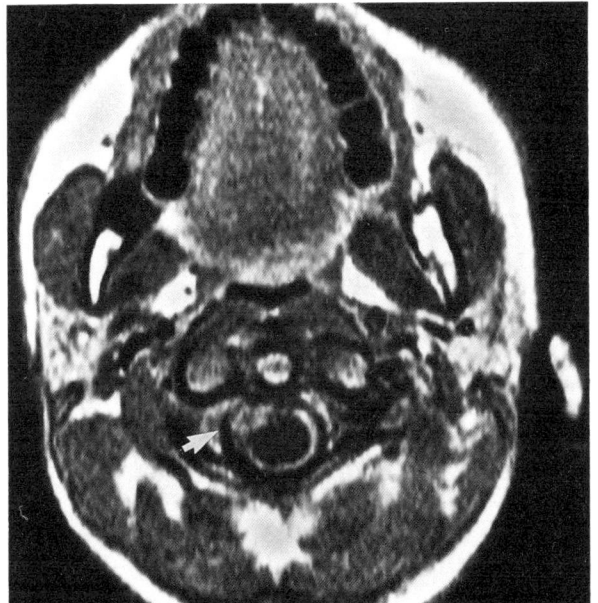
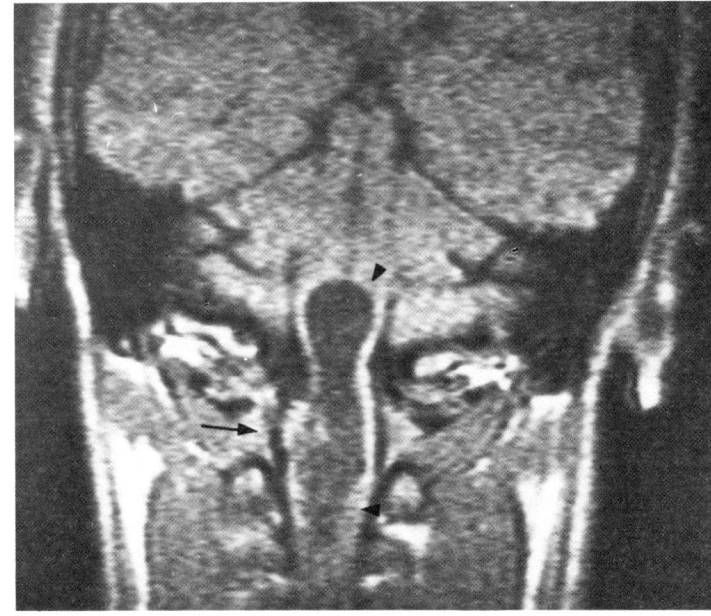
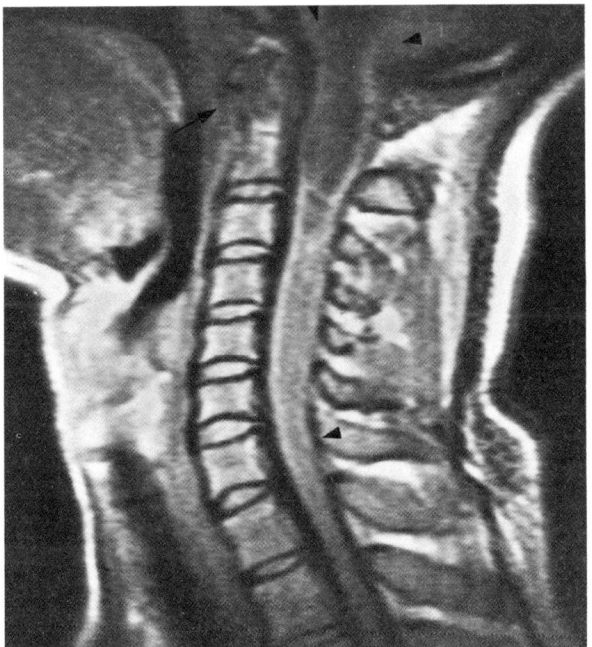

Figure 2.22

Cystic hemangioblastoma. (**a**) Axial image (SE 500/30) reveals cystic dilatation of cervical cord. Note small mural nodule (arrow). (**b**) Coronal image (SE 500/30) of same region shows that cystic component extends to high medulla (arrowhead). Mural nodule again apparent (arrow). (**c**) Sagittal image (SE 500/30) shows slight mass effect due to full extent of tumor extending below level of cyst to lower cervical cord (arrowhead). Soft-tissue change anterior to the mass extends to odontoid (arrow).

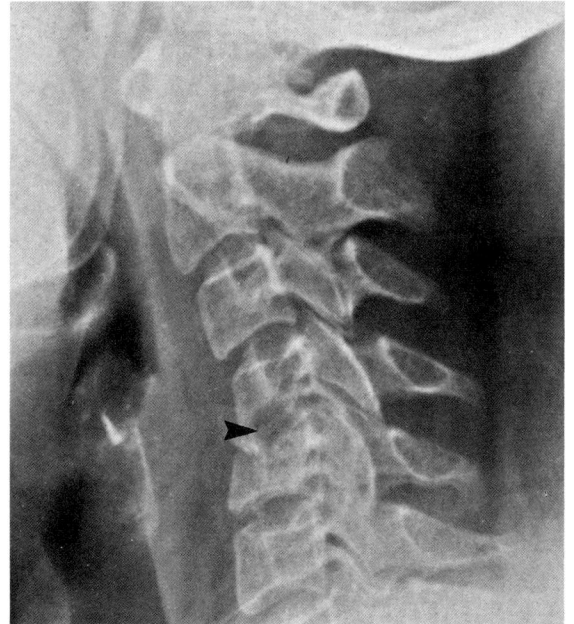
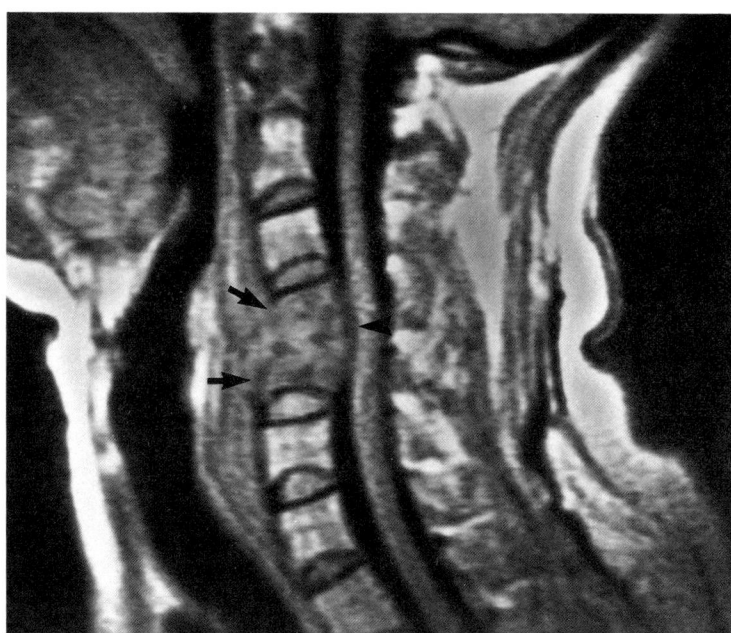
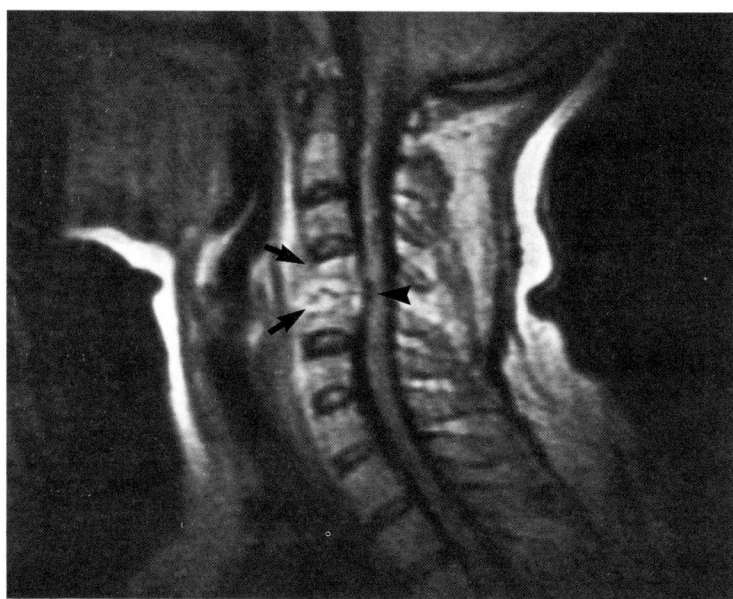

Figure 2.23

Osteomyelitis. (**a**) Lateral radiograph of cervical spine discloses destruction of disk and vertebral end plates at C4–5 level (arrowhead). (**b**) T_1-weighted image (SE 500/28) shows replacement of normal high-signal-intensity marrow and obliteration of cortical margins at C4–5 level (arrow). The process protrudes into the spinal cord and displaces the cord (arrowhead). (**c**) T_2-weighted image (SE 2000/84) shows similar findings, except that disk space is now high in signal intensity, consistent with osteomyelitis.

3 The temporomandibular joint

Cynthia S Sherry and
Steven E Harms

Complaints related to the temporomandibular joint (TMJ) may be found in a significant proportion of the population. Approximately 28 per cent of Americans have some form of TMJ abnormality,[1-3] among which 90 per cent are young women.[4] An interesting correlation with emotional stress, bruxism (an unconscious grinding of the teeth) and other chronic pain syndromes has been described.[5]

The cause of TMJ disease in the majority of cases is unknown. A history of previous trauma may be elicited in about 25 per cent of cases, with orthodontic procedures being the most commonly cited form of trauma.[5,6] An anatomical defect in the joint may predispose to a TMJ disorder. For example, a steep articular eminence and/or a hypermobile joint may result in abnormal joint motion and an increased frequency of TMJ symptoms.[7,8] Some arthritides and metabolic abnormalities may be implicated as rare causes of TMJ pathology.[7,8]

There are primarily five imaging modalities currently utilized in the evaluation of the TMJ. These include plain film radiography, conventional tomography, computed tomography, arthrography and magnetic resonance imaging (MRI).[9,10] Although each of the methods has its relative merits and shortcomings compared to the others in the ability to detect osseous and soft-tissue abnormalities of the TMJ, and in associated patient discomfort, hazard, and cost, MRI has been shown to be an effective imaging modality which combines many of the advantages of the other techniques with few of the disadvantages.[4,11-16] It requires no ionizing radiation, and is not known to produce any biological hazard to young females, the population which most commonly requires TMJ evaluation. Any discomforts related to MRI are limited to claustrophobia, which occurs in only a small percentage of patients, and the necessity of remaining motionless throughout the examination of a painful joint. The major advantages of MRI include improved visualization of the articular disk, retrodiscal laminae, surrounding joint space and adjacent musculature. Bony abnormalities of the mandibular condyle, articular fossa, and temporal bone eminence are also demonstrated. In the near future, cine MRI may provide additional information regarding joint motion and dynamics. The high cost of an MRI examination is perhaps its major limitation.

Technical considerations

The TMJ images displayed in the following pages were obtained at a magnetic field strength of 0.6 tesla. Although higher field strengths may allow an improved signal-to-noise ratio (S/N), there are factors other than field strength that contribute significantly to the quality of the MR image, and that also must be optimized to obtain the greatest possible resolution. For example, the use of surface coils is mandatory.[17] Since the S/N is inversely related to the diameter of the coil, the smallest possible coil which covers the area of interest yields the best image. Moreover, because the TMJ is a superficial structure, the signal detection capability of a small coil will not be hampered by thick overlying tissue. Although a surface coil that permits a bilateral examination greatly reduces the time required for imaging by allowing simultaneous examination of both joints, the resultant increase in the size of the coil also results in a reduced S/N. A surface coil capable of quadrature detection could overcome some of this reduction by improving the S/N by a factor equal to the square root of 2. Alternatively, a surface coil design in which each side is examined by a separate receiver system would have the same S/N as a unilateral coil.

The constituents of the normal TMJ generally have moderate T_1 values and short T_2 values. In the usual situation, a pulse sequence which enhances the inherent soft-tissue T_1-relaxation differences is utilized. A short TR/short TE sequence will maximize the inherent tissue contrast by maximizing the T_1-weighting and enhancing the relative signal of fat compared to muscle and fibrocartilaginous disk. If the short TR does not provide an adequate S/N, the number of excitations may be increased. Lengthening the TR will result in diminished contrast, whereas increasing the number of excitations will maintain contrast, improve the S/N (by a factor equal to the square root of the number of excitations), and reduce motion artifact.

When joint infection, inflammation or neoplasm is suspected, T_2-weighted images will provide additional information. T_2-weighting increases the signal intensity due to edema and/or fluid relative to fat and muscle. Indeed, a long TR/long TE pulse sequence produces the best contrast between fat and tissues with increased water content.

In the future, fast sequences which employ gradient-refocused echoes and partial flip angles may be utilized to obtain images with adequate tissue contrast despite a reduced scan time.[18-20] In addition, multiple images rapidly acquired during opening and closing the mouth may be played back in a cine mode, permitting noninvasive evaluation of joint dynamics and function.

The most commonly utilized acquisition method is the two-dimensional (2D) multislice Fourier transformation technique (Figure 3.1). This method is somewhat limited, however, in its application to TMJ imaging because optimal evaluation of the TMJ requires a large number of thin slice, short TR, heavily T_1-weighted images, and the short TR limits the number of slices which may be obtained. Since the 2D technique defines the slice by a selective excitation, the thinness of the slice is limited by the bandwidth narrowness of the RF pulse and the strength of the slice-selecting gradient. Furthermore, the shape of the slice excitation influences the magnitude of the gap between slices. Thus square-shaped slice excitation RF pulses are desirable in order to minimize the interslice gaps. Finally, the narrowness of the RF bandwidth (and therefore the slice thinness) is limited by the desire to utilize a short TE pulse sequence.

The three-dimensional (3D) technique offers certain advantages over the 2D method (Figure 3.2). In selective 3D imaging, a thick slab is selected and very thin slices are defined by an additional phase-encoding gradient. Because the thinness of the slice is not limited by the narrowness of the bandwidth, wider bandwidths may be utilized, allowing shorter TE sequences with no interslice gaps. Finally as the number of slices is not dictated by the TR, large numbers of highly T_1-weighted images may be obtained. Unfortunately, because of the additional phase-encoding gradient the images obtained by the 3D technique may be severely degraded by motion artifacts. A 3D multislab technique permits both joints to be imaged during the time required to examine a single joint (Figure 3.3).

Anatomy, physiology, and function

The TMJ is composed of the mandibular condyle, temporal bone eminence, and articular disk. Since the disk functions as a separate structure, the TMJ is considered a compound joint. The entire joint is contained within the articular capsular ligament. The condyle resides within the temporal fossa, posterosuperior to the eminence. The articular disk is interposed between the condyle and the temporal bone, and is tightly adherent to the condyle by the highly vascularized and innervated collateral ligaments which attach to the medial and lateral aspects of the condyle (Figure 3.4a–d).

The articular disk is a fibrocartilaginous structure which has a poor neurovascular supply. It has a bowtie-like configuration and is morphologically divided into three segments: thick anterior and posterior bands and a thin intermediate zone. The intermediate zone is situated at the stress-bearing portion of the temporomandibular interface. In the closed mouth position, the posterior band is located at 12 o'clock with respect to the mandibular condyle.

The disk is attached anteriorly to the superior belly of the lateral pterygoid muscle. Posteriorly, the disk is attached by two retrodiscal laminae to the joint capsule. The retrodiscal laminae are richly innervated and vascularized, and provide proprioception. The superior retrodiscal lamina is elastic and opposes the normal tension of the lateral pterygoid muscle. The inferior lamina is composed of collagen and provides passive resistance to movement of the disk. The two retrodiscal laminae are also referred to as the bilaminar zone.

The motion of the TMJ is complex and is composed of two concurrent actions. With normal mouth opening, the articular disk rotates posteriorly on the condyle, constituting one action. Simultaneously, the condyle–disk complex slides anteroinferiorly, translating to a final position inferior to the eminence (Figure 3.5). The process is reversed for mouth closing.

Pathology

The most common abnormalities of the TMJ which require evaluation are internal derangements. Most commonly, the articular disk becomes anteriorly displaced. In mild derangements, the anteriorly displaced disk reduces upon opening the mouth, and a 'click' may be heard when the disk snaps back into the physiologic position. In more severe derangements, the displacement of the disk persists throughout the joint motion. Frequently, the displaced disk acts as a mechanical obstruction to normal condylar translation, and a 'locked' condyle is observed. Examples of anterior displacements with and without reduction are provided in Figures 3.6 and 3.7.

Several anatomic features identifiable by MRI which are known to be associated with the development of internal derangement include a steep articular eminence (Figure 3.8) and a hypermobile joint (Figure 3.9). The predisposition to internal derangements is thought to be related to the abnormal joint motions which result from these anatomic disorders. The development of internal derangement may be presaged by the identification of a thin posterior band (Figure 3.10) or reduced translation (Figure 3.11).

Chronic internal derangements are manifested by osteophyte formation (Figure 3.12), adhesions (Figure 3.13) and endstage ankylosis.

The disk may also be displaced medially (Figure 3.14), laterally (Figure 3.15), or rarely, posteriorly (Figure 3.16).

Enlargement of the retrodiscal tissues may occur in response to either a chronic or an acute process, the distinction of which is an important factor in determining treatment. T_2-weighted images can frequently provide confirmation of the acuteness of a lesion by depicting increased joint fluid and/or edema of the adjacent musculature. Figure 3.17 depicts acute retrodiscal thickening in response to recent arthroscopic lysis of adhesions; this procedure is a common cause of edema and increased joint fluid, which are seen best on T_2-weighted images. Figure 3.18 is an example of a chronic process resulting in retrodiscal hyperplasia in which the T_2-weighted images fail to reveal the abnormal high signal noted in the previous example.

The treatment of TMJ abnormalities includes relatively conservative measures, such as splint therapy, and more aggressive techniques including surgery. If the articular disk is severely distorted or fragmented, a prosthetic disk may be required.[21] An example of a Proplast implant is illustrated in Figure 3.19. Proplast may incite an exuberant giant-cell foreign body reaction (Figure 3.20). The giant-cell reaction may be so destructive as to erode into the middle cranial fossa and temporal lobe (Figure 3.21). A dermal graft is a recent therapeutic alternative to the Proplast implant, and may prove more durable and less toxic (Figure 3.22).

Mandibular trauma is frequently associated with a condylar fracture. Evaluation by MRI provides information regarding the integrity of the articular disk, which may also be injured. Figure 3.23a is an example of a condylar fracture in which the articular disk is intact, while Figure 3.23b demonstrates a condylar dislocation.

Several systemic diseases may rarely involve the TMJ, including rheumatoid arthritis, gout, calcium pyrophosphate dihydrate deposition disease, ankylosing spondylitis, and psoriatic arthritis (Figure 3.24).[8,9]

Conclusions

MRI is a valuable adjunctive or primary imaging modality in the evaluation of TMJ abnormalities. MRI, if properly performed, yields high-quality images of the osseous and soft-tissue structures, furthering the investigation of internal derangements and other disorders.

References

1 GURALNIC W, KABAN LB, MERRIL RG, Temporomandibular joint afflictions, *N Engl J Med* (1978) **229**(3):123–9.

2 SILCHER H, *Oral Anatomy*, (CV Mosby: St Louis 1949).

3 SOLBERG WK, WOO MW, HOUSTON JB, Prevalence of mandibular dysfunction in young adults, *J Am Dent Assoc* (1979) **98**:25–34.

4 HARMS SE, WILK RW, Magnetic resonance of the temporomandibular joint, *Radiographics* (1987) **7**(3):521–42.

5 HELMS CA, RICHARDSON ML, MOON KL et al, Nuclear magnetic resonance of the temporomandibular joint: preliminary observations, *J Craniomand Pract* (1984) **2**(3):219–24.

6 KATZBERG RW, DOLWICK MF, BALES DJ et al, Arthrotomography of the TMJ: new technique and preliminary observations, *AJR* (1979) **132**:949–55.

7 BELL WE, *Clinical Management of Temporomandibular Joint Disorders*, (Year Book: Chicago 1982).

8 WILK RM, HARMS SE, WOLFORD LM, Magnetic resonance imaging of the temporomandibular joint with a surface coil, *J Oral Maxillofac Surg* (1986) **44**:935–43.

9 WESTESSON P, Double-contrast arthrotomography of the temporomandibular joint, *J Oral Maxillofac Surg* (1980) **41**:163–72.

10 HELMS CA, VOGLER JB III, MORRISH RB JR et al, Temporomandibular joint internal derangements: CT diagnosis, *Radiology* (1984) **152**:459–62.

11 CHILES DG, WILK RM, HARMS SE, MRI in the diagnosis of temporomandibular joint disorders with a report of two cases, *J Craniomand Pract* (1986) **4**(4):306–12.

12 FARRAR WB, MCCARTY WL JR, The TMJ dilemma, *J Atla Dent Assoc* (1979) **63**:19.

13 KAPLAN PA, Computed tomography vs. arthrography in the evaluation of the temporomandibular joint, *Radiology* (1984) **152**:825–7.

14 KATZBERG RW, KEITH DA, GURALNICK WC et al, Internal derangements and arthritis of the temporomandibular joint, *Radiology* (1983) **146**:107–12.

15 OKESON JP, *Fundamentals of Occlusion and Temporomandibular Disorders*, (CV Mosby: St Louis 1985).

16 WILK RM, HARMS SE, Multislab 3DFT magnetic resonance imaging in temporomandibular joint imaging. Radiology. In: *Proceedings of Sixth Annual Meeting of the Society of Magnetic Resonance in Medicine*, (Society of Magnetic Resonance in Medicine: New York 1987) 139.

17 HARMS SE, WILK RM, WOLFORD LM et al, The temporomandibular joint: magnetic resonance imaging using surface coils. *Radiology* (1985) **157**:133–6.

18 HASSE A, FRAHM J, MATTHAEI W et al, Rapid images and NMR movies. In: *Proceedings of Fourth Annual Meeting of the Society of Magnetic Resonance in Medicine*, (Society of Magnetic Resonance in Medicine: Berkeley, California 1985) 980:1.

19 FRAHM J, HASSE A, MATTHAEI D, Technical note. Rapid three-dimensional MR imaging using the FLASH technique, *J Comput Assist Tomogr* (1986) **10**:363–8.

20 HASSE A, FRAHM J, MATTHAEI D et al, MR imaging using stimulated echoes (STEAM), *Radiology* (1986) **160**:787–90.

21 TIMMIS DP, ARAGON SB, VAN SICKELS JE et al, Comparative study of alloplastic materials for temporomandibular joint disc replacement in rabbits, *J Oral Maxillofac Surg* (1986) **44**:541–54.

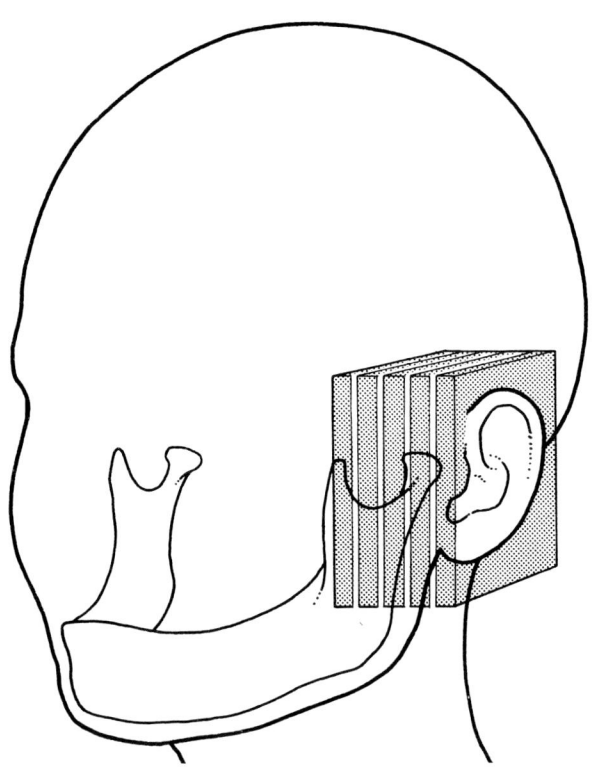

Figure 3.1
2D multislice acquisition. The magnitude of the slice-selecting gradient and the bandwidth of the RF pulse determine the slice thickness. The shape of the RF pulse influences the width of the interslice gap.

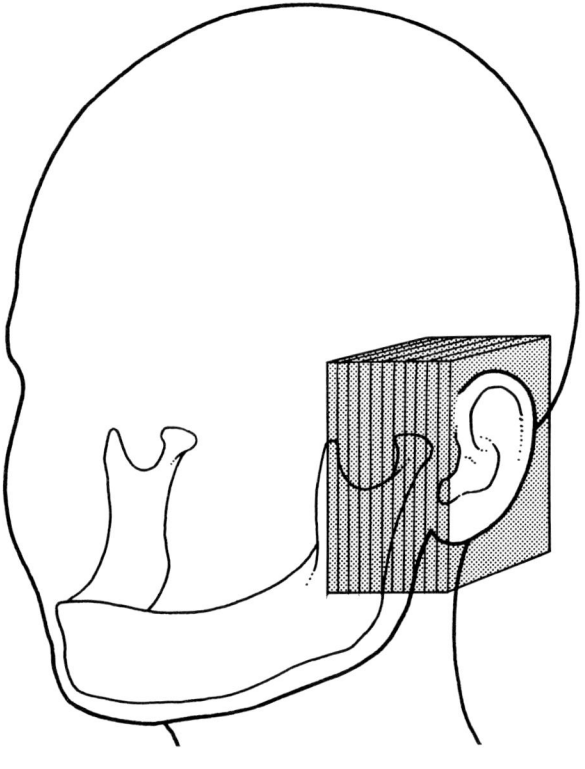

Figure 3.2
Selective 3D acquisition. A thick slab is excited and very thin slices can then be defined by an additional phase-encoding gradient. Because a narrow bandwidth RF pulse is not necessary, a shorter TE sequence may be utilized.

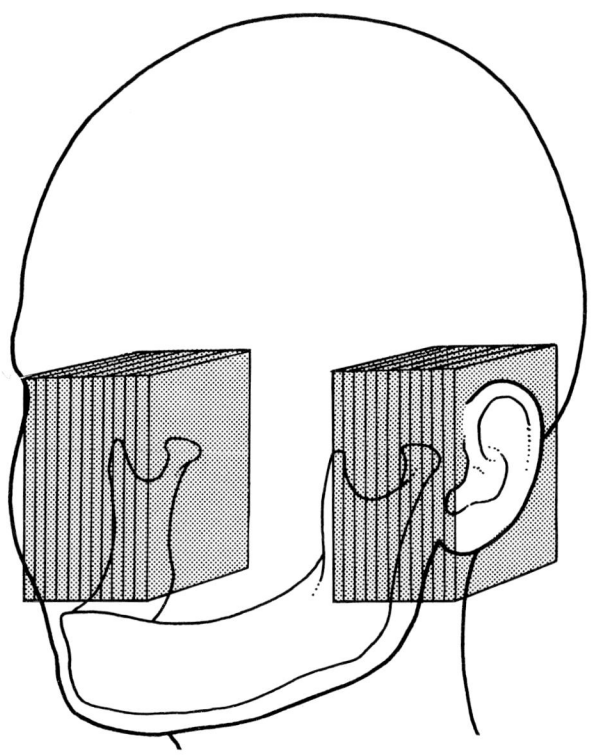

Figure 3.3
3D multislab acquisition. Two thick slabs are sequentially excited (in a manner similar to 2D multislice) and thin slices are defined by a phase-encoding gradient (as in the selective 3D acquisition).

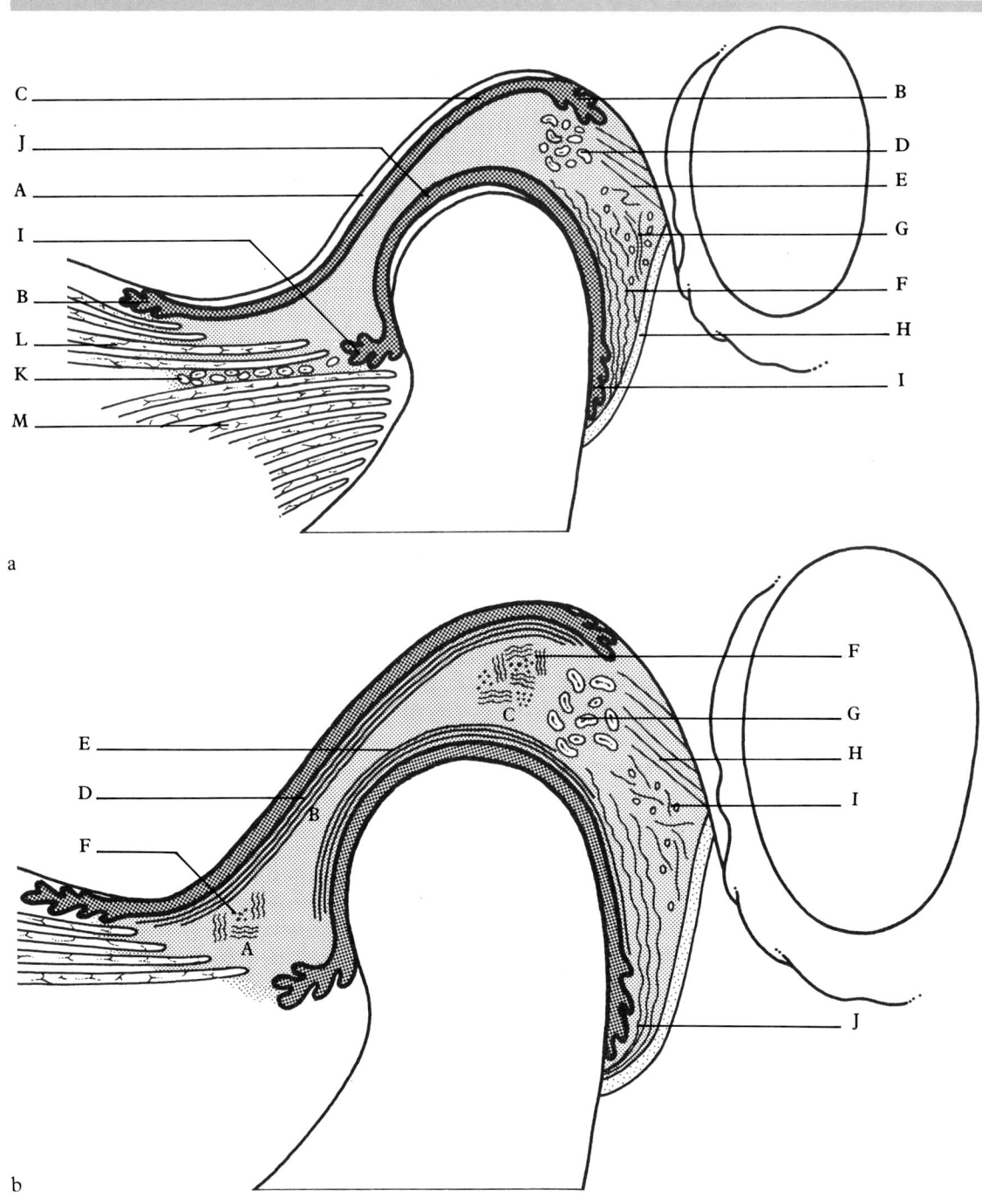

Figure 3.4

(a) TMJ anatomy. A sagittal view of the TMJ shows the following structures: (A) articular surface of eminence and fossa, (B) synovial membrane of superior joint cavity, (C) superior joint cavity, (D) vascular knee, (E) superior retrodiscal lamina, (F) inferior retrodiscal lamina of the bilaminar zone, (G) loose areolar connective tissue, (H) posterior capsule, (I) synovial membrane of inferior cavity, (J) articular surface of condyle, (K) blood vessels, (L) superior belly of the lateral pterygoid muscle, and (M) inferior belly of the lateral pterygoid muscle.
(b) Disk anatomy. A sagittal diagram of disk anatomy shows (A) anterior band, (B) thin or intermediate zone, (C) posterior band, (D) superior joint space, (E) inferior joint space, (F) collagen oriented in all 3 directions, (G) elastic tissue of the superior retrodiscal lamina of the bilaminar zone, (H) superior retrodiscal lamina, (I) loose areolar connective tissue of the bilaminar zone, and (J) inferior retrodiscal lamina of the bilaminar zone.
(c) MRI anatomy, closed. T_1-weighted (SE 250/25) sagittal image of the TMJ in the closed mouth position demonstrates the normal relationship of the mandibular condyle (c) and temporal bone eminence (te), a normal position of the posterior band (pb), anterior band (ab), and the thin zone (tz), and their relationship to the inferior belly of the lateral pterygoid muscle (ilp) and the external auditory meatus (eam). (d) MRI anatomy, open. T_1-weighted sagittal images (SE 250/25) in the open mouth position reveal normal translation of the condyle and a normal position of the articular disk.

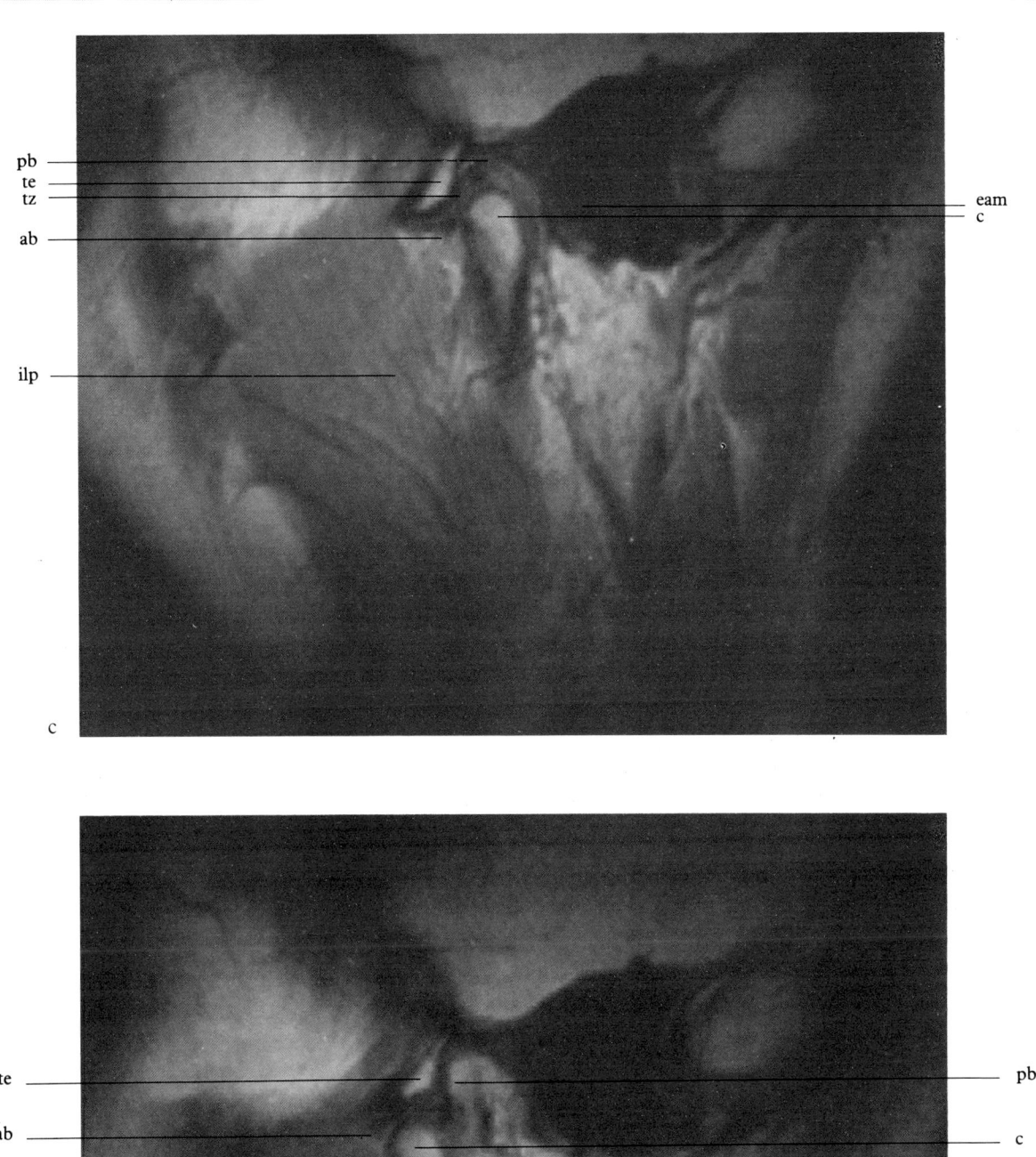

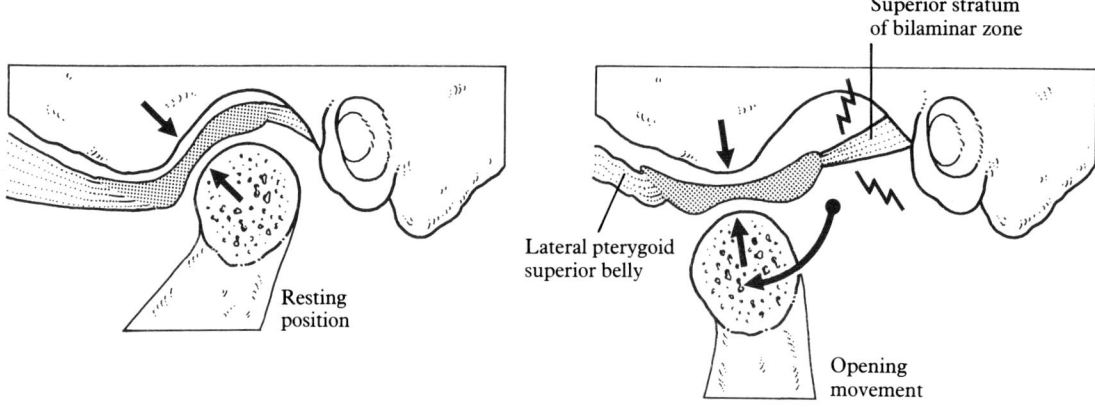
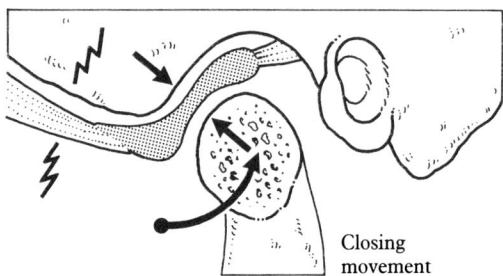

Figure 3.5

Normal TMJ motion. When proceeding from the resting to open mouth position, the disk rotates posteriorly on the condyle, and the condyle–disk complex translates anteroinferiorly.

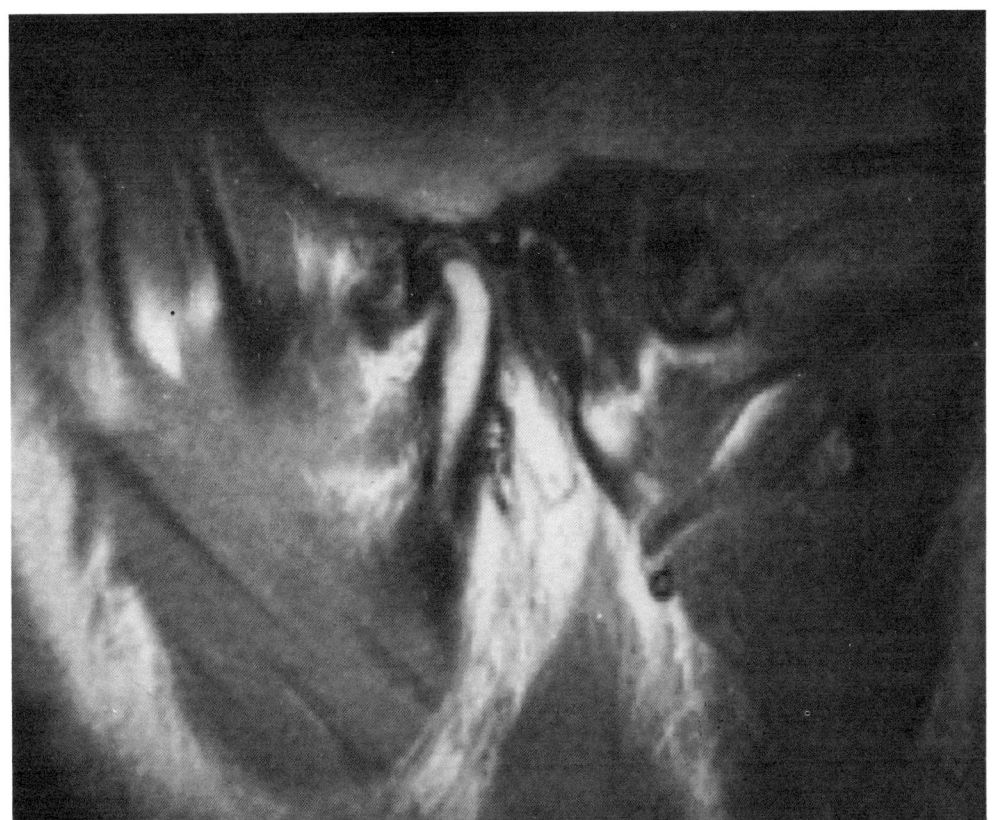

a

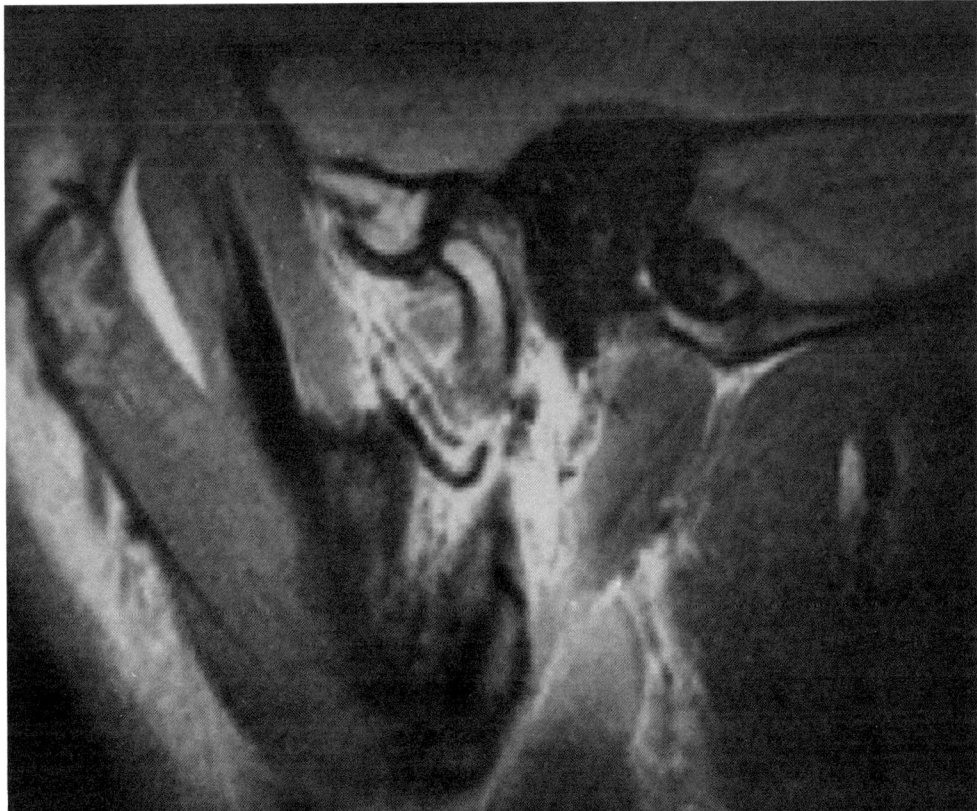

b

Figure 3.6
Anterior disk displacement with reduction. (**a**) The T_1-weighted image (SE 250/25) demonstrates anterior displacement of the articular disk in the closed mouth position. The posterior band lays at about 9 o'clock with respect to the condyle. (**b**) The T_1-weighted image (SE 250/25) shows reduction of the disk in the open mouth position.

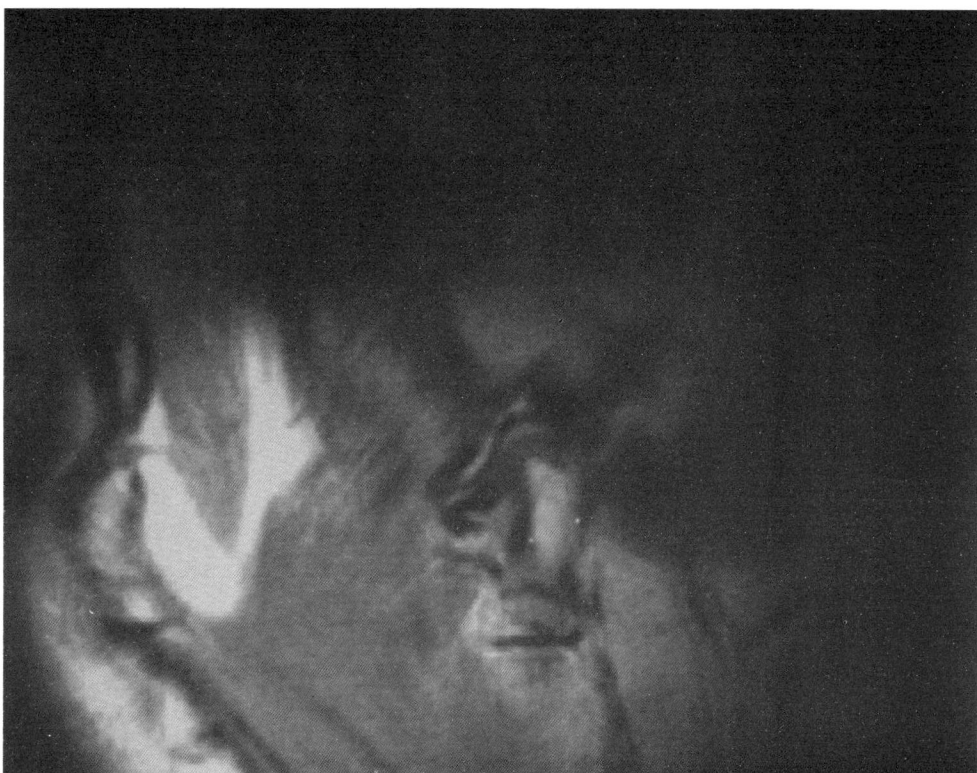

a

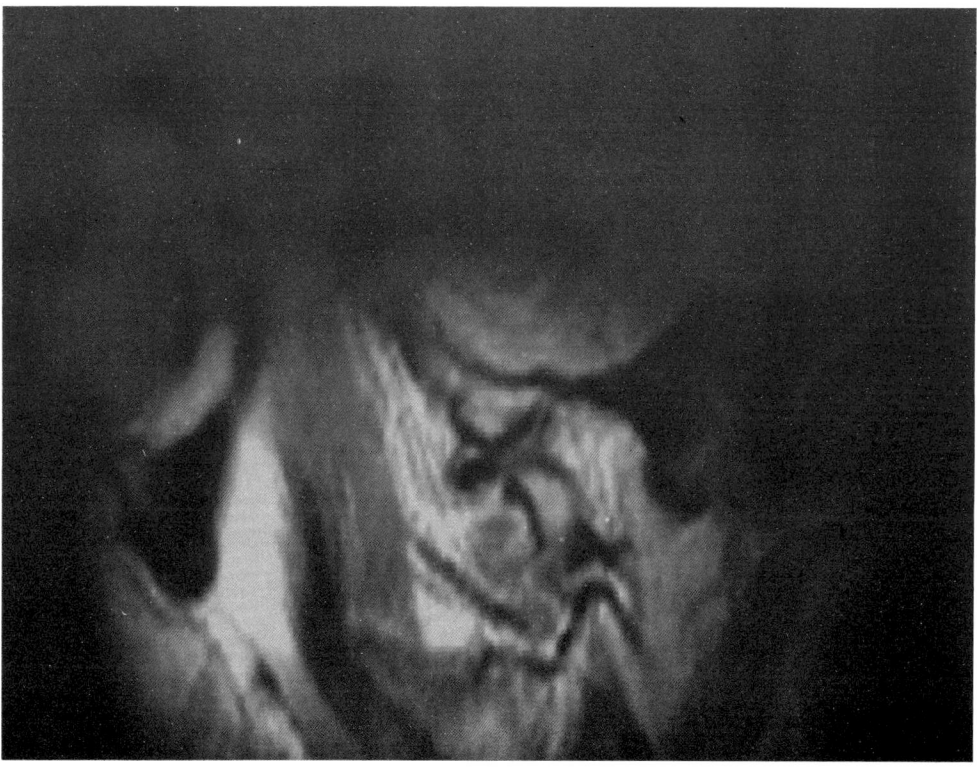

b

Figure 3.7

Anterior disk displacement without reduction. (a) The T_1-weighted image demonstrates anterior displacement of the disk in the closed mouth position with the posterior band at approximately 8 o'clock with respect to the condyle. Upon opening the mouth (b), persistent anterior displacement of the disk is seen. This represents a more severe internal derangement (a and b = SE 250/25).

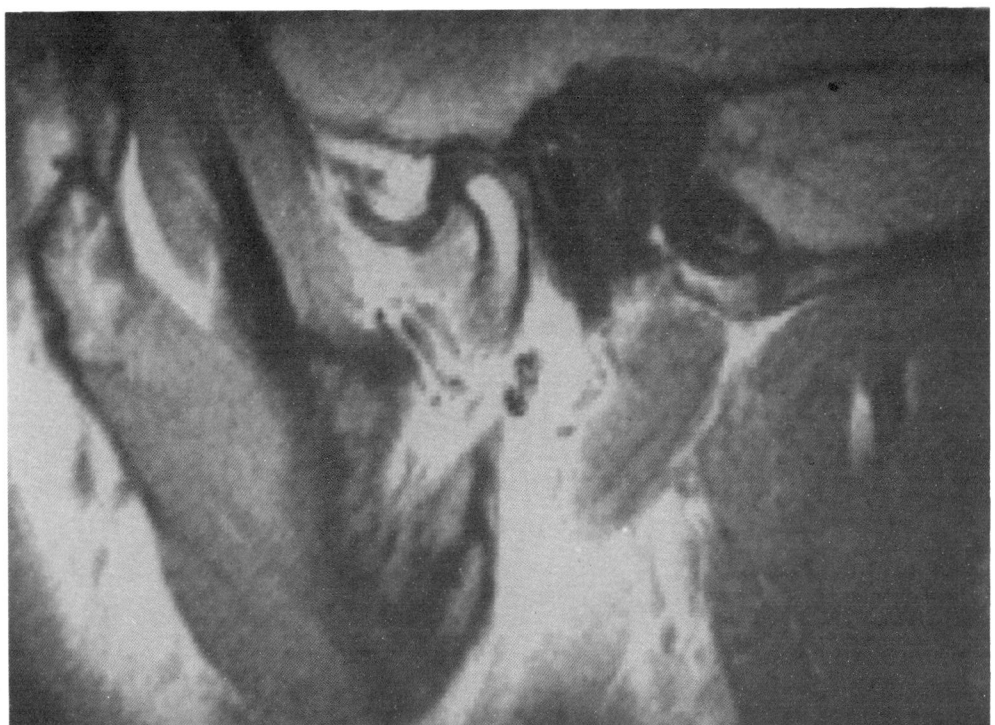

a

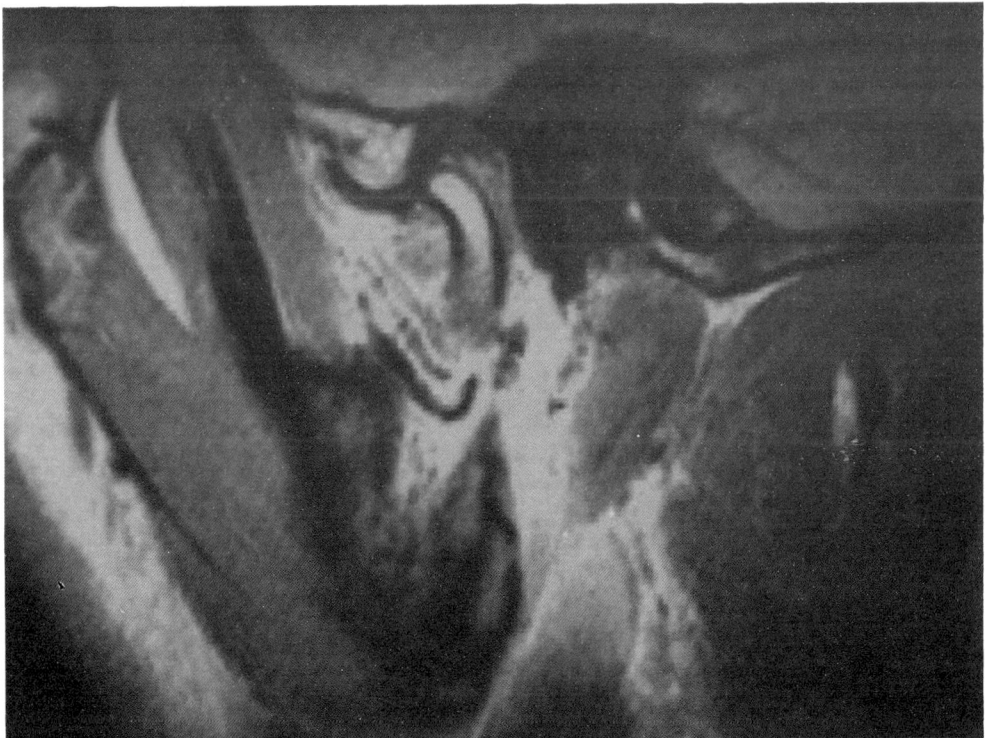

b

Figure 3.8
A steep articular eminence may lead to the development of internal derangement. The closed (**a**) and open (**b**) mouth T_1-weighted images (SE 250/25) demonstrate this anatomic variation.

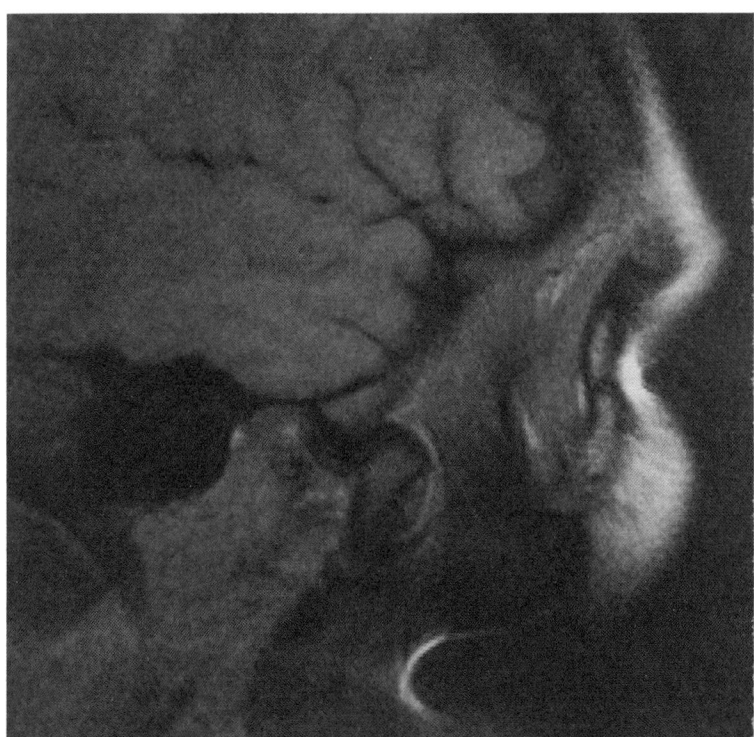

Figure 3.9
A hypermobile joint may predispose to internal derangement. Note the exaggerated translation of the mandibular condyle, with the final position anterior to the eminence (SE 250/25).

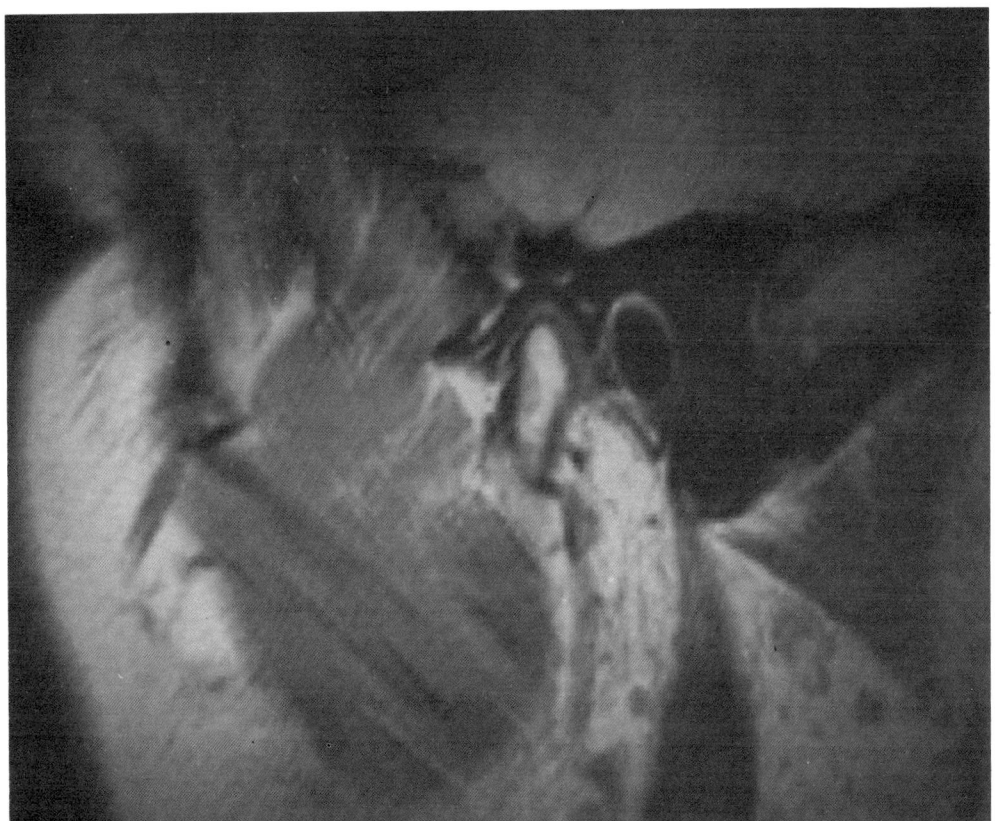

a

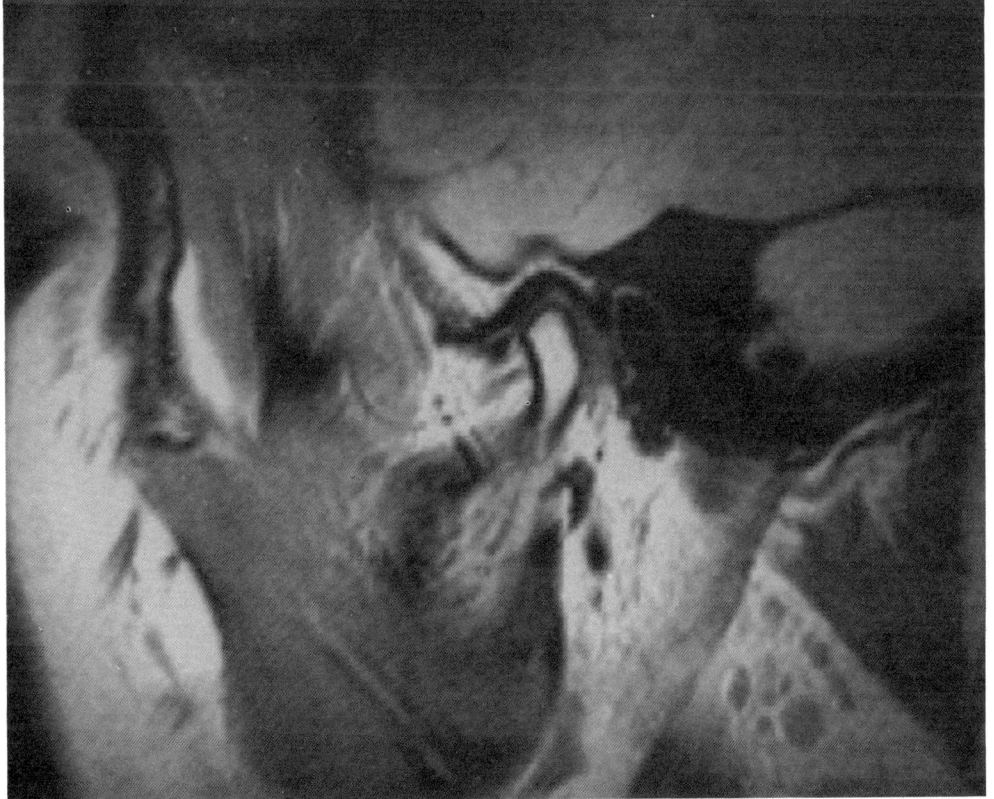

b

Figure 3.10

Thinning of the posterior band may be an early indication of TMJ abnormality. This finding is depicted in the T_1-weighted closed (**a**) and open (**b**) mouth views (SE 250/25).

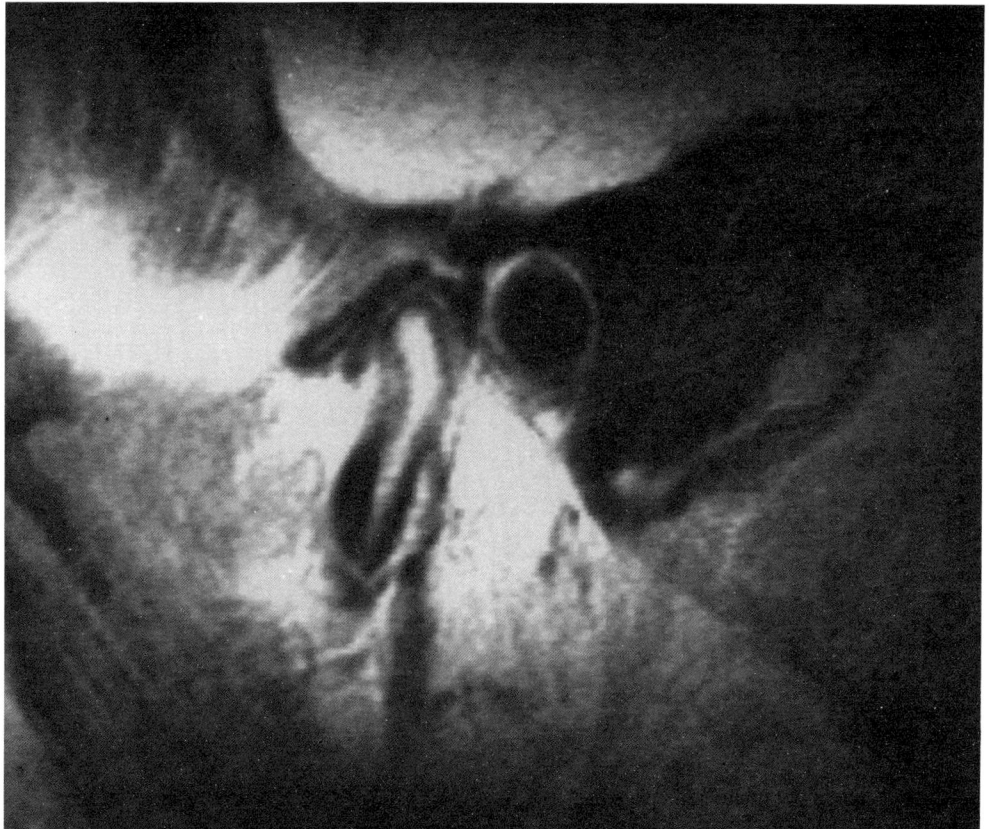

a

Figure 3.11
Reduced translation may be seen as an early indicator of internal derangement. T_1-weighted images in (**a**) open and (**b**) closed mouth positions demonstrate this finding (SE 250/25).

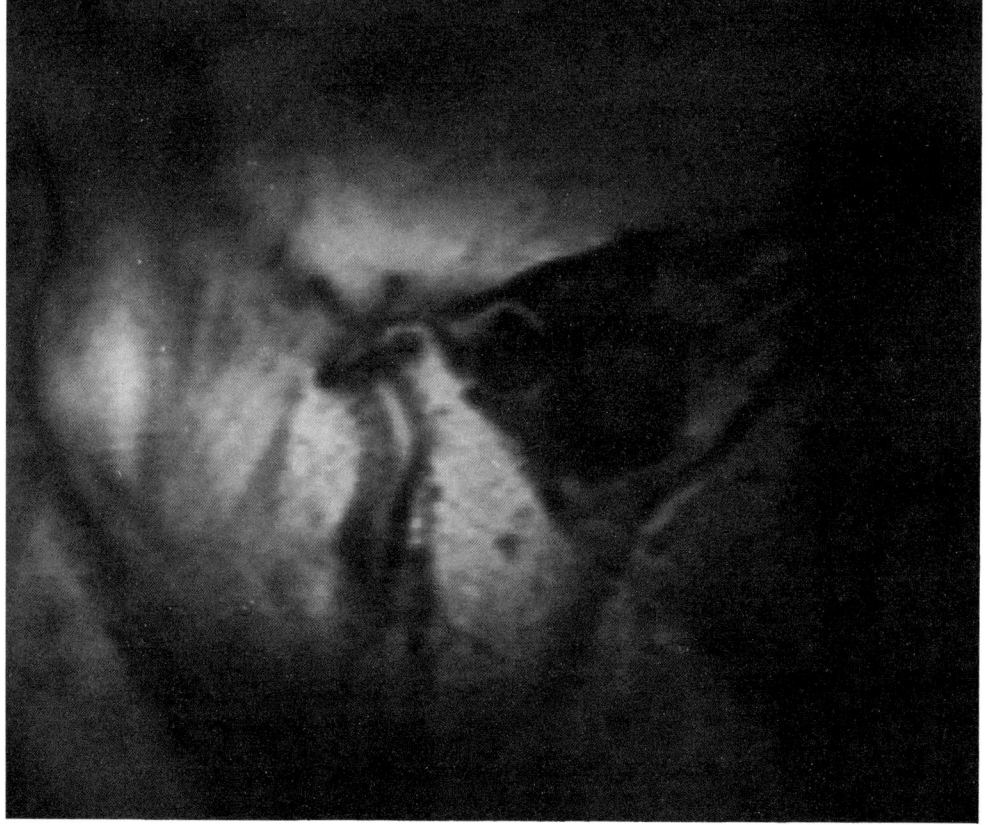

b

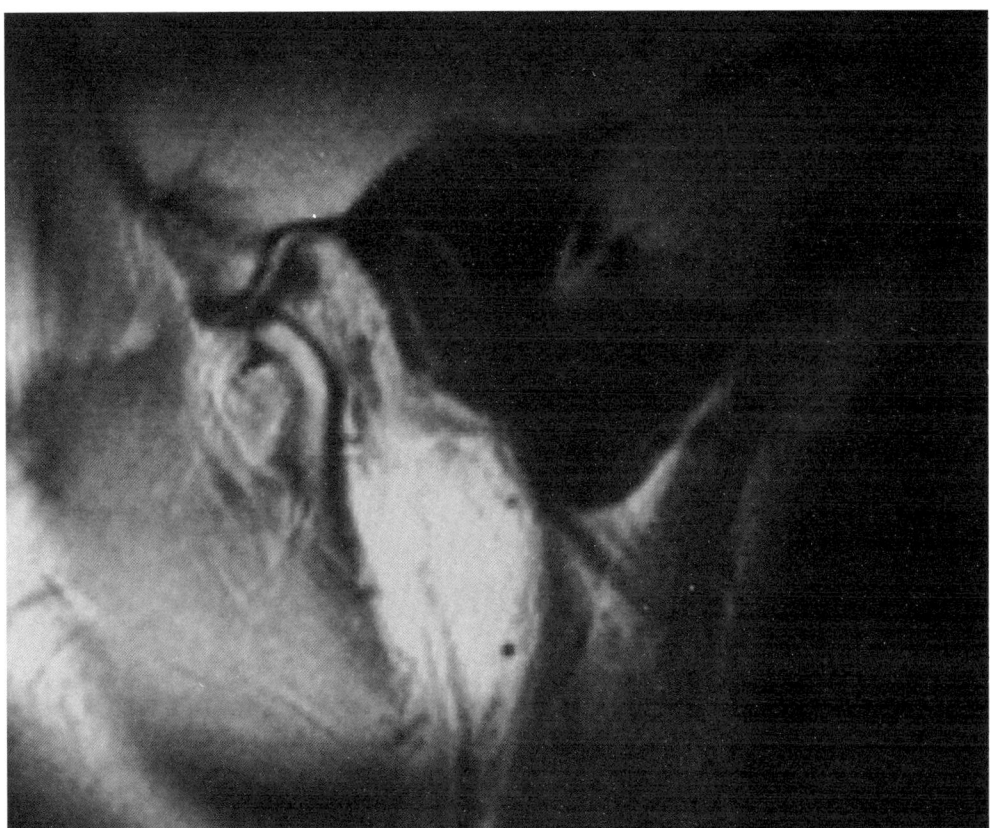

Figure 3.12

Osteophyte formation is evidence of longstanding abnormal stress or motion involving the TMJ. A condylar spur is shown in the open mouth T_1-weighted image (SE 250/25).

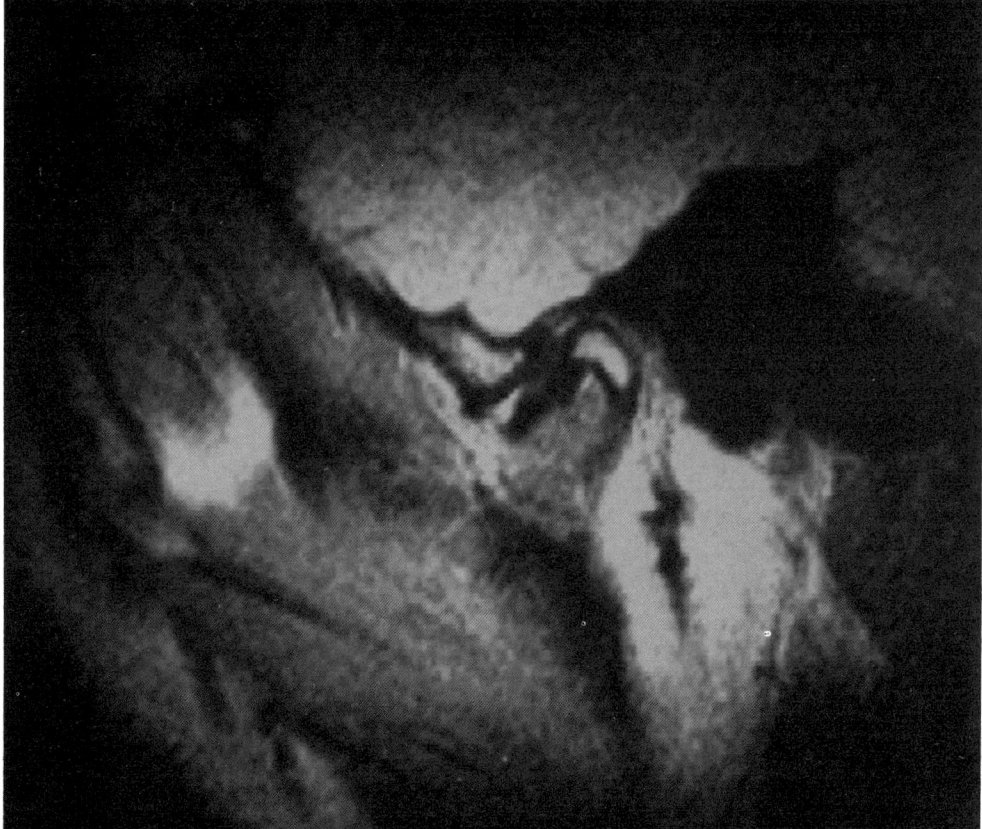

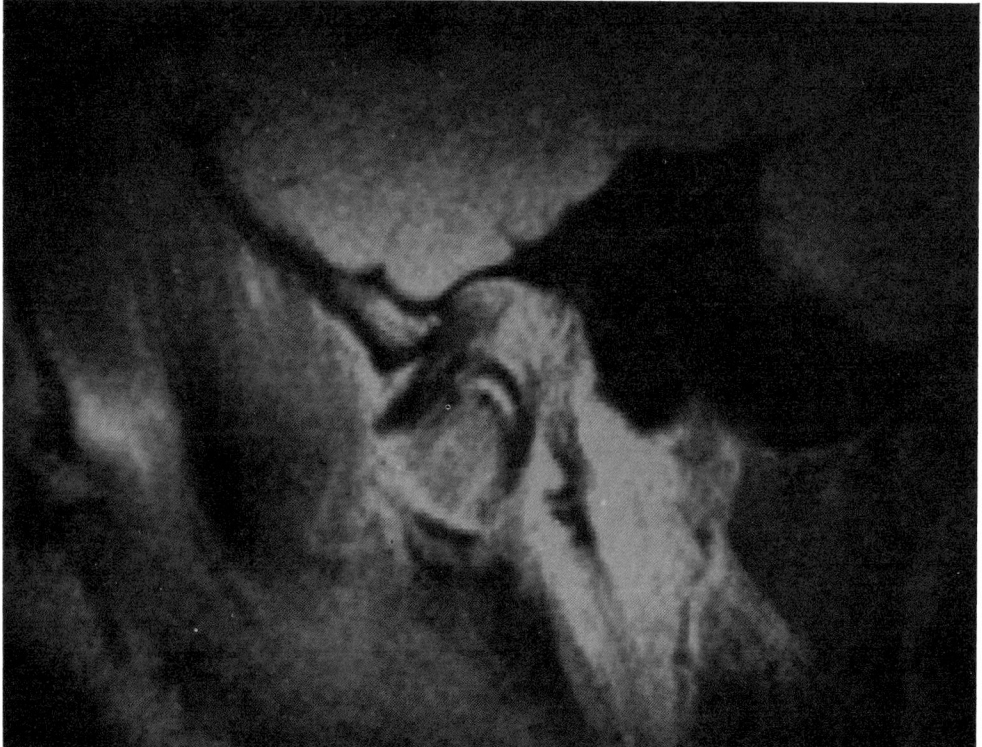

Figure 3.13

Adhesions may develop in response to internal derangement and articular disk disease. This abnormality, characterized by lack of change in the position and shape of the disk throughout the range of TMJ motion, is demonstrated in the closed (**a**) and open (**b**) mouth T_1-weighted images (SE 250/25).

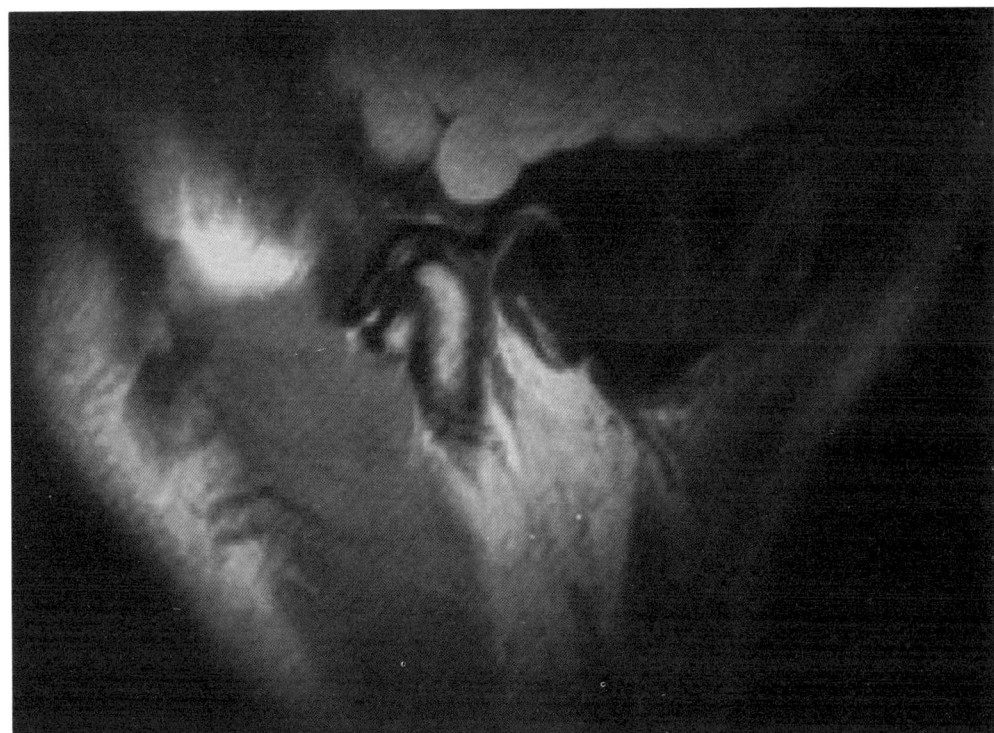

a

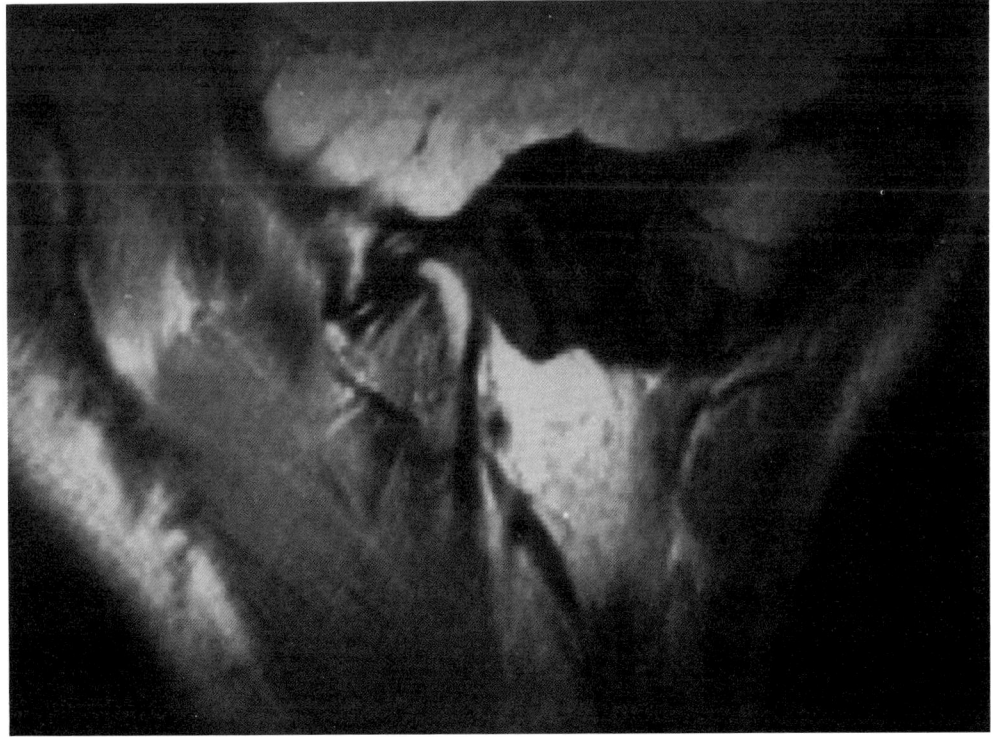

b

Figure 3.14
Medial displacement of the articular disk. T_1-weighted images (SE 250/25). Note the anterior displacement of the disk in the closed mouth position (**a**). In (**b**), the slice is beyond the medial aspect of the condyle, but the thickened and distorted posterior band is clear evidence of medial disk displacement. In the open mouth views, (**c**) and (**d**), although the anterior displacement has been reduced, the medial position of the posterior band persists.

76 *MRI atlas of the musculoskeletal system*

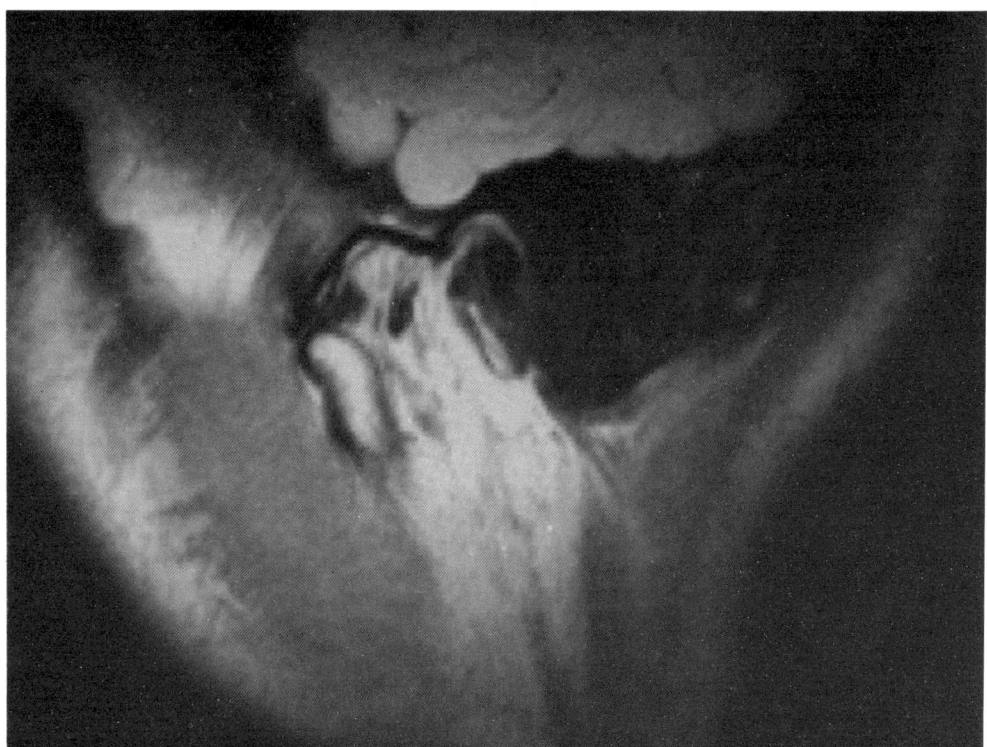

c

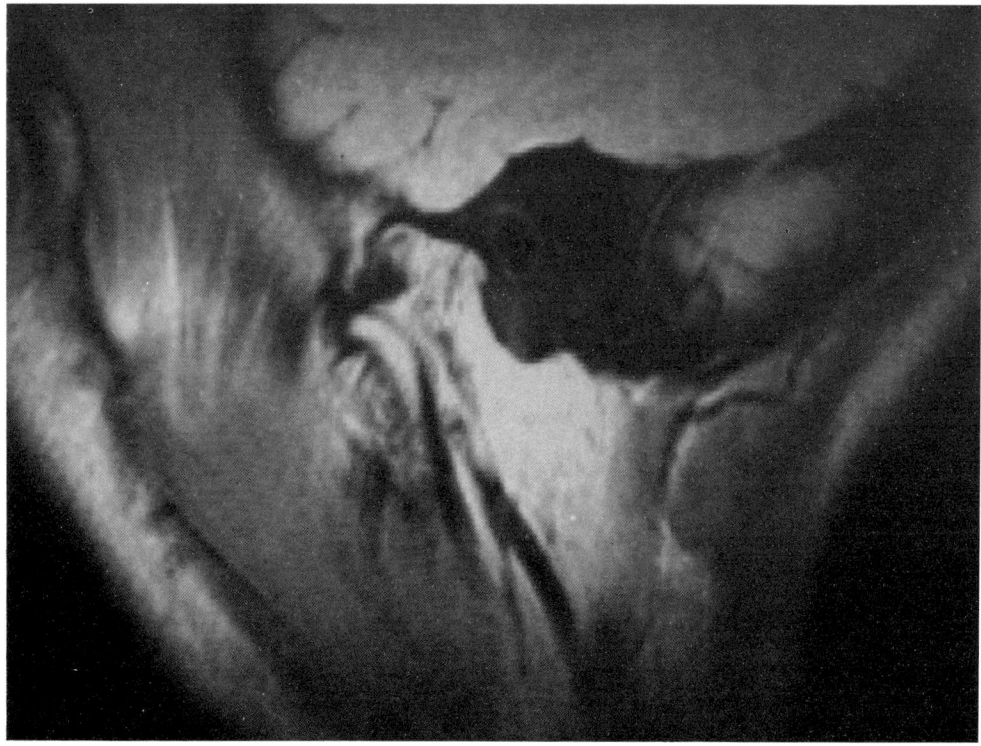

d

Figure 3.14 *continued*

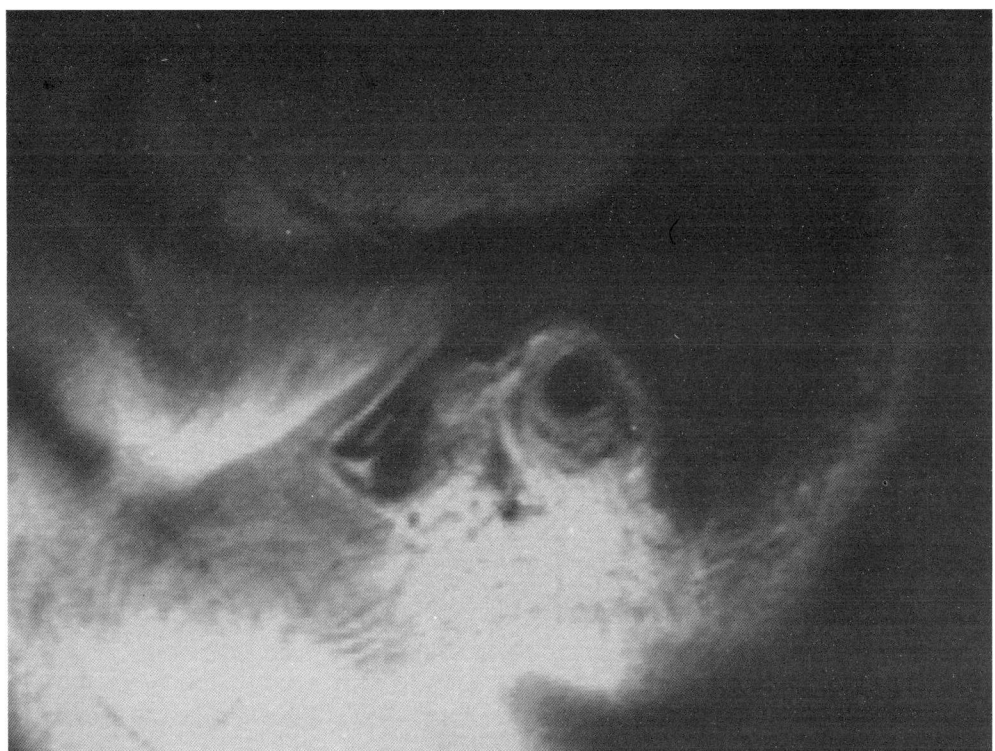

a

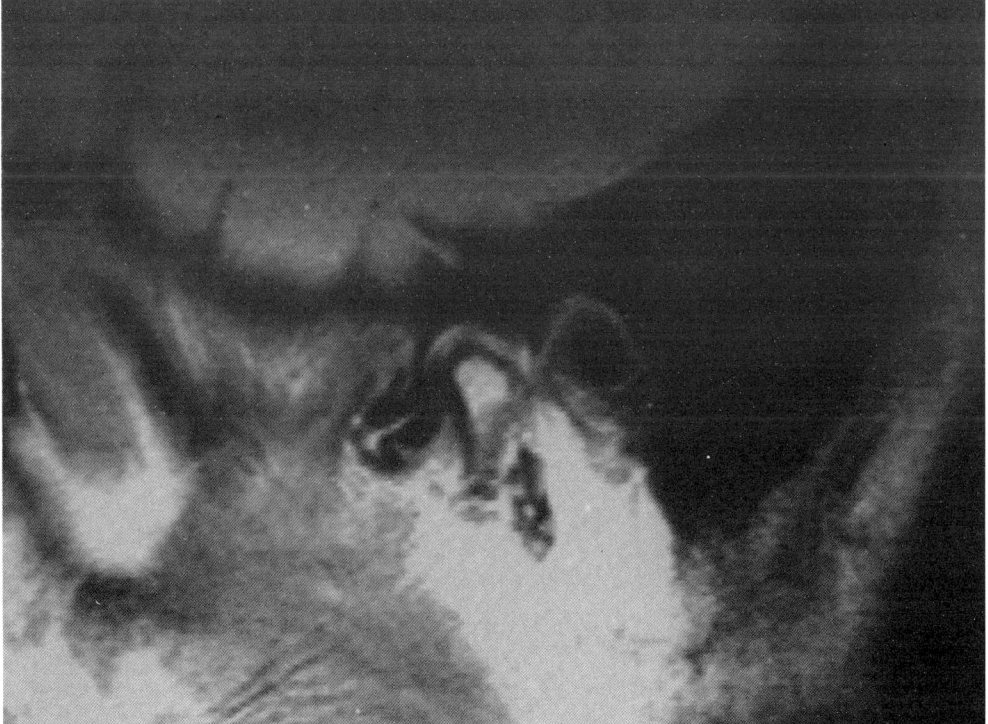

b

Figure 3.15
Lateral displacement of the articular disk. Figures (**a–d**) are T$_1$-weighted images (SE 250/25) obtained in the closed mouth position and progress from the most lateral slice (**a**) to the most medial slice (**d**). Note the presence of low-signal-intensity disk material beneath the high-signal intensity of fat of the eminence (**a**). Slices (**b–d**) include the condyle and the anteriorly displaced disk. The bulk of the articular disk is present in slice (**a**), lateral to the condyle.

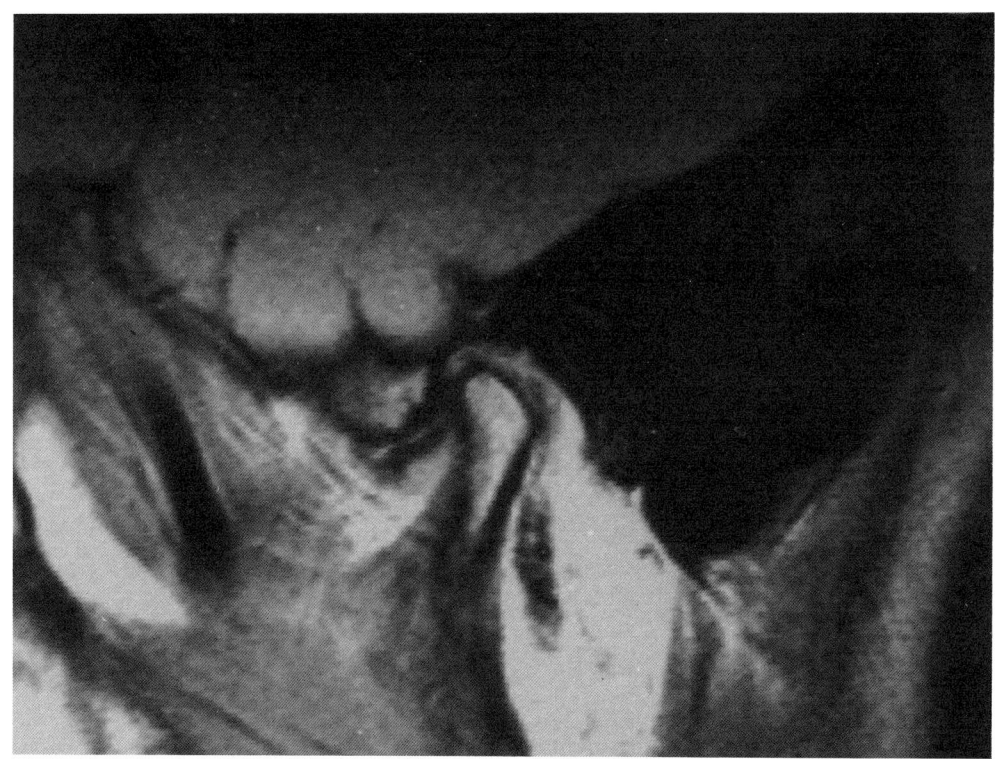

c

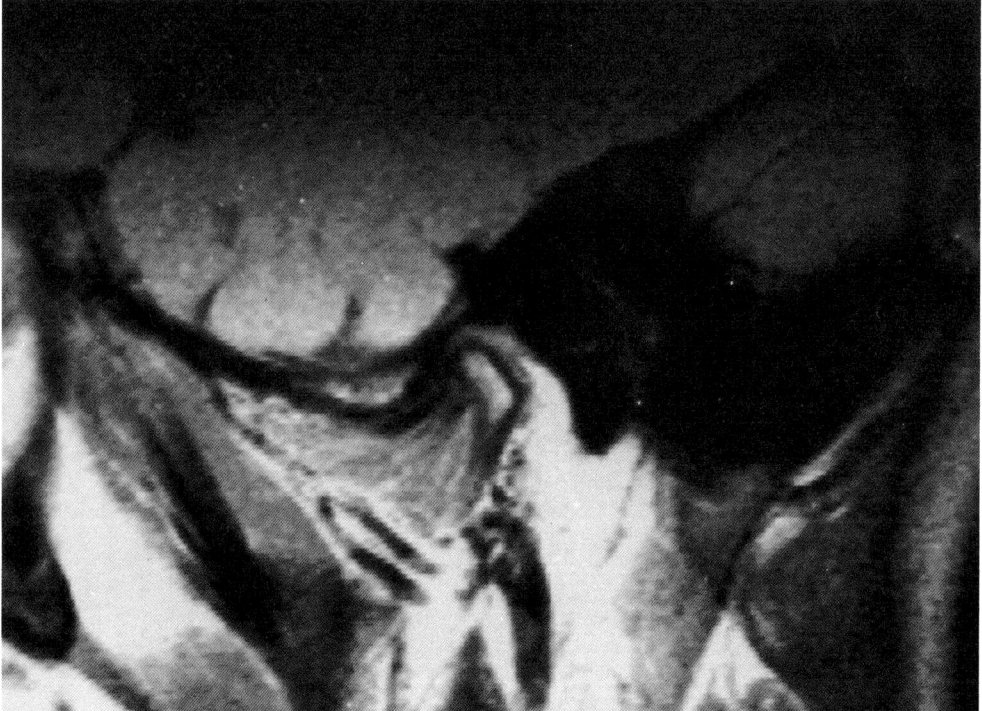

d

Figure 3.15 continued

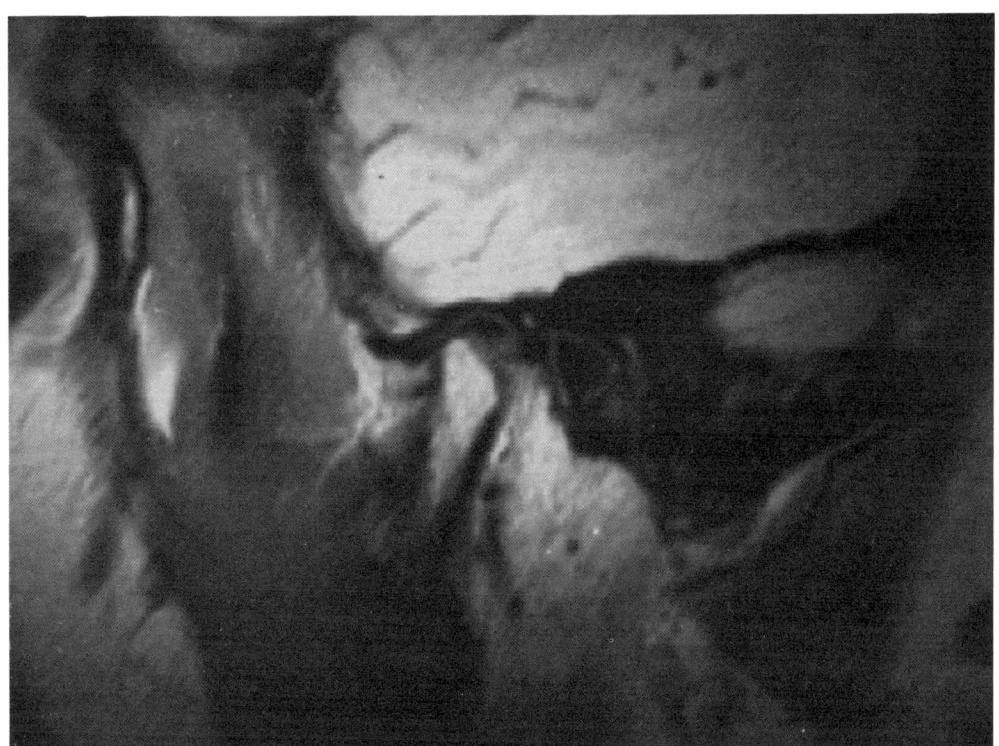

a

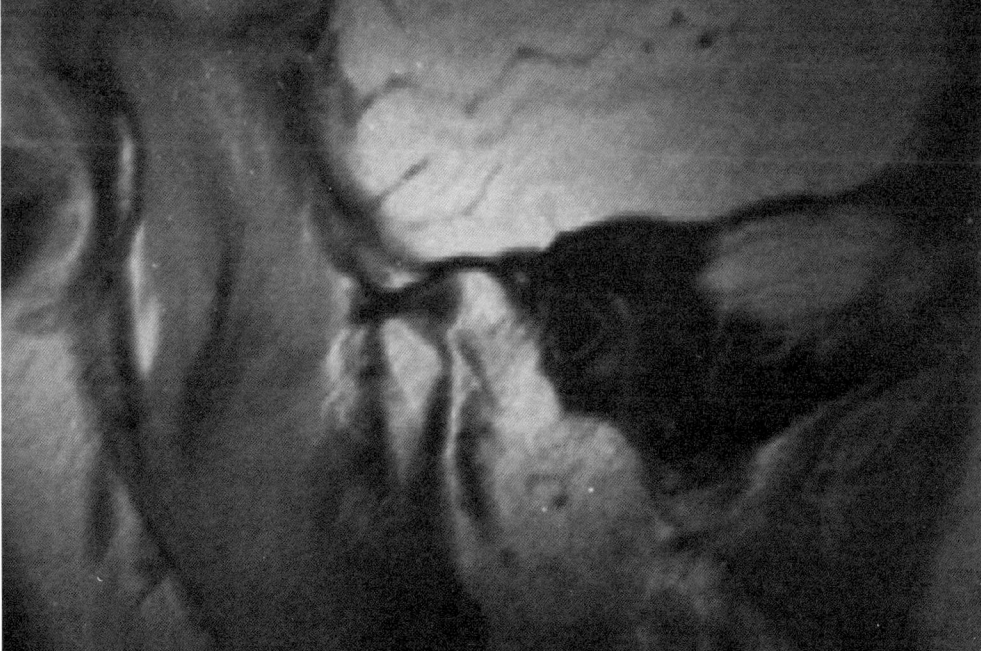

b

Figure 3.16

Posterior displacement of the articular disk is distinctly uncommon, and is shown in the T_1-weighted images (SE 250/25) in the closed (**a**) and open (**b**) mouth positions.

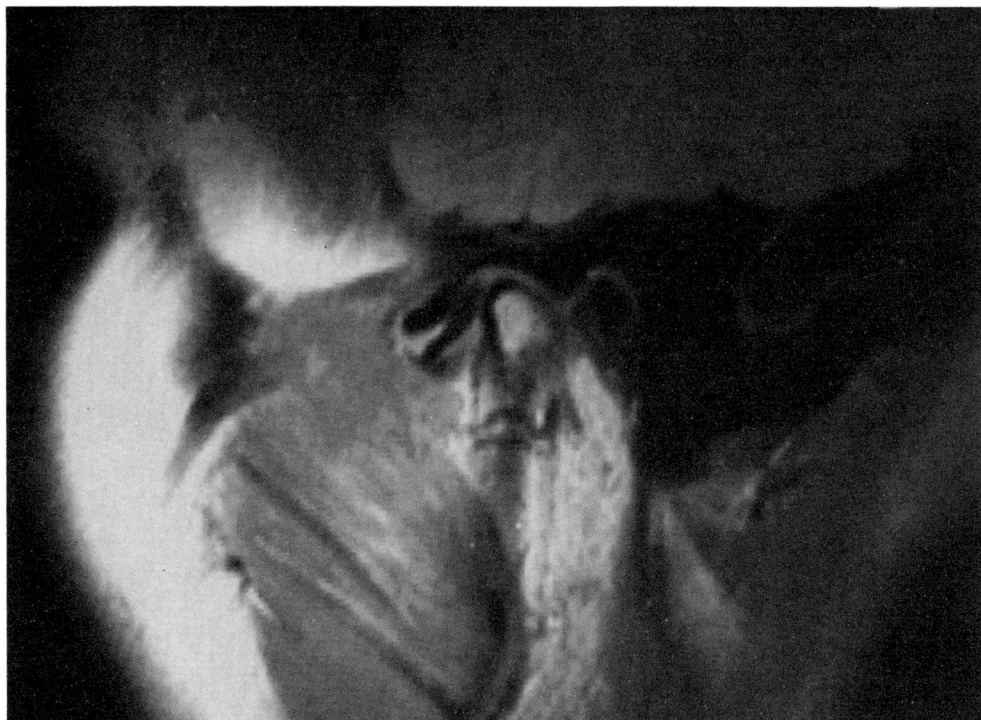

a

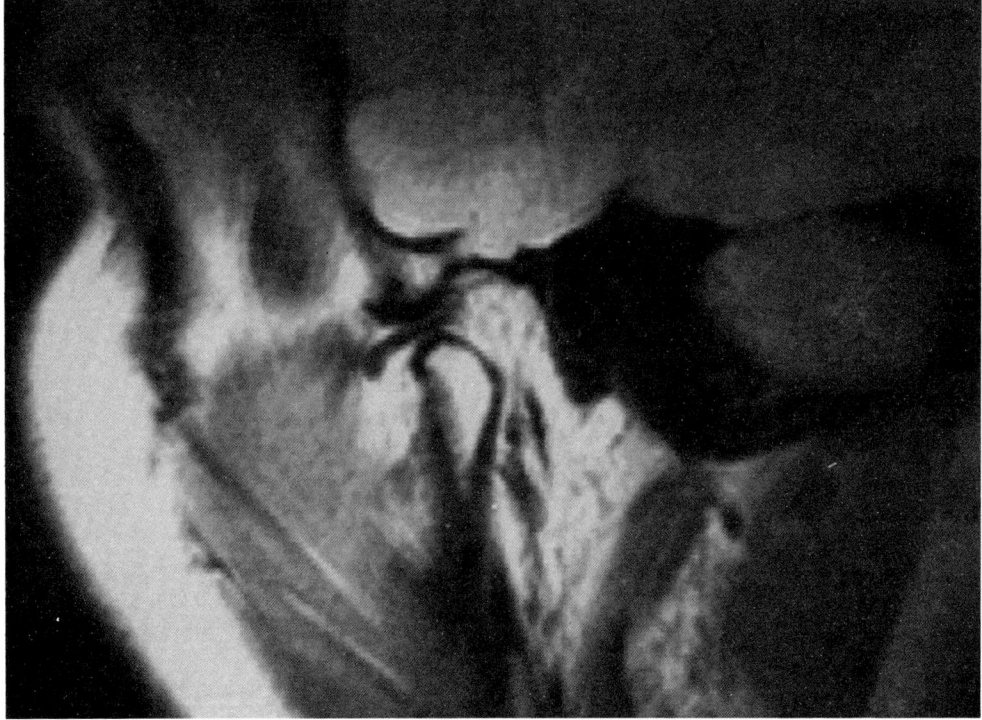

b

Figure 3.17

Acute retrodiscitis. The closed (**a**) and open (**b**) mouth T_1-weighted images (SE 250/25) prior to arthroscopy reveal anterior disk displacement without reduction. Post-arthroscopic closed (**c**) and open (**d**) mouth T_1-weighted images reveal anterior disk displacement and thinning of the posterior band. The T_2-weighted images (SE 2000/90) yield poor anatomic information, but reveal streaks of high-signal intensity within the lateral pterygoid muscle (**e**) and high-signal intensity within the joint space (**f**), evidence of postoperative edema and joint fluid.

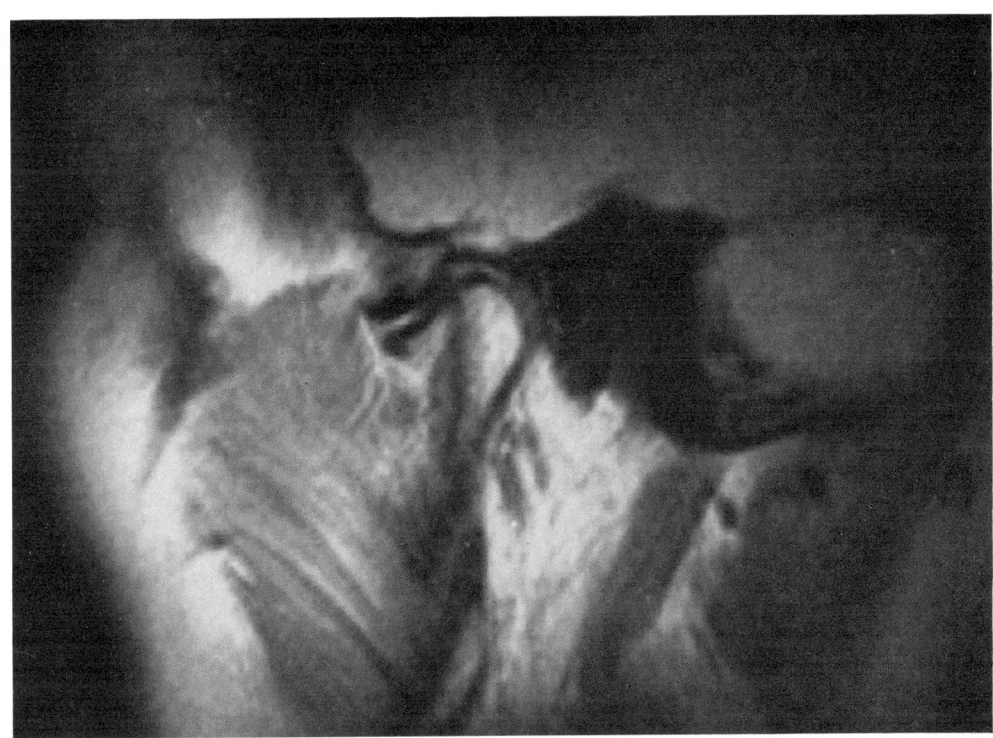

c

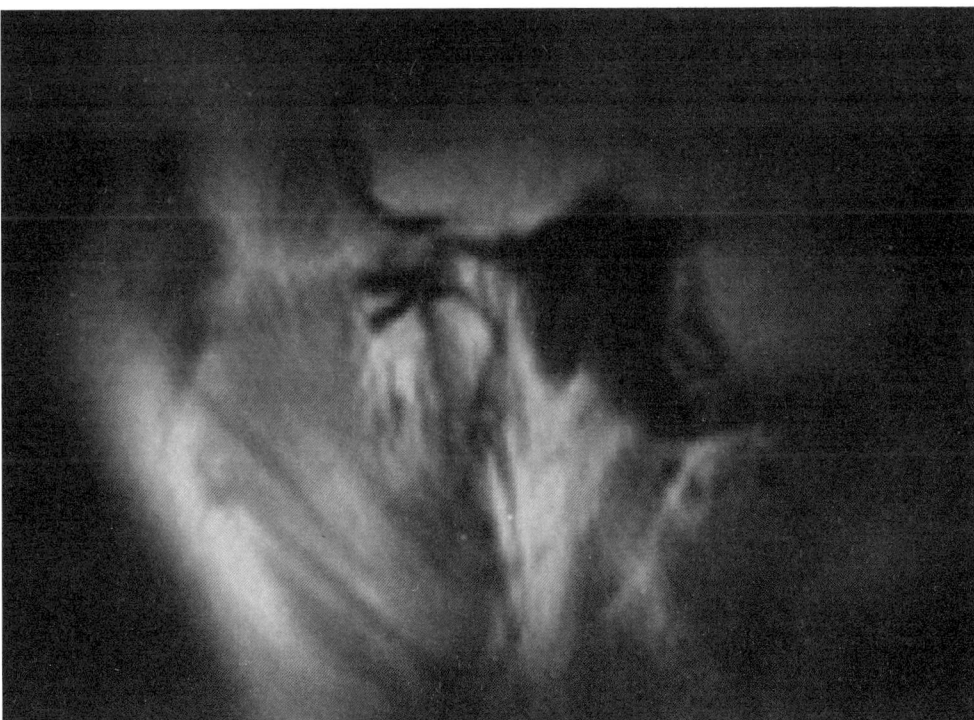

d

Figure 3.17 continued

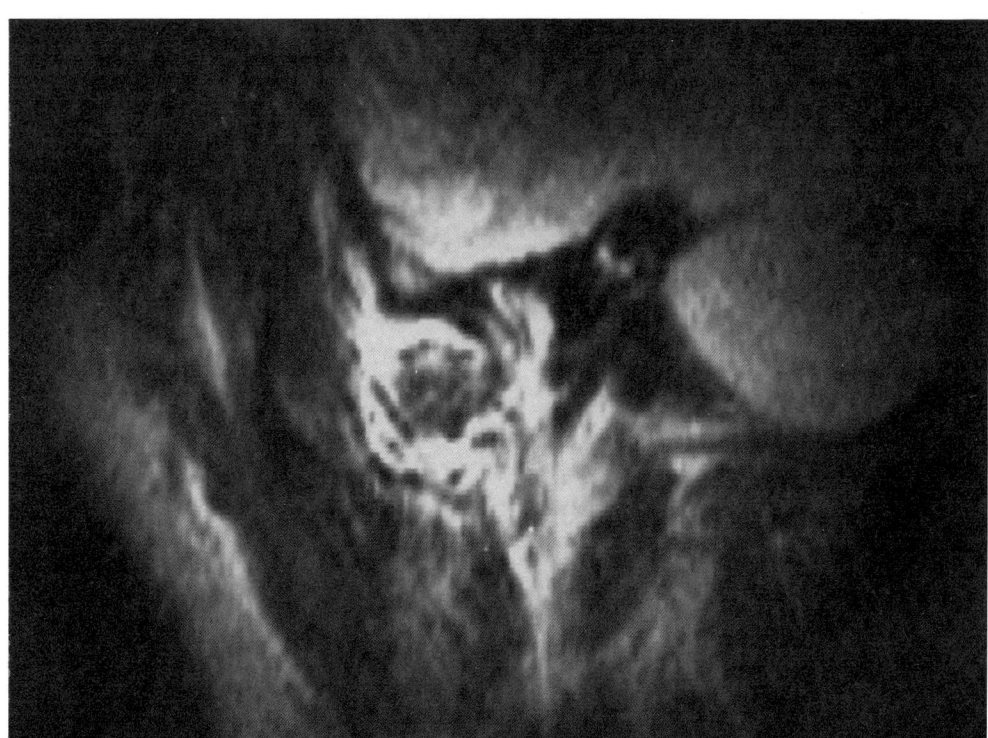

e

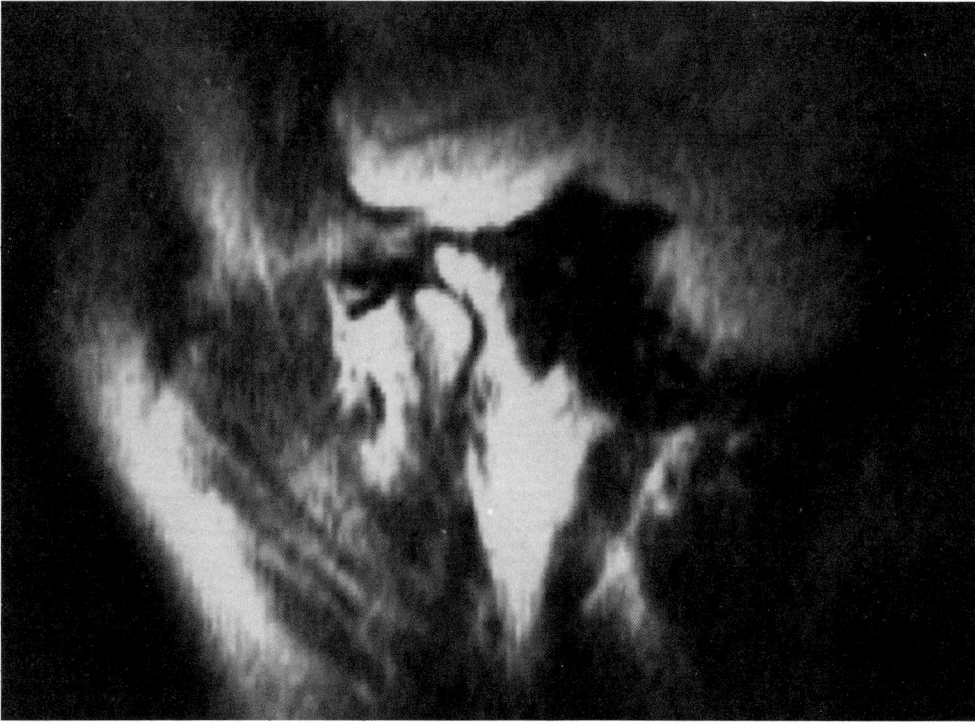

f

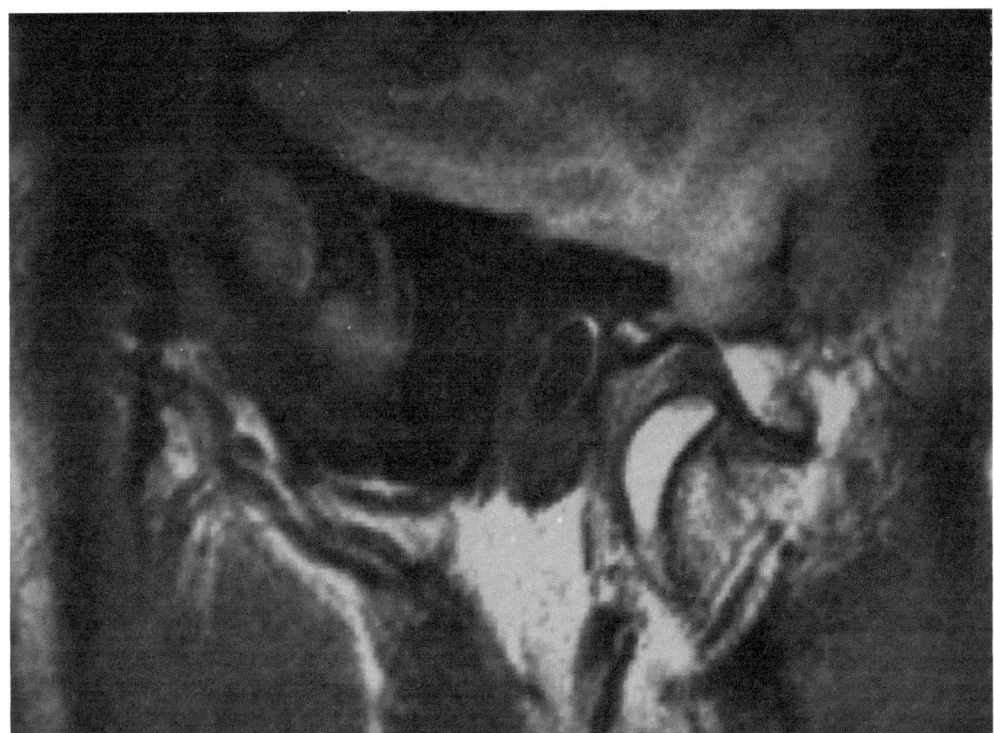

a

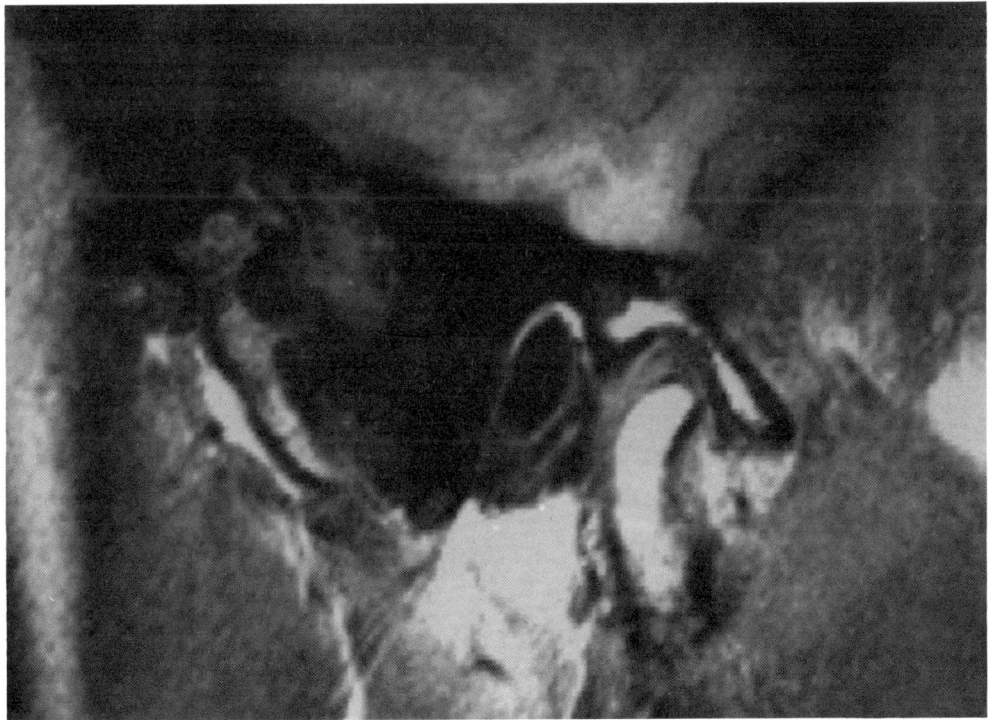

b

Figure 3.18

Retrodiscal hyperplasia. (**a–c**) The closed mouth T$_1$-weighted images (SE 250/25) are very thin and contiguous, and disclose thickening of the retrodiscal tissues and an otherwise normal disk and joint. (**d–f**) The open mouth T$_1$-weighted images (SE 250/25) reveal normal translation and persistently thickened retrodiscal tissues. (**g–i**) T$_2$-weighted images (SE 2000/90) in the open mouth position fail to reveal abnormal high-signal intensity, implying a chronic, longstanding TMJ abnormality, such as may be seen with a chronic malocclusion.

84 *MRI atlas of the musculoskeletal system*

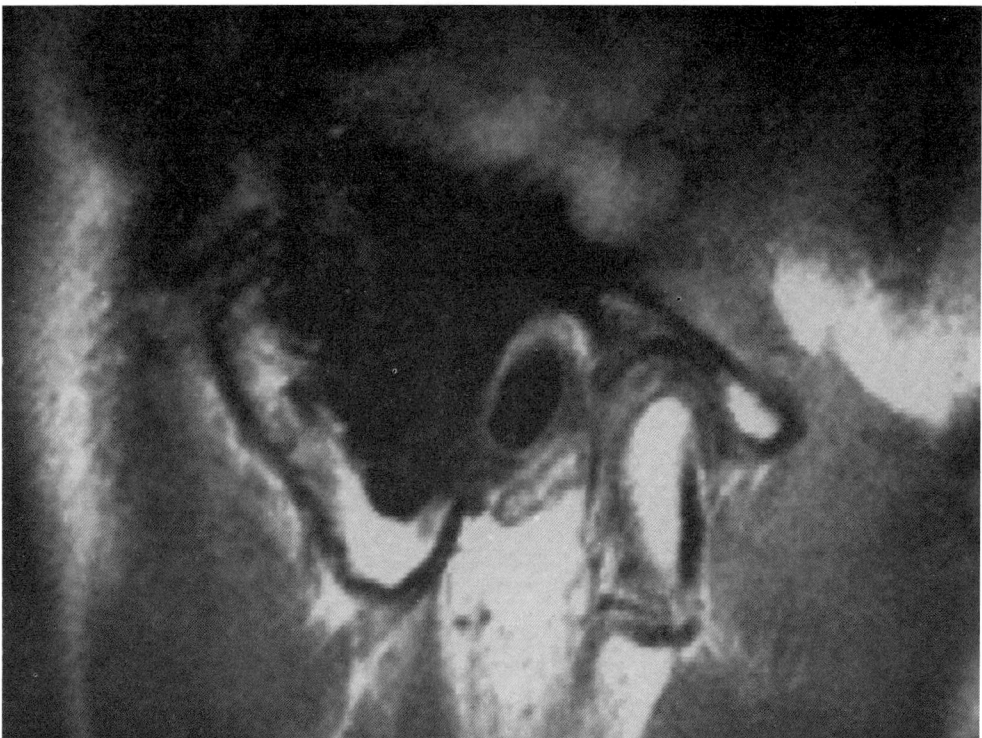

c

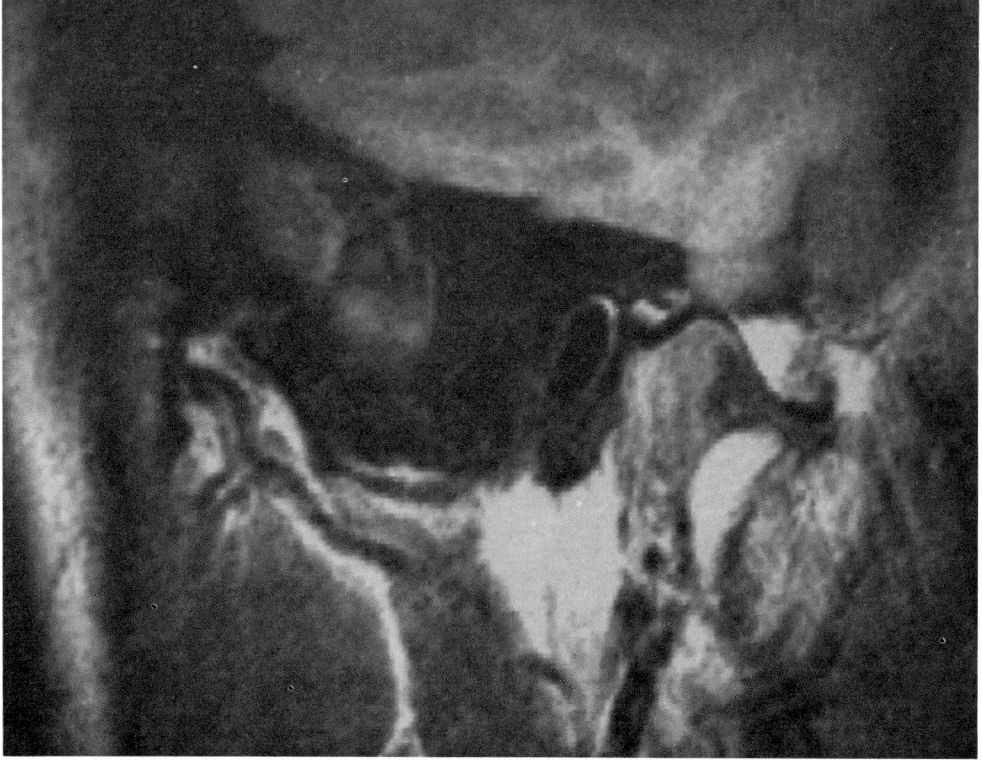

d

Figure 3.18 *continued*

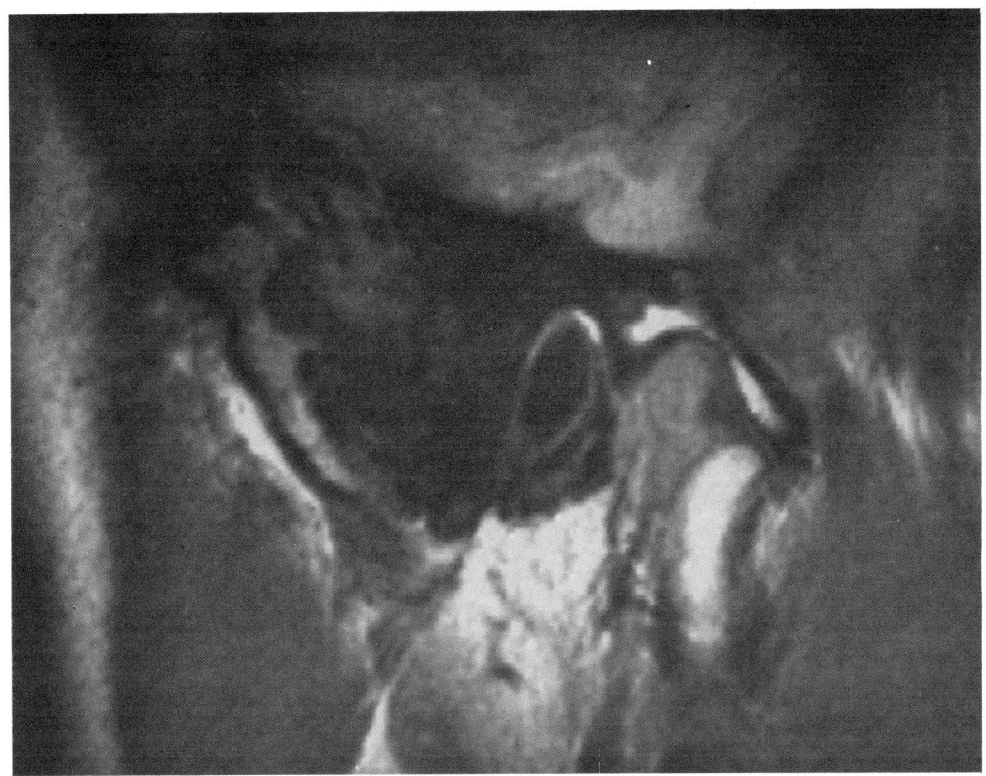
e

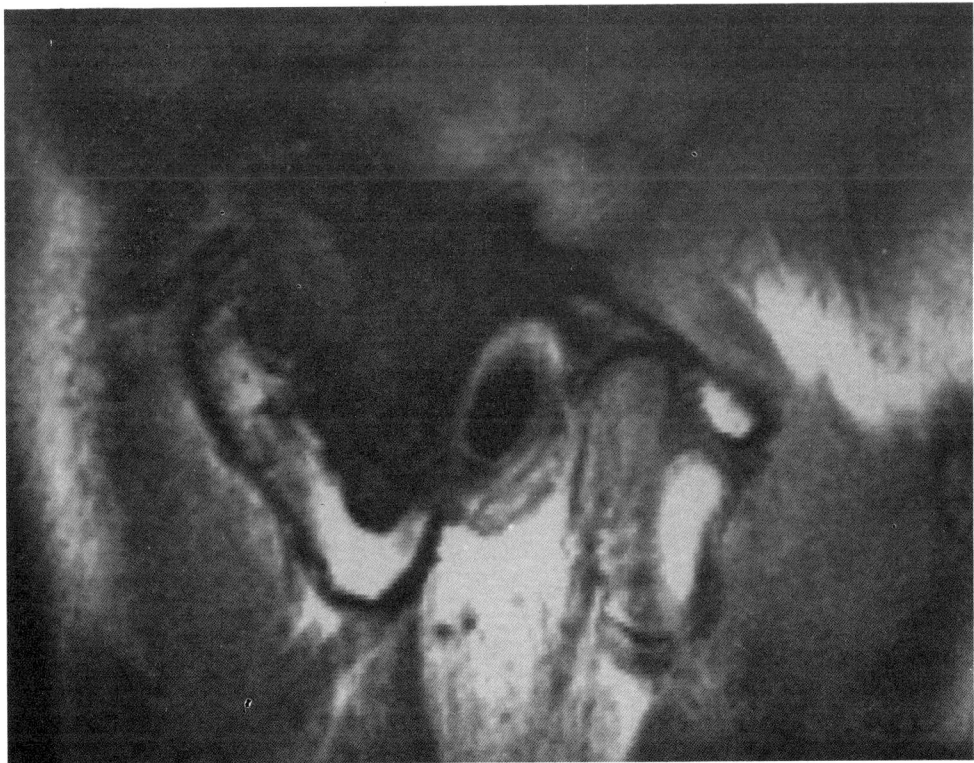
f

86 MRI atlas of the musculoskeletal system

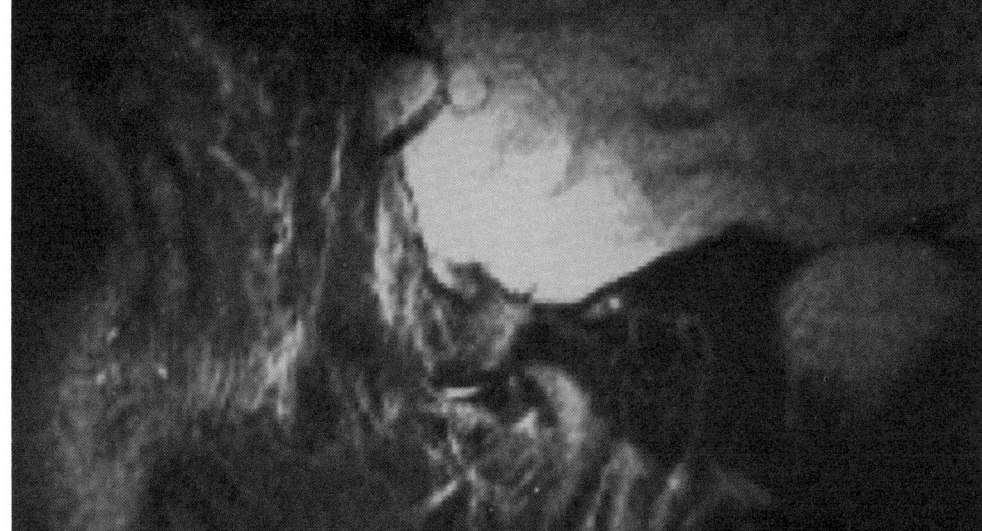

g

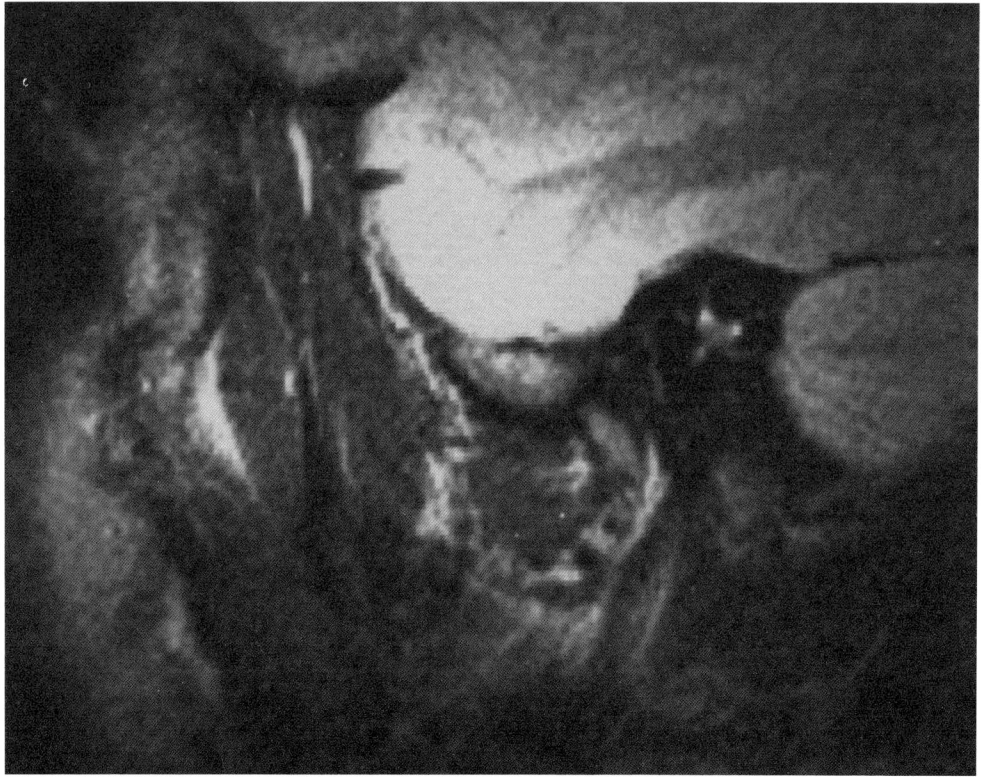

h

Figure 3.18 continued

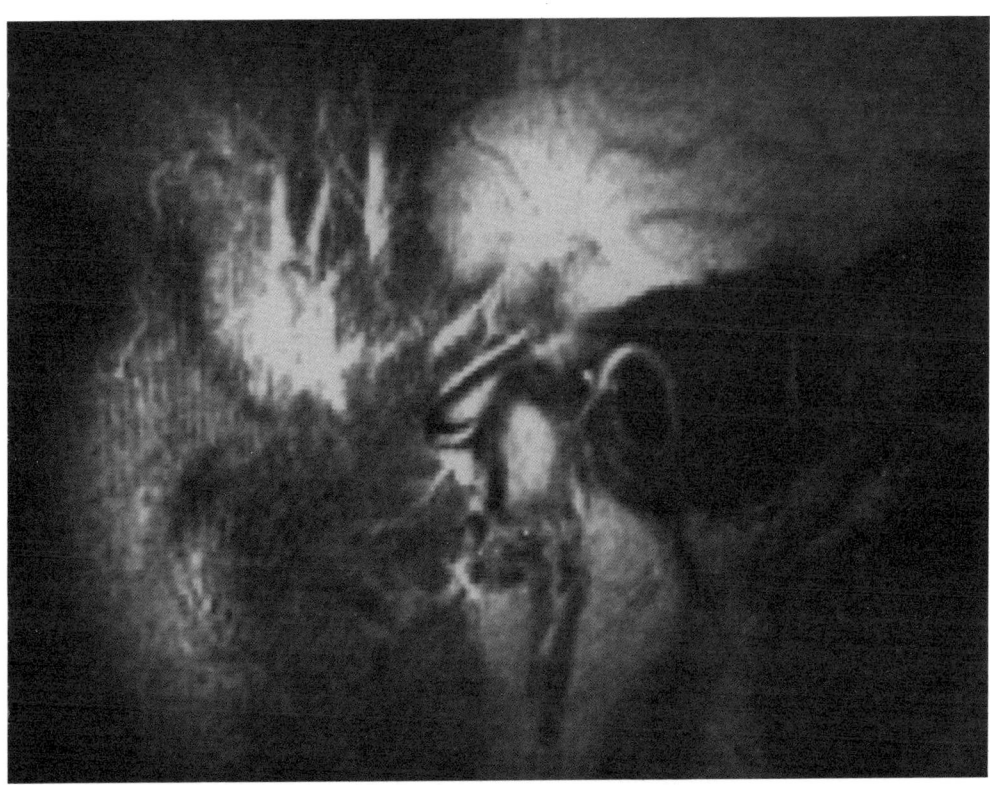

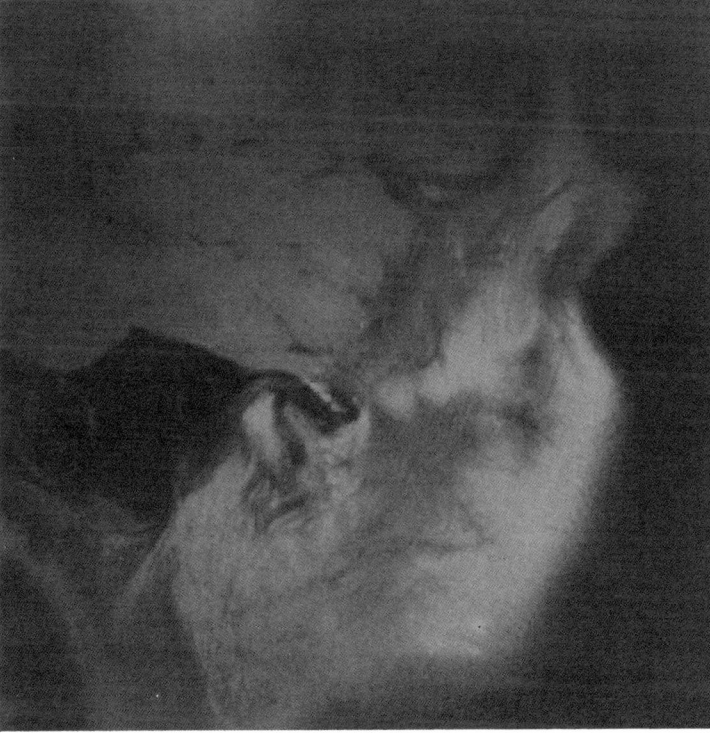

Figure 3.19

Proplast implant. Note the thin band of low-signal intensity which represents the implant in the normal position of the articular disk (SE 250/25).

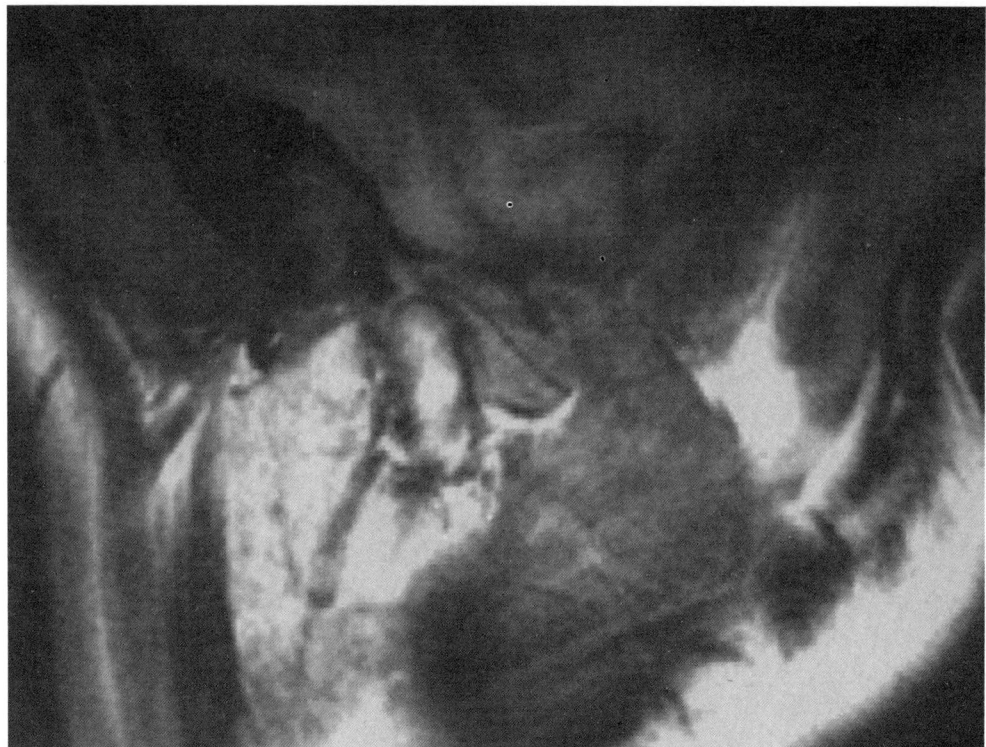

a

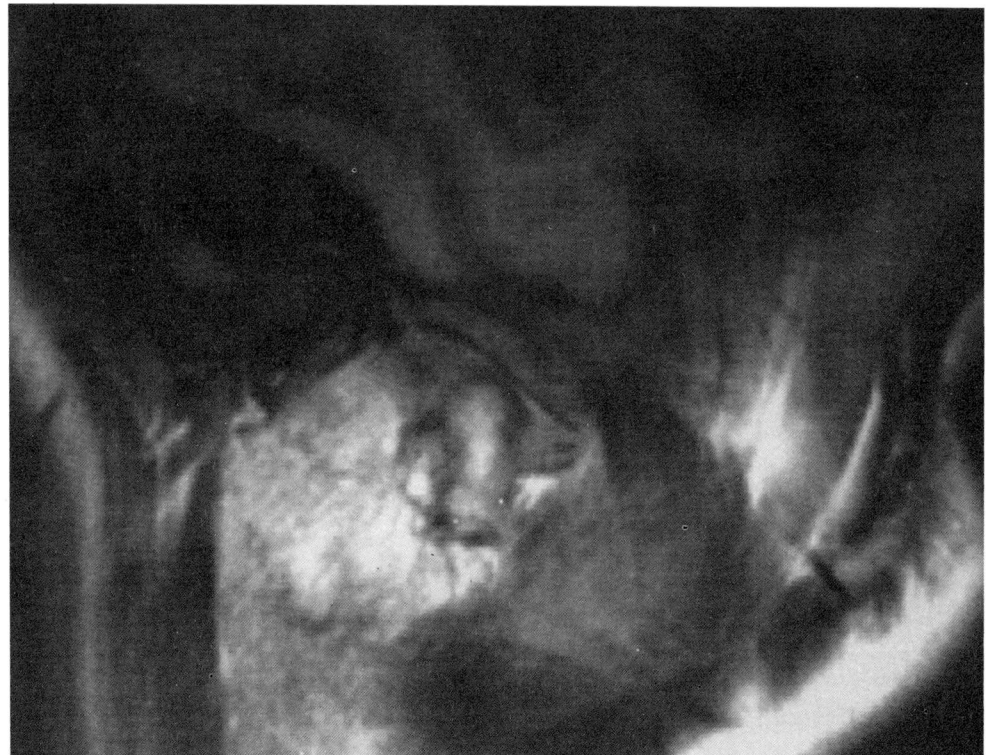

b

Figure 3.20
(**a,b**) T$_1$-weighted images represent a mild giant-cell reaction to a Proplast implant. The thin linear low-intensity signal centrally placed represents the implant. The surrounding intermediate signal intensity is secondary to the giant-cell reaction. (**c–f**) T$_1$-weighted images demonstrate a more exuberant reaction. Although the normal anatomic landmarks are difficult to identify, no destruction or erosion is evident. (SE 250/25.)

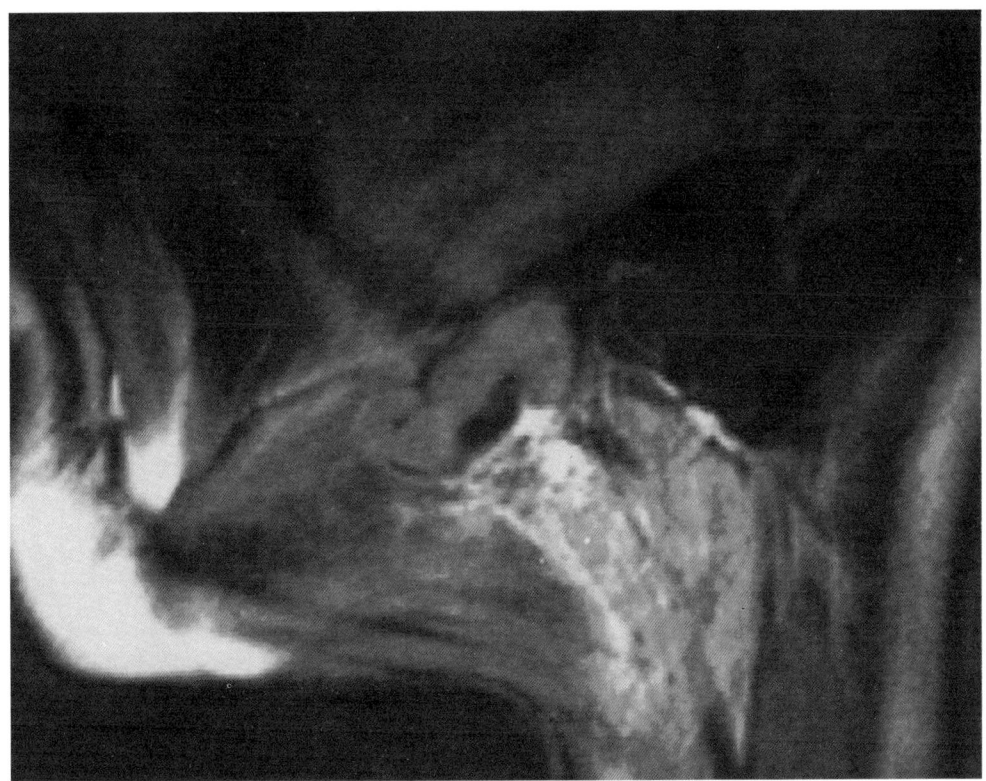

c

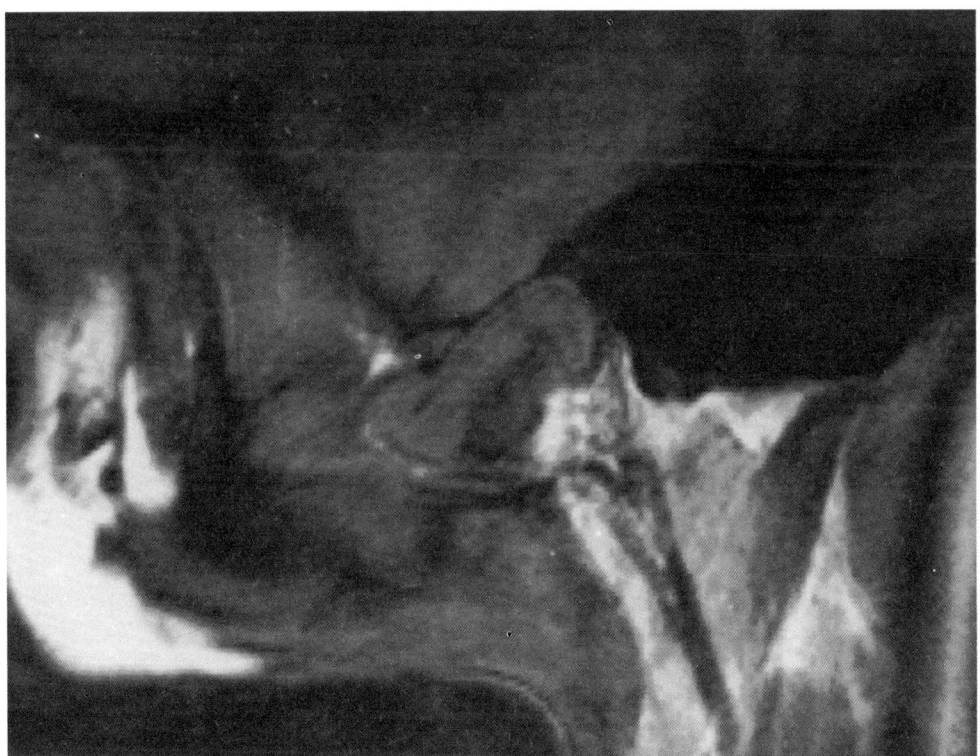

d

Figure 3.20 continued

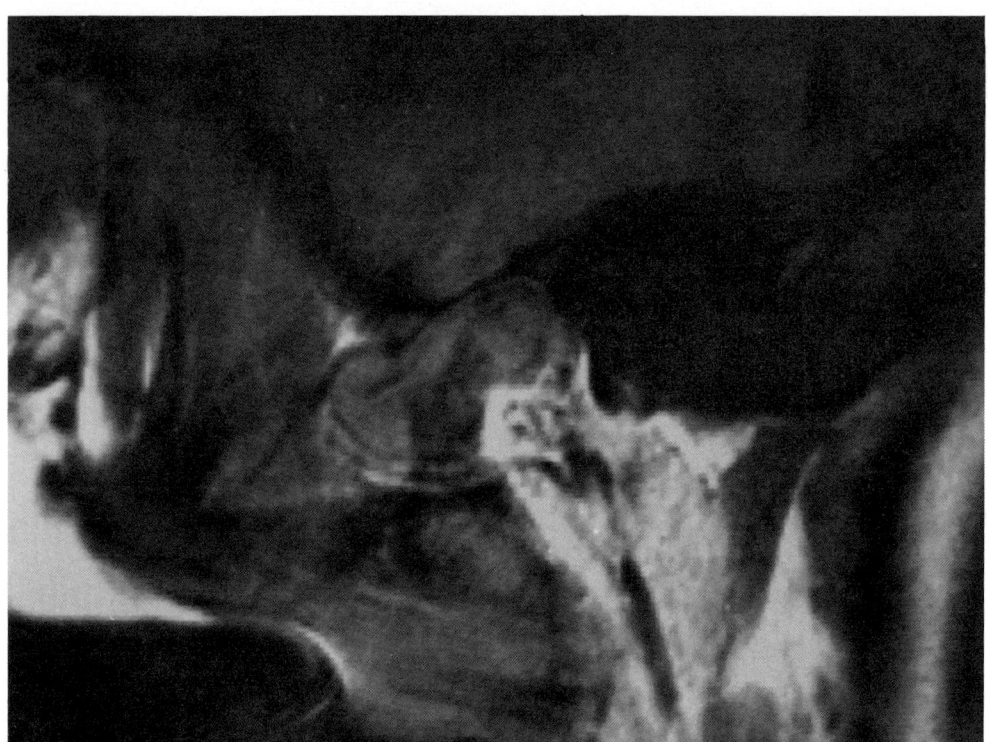

e

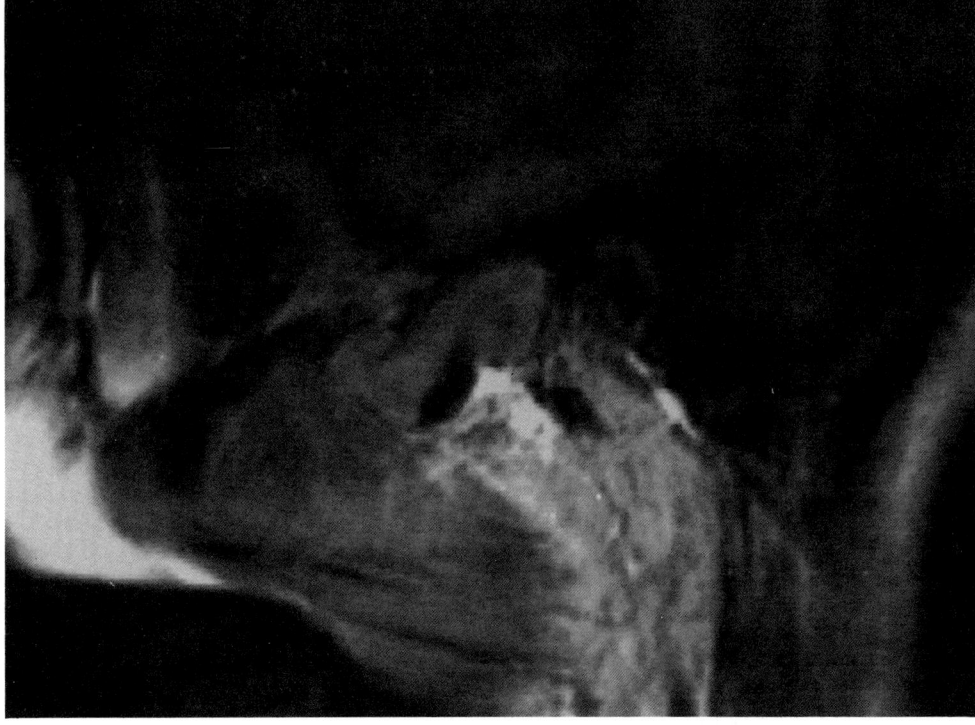

f

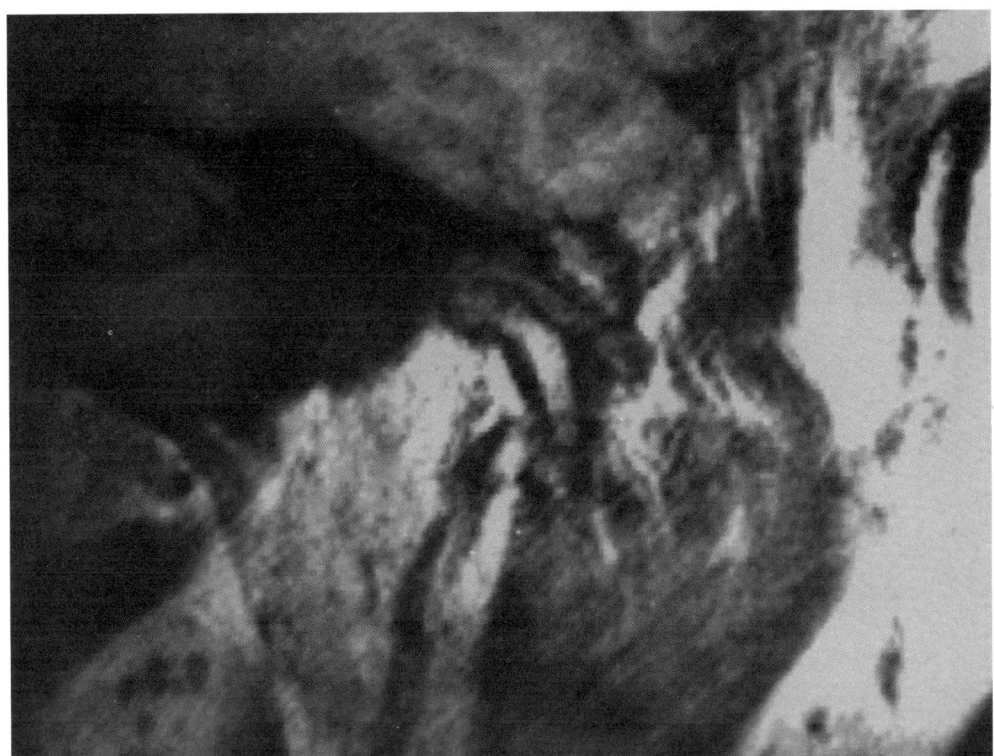

a

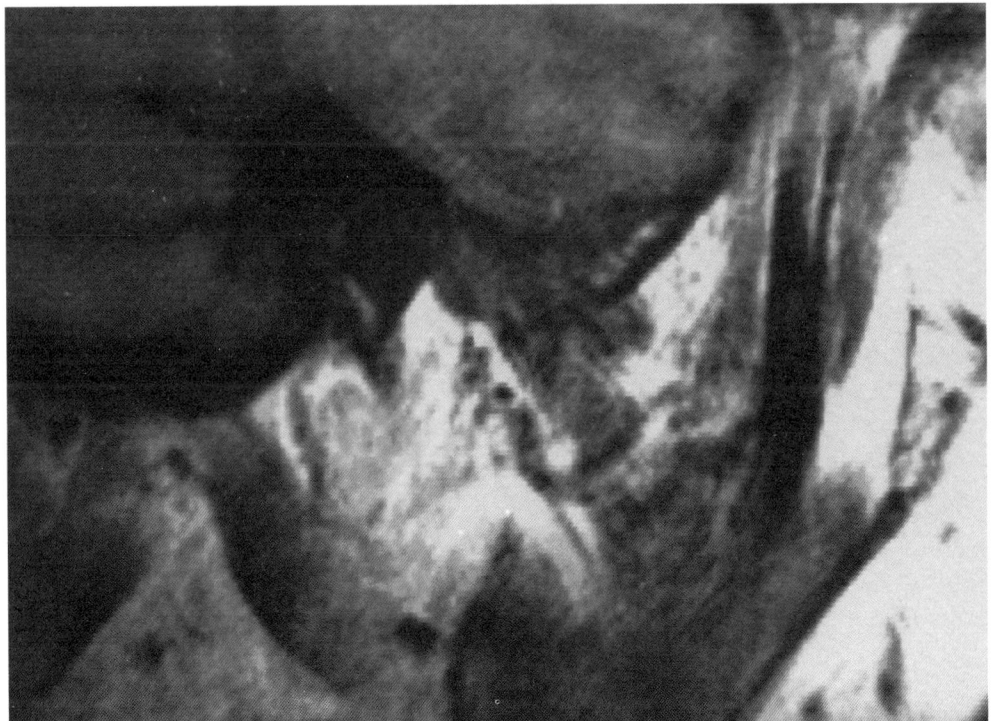

b

Figure 3.21
As the giant-cell reaction progresses, the amount of destruction can increase. In this case, note erosion of the floor of the middle cranial fossa in the T_1-weighted images in the closed (**a,b**) and open (**c,d**) mouth position (SE 250/25). Involvement of the temporal lobe of the CNS is imminent.

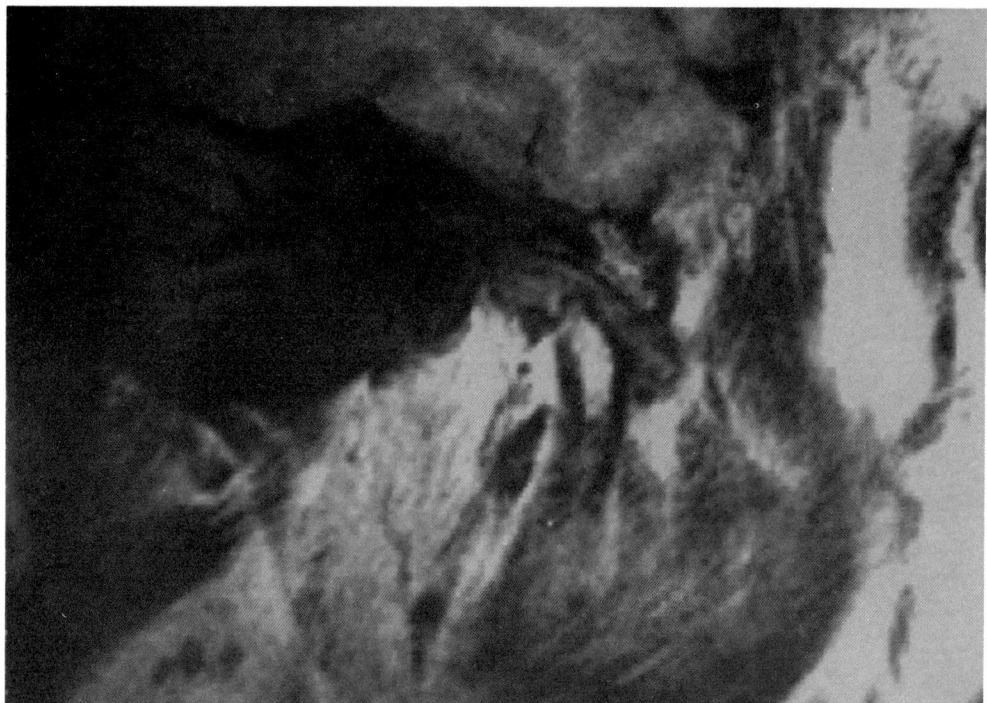

c

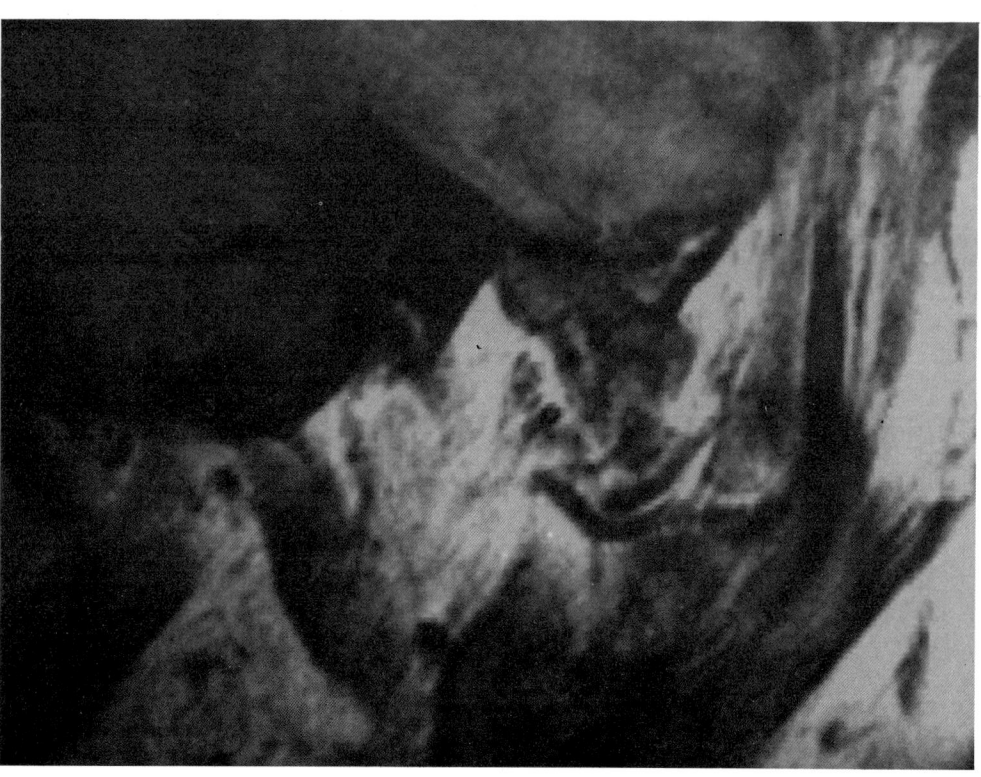

d

Figure 3.21 *continued*

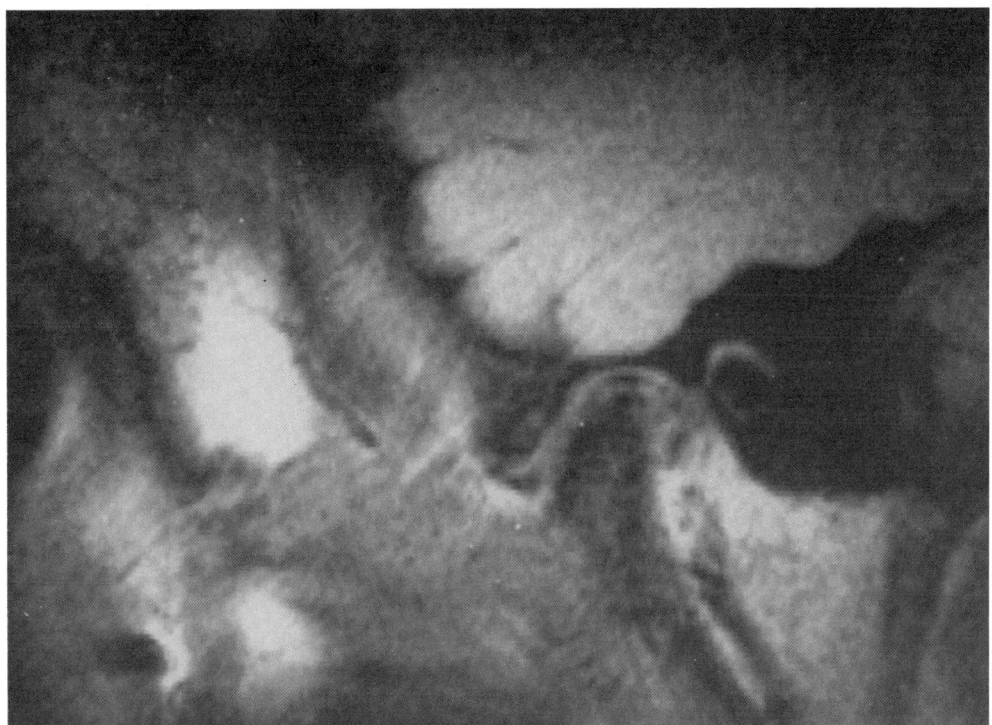

Figure 3.22

A dermal graft is an alternative to a synthetic implant. Closed mouth T_1-weighted image (SE 250/25) exemplifies such a graft. Note the intermediate signal intensity of the tissue interposed between the sclerotic condyle and the fossa.

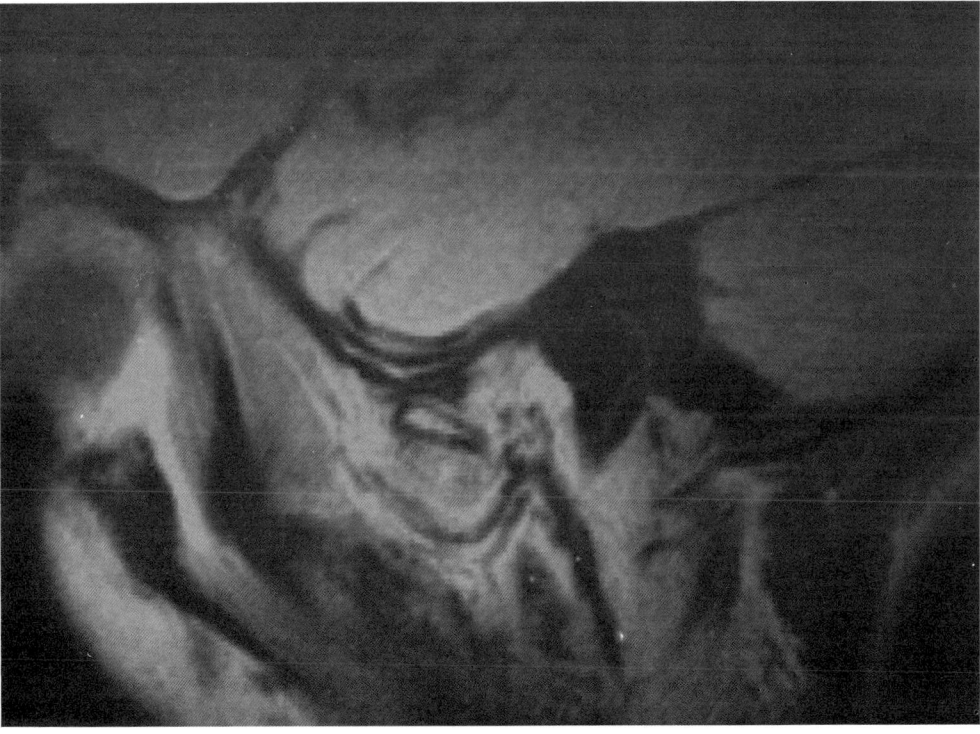

Figure 3.23

Trauma to the mandible may also damage the articular disk. (**a**) The T_1-weighted image reveals a condylar fracture. (**b**) The T_1-weighted image is an example of a TMJ dislocation (SE 250/25.)

a

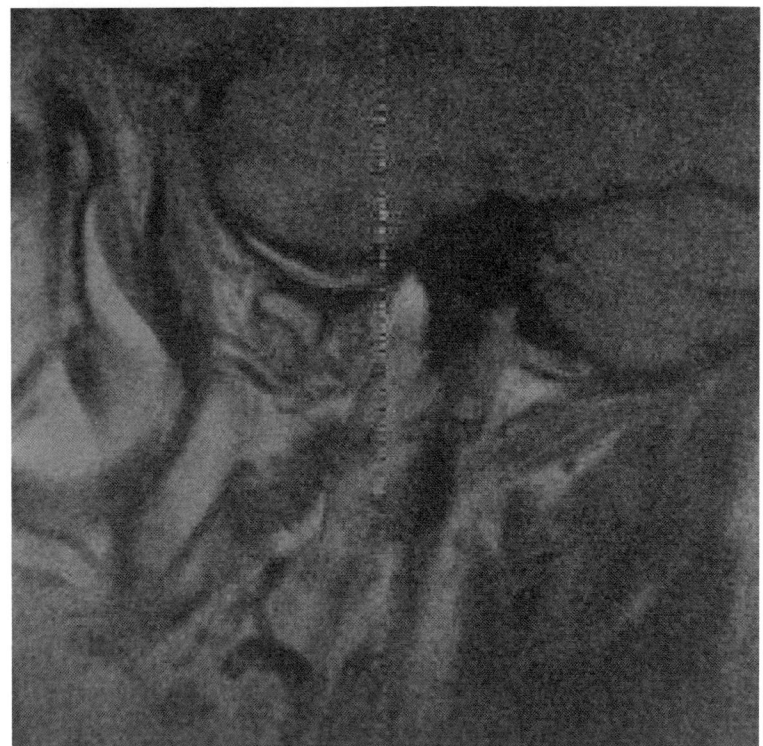

Figure 3.23 *continued*

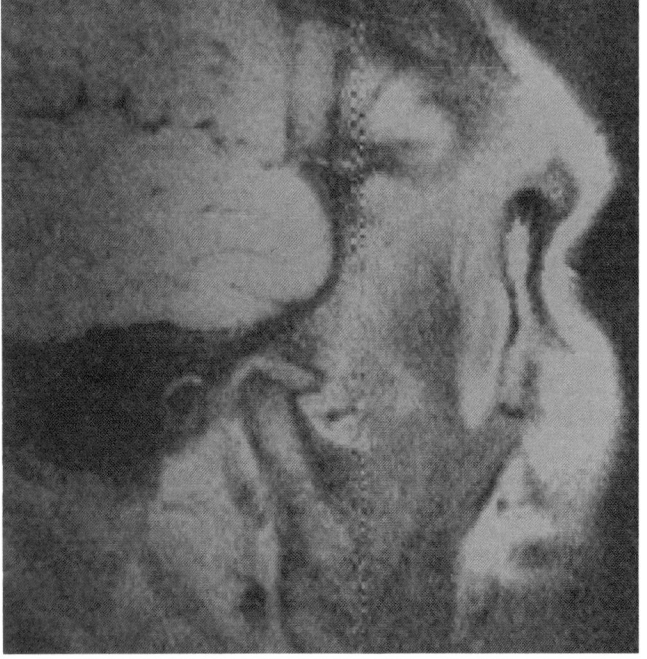

a

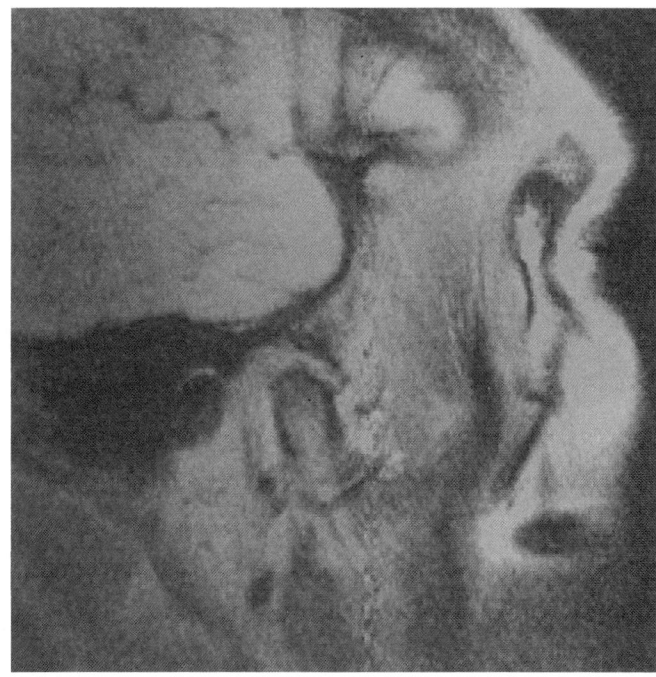

b

Figure 3.24

(**a,b**) The T_1-weighted images depict psoriatic arthritis involving the TMJ, causing deformity and erosion of the condyle (SE 250/25). Ankylosis would represent the end stage of this process.

4
The shoulder

Leanne L Seeger

With the recent surge of public interest in strenuous physical activity, disorders of the shoulder joint have become a prominent topic in the literature of radiology, orthopedics, and sports medicine. Rotator cuff disease is no longer considered a condition afflicting only the elderly or the rheumatoid patient, and disorders such as shoulder impingement syndrome and shoulder instability are now under intensive investigation regarding their mechanism, pathology, and management.

Studies of magnetic resonance imaging (MRI) of the glenohumeral joint have shown that this imaging modality can display accurately the normal anatomy of the shoulder,[1-4] and give promise that it may eventually replace invasive methods for the diagnosis and treatment planning of several forms of shoulder pathology.[5-9] MRI may also permit the earlier detection of conditions which, without expeditious treatment, can lead to chronic pain and disability.

All images shown in this chapter were acquired with a 0.3 tesla magnet, and spin-echo pulse sequences. A 16 to 19.2 field of view was used, with an imaging matrix of 256 × 256 which was interpolated to 512 × 512 for display. Slice thickness was 3 to 5 mm, obtained at 5 to 7 mm intervals. A surface coil was used for all scans.

Imaging modalities for detection of shoulder pathology

The detailed images that are possible with MRI provide the first noninvasive means of evaluating several types of commonly occuring disorders of the shoulder. Plain radiography and computed tomography (CT), although the most appropriate means for depicting soft-tissue calcification, lack the inherent soft-tissue contrast afforded with MR imaging, and therefore often fail to depict the full extent of pathology.

Double-contrast arthrography is an established diagnostic tool for evaluation of rotator cuff disease. However this invasive procedure yields helpful information only in the presence of a cuff tear, and is of no assistance in the diagnosis of impingement syndrome. While double-contrast arthrography combined with CT has a higher diagnostic yield than arthrography alone for suspected labral pathology, CT-arthrography is invasive, and is associated with significant radiation exposure to the neighboring thyroid gland. Moreover, this study requires the almost concomitant availability of both fluoroscopy and CT, which could lead to scheduling problems if, for example, the CT scan was delayed because of difficulties encountered during the preceding arthrogram. Bursography has been advocated for the early detection of shoulder impingement syndrome, but has not gained wide acceptance. Shoulder ultrasonography (US) has been shown to be a useful screening modality for suspected rotator cuff tears, however, it is difficult to perform, and highly operator-dependent. While a negative US examination is evidence that symptomatology is not related to the rotator cuff, and while it permits large tears of the cuff to be diagnosed, US appears insensitive to small foci of tendon pathology, and cannot reliably differentiate a small tear from tendonitis. Moreover, US is not useful for evaluation of the anterior glenoid labrum.

MRI is ideally suited for the evaluation of soft-tissue pathology of the shoulder. The tendinous rotator cuff is well displayed, the cuff muscles can be evaluated for atrophy, and the precise location of the musculotendinous junction can be visualized in order to determine the extent of muscular retraction. Differences in signal intensity allow the synovial tissues and/or fluid to outline clearly the glenoid labrum. The exquisite depiction of marrow afforded by MRI also allows the early detection of ischemic necrosis, primary and metastatic tumor, and infection.

Technical considerations of MRI of the shoulder

MRI of the shoulder poses several difficulties which are not encountered when imaging other parts of the body. First, the shoulder lies in the periphery of the magnetic field. This necessitates significant lateral shift for scan centering, leading to image acquisition in an area where the signal-to-noise ratio is inherently low. The amount of lateral shift required is determined by the size of the patient, but generally falls between 90 and 120 mm toward the shoulder of interest. Secondly, the majority of shoulder disorders require the imaging of relatively small soft-tissue structures. High-resolution scanning techniques and the use of a surface coil can overcome these two problems. The final problem in MRI of the shoulder results from the complex three-dimensional anatomy of this joint. As the musculotendinous and ligamentous structures of the shoulder are oriented along multiple non-orthogonal planes, oblique imaging is needed for their full evaluation.

Because of the great mobility of the shoulder girdle, minor differences in patient position can markedly alter both anatomic relationships and the course of soft-tissue structures. Careful attention to patient positioning will serve to avoid potential confusion, especially when scanning is performed in non-orthogonal planes. We have found the optimal position to be with the shoulder in internal rotation, which is accomplished by placing the forearm across the abdomen. The elbow is then elevated to parallel the humeral head, thus allowing the entire proximal humerus to be depicted on images in the sagittal, coronal, or coronal-to-sagittal oblique planes.

The planar surface coil used for the images in this chapter is positioned obliquely over the shoulder so that its axis is perpendicular to the magnetic field of the scanner.[3] Foam wedges placed between the coil and the patient assist in stabilizing the coil as well as in improving the uniformity of the signal over the entire shoulder girdle. Masking tape is used to secure the coil to the table.

The optimal plane for shoulder MRI is dictated by the type of pathology under investigation. In general, imaging in the axial and frontal oblique planes (along the course of the supraspinatus muscle) constitutes a complete examination. The only common exception to this rule applies to the investigation of ischemic necrosis of the humeral head; with the arm in internal rotation, the early changes of ischemic necrosis are best depicted in the sagittal plane. Although we have not found imaging the shoulder in the coronal plane to be of value in clinical MRI, we have included coronal images in the section on normal anatomy for the sake of completeness.

The choice of the appropriate pulsing sequence is also determined by the suspected lesion. T_1-weighted images are most useful for detailed anatomy and evaluation of the bone marrow. T_2-weighted images are required to depict small collections of fluid, for example within the glenohumeral joint or rotator cuff tendons.

We advise that MR scanning should be delayed a minimum of two weeks after arthrography or corticosteroid injection. MRI is extremely sensitive to changes in the soft tissues, and these invasive procedures may alter their signal intensities, leading to false-positive diagnoses.

Since MRI is insensitive to small foci of soft-tissue calcification, it is essential that plain radiographs be available at the time of scan interpretation.

Normal anatomy

Axial plane (*Figure 4.1a–d*)

Axial imaging is optimal for evaluating the anteroposterior relationship of the glenohumeral joint, the portions of the rotator cuff which course transversely across the joint, and the anterior and posterior glenoid labra.

Regardless of patient position, in the normal state, the humeral head closely articulates with the bony glenoid. With the limited spatial resolution provided by current scanning techniques, the articular surfaces of these two structures cannot be separated in MR images of normal individuals.

The inferior part of the humeral head is oblong, its flattened posterior surface blending into an indentation laterally. This area serves as the site of attachment for some fibers of the infraspinatus and teres minor tendons, and thus lacks hyaline cartilage. More superiorly, the humeral head is rounded and completely surrounded by a signal void representing the cortex of the humeral head and the adjacent cuff tendons. Although the level of transition from oblong to round varies somewhat between individuals, the head should be round at the level of the coracoid process. This concept is useful in the evaluation of a possible Hill–Sachs deformity. Because of partial volume averaging along the curving superior aspect of the humeral head, articular cartilage is not well delineated on high axial scans.

Through mid-adulthood, the bone marrow of the humeral head appears inhomogeneous on axial images. This results primarily from the fact that the obliterated physis (growth plate) has an oblique orientation in an anteroposterior direction, and the metaphysis has an inherently different signal intensity from the epiphysis. With the arm internally rotated, the brighter and more homogeneous signal intensity of the epiphysis is anterior, while the lower and more mottled signal intensity of the metaphysis is posterior. Since the level of the former physis and thus the transition between epiphysis and metaphysis varies greatly between individuals, significant confusion can arise when attempting to evaluate subtle changes in marrow signal intensity with axial imaging. For this reason, diagnoses of ischemic necrosis or marrow infiltrative disorders should not be attempted solely on the basis of axial images.

The anterior and posterior glenoid labra are ideally imaged in the axial plane. These structures are redundant folds of the fibrous joint capsule and glenohumeral ligaments, and are thus imaged as a signal void. Since the labrum is quite pliable, its shape is determined by the position of the humeral head. With the arm internally rotated, the larger anterior labrum becomes sharply pointed, and the smaller posterior labrum is more rounded. A thin rim of intermediate to high intensity signal separating the labra from the surrounding cuff tendons and joint capsule is felt to represent fat-laden folds of synovium which, by their redundancy, allow the shoulder its extreme mobility.

Axial imaging is useful for evaluating the joint capsule and the rotator cuff tendons anteriorly (subscapularis) and posteriorly (infraspinatus and teres minor). Although portions of the supraspinatus tendon can be seen superiorly on axial images, adequate evaluation is impossible due to the curve of the tendon over the humeral head.

The bicipital groove is clearly seen on axial scans. The tendon of the long head of the biceps muscle is seen as a signal void within the groove and, in individuals in whom the synovium of the joint extends into the groove, can be separated from the surrounding cortex and transverse humeral ligament by a thin rim of intermediate to high intensity signal.

Sagittal plane (*Figure 4.2a,b*)

The sagittal plane is useful for evaluating the course of the 'scar' left by the former physis of the humeral head, and avoids the confusion arising from the inhomogeneity seen within this region of the marrow on axial scans. This plane is also well suited for evaluating the acromion, both its general slope with respect to the supraspinatus tendon and its angle anteriorly in the case of an unfused os acromiale. Diagnostic information regarding these two conditions, however, is best acquired in the frontal oblique plane, where the longitudinal extent of the supraspinatus tendon can be evaluated in a single image.

Coronal plane (*Figure 4.3a,b*)

The coronal plane is useful for imaging the two limbs of the coracoclavicular ligament. Although this plane provides information regarding other structures such as the articular cartilage of the humeral head and the acromioclavicular joint, they are also well imaged in the frontal oblique plane, which is superior to the coronal plane for display of several additional structures.

Frontal oblique plane (*Figure 4.4a–d*)

Alignment for frontal oblique scanning is determined from a high axial image which shows the supraspinatus muscle.[3] Cursors aligned along the long axis of this muscle yield images which are oblique along a coronal-to-sagittal plane.

This orientation provides visualization of the supraspinatus muscle and tendon in continuity, and shows the relationship of the supraspinatus tendon to the acromion and acromioclavicular joint. The tendon should be evaluated for signal intensity, anatomic course, and integrity.

Because the angle of the scapula is similar to that of the supraspinatus apparatus, the oblique plane is superior to the coronal plane for evaluation of the superior and inferior portions of the bony glenoid and the glenoid labra.

Pathology

Impingement syndrome/rotator cuff tear (*Figures 4.5–4.23*)

In its classic form, shoulder impingement syndrome refers to entrapment of the supraspinatus tendon and subacromial bursa between the humeral head below and the structures of the coracoacromial arc above.[10-13] A predisposition to impingement may be associated with the presence of spurs arising from the inferior aspect of the anterior acromion or acromioclavicular joint, hypertrophy of the acromioclavicular joint capsule, or a congenitally low-lying anterior acromion.[10,13] Certain movements of the glenohumeral joint reduce the space available for the subacromial bursa and supraspinatus tendon. The resulting repeated mechanical trauma may result in inflammation, fibrosis, and eventual tendon rupture. Neer believes that over 95 per cent of rotator cuff tears are due to the impingement syndrome.[10]

MRI provides the first noninvasive means for evaluating all of the soft-tissue abnormalities associated with the impingement syndrome.[7,8] By imaging in the frontal oblique plane (along the long axis of the supraspinatus muscle and tendon), the relationship of the subacromial bursa and supraspinatus tendon to the osseous structures directly above is readily apparent. Since positioning the arm in internal rotation with elevation of the elbow accentuates mechanical impingement, this position is useful for demonstrating depression of the bursa and tendon in patients with minimal but symptomatic supraspinatus impingement.

MRI not only affords evaluation of the soft-tissue pathology associated with impingement syndrome, but also provides information regarding the offending structures above the bursa and tendon. Small spurs of the anterior acromion or acromioclavicular joint are imaged as regions of signal void (Figure 4.5). Because larger spurs generally contain marrow, they are seen as extensions of the high-intensity signal from the parent bone, surrounded by a cortical signal void (Figure 4.6). Hypertrophy of the acromioclavicular joint capsule appears as a rounded mass of medium-intensity signal surrounding the joint (Figure 4.7). In T_2-weighted images, fluid within the acromioclavicular joint appears as a focus of high-intensity signal (Figure 4.8). Degenerative changes of the acromioclavicular joint lead to a loss of the sharp, smooth margins of the opposing cortical surfaces and replacement by irregular margins, with a resultant intervening irregular, inhomogeneous band of medium signal intensity (Figure 4.9).

The earliest soft-tissue change in patients with the impingement syndrome may be subacromial bursitis (Figures 4.10–4.11). This is manifested as a widening of the high-intensity signal of the bursa medial to the region of depression, and an abrupt cut-off at the site of impingement. The bursa shows high-signal intensity in both T_1- and T_2-weighted images, probably due to the high fat content of its hypertrophied synovium. It is unusual for excessive fluid to be present within the bursa, even in cases of severe, chronic bursitis.

In cases of supraspinatus tendonitis, MRI reveals a medium-intensity signal within the substance of the tendon in T_1-weighted images, which does not become bright with T_2-weighting (Figures 4.12–4.13). The supraspinatus musculotendinous junction retains its normal position beneath the anterior acromion or acromioclavicular joint, and the bulk and signal intensity of the supraspinatus muscle remain normal.

In T_1-weighted images, tendonitis cannot be differentiated from a small or partial tear of the supraspinatus tendon. In T_2-weighted images, however, a well-defined focus of high-intensity signal within the supraspinatus tendon is diagnostic of a tear. The size of the tear, amount of muscle retraction, and evidence of muscle atrophy are all important factors to consider when planning possible surgical correction of rotator cuff tears.

In cases of a small or partial supraspinatus tear, there is no retraction of the musculotendinous junction (Figures 4.14–4.16). These small tears are usually within the distal portion of the tendon, near its insertion on the greater tuberosity, and are therefore best seen in extreme anterior images in the frontal oblique plane. Larger tears can also be seen in more posterior images.

With a massive supraspinatus tear (Figures 4.17–4.21), the muscle retracts medially. Atrophic changes of the muscle are seen in cases of extensive and long-standing rotator cuff disruption. Atrophy is imaged as bands of high-signal intensity within the supraspinatus muscle in T_1-weighted images, indicating fatty replacement of muscle fibers. The mass of the muscle is also diminished in cases of atrophy, but this finding must be evaluated in light of the overall muscular development of each individual patient. With chronic tears of the rotator cuff, the subacromial bursa and supraspinatus tendon are gradually replaced by intermediate-signal intensity scar tissue, and are eventually obliterated by the superiorly migrating humeral head. Although our experience is still limited, MRI occasionally has shown tears of the rotator cuff that were later surgically confirmed in patients with a negative arthrogram (Figure 4.21).

Tears of the infraspinatus portion of the rotator cuff are not uncommon. Although they are usually seen in association with large tears of the supraspinatus tendon (Figure 4.22), they may be isolated (Figure 4.23). The integrity of the infraspinatus musculotendinous unit is best evaluated in the axial plane.

The pathology of the subscapularis portion of the rotator cuff is discussed in the following section on shoulder instability.

Instability (*Figures 4.24–4.33*)

The glenohumeral joint is the most mobile joint in the body, due to the fact that only a small portion of the humeral head is in contact with the bony glenoid at any given time.[14] The trade-off for the exceptional mobility of the glenohumeral joint is its inherent instability. Shoulder instability, either in the form of recurrent subluxation or dislocation, is a common cause of chronic pain and disability.

Although glenohumeral instability often follows trauma, many patients with instability cannot recall a specific traumatic episode. In patients with subclinical instability, the presenting complaint may be vague, including arm numbness, decreased range of motion, or non-specific pain.[15] Diagnosis in this situation may be difficult. There is also a group of patients with 'multidirectional instability' (Figure 4.30), in whom subluxation occurs in both the anterior and posterior directions.[16] If only the more common anterior lesion is repaired, these patients will experience persistent pain and disability due to what will have become a fixed, unidirectional posterior instability.

The cause of non-traumatic instability has long been debated, but is felt to relate to a deficiency in the soft-tissue support of the joint.[17] The so-called 'capsular mechanism' probably provides the majority of natural stability. Anteriorly, this consists of the subscapularis muscle and tendon, the joint capsule and synovial membrane, the three glenohumeral ligaments, and the fibrous glenoid labrum. Persons with a predisposition to instability may be found to have either a relatively medial attachment of the anterior capsule to the scapula, or underdevelopment of one of the anterior glenohumeral ligaments. In the former situation, the subscapularis recess is unusually large, and the humeral head tends to subluxate into the excessively capacious joint space. In the latter situation, the deficiency of ligamentous support allows the joint capsule to be stretched or torn from its attachment to the scapula, leading to repeated subluxation.

Regardless of the mechanism of instability, the resultant pathology is well known. The glenoid labrum generally suffers the earliest and most severe trauma, and may become torn or detached from the bony glenoid. In cases of recurrent subluxation or dislocation, the labrum may become severely attenuated. Additional pathology may include separation of the joint capsule from the scapula, and trauma to the subscapularis tendon or muscle.

Because of the inherent ability of MRI to depict the soft tissues without the injection of intra-articular contrast material, it is well suited to the evaluation of pathology secondary to dislocation or recurrent subluxation of the humeral head. Labral tears are evident as linear regions of increased signal intensity within the normal signal void of the fibrous labrum. The tears show an intermediate-signal intensity in T_1-weighted images, and become bright in T_2-weighted images. With recurrent subluxation or dislocation, the labrum may become severely attenuated (Figure 4.28). High-signal intensity along the base of the labrum indicates labral detachment (Figures 4.29–4.30). In regions of capsular detachment, T_2-weighted images show high-intensity fluid dissecting along the anterior border of the scapula (Figure 4.30). The marrow of the bony glenoid underlying a site of labral detachment or attenuation may show abnormal intermediate-signal intensity, even in cases where an osseous Bankart lesion is not evident radiographically.

Hill–Sachs lesions are depicted on MR images as focal depressions in the contour of the humeral head (Figure 4.31). When subtle, these depressed fractures are best seen in the axial plane. Large lesions may also be evident in the frontal oblique plane. Care must be taken not to mistake the normal posterolateral flattening of the inferior humeral head for a Hill–Sachs deformity. This flattened region of attachment for the fibers of the teres minor and infraspinatus tendons is found low on the humeral head. As most humeral head dislocations are anteroinferior, Hill–Sachs defects are seen on the high posterolateral aspect of the head, at or above the level of the coracoid process in axial images.

Patients with significant trauma to the subscapularis tendon may show medial retraction of the subscapularis musculotendinous junction, indicating tendon rupture (Figures 4.32–4.33). Disproportionate atrophy of the subscapularis muscle in comparison to the adjacent muscles often indicates remote or chronic instability.

Avascular necrosis (AVN) of the humeral head (*Figures 4.34–4.36*)

The humeral head is a common site for avascular necrosis. The exquisitely detailed evaluation of bone marrow that is possible with MRI makes it an ideal means for the diagnosis of early AVN in patients with a suspicious history, and for evaluating the extent of disease in patients who already have radiographic changes.

The sagittal plane is optimal for the evaluation of possible AVN. With the arm in internal rotation, the earliest changes are seen in the posterosuperior aspect of the humeral head. Axial images can be confusing because of the normally mottled signal intensity of the metaphysis, and in the coronal plane small regions of abnormal signal intensity may be inapparent due to partial volume averaging. Sagittal images show the oblique course of the physis tangentially, and clearly depict regions of abnormally diminished signal intensity in the epiphysis.

Osseous trauma (*Figures 4.37–4.38*)

Because of their superior definition of small fracture fragments and trabecular detail, plain radiography and

CT remain the primary imaging modalities for the evaluation of bony trauma. In certain situations, however, MRI may be helpful in the evaluation of fractures. With its ability to image directly in multiple planes, MRI may be useful in determining the relationship of large fracture fragments to adjacent soft tissue. In cases of suspected non-union of a fractured tuberosity, complete osseous union can be documented by depicting the continuity of bone marrow across the fracture line (Figure 4.37). Spin-echo techniques generally will not differentiate non-union from fibrous union, as both cortical bone and fibrous tissue appear as a signal void in both T_1- and T_2-weighted images (Figure 4.38). In the evaluation of a remote fracture, the presence of a band of intermediate- or high-intensity signal separating two regions of signal void along the fracture line may indicate a pseudarthrosis, with the higher intensity signal representing a neo-synovial membrane.

Summary

The use of high-resolution scanning techniques and a surface coil permit the acquisition of detailed MR images of the shoulder. Several forms of soft-tissue pathology can be thoroughly evaluated without radiation exposure or injection of intra-articular contrast material. With the development of faster scanning techniques and a reduction in its cost, MRI may eventually replace arthrography.

MRI is useful in planning the management of several disorders of the shoulder. Because it allows for more appropriate decisions to be made regarding the indications for arthroscopic versus open repair of lesions resulting from impingement syndrome and instability, MRI may eventually replace 'diagnostic' arthroscopy in individuals with elusive shoulder complaints. The use of MRI may also avoid unnecessary surgical procedures that are doomed to failure in patients with chronic, massive rotator cuff tears in whom marked muscle retraction and atrophy are present.

References

1 DOMINIK HJ, SAUTER R, MUELLER E et al, MR imaging of the normal shoulder, *Radiology* (1986) **158**:405–8.
2 KIEFT GJ, BLOEM JL, OBERMANN WR et al, Normal shoulder: MR imaging, *Radiology* (1986) **159**: 741–5.
3 SEEGER LL, RUSZKOWSKI JT, BASSETT LW et al, MR imaging of the normal shoulder: anatomic correlation, *AJR* (1987) **148**:83–91.
4 MIDDLETON WD, KNEELAND JB, CARRERA GF et al, High resolution MR imaging of the normal rotator cuff, *AJR* (1987) **148**:559–64.
5 KIEFT GJ, SARTORIS DJ, BLOEM JL et al, Magnetic resonance imaging of glenohumeral joint diseases, *Skeletal Radiol* (1987) **16**:285–90.
6 KNEELAND JB, MIDDLETON WD, CARRERA GF et al, MR imaging of the shoulder: diagnosis of rotator cuff tears, *AJR* (1987) **149**:333–7.
7 SEEGER LL, GOLD RH, BASSETT LW et al, Shoulder impingement syndrome: MR findings in 53 shoulders, *AJR* (1988) **150**:343–57.
8 KIEFT GJ, BLOEM JL, ROZING PM et al, Rotator cuff impingement syndrome: MR imaging, *Radiology* (1988) **166**:211–14.
9 SEEGER LL, GOLD RH, BASSETT LW, MR imaging of shoulder instability, *Radiology* (1988) **168**:695–7.
10 NEER CS II, Impingement lesions, *Clin Orthop* (1983) **173**: 70–7.
11 PENNY JN, WELSH MB, Shoulder impingement syndromes in athletes and their surgical management, *Am J Sports Med* (1981) **9**:11–15.
12 HAWKINS RJ, KENNEDY JC, Impingement syndrome in athletes, *Am J Sports Med* (1980) **8**:151–7.
13 NEER CS II, Anterior acromioplasty for the chronic impingement syndrome in the shoulder, *J Bone Joint Surg [Am]* (1972) **54A**:41–50.
14 ROTHMAN RH, MARVEL JP, HAPPENSTALL RB, Recurrent anterior dislocation of the shoulder, *Orthop Clin North Am* (1975) **6**:415–22.
15 ROWE CR, ZARINS B, Recurrent transient subluxation of the shoulder, *J Bone Joint Surg [Am]* (1981) **62A**:863–71.
16 NEER CS II, FOSTER CR, Inferior capsular shift for involuntary and multidirectional instability of the shoulder, *J Bone Joint Surg [Am]* (1980) **62A**:897–908.
17 MOSLEY HF, OVERGAARD B, The anterior capsular mechanism in recurrent anterior dislocation of the shoulder, *J Bone Joint Surg [Am]* (1962) **44A**:913–27.

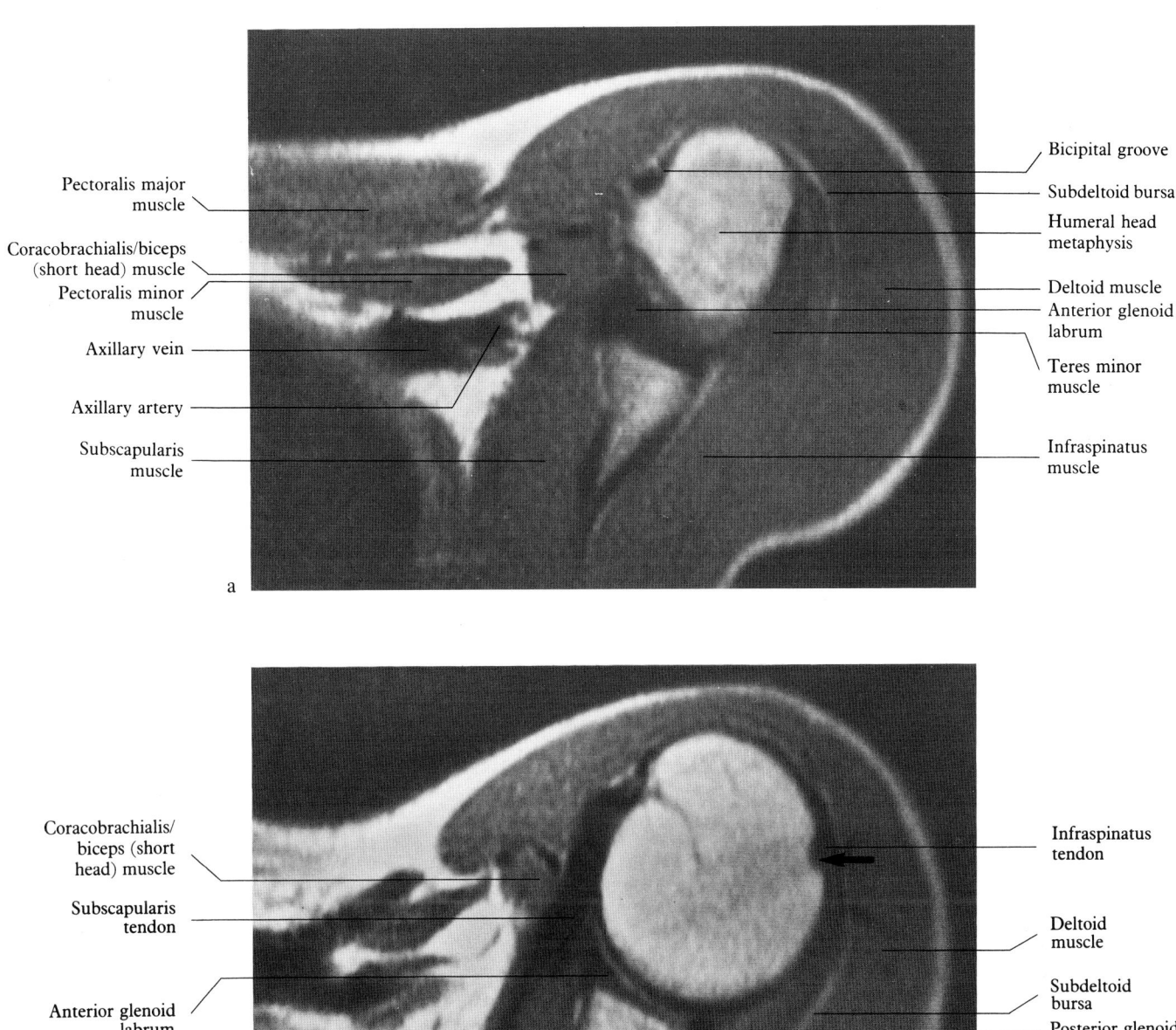

Figure 4.1

Normal anatomy: axial plane (SE 500/28). (**a**) Level of inferior humeral head. The head is oblong, and flattened posteriorly. (**b**) Level of mid-humeral head. The indentation in head laterally (arrow) is site of attachment for some fibers of the infraspinatus muscle, and should not be confused with Hill–Sachs deformity. The anterior glenoid labrum is larger and more sharply pointed than the smaller, more rounded posterior labrum. (**c**) Level of coracoid process. The humeral head is surrounded by signal void, which represents both cortical bone and tendons of rotator cuff. (**d**) Level of supraspinatus muscle.

Figure 4.1 continued

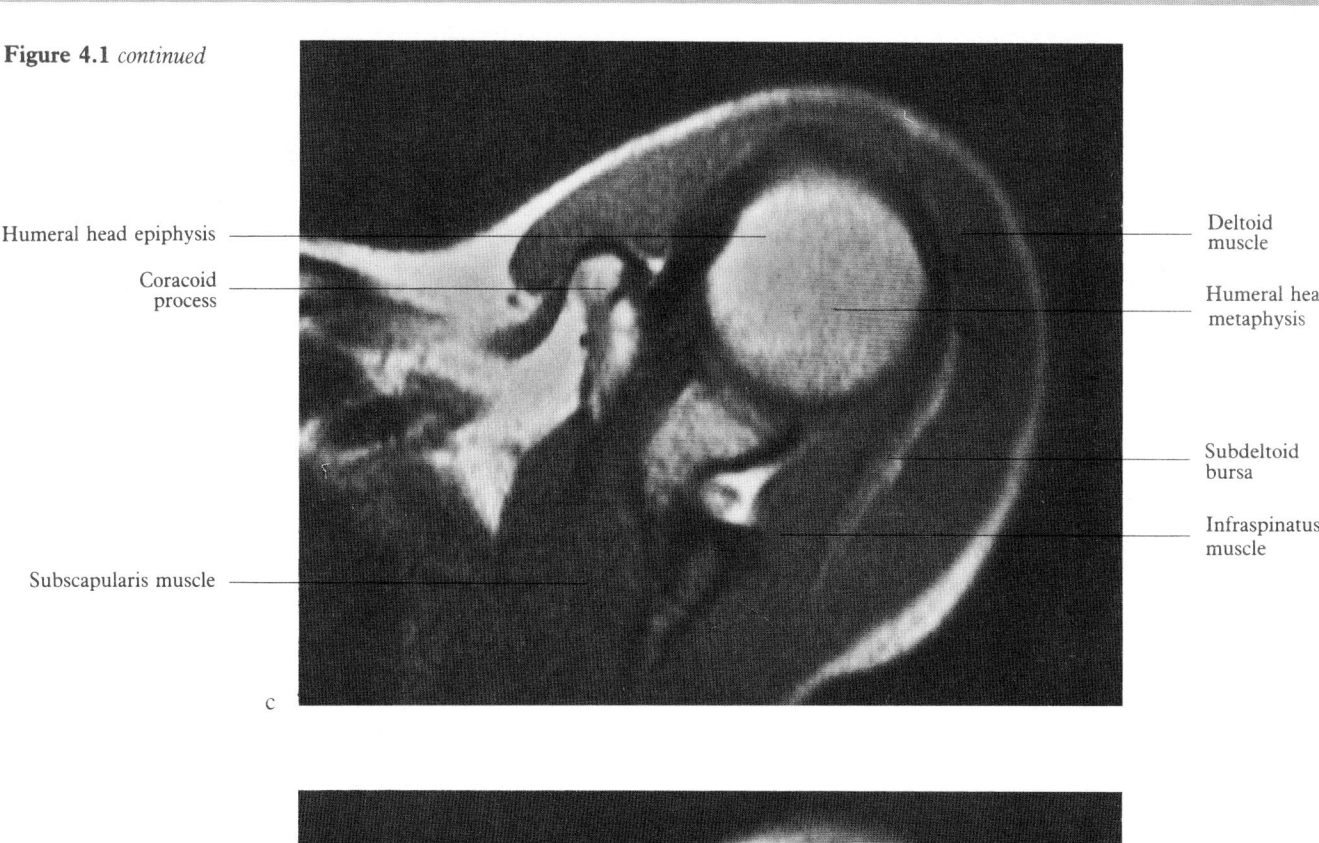

- Humeral head epiphysis
- Coracoid process
- Subscapularis muscle
- Deltoid muscle
- Humeral head metaphysis
- Subdeltoid bursa
- Infraspinatus muscle

c

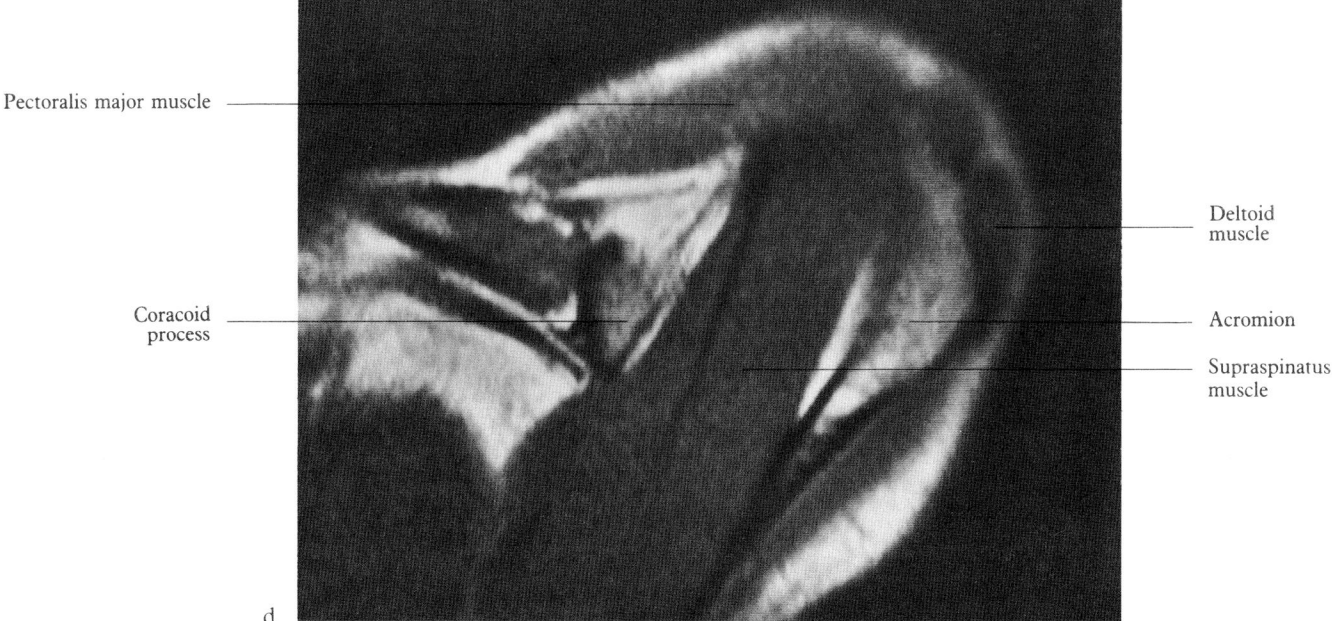

- Pectoralis major muscle
- Coracoid process
- Deltoid muscle
- Acromion
- Supraspinatus muscle

d

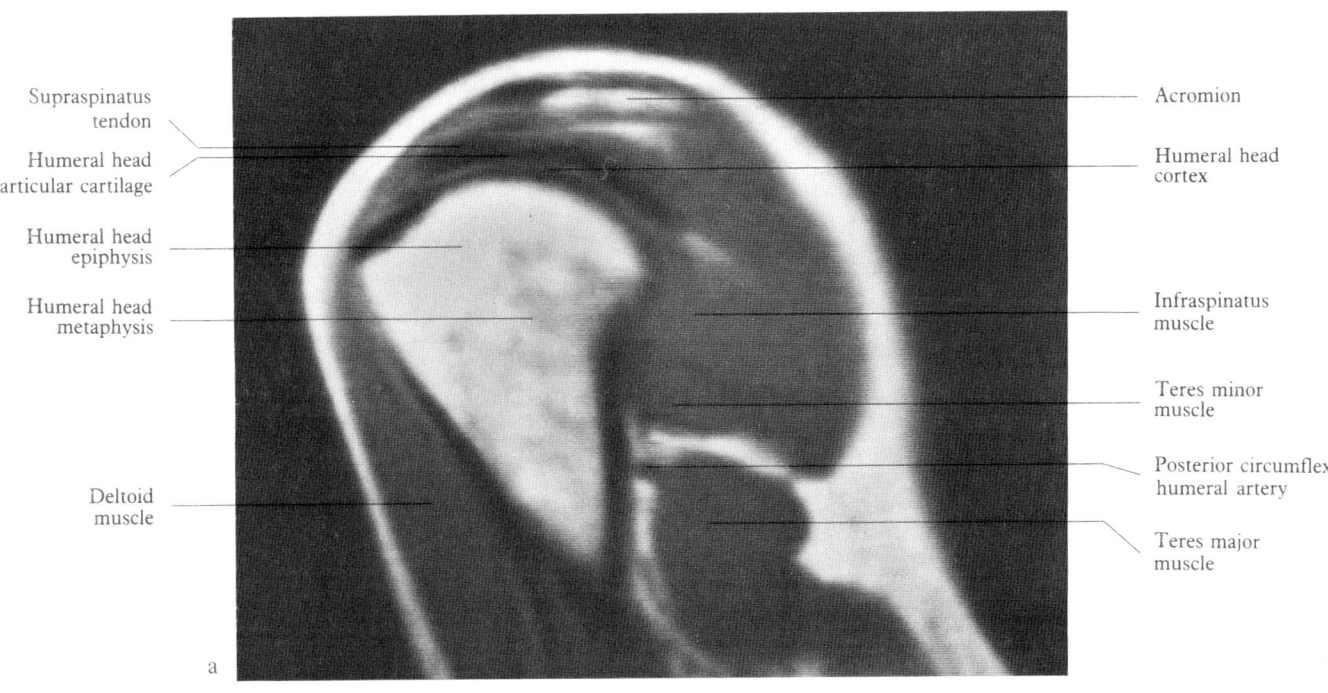

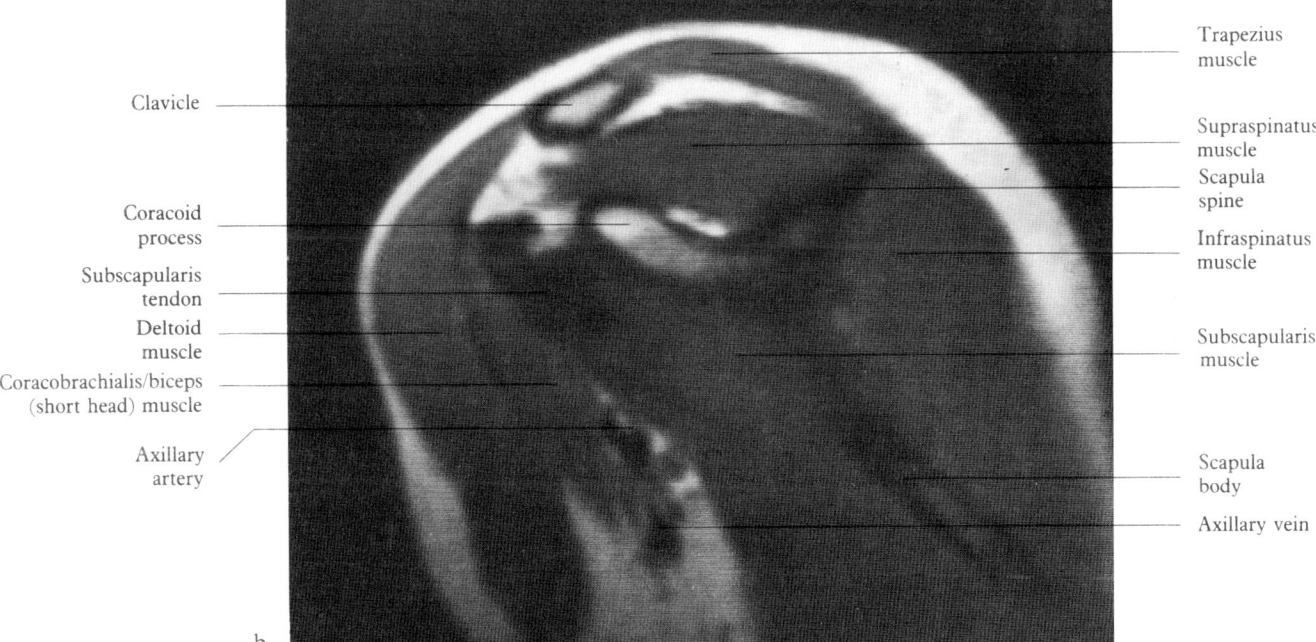

Figure 4.2

Normal anatomy: sagittal plane (SE 500/28). (**a**) Level of mid-humeral head. (**b**) Level of coracoid process (medial to humeral head).

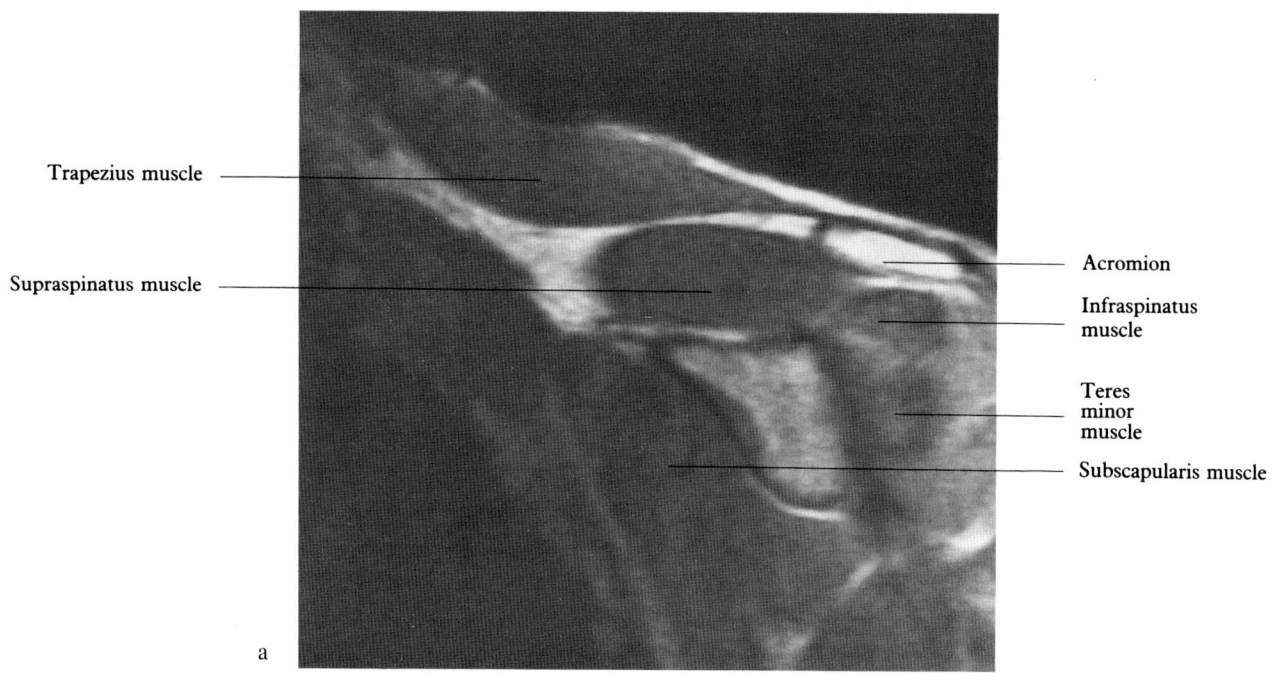

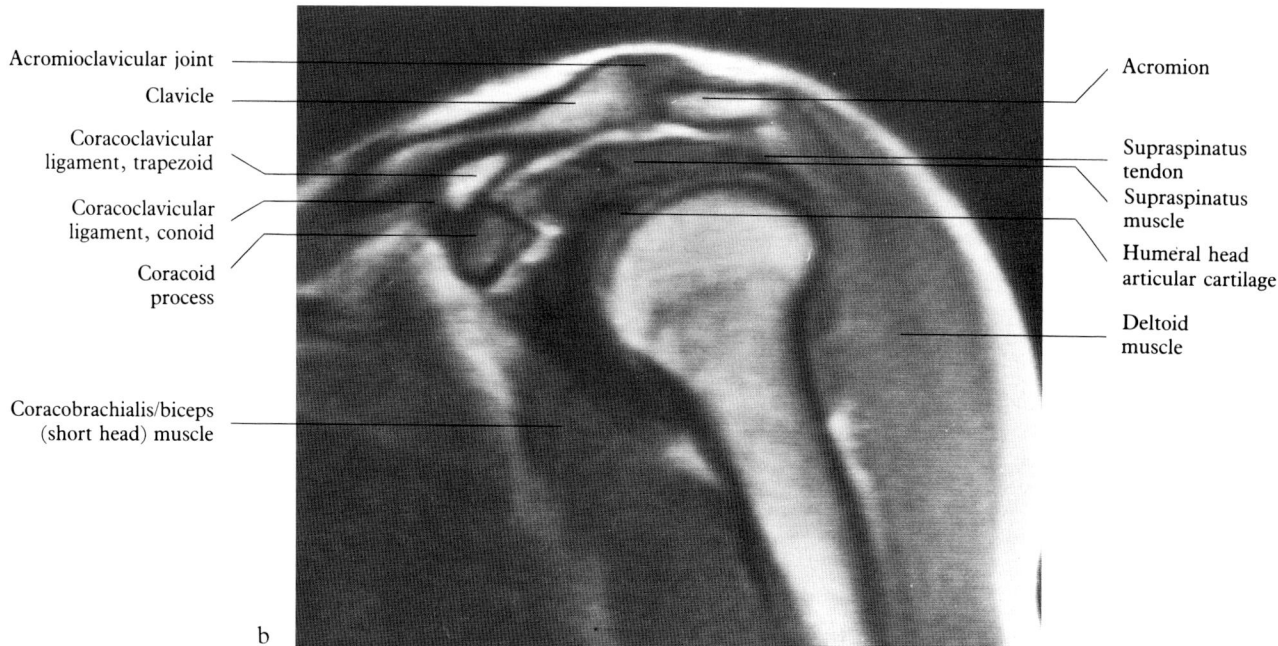

Figure 4.3

Normal anatomy: coronal plane (SE 500/28). (**a**) Posterior to humeral head. (**b**) Level of mid-humeral head. The 2 limbs of the coracoclavicular ligament are clearly visualized.

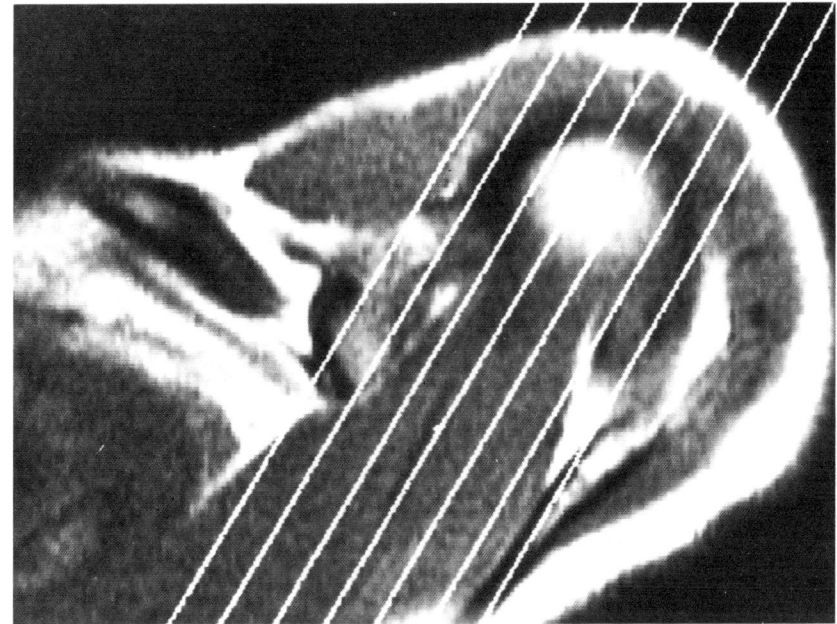

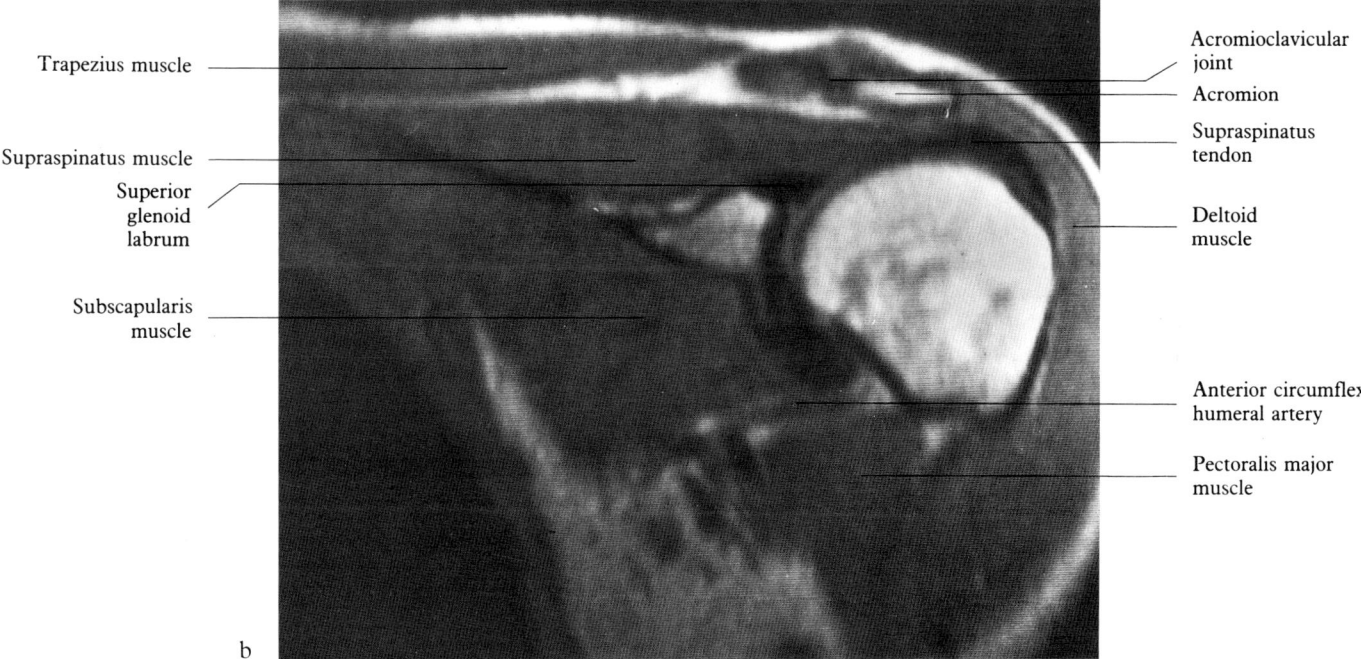

Figure 4.4

Normal anatomy: frontal oblique plane (SE 500/28). (**a**) Axial image showing alignment for frontal oblique imaging. The cursors are placed along long axis of supraspinatus muscle belly. (**b**) Level of mid-acromioclavicular joint. The supraspinatus muscle and tendon are seen in continuity, and should course beneath the acromioclavicular joint and anterior acromion in a straight line, without focal depression. A thin line of bright signal intensity below the anterior acromion represents the subacromial bursa. Supraspinatus tendon is a homogeneous signal void. (**c**) Level of anterior coracoid process. Coracoclavicular ligament is clearly seen. Long head of biceps tendon is evident as it exits from the bicipital groove and travels medially within the joint space to insert above superior glenoid labrum. (**d**) Level of bicipital groove. Signal void surrounding superior and medial humeral head represents rotator cuff tendons: supraspinatus above, and subscapularis inferomedial.

106 *MRI atlas of the musculoskeletal system*

Figure 4.4 *continued*

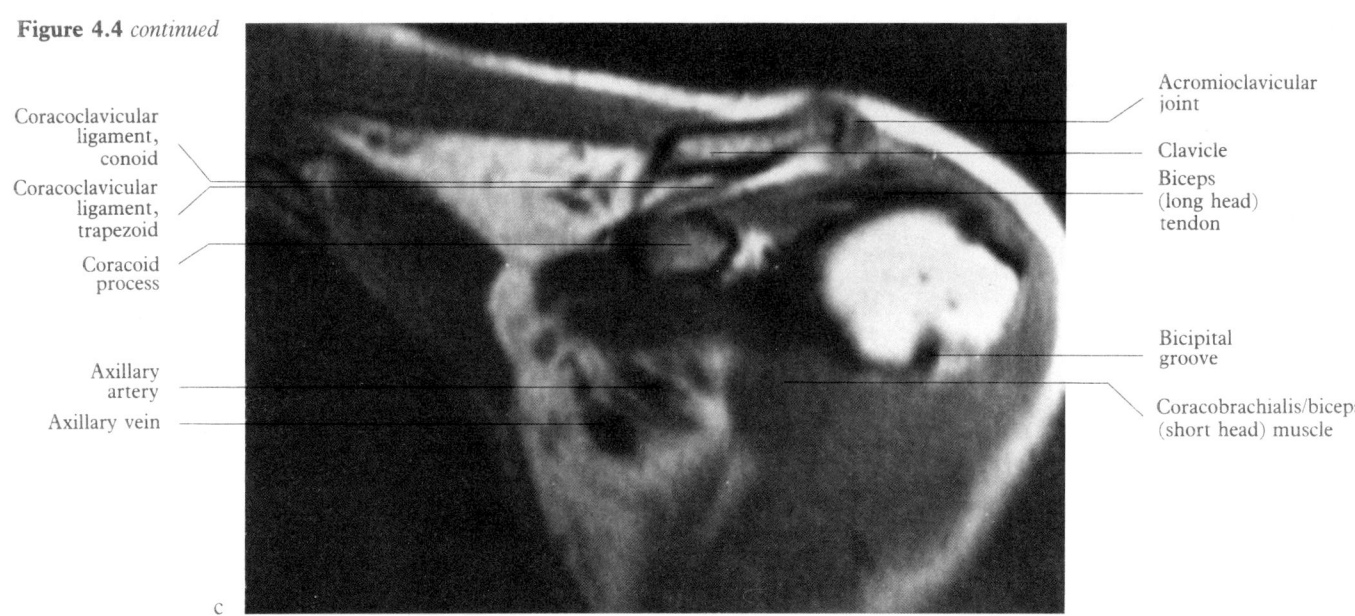

- Coracoclavicular ligament, conoid
- Coracoclavicular ligament, trapezoid
- Coracoid process
- Axillary artery
- Axillary vein
- Acromioclavicular joint
- Clavicle
- Biceps (long head) tendon
- Bicipital groove
- Coracobrachialis/biceps (short head) muscle

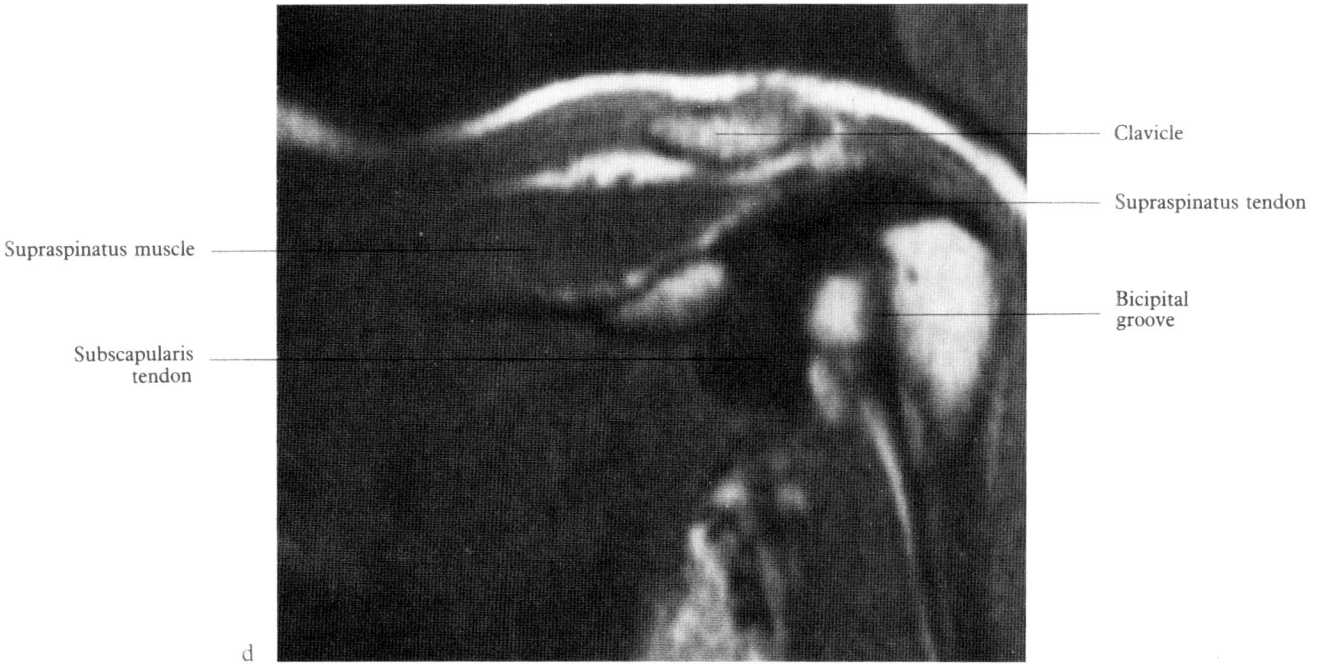

- Supraspinatus muscle
- Subscapularis tendon
- Clavicle
- Supraspinatus tendon
- Bicipital groove

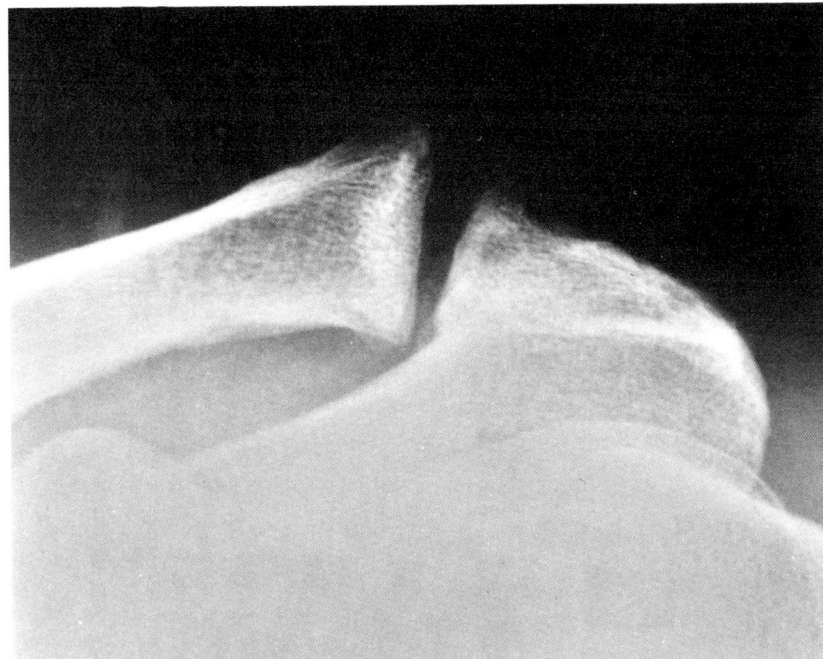

a

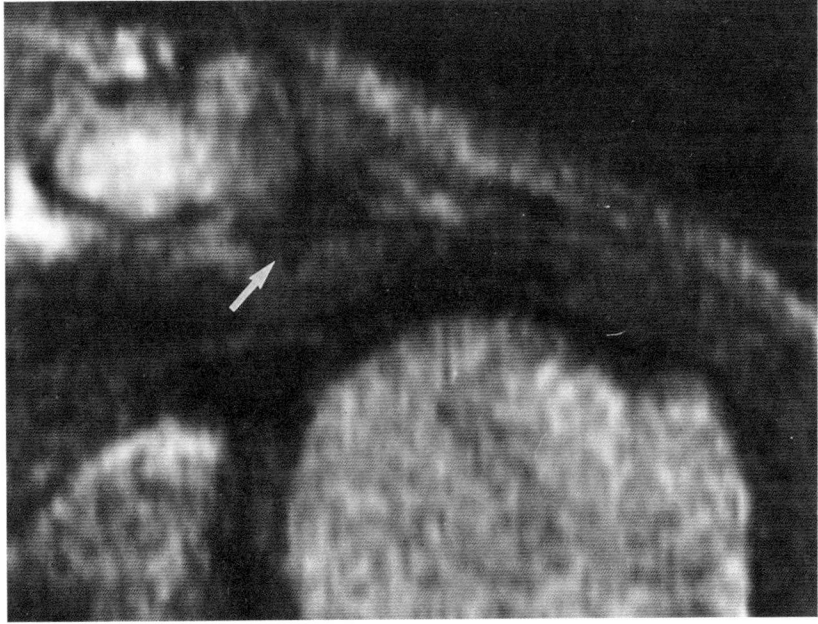

b

Figure 4.5

Small subacromial spur. A 44-year-old man who suffered repeated trauma to shoulder over several years through body building and boxing. (**a**) Caudal oblique anteroposterior radiograph shows small inferior spur off acromioclavicular joint. (**b**) Frontal oblique image (SE 500/28). Spur is represented by small area of signal void (arrow) inferior to acromioclavicular joint.

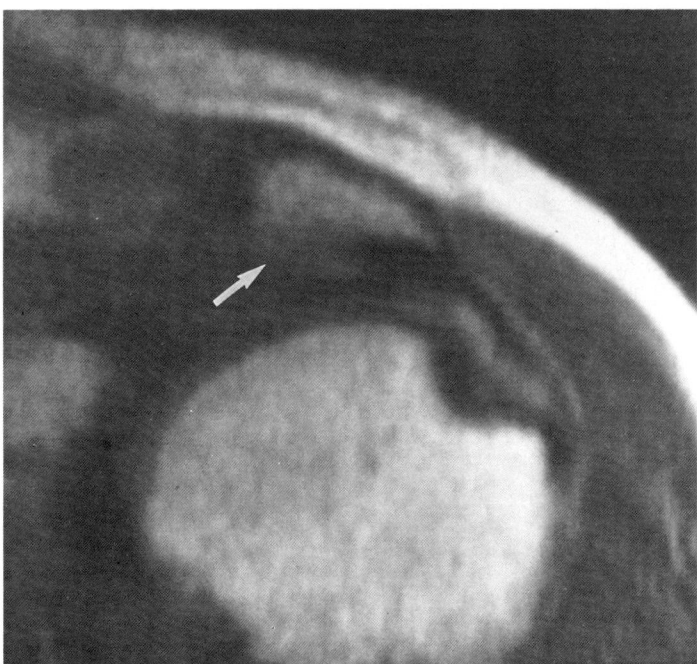

Figure 4.6

Large subacromial spur. A 54-year-old man with shoulder pain and decreased range of motion. Frontal oblique MR image (SE 700/28). Large subacromial spur appears as area of high-signal intensity extending from anterior acromion (arrow). Underlying supraspinatus tendon signal intensity is abnormal.

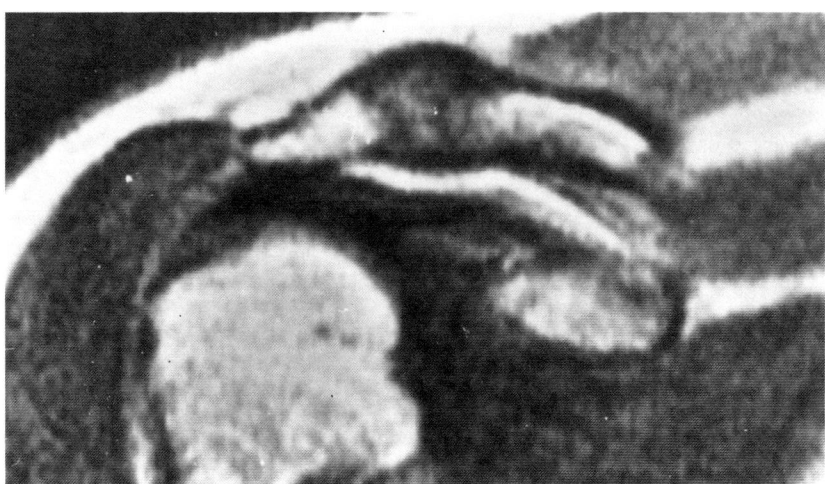

Figure 4.7

Acromioclavicular joint capsule hypertrophy. A 28-year-old man complained of shoulder pain. Frontal oblique T_1-weighted image (SE 500/28) shows rounded mass of intermediate-signal intensity around acromioclavicular joint. This indicates hypertrophy of acromioclavicular joint capsule.

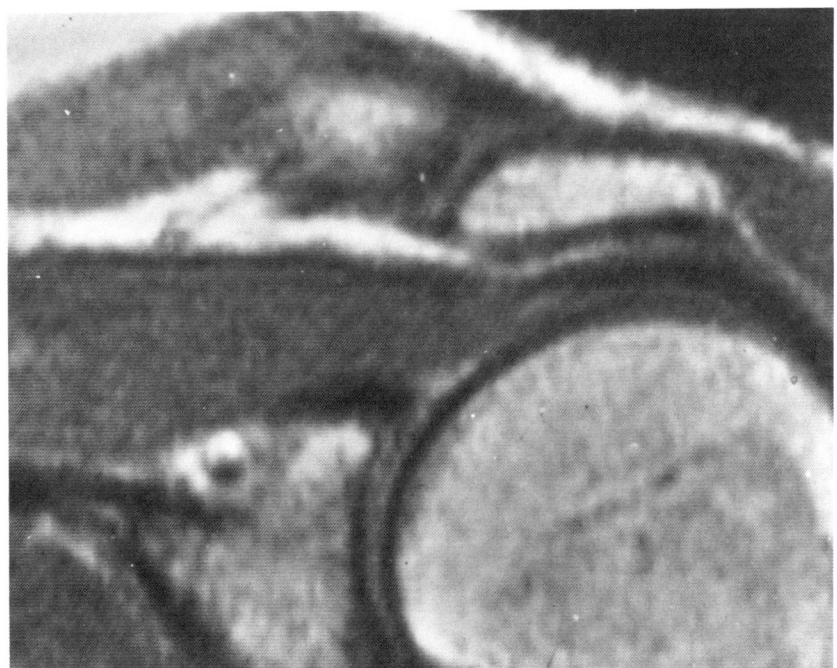

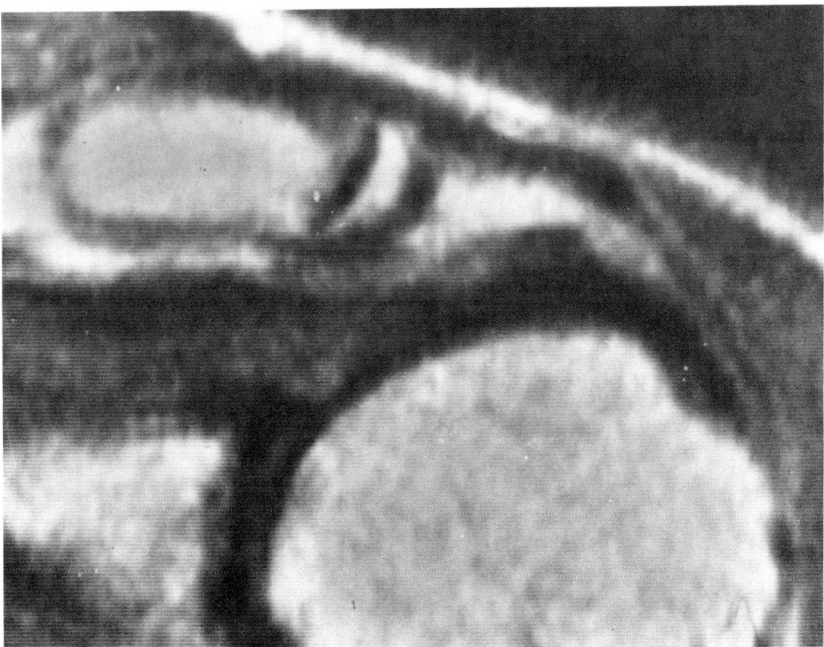

Figure 4.8

Acromioclavicular joint fluid. A 37-year-old man with shoulder pain. (**a**) T_1-weighted frontal oblique image (SE 500/28) shows medium-signal intensity in region of acromioclavicular joint. (**b**) T_2-weighted image (SE 1500/56) at the same level shows high-signal intensity fluid within joint. Note that acromion lies too low with respect to distal clavicle.

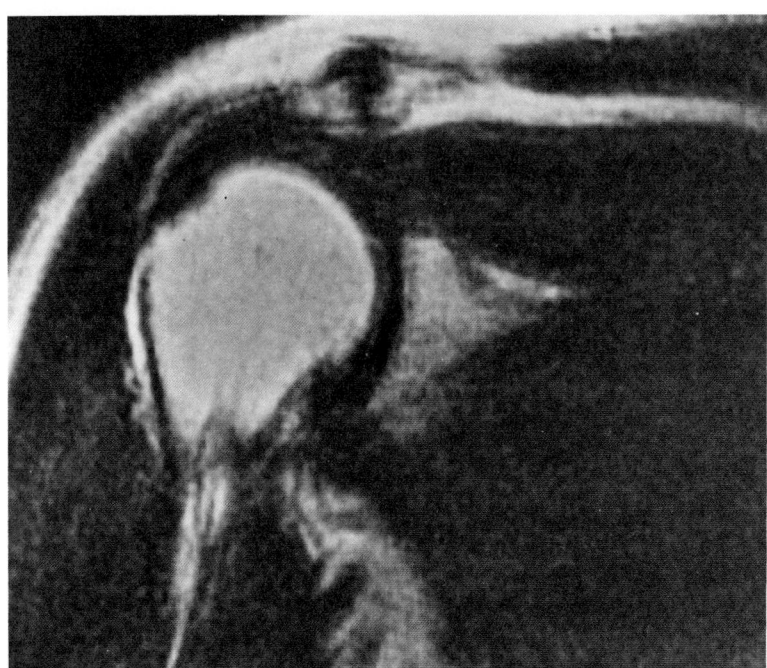

Figure 4.9

Acromioclavicular degenerative joint disease. A 38-year-old man with shoulder pain. Frontal oblique image (SE 2000/85) shows loss of normal well-defined signal void of 2 cortices at acromioclavicular joint margin. Instead, joint margins are indistinct.

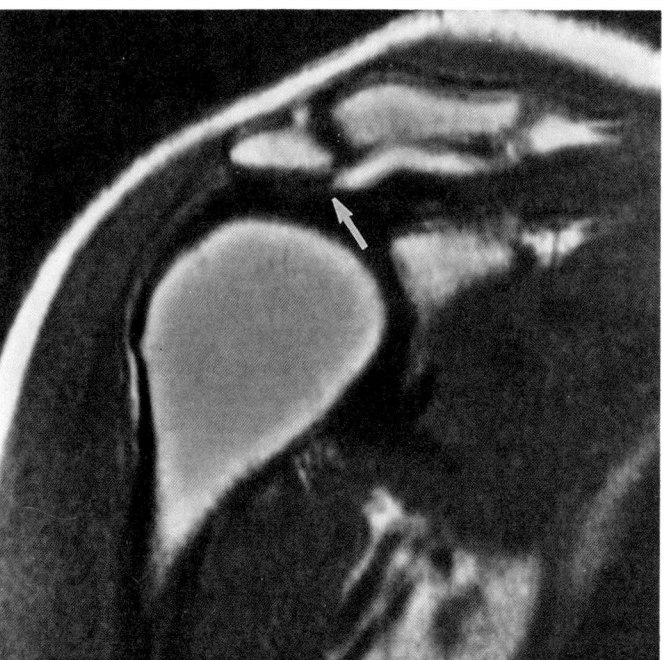

Figure 4.10

Subacromial bursitis. A 29-year-old male competitive swimmer with shoulder pain and decreased range of motion. Frontal oblique image (SE 500/28). Bright signal intensity of subacromial bursa is thickened below distal clavicle and acromioclavicular joint, and is abruptly cut off below anterior acromion (arrow). Anterior acromion lies too low with respect to clavicle. Signal intensity of supraspinatus tendon is normal.

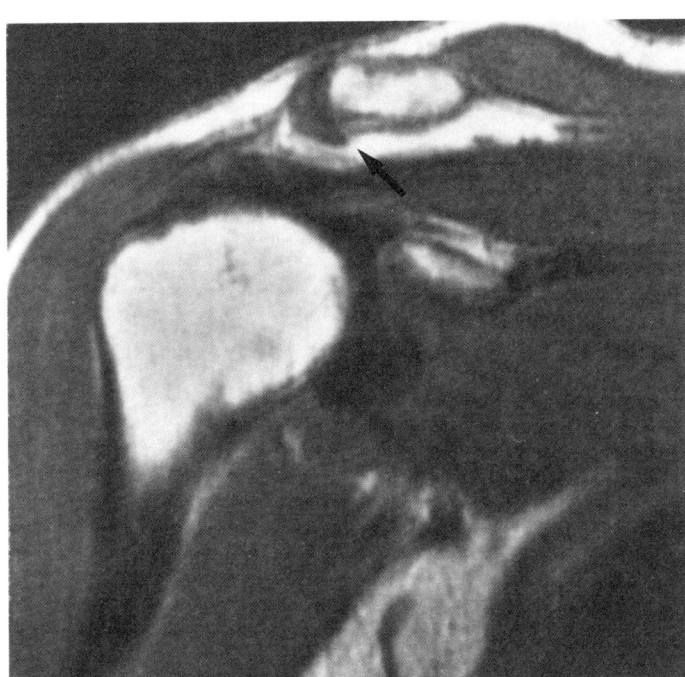

a

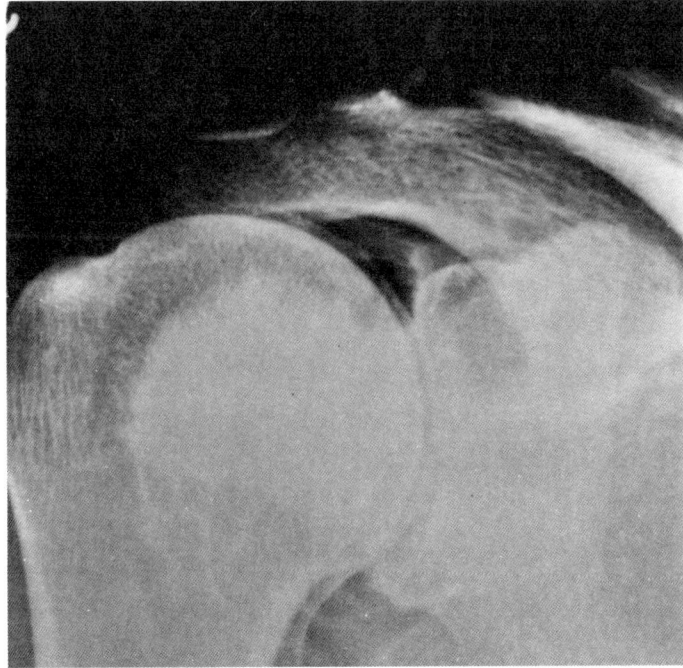

b

Figure 4.11

Subacromial bursitis. A 36-year-old man with shoulder pain. (**a**) Frontal oblique image (SE 500/28). The subacromial bursa is thickened (arrow), and indented by small spur of distal clavicle. Signal intensity of supraspinatus tendon is normal. (**b**) Arthrogram confirms integrity of rotator cuff.

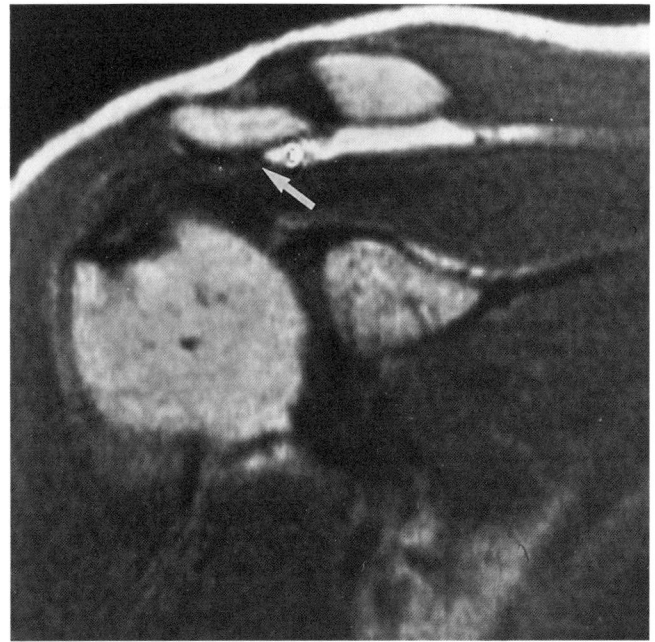

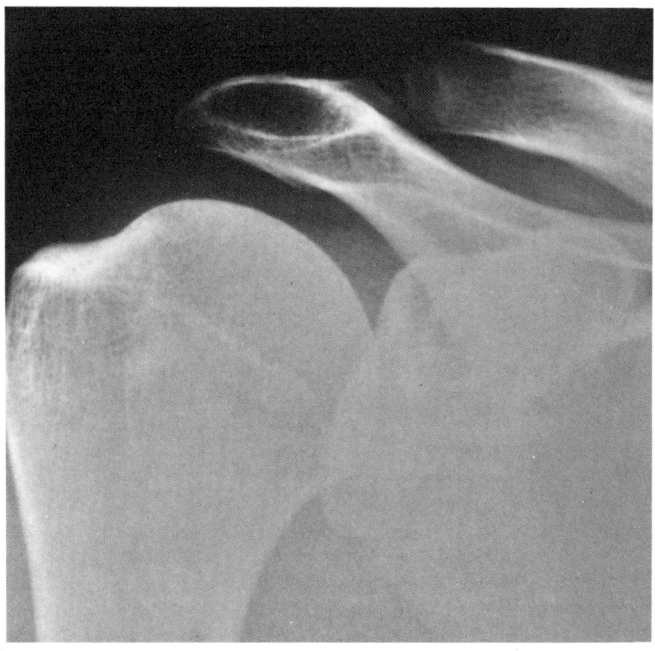

Figure 4.12

Supraspinatus tendonitis due to low-lying anterior acromion. A 35-year-old male tennis player with shoulder pain. (**a**) Frontal oblique image (SE 500/28). The subacromial bursa is thickened, and abruptly cut off at level of anterior acromion (arrow). The signal intensity of the supraspinatus tendon is diffusely abnormal. On this T_1-weighted image, a small tear of the tendon cannot be excluded. Normal position of supraspinatus musculotendinous junction excludes massive tear. (**b**) Anteroposterior radiograph shows sclerosis of greater tuberosity. This finding may be normal variant. (**c**) Axial image through level of coracoid process (SE 500/28). The signal intensity of bone marrow of greater tuberosity is abnormally low and inhomogeneous. This finding is often seen in patients with impingement syndrome, and is felt to be due to repetitive trauma as the humeral head is forced against the anterior acromion with abduction and elevation of the arm. Although this is also evident on the frontal oblique image, the inhomogeneous nature of marrow change is best appreciated in axial images.

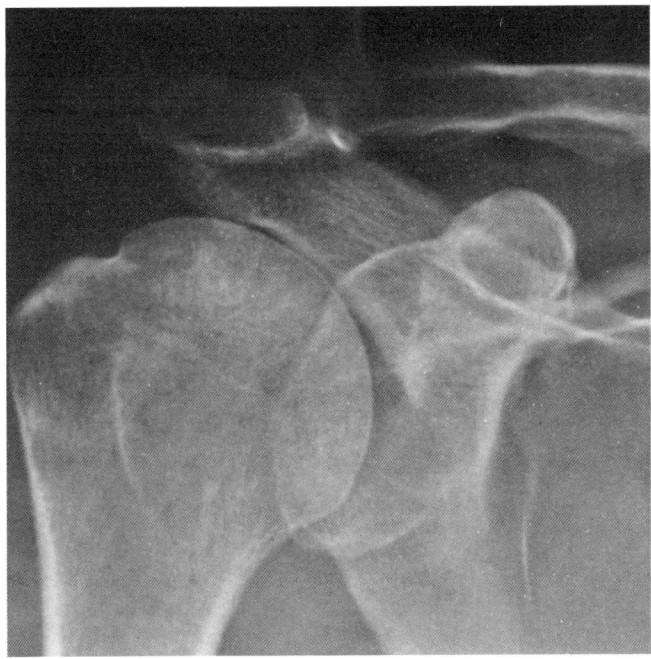
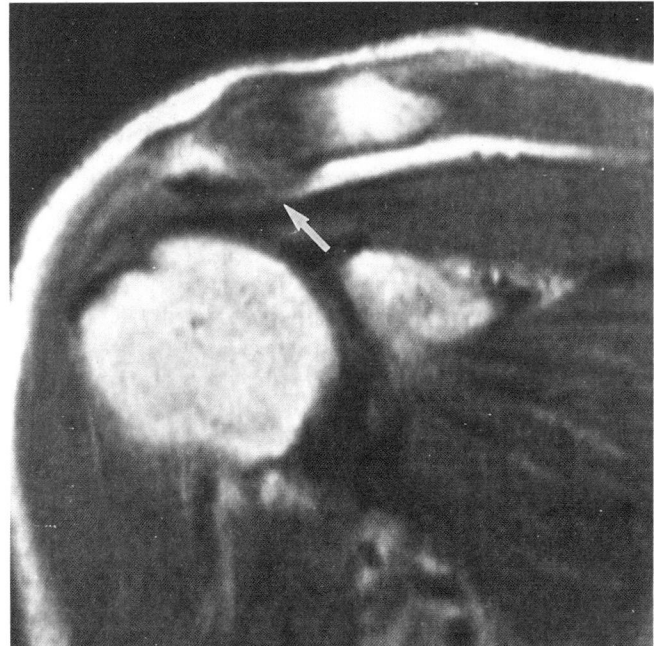

a b

Figure 4.13

Supraspinatus tendonitis due to hypertrophy of the acromioclavicular joint capsule. A 63-year-old man with shoulder pain. (**a**) Anteroposterior radiograph shows only mild degenerative changes of acromioclavicular joint. (**b**) Frontal oblique image (SE 500/28) shows thickening and abrupt cut off of subacromial bursa (subacromial bursitis), and diffuse abnormal signal intensity of supraspinatus tendon (supraspinatus tendonitis). Supraspinatus musculotendinous junction is depressed by marked hypertrophic changes of the acromioclavicular joint capsule (arrow). Cortical margins of the acromioclavicular joint are irregular and poorly defined, indicating degenerative joint disease.

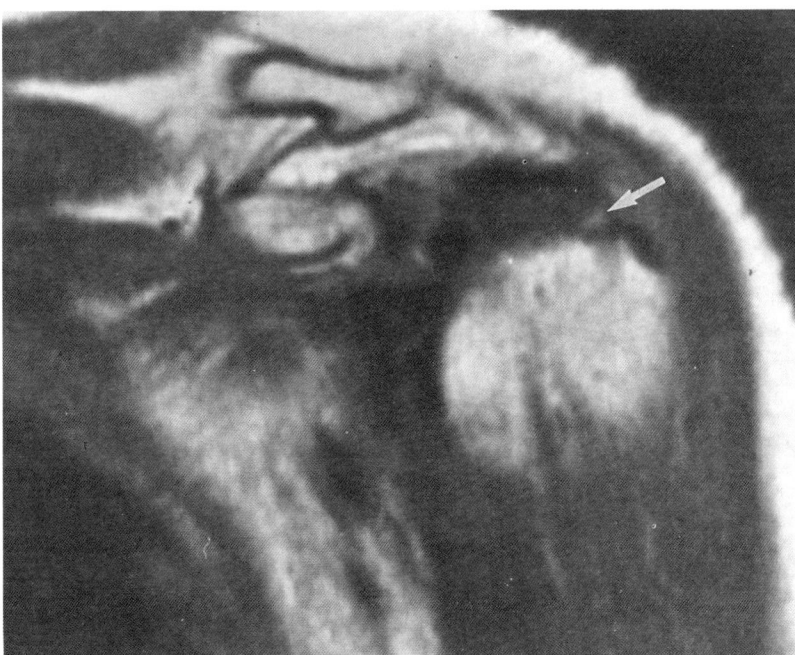

Figure 4.14

Small supraspinatus tendon tear. A 65-year-old woman with persistent pain for 6 months, after lifting heavy object. T_2-weighted frontal oblique image (SE 2000/56) shows small focus of high-intensity signal within tendon (arrow), indicating rotator cuff tear. Muscle is not retracted.

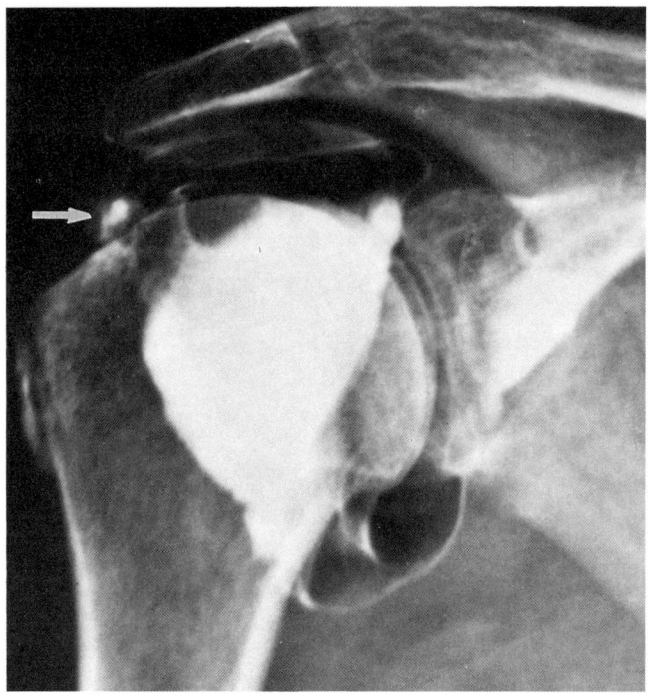
a

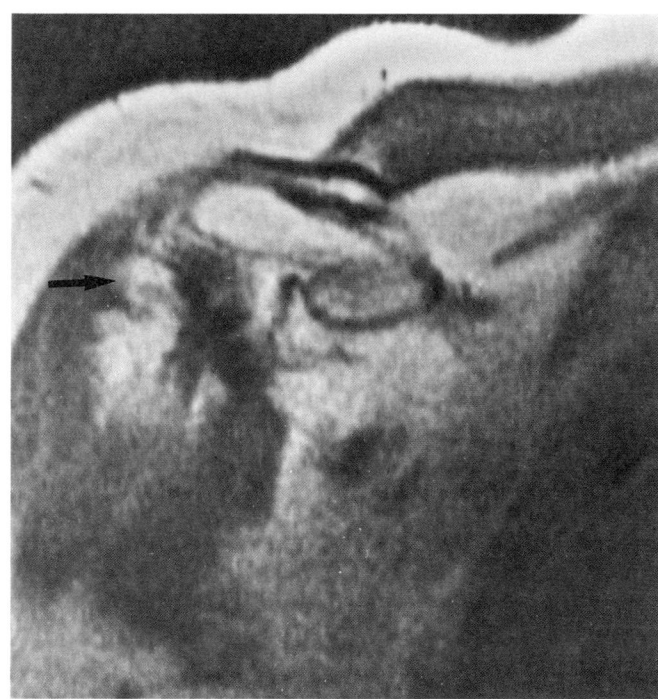
b

Figure 4.15

Small supraspinatus tendon tear. A 54-year-old woman with a 2-year history of pain and decreased range of motion. There was no history of trauma. (**a**) Arthrogram reveals small collection of contrast material within supraspinatus tendon near its insertion onto greater tuberosity (arrow). (**b**) Extreme anterior T_2-weighted frontal oblique image (SE 1500/56) reveals focus of high-intensity signal in corresponding region of tendon, implying tear (arrow).

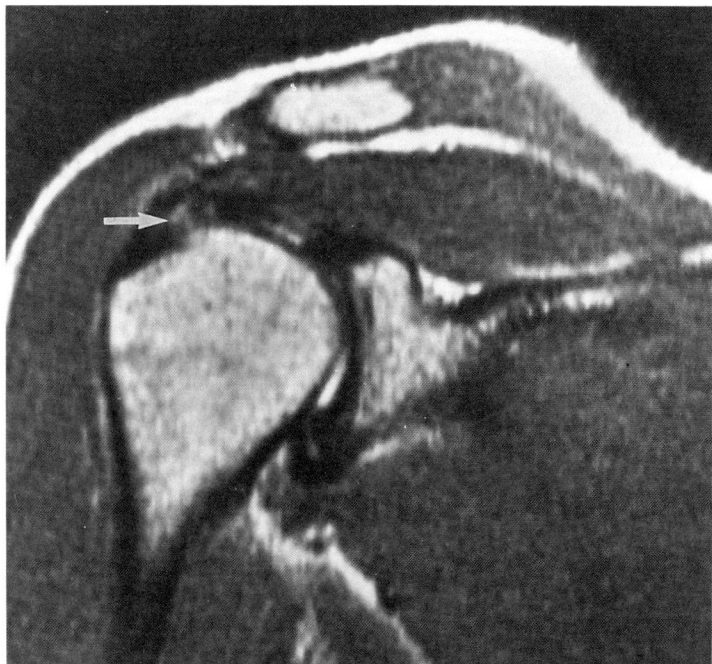

Figure 4.16

Small supraspinatus tendon tear. A 22-year-old female softball player with 6-month history of shoulder pain. Frontal oblique T_2-weighted image (SE 2000/56) discloses high-intensity signal within supraspinatus tendon (arrow). The musculotendinous junction is in the normal position, and there are no changes of atrophy of the supraspinatus muscle.

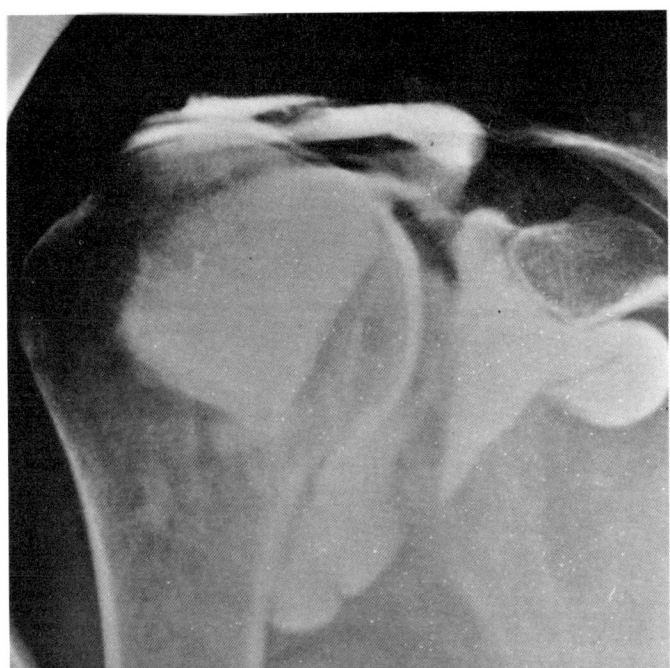

Figure 4.17

Massive supraspinatus rotator cuff tear. A 65-year-old man who complained of decreased range of motion after fall 6 months previously.
(a) Arthrogram shows large tear of supraspinatus tendon, with contrast material flooding subacromial space. (b) Frontal oblique T_1-weighted image (SE 500/28) through mid-humeral head. Supraspinatus tendon is replaced by medium-intensity signal, and there is marked retraction of the musculotendinous junction (arrow). The acromiohumeral distance is normal, and there are no changes of atrophy within supraspinatus muscle belly. These findings reflect the relatively recent nature of the tear.

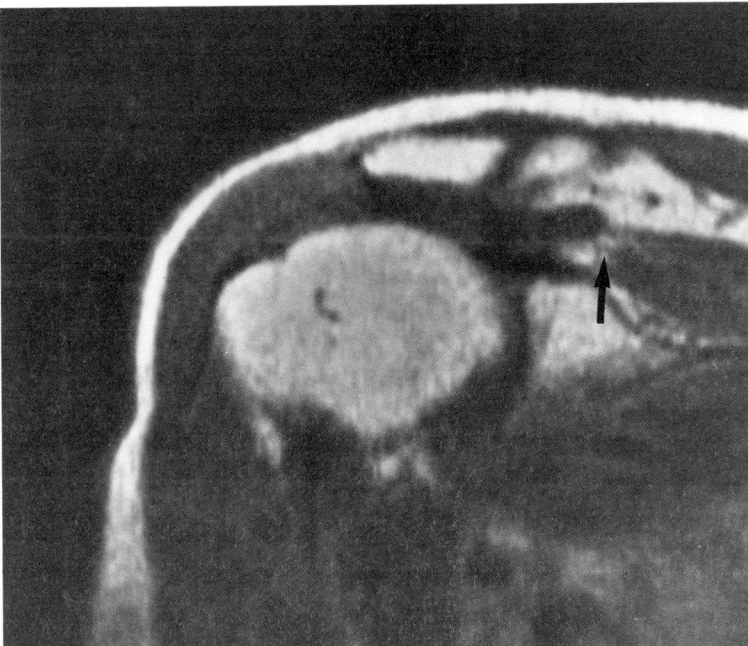

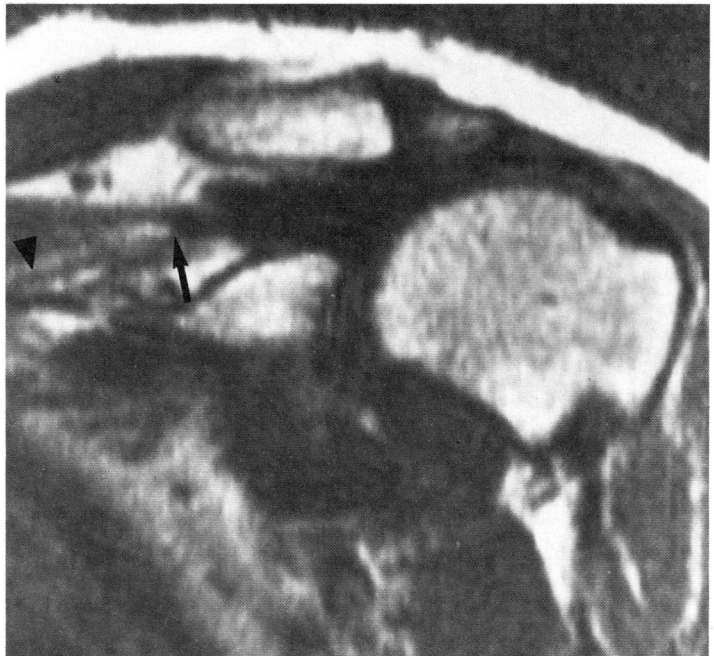

a

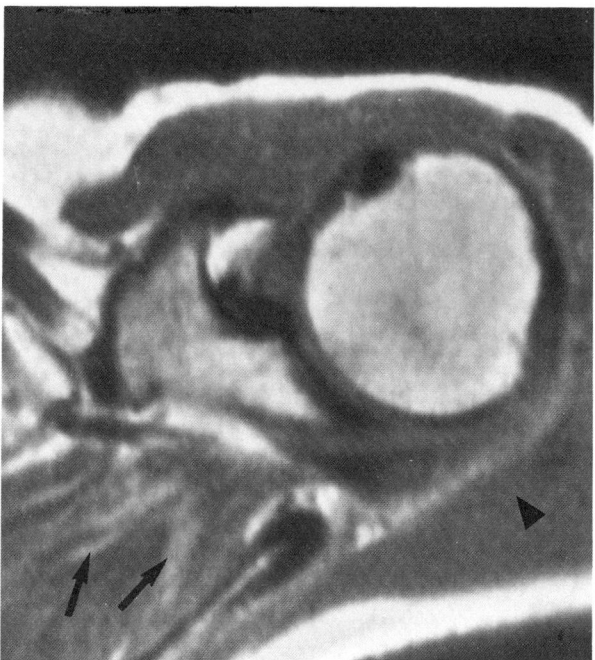

b

Figure 4.18

Chronic, massive supraspinatus rotator cuff tear. A 63-year-old man who had fallen 5 years previously while having a myocardial infarct. He had experienced shoulder pain since the fall, and was forced to stop playing golf due to limited range of motion. (**a**) Frontal oblique T_1-weighted image (SE 500/28) through acromioclavicular joint. Supraspinatus musculotendinous junction is markedly retracted (arrow). Atrophy of supraspinatus muscle is characterized by bands of high-signal intensity fat replacing muscle fibers (arrowhead). (**b**) Axial T_1-weighted image (SE 500/28) at level of coracoid process confirms atrophy of supraspinatus muscle (arrow). There are no atrophic changes in deltoid muscle (arrowhead).

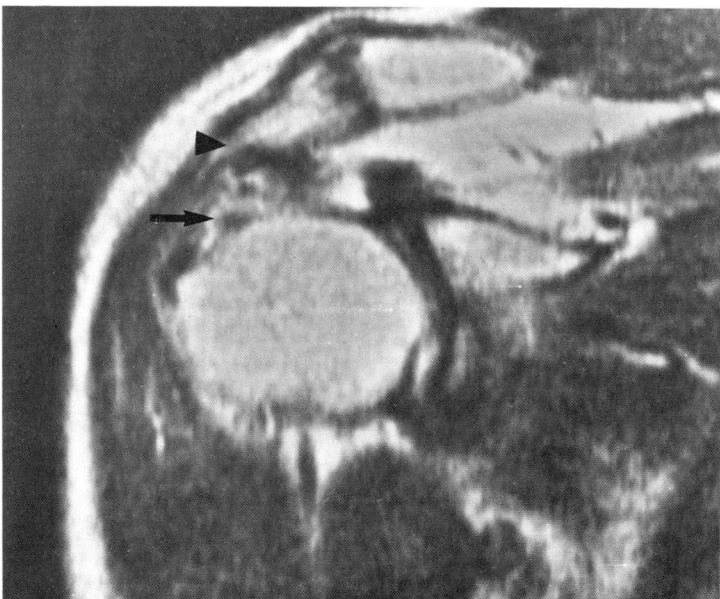

Figure 4.19

Massive supraspinatus rotator cuff tear. A 36-year-old dock worker experienced persistent shoulder pain and limited range of motion 4 months after fall on shoulder. Frontal oblique T_2-weighted image (SE 1500/56) through acromioclavicular joint. Multiple linear foci of bright signal intensity are present within supraspinatus tendon (arrow). The anterior acromion lies low, and its lateral margin slopes inferiorly into large spur (arrowhead). There are degenerative changes of the acromioclavicular joint.

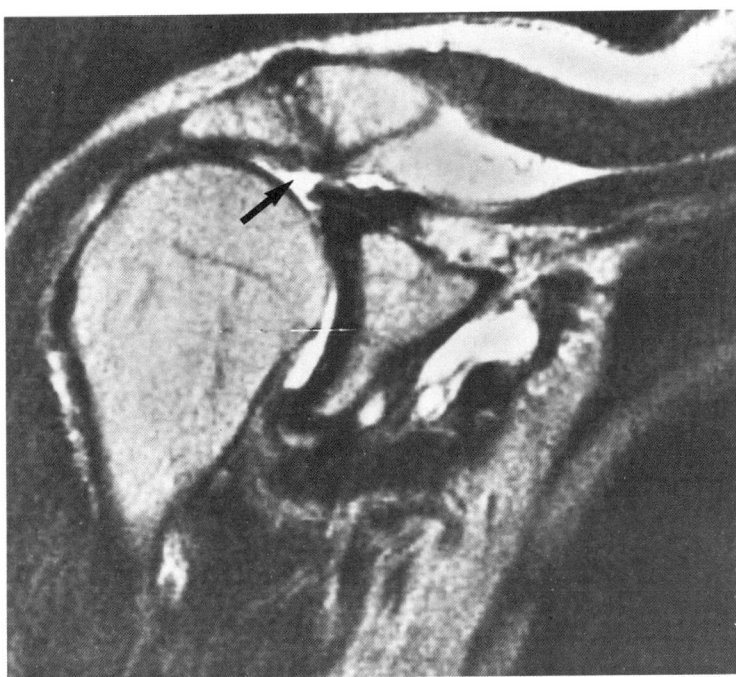

Figure 4.20

Massive supraspinatus rotator cuff tear. A 73-year-old man with several-year history of shoulder pain and limited range of motion. Frontal oblique T_2-weighted image (SE 2000/84) through acromioclavicular joint. High-signal intensity fluid is seen communicating between glenohumeral joint and subacromial space (arrow). There is no supraspinatus tendon tissue present, and the acromiohumeral distance is diminished. Acromioclavicular joint shows degenerative changes and small amount of fluid.

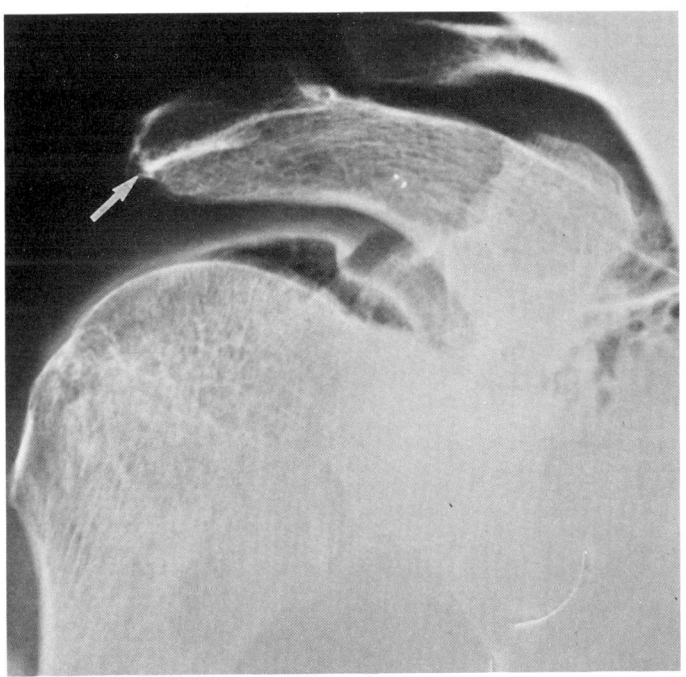

a

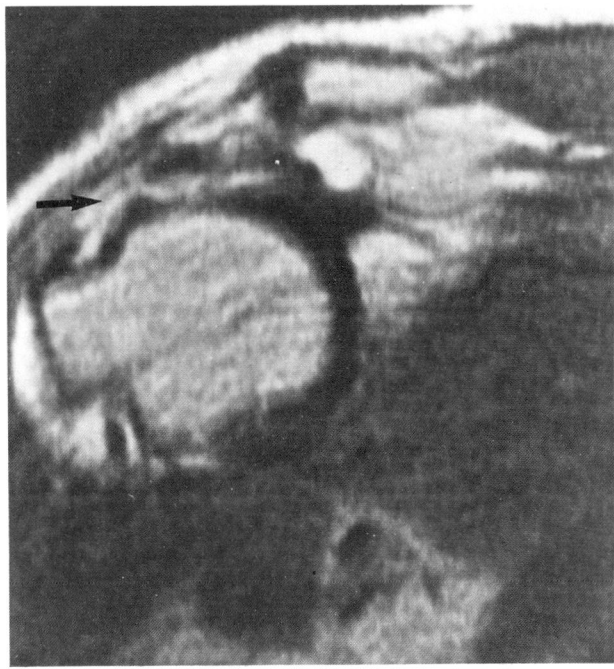

b

Figure 4.21

Large supraspinatus tendon tear with negative arthrogram. A 64-year-old man with chronic shoulder pain. (a) Double contrast arthrogram was negative for rotator cuff tear. Inferior margin of supraspinatus tendon is smooth. A small subacromial spur is present (arrow). (b) Anterior T_2-weighted coronal oblique image (SE 2000/56) obtained 5 weeks after arthrography. There is high-signal intensity fluid within supraspinatus tendon (arrow). The anterior acromion lies relatively low with respect to the distal clavicle, and the subacromial spur is evident as signal void. Arthrography was repeated 3 weeks after MRI, and was again interpreted as normal. Because of continued complaints of severe pain, the patient underwent surgery. A large rotator cuff tear was found.

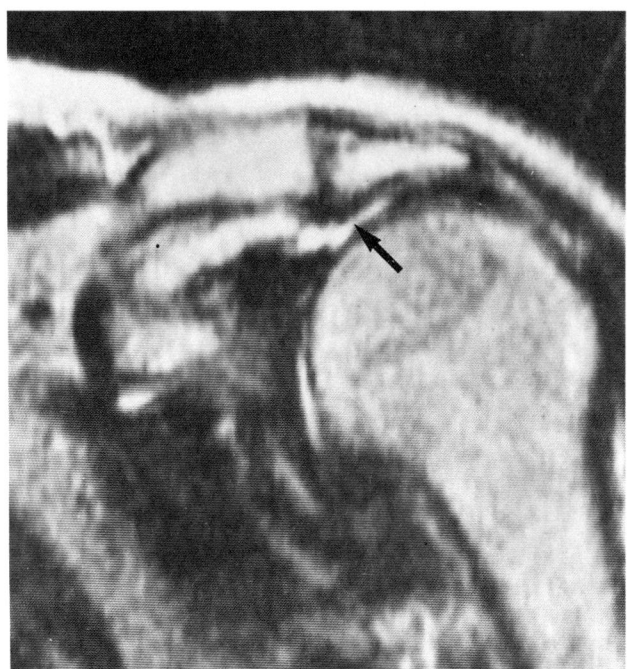
a

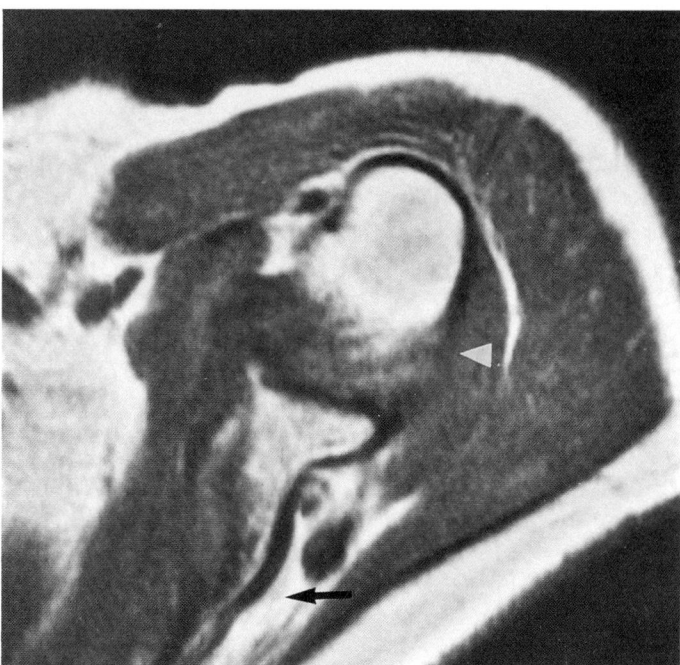
b

Figure 4.22

Supraspinatus and infraspinatus tendon tears. A 67-year-old man with shoulder pain and limited range of motion for 1½ years. (**a**) Frontal oblique image (SE 2000/84) shows marked reduction of acromiohumeral space, and high-signal-intensity fluid communicating between glenohumeral joint and subacromial bursa (arrow). There is no identifiable supraspinatus tendon tissue. (**b**) Axial T_1-weighted image (SE 500/28) through surgical neck of humerus. Infraspinatus muscle is totally replaced by high-signal-intensity fat (arrow). Laterally, normal teres minor muscle is seen (arrowhead).

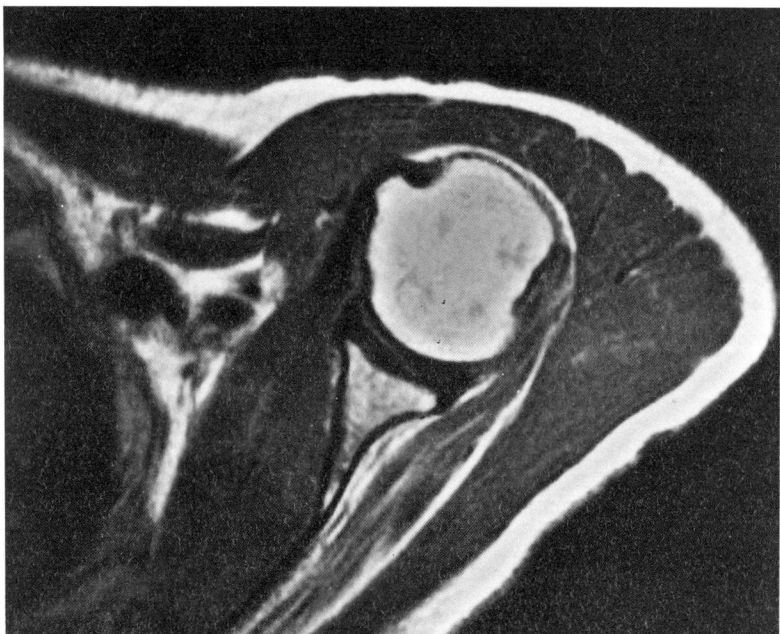

Figure 4.23

Isolated infraspinatus tendon tear. A 61-year-old man with decreased strength in shoulder for 3 years. Axial T_1-weighted image shows abnormal bands of high-signal-intensity fat confined to infraspinatus muscle, indicating chronic disuse. Selective atrophy of this muscle could only be due to a tendon tear. Site of tear is not seen without T_2-weighted imaging. Supraspinatus musculotendinous unit was normal on frontal oblique images.

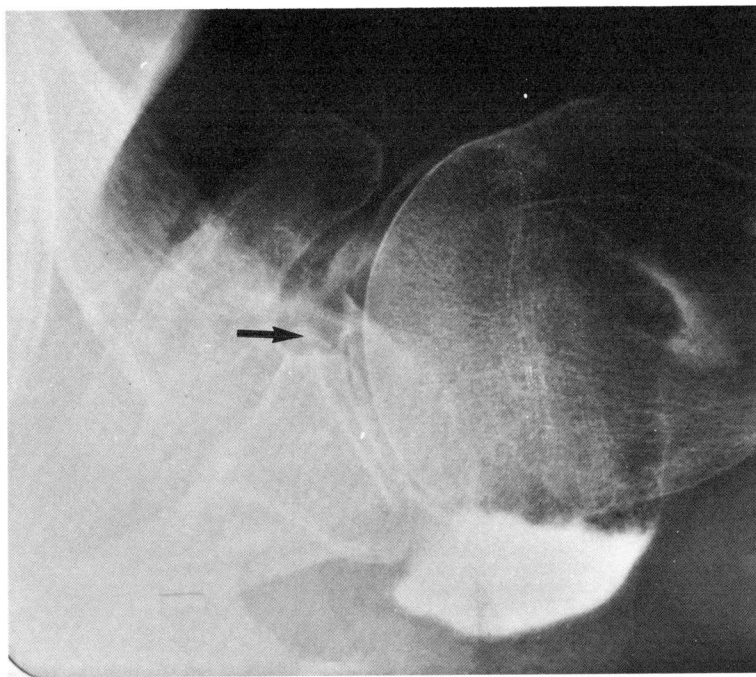

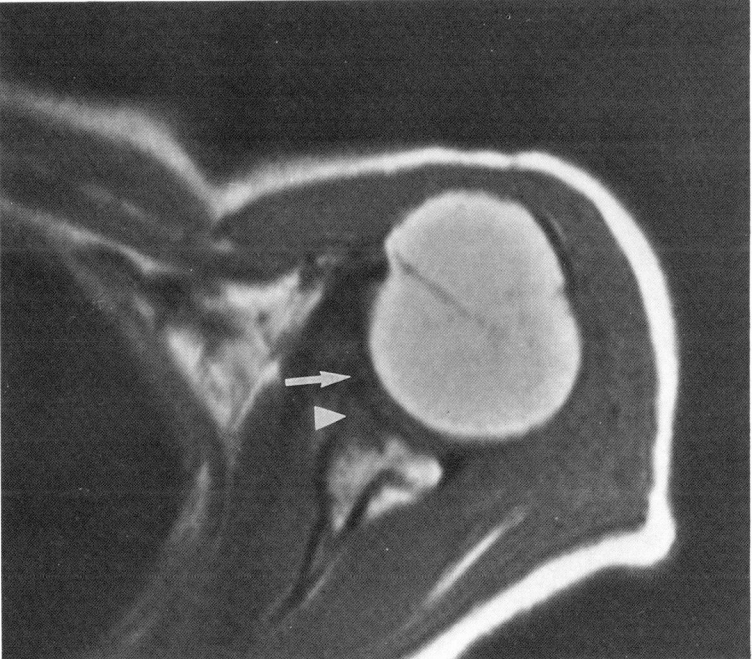

Figure 4.24

Anterior labral tear. A 65-year-old woman with persistent pain after falling down stairs 10 days earlier. (**a**) Supine axillary view from double-contrast arthrogram shows collection of contrast material in anterior labrum, and irregularity of anterior bony glenoid rim (arrow). Posterior labrum is normal, and rotator cuff was intact. (**b**) Axial T_1-weighted image (SE 500/28) reveals abnormal intermediate signal intensity within anterior labrum (arrow) and in marrow of anterior bony glenoid (arrowhead). Posterior labrum has normal homogeneous signal void.

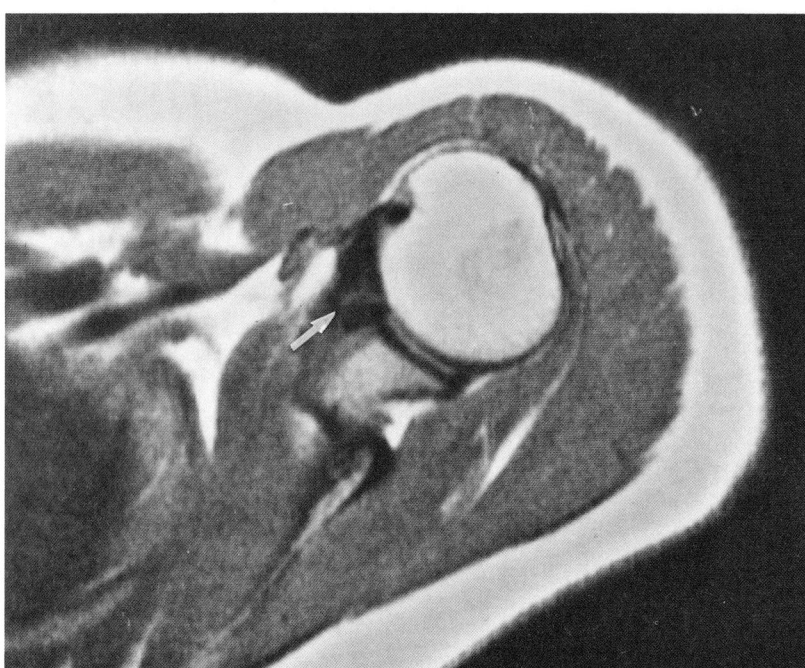

a

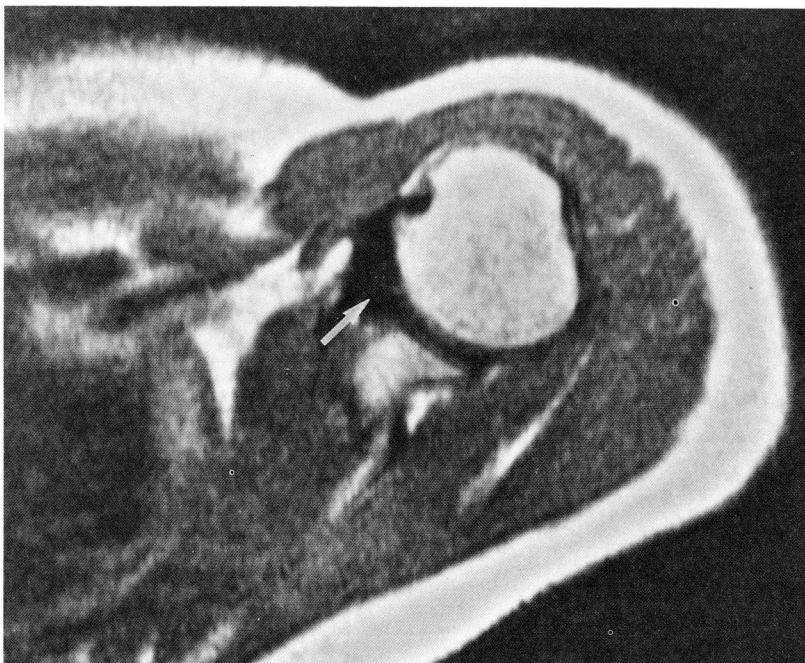

b

Figure 4.25

Anterior labral tear. A 19-year-old female blacksmith who complained of shoulder pain after a horse fell on her 2 months previously. (**a**) T_1-weighted axial image (SE 500/28) through mid-humeral head shows band of medium-signal intensity traversing anterior labrum (arrow). The posterior labrum is small but normal in contour and signal intensity. (**b**) T_2-weighted image (SE 1500/56) at same level displaying fluid within labral tear as band of high-intensity signal (arrow).

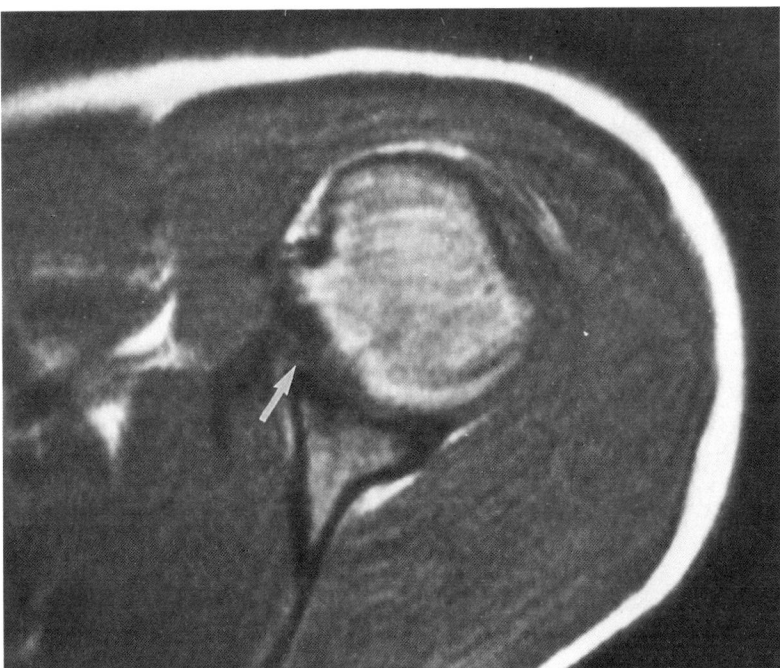

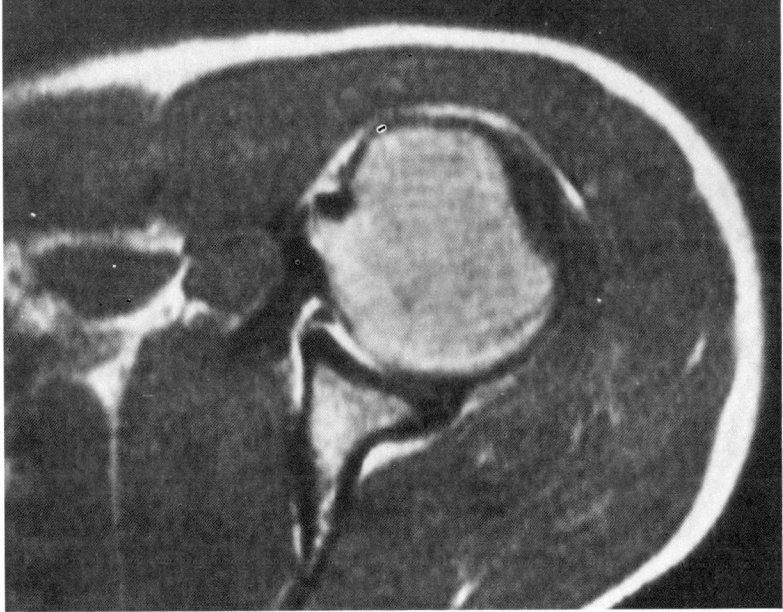

Figure 4.26

Anterior labral tear. A 22-year-old male baseball pitcher with a history of multiple anterior subluxations over past 4 months. (**a**) Axial T_1-weighted image (SE 500/28) reveals band of abnormal intermediate signal intensity in anterior labrum (arrow). (**b**) With T_2-weighted imaging (SE 2000/56), tear shows high-intensity signal. At surgery, bucket-handle-type tear was discovered.

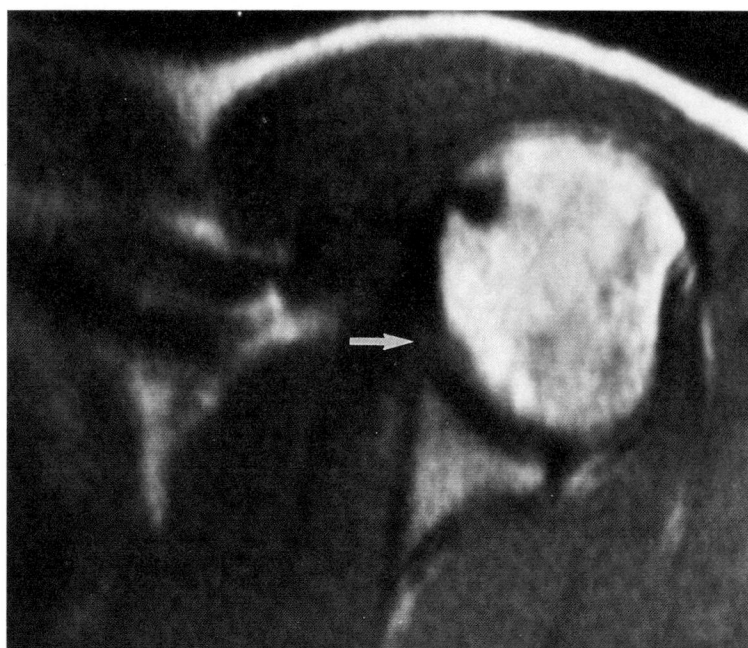

Figure 4.27

Recurrent anterior dislocation. A 33-year-old man with history of over 20 anterior dislocations. T_1-weighted axial image (SE 500/28) through mid-humeral head. The anterior labrum is markedly attenuated, and its border is ill-defined (arrow). Posterior labrum is normal, as are subscapularis muscle and tendon. At surgery, anterior labrum was nearly absent, and its margin was frayed.

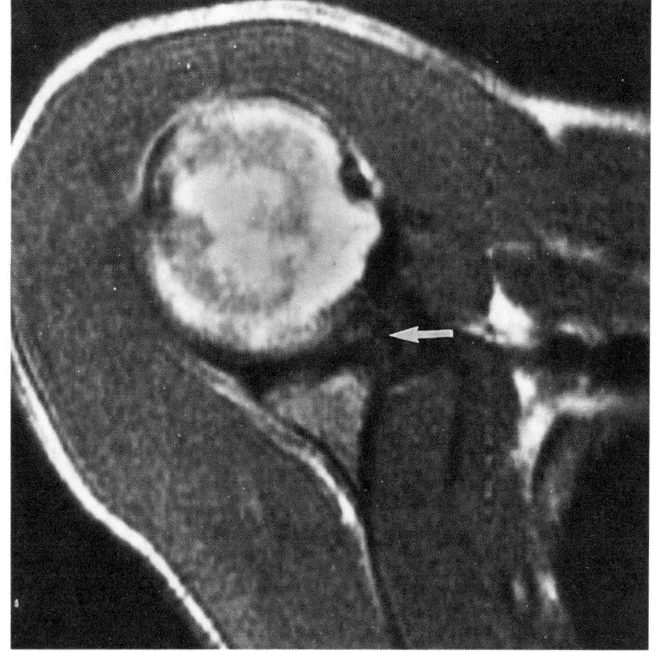
a

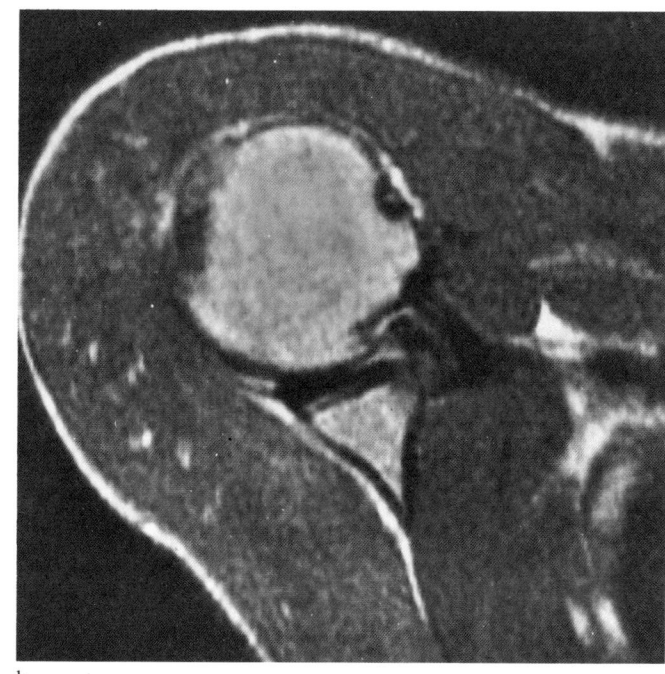
b

Figure 4.28

Anterior labral tear. A 26-year-old male racquetball player with history of multiple anterior dislocations. (**a**) T_1-weighted axial image (SE 500/28) through the mid-humeral head shows curvilinear region of medium-signal intensity crossing anterior labrum (arrow). Posterior labrum is normal. (**b**) T_2-weighted image (SE 2000/56) at same level indicates high-signal-intensity fluid within tear.

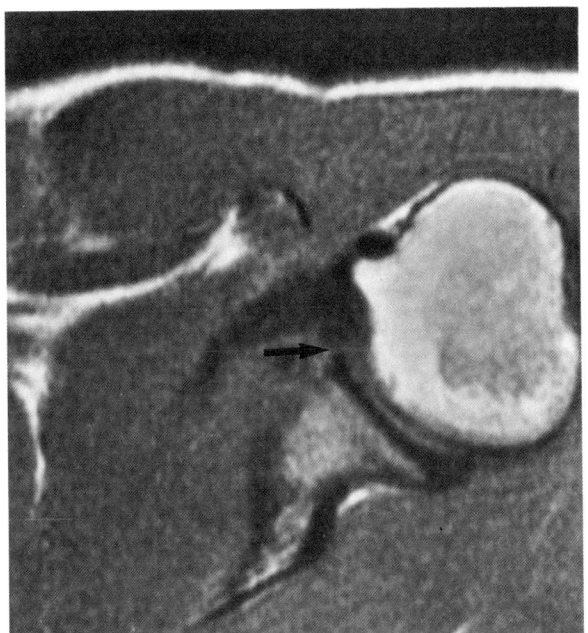
a

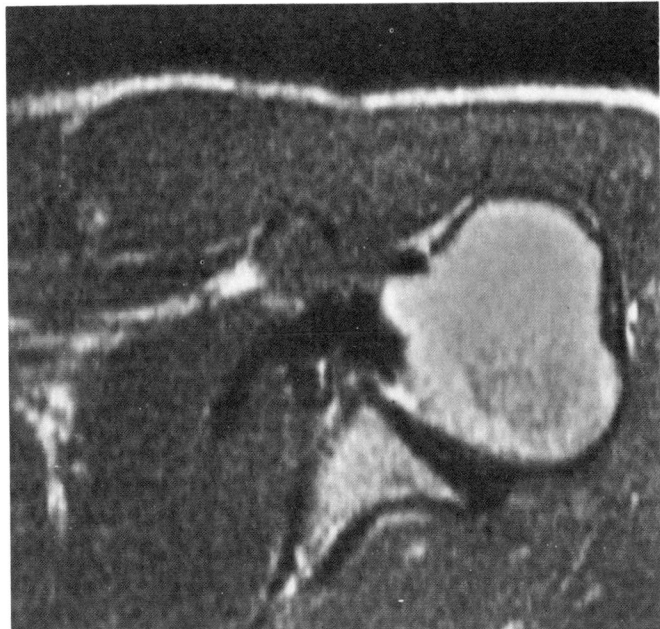
b

Figure 4.29

Anterior labral tear. A 23-year-old man with history of recurrent anterior dislocation. (**a**) T$_1$-weighted axial image (SE 500/28) through lower humeral head shows medium-intensity signal completely traversing mid-substance of anterior labrum (arrow). (**b**) T$_2$-weighted image (SE 2000/84) at same level highlights fluid that separates labral fragment from glenoid.

Figure 4.30

Multidirectional instability. A 42-year-old man with shoulder pain, who denied history of glenohumeral subluxation or dislocation. (**a**) Axial T$_1$-weighted image (SE 500/28) shows marked attenuation of anterior labrum, and abnormal intermediate-signal intensity around base of both anterior and posterior labra. (**b**) T$_2$-weighted image (SE 2000/56). There is no identifiable anterior labral tissue. High-signal-intensity fluid surrounds both labra and extends along anterior border of scapula. At surgery, both anterior and posterior Bankart lesions were found.

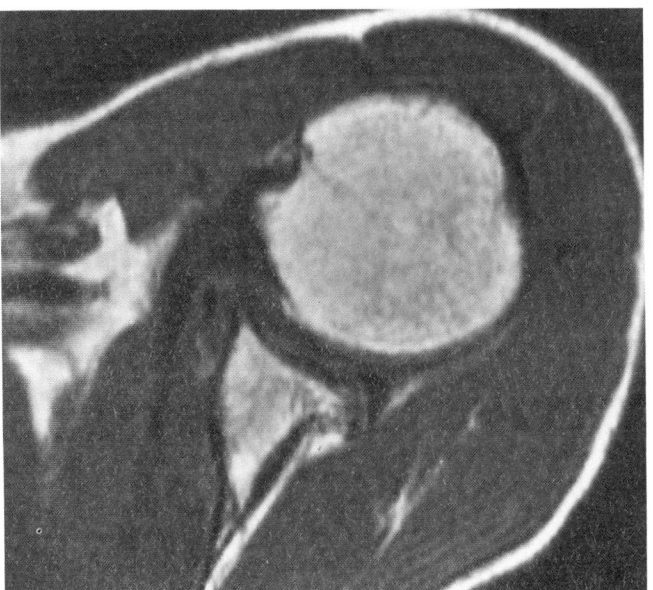
a

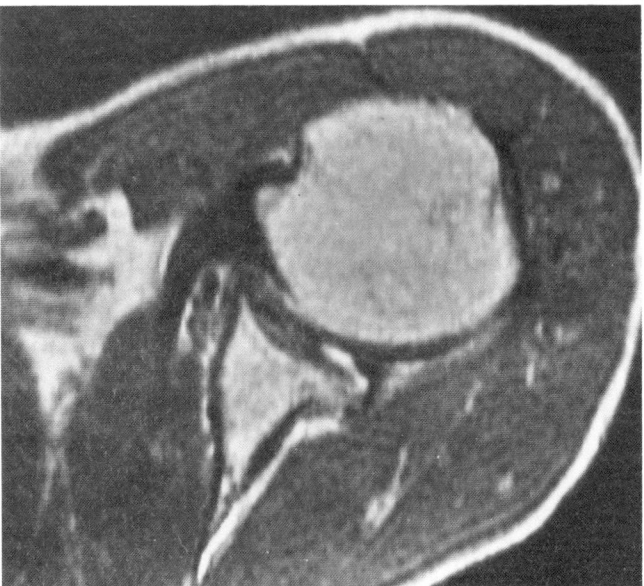
b

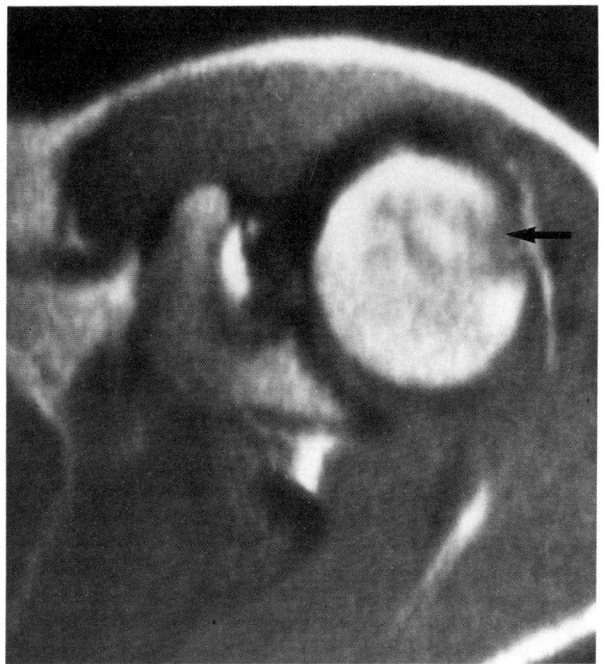 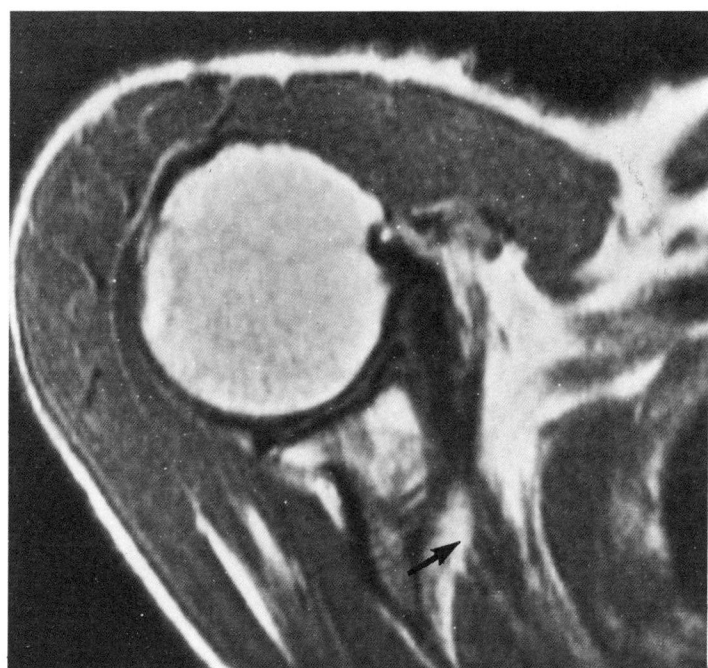

Figure 4.31

Hill–Sachs lesion. A 33-year-old man with history of multiple anterior dislocations. Axial image (SE 500/28) at level of coracoid process shows depression of posterolateral aspect of humeral head (arrow). The high humeral head should normally be round.

Figure 4.32

Subscapularis tendon rupture. A 65-year-old man with limited range of motion and pain for 5 years. Axial T_1-weighted image (SE 500/28) shows medial retraction of subscapularis musculotendinous junction (arrow), and fatty replacement of muscle fibers along border of scapula.

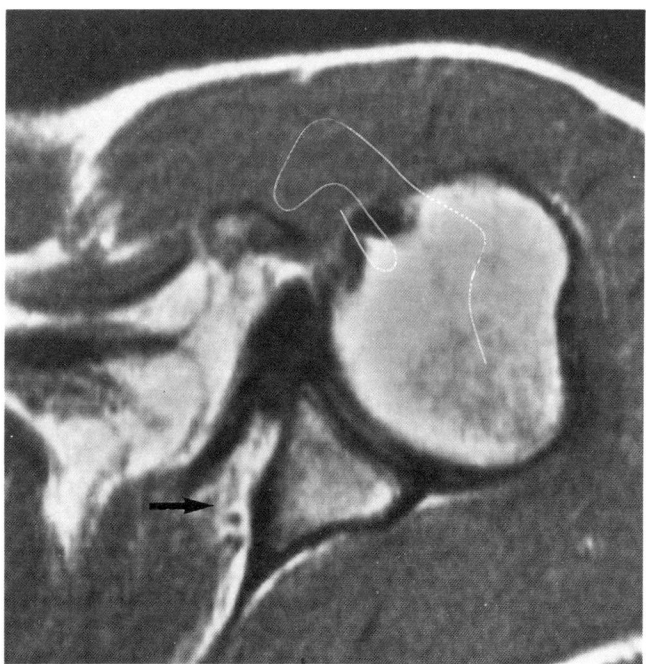

Figure 4.33

Subscapularis tendon rupture. A 37-year-old man with history of anterior glenohumeral dislocation 20 years ago. Since initial injury, he had experienced recurrent multidirectional subluxations. T_1-weighted axial image (SE 500/28) through level of inferior coracoid process shows marked retraction of subscapularis musculotendinous junction (arrow). There is fatty replacement of muscle at musculotendinous junction and along muscle origin from scapula. Both anterior and posterior labra are severely attenuated.

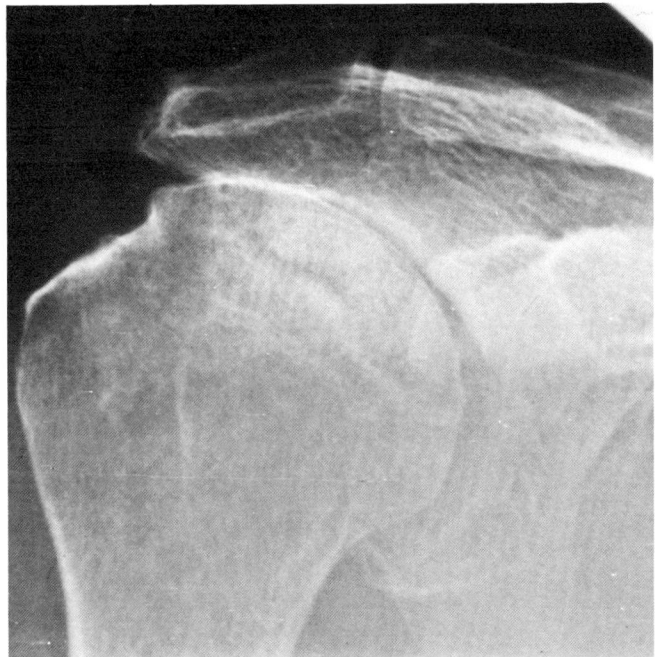

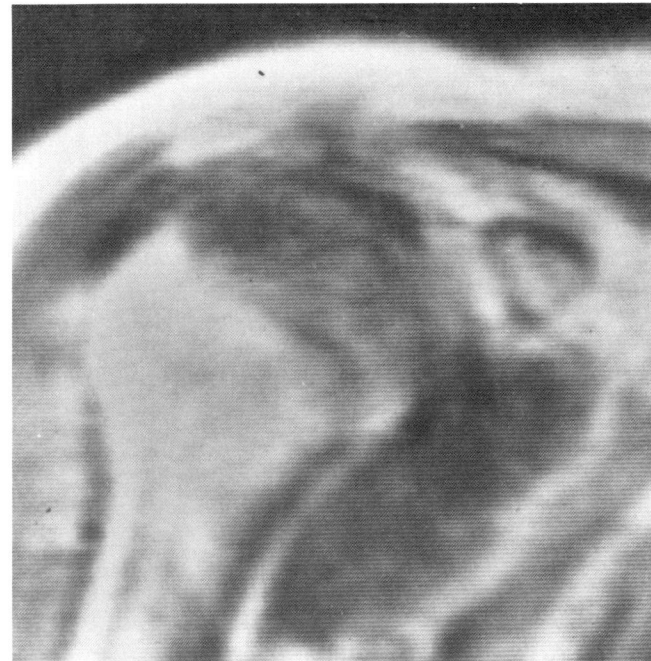

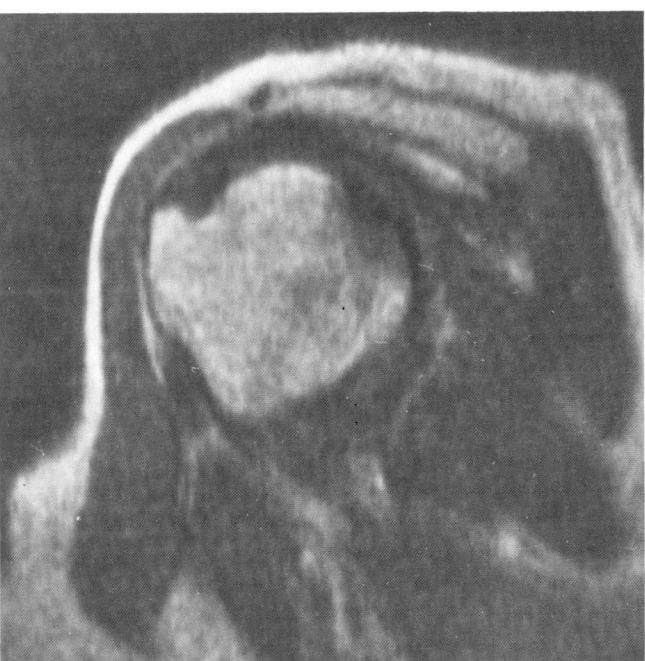

Figure 4.34

Avascular necrosis of humeral head. A 70-year-old man who complained of bilateral shoulder pain. He had been on corticosteroid therapy for asthma for several years. (**a**) Anteroposterior radiograph of right shoulder shows fracture of humeral head with increased density of fragment, typical of avascular necrosis. (**b**) Coronal image (SE 500/28) of right shoulder confirms presence of AVN, with intermediate-intensity signal corresponding to region of increased density which was seen in radiograph. (**c**) Radiographs of left shoulder were normal, and no abnormalities were detected on coronal MR images. Sagittal MR image (SE 500/28) revealed focus of low-signal intensity in posterosuperior aspect of humeral head, which was proven by core biopsy to represent AVN. This abnormality was probably not appreciated in coronal images because of partial volume averaging through round humeral head.

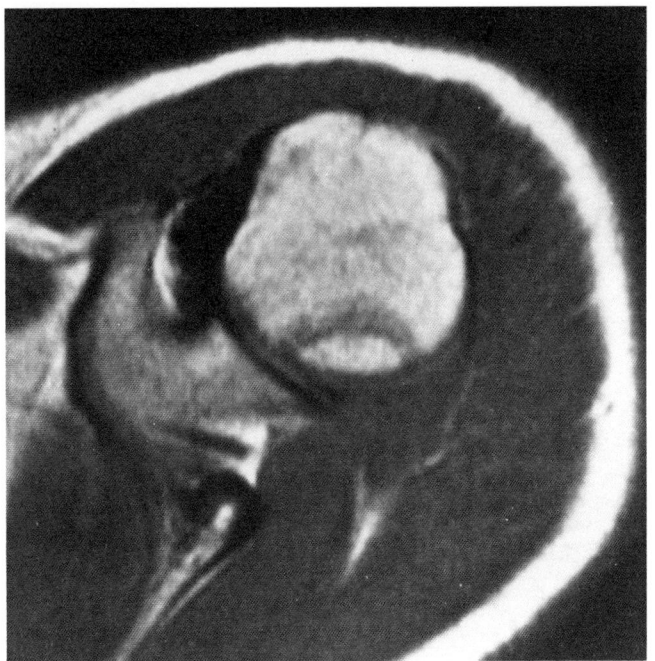 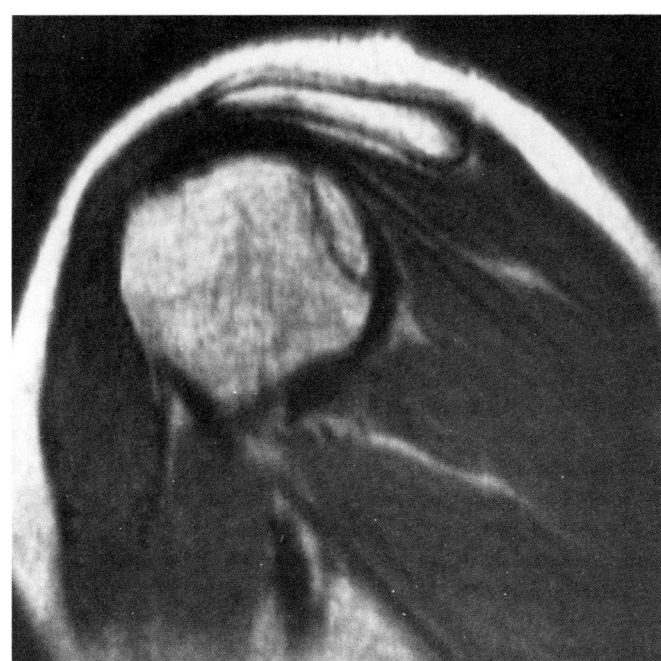

a b

Figure 4.35

Avascular necrosis of the humeral head. A 23-year-old man who complained of pain in left shoulder. He had undergone cadaveric renal transplant for end-stage renal disease. Radiographs were normal. (**a**) Axial T_1-weighted image (SE 500/28) reveals curvilinear band of medium-signal intensity along posterior aspect of humeral head, consistent with Waldenstrom crescent. (**b**) Sagittal T_1-weighted image (SE 500/28) through mid-humeral head confirms that crescent lies in posterosuperior aspect of the humeral head, and involves epiphysis.

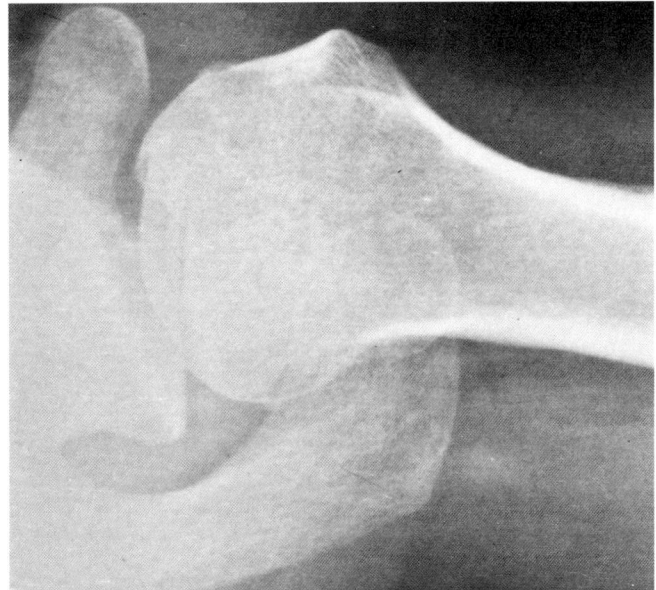

a

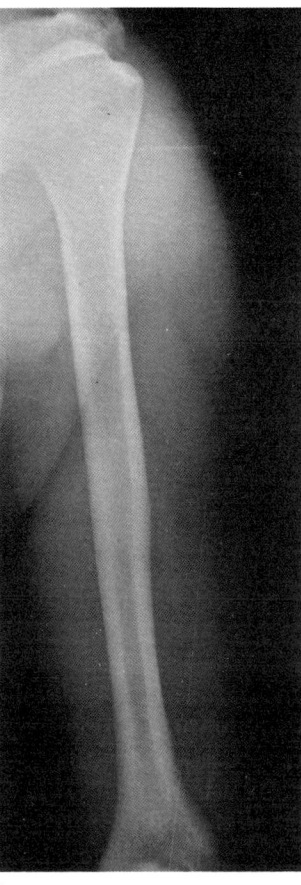

b

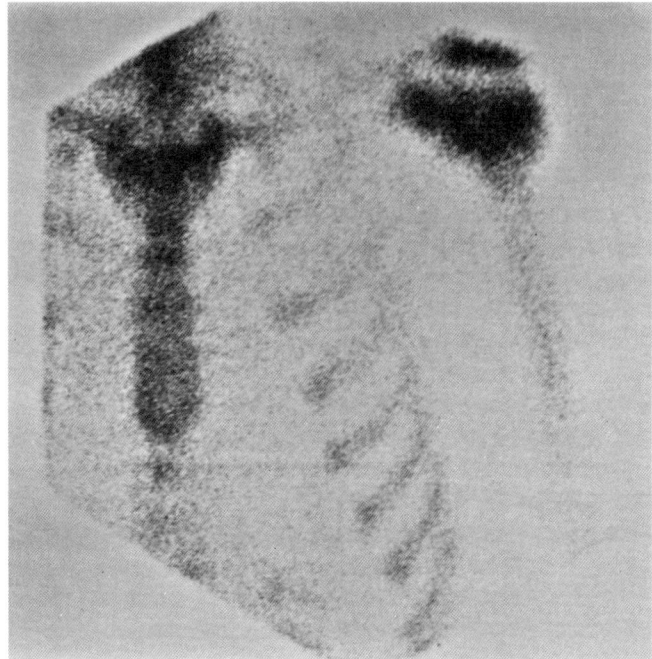

c

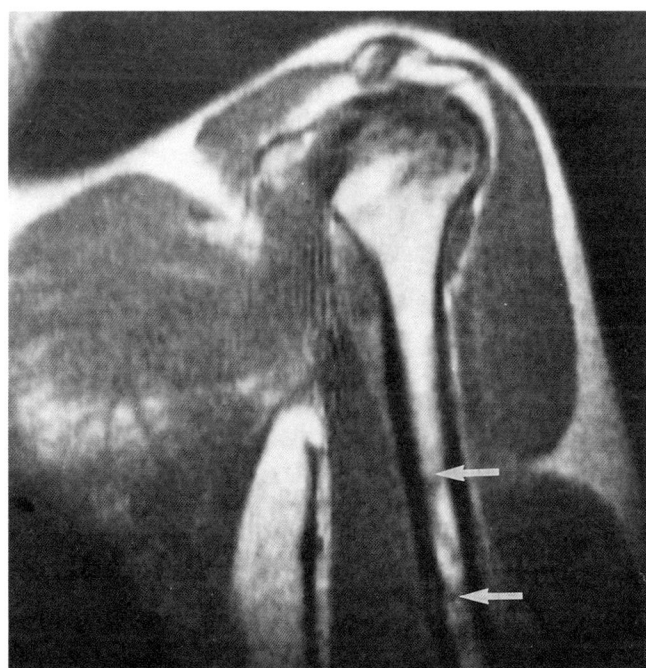

d

Figure 4.36
Avascular necrosis of humeral head and medullary cavity. A 27-year-old deep-sea diver who complained of pain in right shoulder and mid-humeral shaft. (**a**) Axillary radiograph of humeral head shows region of collapse, indicating AVN. (**b**) Radiograph of humeral shaft is normal. (**c**) Radionuclide bone scan shows abnormal tracer accumulation in humeral head, but uptake in shaft is normal. (**d**) Coronal T_1-weighted image (SE 500/28) through shoulder and humerus. This image was acquired with body coil in order to image humeral shaft. Abnormal signal intensity is present within humeral head, consistent with AVN. There are also foci of abnormally low signal intensity within humeral shaft (arrows), indicating diaphyseal medullary infarcts.

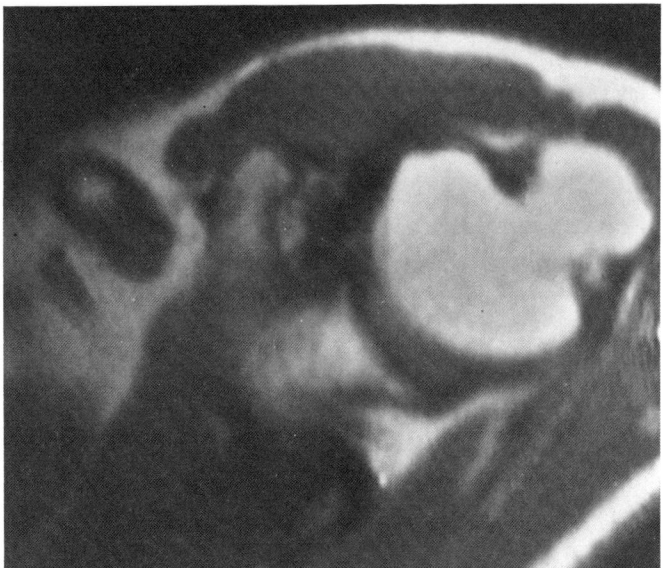

Figure 4.37

Osseous trauma: fracture-dislocation. A 49-year-old airline pilot who had suffered fracture of greater tuberosity, and complained of persistent pain and decreased function. Arthrography was negative for rotator cuff tear, and the question of possible fracture non-union was raised. T_1-weighted axial image (SE 500/28) reveals that marrow completely bridges fracture site, indicating osseous union. Malunion of greater tuberosity resulted in abnormal angulation of rotator cuff tendons, and was clearly depicted with MRI.

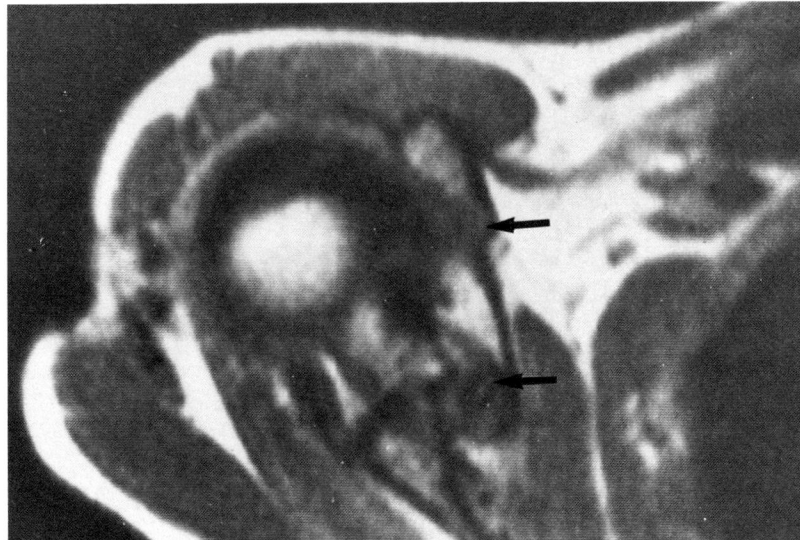

a

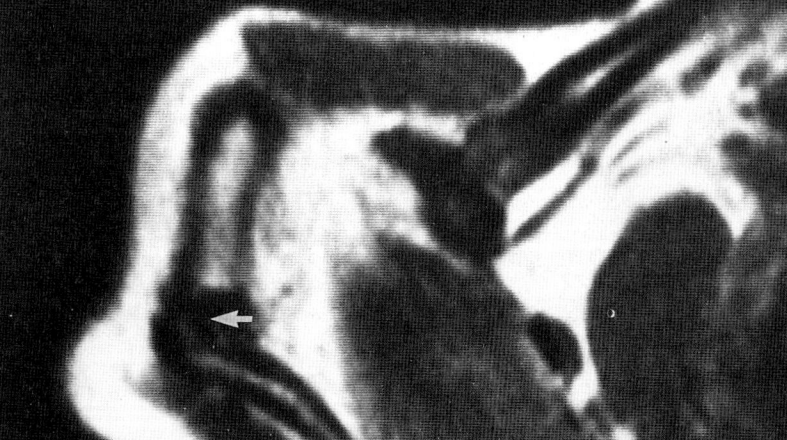

b

Figure 4.38

Osseous trauma. A 39-year-old woman who complained of persistent severe shoulder pain several months after being struck by car. At time of accident, she had suffered comminuted fracture of scapula. (**a**) Axial T_1-weighted image (SE 500/28) through level of coracoid process reveals wide distraction of ununited fracture of base of coracoid process (arrows). Skin defect is secondary to initial trauma. (**b**) Axial T_1-weighted image (SE 500/28) through the acromion indicates non-union of acromion fracture (arrow), but fragments are in near-anatomic alignment. At surgery, this represented unstable fibrous union.

5

The elbow

Daniel H Bunnell and
Lawrence W Bassett

Magnetic resonance imaging's high soft-tissue contrast and multiplanar imaging capabilities have established it as a very useful tool in the examination of the musculoskeletal system.[1-4] Improved solenoid surface coil technology and updated computer software allows thinner section images with higher resolution in non-orthogonal planes.[4-10] These advances have allowed detailed depiction of the soft-tissue structures of the joints for the first time. A further technical advance, lateral decentering, allows imaging of the appendicular joints beyond the center of the magnetic field.[4] These improvements allow positioning with the patient's arm at his or her side, an important advantage in the evaluation of a painful or injured elbow.

Imaging technique

We perform routine MR images of the elbow joint with the elbow positioned in pronation at the subject's side, with the palm of the hand against the thigh.[6] This is generally a very comfortable position that is easily maintained without motion during the time of scanning. The elbow is elevated slightly with foam padding to allow positioning closer to the center of the magnetic field. A wraparound solenoid surface coil designed especially for the elbow enhances signal detection to improve the signal-to-noise ratio. A software modification in our 0.3 tesla permanent magnet imaging system automatically directs the data to the center of the magnetic field. This software tells the computer to establish a new center at any predetermined lateral shift (lateral decentering) without loss of signal or spatial resolution. Depending on the size of the patient, an average of 170 mm of lateral shift is required for elbow imaging.

Axial images are first obtained, one of which serves as a scout for the precise cursor placement required for imaging in non-orthogonal planes (Figure 5.1). Cursors are aligned and images are obtained in non-orthogonal planes both parallel and perpendicular to a line drawn between the lateral and medial humeral epicondyles. Scans are obtained with a section thickness of 5 mm, the sections being at 7 mm intervals. The time of examination averages 10–12 minutes for seven images.

A spin-echo pulse sequence is routinely used with relative T_1-weighting (TE = 28, TR = 500) and four averages. In our experience, this pulse sequence has provided the best depiction of soft tissues, articular cartilage and bone marrow.

Imaging of the normal elbow joint

MR images of the normal elbow joint in axial, sagittal, and coronal planes are shown in Figures 5.2–5.4. All images utilize a spin-echo sequence with TE = 28 and TR = 500. The intermediate signal intensity of articular cartilage, the moderately low signal intensity of muscular bundles and the even lower signal intensity of tendinous insertions are well delineated.

Muscles of the elbow region can be divided into four groups: posterior, anterior, medial and lateral. The posterior group consists of the anconeus and triceps muscles while the anterior group includes the biceps brachii and brachialis muscles. The lateral group includes the supinator, brachioradialis, and the extensor muscles of the wrist while the medial group includes the pronator teres, the palmaris longus and the flexor muscles of the wrist.

On images in the axial plane, the ulnar nerve is seen as a dot of relatively low signal intensity just posterior to the medial humeral epicondyle, surrounded by high-intensity perineural fat (Figure 5.2b). The cephalic and basilic venous systems are delineated by their intermediate signal intensities owing to the paradoxical enhancement of slow bloodflow. The brachial artery is seen as a signal-void structure running just superficial to the brachialis muscle and medial to the biceps brachii muscle and tendon. It divides into the radial and ulnar arteries at a variable distance, usually about 1–2 cm distal to the elbow joint. We have found that because these longitudinally oriented neurovascular and musculotendinous bundles do not always travel in strict sagittal or coronal planes, they are usually best evaluated in the axial plane.

The sagittal and coronal imaging planes appear to be the most effective for evaluation of the joint space and articular surfaces of the elbow joint (Figures 5.3, 5.4). The elbow is a hinge joint, and optimal axial and coronal images depend on the ability of the patient to extend the joint fully. In the sagittal plane, anatomic structures can be more easily identified when the elbow is in varying degrees of flexion; this is an important advantage in the evaluation of an injured or painful joint in which full extension may be impossible.

In the sagittal plane, the anterior and posterior fat pads of the elbow joint are depicted as high-intensity bands along the anterior and posterior surfaces of the humerus, in the coronoid and olecranon fossae respectively. The normal high signal intensity of the bone marrow in the radial head, olecranon, and distal humerus is also well visualized (Figure 5.3). The intermediate-intensity articular cartilage at the trochlear-olecranon (Figure 5.3a) and the radial head-capitellum (Figure 5.3b,c) articulations are well delineated. The intermediate-intensity biceps muscle is seen coursing anteriorly over the elbow joint, to become the low-signal-intensity biceps tendon, which inserts on the radial tuberosity (Figure 5.3b).

In the coronal plane, the high-intensity signal of the bone marrow within the radial head and capitellum and their respective articular surfaces are again well depicted (Figure 5.4a). Also well seen are the longitudinally oriented, intermediate-signal-intensity muscle bundles (Figure 5.4). The ulnar nerve, which courses immediately posterior to the medial epicondyle of the humerus, is depicted as a band of relatively low-signal intensity surrounded by the higher-intensity signal of perineural fat (Figure 5.4).

Imaging of the abnormal elbow joint

MR imaging of the elbow joint has many potential roles, most of which have not yet been fully explored. These include evaluation of inflammatory, traumatic or degenerative joint disorders; investigations of bone marrow disorders involving the elbow region; and the study of abnormalities of the neurovascular structures which traverse the joint. Examples of pathologic entities involving the elbow joint are shown in Figures 5.5–5.7.

References

1 MOON KL, GENANT HK, DAVIS PL et al, Nuclear magnetic resonance imaging in orthopedics: principles and applications, *J Orthop Res* (1983) **1**:101–14.
2 REICHER MA, RAUSCHNING W, GOLD RH, BASSETT LW et al, High-resolution magnetic resonance imaging of the knee joint: normal anatomy, *AJR* (1985) **145**:895–902.
3 DAFFNER RH, LUPETIN AR, DASH N et al, MRI in the detection of malignant infiltration of bone marrow, *AJR* (1986) **146**:353–8.
4 SEEGER LL, RUZKOWSI JR, BASSETT LW et al, MR imaging of the normal shoulder: anatomic correlation, *AJR* (1987) **148**:83–91.
5 MIDDLETON WD, MACRANDER S, KNEELAND JB et al, MR imaging of the normal elbow: Anatomic correlation, *AJR* (1987) **149**:543–7.
6 BUNNELL DH, FISHER DA, BASSETT LW et al, Elbow joint: Normal anatomy on MR images, *Radiology* (1987) **165**(2): 527–31.
7 HUBER DJ, MUELLER E, HUEBES P, Oblique magnetic resonance imaging of normal structures, *AJR* (1985) **145**:843–6.
8 EDELMAN RR, STARK DD, SAINI S et al, Oblique planes of section in MR imaging, *Radiology* (1986) **159**:807–10.
9 FISHER MR, BARKER B, AMPARO EG et al, MR imaging using specialized coils, *Radiology* (1985) **157**:443–7.
10 EHMAN RL, MR imaging with surface coils, *Radiology* (1985) **157**:549–50.

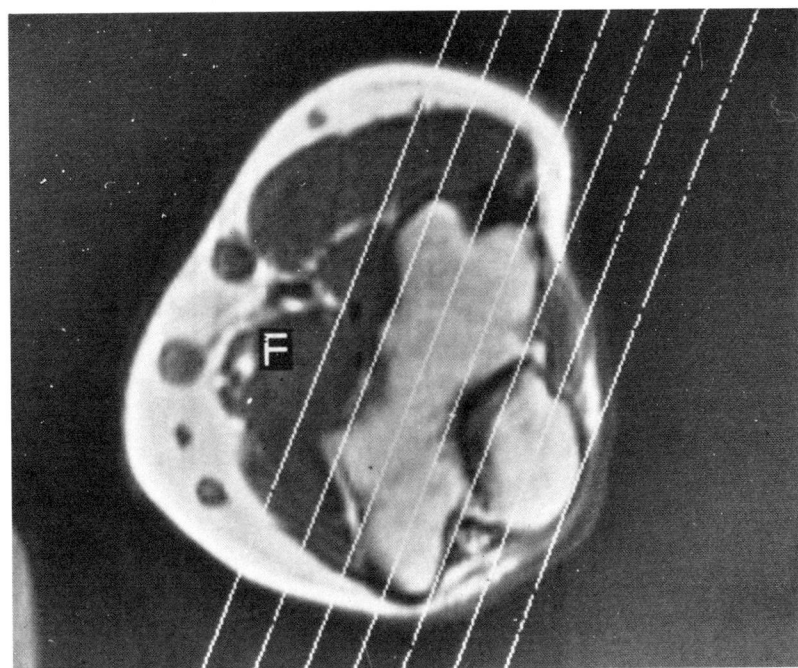

Figure 5.1

Axial scout image with cursors demonstrating oblique plane used for coronal images (SE 500/28). Coronal images are obtained in a plane parallel to a line drawn between the lateral and medical epicondyle. Sagittal images are performed in a plane perpendicular to this line.

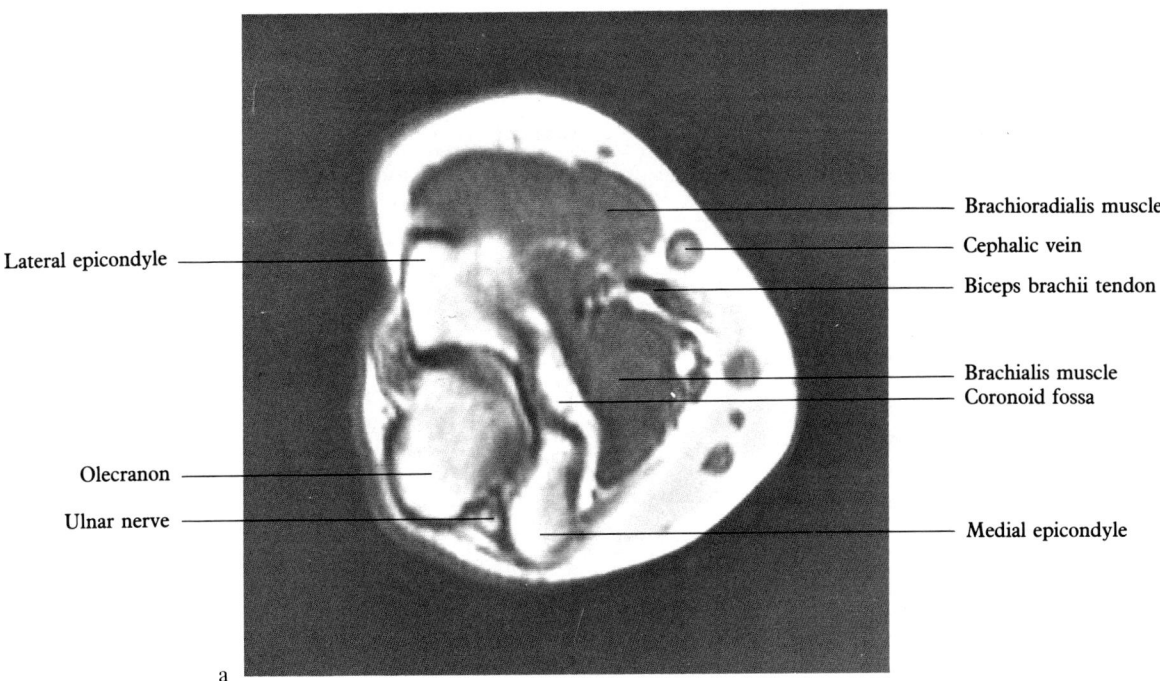

a

Figure 5.2

Normal anatomy, axial. (**a–d**) Serial MR images (SE 500/28) of the normal elbow joint in the axial plane, from superior to inferior.

Figure 5.2 *continued*

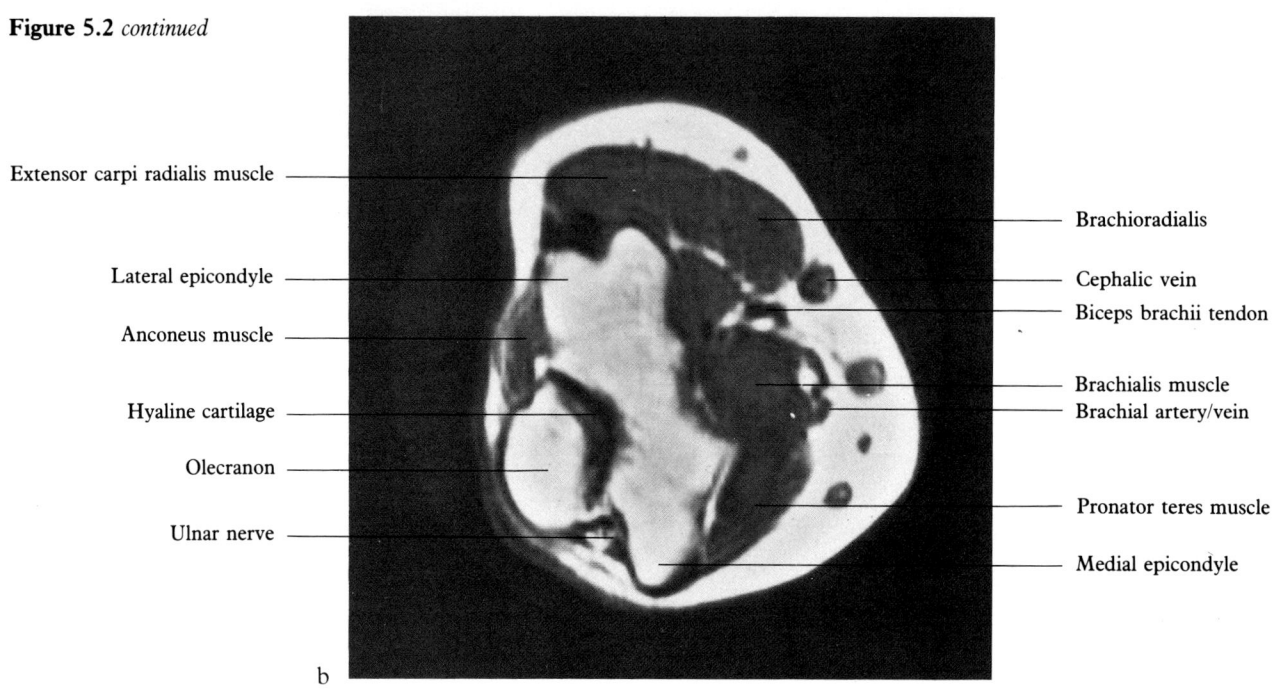

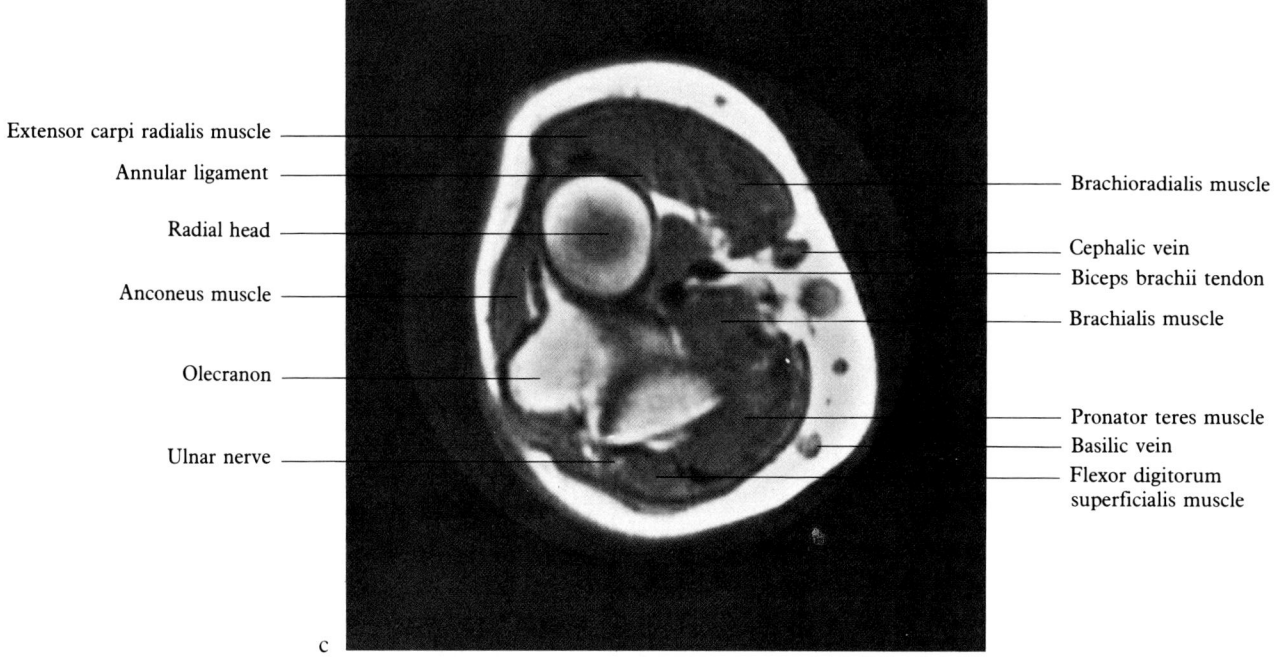

The elbow 133

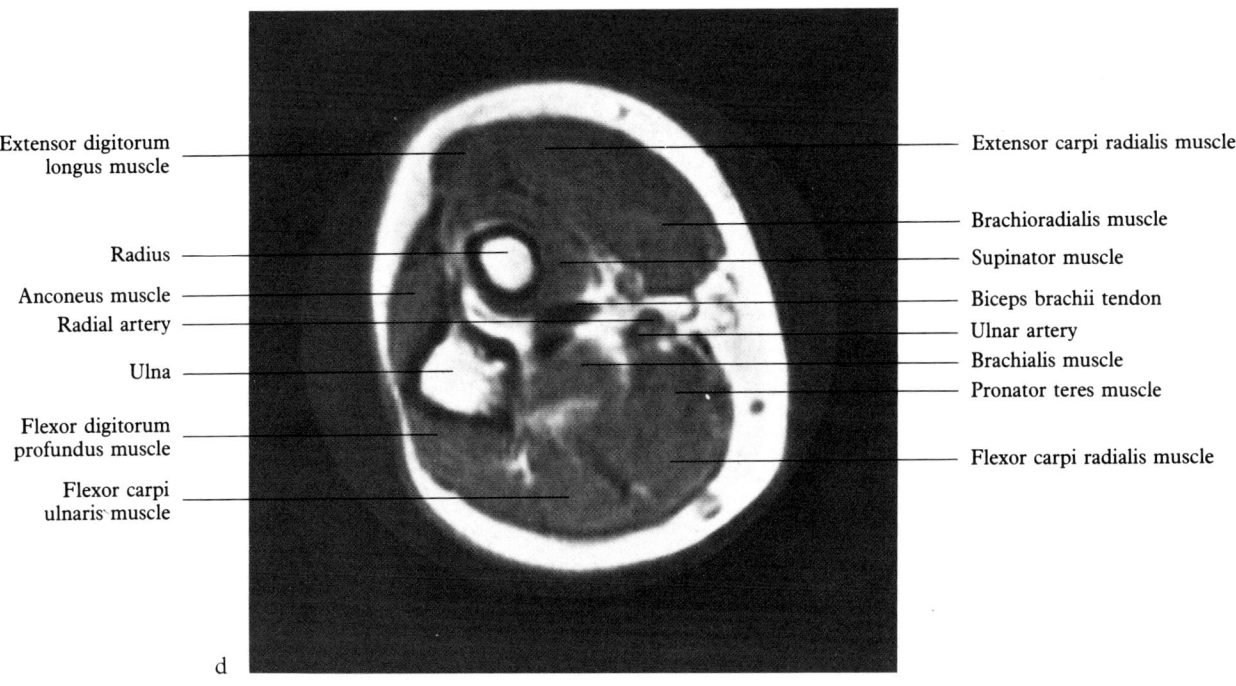

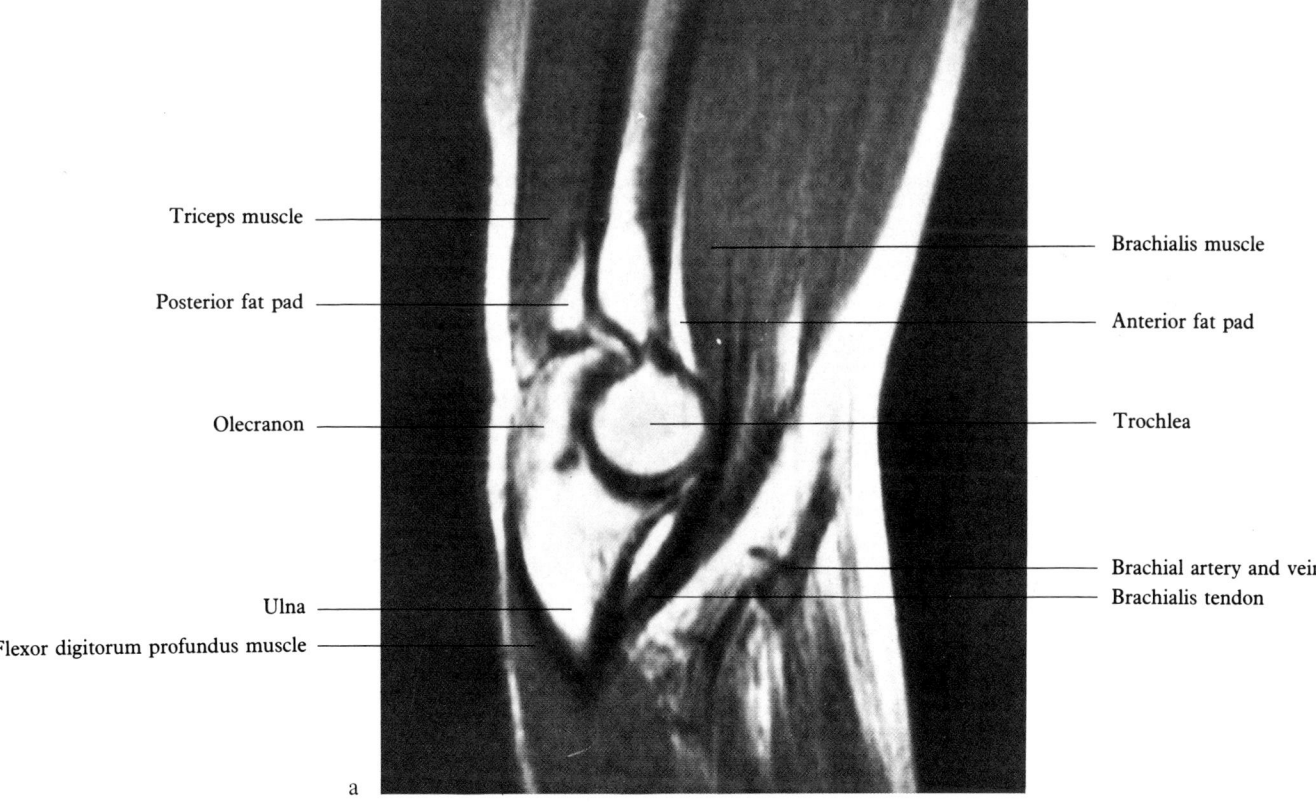

Figure 5.3

Normal anatomy, sagittal. (**a–c**) Serial MR images (SE 500/28) of the normal elbow joint in the sagittal plane, from medial to lateral.

Figure 5.3 *continued*

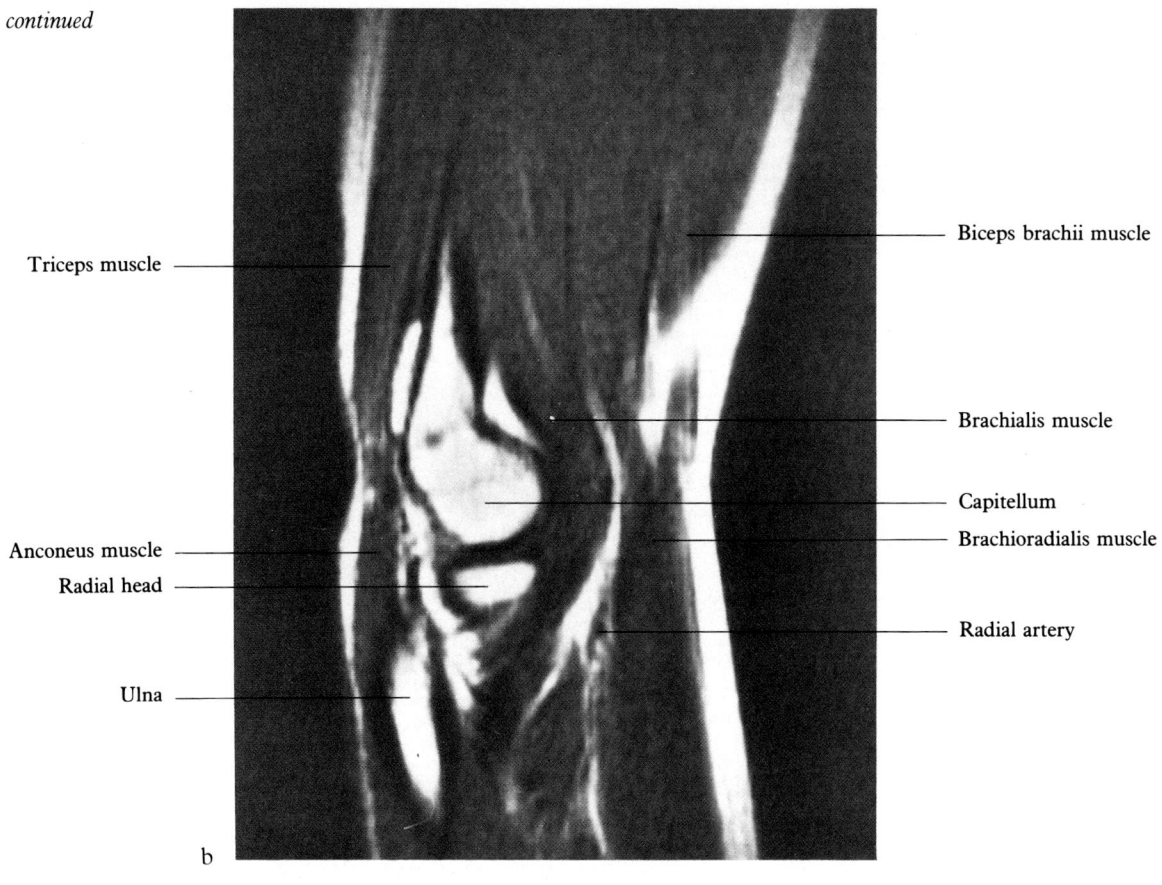

b

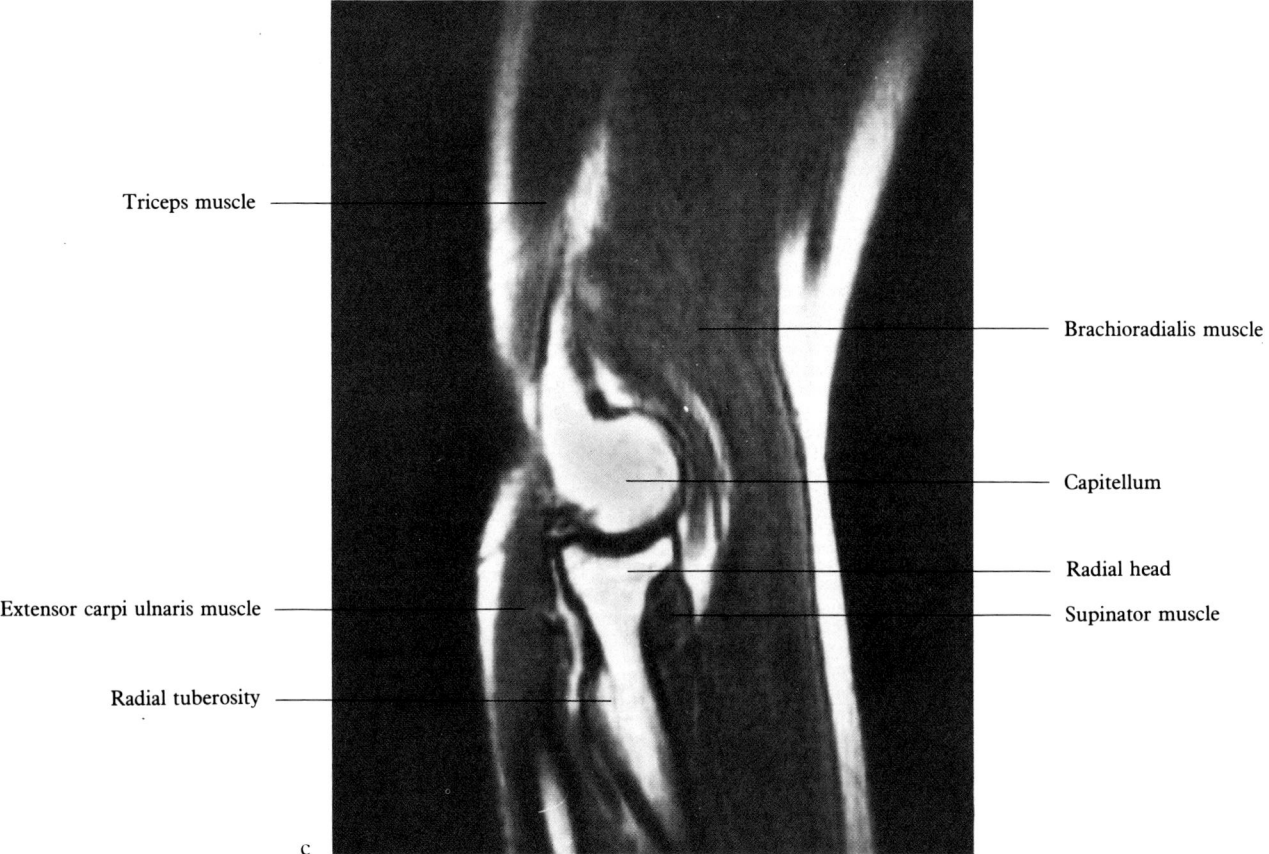

c

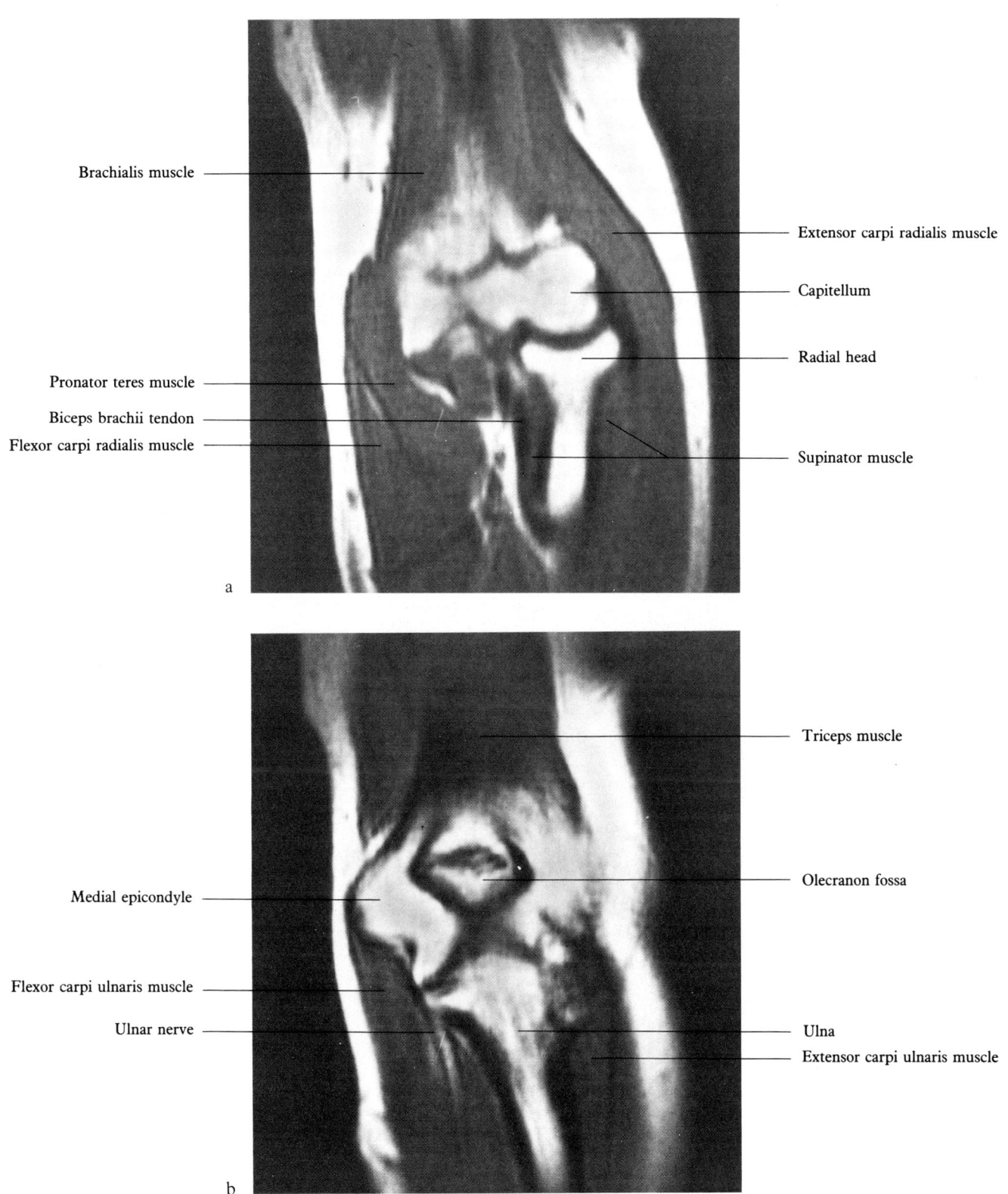

Figure 5.4

Normal anatomy, coronal. (a–c) Serial MR images (SE 500/28) of the normal elbow joint in the coronal plane, from anterior to posterior.

Figure 5.4 continued

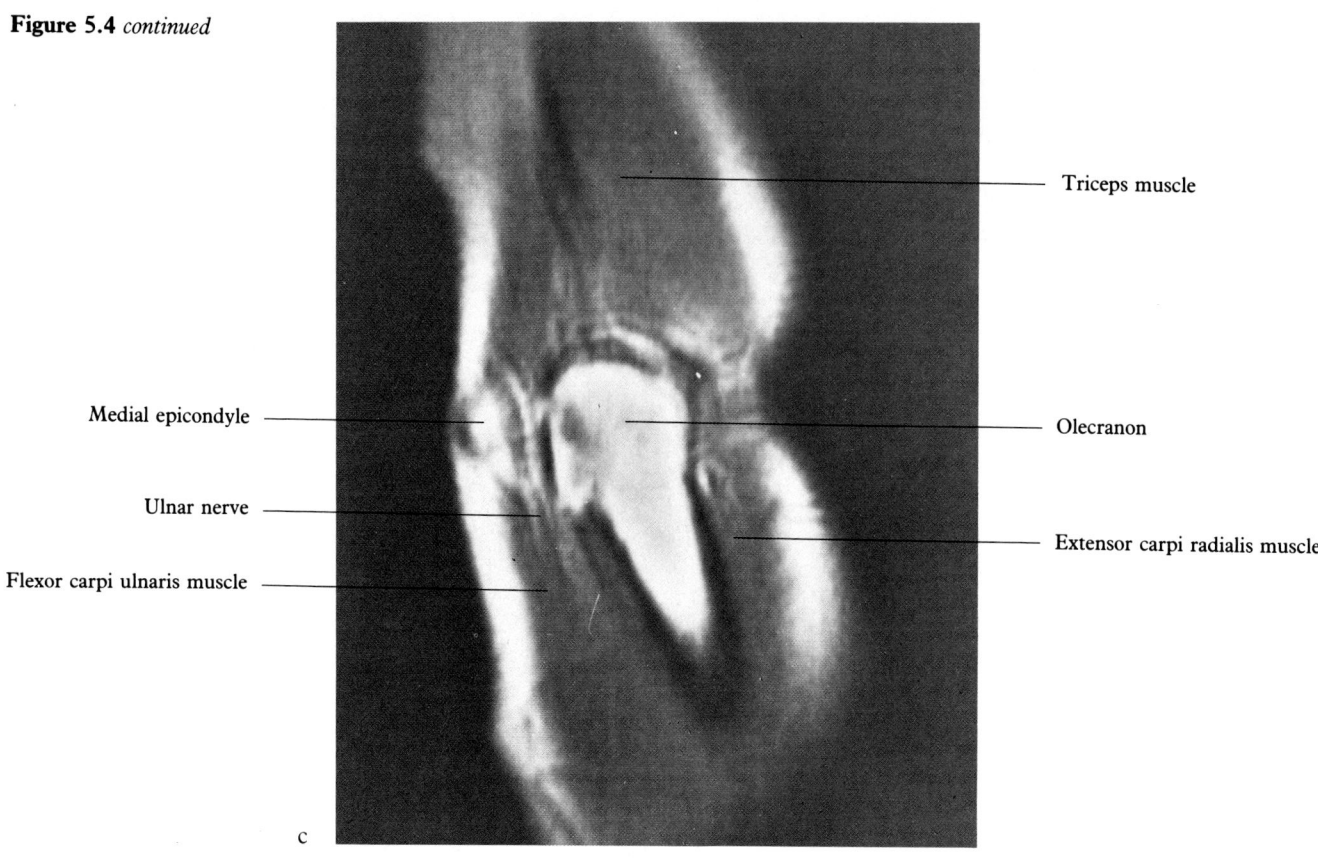

- Triceps muscle
- Medial epicondyle
- Ulnar nerve
- Flexor carpi ulnaris muscle
- Olecranon
- Extensor carpi radialis muscle

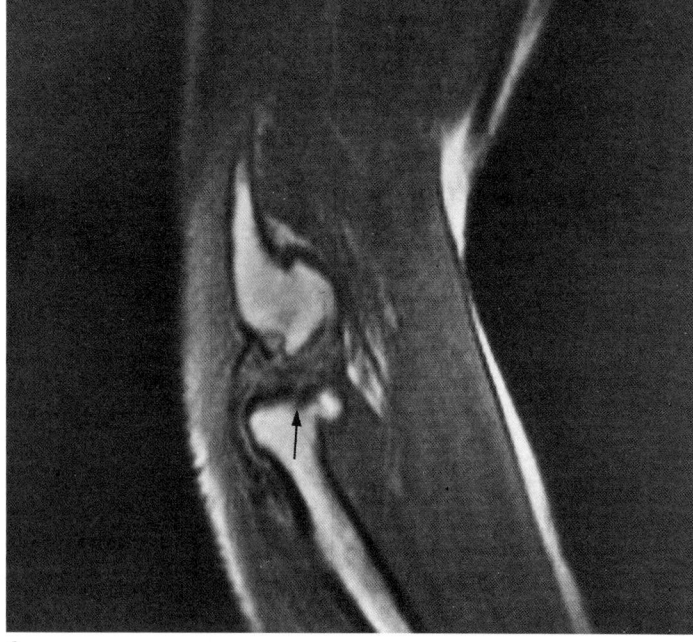

Figure 5.5

Sagittal MR images (SE 500/28) of a patient with complaints of 'locking' in extension after trauma. A radial head cortical step-off, consistent with radial head fracture, is well seen (**a,b**). Joint effusion is especially prominent on (**c**), where it displaces anterior and posterior fatpads (arrows).

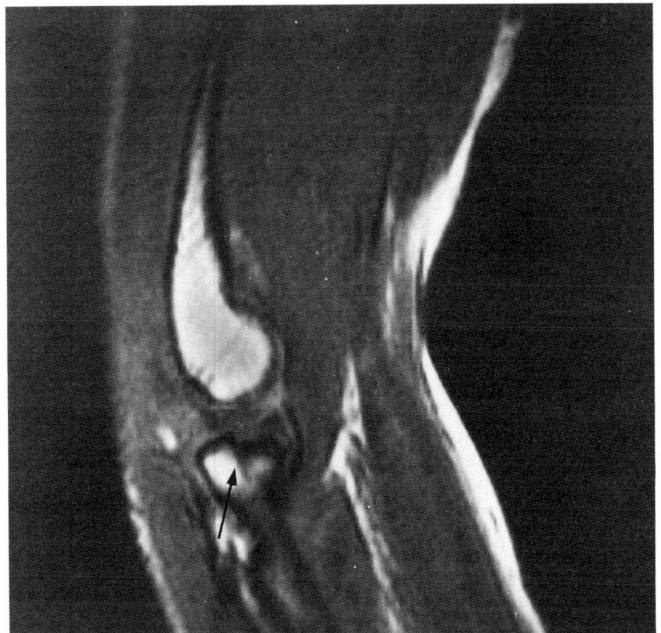

b

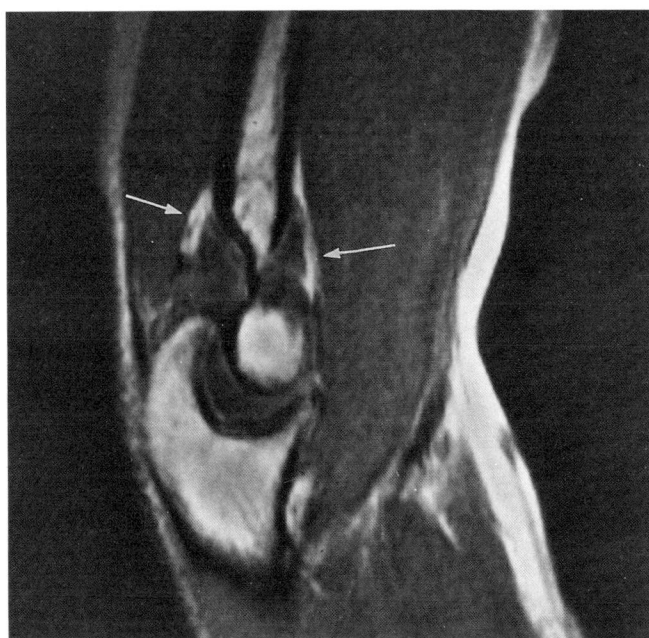

c

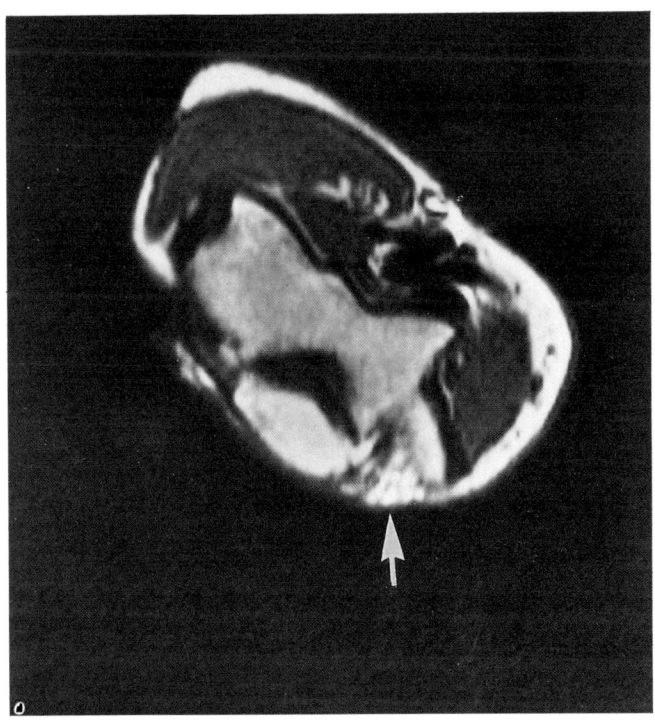

Figure 5.6

T_1-weighted axial MR image (SE 500/28) through the distal humerus in a 62-year-old man with a history of ulnar nerve palsy after an old shrapnel injury. A region of heterogeneously increased signal posterior to the medial humeral epicondyle is seen in the usual course of the ulnar nerve (arrow). This is consistent with increased fat and scar-tissue deposition along the course of the ulnar nerve. A discrete ulnar nerve could not be visualized, even with interleaving sections and multiple planes.

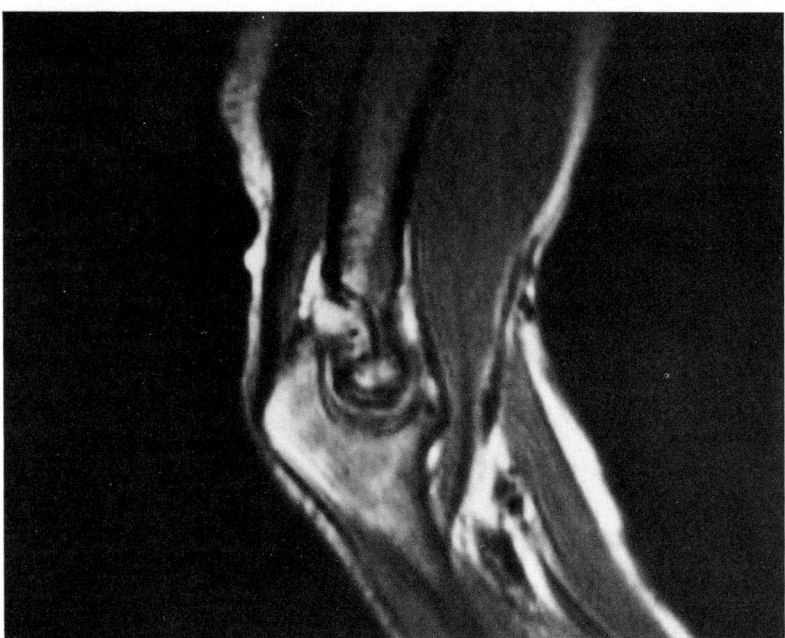

Figure 5.7

Sagittal MR image (SE 500/28) of a 35-year-old man with Gaucher disease diagnosed at 4 years of age. Gaucher disease is a hereditary metabolic disturbance characterized by the accumulation of cerebroside in reticuloendothelial cells (Gaucher cells). Note the heterogeneously decreased signal of the marrow consistent with infiltration by Gaucher cells combined with ischemic necrosis. Note also the bowing deformity of the distal humerus and expansion of the proximal ulna.

6
The wrist and hand

Mamed Mesgarzadeh
Carson D Schneck and
Akbar Bonakdarpour

Radiological examination of the painful wrist has been largely confined to the bony structures and has contributed little to the diagnosis of soft-tissue-related disorders such as carpal tunnel syndrome (CTS), ligament injuries and tendon disease. Even CT has had limited application to the evaluation of the soft tissues of the wrist.[1-3] The unique ability of MRI to demonstrate soft-tissue abnormalities and bone marrow disorders, coupled with its non-ionizing nature and multiplanar imaging capabilities, has brought new hope for solving the mystery of the painful wrist.[4-9] The preliminary findings reported here will require more extensive clinical investigation and further advances in MRI technology to realize the full potential of this modality in imaging the painful wrist.

Anatomy of the wrist

A comprehensive understanding of the complex soft-tissue anatomy of the wrist is required to provide a basis for interpreting MR images of the wrist. To provide a three-dimensional orientation to some of the radiologically unfamiliar soft-tissue structures of the wrist depicted by MR, some dissected anatomic specimens are used for demonstration (Figures 6.1–6.5). The sectional appearance of these structures is demonstrated by matched pairs of T_1-weighted MR images and anatomical sections in axial, coronal and sagittal planes (Figures 6.6–6.20).

Axial plane

The anatomy of the wrist in the axial plane is presented at six levels: distal radioulnar joint, proximal carpal tunnel, intermediate carpal tunnel, distal carpal tunnel, metacarpal base, and metacarpal shaft (Figures 6.6–6.16).

Distal radioulnar joint level (Figure 6.6)

The pronator quadratus muscle overlies the palmar aspect of the distal radius and ulna where it is separated from the overlying flexor tendons by a fat-filled (Parona's) space, which can be a pathway for the spread of inflammatory processes from the hand to the forearm.[10] The axial section of the median nerve is round or oval in shape and shows moderate signal intensity. It is partly surrounded by fat in some cases, but is consistently delimited by specifically identifiable signal-void tendons: superficially the palmaris longus, radially the flexor carpi radialis, deeply and radially the flexor pollicis longus, ulnarly the flexor digitorum profundus.

The radial artery is radial to the flexor carpi radialis. Locating the flexor carpi ulnaris permits identification of the moderate signal intensity of the ulnar nerve deep to it and the ulnar artery radial to it. The dorsal tendons can be identified by their typical relationships to the radius and ulna. From radial to ulnar they include the: abductor pollicis longus and extensor pollicis brevis, radial to the radius; extensor carpi radialis longus and brevis, just radial to the radial (Lister's) tubercle; extensor pollicis longus, just ulnar to the radial tubercle; extensor digitorum and extensor indicis, dorsal to the ulnar side of the radius; extensor digiti minimi, dorsal to the radioulnar joint; and extensor carpi ulnaris, dorsal to the ulna. At this level each of these tendons is surrounded by a lubricating synovial tendon sheath which can be involved in tenosynovitis. The cephalic vein overlies the anatomic snuffbox which is bounded by the abductor pollicis longus and extensor pollicis brevis palmarly, and extensor pollicis longus dorsally.

Proximal carpal tunnel (Figures 6.7–6.9)

At this level the walls of the proximal carpal tunnel include the trapezium and scaphoid radially, the capitate, hamate and triquetrum deeply, the pisiform ulnarly, and the flexor retinaculum superficially. The low-signal-intensity flexor retinaculum extends from the pisiform to the scaphoid and trapezium. The median nerve is just deep to the flexor retinaculum and palmaris longus tendon, between the flexor pollicis longus deeply and radially and the flexor digitorum superficialis tendons deeply and ulnarly. The median nerve is slightly flattened by the adjacent tendons. The flexor digitorum profundus tendons occupy the deep carpal tunnel. Synovial tendon sheaths surround the flexor pollicis longus, flexor digitorum superficialis and flexor digitorum profundus tendons as they traverse the carpal tunnel. The flexor carpi radialis lies in its groove on the trapezium (Figure 6.9). The low-signal-intensity palmar scaphoid-triquetral (Figures 6.8–6.9) and palmar lunate-triquetral (Figure 6.7) carpal ligaments form the deep walls of the carpal tunnel at this level. On the dorsal surface of the carpal bones the dorsal scaphoid-triquetral (Figure 6.7) and dorsal scaphoid-hamate (Figure 6.8) carpal ligaments are visible as thin bands of low-signal intensity. The ulnar nerve is situated medial to the ulnar artery and both are well delimited by the fat of the ulnar (Guyon's) canal, a potential ulnar nerve entrapment site.[11-13] At this level the structures bounding Guyon's canal include the flexor retinaculum deeply and the pisiform bone ulnarly. As the ulnar nerve and artery pass distally in Guyon's canal they divide into their superficial and deep branches (Figure 6.9). Dorsally the radial artery passes deep to the abductor pollicis longus and extensor pollicis brevis to enter the anatomic snuffbox. The extensor pollicis longus courses radially, superficial to the extensor carpi radialis longus and brevis tendons.

Intermediate part of the carpal tunnel (Figures 6.10–6.12)

The palmar capitate-trapezium (Figures 6.11–6.12) and the palmar capitate-scaphoid (Figure 6.10) ligaments contribute to the deep wall of this portion of the carpal tunnel. At this level the flexor retinaculum is well visualized, extending from the hamulus of the hamate to the tubercle of the trapezium. The thenar muscle mass originates from the flexor retinaculum and trapezium. Generally, the median and ulnar nerves maintain the same relationships as in proximal sections (Figure 6.11). At times, however, the position of the median nerve is atypical. For example, it can be situated more deeply within the carpal tunnel (Figures 6.10–6.12).

Distal carpal tunnel (Figure 6.13)

At this level the hook of the hamate forms the ulnar wall of the carpal tunnel and the deep wall of Guyon's canal. The branches of the ulnar nerve and ulnar artery are superficial to the hook of the hamate. The flexor retinaculum attaches to the hook of the hamate as does the hypothenar muscle mass. The flattened median nerve maintains a typical relationship with its tendinous boundaries. Deep in the carpal tunnel a fat plane develops on the anterior aspect of the palmar capitate-trapezium ligament.

Metacarpal base level (Figures 6.14–6.15)

The flexor retinaculum becomes continuous with the palmar aponeurosis. Branches of the median nerve exit from the distal end of the carpal tunnel. The three common digital branches of the median nerve are well delineated by the fat plane deep to the palmar aponeurosis (Figure 6.15). The palmar metacarpal ligaments are visible (Figure 6.14), attaching to the palmar aspect of the metacarpal bases.

Metacarpal shaft level (Figure 6.16)

At this level the common digital branches of the median nerve lie superficial to the spaces between the paired flexor digitorum superficialis and profundus tendons to each finger. The common digital branch of the ulnar nerve is deep to the superficial and deep branches of the ulnar artery. A deep fat plane occupies the spaces between the flexor digitorum profundus tendons with their attaching lumbricals and the more deeply lying metacarpals, interossei and adductor pollicis. The deep palmar branch of the radial artery passes palmarward through the space between the first and second metacarpals.

Summary of MR anatomy

Axial plane

Axial images are most informative and permit identification of all ligaments, tendons, nerves and vessels of the wrist. The most logical algorithm for identifying the soft-tissue structures is to use the bones to locate the ligaments, including the flexor retinaculum. Then the bones and ligaments can serve as landmarks to identify all of the flexor and extensor tendons. Finally, the tendons can be utilized to locate the important nerves and vessels.

Axial images provide the best delineation of the walls of the carpal tunnel, including the palmar carpal ligaments deeply and the flexor retinaculum superficially. The median nerve is identifiable as a structure of intermediate-signal intensity surrounded in some cases by high-signal-intensity fat, but relatively consistently delimited by low-signal-intensity tendons including the palmaris longus superficially, the flexor carpi radialis radially, the flexor pollicis longus deeply and radially, the flexor digitorum superficialis ulnarly and the flexor digitorum profundus deeply and ulnarly. The median nerve can be traced through the carpal tunnel and into the hand, where even its small terminal branches can be discriminated. Axial images clearly visualize the intermediate-signal intensity of the ulnar nerve and the low-signal intensity of the ulnar artery as they traverse Guyon's canal to their division into superficial and deep branches. The ability of MRI to identify and follow the median and ulnar nerves through their potential entrapment sites at the wrist is useful in evaluating carpal tunnel syndromes and similarly may be helpful in diagnosing Guyon's canal syndromes.

Axial images also permit evaluation of each of the flexor and extensor tendons at the wrist. Recognition of these tendons should permit identification of tenosynovitis involving their surrounding synovial tendon sheaths. The palmar carpal ligaments, which are well demonstrated on axial views, include the scaphoid-triquetral, lunate-triquetral, capitate-scaphoid, capitate-trapezium and palmar metacarpal ligaments. The dorsal scaphoid-triquetral and scaphoid hamate ligaments are also identifiable. The capability of identifying specific carpal ligaments may permit the diagnosis of specific ligamentous injuries.

Coronal and sagittal planes

Three pairs of coronal and one pair of sagittal sections are presented in Figures 6.17–6.20.

The coronal images are most helpful in identifying each of the carpal bones, since they provide a simultaneous panoramic view of all the carpal bones. The triangular fibrocartilage is seen as a signal-free area in the ulnocarpal space. The radial collateral ligament, which attaches the radial styloid to the scaphoid, and the ulnar collateral ligament, that extends from ulnar styloid to the triquetrum, are also identifiable in coronal images (Figures 6.17–6.18). The radial artery courses obliquely and dorsally through the anatomic snuffbox and is surrounded by fat (navicular fat pad). This fat pad separates the radial collateral ligament from the abductor pollicis longus (Figure 6.17).

Both coronal and sagittal images also provide longitudinal views of the flexor and extensor tendons. Sagittal images can demonstrate the palmar radiocapitate and dorsal radiolunate and capitate-lunate ligaments (Figures 6.19–6.20).

In summary, coronal images yield the best visualization of the radial and ulnar collateral ligaments and the triangular fibrocartilage. The carpal bones are also more easily identifiable on coronal images. Long views of the tendons are provided by both coronal and sagittal images. Sagittal images also provide visualization of the palmar and dorsal carpal ligaments.

Technique

Positioning

The anatomical structures of the wrist are very small and the resolution of the commercially available MR units is limited. Therefore, all efforts should be made to increase resolution and decrease motion artifacts. For this purpose a small surface coil should be used, and the wrist positioned in the middle of the magnetic bore where the magnetic field is most homogeneous. A small field of view should be chosen. In order to eliminate even minor motion artifacts, the entire upper extremity should be positioned comfortably. Most wrist MR images acquired to date have been obtained by positioning the wrist transversely above the supine patient's head. This is done by abducting the arm alongside the head and flexing the elbow. Placing the upper limb straight overhead is usually poorly tolerated by the patient because of the strain placed on the shoulder. An alternative is to place the wrist over the pubis.

We have designed a special fixture consisting of a fiberglass tray that can be adjusted according to the patient's size. The surface of the tray is covered by a velcro-type material allowing the surface coil and patient's forearm to be immobilized over the tray by several velcro straps. The tray, which eliminates transmission of the respiratory motions to the coil, is positioned over the lower thorax or upper abdomen of the supine patient.

Image strategy

The selection of the optimum imaging plane will be partly dependent on the suspect anatomical structure. After acquiring a T_1-weighted scout series, T_1- and T_2-weighted images are obtained in the optimum plane.

In most instances the axial images are preferred, since most of the soft-tissue structures of the wrist and carpal tunnel are best demonstrated and can be most easily identified in the axial plane. However, evaluation of the carpal bones, radial and ulnar collateral ligaments, and triangular-fibrocartilage is better accomplished with coronal images. Nevertheless, the sagittal or coronal images are equally useful for evaluation of the radiocarpal and intercarpal joints. For example, to determine the intra-articular extension of an intra-osseous tumor such as giant cell tumor, coronal or sagittal images are preferred. To decrease the acquisition time, thus minimizing motion artifact, an attempt should be made to use a relatively short repetition time. Typical pulse sequences are TE = 16 msec, TR = 400 msec, and TE = 84 msec, TR = 1200–1500 msec for T_1- and T_2-weighted imaging, respectively.

Indications

MRI can be used for evaluation of painful wrists due to various pathologies. Examples are: CTS, soft-tissue neoplasms, and any soft-tissue-related abnormality such as ganglion cyst, adhesive tendonitis (Figure 6.27) and incisional neuroma[14] (Figure 6.25). Since MRI is very sensitive in demonstrating marrow abnormalities, it can also be utilized for early detection of ischemic necrosis in a fractured scaphoid (Figure 6.24) or Kienbock's disease of the lunate (Figure 6.23). For the same reason, MRI will aid the early detection of a marrow-infiltrative process such as metastatic tumor. For evaluation of primary bone tumors, the main indications for MRI are to determine the extent of marrow involvement, and to determine the presence or absence of soft-tissue or intra-articular involvement (Figure 6.22).[5-19] Occasionally the signal characteristics of the lesion represent a clue regarding its tissue content. For example, fatty tissues and organized hematomas have bright signal intensity in both T_1- and T_2-weighted images.[20,21] Cellular tumors have prolonged T_1- and T_2-relaxation times. Regardless of their aggressiveness, less cellular tumors exhibit low-signal intensity in T_1- and T_2-weighted images because of their higher collagen content.[22-24] Pigmented villonodular synovitis produces intermediate-signal intensity on T_1-weighted images, and a mixture of low- and high-signal intensities on T_2-weighted images, the high-intensity signal resulting from increased joint fluid and congested synovium, and the low-intensity signal from deposits of hemosiderin within the synovium.

Carpal tunnel syndrome

Definition

Carpal tunnel syndrome is a relatively common chronic, disabling condition characterized by nocturnal hand discomfort, finger paresthesias in the median nerve distribution and thenar muscle atrophy. It is most frequently caused by compression of the median nerve in the carpal tunnel. Most often it occurs between ages 30 and 60, and is 2 to 5 times more common in women than men.[25-27] The dominant hand is affected more frequently, and in 32 to 50 per cent of cases the involvement is bilateral.[25,27,28]

Clinical

Numbness and tingling of the tip of the middle finger is an early manifestation. In well established cases the main complaint is a poorly described burning and aching discomfort characteristically occurring during the night, most commonly 3 to 4 hours after retiring. The fifth finger is usually spared. The typical objective findings are Tinel's and Phalen's signs. Tinel's sign is a tingling sensation in the median nerve distribution, elicited in the digits by tapping the nerve at the wrist. Phalen's sign is the exacerbation of the symptoms by acute flexion of the wrist.

Etiology

Multiple etiologies have been reported for CTS.[28] For many of these, the cause for symptoms is compression of the median nerve. In other instances, the cause of symptoms is poorly understood. For an easier understanding of the various etiologies, we have classified them into three main groups.

1. Local entrapment of the median nerve within the carpal tunnel, which we subclassify into three groups:
   (a) Decrease in size of the carpal tunnel due to bony or soft-tissue changes such as malalignment of the carpal bones, displaced fractures, callus formation, hypertrophic osteophytes[29] and fibrous scarring, for example, from a third-degree burn. Colles' fracture is considered the most common traumatic etiology of the carpal tunnel syndrome.[26,30]
   (b) Increase in volume of the normal content of the carpal tunnel. This can be due to occupational hypertrophy of the muscles and tendons in the carpal tunnel, and has been reported in dentists, disabled patients using a wheel chair, typists, tennis and golf players, and we have even noted it in a barber. Other causes include an aberrant lubrical muscle,[31] synovial proliferation in arthritis, tenosynovitis,[26] edema of congestive heart failure, diffuse deposition of various materials such as amyloid in patients on dialysis,[32] fat in obesity, or other deposits secondary to various endocrinopathies such as Graves' disease, myxedema and acromegaly.[33]
   (c) Space-occupying lesions such as hemangioma, lipoma, ganglion cyst,[26,29] gouty tophus[34] and hematoma.
2. Systemic diseases causing neuritis, of which diabetes is the best known and is more common in patients with CTS than in the general population. Seven per cent of patients with CTS have diabetes, while in the overall population the incidence of diabetes is 1.7 per cent.[27]
3. Idiopathic: in approximately 50 per cent of patients the etiology and mechanism of the syndrome are not well understood. Examples are CTS associated with menopause, late trimester of pregnancy,[26,29] and osteopetrosis.[35]

Entrapment of the median nerve or its segmental spinal nerve contribution proximal to the carpal tunnel is most frequently due to cervical spondylosis, which can produce symptoms identical to entrapment of the nerve within the carpal tunnel. This is the cause of the syndrome in 11 per cent of patients and in whom bilateral involvement occurs in 41 per cent.[26,27]

Diagnosis and treatment

The diagnosis of CTS has been mainly empiric. Present diagnostic parameters are limited to clinical history, clinical signs, and nerve conduction studies which are on occasion equivocal. Imaging modalities prior to MRI have been largely non-contributory to the diagnosis and treatment of CTS except in a limited number of cases of spurs of malalignments of carpal bones. On CT examination no significant difference has been found between the cross-sectional area of the carpal tunnels of symptomatic patients and of a control population.[36] To date, except for flattening of the median nerve under the flexor retinaculum no other objective imaging signs of increased pressure in the carpal tunnel have been described.[8]

Likewise, the choice of conservative management (splinting, injection of corticosteroid, or administration of pyridoxine) or surgical treatment is largely empiric. The reason for success or failure of conservative treatment is poorly understood, since there are many possible etiologies, and one is rarely established prior to treatment. The reasons for the failure of surgical treatment or later recurrence could be due to either inappropriate diagnosis, Wallerian degeneration of the nerve due to delayed treatment, inadequate incision of the flexor retinaculum, postoperative formation of scar or neuroma, or existence of a growing space-occupying lesion within the carpal tunnel.

In a preliminary MRI study, we noted two classes of MRI findings which may prove to be useful diagnostic descriptors of CTS: general findings that were seen regardless of the etiology, and etiologically specific findings. The general findings consist of palmar bowing of the retinaculum, swelling of the median nerve proximal to the carpal tunnel (pseudoganglion) and constriction of the nerve within the carpal tunnel (Figure 6.22). The etiologically specific findings include edema around or fluid within a tendon sheath (tenosynovitis) (Figure 6.23), ischemic necrosis of bone (Figure 6.24), abnormal location of the median nerve, and incisional neuroma (Figure 6.25).[18] In one case of recurrence after surgery we were able to show an excessive amount of fat within the carpal tunnel, which may or may not have been etiologically significant (Figure 6.27).

Our preliminary findings in conjunction with other reports indicate that the present largely empirical diagnosis and treatment of CTS could be made substantially more objective if the structural parameters of CTS and its multiple etiologies could be specifically identified by MRI. Imaged structural parameters localizable to the carpal tunnel could differentiate CTS from higher level nerve entrapment. Etiology-specific MRI findings could permit a more rational choice of treatment options directed toward that etiology. Furthermore, MRI localization of an atypically located median nerve could help avoid median nerve injury during surgery or injection.

Finally, post-treatment MRI findings could serve as an early predictor of the success or failure of a given mode of treatment. For example, demonstration of the completeness (Figure 6.22), or incompleteness (Figure 6.28), of incision of the flexor retinaculum in cases treated surgically could predict the probability of future symptom recurrence, or establish the reason for an already clinically apparent treatment failure. In cases that are treated conservatively, the early prediction of therapeutic success or failure by follow-up MRI could be important in deciding upon continuation or replacement of the selected therapy, before irreversible nerve degeneration occurs. For example, clearance of peritendinous edema and fluid in cases of tenosynovitis, decrease in bowing of the flexor retinaculum, and diminished swelling of the median nerve might be encouraging findings favoring the continuation of conservative treatment.

References

1. CONE RD, RESNICK D, GELBERMAN R et al, Computed tomography of the normal soft tissues of the wrist, *Invest Radiol* (1985) **18**:546–50.
2. ZUCKER-PINCHOFF B, HERMAN G, STRINIVASAN R, Computed tomography of the carpal tunnel: a radioanatomical study, *J Comput Assist Tomogr* (1981) **5**:525–8.
3. WEEKS PM, VANNIER MW, STEVENS G et al, Three dimensional imaging of the wrist, *J Hand Surg* (1985) **10A**:32–9.
4. MESGARZADEH M, SCHNECK CD, BONAKDARPOUR A, Magnetic resonance imaging of wrist and carpal tunnel: correlation with normal anatomy, *Radiology* (1985) **157**(P):30.
5. MESGARZADEH M, SCHNECK CD, BONAKDARPOUR A, MR imaging of the knee and wrist: correlation with normal anatomy, *Radiology* (1985) **157**(P):362.
6. WEISS KL, BELTRAN J, SHAMON OM et al, High field MR surface-coil imaging of the hand and wrist. Part I. Normal anatomy, *Radiology* (1986) **160**:143–6.
7. WEISS KL, BELTRAN J, LUBBERS LM, High-field MR surface-coil imaging of the hand. Part II. Pathologic correlations and clinical relevance, *Radiology* (1986) **160**:147–52.
8. MIDDLETON WD, KNEELAND JB, KELLMAN GM et al, MR imaging of the carpal tunnel: normal anatomy and preliminary findings in the carpal tunnel syndrome, *AJR* (1987) **148**:307–16.
9. KOENIG H, LUCAS D, MEISNER R et al, The wrist: a preliminary report on high-resolution MR imaging, *Radiology* (1986) **160**:463–7.
10. LAMPE EW, Surgical anatomy of the hand, *Clin Symp* (1957) **9**:41.
11. SHEA JD, McCLAIN E, Ulnar-nerve compression syndromes at and below wrist, *J Bone Joint Surg* (1969) **51A**:1095–1103.
12. KLEINET HE, HAYES JE, The ulnar tunnel syndrome. *Plast Reconstr Surg* (1971) **47**:21–4.
13. SUNDERLAND S, Ulnar nerve lesions. In *Nerves and nerve injuries*, 2nd edn. (Churchill Livingstone: Edinburgh 1981) 750–79.
14. SINGSON RD, FELDMAN F, SLIPMAN CW et al, Postamputation neuroma and other symptomatic stump abnormalities: detection with CT, *Radiology* (1987) **162**:743–5.
15. REINUS WR, CONWAY WF, TOTTY WG et al, Carpal avascular necrosis: MR imaging, *Radiology* (1986) **160**:689–93.
16. McKINSTRY CS, STEINER RE, YOUNG AT et al, Bone marrow in leukemia and aplastic anemia: MR imaging before, during and after treatment, *Radiology* (1987) **162**:701–7.
17. MOORE SG, GOODING CA, BRASCH RC et al, Bone marrow in children with acute lymphocytic leukemia: MR relaxation times, *Radiology* (1986) **160**:237–40.
18. RAO VM, FISHMAN M, MITCHEL DG et al, Painful sickle cell crisis: bone marrow patterns observed with MR imaging, *Radiology* (1986) **161**:211–15.
19. HERMAN SD, MESGARZADEH M, BONAKDARPOUR A et al, The role of magnetic resonance imaging in giant cell tumor of bone, *Skeletal Radiol* (1987) **16**:635–43.
20. DOOMS GC, KRICAK H, SOLLOTTO RA et al, Lipomatous tumors and tumors with fatty component: MR imaging potential and comparison of MR and CT results, *Radiology* (1985) **157**:479–83.
21. RUBIN JI, GOMORI JM, GROSSMAN RI et al, High-field MR imaging of extracranial hematomas, *AJR* (1987) **148**:813–17.
22. SUNDARAM M, McGUIRE MH, SCHAJOWICZ F, Soft tissue masses: histologic basis for decreased signal (short T_2) on T_2-weighted images, *AJR* (1987) **148**:1247–50.
23. PETASNICK JP, TURNER DA, CHARTERS JR et al, Soft-tissue masses of the locomotor system: comparison of MR imaging with CT, *Radiology* (1986) **160**:135–41.
24. TOTTY WG, MURPHY AM, LEE JK, Soft-tissue tumors: MR imaging, *Radiology* (1986) **160**:135–41.
25. TANZER RC, Compression neuropathies. In Flynn JE, ed. *Hand surgery*, 2nd edn. (Williams and Wilkins: Baltimore 1975) 317–21.
26. MILFORD L, The hand, Chapter 14, Carpal tunnel and ulnar tunnel syndromes and stenosing tenosynovitis. In Edmonson AS, Chrenshaw AH, eds. *Campbell's Operative Orthopedics*, 6th edn. (Mosby: St Louis 1982) 282–285.
27. HURST LC, WEISSBERG D, CARROLL RE, The relationship of the double crush to carpal tunnel syndrome (analysis of 1000 cases of carpal tunnel syndrome), *J Hand Surg* (1985) **10B**:202–4.
28. ARMINIO JA, Etiology of carpal tunnel syndrome, *Del Med J* (1986) **58**:189–92.
29. KESSLER FB, Complication of the management of carpal tunnel syndrome, *Hand Clin* (1986) **2**:401–6.
30. REIYZ KA, ONNE L, Analysis of sixty-five operated cases of carpal tunnel syndrome, *Acta Chir Scand* (1967) **133**:443–7.
31. ASAI M, WONG ACW, MATSUNAGA T et al, Carpal tunnel syndrome caused by aberrant lumbrical muscles associated with cystic degeneration of tenosynovium: a case report, *J Hand Surg* (1986) **11A**:218–21.
32. FENVES AZ, EMMET M, WHITE MG et al, Carpal tunnel syndrome with cystic bone lesions secondary to amyloidosis in chronic hemodialysis patients. *Am J Kidney Dis* (1986) **7**:130–34.
33. BEARD A, KUMAR A, ESTEP HL, Carpal tunnel syndrome caused by Graves' disease, *Arch Intern Med* (1985) **145**:345–6.
34. HYOT RE, Carpal tunnel syndrome and gout: case report, *Va Med* (1986) **113**:407–9.
35. RAKIC M, ELHOSSEINY A, RAMADAN F et al, Adult-type osteopetrosis representing as carpal tunnel syndrome, *Arthritis Rheum* (1986) **29**:926–8.
36. MERHAR GI, CLARK RA, SCHNEIDER HJ, High resolution CT scans of the wrist in patients with carpal tunnel syndrome, *Radiology* (1985) **157**(P):30.

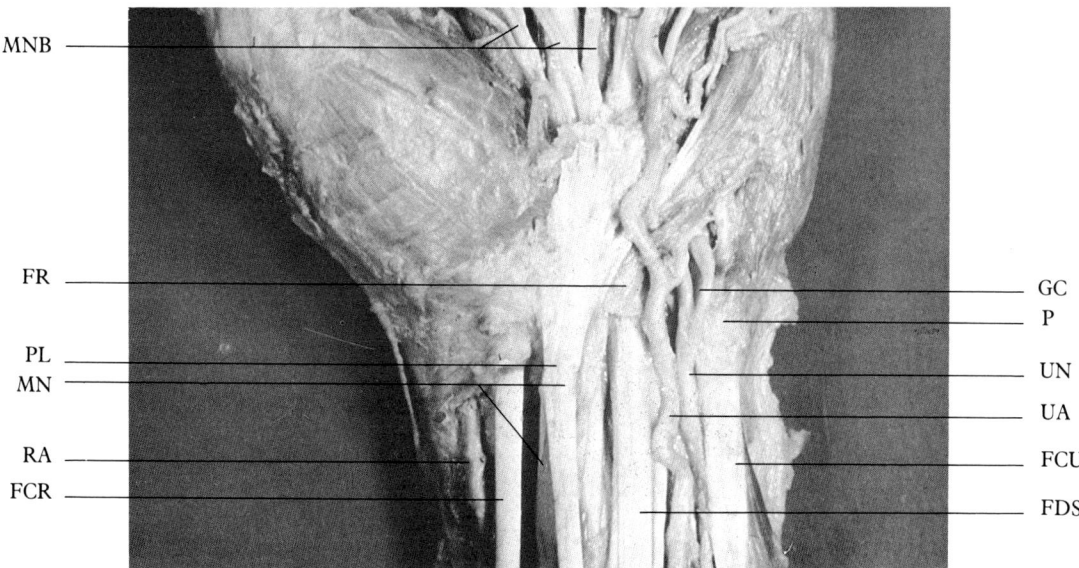

Figure 6.1

Anterior view of wrist demonstrates median nerve (MN) just proximal to flexor retinaculum (FR) where it lies deep to palmaris longus tendon (PL), with flexor carpi radialis tendon (FCR) radially and flexor digitorum superficialis tendons (FDS) ulnarly. The median nerve passes deep to flexor retinaculum within carpal tunnel. As it emerges from distal end of carpal tunnel, it divides into terminal median nerve branches (MNB). Radial artery (RA) is located radial to flexor carpi radialis tendon. Proximal to flexor retinaculum, ulnar nerve (UN) and artery (UA) lie between flexor carpi ulnaris tendon (FCU) ulnarly and flexor digitorum superficialis tendons radially. They pass superficial to the flexor retinaculum where they lie just radial to the pisiform (P) within Guyon's canal (GC) where they divide into superficial and deep branches.

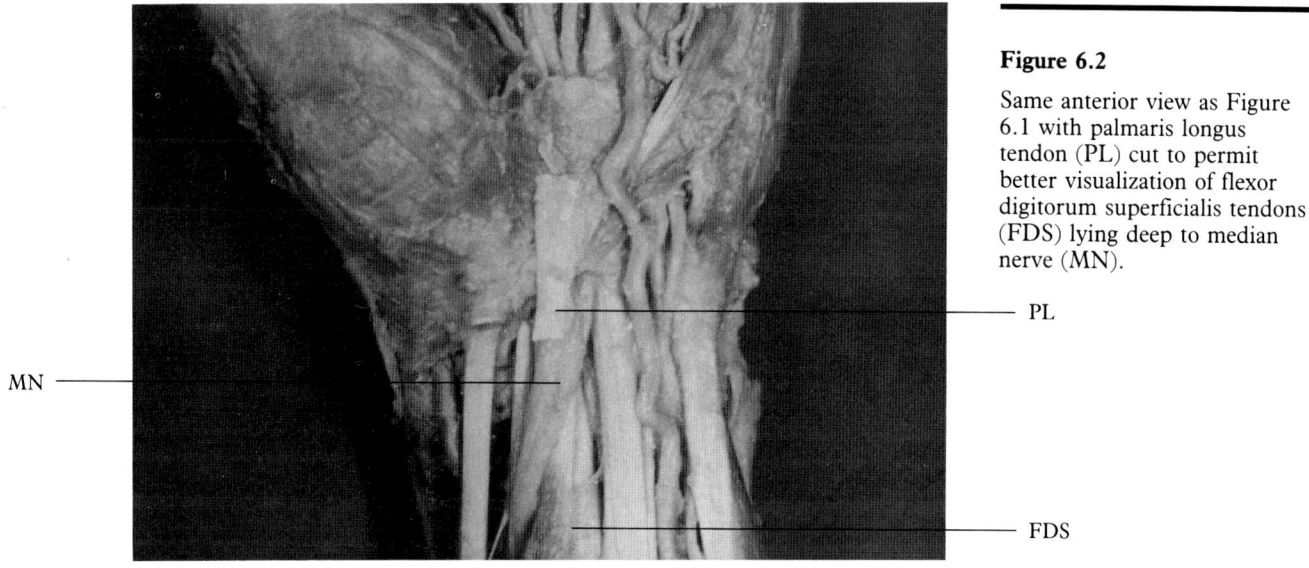

Figure 6.2

Same anterior view as Figure 6.1 with palmaris longus tendon (PL) cut to permit better visualization of flexor digitorum superficialis tendons (FDS) lying deep to median nerve (MN).

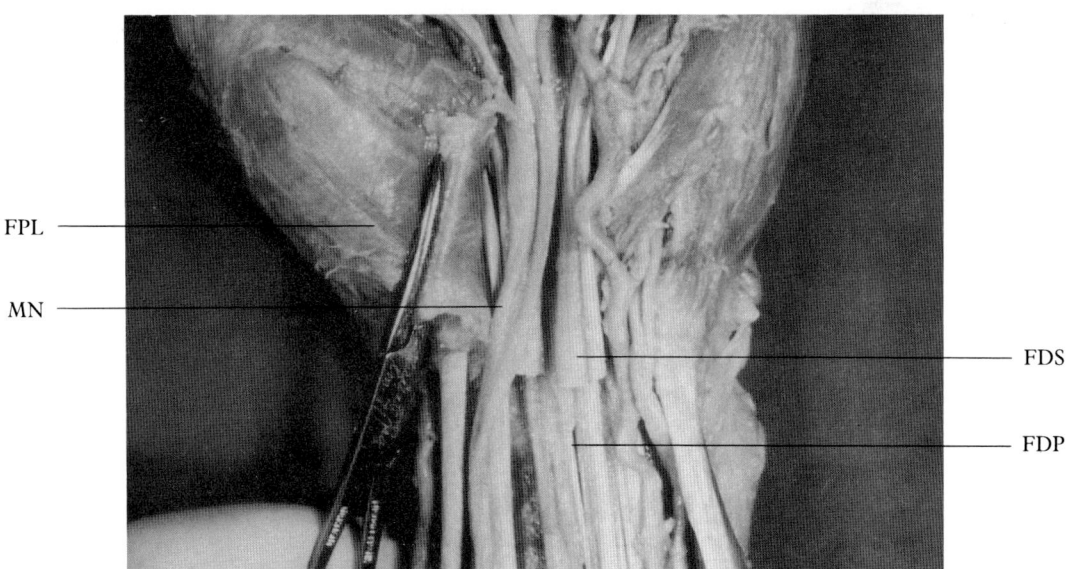

Figure 6.3

Flexor retinaculum has been cut, clamped, and reflected radially to reveal median nerve within carpal tunnel. Flexor digitorum superficialis tendons (FDS) have been cut just proximal to carpal tunnel to improve visualization of more deeply lying flexor pollicis longus (FPL) and flexor digitorum profundus (FDP) tendons. Median nerve (MN) is bounded by flexor pollicis longus tendon deeply and radially, flexor digitorum superficialis tendons ulnarly, and flexor digitorum profundus tendons deeply and ulnarly.

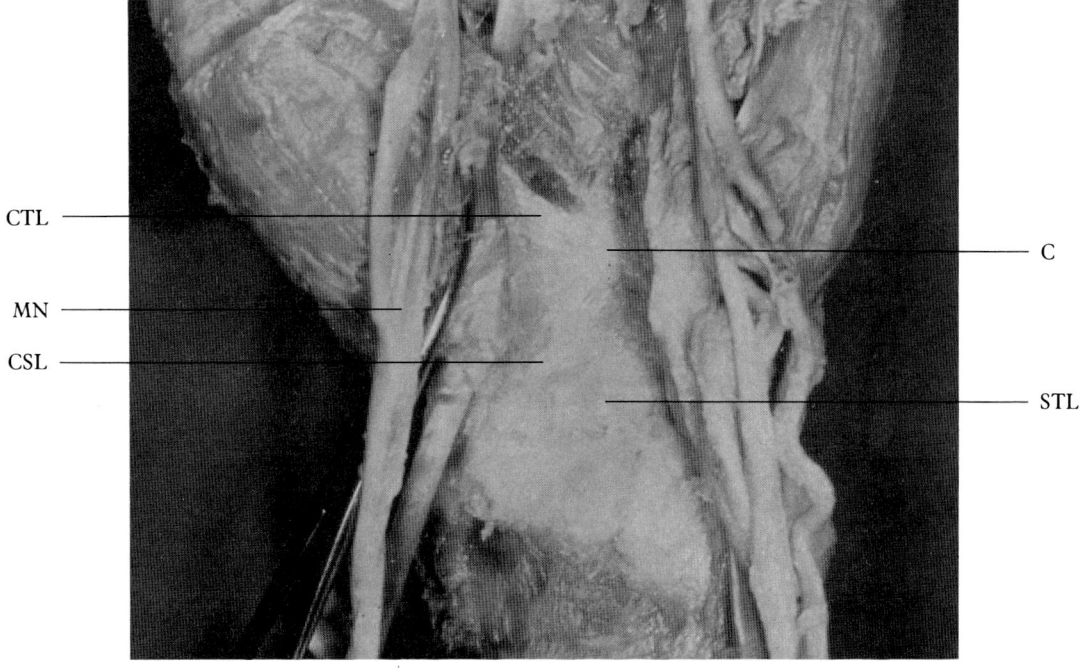

Figure 6.4

Anterior view of floor of carpal tunnel with median nerve (MN) retracted radially and all tendons in carpal tunnel removed. Head of capitate (C) is visualized along with capitate-scaphoid (CSL) and capitate-triquetral (CTL) ligaments. Less well defined scaphoid-triquetral ligament (STL) is situated more proximally.

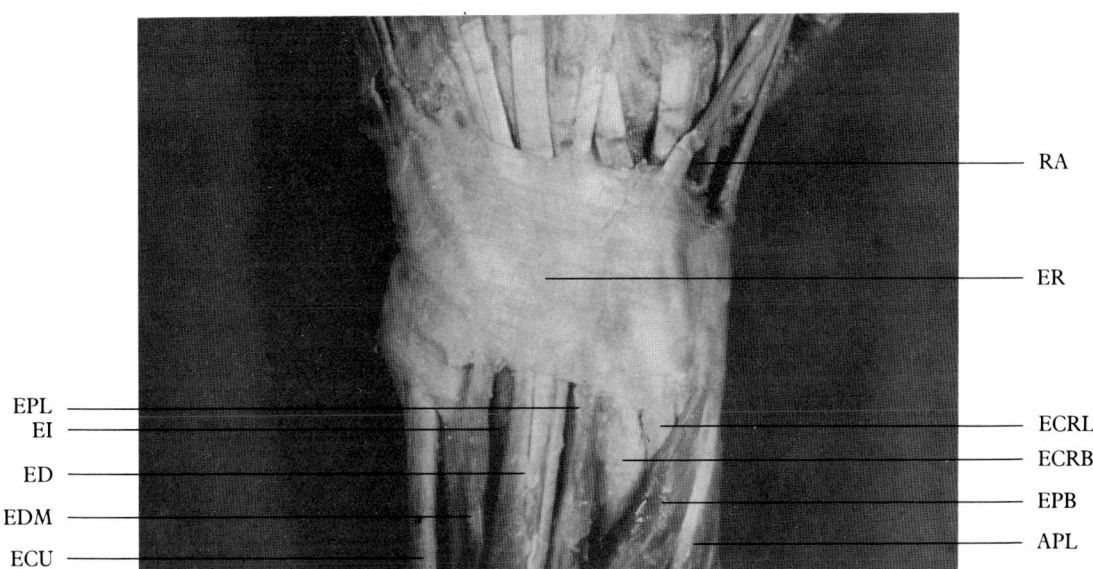

Figure 6.5

Dorsal view of the wrist demonstrating extensor retinaculum (ER) and 6 sets of tendons which run deep to it. From radial to ulnar they are (1) abductor pollicis longus (APL) and extensor pollicis brevis (EPB), (2) extensor carpi radialis longus and brevis (ECRL and ECRB), (3) extensor pollicis longus (EPL), (4) extensor digitorum (ED) and extensor indicis (EI), (5) extensor digiti minimi (EDM), and (6) extensor carpi ulnaris (ECU). By distal end of extensor retinaculum, extensor pollicis longus has crossed over extensor carpi radialis tendons to form dorsal border of anatomic snuffbox. Abductor pollicis longus and extensor pollicis brevis tendons form anterior border of snuffbox which has radial artery (RA) running through it at this level.

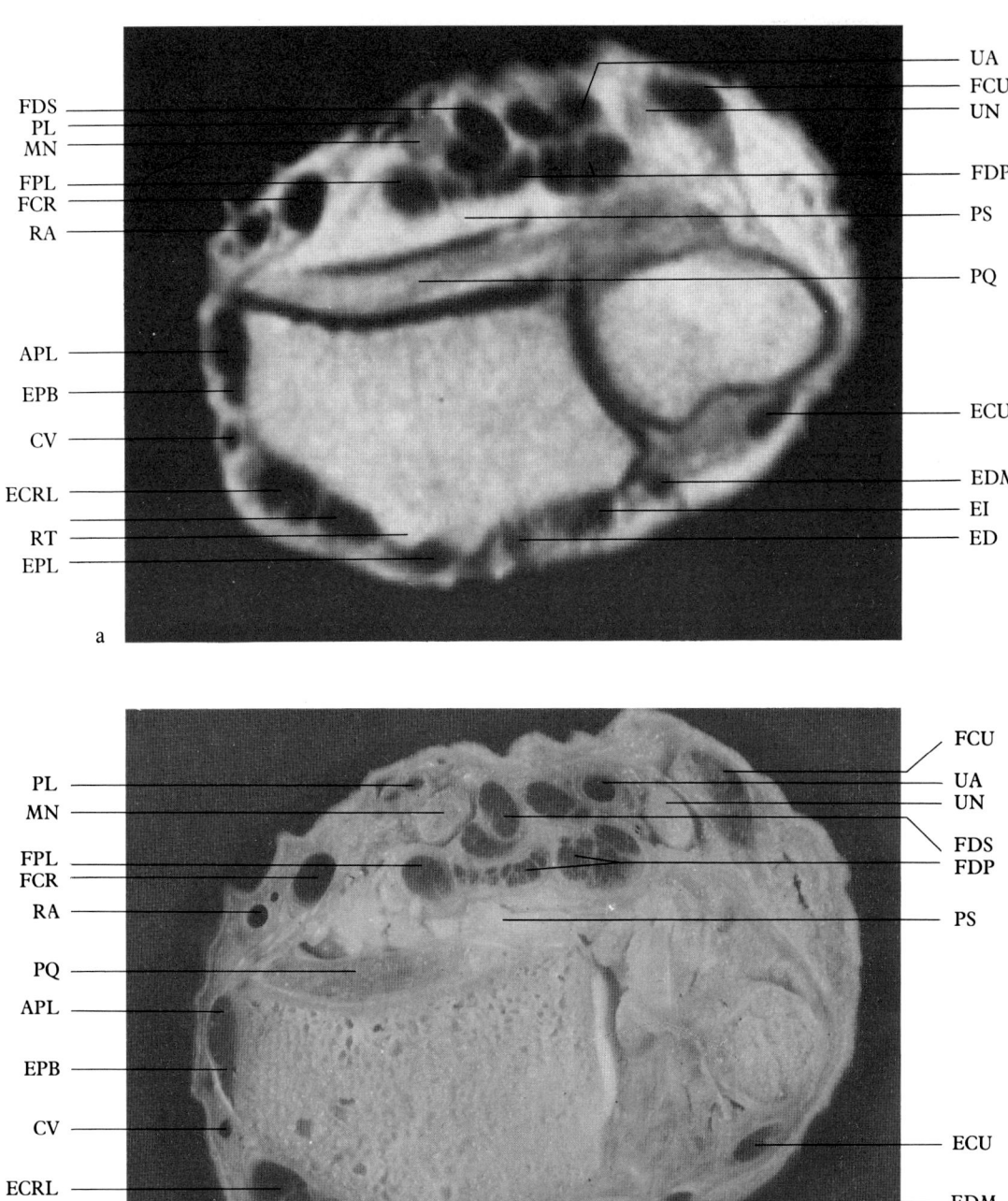

Figure 6.6

Section through the distal radioulnar joint shows: pronator quadratus muscle (PQ), Parona's space (PS), median (MN) and ulnar nerves (UN), palmaris longus (PL), flexor digitorum superficialis (FDS) and profundus (FDP), flexor pollicis longus (FPL), flexor carpi radialis (FCR) and ulnaris (FCU), radial (RA) and ulnar (UA) arteries, abductor pollicis longus (APL), extensor pollicis brevis (EPB) and longus (EPL), cephalic vein (CV), extensor carpi radialis longus (ECRL) and brevis (ECRB), radial tubercle (RT), extensor digitorum (ED), extensor indicis (EI), extensor digiti minimi (EDM) and extensor carpi ulnaris (ECU).

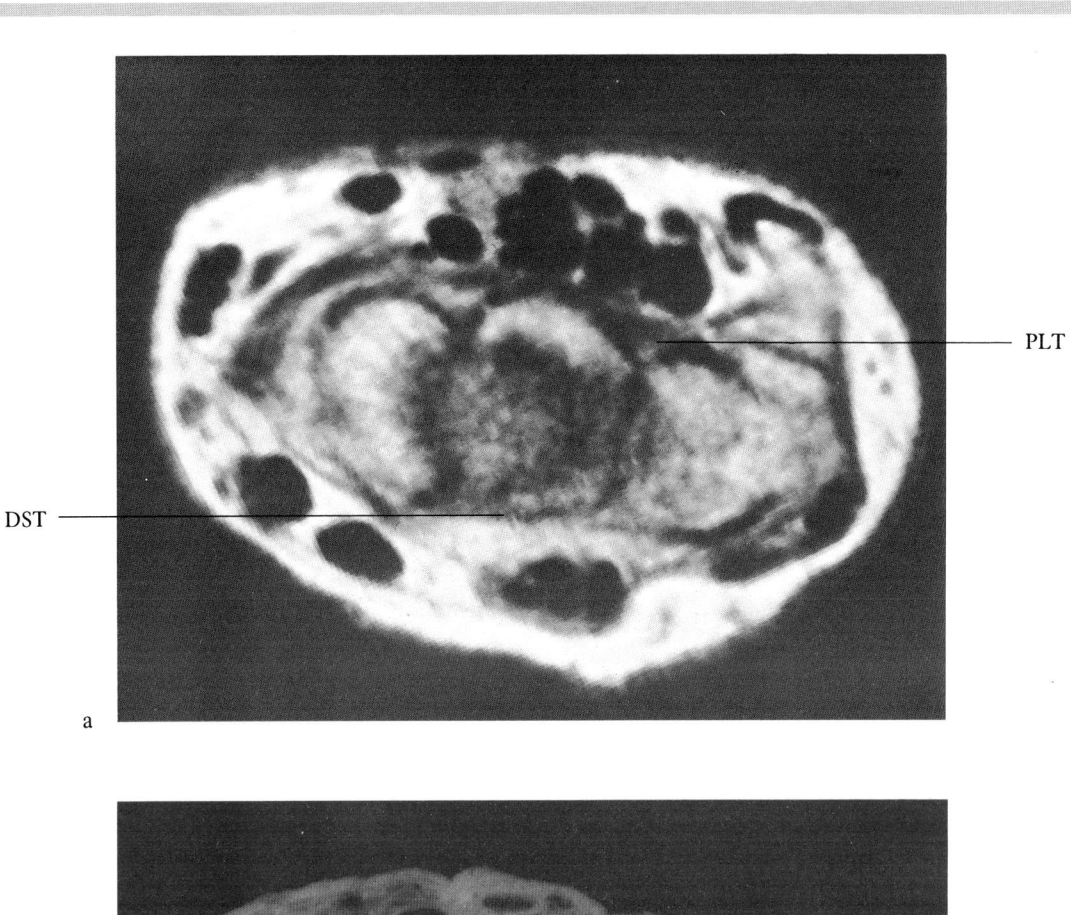

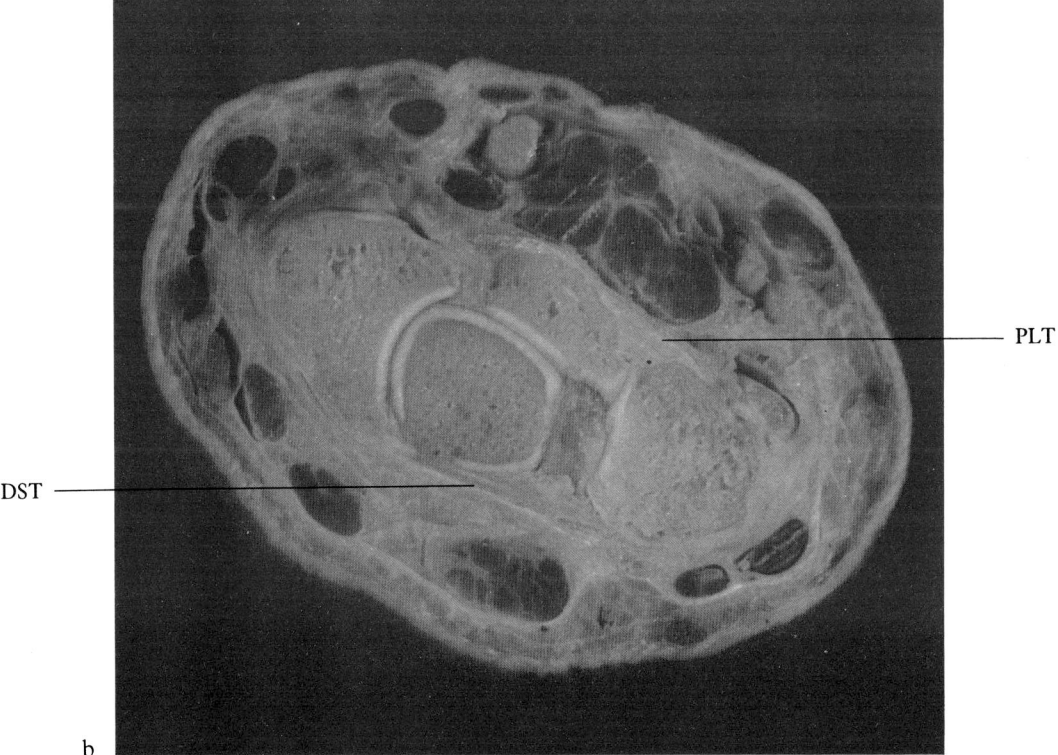

Figure 6.7

Section through the proximal carpal tunnel shows palmar lunate-triquetral ligament (PLT) and dorsal scaphoid-triquetral ligament (DST).

150 MRI atlas of the musculoskeletal system

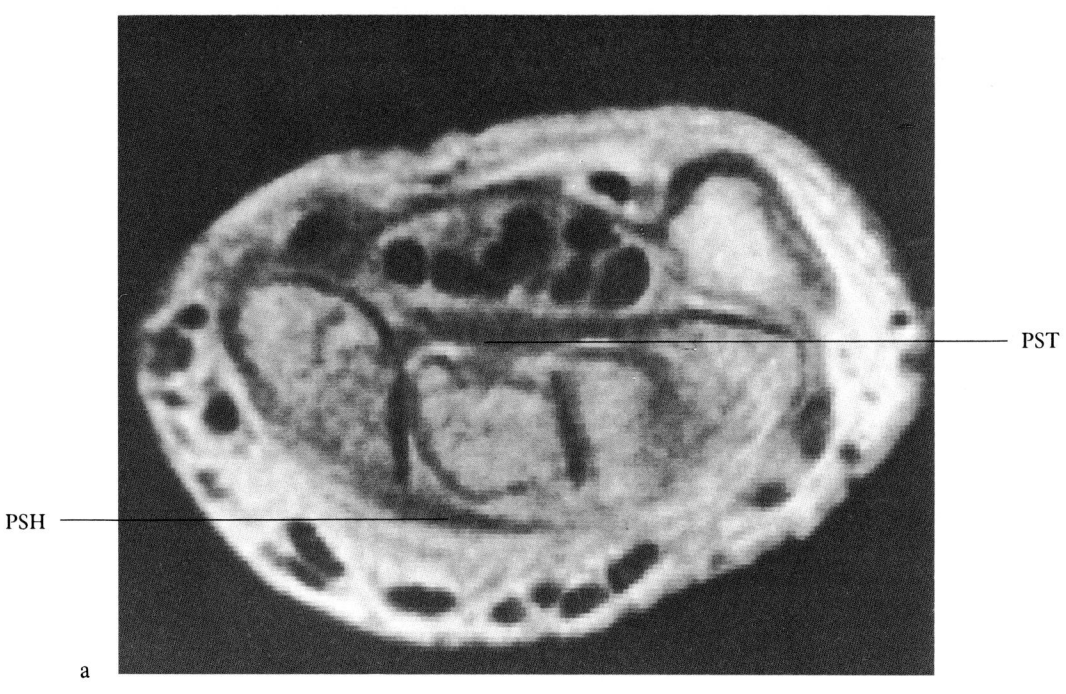

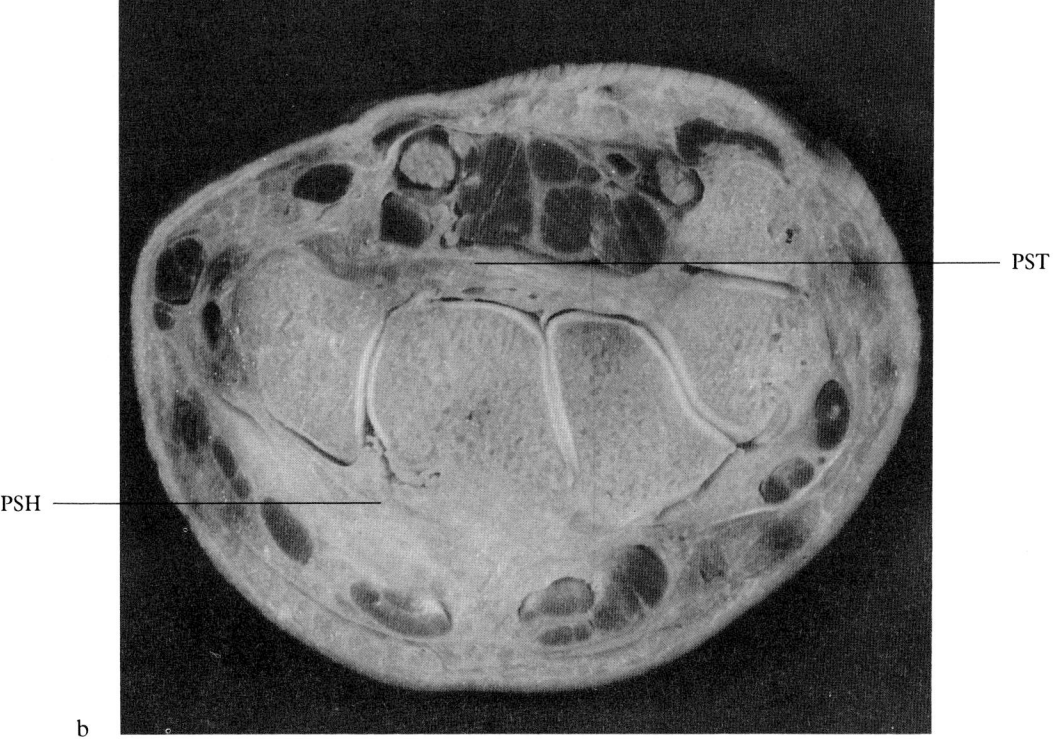

Figure 6.8

Palmar scaphoid-triquetral ligament (PST) forms deep wall of carpal tunnel. Dorsal scaphoid-hamate (DSH) ligament is visualized on dorsum of these bones.

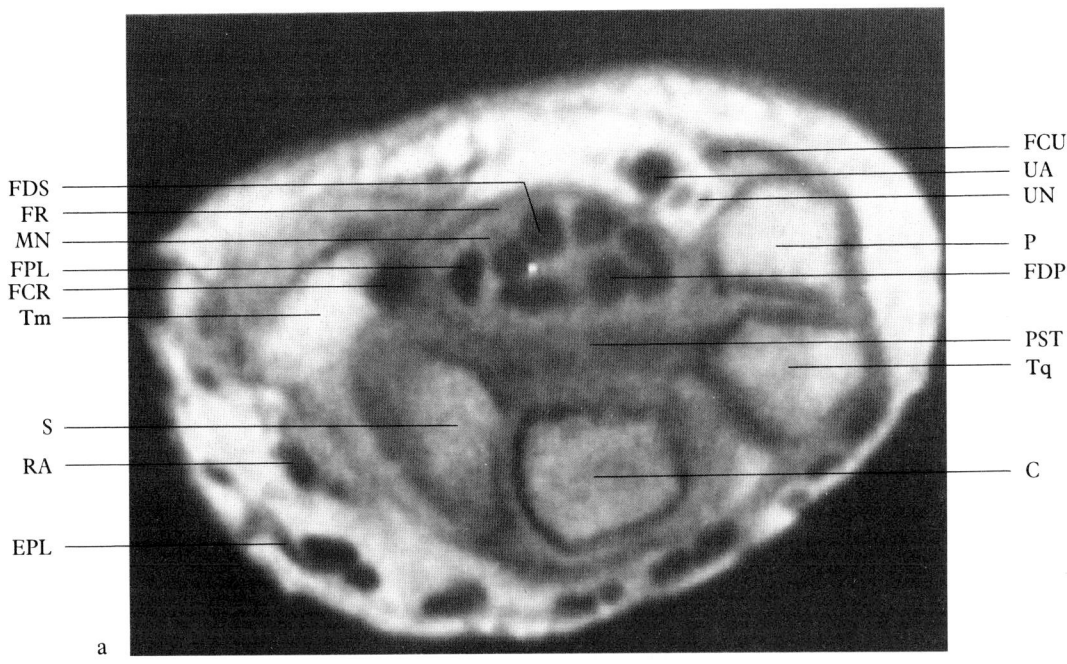

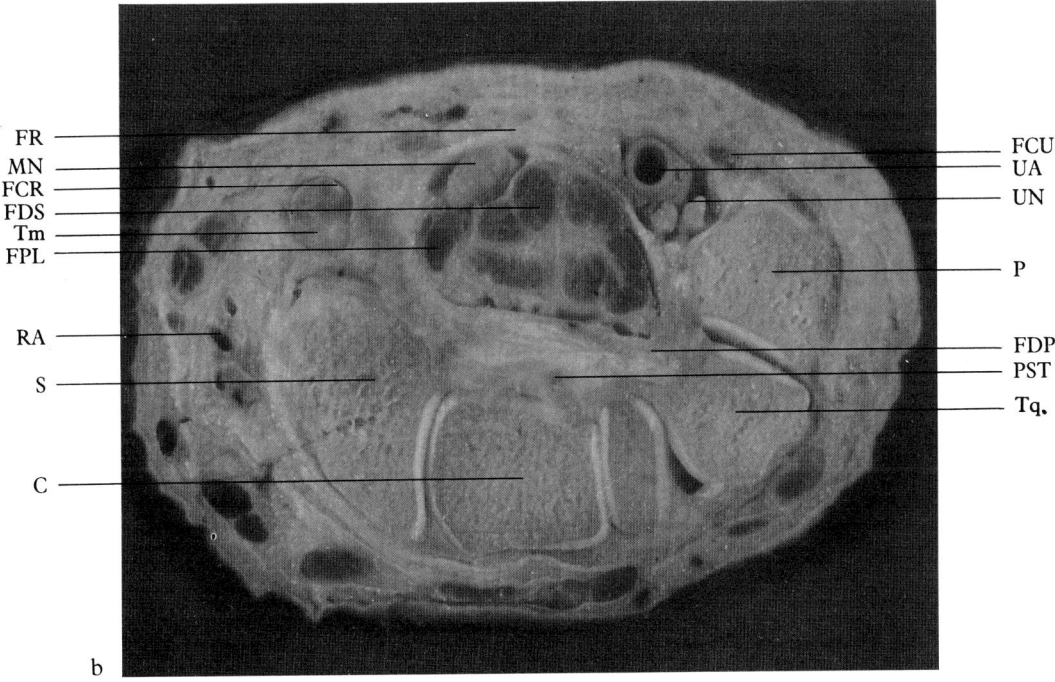

Figure 6.9

Section through proximal carpal tunnel demonstrates: trapezium (Tm), scaphoid (S), capitate (C), triquetrum (Tq), pisiform (P), palmar scaphoid-triquetral ligament (PST), flexor carpi radialis (FCR) and ulnaris (FCU), median (MN) and ulnar (UN) nerves, radial (RA) and ulnar (UA) arteries, flexor pollicis longus (FPL), flexor digitorum superficialis (FDS) and profundus (FDP), flexor retinaculum (FR) and extensor pollicis longus (EPL).

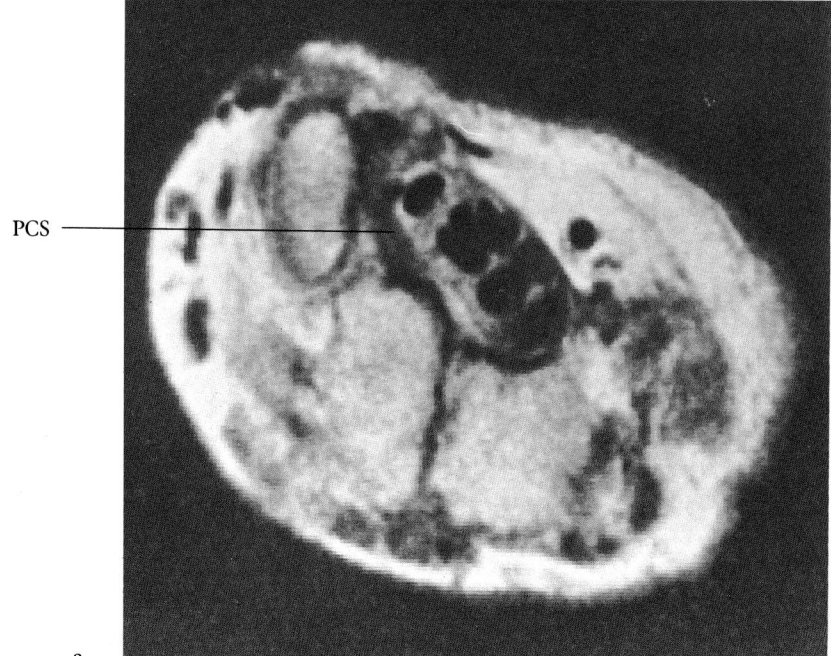

PCS

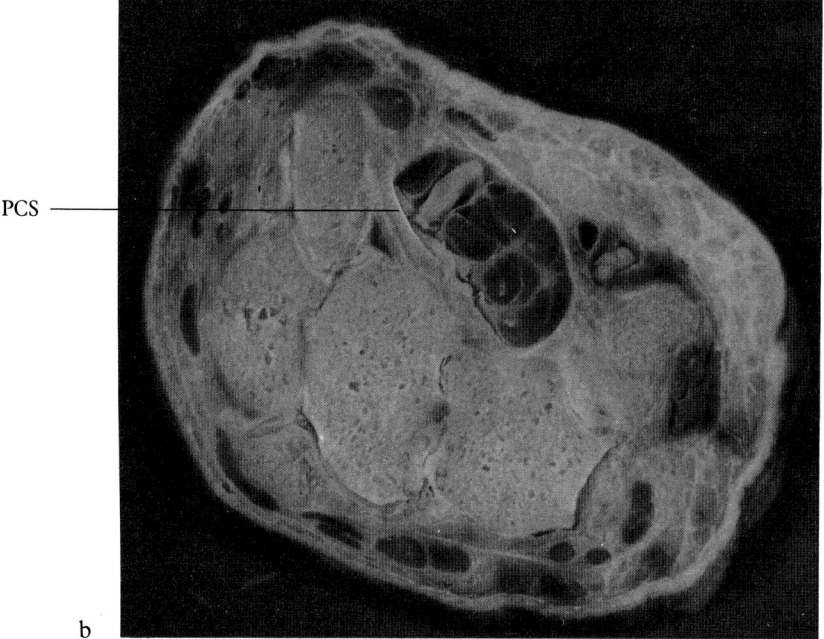

PCS

Figure 6.10

Section through intermediate carpal tunnel demonstrates the palmar capitate-scaphoid ligament (PCS).

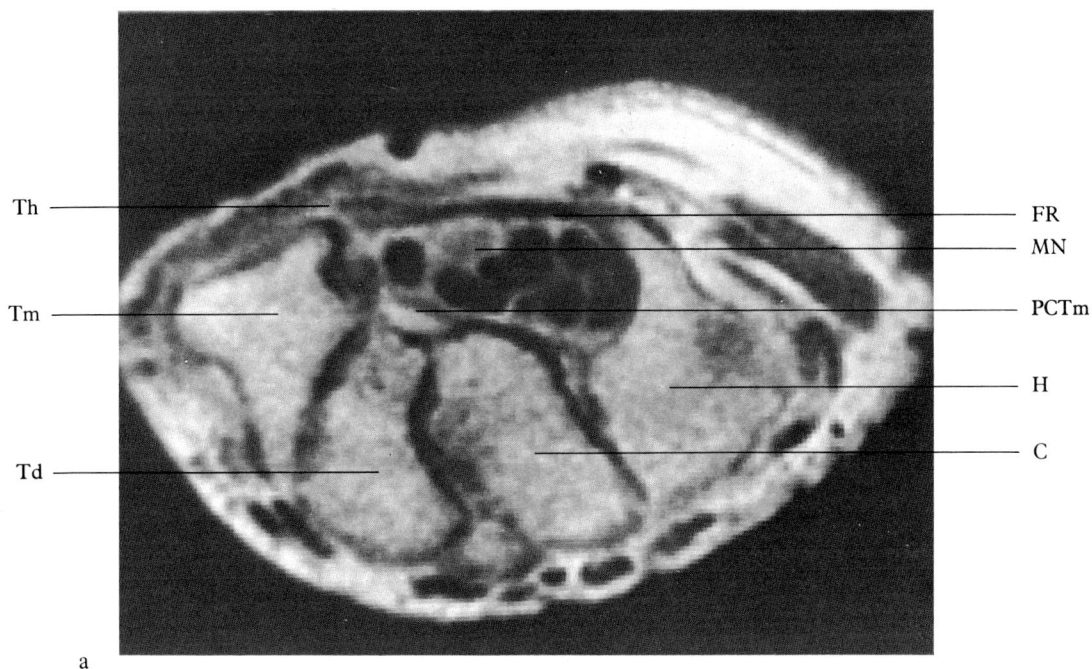

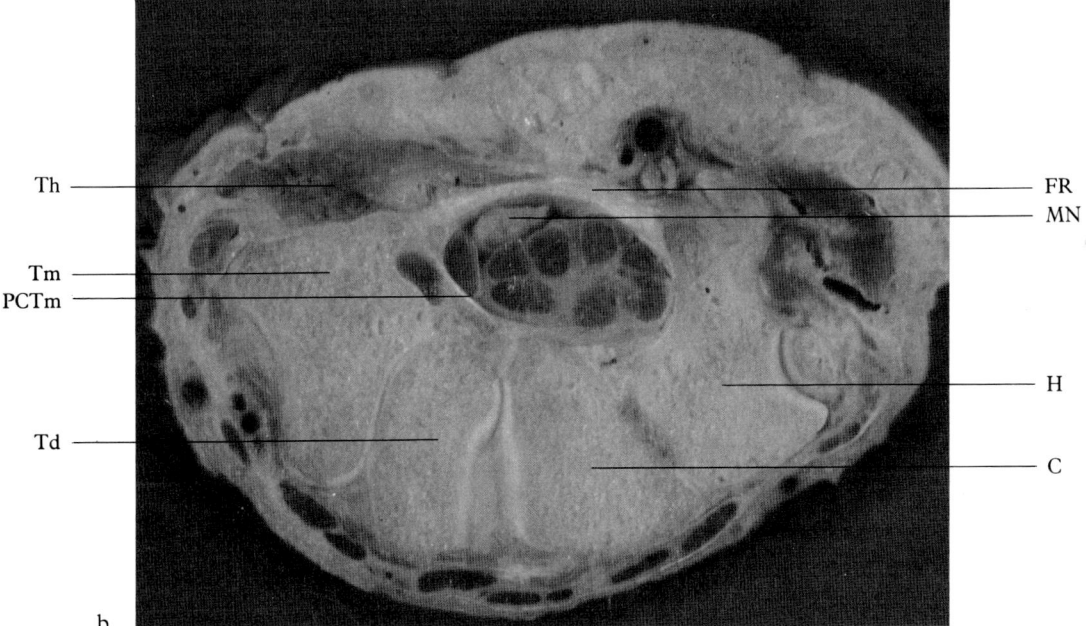

Figure 6.11

Section through the intermediate part of the carpal tunnel. Visualized are: trapezoid (Td), hamate (H), trapezium (Tm), capitate (C), palmar capitate-trapezium ligament (PCTm), median nerve (MN), flexor retinaculum (FR), and thenar muscles (Th).

154 *MRI atlas of the musculoskeletal system*

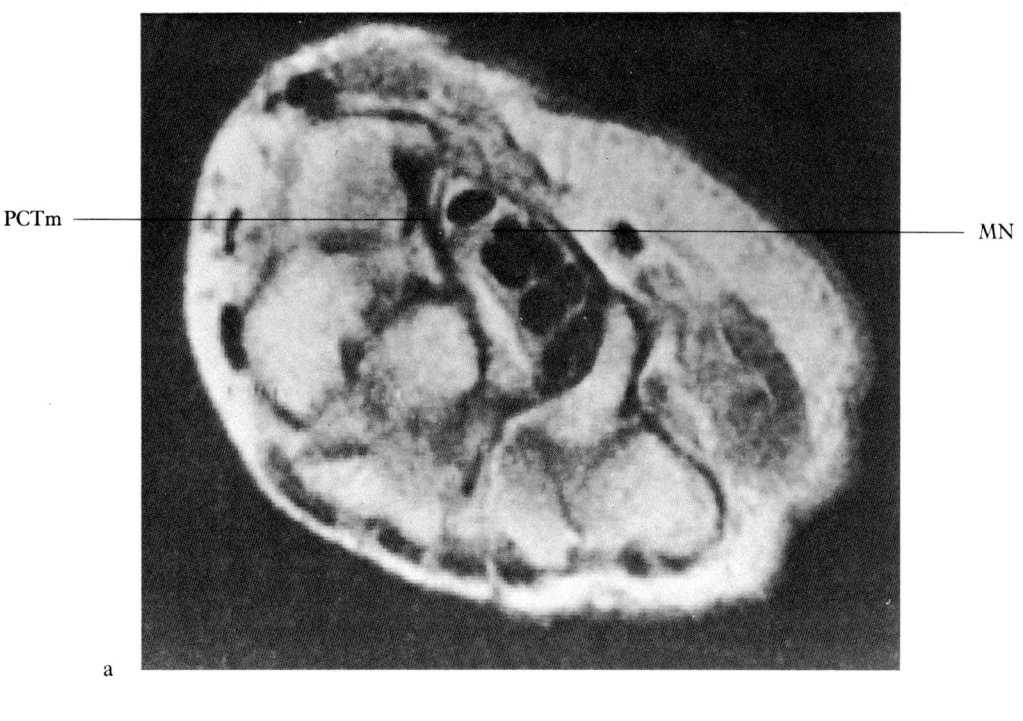

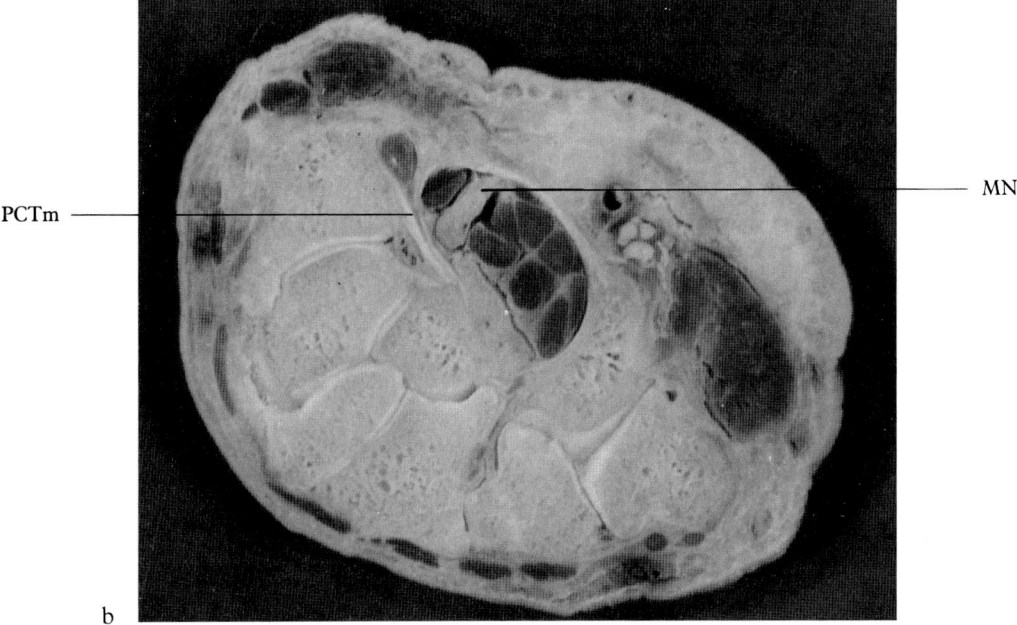

Figure 6.12

Section through intermediate carpal tunnel provides good visualization of palmar capitate-trapezium ligament (PCTm). The median nerve (MN), even though it is unusually deep, still can be identified by its moderate signal intensity and its consistently adjacent tendons.

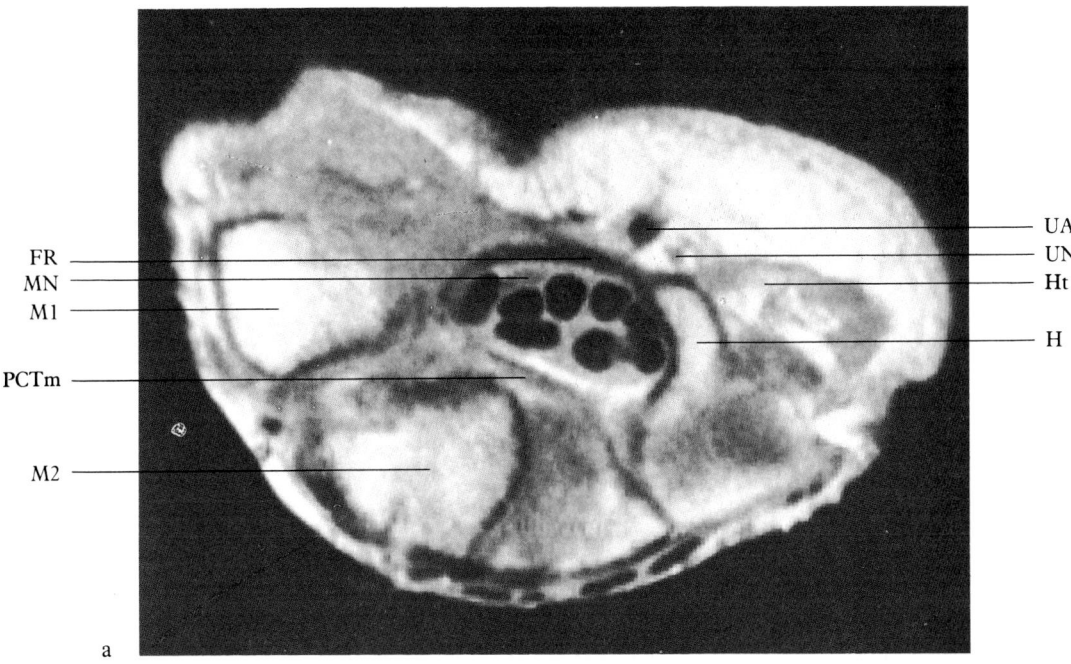

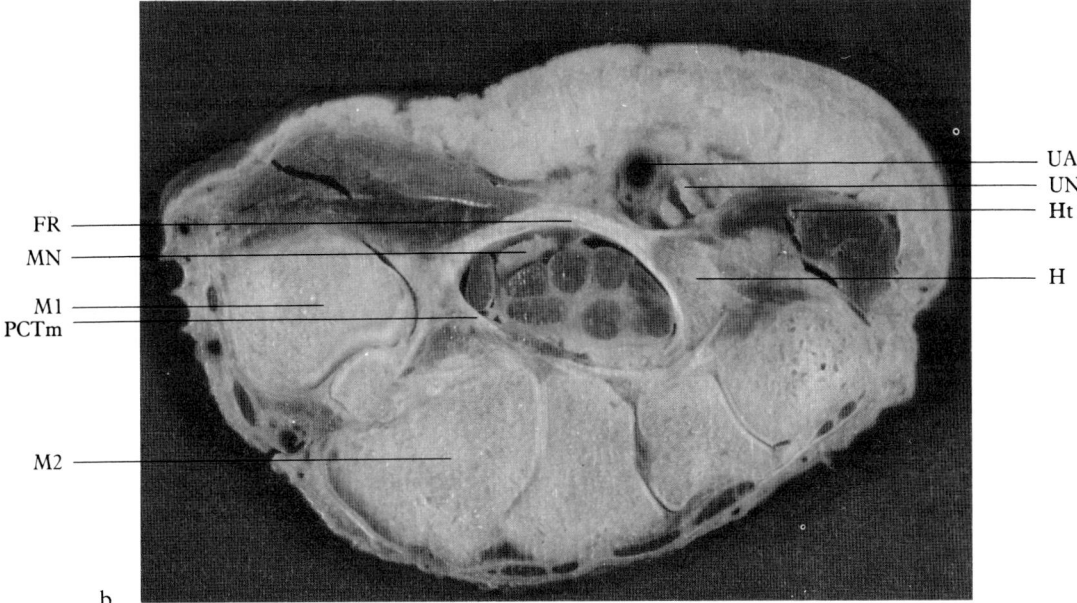

Figure 6.13

Section through distal carpal tunnel shows hook of hamate (H), first (M1) and second (M2) metacarpal, ulnar artery (UA), branches of ulnar nerve (UN), flexor retinaculum (FR), median nerve (MN), palmar capitate-trapezium ligament (PCTm) and hypothenar muscle (Ht).

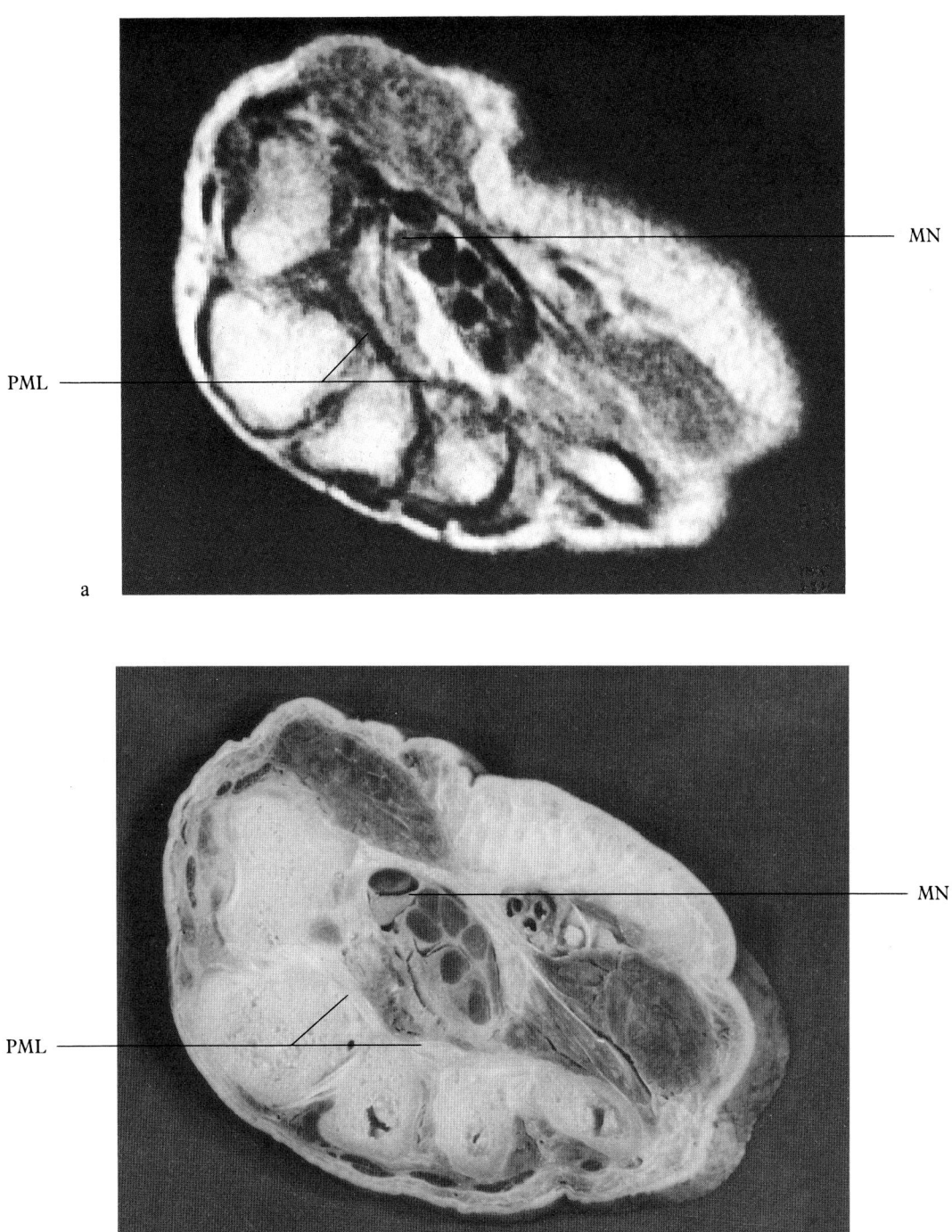

Figure 6.14

Section through base of metacarpals shows palmar metacarpal ligaments (PML). Specimen also demonstrates an unusually deep median nerve (MN).

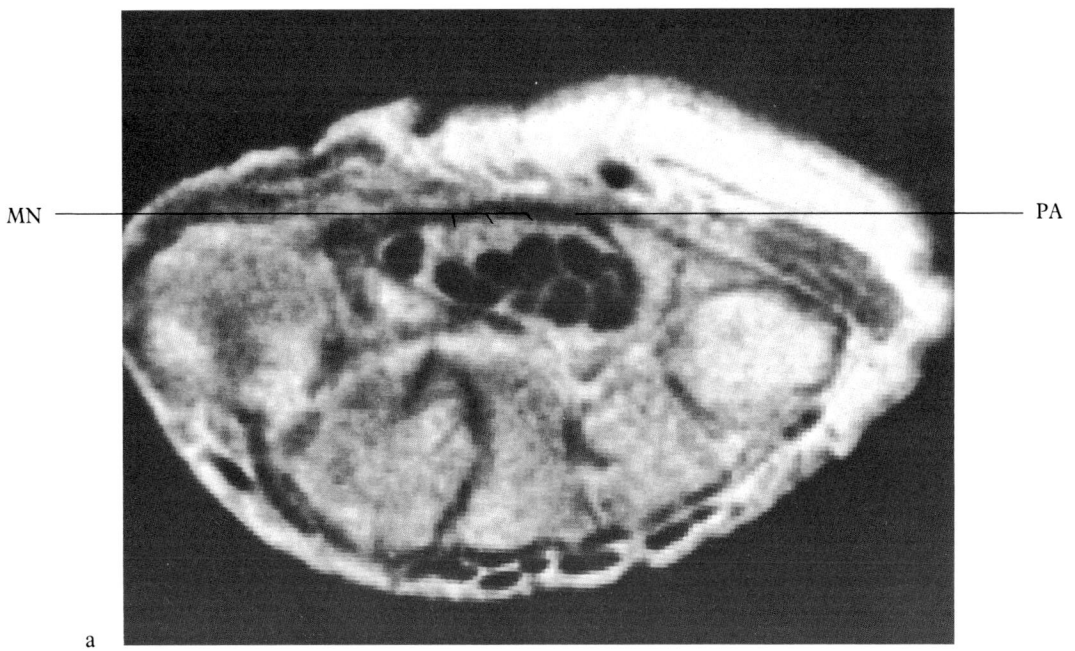

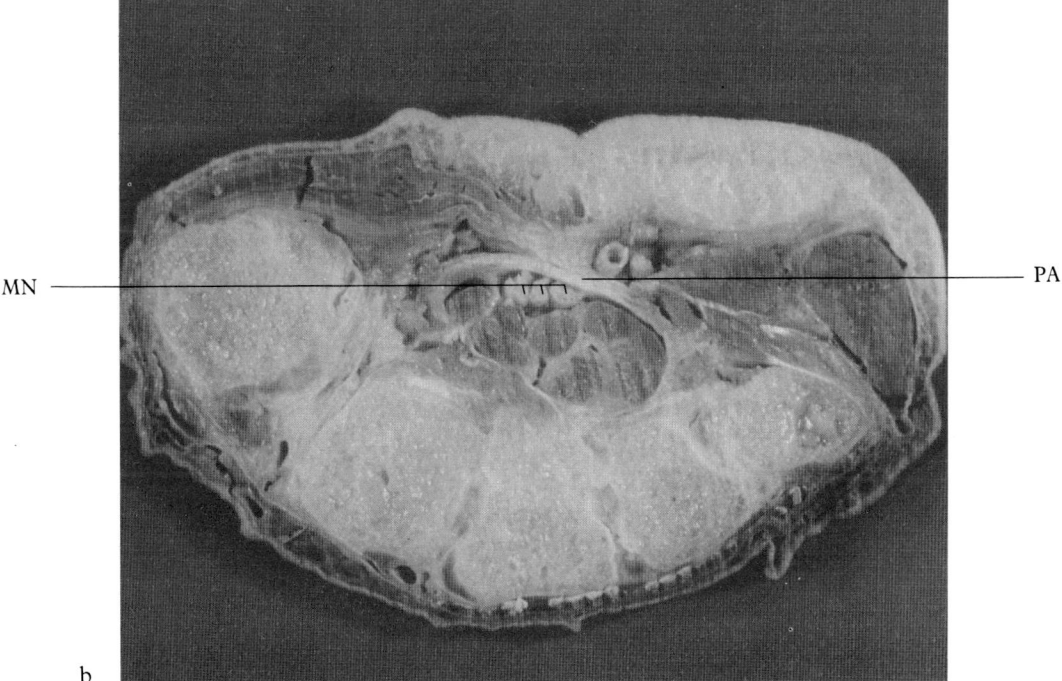

Figure 6.15

Section through base of metacarpals. Three branches of median nerve (MN) and palmar aponeurosis (PA) replace flexor retinaculum.

158 MRI atlas of the musculoskeletal system

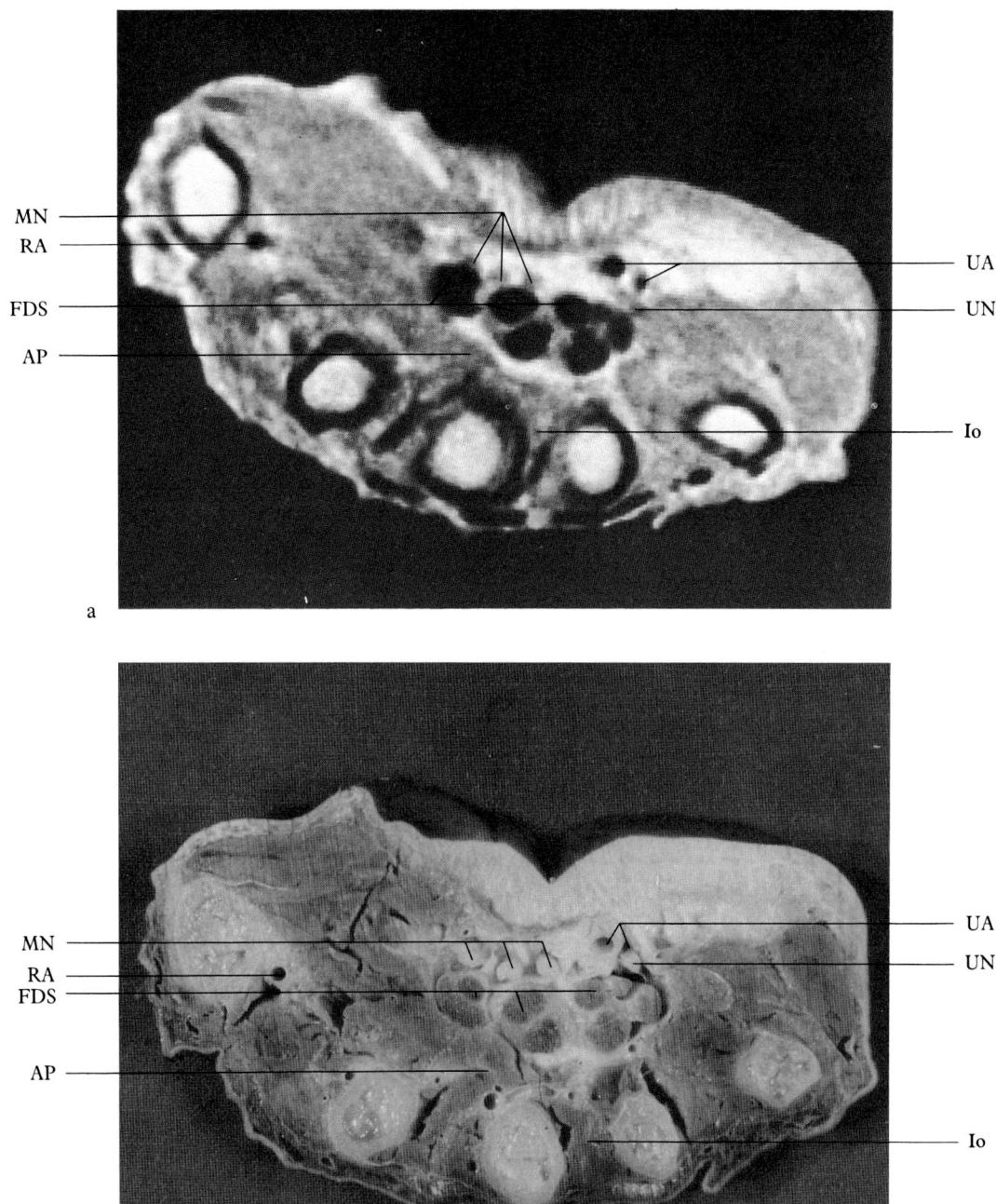

Figure 6.16

Section through metacarpal shafts shows 3 common digital branches of median nerve (MN) overlying the spaces between paired flexor digitorum superficialis (FDS) and profundus tendons to each finger. The common digital branch of ulnar nerve (UN), deep palmar branch of radial artery (RA), abductor pollicis (AP) and interossei (Io) are well seen.

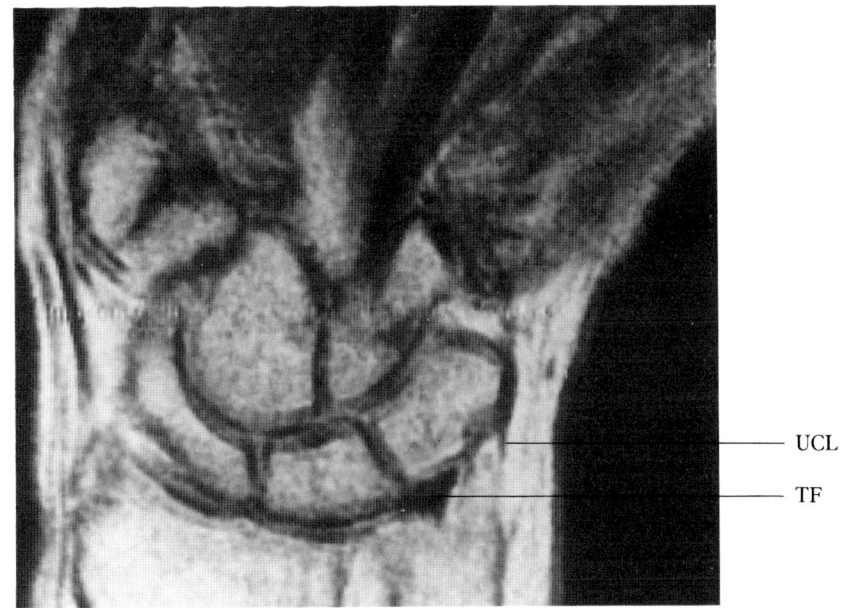

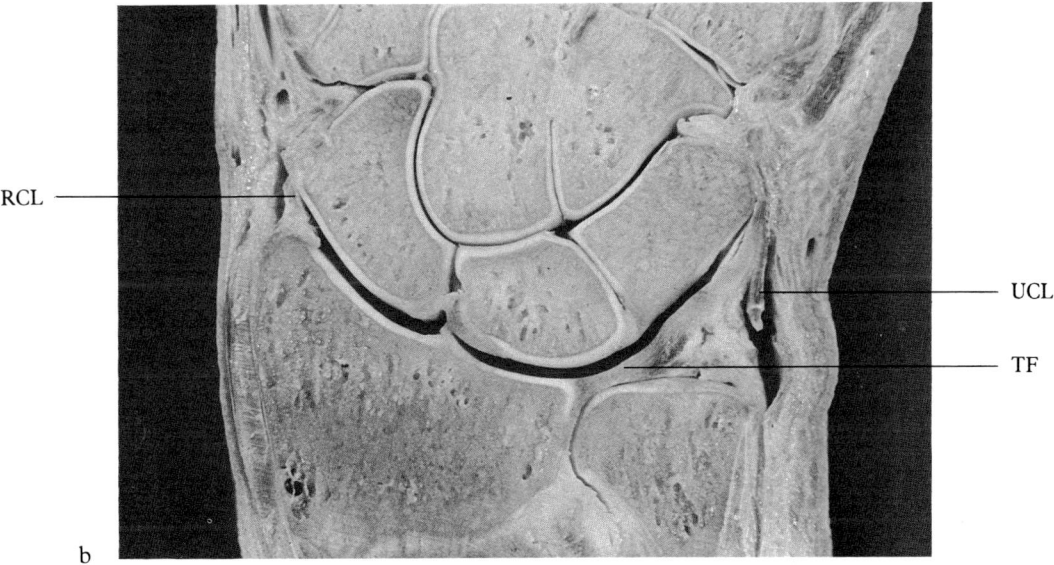

Figure 6.17

Posterior coronal section demonstrates carpal bones, triangular fibrocartilage (TF) radial collateral ligament (RCL) and distal end of ulnar collateral ligament (UCL).

160 *MRI atlas of the musculoskeletal system*

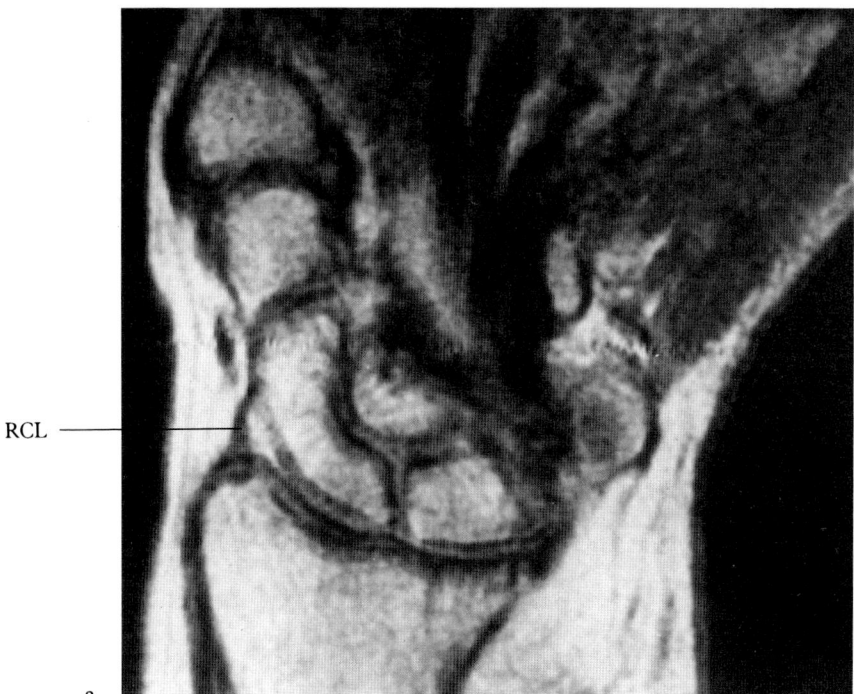

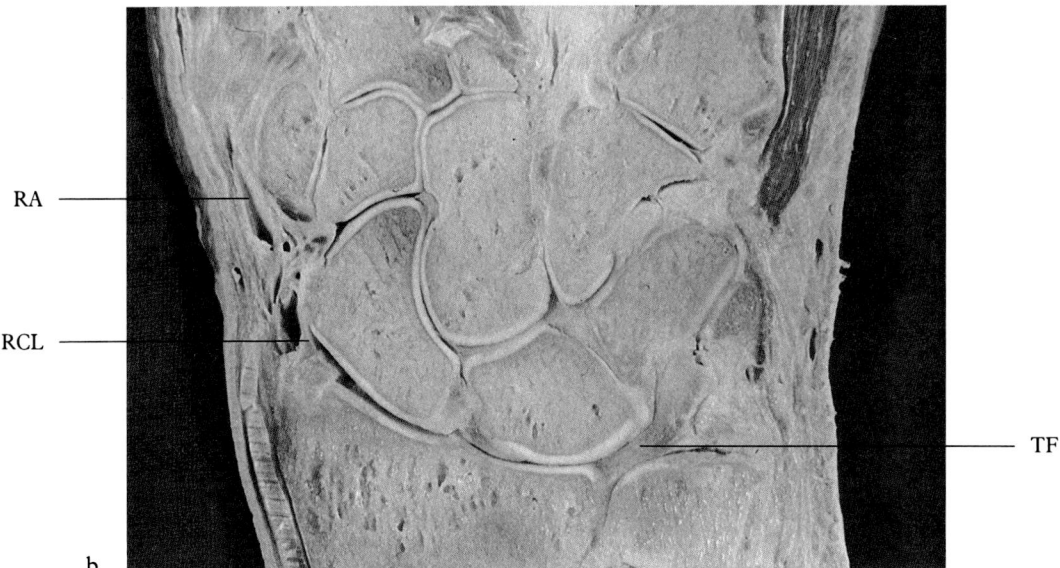

Figure 6.18

Coronal section slightly palmar to Figure 6.17 shows radial collateral liagment (RCL), radial artery (RA) and triangular fibrocartilage (TF).

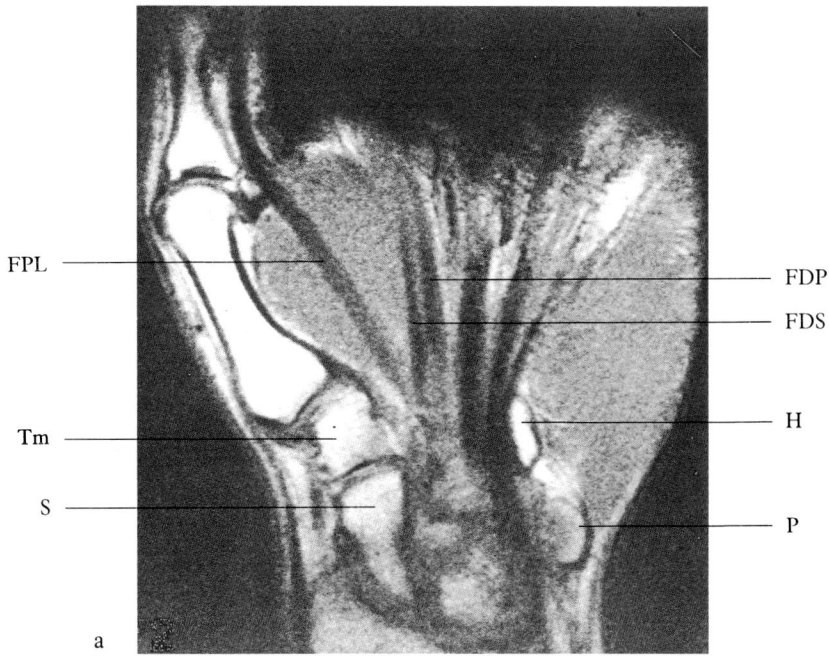
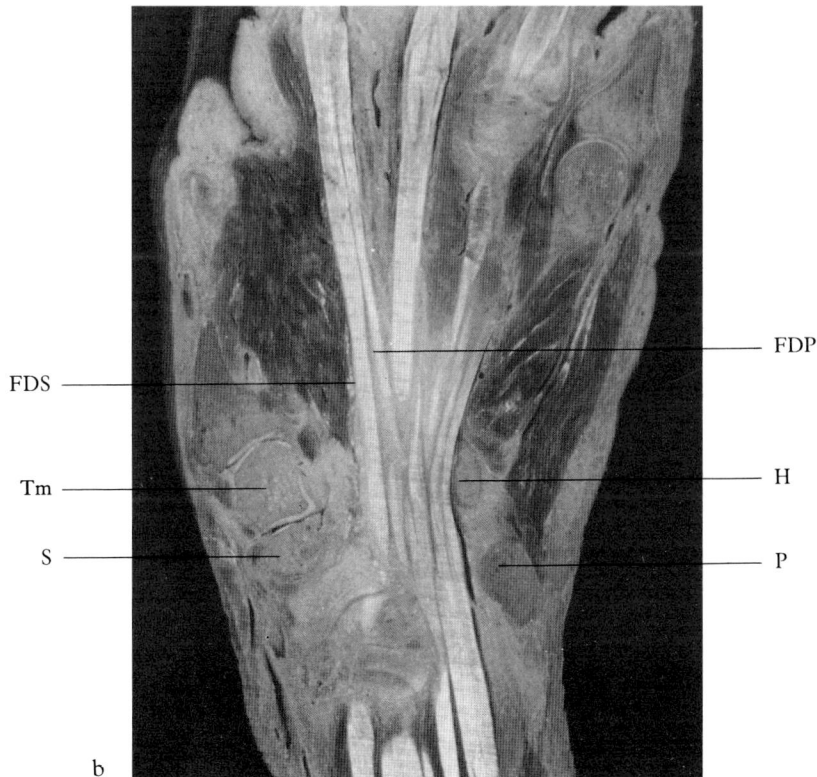

Figure 6.19

Coronal section through carpal tunnel displays trapezium (Tm), scaphoid (S), hook of hamate (H), pisiform (P), and provides longitudinal view of flexor pollicis longus (FPL) and flexor digitorum superficialis (FDS) and profundus (FDP) tendons.

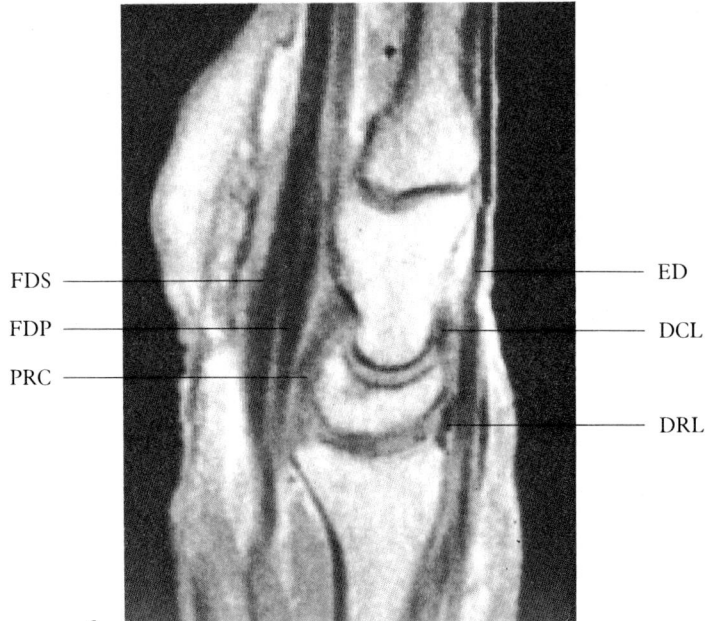

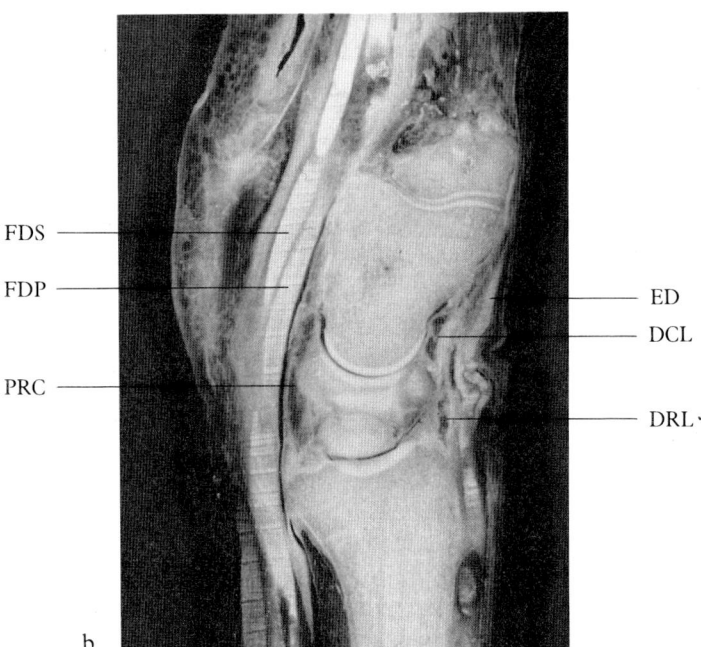

Figure 6.20

Sagittal section through radius, lunate, capitate and 3rd metacarpal shows flexor digitorum superficialis (FDS) and profundus (FDP) and extensor digitorum (ED) tendons to middle finger. Palmar radiocapitate (PRC), dorsal radiolunate (DRL) and dorsal capitate-lunate (DCL) ligaments are also seen.

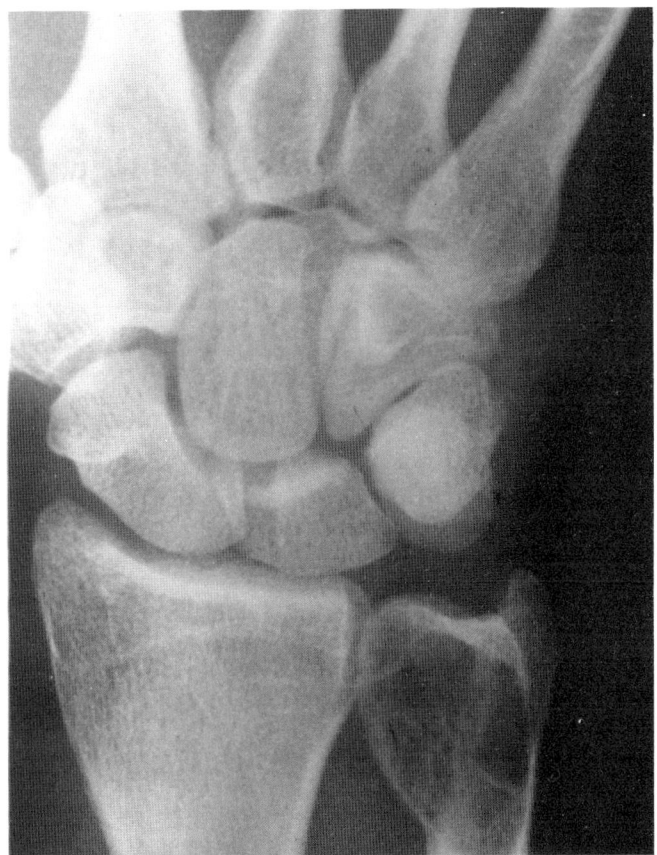
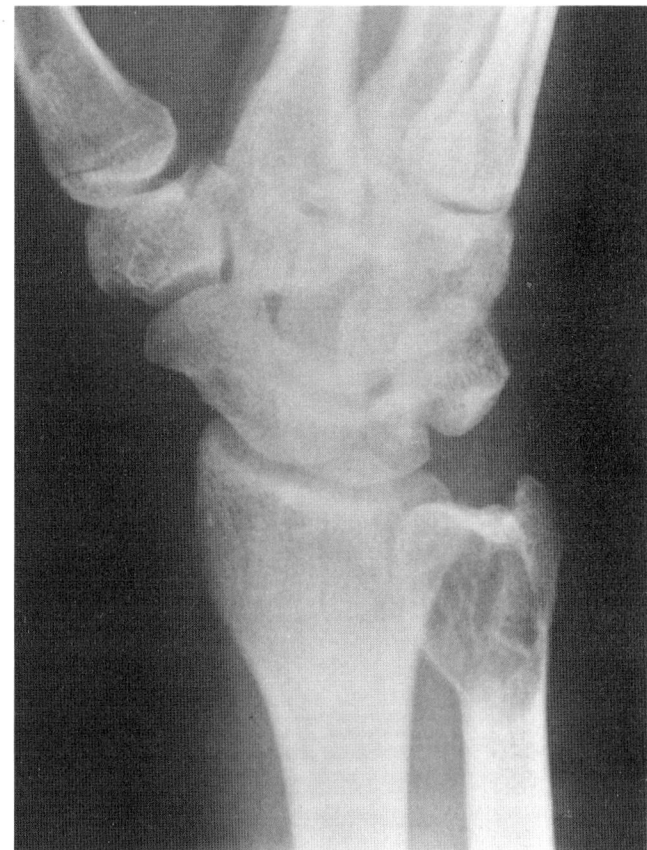
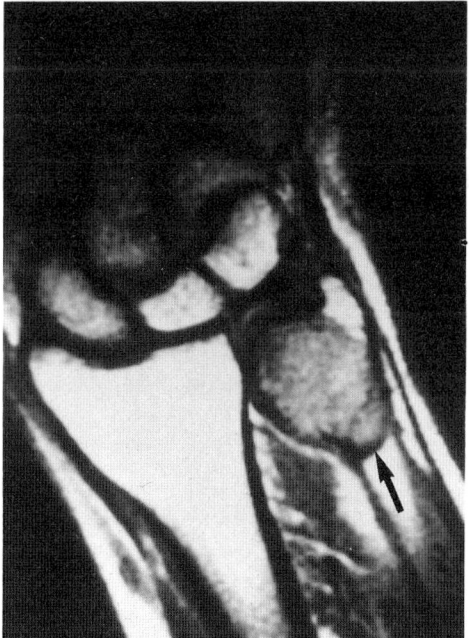
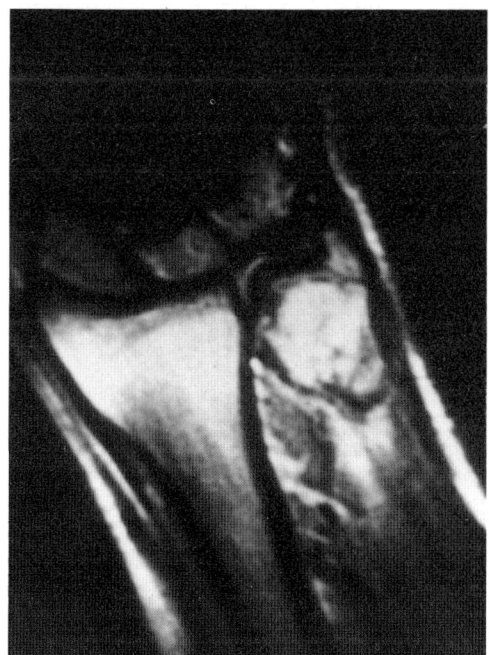

a

b

c

d

Figure 6.21

A 38-year-old man with giant cell tumor of distal ulna. (**a,b**) Plain radiographs show typical appearance of giant cell tumor of distal ulna, which extends to subchondral bone. (**c**) T_1- (SE 400/25) and (**d**) T_2- (SE 2000/80) weighted images show lesion is well confined to bone. The intramedullary extent of lesion is best appreciated in the T_1-weighted image, which also demonstrates that the ulnar styloid has been spared. The lesion is well demarcated by a rim of low-signal intensity (arrow) which corresponds to sclerotic margin. If it were present, intra-articular invasion would be easy to detect in the T_2-weighted image. The high T_2 signal intensity of the tumor contrasts with the low-signal-intensity triangular fibrocartilage, which remains uninvolved.

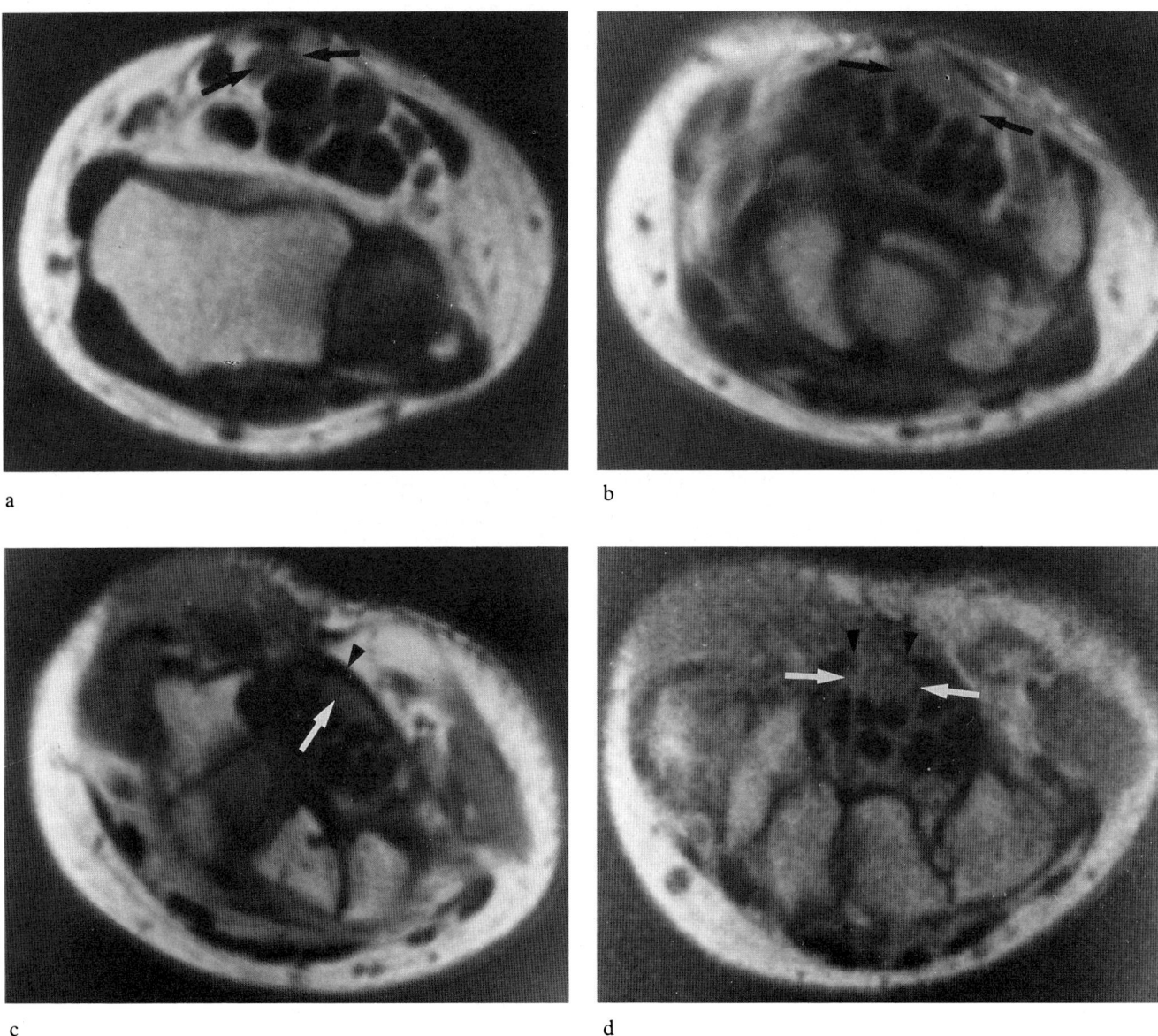

Figure 6.22

A 42-year-old female typist with clinical idiopathic carpal tunnel syndrome, underwent successful surgical release. All images are T_1-weighted (SE 500–800/28). (**a**) Preoperative image at level of radioulnar joint reveals normal size median nerve (arrows) deep to palmaris longus. (**b**) Preoperative image at proximal edge of carpal tunnel shows palmaris longus blending with flexor retinaculum. Size of median nerve (arrows) has almost doubled in comparison to its size more proximally, instead of gradual distal tapering. (**c**) Preoperative image at mid-carpal level where median nerve becomes smaller and flattened (arrow). Palmar bowing of flexor retinaculum (arrowhead) indicates increased pressure within carpal tunnel. (**d**) Two weeks postoperative, image at mid-carpal level shows satisfactory incision of flexor retinaculum (arrowheads). Contents of carpal tunnel have expanded palmarly. Decompressed median nerve (arrows) is no longer flattened and appears to have enlarged from its preoperative size.

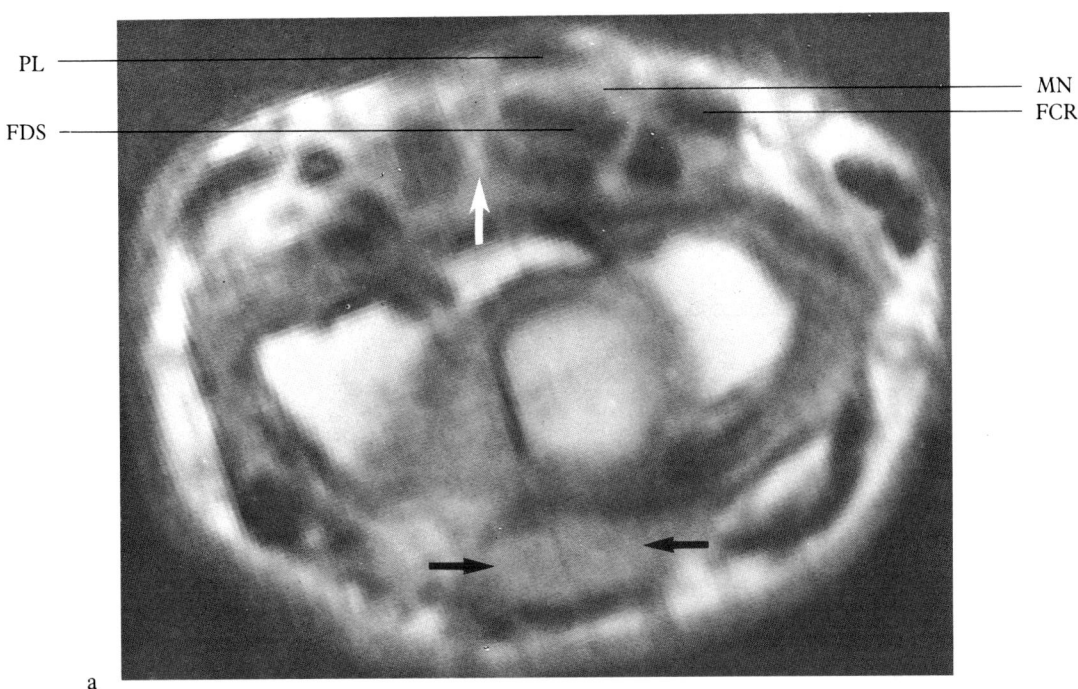

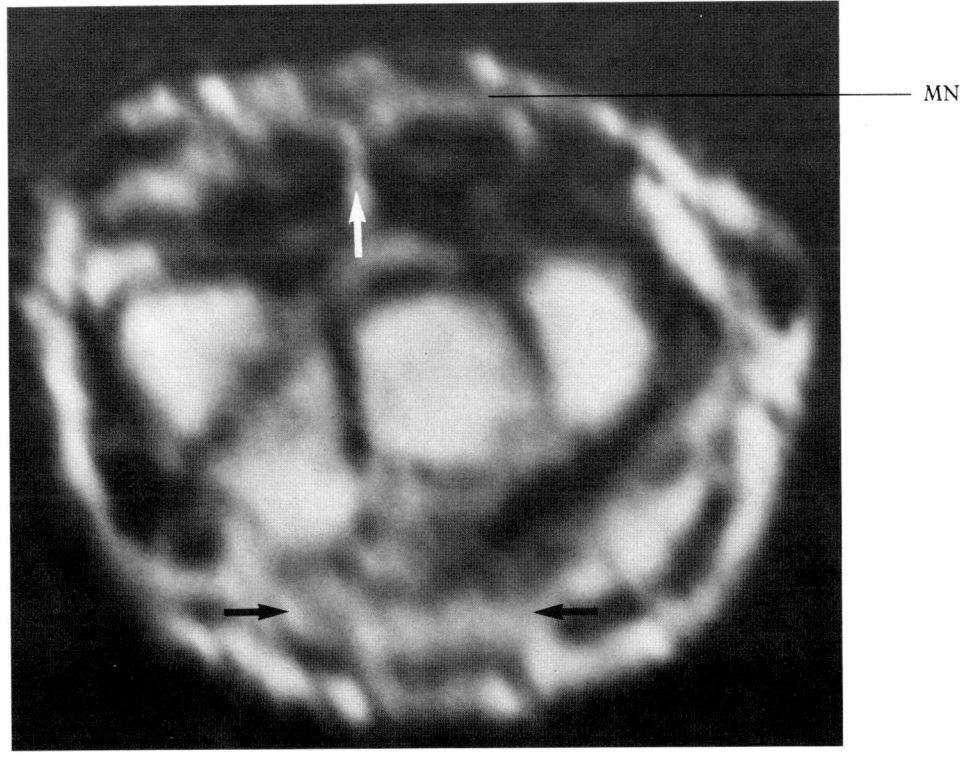

Figure 6.23

A 42-year-old man with carpal tunnel syndrome due to tenosynovitis. (a) T_1- (SE 500/28) and (b) T_2- (SE 1500/84) weighted images at proximal wrist before entering the carpal tunnel show normal size median nerve (MN) in normal location deep and slightly radial to palmaris longus (PL) tendon between flexor carpi radialis (FCR) and flexor digitorum superficialis (FDS). Due to peritendonitis, margins of tendons are not sharp. Fluid in tendon sheath of flexor digitorum superficialis and profundus produces intermediate-signal intensity with T_1-weighting (white arrow) and high-signal intensity with T_2-weighting (white arrow). On dorsal aspect of wrist, deep soft tissues between extensor digitorum tendons and dorsal scaphoid lunate ligament are also edematous (black arrows) and the signal intensity increases with T_2-weighting (black arrows).

MN —

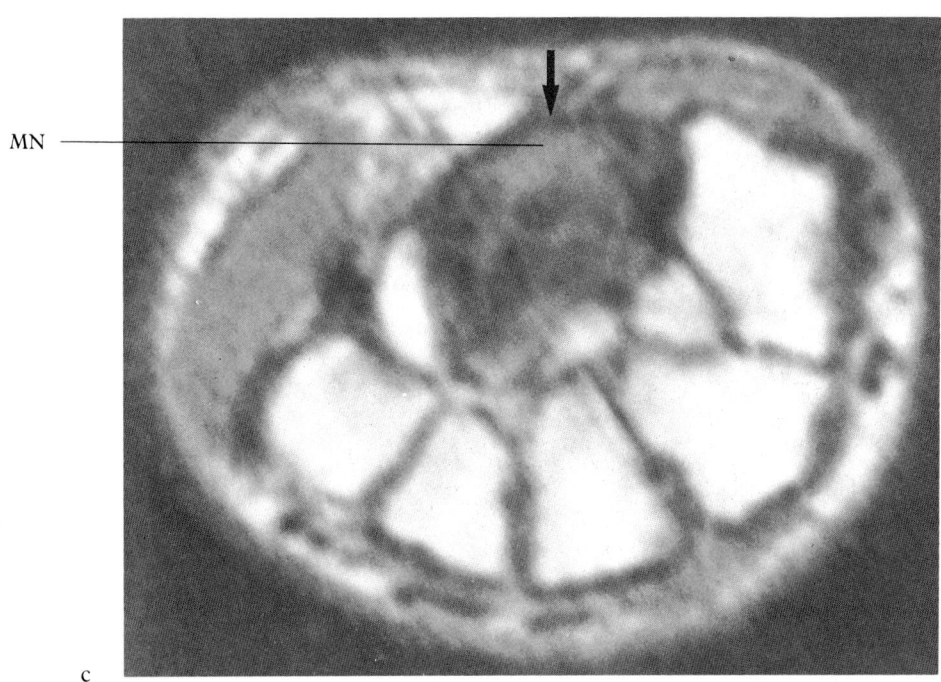

c

MN —

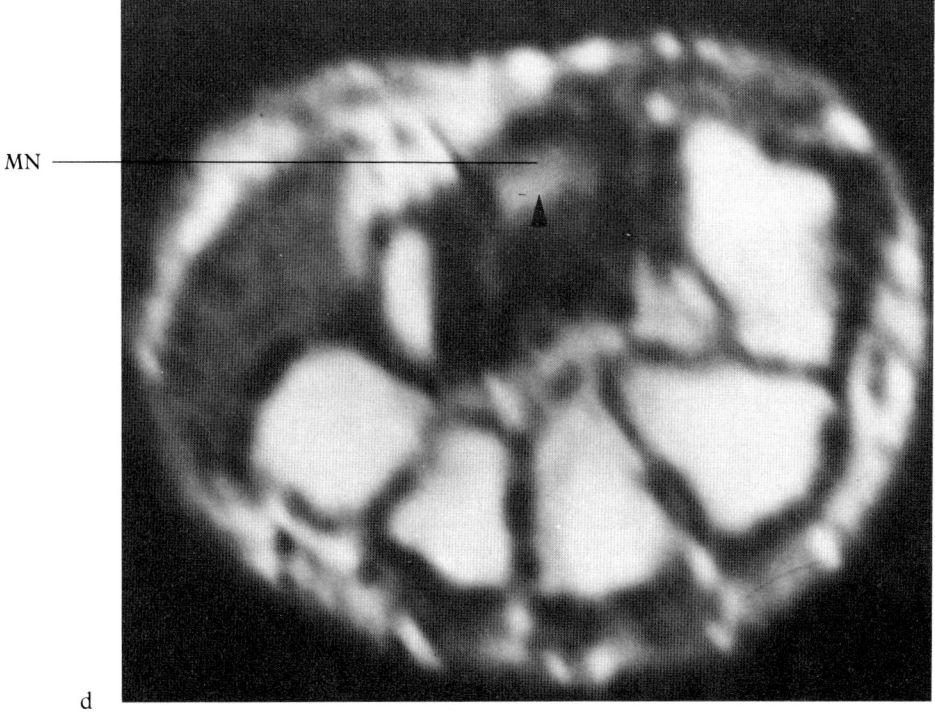

d

Figure 6.23 *continued*

(**c**) T_1- (SE 500/28) and (**d**) T_2- (SE 1500/84) weighted images at midcarpal level show poor definition of tendons due to peritendinous edema. For same reason the median nerve (MN) is not well defined and appears enlarged. Signal intensity from region of median nerve increases with T_2-weighting (arrowhead). Flexor retinaculum (arrow) is bowing palmarly due to increased pressure within carpal tunnel.

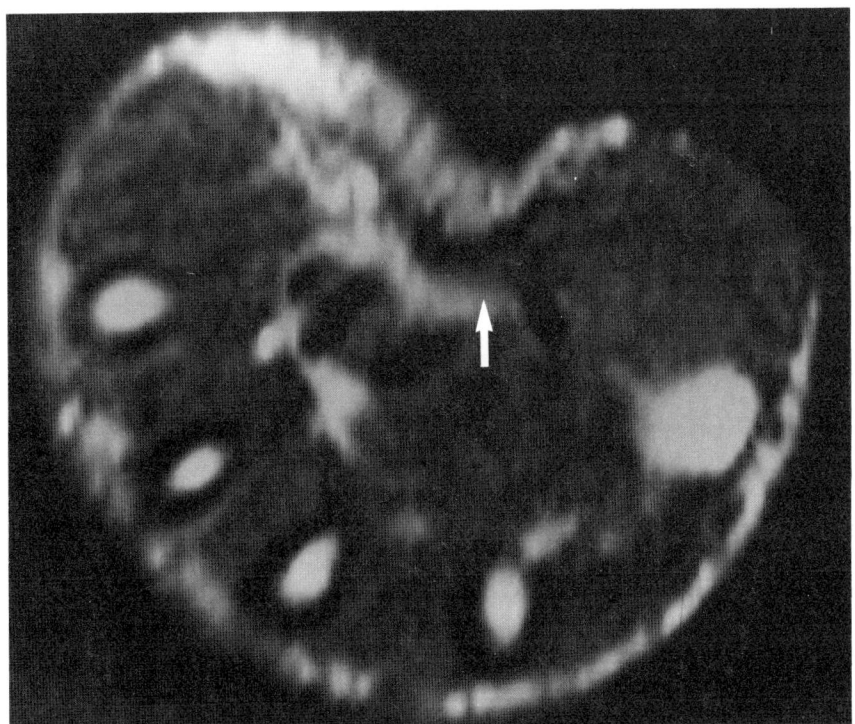

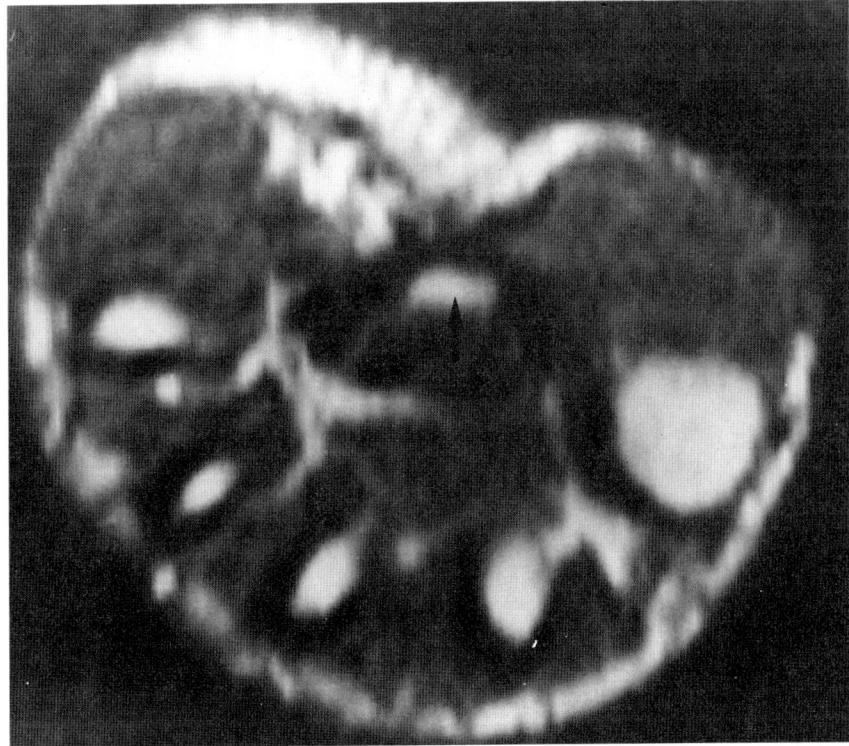

(**e**) T_1- and (**f**) T_2-weighted images through metacarpal shafts show distal extension of abnormality. Fluid and edema around tendons cause poor definition of their margins on T_1-weighted images. They gain in signal intensity with T_2-weighting. Just deep to palmar aponeurosis an ovoid area of intermediate signal intensity (white arrow) on T_1-weighted image becomes hyperintense with T_2-weighting (black arrow). Differential diagnosis could include focal edema and fluid, swelling of median nerve, synovial cyst or soft-tissue neoplasm. It is not likely to be swelling of the median nerve, since by mid-metacarpal shaft level the median nerve has divided into multiple small terminal branches (see Figure 6.16).

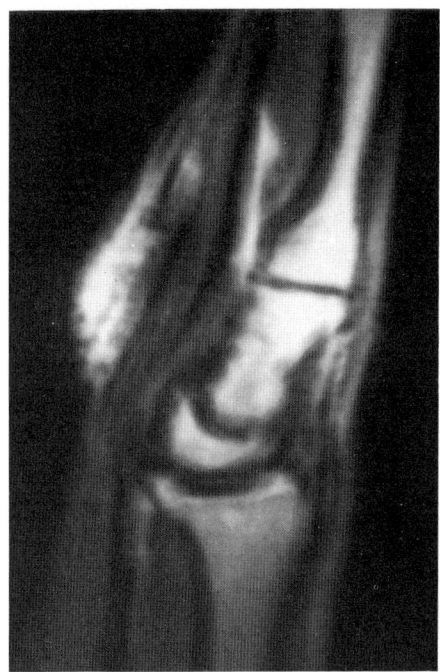
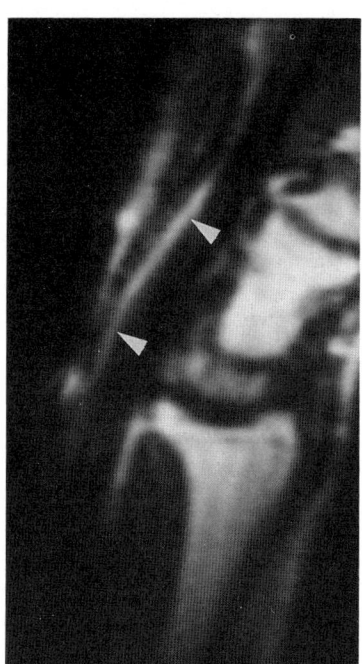

g h

Figure 6.23 *continued*
(**g**) T_1- (SE 500/28) and (**h**) T_2- (SE 1550/84) weighted images in sagittal plain show longitudinal extent of abnormality (arrowheads) and constant relationship to tendons which favors diagnosis of fluid accumulation within their sheaths consistent with tenosynovitis.

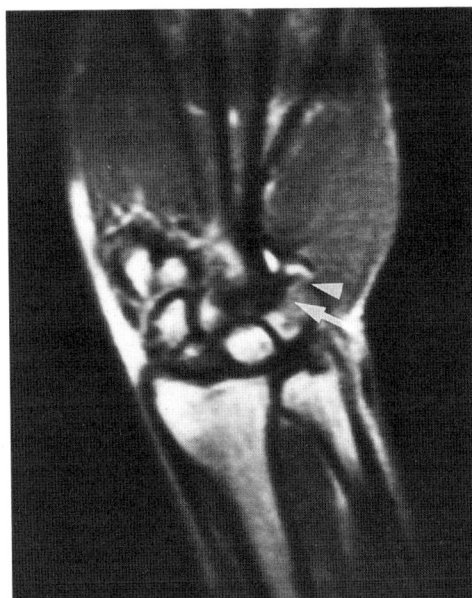

Figure 6.24

A T_1-weighted (SE 500/28) coronal image of wrist shows decreased signal in proximal pole scaphoid due to ischemic necrosis (arrow) secondary to fracture (arrowhead). (*Courtesy of Lawrence W Bassett, UCLA School of Medicine*)

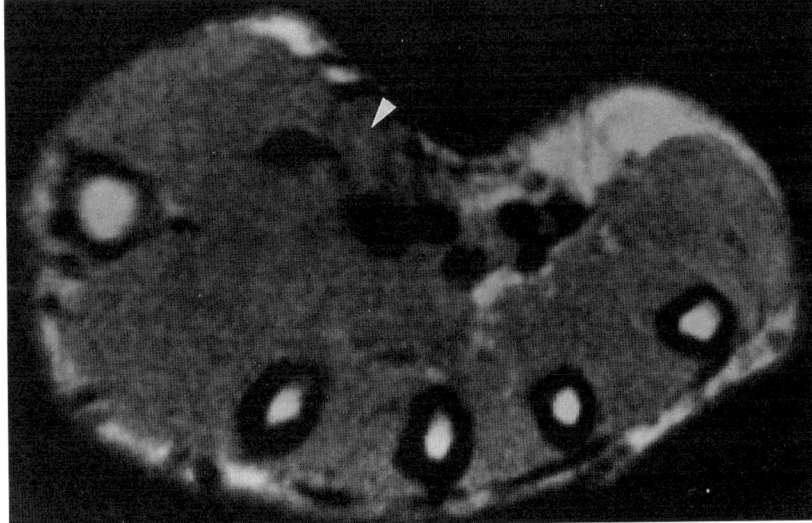

a

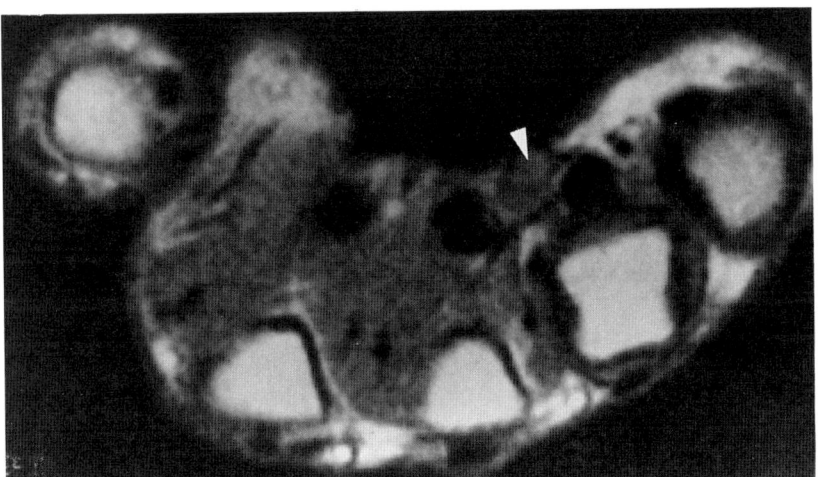

b

Figure 6.25

A 39-year-old man with history of laceration of hand at distal metacarpal level, and who later developed discomfort. On clinical examination, nodular indurations were palpable on palmar aspect of hand, which corresponded to postincisional neuromas. Axial T_1-weighted images (SE 500/28) at (**a**) metacarpal shaft and (**b**) distal metacarpal levels are shown. Proximally (**a**), lobulated soft-tissue masses (arrowhead) lie palmar to the space between flexor pollicis longus and flexor tendons of index finger. More distally (**b**), similar masses (arrowhead) are situated palmar to the space between flexor tendons of middle and ring fingers. Signal intensity of masses is similar to normal nerve and equal to muscle. However, as position of masses corresponds with position of branches of median nerve, neural origin is strongly suggested.

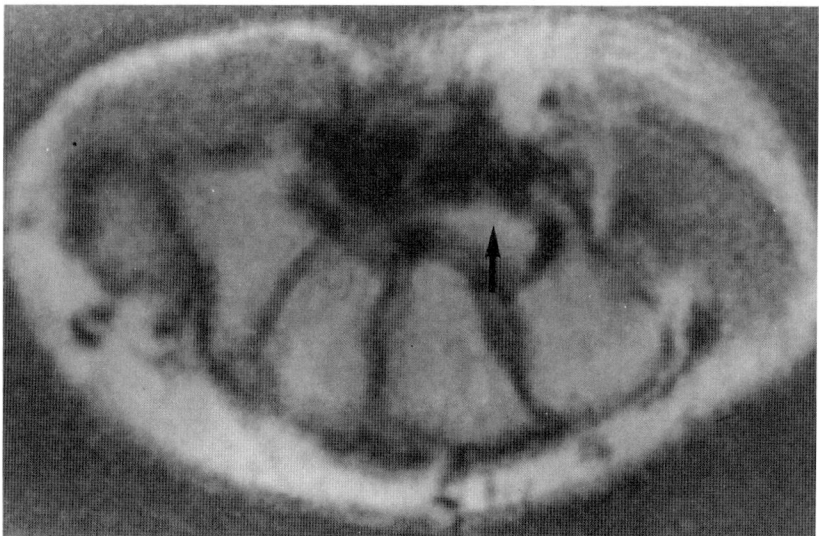

a

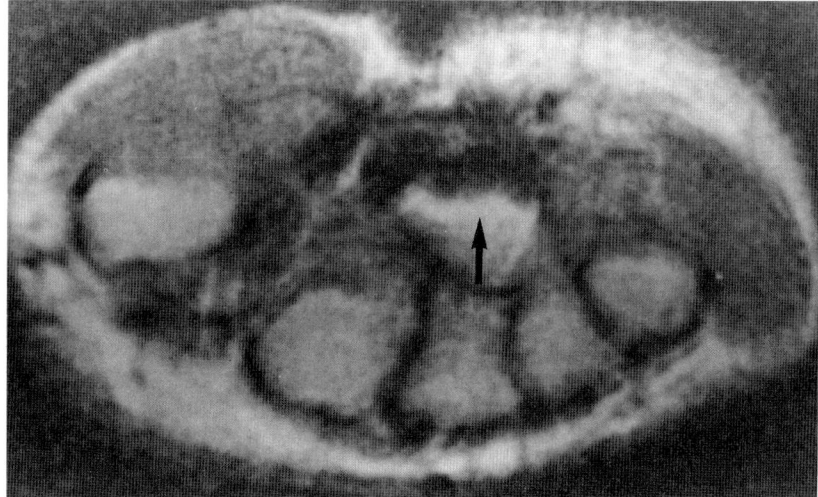

b

Figure 6.26
A 32-year-old man who had recurrence of symptoms after surgical therapy for carpal tunnel syndrome which had probably resulted from excessive fat within the carpal tunnel. (**a**) Axial T_1-weighted (SE 500/28) image, although degraded by motion artifact, clearly shows hook of hamate and tubercle of the trapezium where the flexor retinaculum attaches. Flexor retinaculum is no longer visible, and contents of carpal tunnel have moved palmarly beyond imaginary line drawn between hook of hamate and tubercle of trapezium which flexor retinaculum should approximate. There is excessive amount of fat (arrow) behind flexor digitorum profundus and in front of hamate and capitate.
(**b**) Another T_1-weighted axial image (SE 500/28) at metacarpal bases again shows excessive fat (arrow) in front of the capitate.

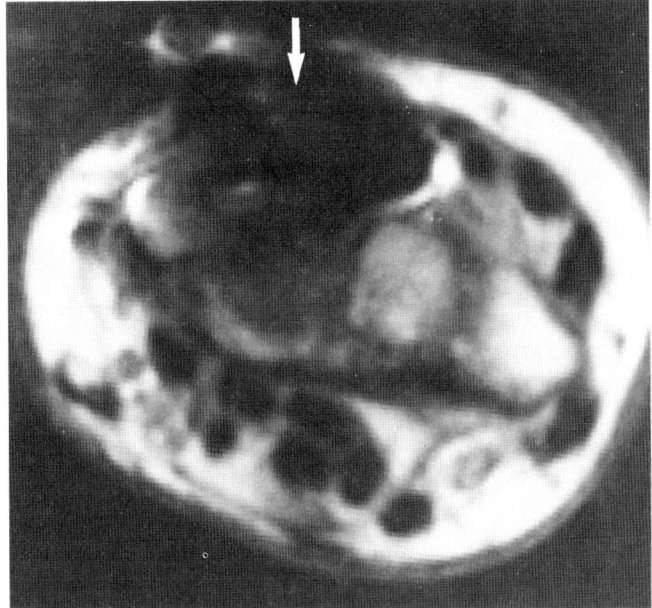

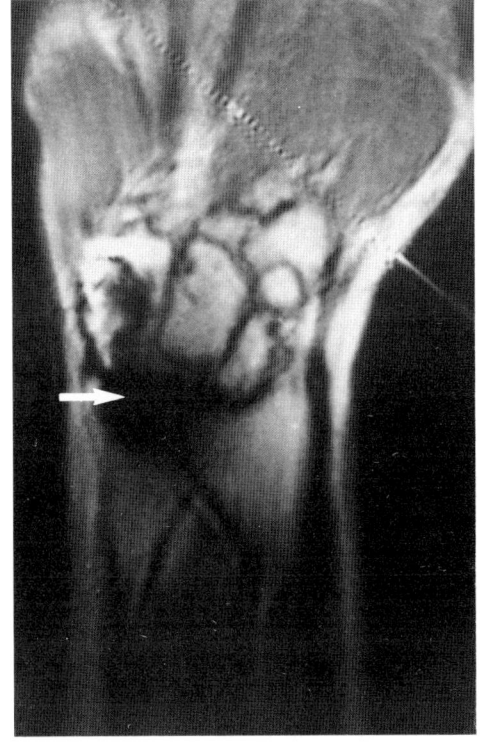

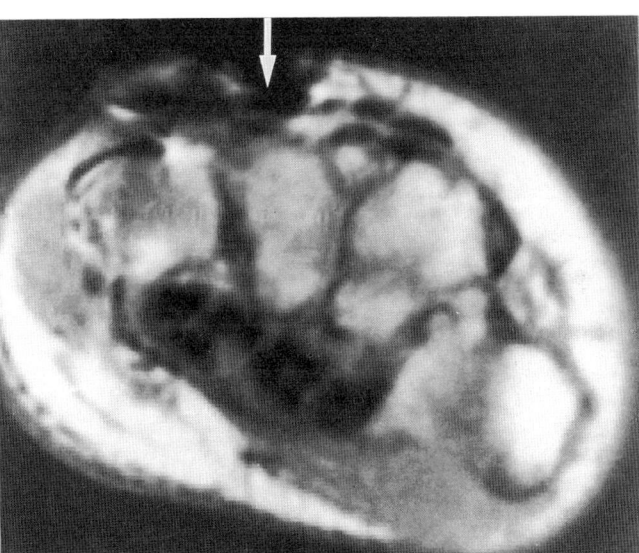

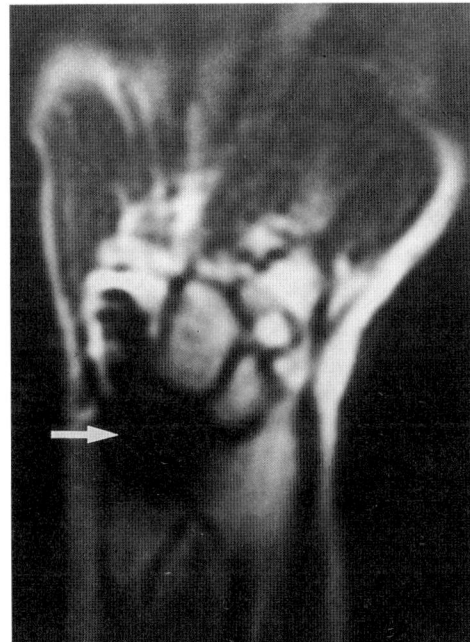

Figure 6.27

A 24-year-old woman with history of previous laceration of dorsal aspect of wrist. She developed symptoms because of post-traumatic fibrosis and adhesive tendonitis. Plain radiographs were normal. Axial T_1-weighted (SE 500/28) images from (**a**) proximal and (**b**) distal wrist show large area of decreased signal intensity due to fibrosis (arrow) which is replacing subcutaneous fat. All extensor tendons on ulnar side of wrist (extensor carpi ulnaris, extensor digiti minimi and the extensors of the 3rd and 4th fingers) are adherent to each other, and no individual tendons can be separately identified. (**c**) Coronal T_1-weighted (SE 500/28) and (**d**) T_2-weighted (SE 1500/84) images show no change in signal intensity of fibrosis (arrow) over dorsal aspect of ulnar side of wrist. Large area of decreased signal in lunate and triquetrum is due to partial volume averaging and not avascular necrosis.

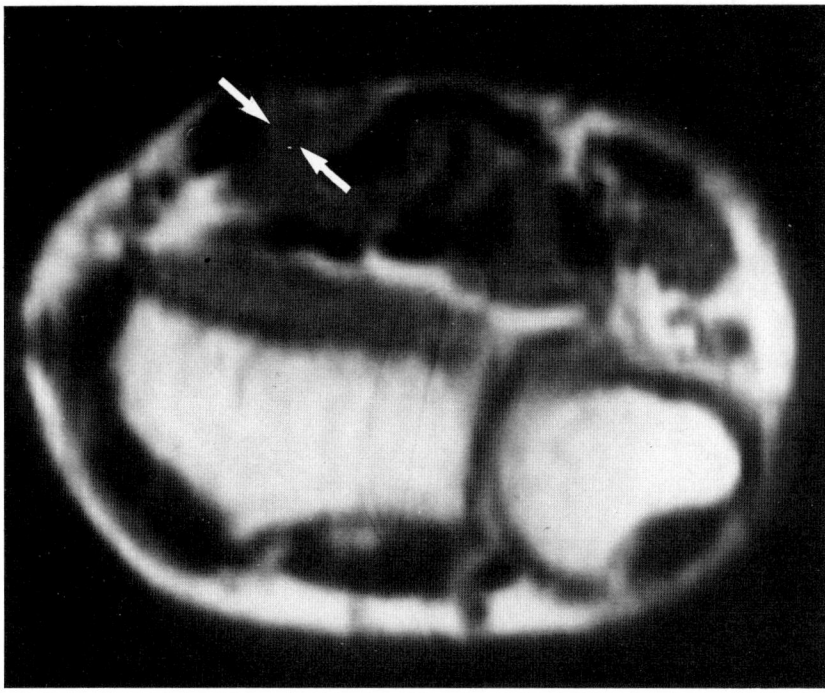

a

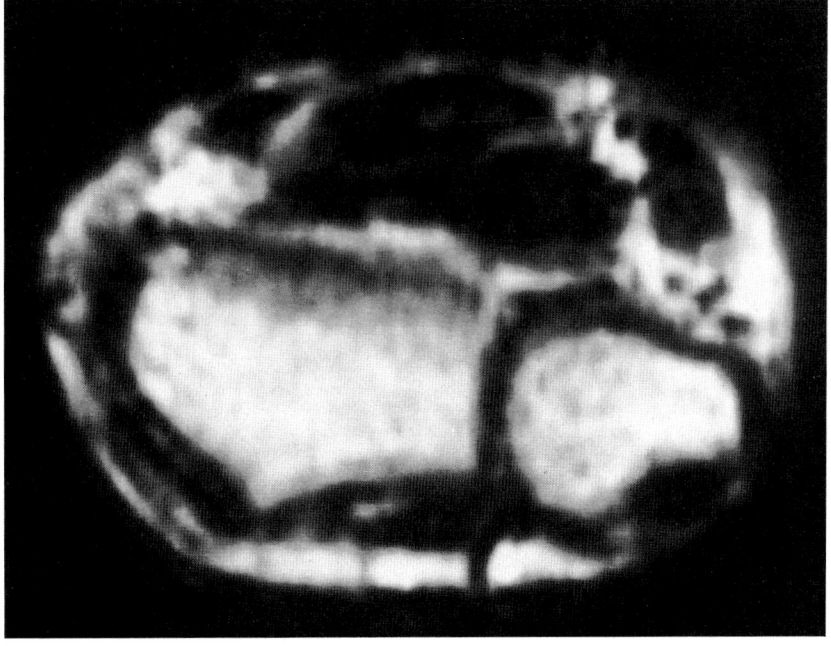

b

Figure 6.28

A 33-year-old woman with carpal tunnel syndrome who had only minimal relief of symptoms immediately after surgery, and who currently is symptomatic. Axial T_1- (SE 400/16) and T_2- (SE 1800/84) weighted images at (**a,b**) radioulnar joint, (**c,d**) proximal carpal tunnel and (**e,f**) distal carpal tunnel levels. At radioulnar joint (**a,b**), median nerve (arrows) is slightly large (compared to opposite side, not shown), but its signal intensity does not change with T_2-weighting. In proximal carpal tunnel (**c,d**), ulnar part of flexor retinaculum is visualized, but radially and anterior to median nerve it is not visible, due to previous incision. In distal part of carpal tunnel, flexor retinaculum is intact from hook of hamate to tubercle of trapezium, indicating an incomplete incision. Further evidence that this is the cause of postoperative symptoms is provided by forward bowing of flexor retinaculum and edema anterior to median nerve, which appears to flatten nerve against underlying flexor tendons. Note the increased signal intensity of the median nerve on T_2-weighted image.

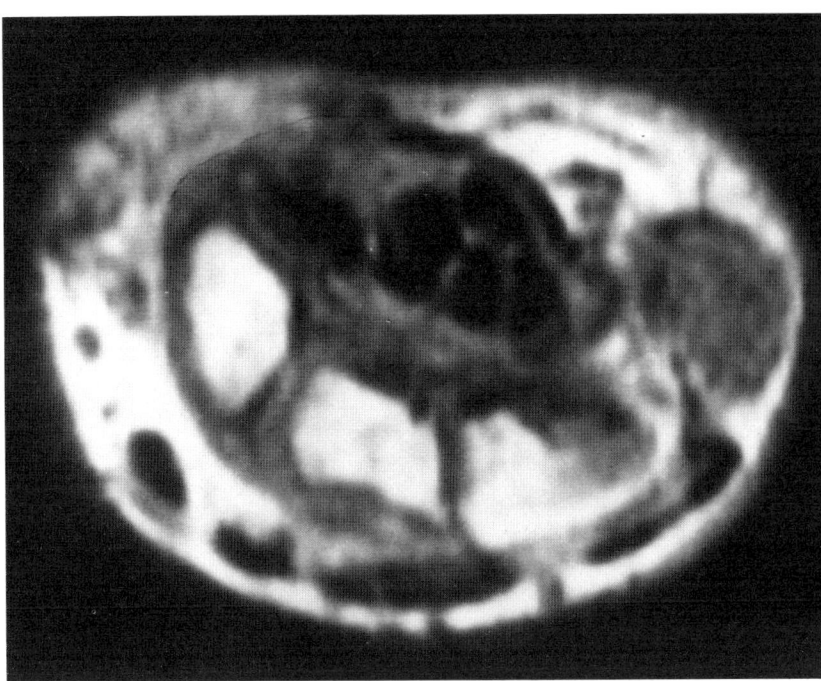
c

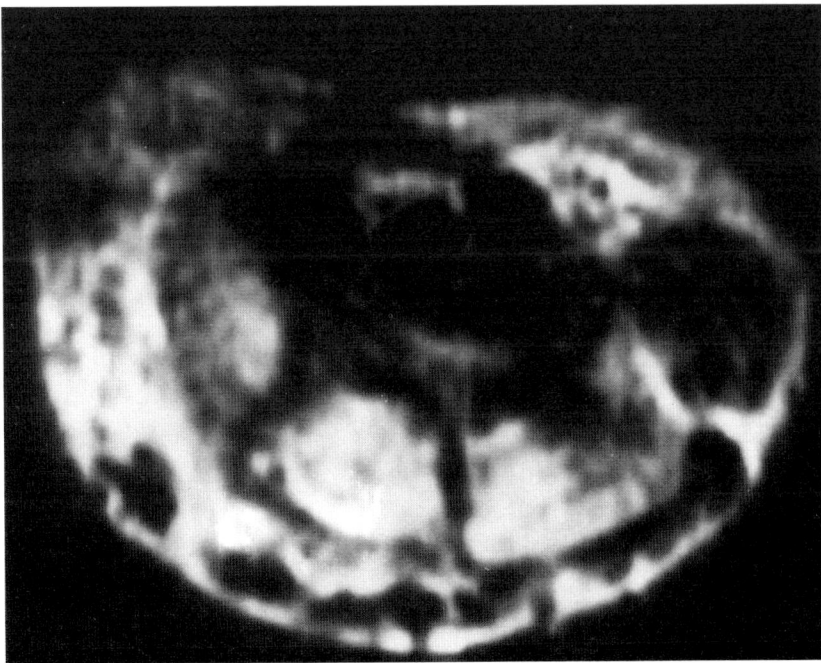
d

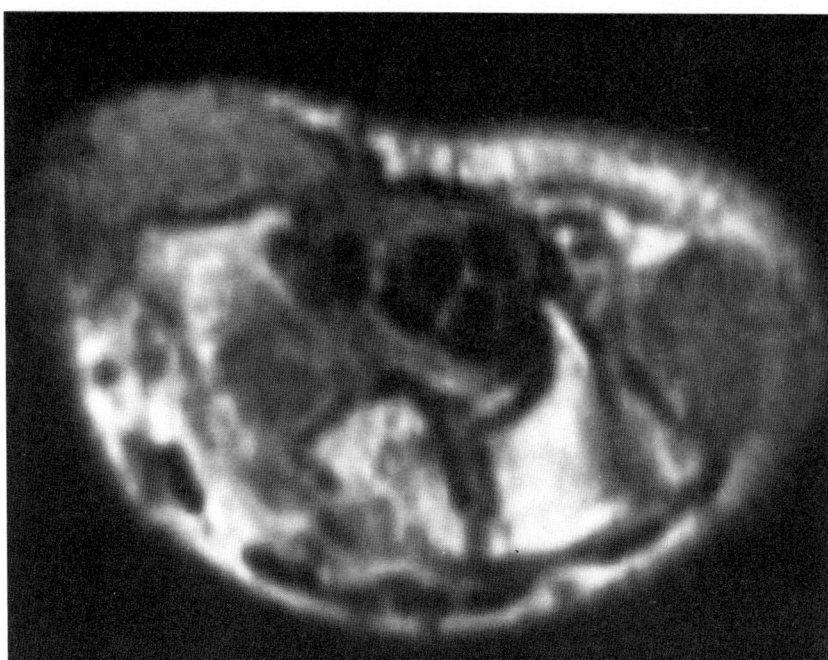

e

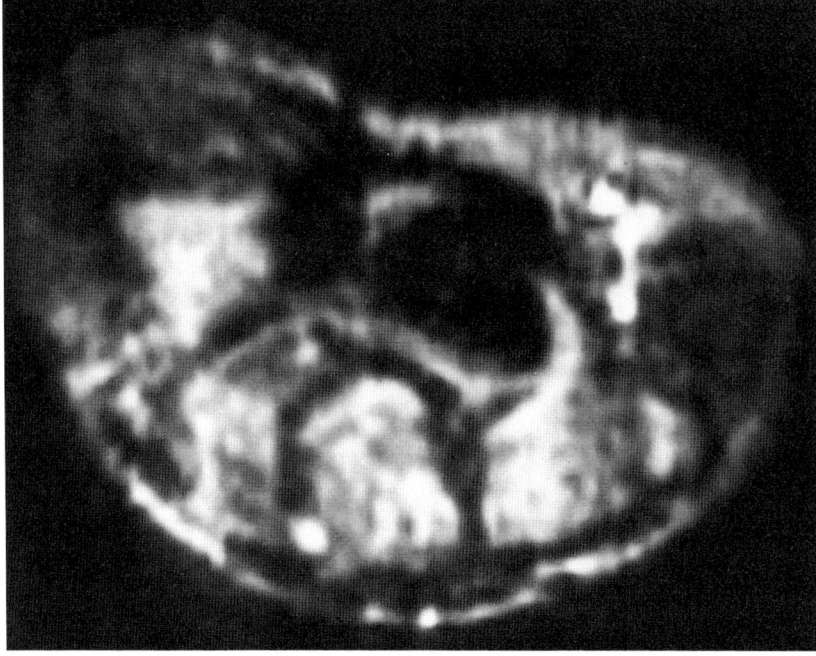

f

Figure 6.28 *continued*

At the level of mid-carpal tunnel (**e,f**), the median nerve is enlarged, flattened, and emits a higher signal with T_2-weighting, presumably due to edema.

7
The hip

Charles T McGlade and Lawrence W Bassett

Imaging of the normal hip

T_1-weighted MR images of the normal hip and surrounding muscles are presented in Figures 7.1–7.3. The normal femoral head contains high-signal bone marrow surrounded by thin cortical signal void. Within the bone marrow, there are normally three regions of signal lower than bone marrow: the fovea capitis medially; a horizontal low-signal band representing the growth plate remnant; and trabeculae oriented along the lines of stress. The capsule of the hip is a low-signal region surrounding the femoral head and inserting on the intertrochanteric line. Stress trabeculae may become unusually prominent when increased longitudinal forces are exerted on the hip (Figure 7.4).

Since red marrow has a lower MR signal than yellow marrow, it is important to understand the normal distribution of these marrow tissues in the skeleton. Just before birth when all the marrow is hematopoietic, conversion from red hematopoietic to yellow fatty marrow begins in the terminal phalanges of the feet and probably the hands. This conversion gradually progresses from the distal to proximal skeleton. Fat can be detected in the distal epiphyses of long bones by seven years of age and in the midshaft of all long bones by 12–14 years.[1] In MRI, in children and young adults the red-marrow femoral neck has a lower signal intensity than the epiphyseal region, where there is no red marrow (Figure 7.5).[2] The femoral capital epiphysis and greater trochanter both contain fatty marrow and have high-signal intensity. Hematopoietic bone marrow is not observed in the femoral capital epiphysis or greater trochanter at any age. The presence of fatty marrow within the femoral neck and intertrochanteric region varies with age. In patients less than 50 years of age, the majority have some hematopoietic marrow within the femoral neck and/or trochanteric region. However, the majority of patients over 50 have predominantly fatty marrow throughout the femur. The distribution of the fatty and hematopoietic marrow is generally bilaterally symmetrical. In disorders such as chronic anemia and marrow replacement disorders such as myelofibrosis, myeloma, infarcts and others, the body responds to the need for additional hematopoiesis by marrow reconversion and local hyperplasia of existing hematopoietic marrow. The spine and flat bones respond more rapidly than other bones to marrow hyperplasia.

Imaging of ischemic necrosis

The most common indication for MR imaging of the hip is the evaluation of suspected ischemic necrosis.[3-7] Ischemic necrosis, aseptic necrosis, avascular necrosis and osteonecrosis are synonyms for bone death resulting from a compromised blood supply. The femoral head is particularly vulnerable to ischemia, and weight bearing predisposes its necrotic bone to irreversible structural collapse. Conservative management and surgical measures designed to limit the progression of the disease and salvage the femoral head depend on early detection.

Early detection of ischemic necrosis of the femoral head implies diagnosis before osseous destruction has advanced to the stage of structural collapse.[8] Symptomatology and physical examination are not dependable for early diagnosis. An important key to early diagnosis is recognition of those patients at high risk. Hip trauma, corticosteroid therapy, alcoholism, chronic pancreatitis, sickle-cell disease, Gaucher disease, and dysbaric atmospheric conditions all predispose to ischemic necrosis.[9-12] The staging of ischemic necrosis of the femoral head is based on plain film radiographic findings.[13] In Stage I, the hip is radiographically normal. Stage II is characterized by mixed sclerosis and porosis of the femoral head. Stage III features collapse of the femoral head, but the joint space remains normal. Stage IV is marked by collapse of the femoral head and a compromised joint space. A preclinical stage, Stage 0, was proposed by Ficat for a femoral head that is still asymptomatic and radiographically normal in the presence of contralateral ischemic necrosis, in a patient at high risk for bilateral disease.[8] A subchondral radiolucent (Waldenström)

crescent is considered by Ficat to reflect an intermediate stage between Stages II and III.

Radionuclide scanning, bone marrow pressure determinations, intraosseous venography, and core biopsy study are diagnostic tools currently in use to detect preradiographic ischemic necrosis.[8,14-17] In the past, bone scanning was the primary modality for the evaluation of ischemic necrosis. However, for the detection of early ischemic necrosis, MRI has proved to be a more sensitive and specific method (Tables 7.1 and 7.2).

Although the MR images of ischemic necrosis are subject to some variability, in all cases the abnormality is characterized on T_1-weighted images by a decrease in the normally high-intensity signal of the femoral head bone marrow (Figure 7.6). On the MR examination, there may be a band or ring-like area (Figures 7.7–7.8) of diminished signal or homogeneous or inhomogeneous focal regions of diminished signal in the subarticular region of the head (Figures 7.9–7.13). A pattern of diffuse decreased signal may also be seen (Figure 7.14). The reasons for this diminished signal have not yet been explained satisfactorily, neither have the pathophysiological mechanisms behind the various MRI changes been proved, but replacement of marrow fat by a histiocytic, fibrous connective tissue response in combination with new bone proliferation undoubtedly plays a contributing role. In those cases where a low-signal band surrounds the lesion, we believe the low-signal zone represents a healing sclerotic interface between viable fibrovascular tissue and necrotic bone. This has been confirmed by MR-histologic correlation in one case (Figure 7.15).[18] A double line sign in T_2-weighted images, consisting of a peripheral low-signal margin with high-signal just inside it, could also be explained by a zone of peripheral sclerosis with a rim of granulation tissue, hyperemia and/or chondroid metaplasia within.[19,20] In T_2-weighted images, fluid may be observed in the ischemic hip joint surrounding the femoral neck or distending capsular recesses. This is represented on the MR images by a bright signal surrounding the cortical bone of the head (Figure 7.16).[18,19] T_2-weighted images show a variety of changes in the ischemic femoral head, depending on the stage of disease and the presence or absence of fluid and fibrous tissue (Figure 7.17).[21] In advanced cases, joint space narrowing, femoral head collapse and loose bodies may be observed on the MR examination.

Pathologic fracture secondary to ischemic necrosis is a rare complication that may be observed on MR (Figure 7.18).

Following core biopsy of the ischemic hip, a cylindrical, low-signal region is observed extending the full length of the trephine defect. In our experience, the defect has not interfered with evaluation of the surrounding marrow (Figure 7.19).

Table 7.1 Forty-seven hips with positive MRI for ischemic necrosis

Stage	Number	Number positive Diphosphonate	Sulfur colloid
0	7	1(7)	3(5)
I	11	7(11)	6(6)
II	14	11(14)	2(7)
III	11	11(11)	8(8)
IV	4	4(4)	3(3)
Totals	47	34(47)	22(29)

( ) = Number of radionuclide studies.
Of the 47 positive MRI cases, 38 were corroborated by pathology and 6 by serial radiography. Of the remaining 3, one refused core bx, one died without autopsy and the last, initially Stage 0, has recently become symptomatic.

All the normal hips on MRI have remained asymptomatic with follow-up times ranging from 6–21 months.

Table 7.2 MRI vs. SPECT in 17 patients (8 hips proven ON)

SPECT	+ON	−ON
Positive	7	3
Negative	1	23

Sensitivity = 87%; specificity = 88%

MRI	+ON	−ON
Positive	8	2
Negative	0	24

Sensitivity = 100%; specificity = 92%

MRI in other hip conditions

Different MR patterns are seen with other diseases. For example, osteoarthritis may present as one of two patterns (Figures 7.20–7.22). If the plain films show sclerosis of the subchondral bone, the MRI will show diminished signal in the subchondral bone marrow. Unlike the findings in ischemic necrosis, this diminished signal is seen both in the femoral head and in the subjacent acetabulum. If the plain films show large osteophytes of the femoral head and/or acetabulum, the MR may show normally high signal of the subarticular bone as well as high signal within the osteophytes, since they are also filled with fatty marrow.

Sickle-cell disease is characterized by bone infarcts throughout the pelvis and femur. These do not necessarily have the same features as localized ischemic necrosis of the hip. In addition, there is proliferation of red marrow secondary to the anemia which also causes decreased signal intensity (Figure 7.23).

Infection is characterized by regions of low signal in the femoral head and adjacent acetabulum on the T_1-weighted sequence. On T_2-weighted sequences, these same regions may show super-high signal intensity, and fluid is identified within the joint (Figure 7.24).

Transient regional osteoporosis of the hip may mimic ischemic necrosis of the hip. In plain films there is severe osteoporosis. The radionuclide bone scan will show intense isotope accumulation in the femoral head. The T_1-weighted MRI shows low to no signal throughout the femoral head. The diminished signal is more uniform than usually seen in ischemic necrosis. A T_2-weighted image should result in conversion to high-signal intensity throughout the femoral head. In ischemic necrosis edematous portions of the head will give a higher signal intensity on T_2-weighted images, but signal voids should remain at the site of new bone formation and fibrovascular tissue proliferation (Figures 7.25, 7.26).

Paget's disease is manifested by thickened trabeculae, which result in linear signal deficits throughout the involved bone (Figure 7.27).

Metastatic disease to bone causes marrow replacement leading to low-signal regions in T_1-weighted images (Figure 7.28). Tumors may arise within the femoral head and acetabulum (Figure 7.29).

Loose bodies within the hip joint would be expected to be low-signal. However, osseous fragments containing bone marrow have high-signal (Figure 7.30).

A metal hip replacement is not a contraindication to MRI examination. Studies of different types of prostheses have shown no adverse effects on the metals employed in these prostheses and there has been no evidence of heating of the metal. However, the metal does produce artifacts. Unlike CT, the metal artifacts of MRI are limited to an area directly at and adjacent to the site of the metal prosthesis. These artifacts are usually signal voids; however, occasionally the metal produces high-signal artifacts as well (Figures 7.31–7.33).[22]

References

1. KRICUN ME, Red-yellow marrow conversion: its effect on the location of some solitary bone lesions, *Skeletal Radiol* (1985) **14**:10–19.
2. MITCHELL DG, RAO VM, DALINKA M et al, Hematopoietic and fatty bone marrow distribution in the normal and ischemic hip: new observations with 1.5-T MR imaging, *Radiology* (1986) **161**:199–202.
3. MITCHELL MD, KUNDEL HL, STEINBERG ME et al, Avascular necrosis of the hip: comparison of MR, CT, and scintigraphy, *AJR* (1986) **147**:67–71.
4. TOTTY WC, MURPHY WA, GANZ WI et al, Magnetic resonance imaging of the normal and ischemic femoral head, *AJR* (1984) **143**:1273–80.
5. THICKMAN D, AXEL L, KRESSEL HV et al, Magnetic resonance imaging of avascular necrosis of the femoral head, *Skeletal Radiol* (1986) **15**:133–40.
6. BASSETT LW, GOLD RH, REICHER M et al, Magnetic resonance imaging in the early diagnosis of ischemic necrosis of the femoral head, *Clin Orthop* (1987) **214**:237–48.
7. MARKISZ JA, KNOWLES RJR, ALTCHEK DW et al, Segmental patterns of avascular necrosis of the femoral heads: early detection with MR imaging, *Radiology* (1987) **162**:717–20.
8. FICAT RP, Idiopathic necrosis of the femoral head. Early diagnosis and treatment. *J Bone Joint Surg* (1985) **67B**:3–9.
9. CATTO M, The histological appearances of late segmental collapse of the femoral head after transcervical fracture, *J Bone Joint Surg* (1965) **47B**:777–91.
10. FICAT RP, ARLET J, Bone necrosis of known etiology, In: Hungerford D (ed), *Ischemia and Necrosis of Bone*, (Williams and Wilkins: Baltimore, 1980) 111–30.
11. JONES JP JR, JAMESON RP, ENGLEMAN EP, Alcoholism, fat embolism, and avascular necrosis, *J Bone Joint Surg* (1968) **50A**:1065(ab).
12. KLIPPER AR, STEVENS MB, ZIZAC TM et al, Ischemic necrosis of bone in systemic lupus erythematosus, *Medicine* (1976) **55**:251–7.
13. ARLET J, FICAT P, Diagnosis of primary femur head osteonecrosis at Stage 1, *Rev Chir Orthop* (1968) **54**:637–48.
14. ALAVI A, McCLOSKEY JR, STEINBERG ME, Early detection of avascular necrosis of the femoral head by 99m-technetium diphosphonate bone scan: preliminary report, *Clin Orthop* (1977) **127**:137–41.
15. CONKLIN JJ, ALDERSON PO, ZIZAC TM et al, Comparison of bone scan and radiograph sensitivity in the detection of steroid-induced ischemic necrosis of bone, *Radiology* (1983) **147**:221–6.
16. MEYERS MH, TELFER N, MOORE TM, Determination of the vascularity of the femoral head with technetium 99m-sulphur-colloid, *J Bone Joint Surg* (1977) **59A**:658–64.
17. WEBBER MM, WAGNER J, CRAGIN MD, Radionuclide patterns of femoral head disease, *Int J Nucl Med Biol* (1977) **4**:167–7.

18 BASSETT LW, MIRRA JM, CRACCHIOLO A et al, Ischemic necrosis of the femoral head: Correlation of magnetic resonance imaging and histologic sections, *Clin Orthop* (1987) **223**: 181–7.

19 MITCHELL DG, RAO BM, DALINKA M et al, Femoral head avascular necrosis: Correlation with MR imaging. Radiographic staging, radionuclide imaging, and clinical findings, *Radiology* (1987) **162**:709–15.

20 MITCHELL DG, JOSEPH PM, FALLON M et al, Chemical-shift MR imaging of the femoral head: an in vitro study of normal hips and hips with avascular necrosis, *AJR* (1987) **148**:1159–64.

21 MITCHELL DG, RAO V, DALINKA M et al, MRI of joint fluid in the normal and ischemic hip, *AJR* (1986) **146**:1215–18.

22 RICHARDSON ML, HELMS CA, Artifacts, normal variants, and imaging pitfalls of musculoskeletal magnetic resonance imaging, *Radiol Clin North Am* (1986) **24**:145–77.

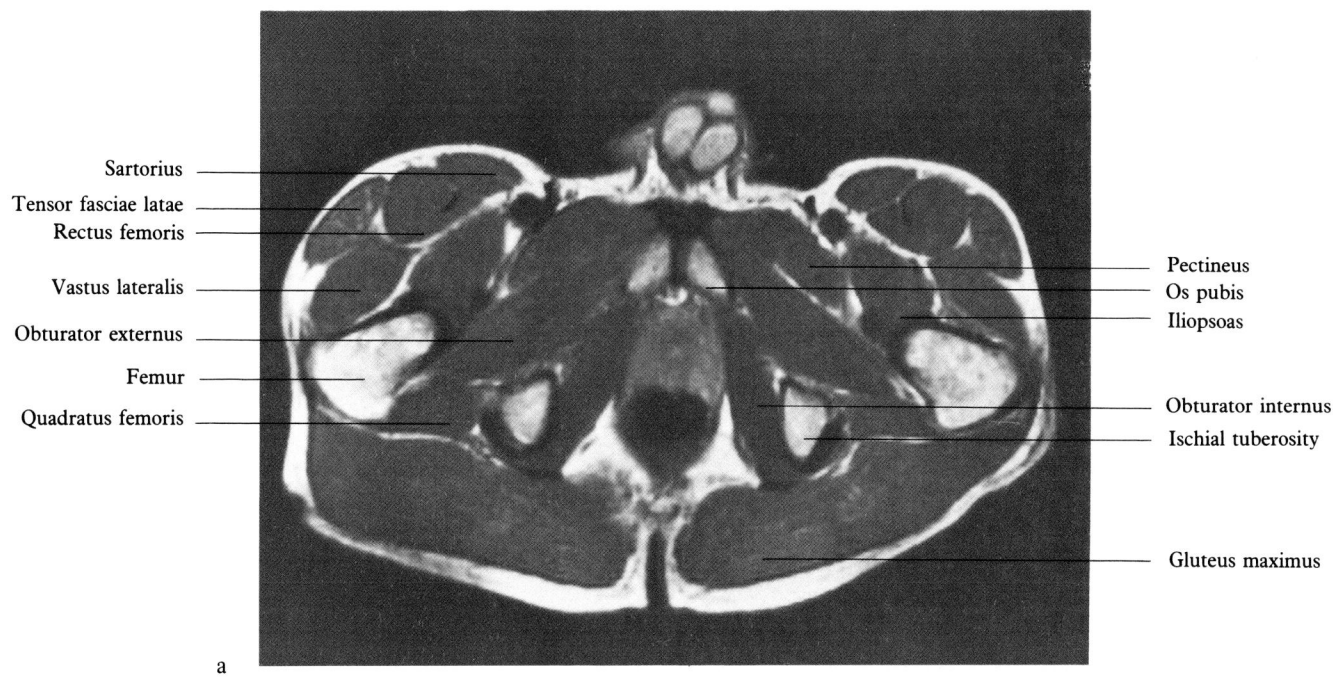

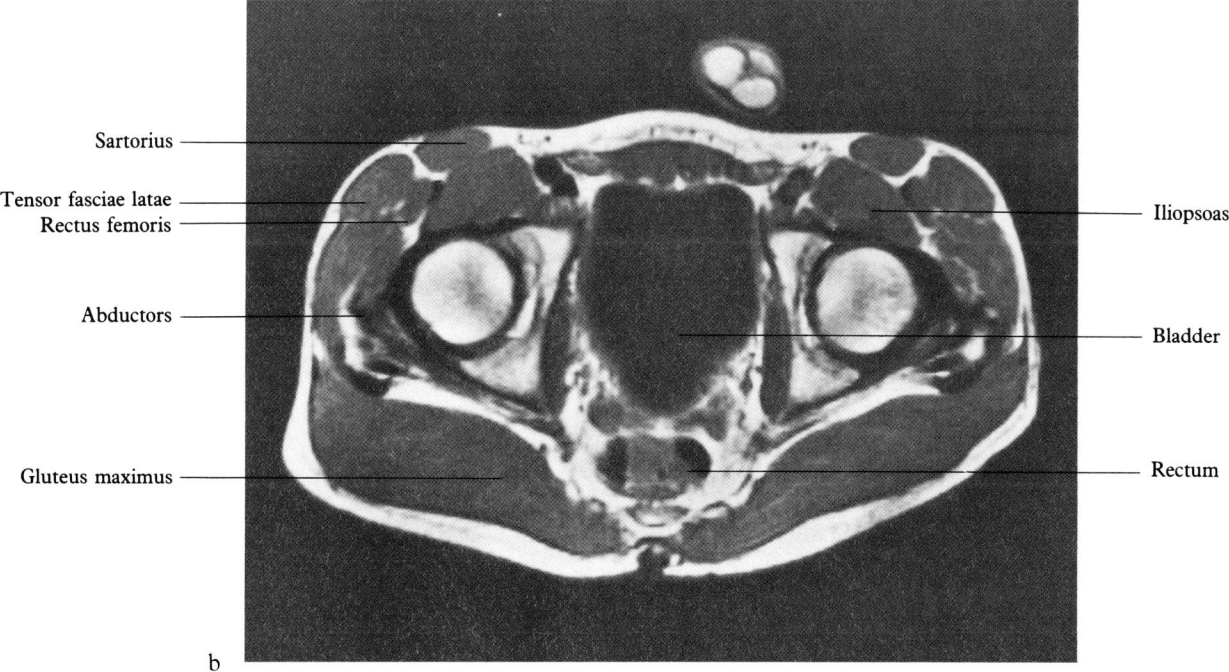

Figure 7.1

Normal axial anatomy (SE 500/28).

180 MRI atlas of the musculoskeletal system

Figure 7.1 continued

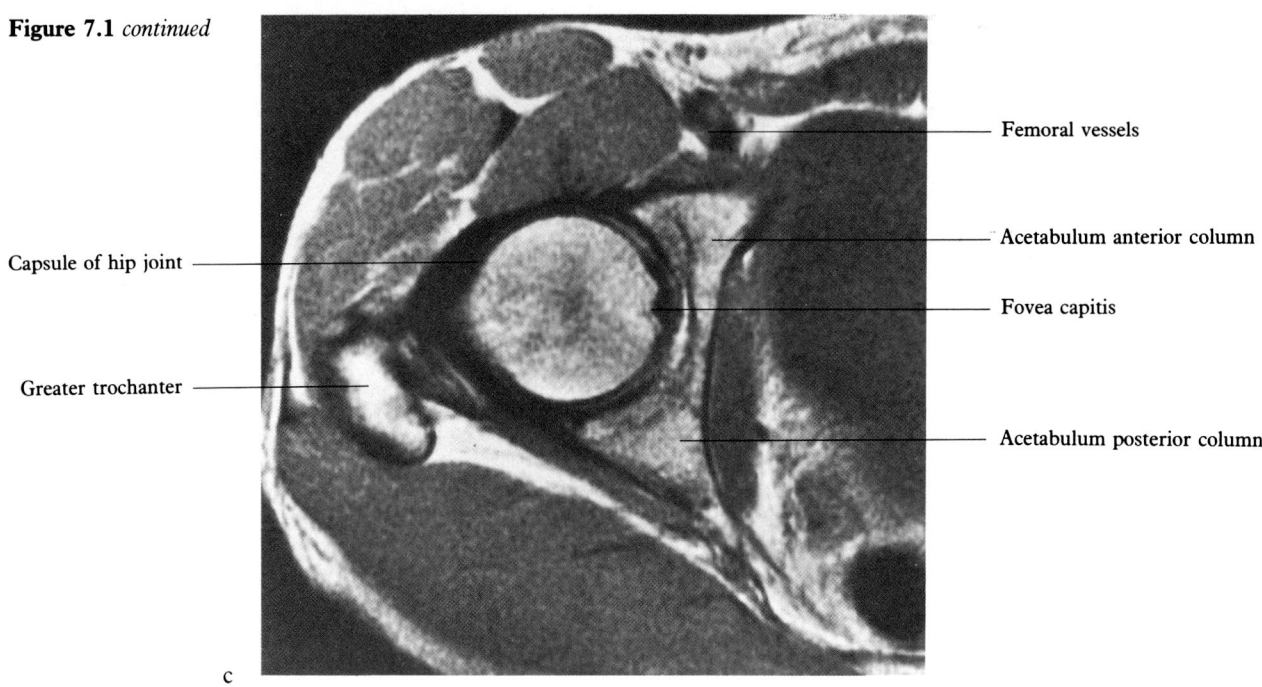

Figure 7.2

Normal coronal anatomy (SE 500/28).

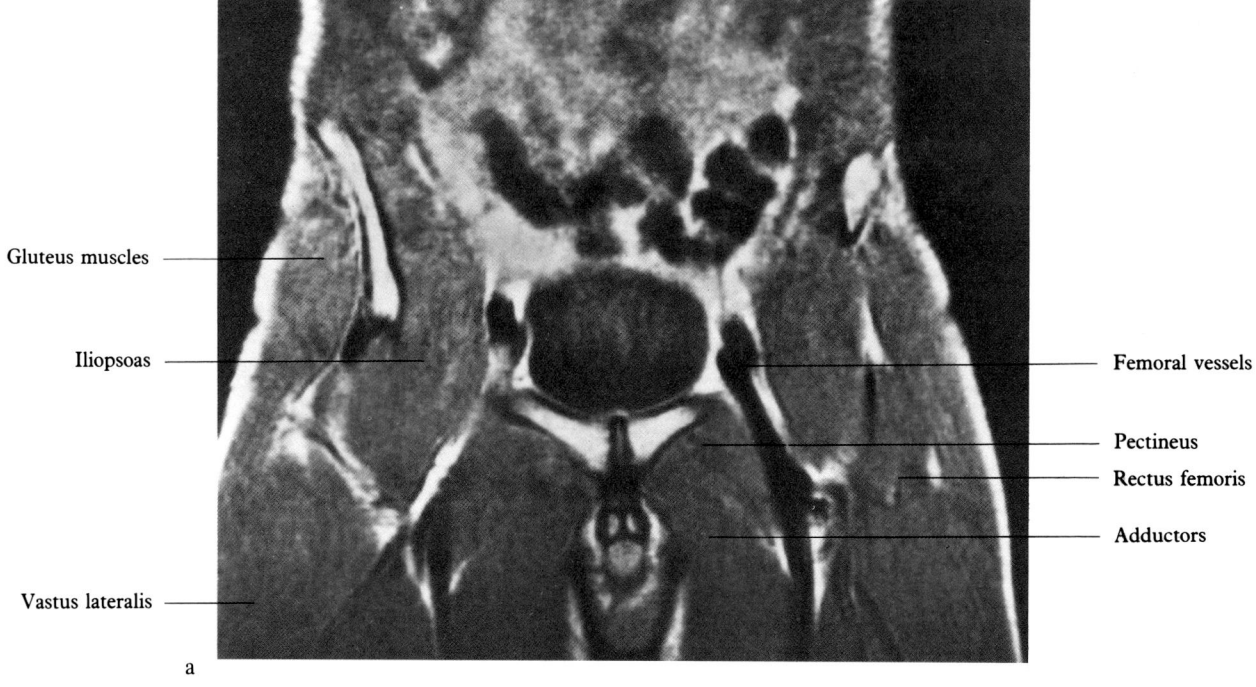

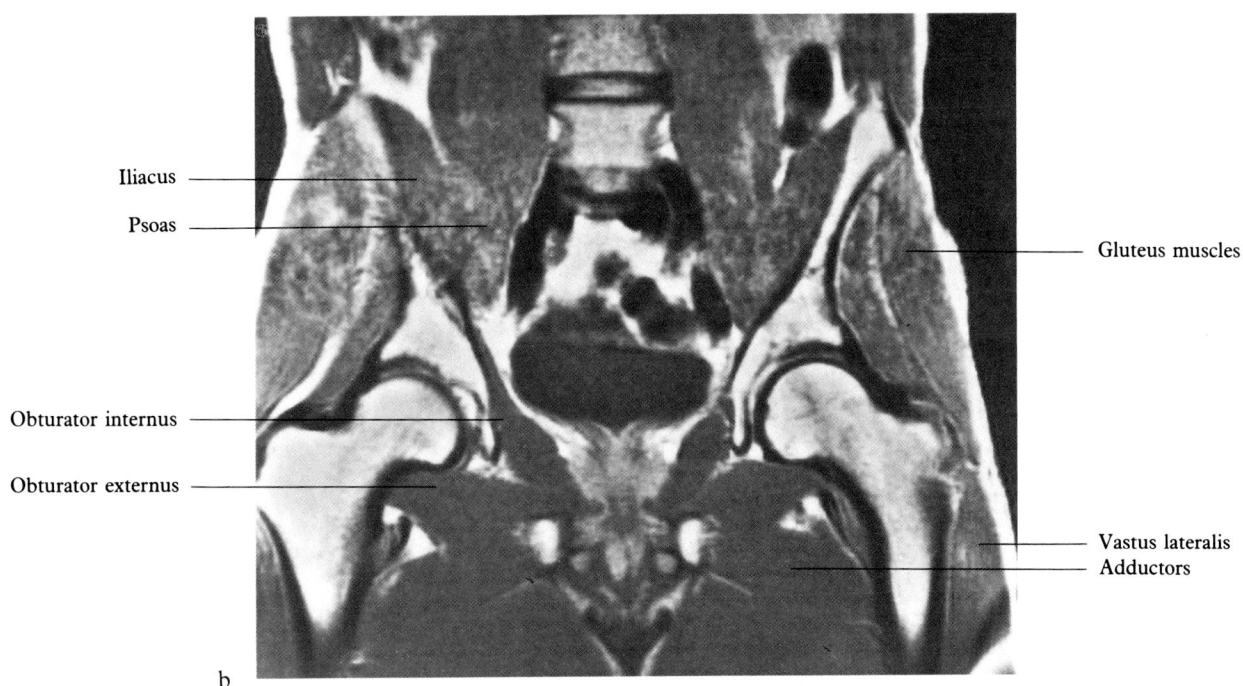

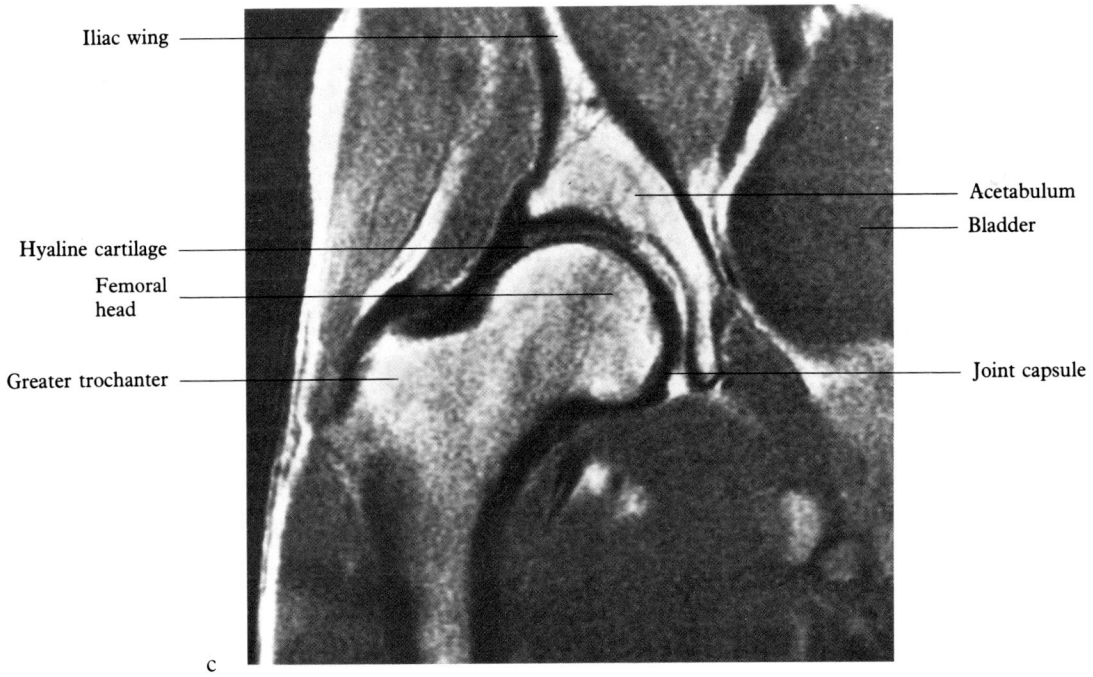

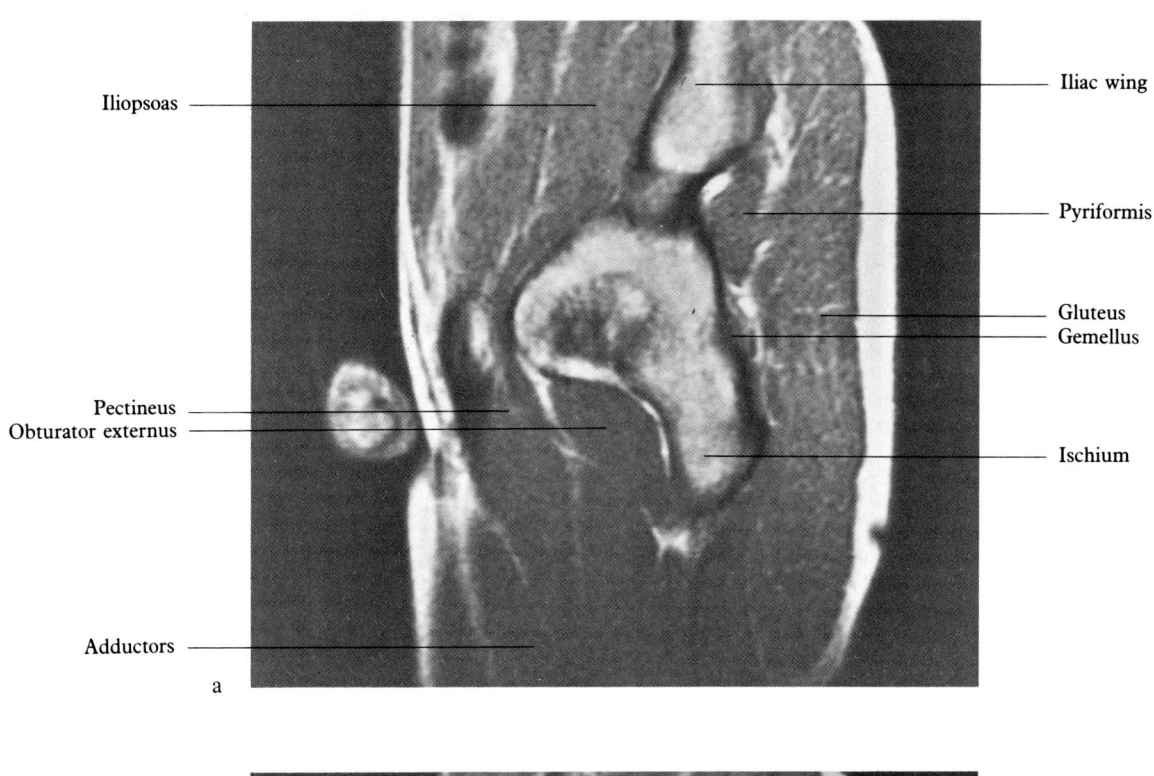

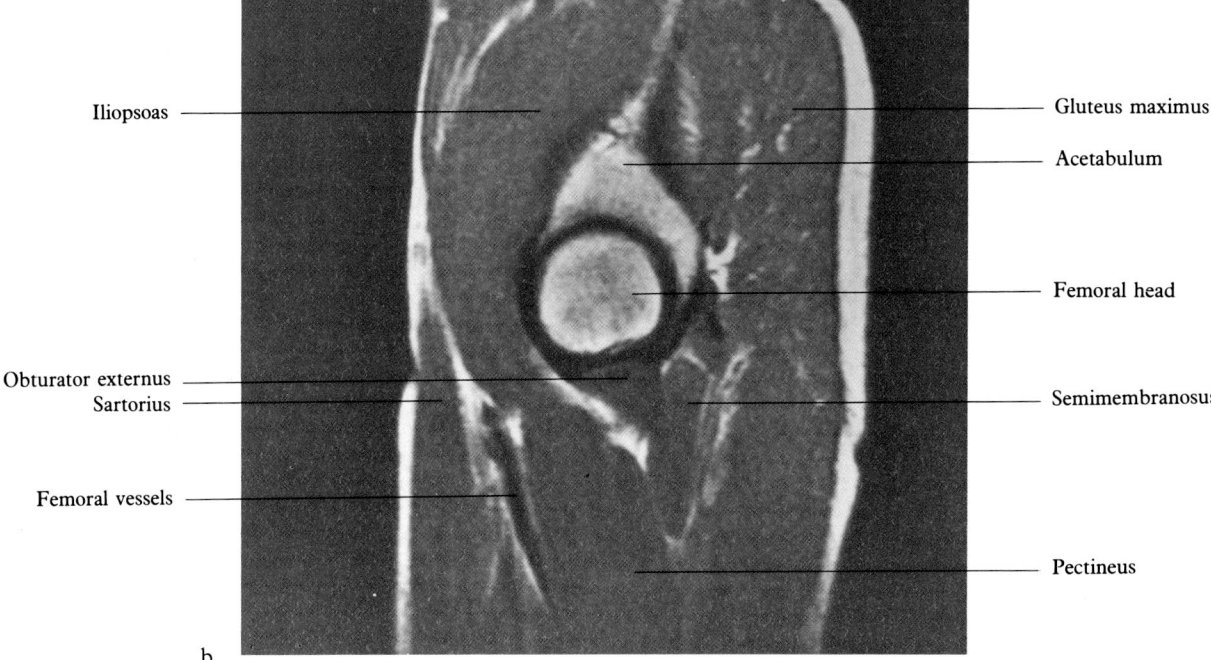

Figure 7.3

Normal sagittal anatomy (SE 500/28).

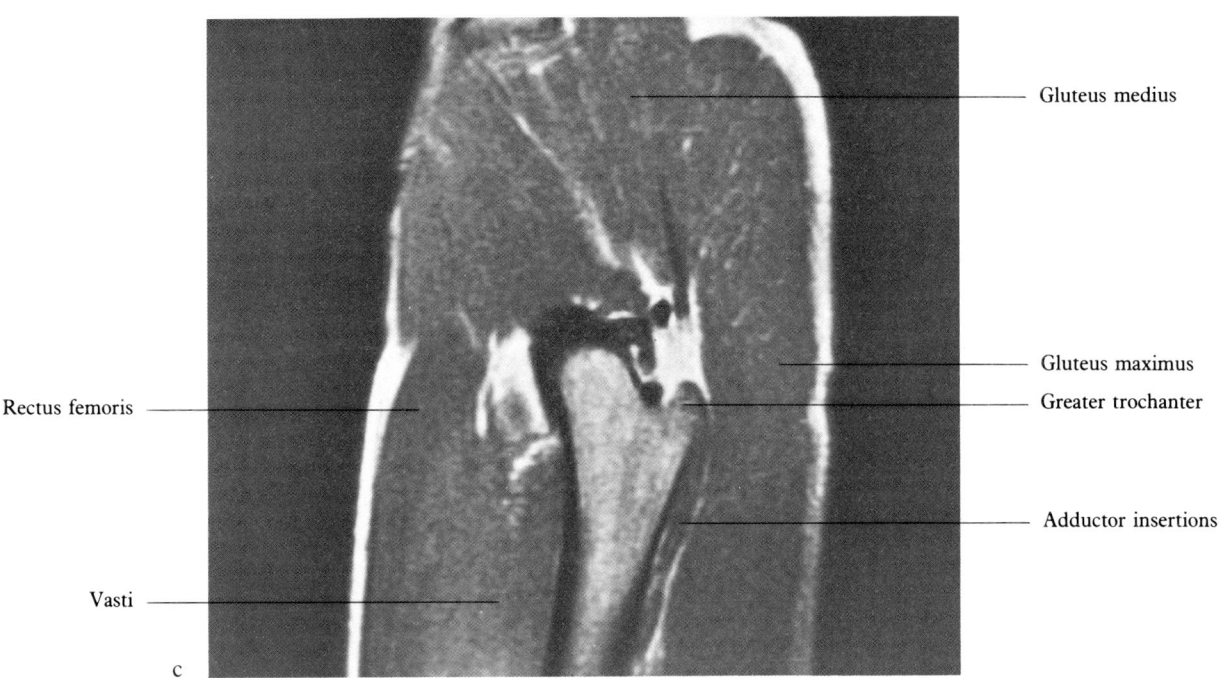

Gluteus medius

Gluteus maximus
Greater trochanter

Adductor insertions

Rectus femoris

Vasti

c

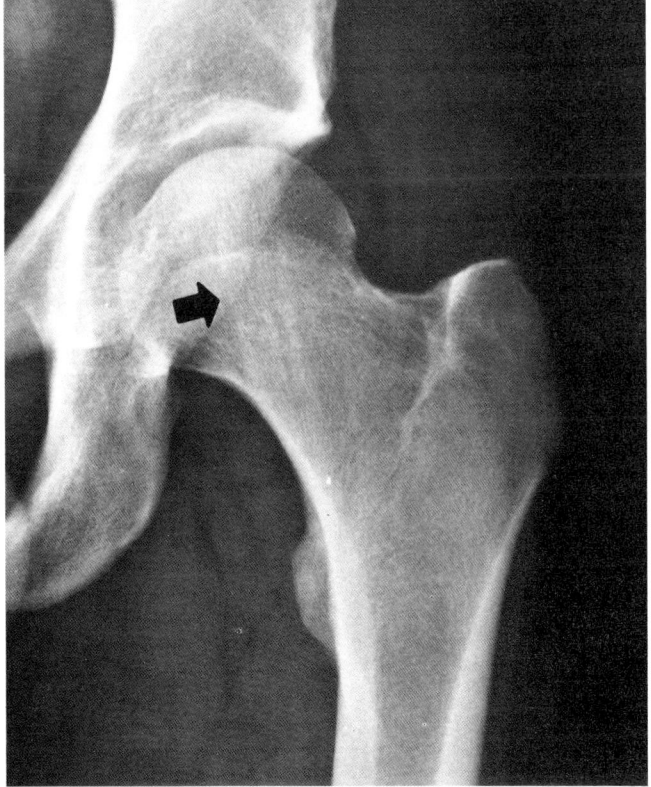

a

Figure 7.4

Prominent stress trabeculae. A 32-year-old weight lifter had thigh pain which resolved. Note the prominent trabeculae along lines of stress in the plain film (**a**), which are manifested by signal deficits in the coronal (**b**) and sagittal (**c**) MR images (SE 500/28).

184 MRI atlas of the musculoskeletal system

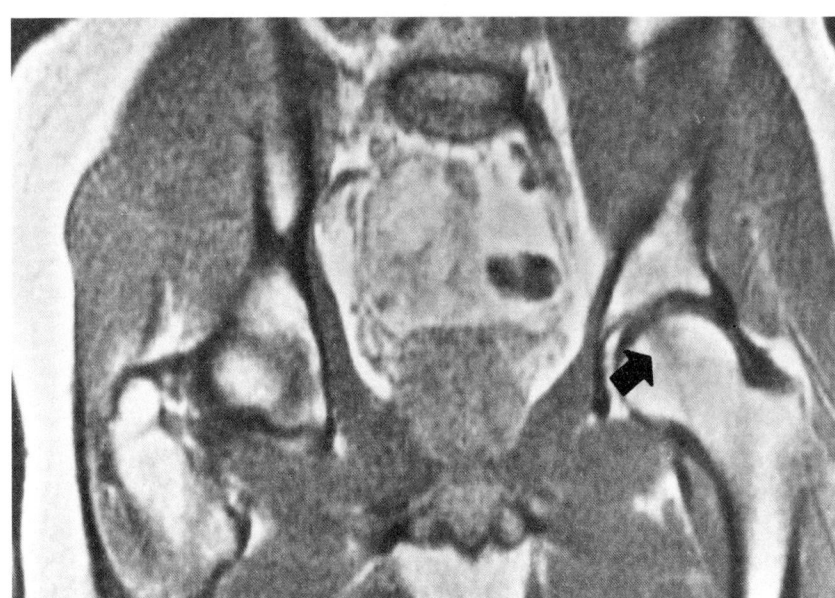

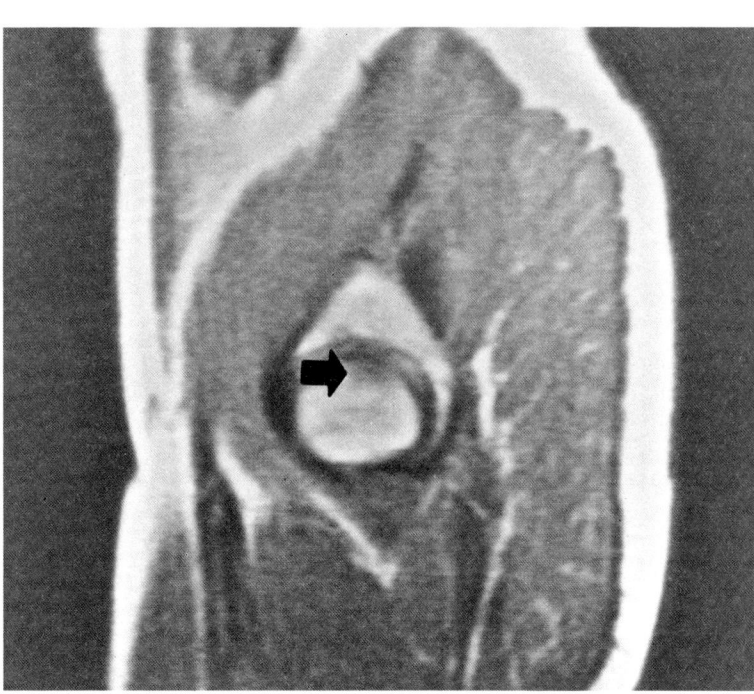

Figure 7.4 continued

b

c

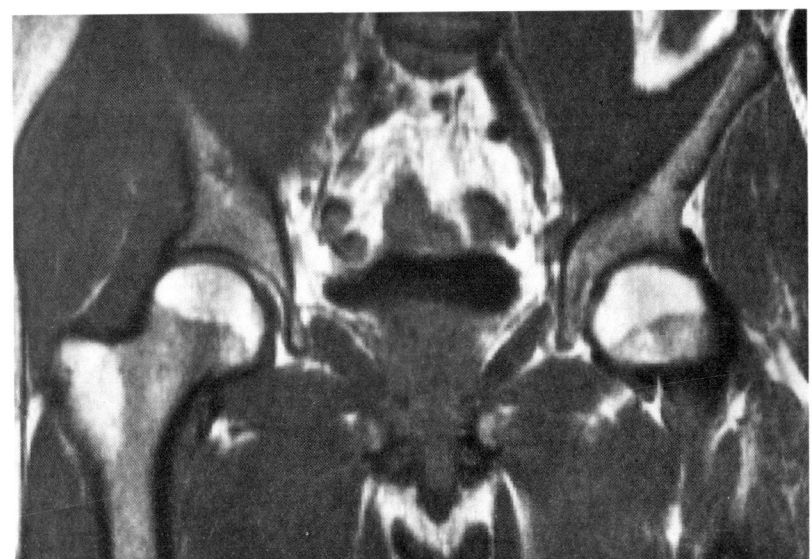

Figure 7.5

Low signal of hematopoietic marrow. A 19-year-old man with transient hip pain. The femoral head epiphysis and trochanteric apophysis contain high-signal fatty marrow while the surrounding actively hematopoietic marrow gives a lower signal. In young patients this may be a normal variant, while in older patients it reflects abnormal hematopoietic activity (SE 500/28).

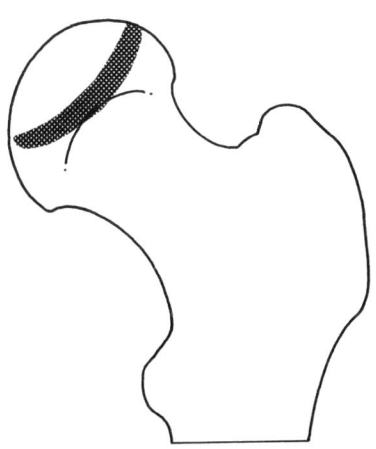
a

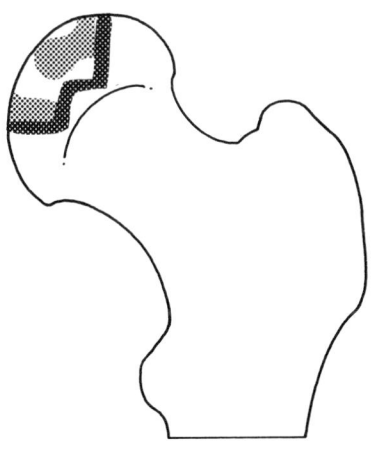
b

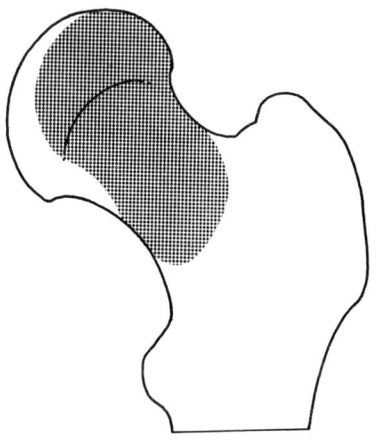
c

Figure 7.6

MR patterns of ischemic necrosis of the hip. (**a**) Band or ring pattern of decreased signal. (**b**) Subchondral region of inhomogeneous decreased signal. (**c**) Homogeneous or diffuse diminished signal intensity.

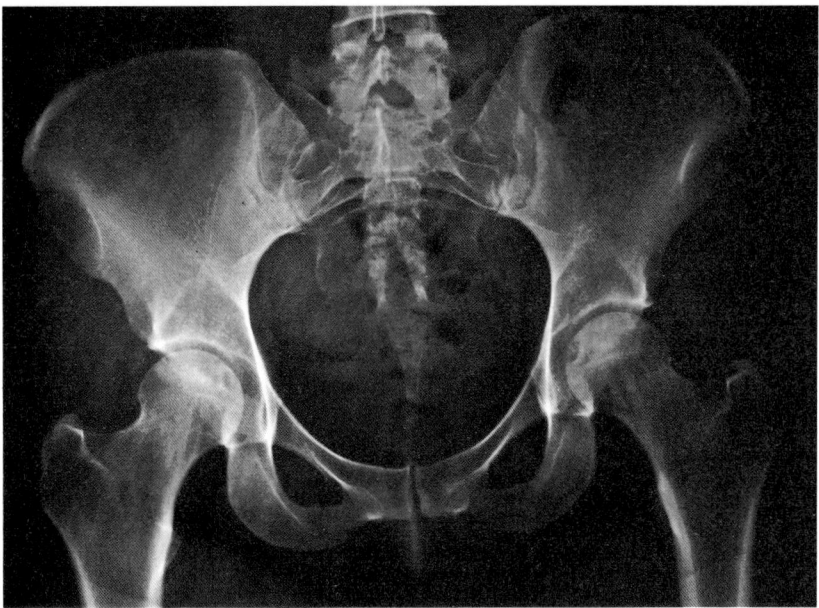

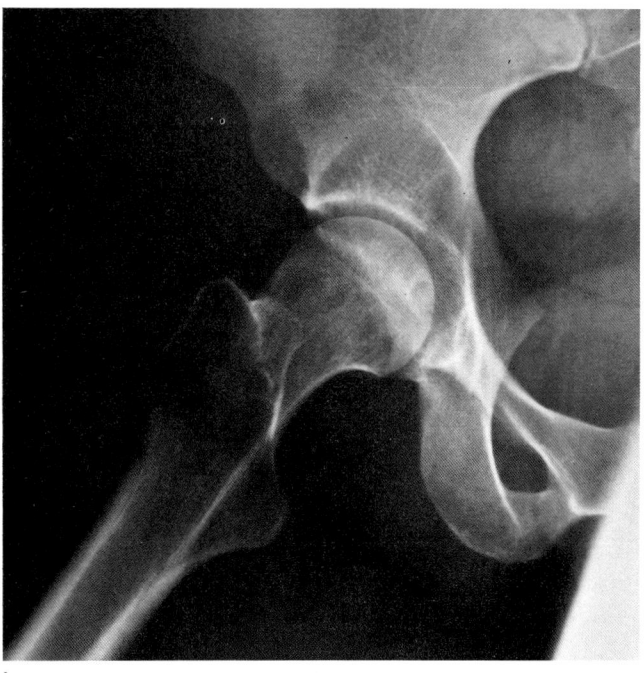

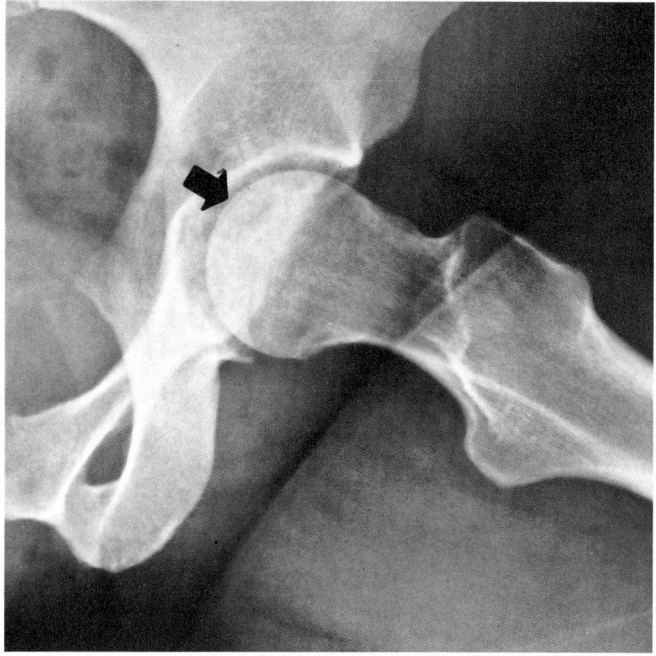

Figure 7.7

Bilateral hip and knee pain in a 29-year-old woman with lupus erythematosis treated with corticosteroids for many years. (a) Anteroposterior radiograph of the pelvis is normal. (b) Right frog lateral radiograph is normal. (c) Left frog lateral radiograph discloses a subtle subchondral radiolucent crescent (arrow). (d) Diphosphonate scan reveals increased activity in the left hip. (e) Sulfur colloid scan discloses a normal bilaterally symmetric uptake in the femoral heads. (f) T_1-weighted coronal MRI (SE 500/28) shows bilateral diminished signal intensity in the femoral heads, focal inhomogeneous on right and band-like on left. (g) Axial MR image confirms the findings in (f). (h) Coronal MR image of the knees (SE 500/28) reveals bilateral infarcts manifested as ring-like signal deficits surrounding normally high-signal bone marrow. Core biopsies revealed bilateral ischemic necrosis.

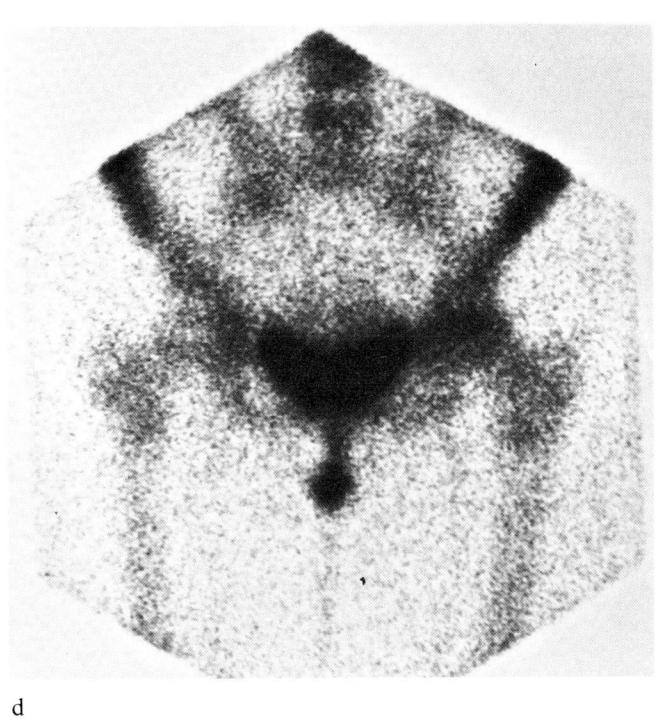

d

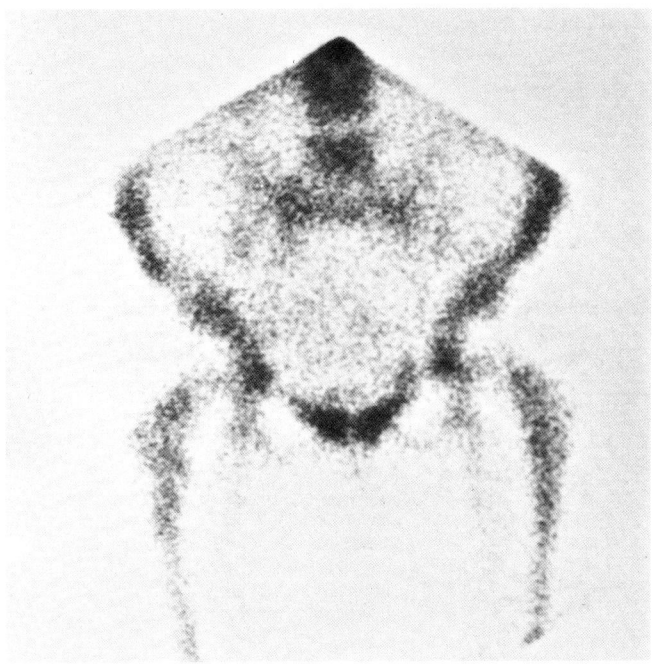

e

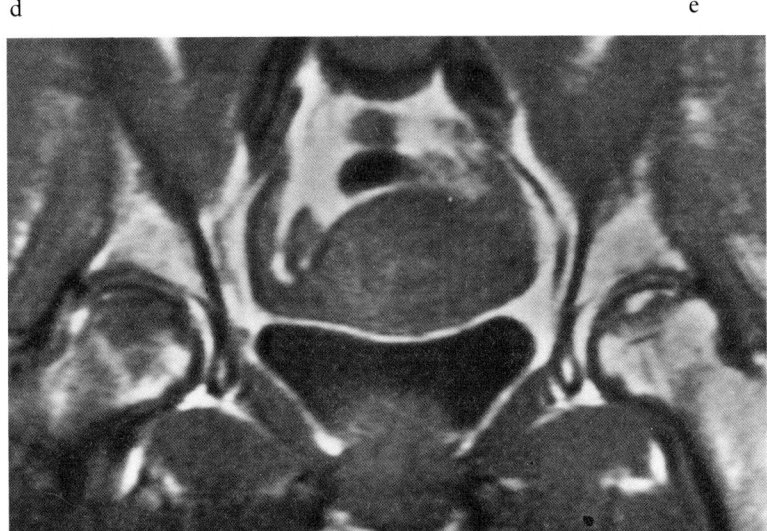

f

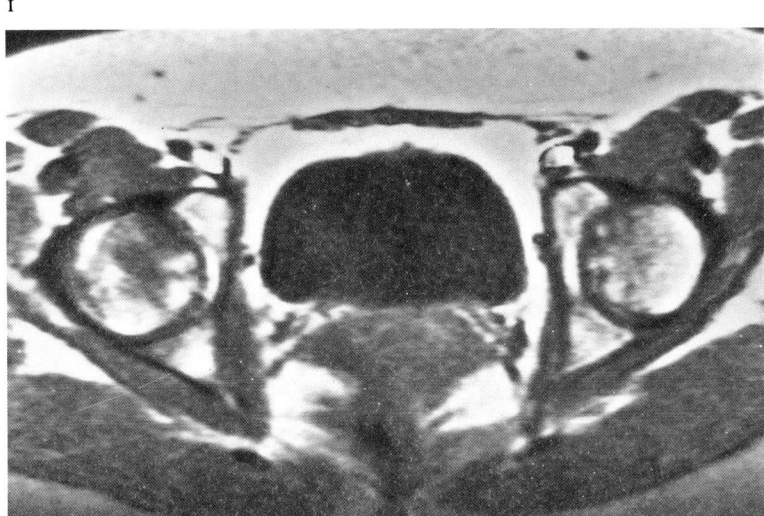

g

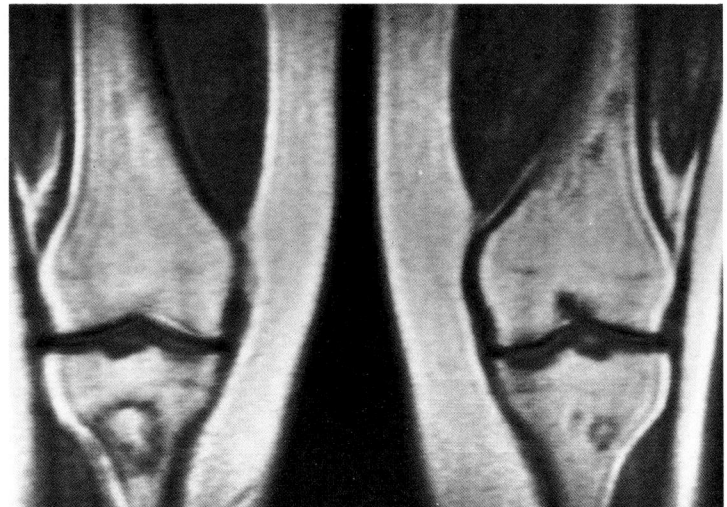

h

Figure 7.7 continued

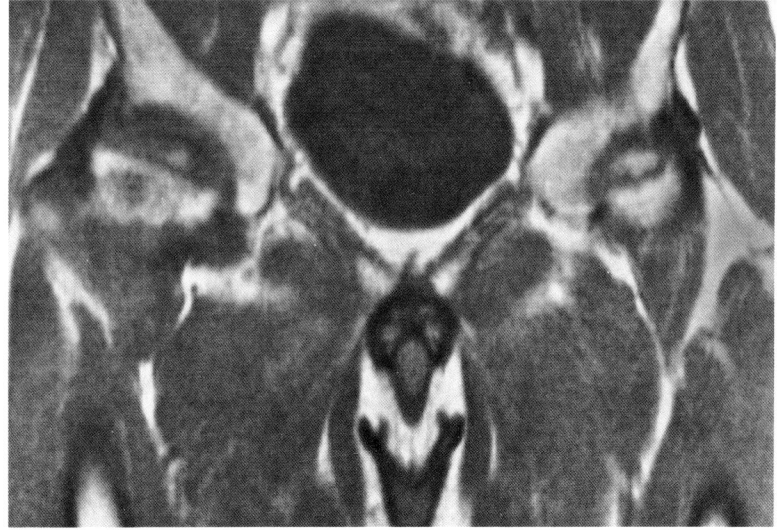

a

Figure 7.8

A 56-year-old male bartender complained of pain in both hips. T_1-weighted MRI (SE 500/28) demonstrates bilateral ring-like signal deficits in coronal (**a**) and left sagittal (**b**) images. Core biopsies showed bilateral ischemic necrosis.

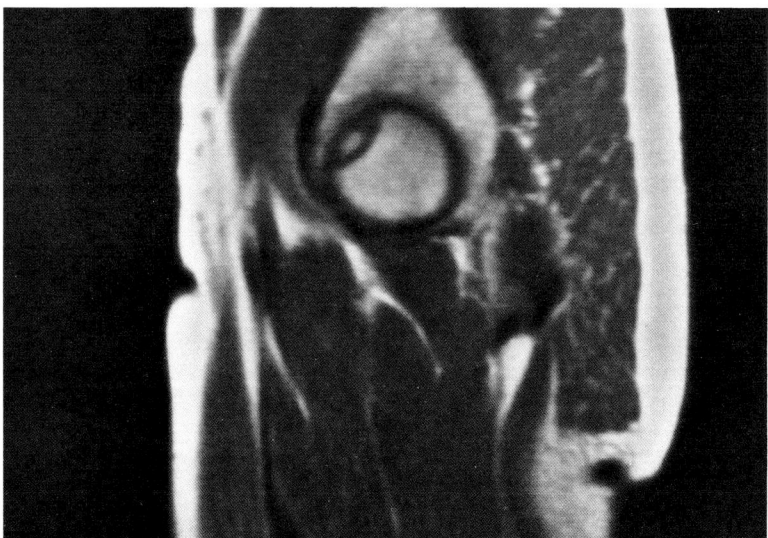

b

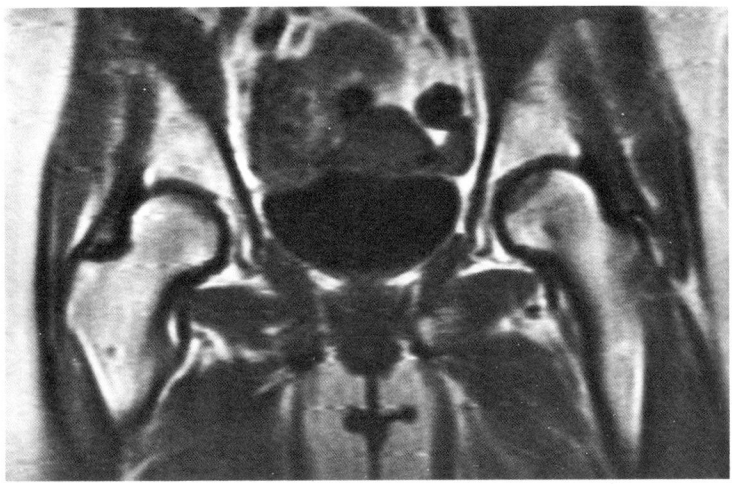

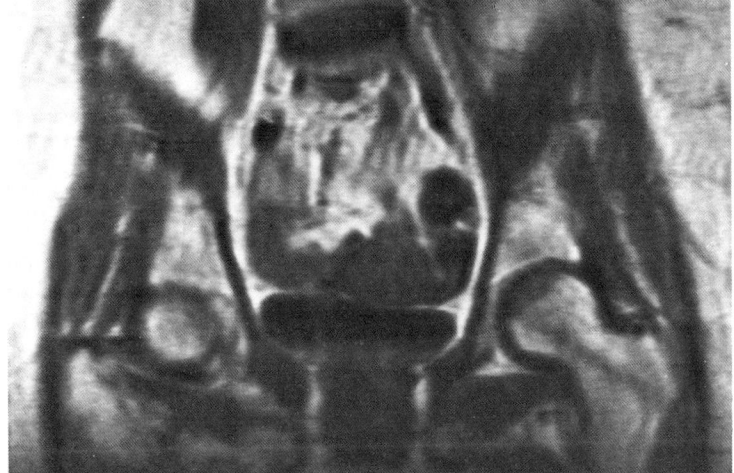

Figure 7.9

A 22-year-old woman receiving corticosteroid therapy for systemic lupus, complained of left hip pain. (**a**) T_1-weighted MR image (SE 500/28) reveals focal subchondral inhomogeneous signal deficits in both femoral heads. The patient subsequently underwent coring procedures which documented bilateral ischemic necrosis. (**b**) MR image (SE 500/28) 6 months post-coring shows the cylindral coring defect; the surrounding bone marrow is well seen.

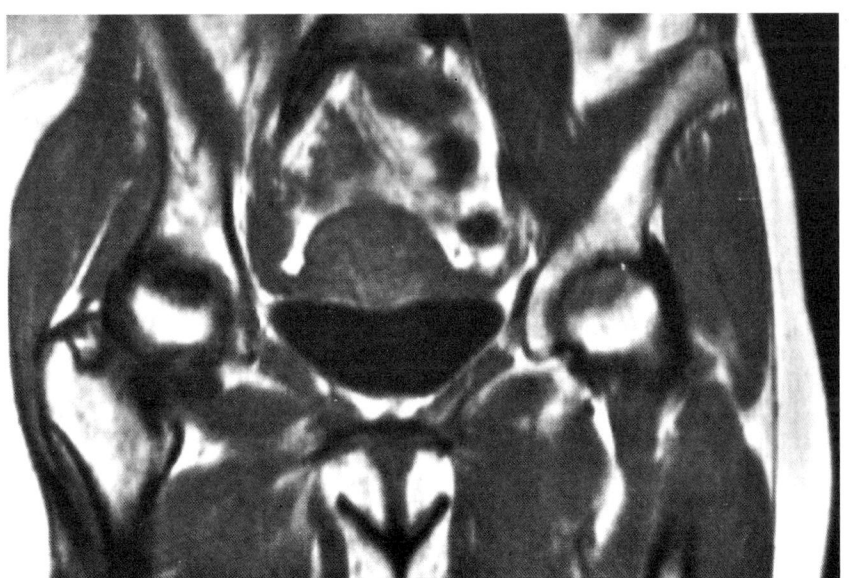

Figure 7.10

A 34-year-old woman treated with corticosteroids for lupus complained of severe hip pain. T_1-weighted coronal MR image (SE 500/28) shows bilateral inhomogeneous subchondral femoral head focal defects. A core biopsy confirmed the diagnosis of ischemic necrosis.

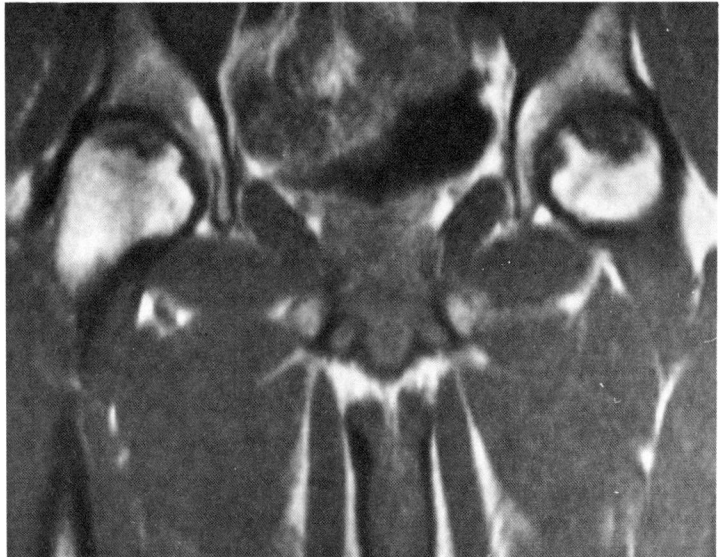
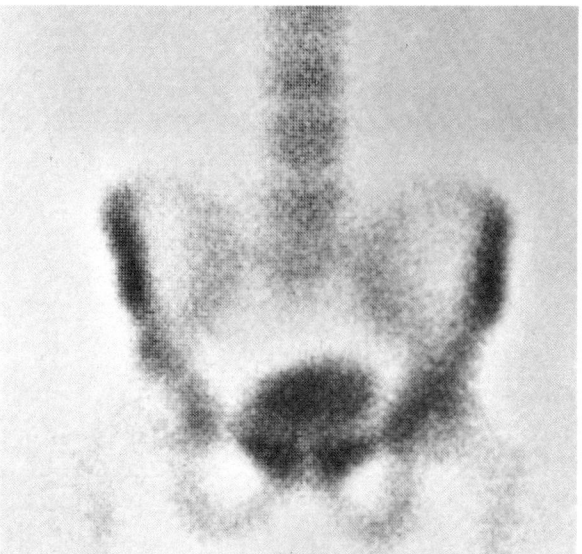

Figure 7.11

A 32-year-old man treated with corticosteroids for a dermatologic condition. (a) T_1-weighted coronal MR image (SE 500/28) shows patchy signal deficits in subchondral regions of both femoral heads. (b) Technetium diphosphonate radionuclide scan demonstrates abnormal accumulation only in the left hip. Core biopsies revealed bilateral ischemic necrosis.

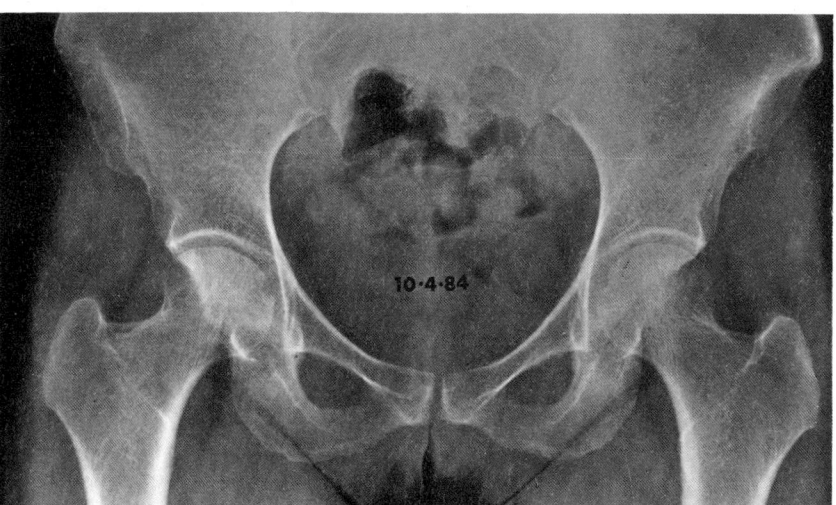

Figure 7.12

A 40-year-old woman was treated 2 years previously with corticosteroids for a brain abscess. In the last year she had increasing left hip pain. (a) Radiograph of the pelvis shows increased density of the right femoral head. (b) Frog lateral radiograph of the right hip reveals slight loss of sphericity and mottled increased density. (c) Frog lateral radiograph of the left hip is normal. (d) Diphosphonate bone scan discloses increased activity in the right femoral head. (e) Sulfur colloid scan reveals diminished uptake in the right femoral head and neck compared to the left. (f) Intraosseous venograph of left femur immediately after contrast injection shows normal rapid clearing; however, reflux into diaphysis (arrow) was considered abnormal. (g) Intraosseous venography 5 minutes post-injection: the contrast medium has completely disappeared. (h) T_1-weighted coronal MR image (SE 500/28) discloses diminished signal from the right femoral head and normal signal from the left femoral head. Core biopsies revealed ischemic necrosis of the right hip and normal left hip.

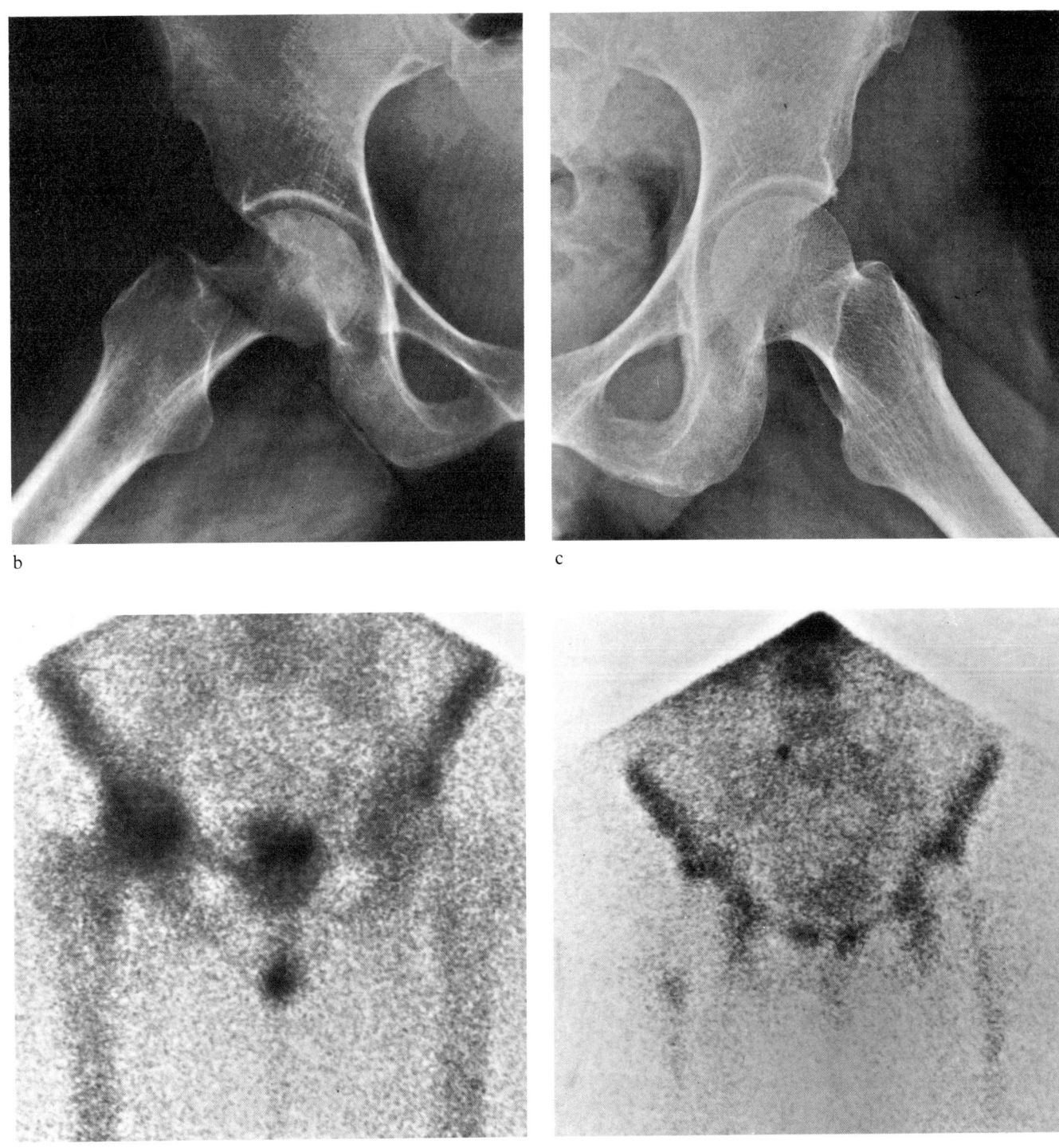

b

c

d

e

192 MRI atlas of the musculoskeletal system

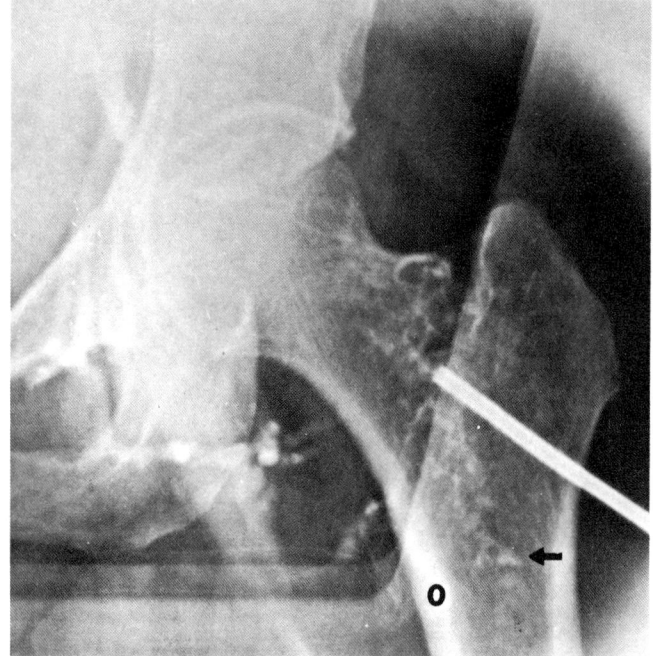

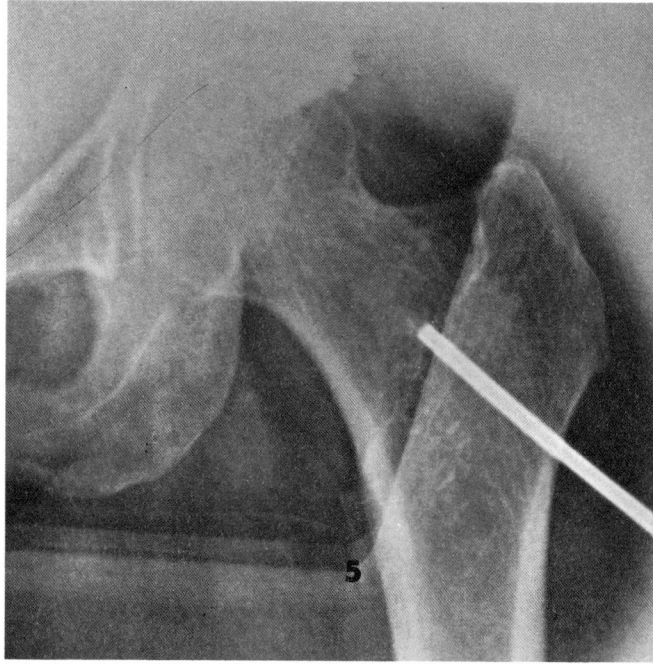

Figure 7.12 continued

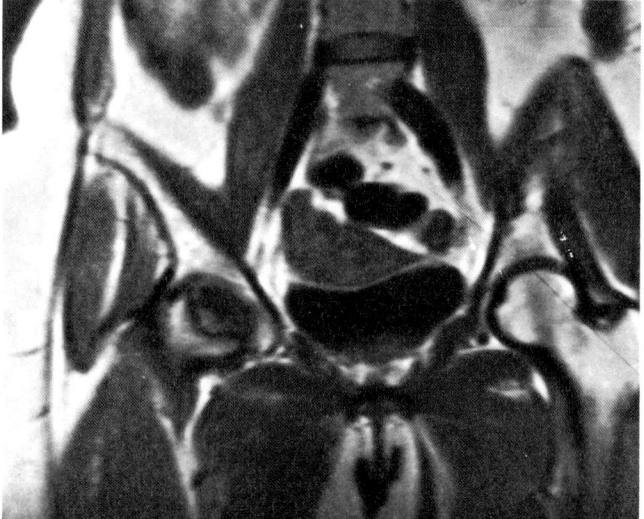

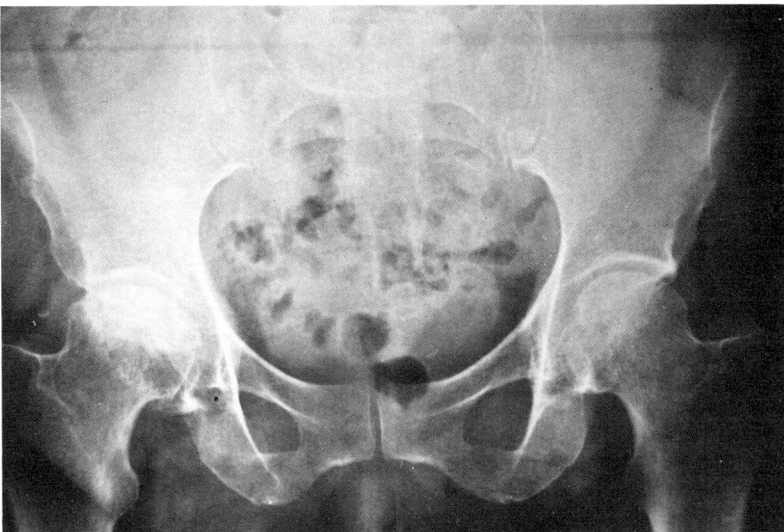
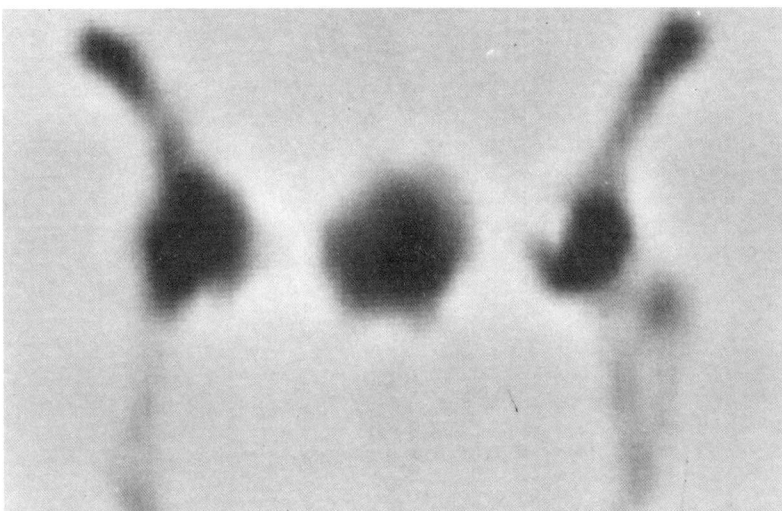
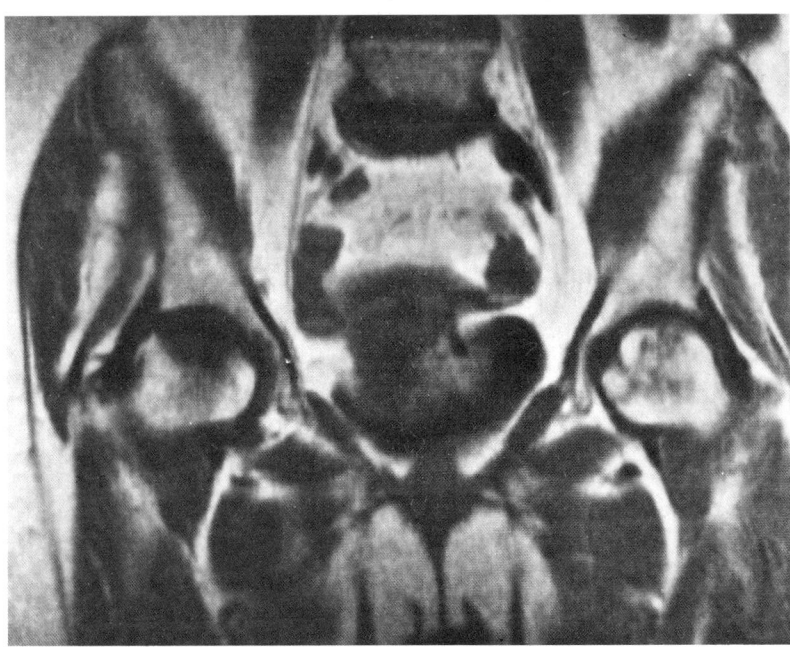

Figure 7.13

A 70-year-old woman with giant cell arteritis treated with steroids and right hip pain. (**a**) Radiograph shows Stage III–IV osteonecrosis on the right and a normal left hip. (**b**) SPECT scan is consistent with bilateral osteonecrosis with intense uptake in the right femoral head and acetabulum and a photopenic defect on the left surrounded by increased isotope accumulation. (**c**) T_1-weighted coronal MR image (SE 500/28) demonstrates a subchondral signal deficit on the right and an inhomogeneous band of decreased signal surrounding a subchondral region of high signal on the left.

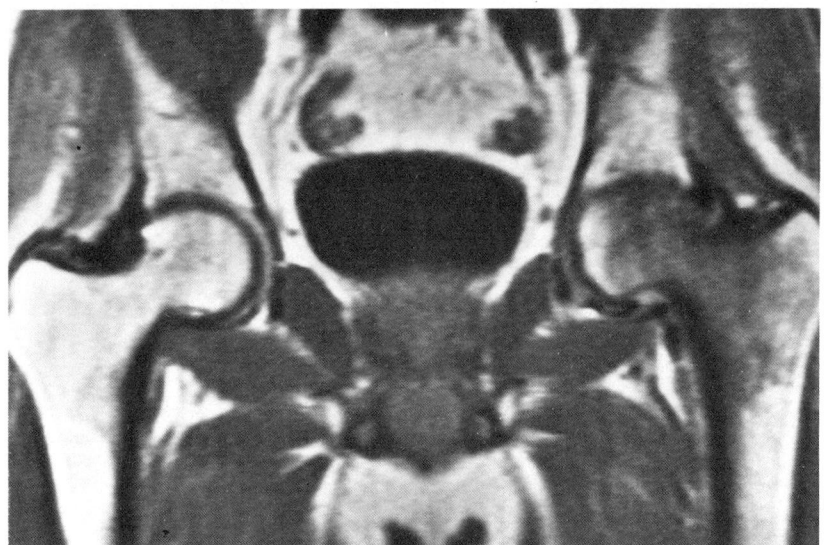

a

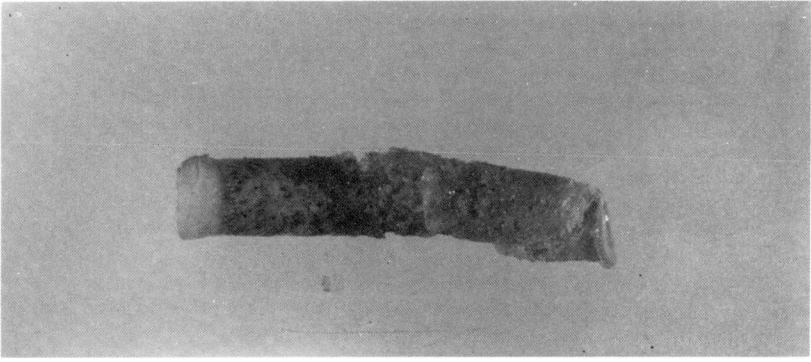

b

Figure 7.14

A 26-year-old man complained of pain in the left hip. (**a**) MR image (SE 500/28) disclosed diffuse decreased signal. (**b**) Core biopsy yielded a chalky fragmented specimen consistent with ischemic necrosis on histologic examination.

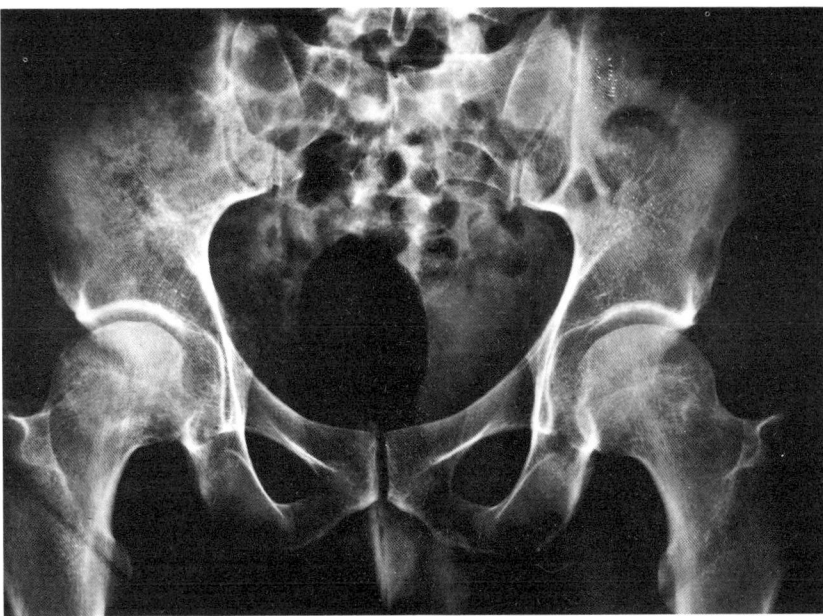

a

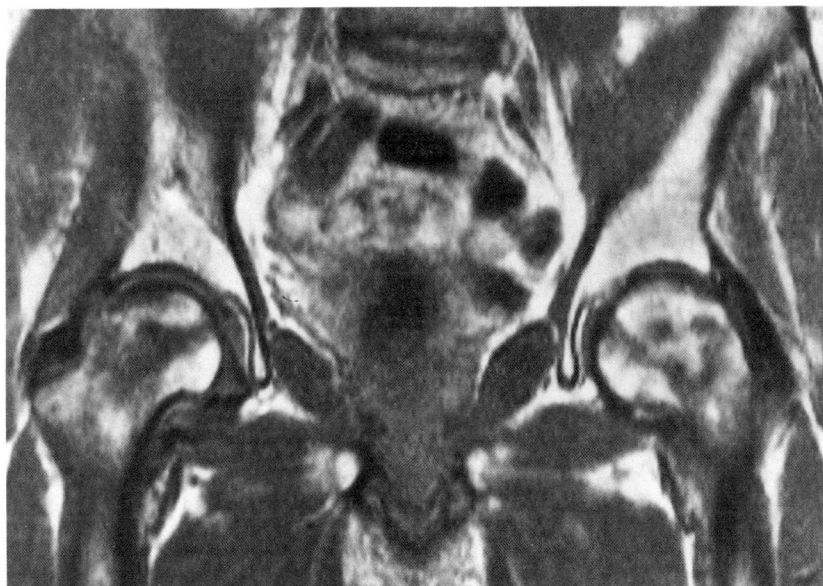

b

Figure 7.15

A 41-year-old man with a painful right hip and mild discomfort of the left hip, treated with corticosteroids following removal of a pituitary adenoma. (**a**) AP radiograph of the pelvis reveals Stage III ischemic necrosis of the right hip and Stage II of the left. (**b**) T_1-weighted MR image (SE 500/28) before total right hip arthroplasty shows alternating regions of normal and diminished signal intensity in both femoral heads. (**c**) Coronal whole-organ gross slice of resected right femoral head and neck. (**d**) T_1-weighted coronal MR image (SE 500/28) of section in (**c**). (**e**) Corresponding histologic section following further fixation, whole-organ slicing with microtome, and H&E staining. (**f**) *Inset 1*: Interface between the region in which marrow has been replaced by calcium soaps—saponified fat (A)—which gives a diminished signal on MR, and the region of mummified anuclear dead fat cells (B), which maintained a normal signal (H&E × 40). (**g**) High-power detail view of Inset 1, showing interface between dark granular saponified fat (A) and mummified fat (B). (**h**) *Inset 2*: Delicate horizontal trabeculae (arrows), at the growth plate also yielded a low signal on the MR image (H&E × 40). (**i**) *Inset 3*: A reparative transition zone composed of fibrovascular tissue (A) caused a diminished signal on the MR image, and lies between infarcted marrow (B) and normal marrow (C).

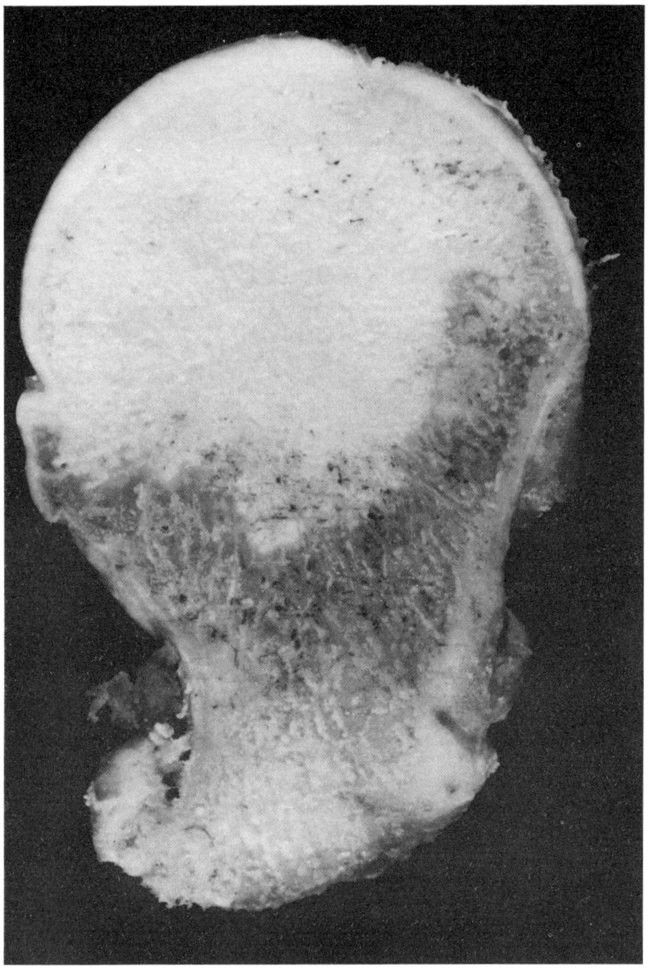

196 *MRI atlas of the musculoskeletal system*

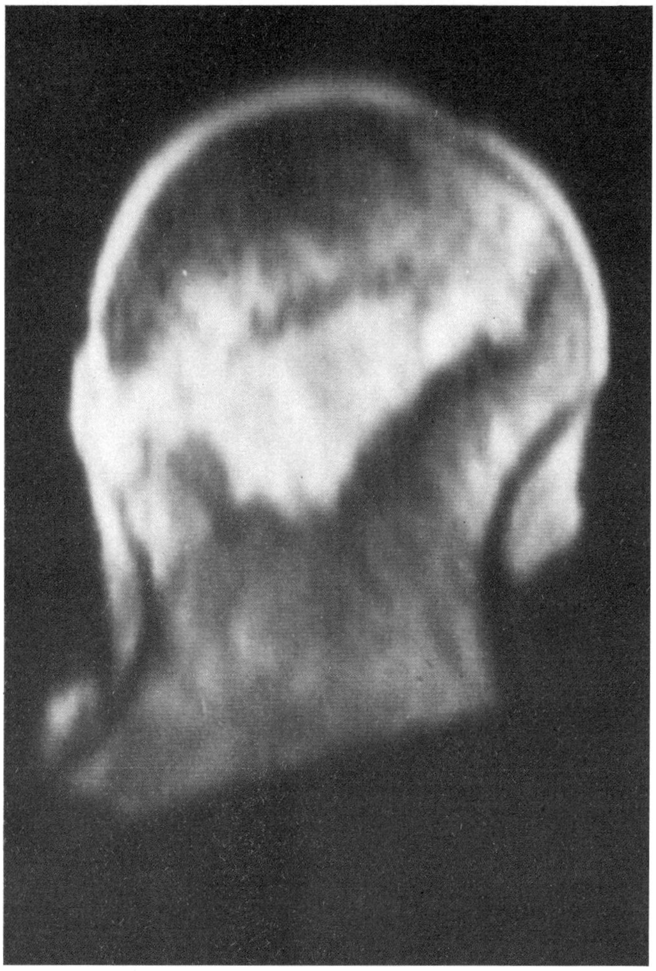

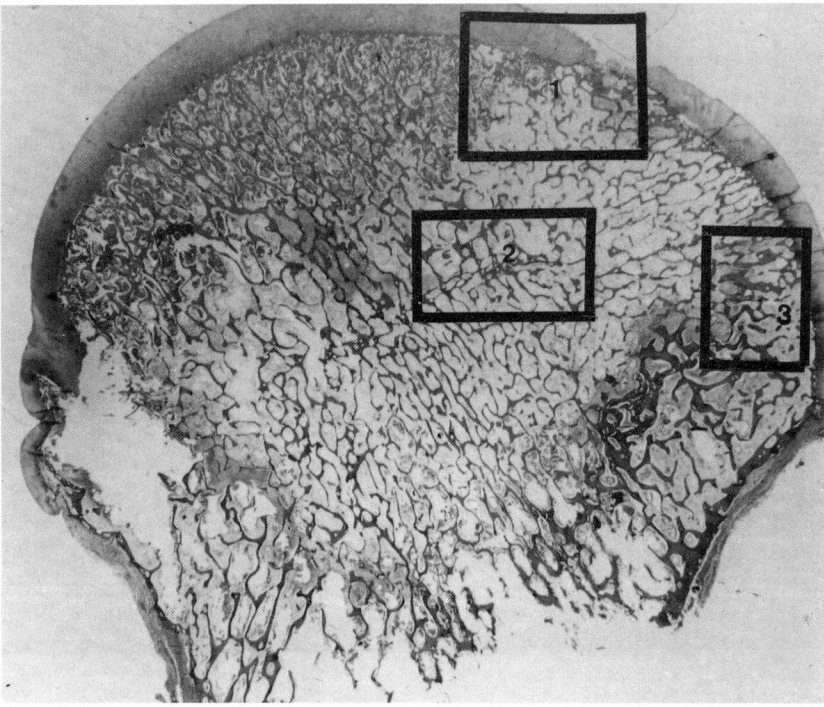

Figure 7.15 *continued*

c

d

e

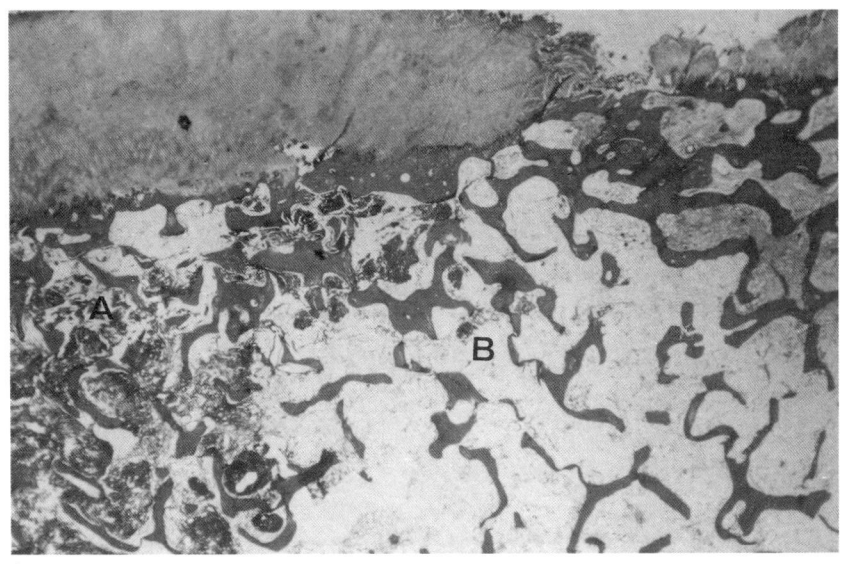
f

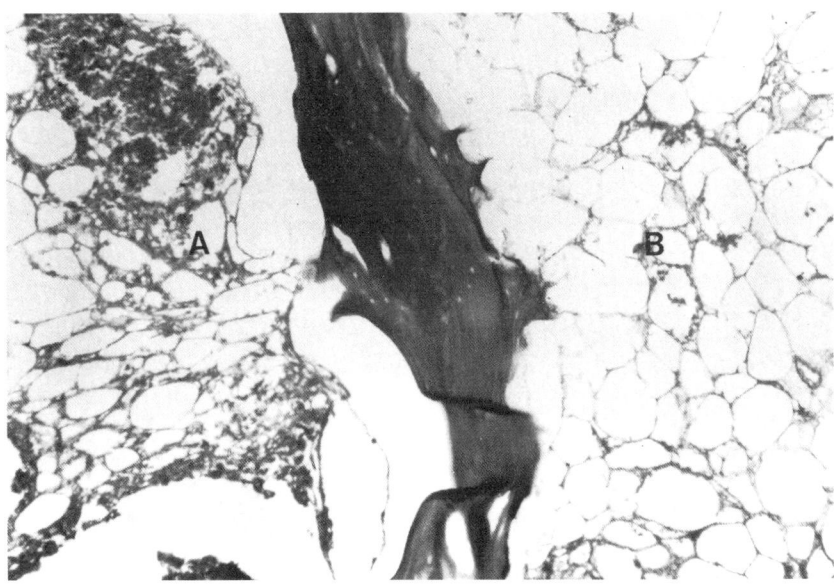
g

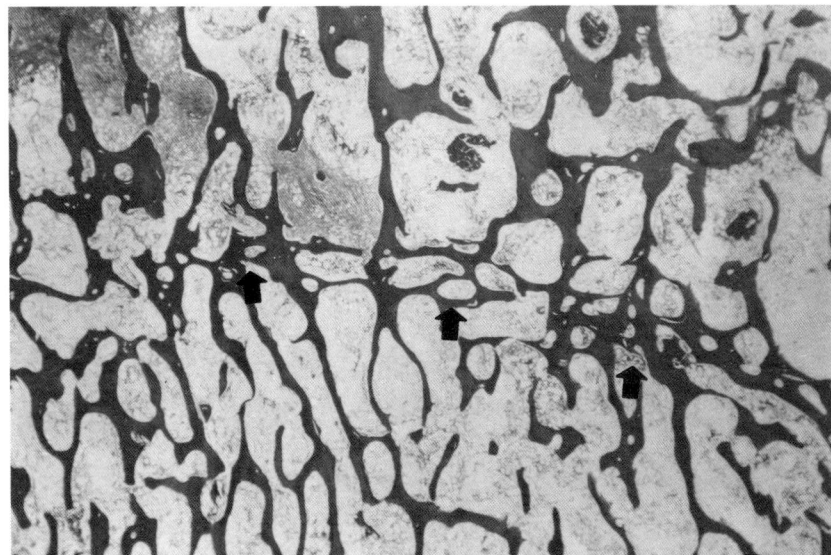
h

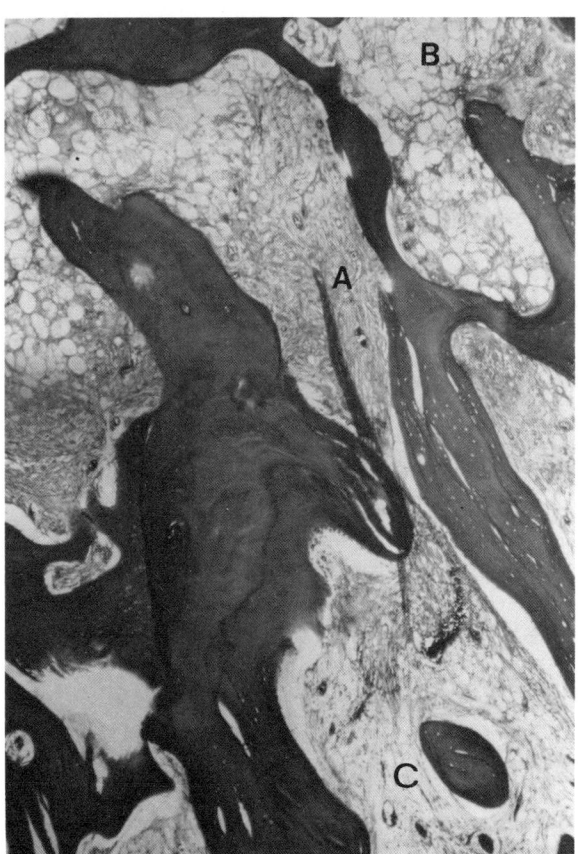

Figure 7.15 continued

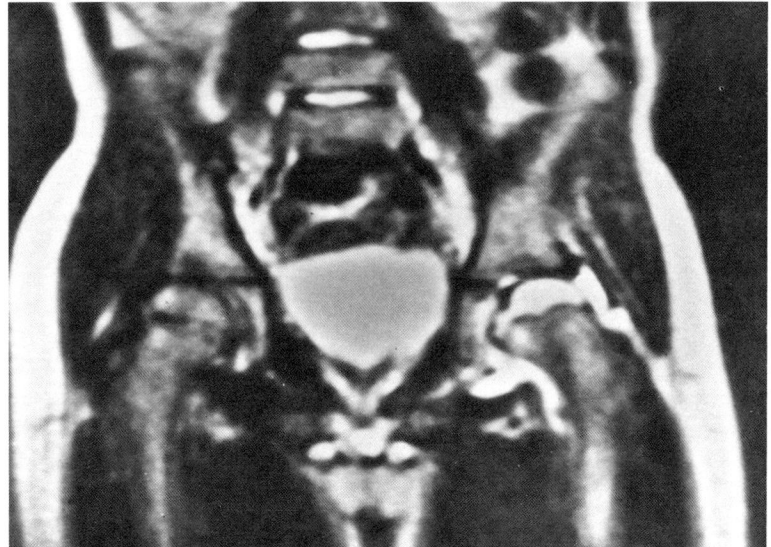

Figure 7.16

A 10-year-old boy with left hip pain. T_2-weighted coronal MR image (SE 2000/85) shows very high signal intensity of the edematous femoral head epiphysis and hip effusion, consistent with Legg-Perthes disease.

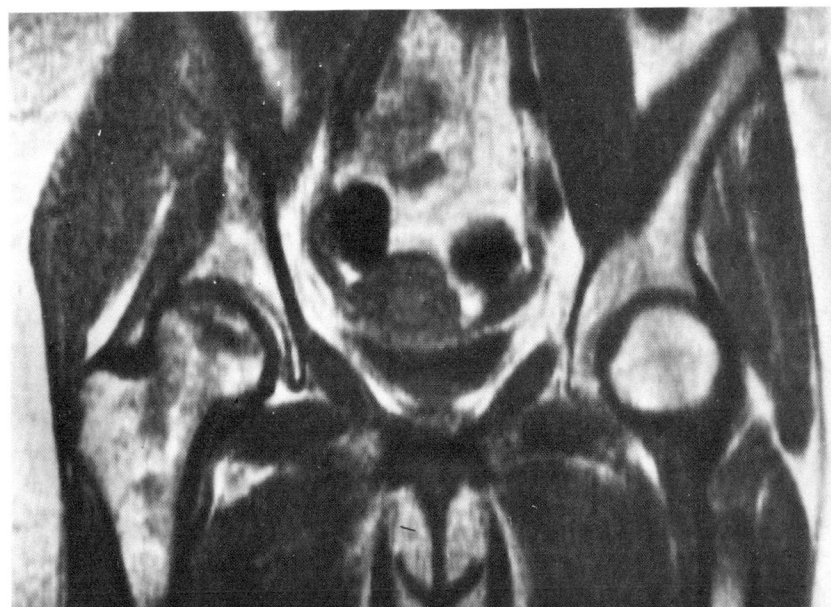

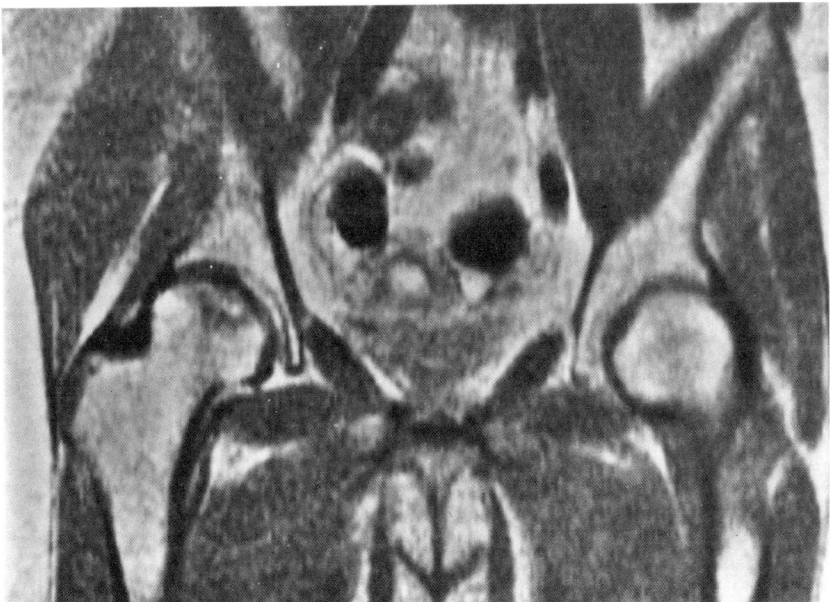

Figure 7.17

(a) T_1-weighted MR image (SE 500/28) shows a ring-shaped subchondral signal deficit in the right femoral head. In addition, a region of decreased signal is seen in the femoral neck and intertrochanteric area. (b) T_2-weighted MR image (SE 2000/85) shows only the ring-like region of diminished signal.

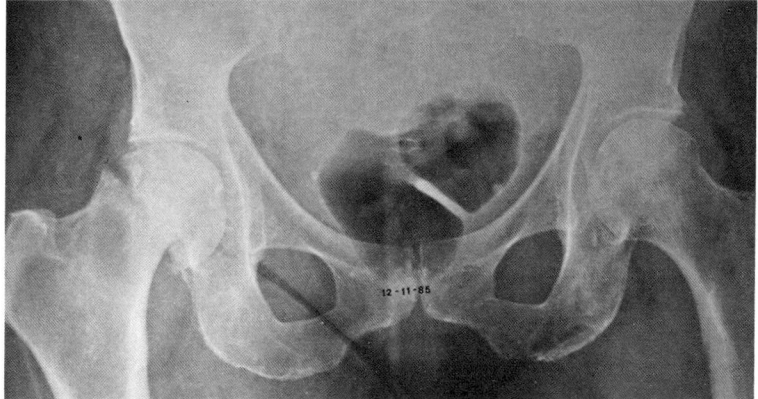

Figure 7.18

A 66-year-old woman receiving corticosteroids, presented with sudden pain in right hip. (**a**) Radiograph shows a right subcapital fracture. (**b**) T_1-weighted MR image (SE 500/28) reveals a diminished signal in both femoral heads, signifying bilateral ischemic necrosis which led to a pathologic fracture.

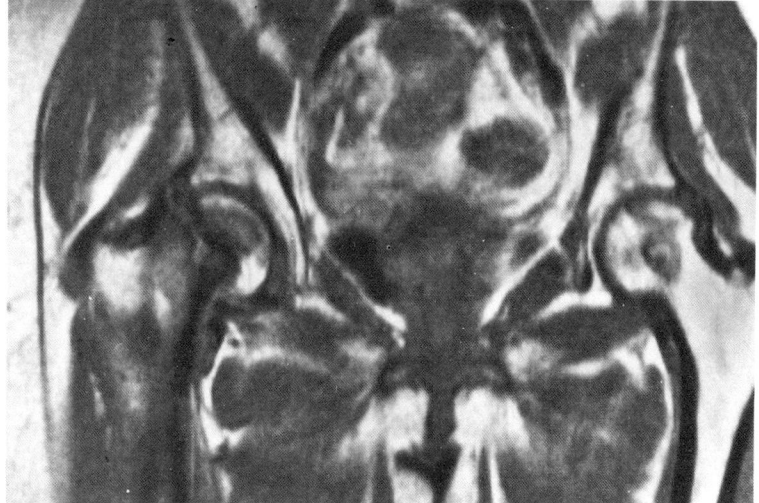

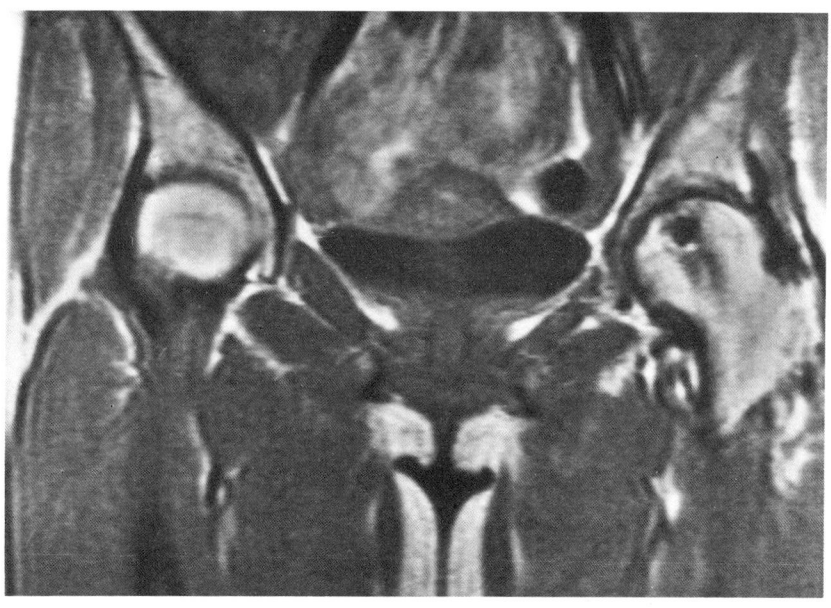

Figure 7.19

A 43-year-old woman had a Phemister procedure to relieve pain from ischemic necrosis involving the left femoral head. T_1-weighted MR image (SE 500/28) reveals the cylindrical coring defect and signal deficit at the site of the bone graft plug.

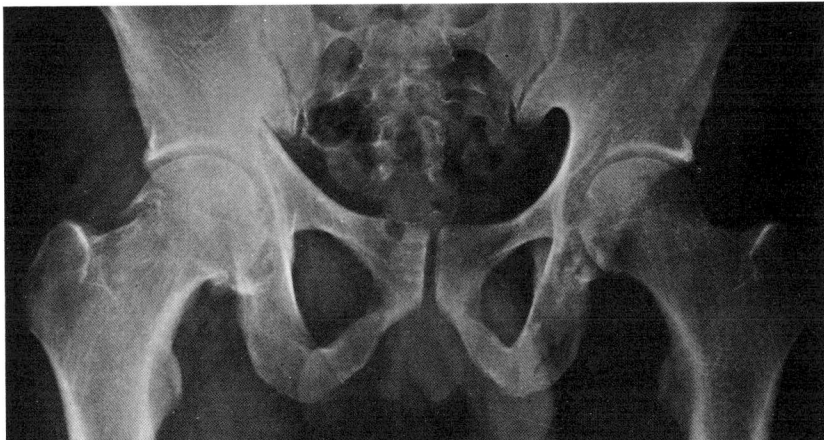

Figure 7.20

Osteoarthritis of the right hip. A 36-year-old man with right hip pain of 7 years' duration. (a) Radiograph of the pelvis shows osteoarthritis on the right, manifested by osteophytes along the inferomedial and superolateral margins of the femoral head, and butressing along the medial margin of the femoral neck. (b) Close-up of the right hip. (c) T_1-weighted MR image (SE 500/28): high-signal intensity is seen in marrow of osteophytes.

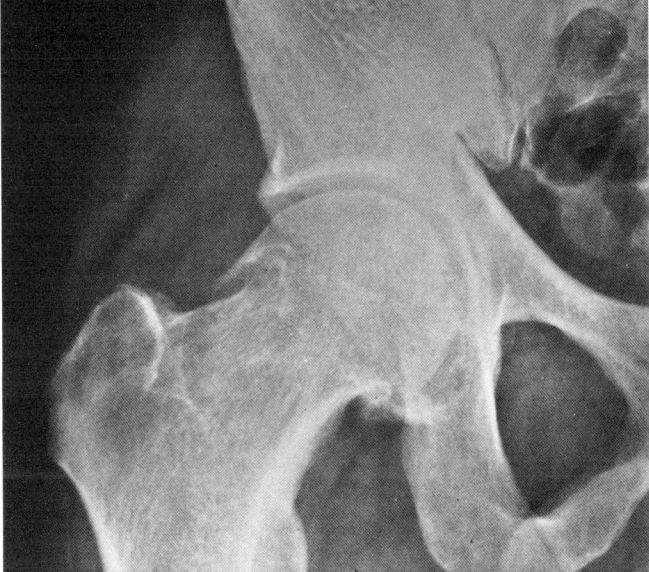

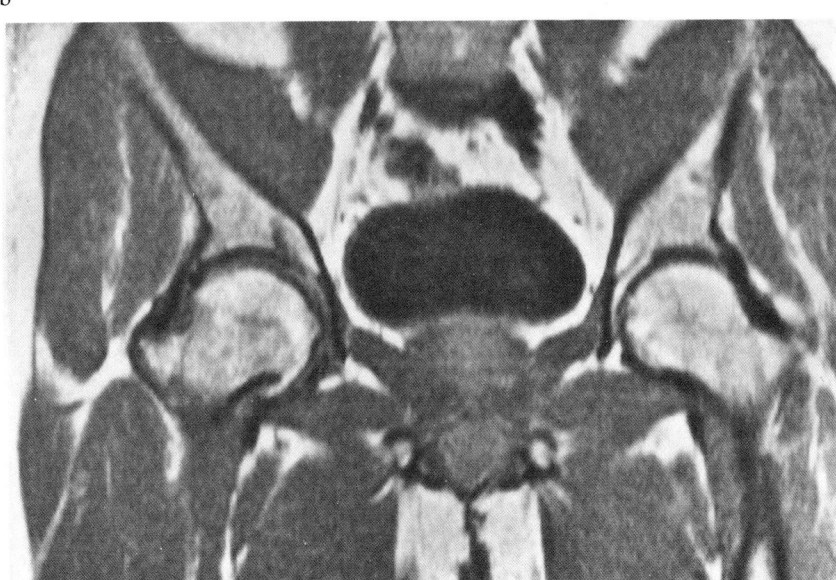

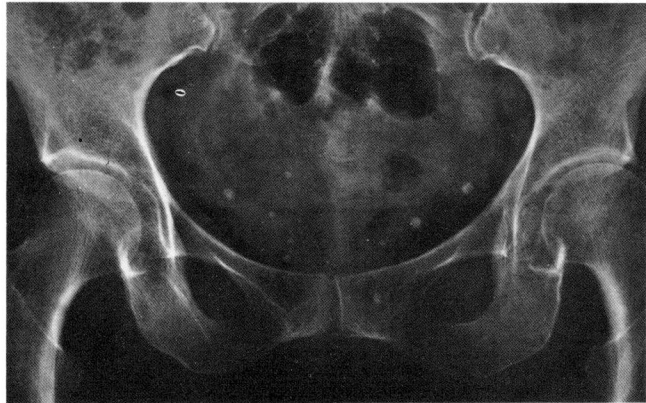

Figure 7.21

Osteoarthritis of the left hip. A 65-year-old woman complained of pain in the left hip. MRI was performed to rule out osteonecrosis. (**a**) Radiograph of the pelvis shows lateral migration of the left femoral head. (**b**) Diphosphonate scan shows intense uptake in the left femoral head. (**c**) T_1-weighted coronal and (**d**) sagittal MR images (SE 500/28) revealed focal signal deficits in the acetabulum and femoral head. (**e**) Radiograph of the left hip 1 year later. There is further lateral migration of the femoral head, superior joint space narrowing, and a large osteophyte medially (arrow). Histologic diagnosis at the time of hip replacement was osteoarthritis.

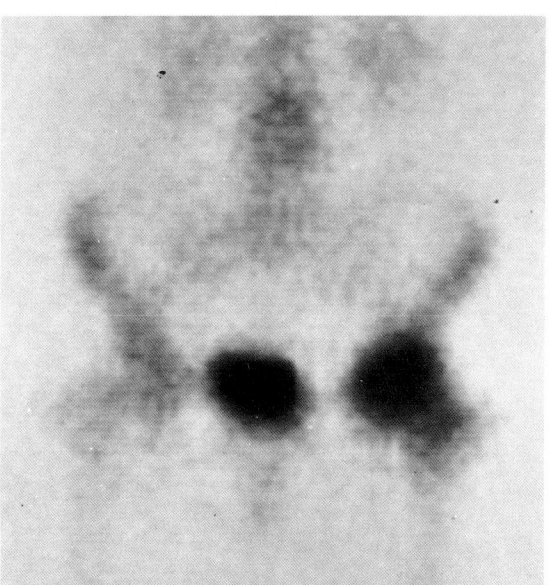
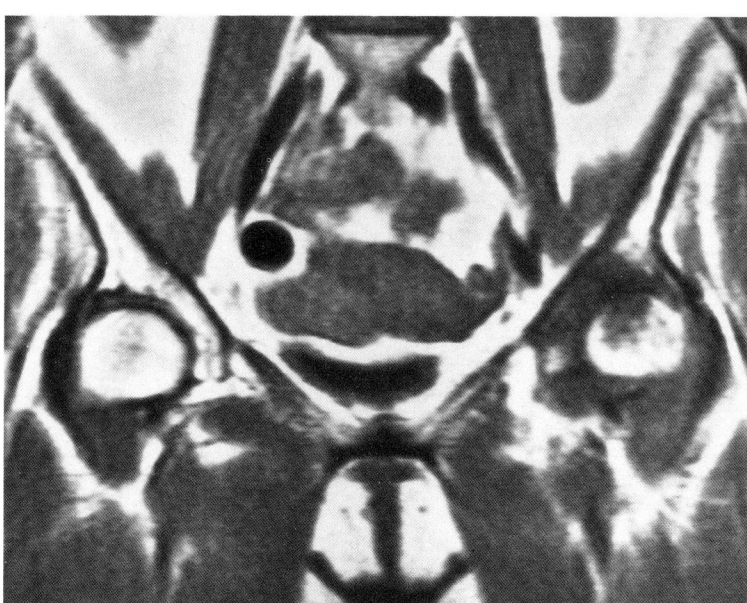
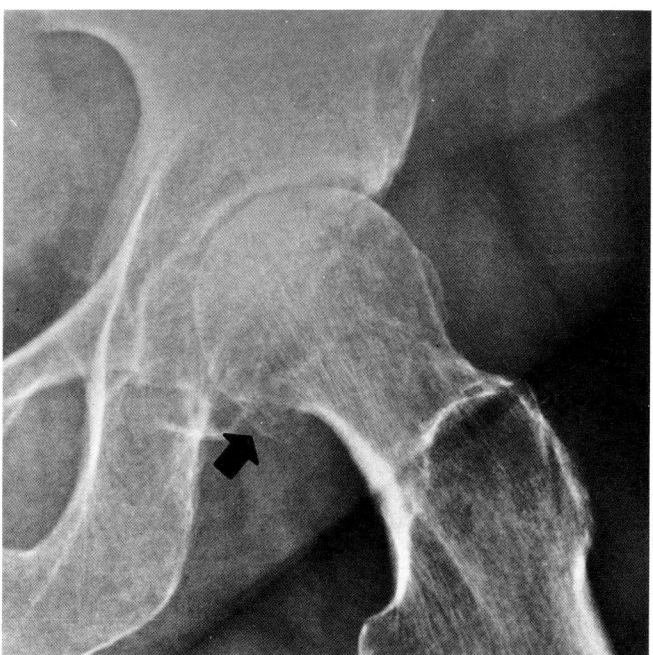

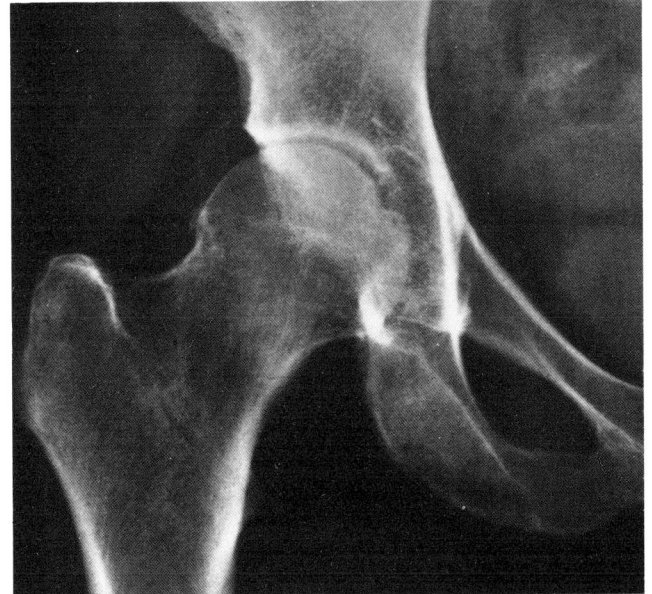

a

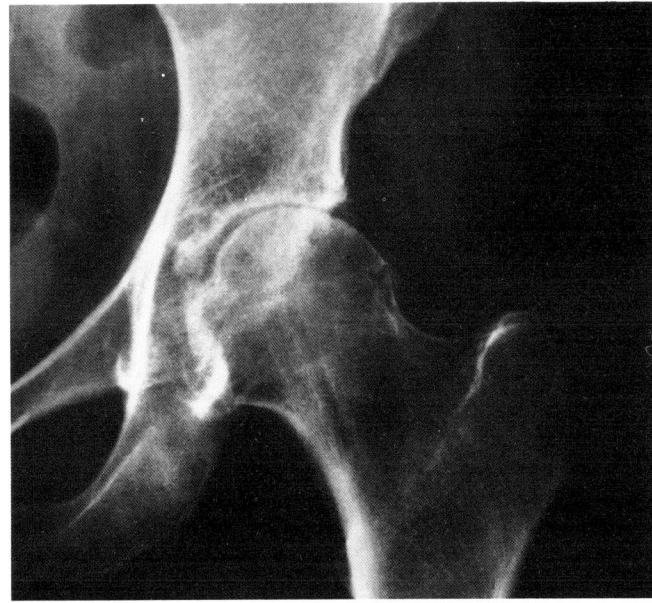

b

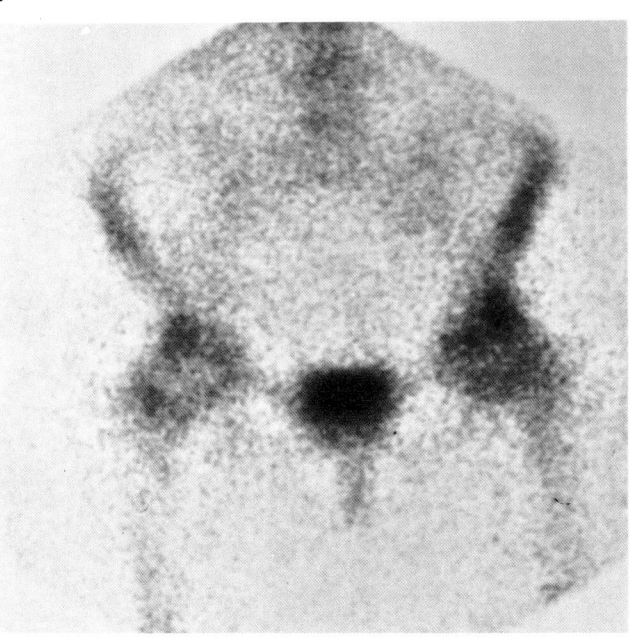

c

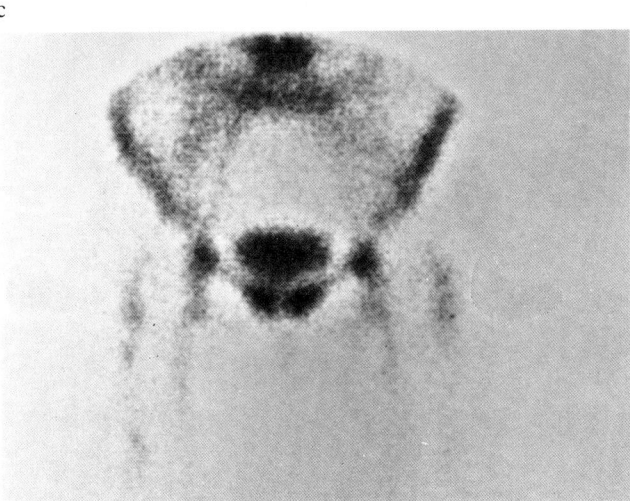

d

Figure 7.22

A 50-year-old woman complained of severe bilateral hip pain. Radiographs of (**a**) right and (**b**) left hips revealed bilateral superior joint space narrowing and lateral femoral head subluxations. (**c,d**) Diphosphonate scan showed bilateral increased isotope accumulations at the superolateral aspects of both hips. Sulfur colloid scans showed symmetrical isotope uptake in the femoral necks. Free technetium is seen in the femoral arteries. (**e**) T_1-weighted coronal MR image (SE 500/28) discloses focal subchondral signal deficits in both femoral heads which might have been confused with ischemic necrosis if there had not been radiographic correlation. (**f**) Sagittal MR image (SE 500/28) of the right hip shows signal deficits of both the femoral head and the acetabulum. At surgery there was no evidence of ischemic necrosis and histologic sections confirmed the diagnosis of bilateral osteoarthritis.

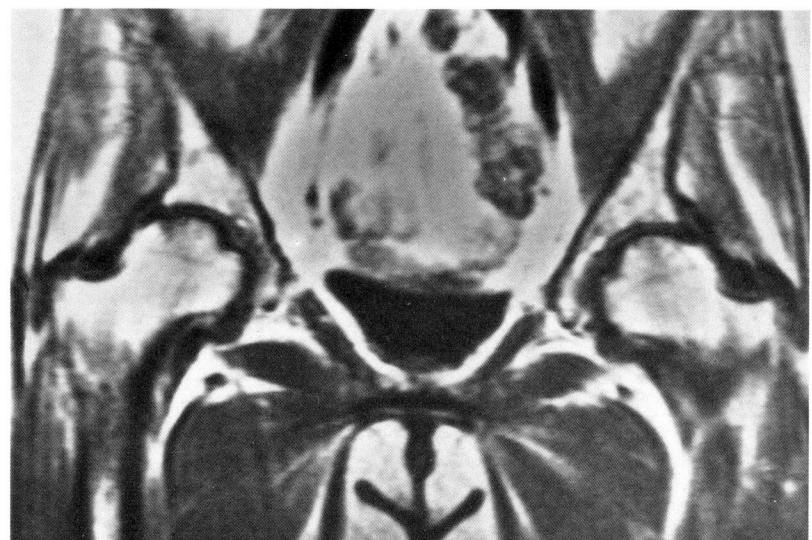

Figure 7.22 continued

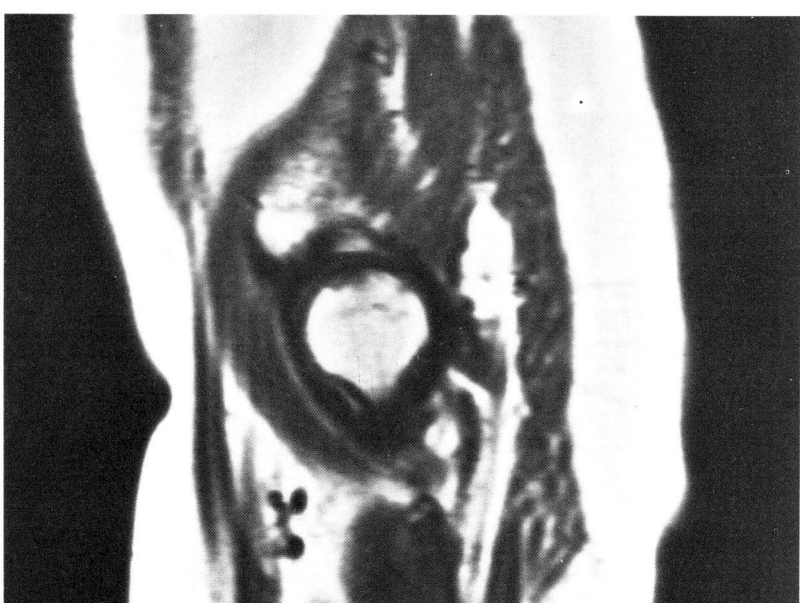

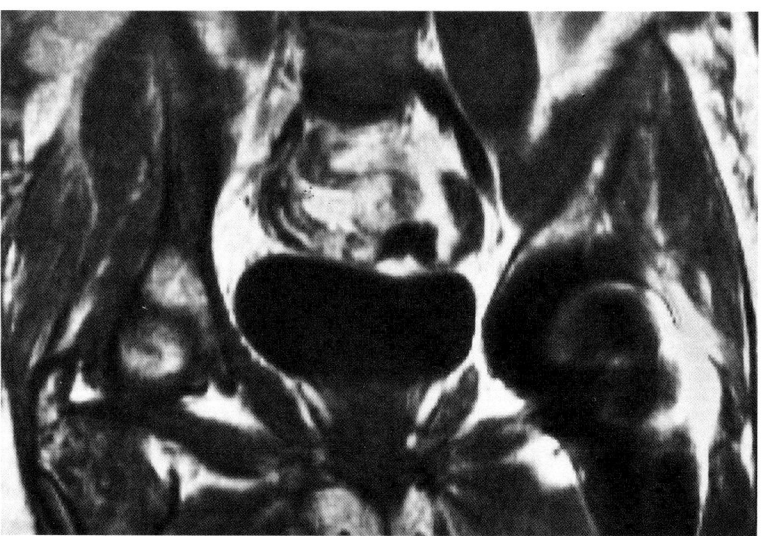

Figure 7.23

Sickle-cell disease. A 55-year-old woman with recurrent painful crises. Foci of diminished signal intensity on T_1-weighted MR image (SE 500/28) represent medullary infarcts and hematopoietic proliferation in the right iliac wing and intertrochanteric region of the right femur. Left hemiarthroplasty was present.

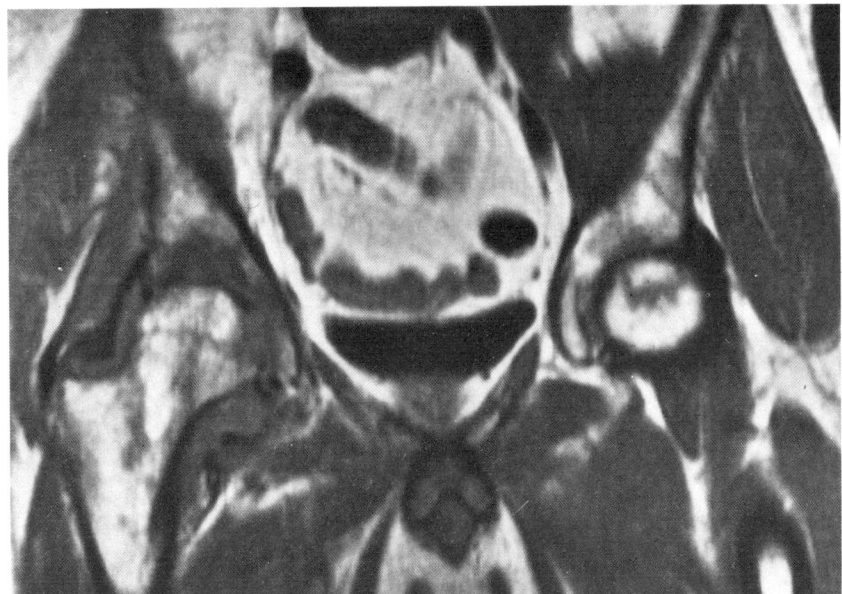

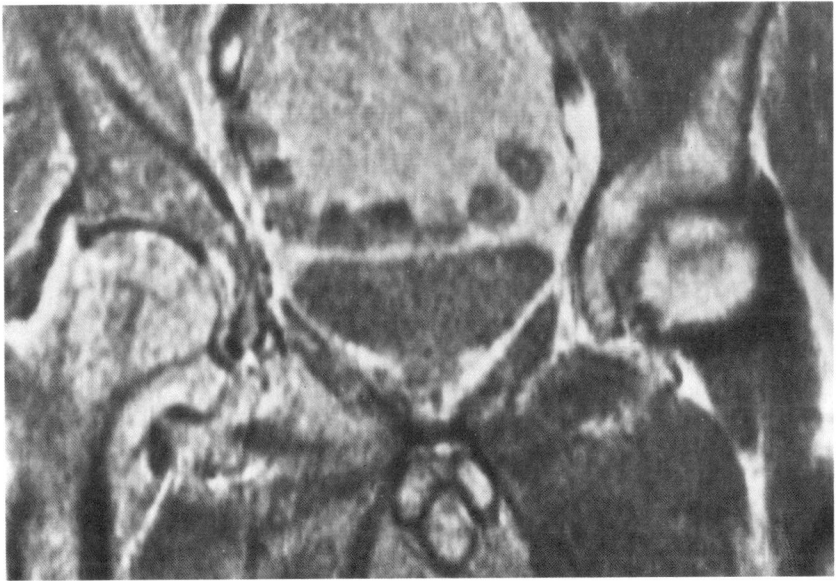

Figure 7.24

Septic arthritis. A 45-year-old man with aplastic anemia and a painful right hip. (**a**) T_1-weighted MR image (SE 500/28) depicts edematous soft tissue of the right hip. The left hip shows ischemic necrosis secondary to corticosteroid medication. (**b**) Signal from edematous soft tissue is enhanced on T_2-weighted image (SE 2000/85). Ring of decreased signal, evidence of ischemic necrosis in the left hip, is unchanged.

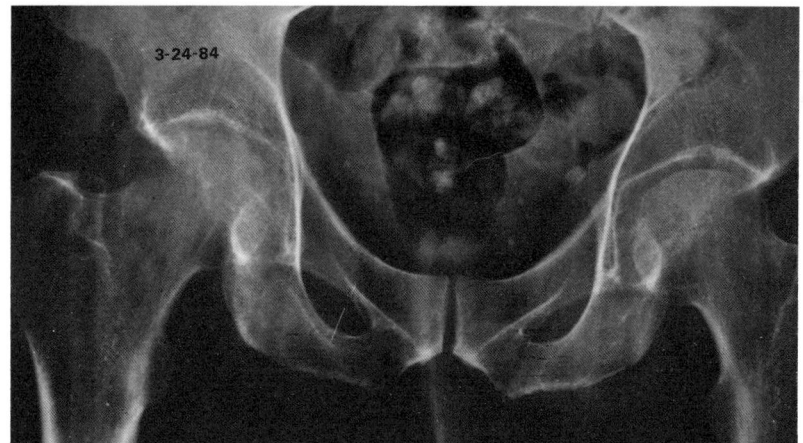

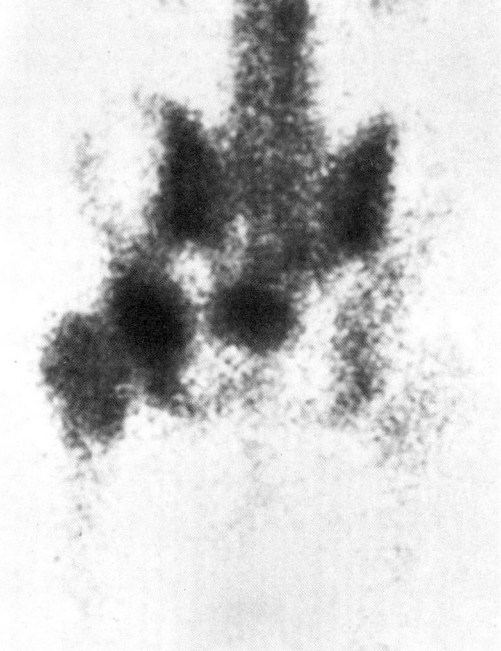

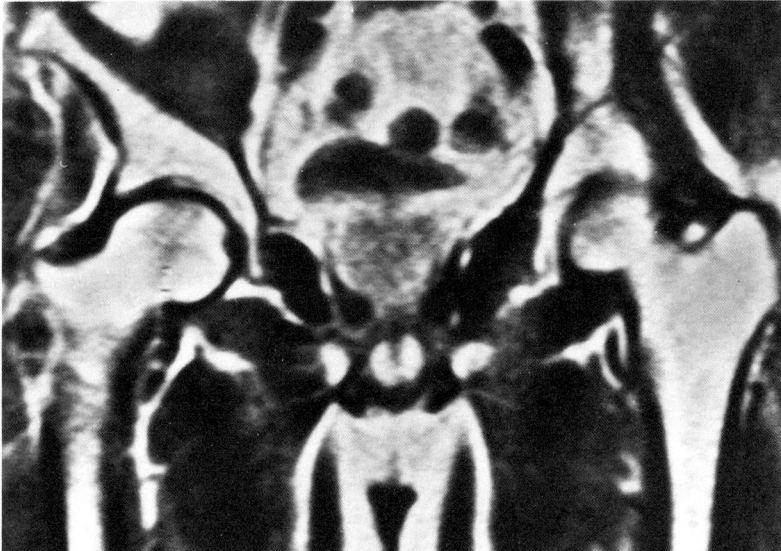

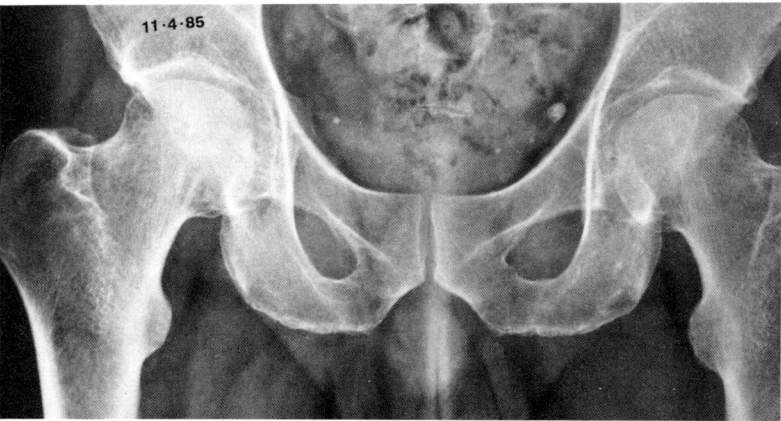

Figure 7.25

Transient regional osteoporosis. A 47-year-old man complained of right hip pain. Cultures and cytology of joint aspirates were negative. (**a**) Radiograph at time of MR scan shows osteoporosis of the right femoral head. (**b**) T_1-weighted MR image (SE 500/28) demonstrates normal signal intensity. (**c**) Diphosphonate bone scan shows markedly increased uptake in the proximal femur.
(**d**) Radiograph 7 months later, when the patient was asymptomatic, shows normal bone density.

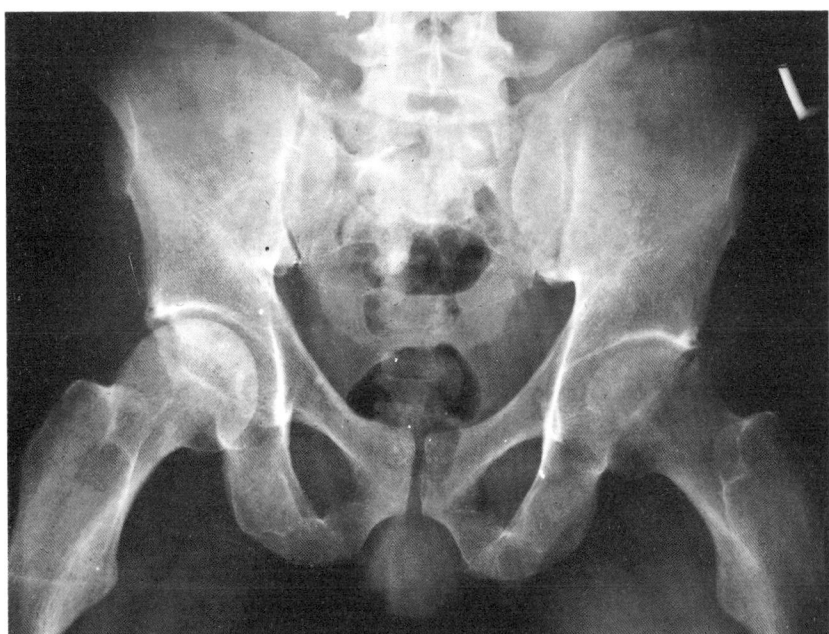

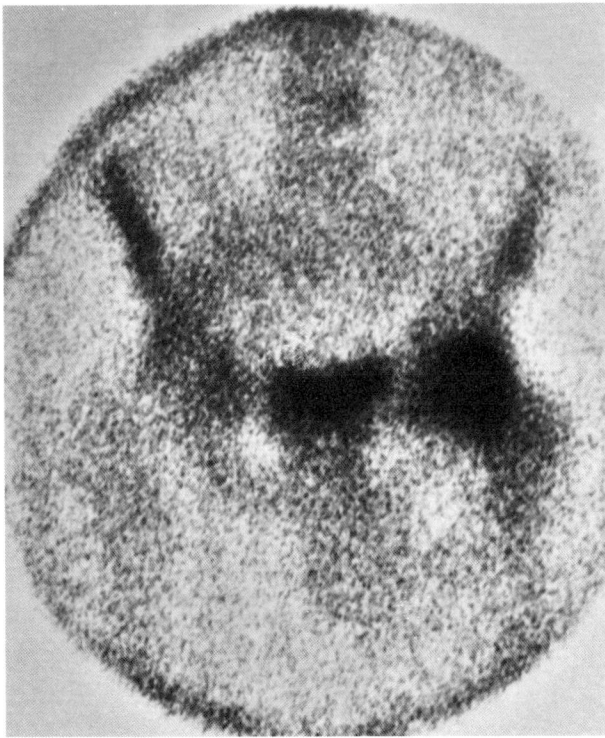

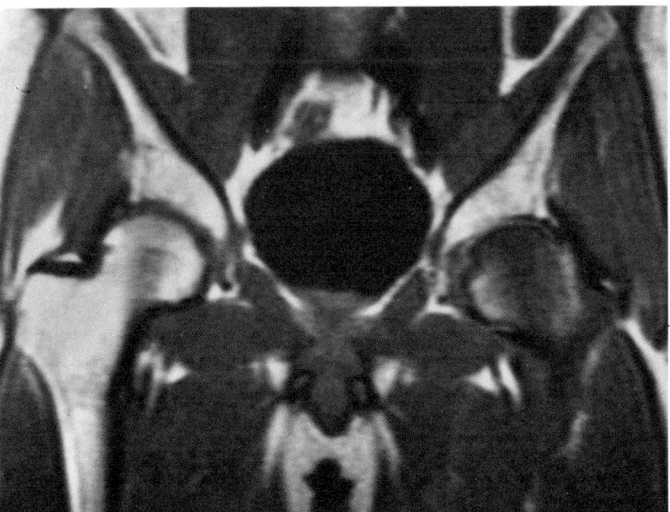

Figure 7.26

Transient regional osteoporosis. A 46-year-old man complained of left hip pain. (**a**) Radiographs showed marked osteoporosis of the left hip. (**b**) Diphosphonate scan revealed intense radionuclide accumulation in the left femoral head. (**c**) T_1-weighted MR image (SE 500/28) demonstrated diffuse decreased signal intensity in the left proximal femur. Core biopsy revealed no evidence of ischemic necrosis or other abnormal processes. (**d**) Radiograph 16 months post-coring reveals cylindrical core defect and normal bone density. (**e**) T_1-weighted MR image (SE 500/28) at the time of (**d**) demonstrated normal intensity bone marrow surrounding the core defect.

208 MRI atlas of the musculoskeletal system

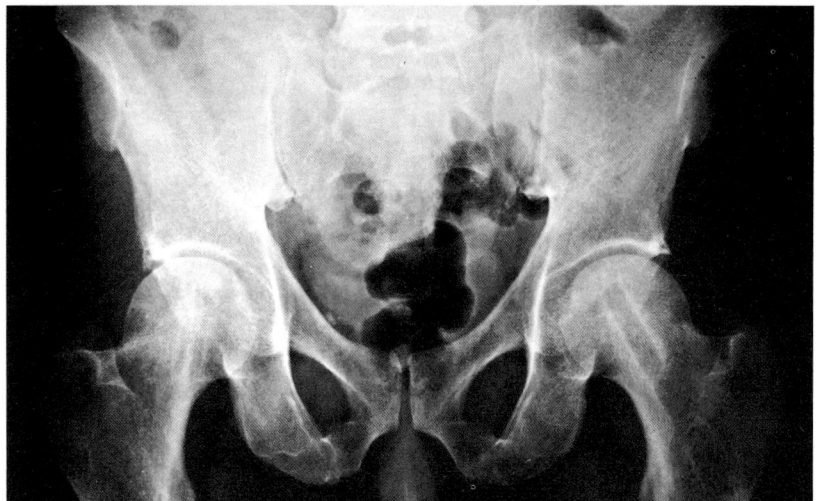

d

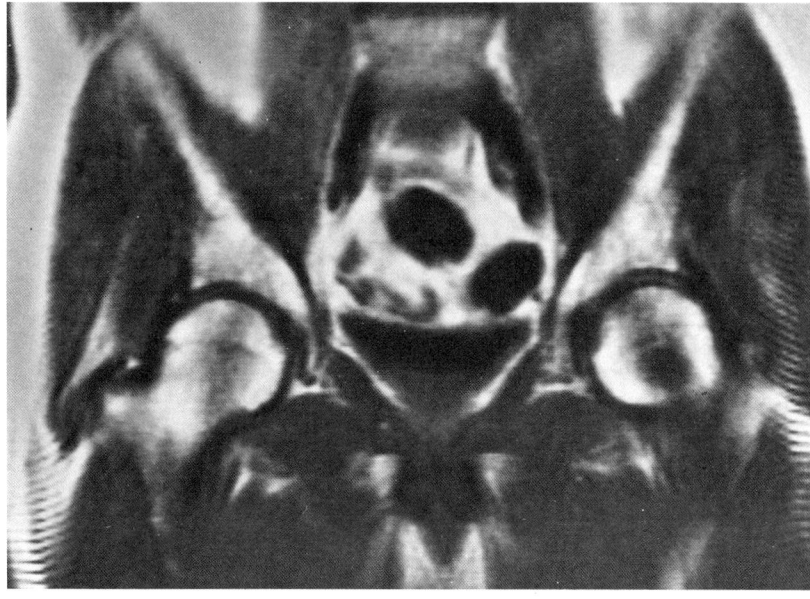

e

Figure 7.26 continued

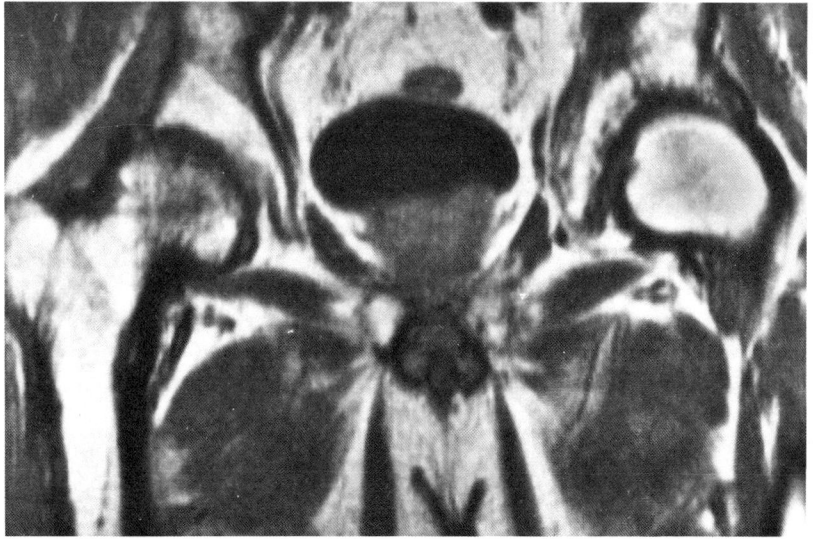

Figure 7.27

Paget's disease. A 70-year-old man with pain in the right hip. Thickened trabeculae and cortex produced no signal on T_1-weighted MR image (SE 500/28).

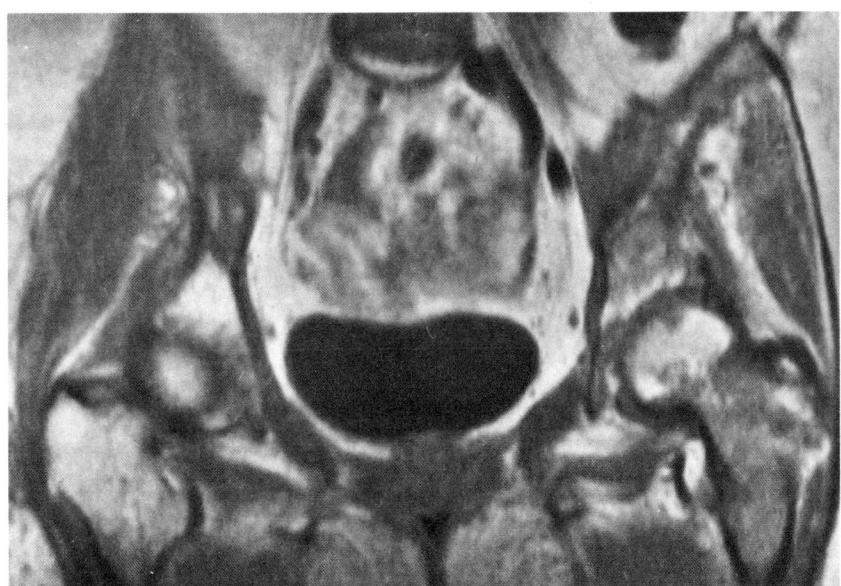

Figure 7.28

Metastatic disease. A 75-year-old woman with metastatic breast carcinoma and left hip pain. T_1-weighted MR image (SE 500/28) reveals diminished signal in the acetabulum, iliac wing and femoral neck on the left and subtrochanteric region of the right femur, consistent with marrow replacement by metastases.

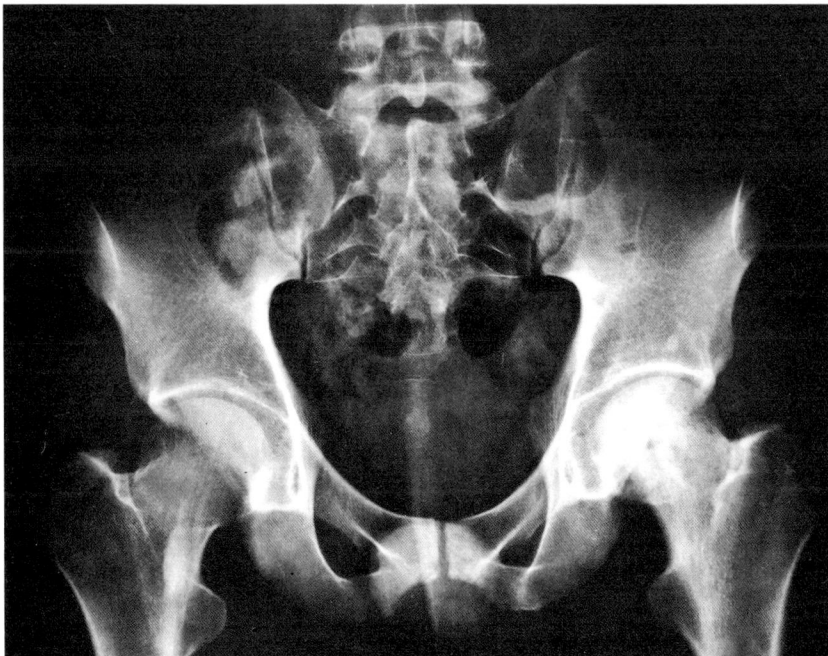

Figure 7.29

(a) Plain radiograph demonstrates increased density in the left femoral head. Ischemic necrosis was suspected. A large bone island is seen in the intertrochanteric region. T_1-weighted (b) coronal and (c) axial MR images (SE 500/28) show extensive signal deficit throughout the left proximal femur. (d) T_2-weighted axial MR image (SE 2000/85) shows inhomogeneous signal intensity in the left femoral head with surrounding high signal consistent with joint fluid. (e) Follow-up radiograph reveals regions of sclerosis and radiolucent mottling. Biopsy revealed a lymphoma.

a

210 *MRI atlas of the musculoskeletal system*

Figure 7.29 *continued*

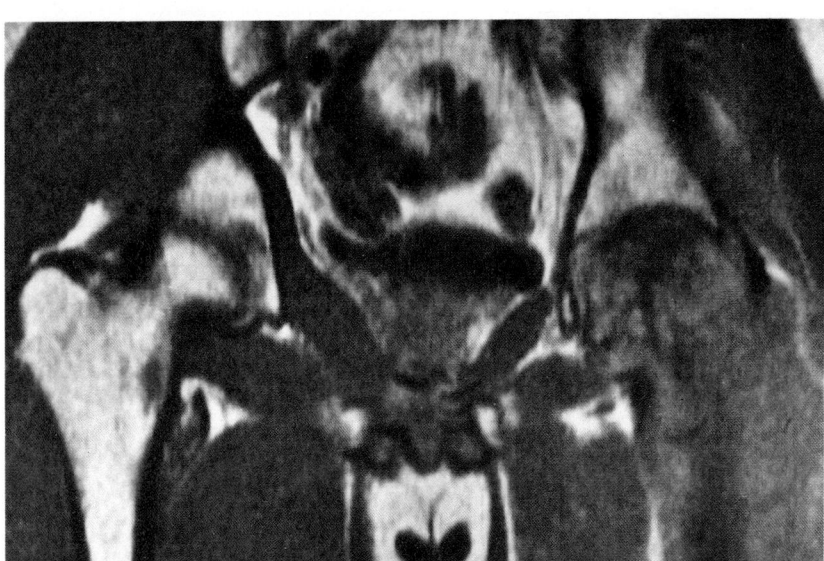
b

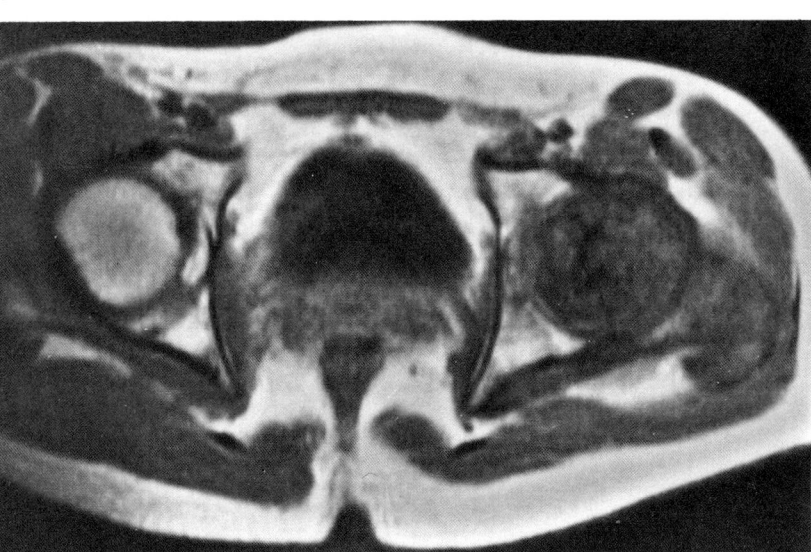
c

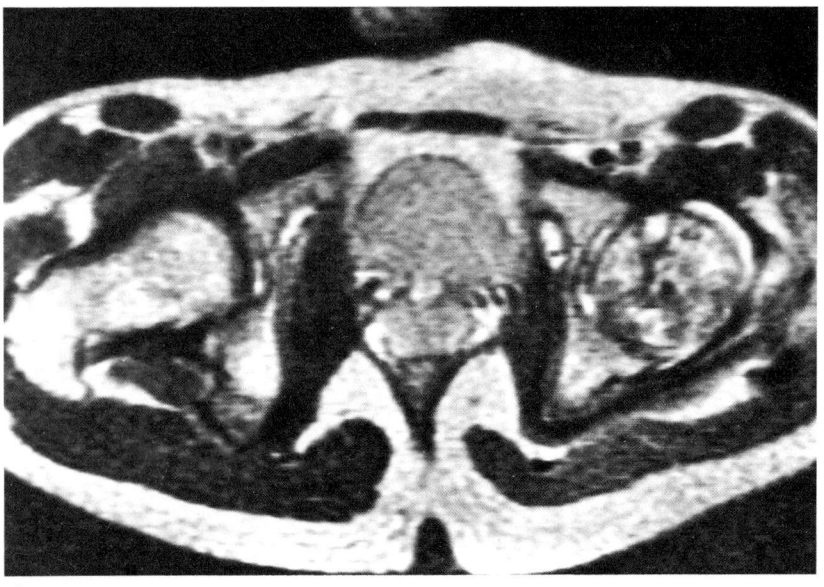
d

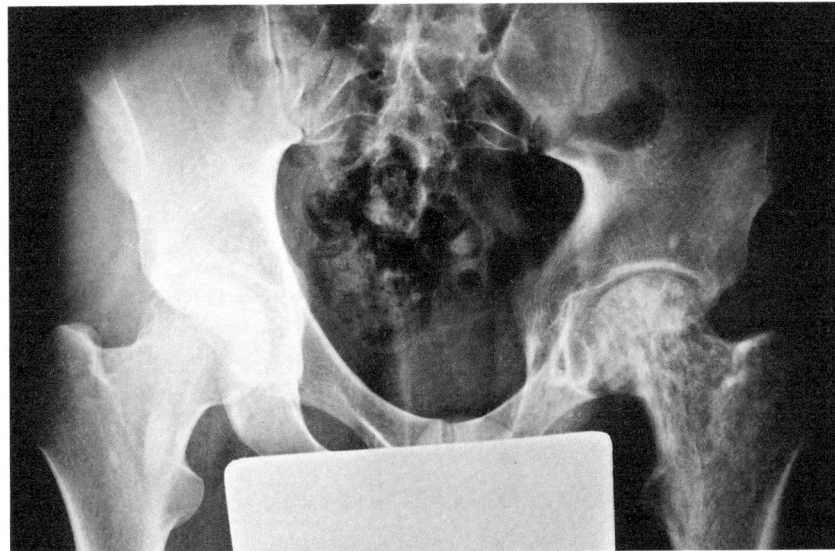

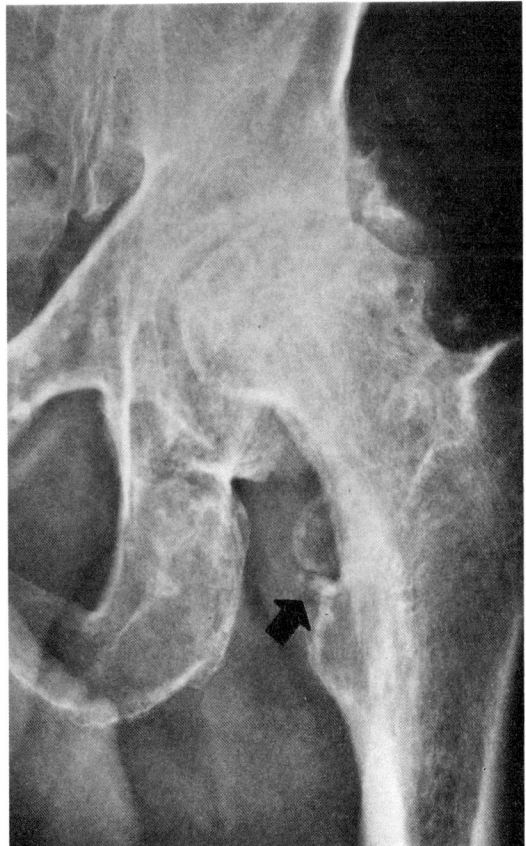

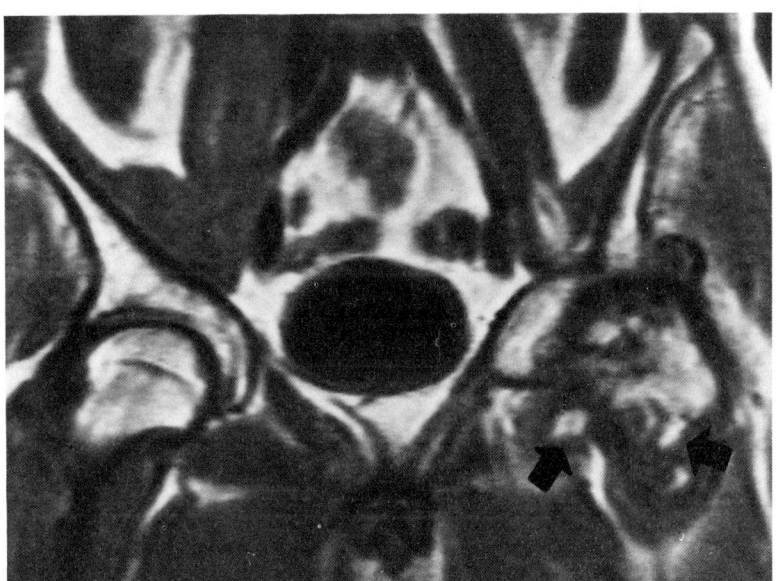

Figure 7.30

Secondary osteochondromatosis. A 36-year-old man with severe osteoarthritis of the left hip, confirmed at surgery. (a) Plain radiograph. (b) T_1-weighted MR image (SE 500/28) depicts osteochondromata (arrows) within joint.

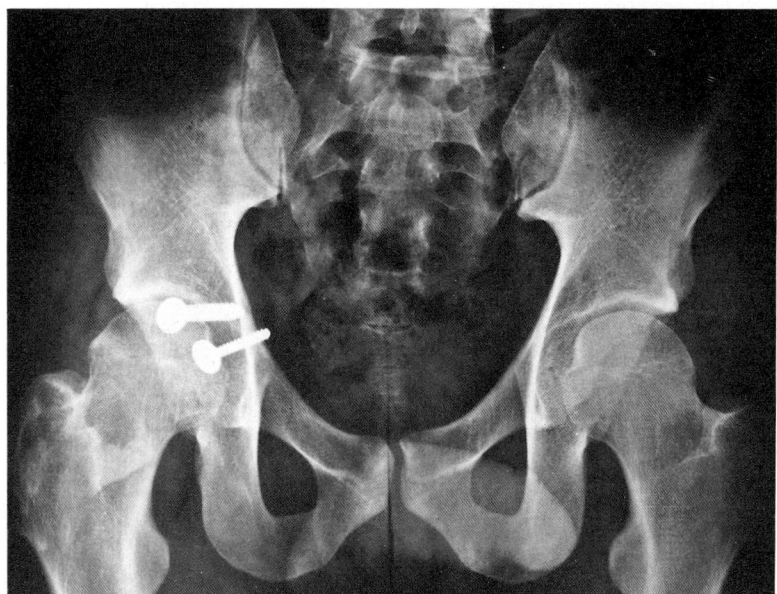

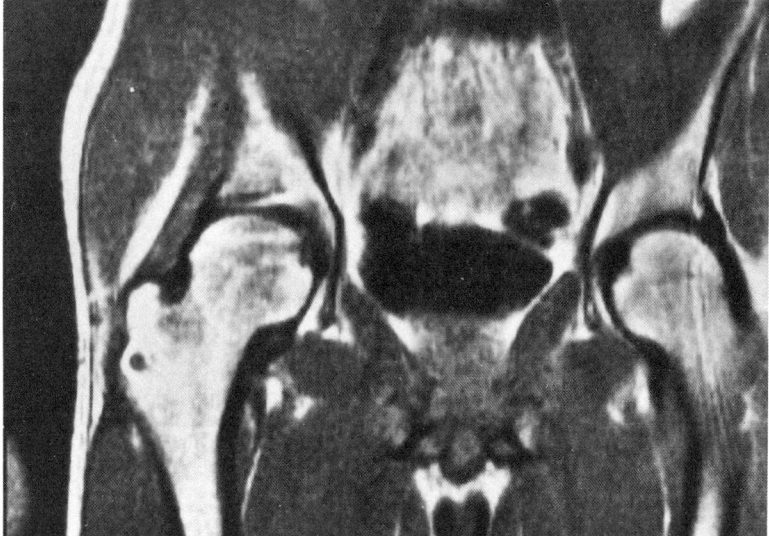

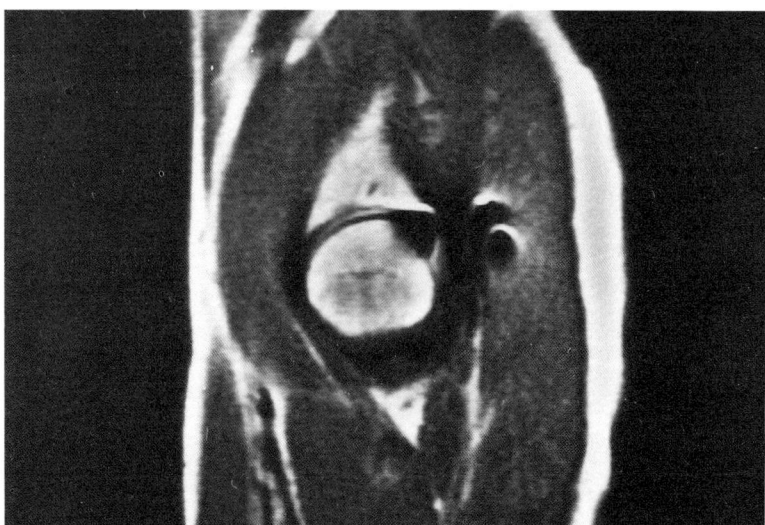

Figure 7.31

Metal artifact. A 40-year-old man had internal reduction of an acetabular fracture/femoral head dislocation. He later experienced hip pain. (**a**) Radiograph shows location of 2 metallic screws in the femoral head. (**b**) T_1-weighted coronal and (**c**) sagittal MR images (SE 500/28) depict normal intensity of the femoral head bone marrow, despite the presence of the adjacent metal artifacts.

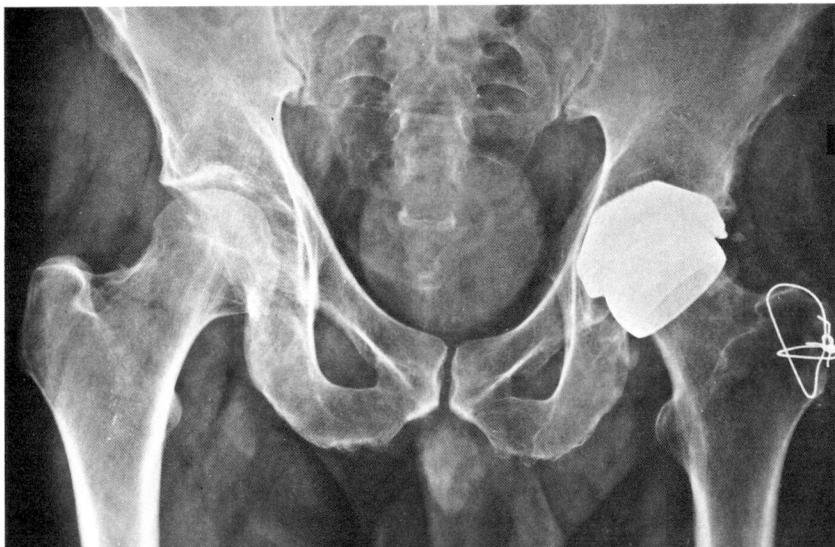

a

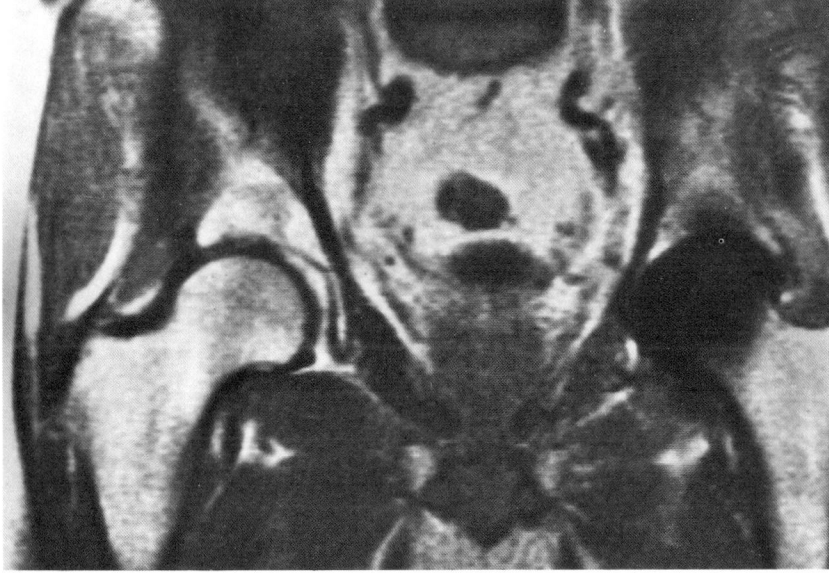

b

Figure 7.32

Total surface hip arthroplasty. A 60-year-old man with a history of left hip arthroplasty for ischemic necrosis complained of right hip pain. (**a**) Radiograph shows left total-surface hip arthroplasty and normal right hip. (**b**) T_1-weighted MR image (SE 500/28). Metallic surface replacement causes local signal void and artifacts around the left hip but does not interfere with the depiction of normal bone marrow in the right hip.

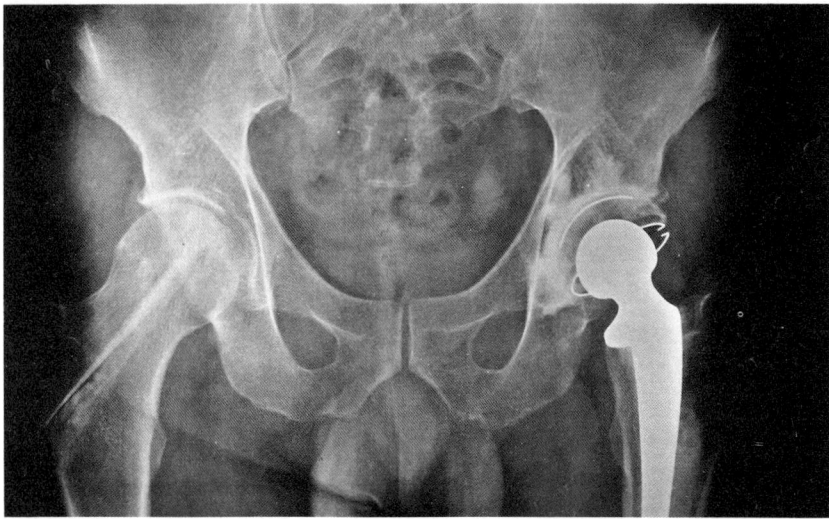

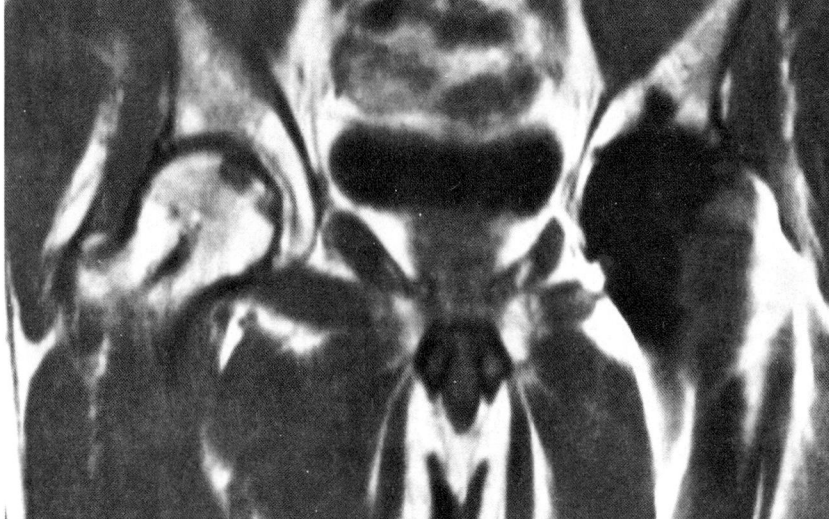

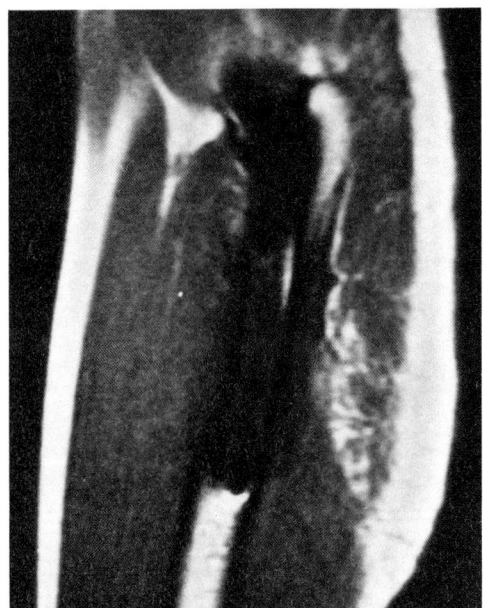

Figure 7.33

Stemmed left total hip arthroplasty. A 65-year-old man had a left total hip replacement and core biopsy with bone grafting of the right hip. (**a**) AP radiograph. T_1-weighted (**b**) coronal and (**c**) sagittal MR images (SE 500/28).

8
The knee

Steven Hartzman and
Richard H Gold

Many recent reports have demonstrated a valuable role for MRI in the evaluation of a wide spectrum of acute and chronic knee disorders.[1-11] A variety of imaging modalities are currently used to evaluate knee abnormalities, including standard radiography, scintigraphy, computed tomography, planar tomography, and arthrography. In many instances, more than one modality is necessary to determine the presence and extent of disease. The proposed advantages of MRI over these modalities include excellent inherent soft-tissue contrast (obviating the need for intra-articular contrast injection), lack of beam hardening artifact, multiplanar imaging capabilities, lack of ionizing radiation, and the ability to depict the periarticular structures including the ligaments, tendons, articular cartilage, and menisci. MRI does not require manipulation of the knee joint and is painless. At our institution and at a growing number of other large centers, MRI has become the primary nonsurgical imaging modality of the knee joint.

Normal anatomy of the knee joint (*Figures 8.1–8.3*)

In the normal knee, a thin layer of articular hyaline cartilage, 2–3 mm thick covers the articular surfaces of the femoral condyles, tibial plateau, and patella. On T_1- and T_2-weighted spin-echo images, the articular cartilage has intermediate signal intensity, less than that of fat but greater than the signal intensity of adjacent cortical bone and menisci, both of which have low-signal intensity and appear black. The tibiofemoral articular cartilage is well seen in both sagittal and coronal planes, while the patellar cartilage is best seen in the sagittal and axial planes.

The menisci are composed of fibrocartilage, and have low intensity signals that appear black on both T_1- and T_2-weighted spin echo images. The menisci are best evaluated in the sagittal plane, where they appear as triangular structures. In sagittal images of the more peripheral regions of the knee, the posterior and anterior horns of each meniscus unite to form a characteristic bow-tie appearance. The junction between each meniscus and the fibrous joint capsule is identified as a thin layer of increased signal intensity. The popliteus tendon and its sheath are best seen in the sagittal plane, where they cross obliquely along the periphery of the posterior horn of the lateral meniscus. The tendon appears as a thin, low intensity (black) band surrounded by the higher signal intensity of the sheath, and can be falsely interpreted as a meniscal tear.

The cruciate ligaments are best evaluated in the sagittal plane. The posterior cruciate ligament (PCL) appears as a thick curvilinear band of low-signal intensity coursing downwards from the posterior femoral condyle to attach on to the posterior aspect of the tibial plateau. The anterior cruciate ligament (ACL) is not as thick as the PCL, and usually manifests slightly higher signal intensity. The ACL extends obliquely from the lateral femoral condyle to attach on the medial aspect of the tibial plateau. With the knee externally rotated 15 degrees, the entire ACL can usually be seen on a single sagittal image.

The lateral collateral and medial collateral ligaments are best evaluated on coronal images, where they appear as low-intensity (black) bands extending between the femur, tibia, and fibula. The patellar tendon appears as a thick band of low-signal intensity anterior to Hoffa's fat pad, coursing inferiorly from the lower pole of the patella to its insertion at the tibial tuberosity. The patellar tendon (ligament) is best seen on sagittal and axial images. The popliteal artery and vein are readily identified in the sagittal and axial planes.

Meniscal injuries (*Figures 8.4–8.18, 8.21–8.23, 8.25*)

Many clinical and experimental studies have shown MRI capable of identifying meniscal abnormalities. Reicher et al were the first authors to present detailed descriptions of meniscal anatomy and abnormalities including meniscal tears.[11,12] Subsequent reported studies have shown MRI capable of detecting both meniscal degeneration and frank tears.[1-10] Crues et al showed MRI to be highly accurate in detecting tears when compared with surgery.[7] MRI has been shown to have low rates of false-positive and false-negative diagnoses. Another great advantage is its ability to detect parameniscal injuries, which may accompany meniscal tears. These include cruciate and collateral ligament tears, meniscal and parameniscal cysts, and popliteal cysts.

Necessary fine detail of the menisci can be obtained only with the use of well designed surface coils, which are currently available from most MR manufacturers. The menisci are best evaluated in the sagittal plane. For scanners not capable of producing contiguous images, interleaving may be necessary to evaluate fully the entire menisci. Slight external rotation (approximately 15 degrees) of the knee allows evaluation of the cruciate ligaments as well as the menisci. Coronal images may be used to supplement the sagittal images, but are not routinely obtained at our institution unless collateral ligament injury is also a consideration. T_1-weighted spin-echo sequences have been shown thus far to be more sensitive than T_2-weighted sequences in the detection of meniscal tears. However, this may soon change owing to the improved spatial resolution of T_2-weighted images available with the newest scanners. In the presence of a joint effusion, T_2-weighted images may provide an 'arthrogram effect', and better delineate the site of a fluid-filled meniscal tear.

Meniscal tears are characterized on both T_1- and T_2-weighted images as defects of high-signal intensity situated within the normal homogeneous low-intensity signal of the menisci. It is thought that this high-intensity signal represents synovial fluid embedded within intrameniscal crevices. However, not all foci of increased signal within the menisci represent frank tears. Menisci frequently contain areas of increased signal which histologically have been shown to correlate with areas of myxoid degeneration.[9] Stoller et al[9] showed that intrameniscal punctate or linear bands of increased signal intensity that did not extend to the articular margins usually proved to be areas of degeneration at pathologic examination. The bands of increased signal intensity that extended to meniscal surfaces were, however, all tears. Stoller et al demonstrated 100 per cent correlation between the MRI and pathologic findings.

It is thought that most meniscal tears are secondary to chronic trauma and secondary degenerative changes within the menisci. In a histologic study Tobler[13] showed that degenerative changes frequently preceded the development of meniscal tears. The corollary of this may be that patients with an intrameniscal high-intensity signal that does not communicate with a meniscal surface are at increased risk of developing frank tears in the future. This hypothesis will require validation through prospective studies. Although the existence of degenerative intrameniscal lesions may explain some false-positive tears and may ultimately evolve into frank tears, acute injuries in nondegenerated menisci do occur, predominantly in young athletes.

At least two systems for grading meniscal abnormalities detected by MRI have been reported.[4,9] Early in our experience, we used a grading system proposed by Reicher et al,[4] but noted a high false-positive rate. We have subsequently adopted a scheme containing features of both Reicher et al and Stoller et al,[9] with a resultant improved accuracy in interpretation. The grading system that we now use is as follows (see Figure 8.4):

Grade 1 homogeneous low-intensity (black) meniscus (no tear)

Grade 2 nonlinear region of increased signal intensity within the meniscus not extending to the articular margins; usually not present on two adjacent scans (tear unlikely)

Grade 3 linear region of increased signal intensity within the meniscus, not extending to the articular margins; may extend to meniscal capsular junction, which is not considered an articular margin (intrameniscal tear, unlikely to be seen at arthroscopy)

Grade 4 gross distortion of normal meniscal shape, truncation of meniscus or large linear or nonlinear region of increased signal intensity which extends to or abuts the articular margin (definite tear).

There are many potential pitfalls in the evaluation of meniscal injuries. Motion artifact can degrade resolution in the normally small menisci. We have been able to diminish this problem greatly by using a leg brace to hold the knee in proper alignment. As noted by Stoller et al, small areas of fraying of the free edge of the meniscus, which are detected at surgery, may not be depicted on MR images. The transverse meniscal ligament, connecting the anterior horns of the medial and lateral menisci, may simulate a frank tear of the anterior horn of the lateral meniscus, and has been noted in 30 per cent of knee studies.[9] The popliteus tendon sheath which runs obliquely across the posterior horn of the lateral meniscus is frequently visualized and may also be mistaken for a meniscal tear.

Cruciate ligament tears (*Figures 8.15, 8.19–8.26, 8.31, 8.32*)

Many reports have shown MRI to be useful in the detection of cruciate ligament tears.[5,11,14,15,16] On MR images, the normal continuity of both the anterior and posterior cruciate ligaments are best demonstrated in the sagittal plane with the knee externally rotated 15 degrees. The PCL appears as a thick, low-intensity curvilinear band, convex posteriorly, extending from its femoral attachment to its attachment on the posterior intercondylar fossa of the tibia. On sagittal MR images, the ACL is imaged slightly lateral to the PCL, and appears as a thin sagittal band of low-signal intensity (but greater than that of the PCL), extending from its attachment on the inner aspect of the lateral femoral condyle to its attachment anterior to the intercondylar process of the tibia. Although we routinely image the cruciate ligaments with the knee extended, some investigators advocate slight flexion of the knee (15–30 degrees), which relaxes the ACL, allowing detection of some normal ACLs not well seen with the knee fully extended.[15] For further evaluation of ligament integrity, other authors recommend obtaining nonorthogonal views along the course of the ACL.

The ACL and PCL are considered to be torn when the normal continuity of the ligament cannot be demonstrated on overlapping, interleaved sagittal images. Complete tears of the PCL are readily identified on both T_1- and T_2-weighted spin-echo images. Incomplete tears of the PCL are also easily detected as areas of increased signal intensity within the substance of the normal low-intensity (black) ligament. We were once fooled by a case of a complete PCL tear, which on MR appeared to be only a partial tear. At surgery, the ends of the torn ligament were in close apposition, accounting for the false-negative interpretation. A stressed flexion view of the PCL may be helpful to separate any fragments and disclose more fully any avulsion from the tibia.

Accurate detection and characterization of ACL tears has proved to be more difficult than for tears of the PCL. Joint hemorrhage or effusion, present in nearly all cases of ligament injury, frequently have an intermediate signal intensity similar to that of the ACL on T_1-weighted images, resulting in obscuration of the ACL, even when it is intact. In many cases, this potential pitfall can be overcome by obtaining a sequence with longer TR, allowing the ligament to be seen, which then has a lower signal intensity than the fluid. Partial tears of the ACL with surrounding hemorrhage and ligamentous edema cannot be reliably differentiated from complete tears. Turner et al suggested obtaining stress images of the ACL with a traction device and the knee slightly flexed.[15] The usefulness of this technique has not, however, been fully investigated.

MRI is also capable of imaging ACL grafts, which are being used in increasing numbers. Although to date we have imaged only intact biologic grafts, we believe that MRI may be capable of identifying tears of both biologic and synthetic graft material.

Collateral ligament tears (*Figures 8.23, 8.24, 8.27, 8.28, 8.32*)

Injuries to the collateral ligaments are best assessed on images obtained in the coronal plane. A combination of T_1- and T_2-weighted image sequences afford optimal evaluation. Tears of the medial or lateral collateral ligaments are considered to be present when there is a disruption of the normally low-signal intensity emanating from those structures. The exact site of tear may be seen on T_2-weighted images as a focus of high signal within the ligament. Usually, however, the precise point of disruption cannot be determined. Hemorrhage and/or edema in the adjacent tissues always accompanies collateral ligament tears, resulting in an increased distance between the subcutaneous fat and underlying bone. Fragments of the torn ligament may be seen outlined by hemorrhage and edema on T_2-weighted images. Gross separation of the medial collateral ligament from the medial meniscus to which it is normally attached may also be seen.

Patellar tendon injuries (*Figures 8.24, 8.29*)

MRI is capable of detecting injuries to the patellar tendon, which is normally depicted on sagittal images as a band of low-intensity signal extending from the lower pole of the patella to the anterior aspect of the tibial tubercle. The tendon is also well seen on axial images, where it appears as an oval, low-intensity region anterior to the higher signal intensity of the infrapatellar fat pad. A tear is diagnosed when there is a disruption in the normal continuity of the ligament. Partial tears or trauma-related inflammation are displayed as foci of increased signal intensity within the normal low-intensity ligament. In most cases, trauma to the patellar ligament is accompanied by adjacent hemorrhage represented by ill-defined regions of intermediate-to-high signal intensity anterior and posterior to the ligament. Injury in this region may also be accompanied by a fluid collection in the infrapatellar bursa, which produces a focus of moderate signal intensity posterior to the distal portion of the patellar tendon and anterior to the infrapatellar fat pad.

Joint effusions (*Figures 8.15, 8.16, 8.23, 8.24, 8.31–8.35*)

Synovial fluid within the knee joint is readily identified on both sagittal and axial images. Similar to other fluids within the body, a joint effusion is characterized by relatively long T_1 and T_2 relaxation values. As a result, effusions appear as regions of low-to-moderate signal intensity on T_1-weighted images and as regions of high-signal intensity on T_2-weighted images. The T_1 elongation of joint effusions may, however, not be as long as other body fluids, thereby producing a signal of intermediate intensity. Locations for knee effusions include the suprapatellar bursa, infrapatellar bursa, intercondylar notch, and posterior recess of the knee joint. On T_2-weighted images, the high signal intensity of the fluid allows delineation of low-signal intensity structures such as cruciate ligaments, menisci, and articular cartilage. As yet, MRI is incapable of differentiating types of joint fluid. Beltran et al showed that experimental joint distention caused by intracapsular injection of fresh blood or saline resulted in similar signal intensities for both fluids.[17] In clinical situations, we have been unable to differentiate between hemorrhagic versus nonhemorrhagic effusions.

Synovial cysts (*Figure 8.36*)

Fluid within synovial cysts has MR imaging characteristics identical to those of joint effusions, having low-signal intensity on T_1-weighted images and high-signal intensity on T_2-weighted images. Popliteal cysts typically appear as posteromedial oval-to-round fluid collections between the medial head of the gastrocnemius and semimembranosus muscles, and are best visualized in the sagittal or axial planes.

Cartilage disorders

Many imaging modalities are currently used for evaluation of cartilaginous lesions involving the knee joint. Plain radiographs and planar tomography indirectly assess cartilage loss by showing joint space narrowing and associated subchondral bony changes. Direct assessment of articular cartilage can be achieved with invasive procedures such as double contrast arthrography, CT arthrotomography, and arthroscopy. MRI has been shown to be capable of detecting cartilaginous defects, and holds the promise of being able to detect early cartilaginous degeneration.

On high resolution spin-echo MR images using a surface coil, the tibial and femoral articular cartilage appear as two opposing layers of intermediate signal intensity, with a combined thickness of 2–3 mm, adjacent to the low-signal intensity of the underlying subchondral cortical bone. The articular cartilage maintains intermediate signal intensity on both T_1-and T_2-weighted images. The posterior patellar cartilage conforms to the shape of the patella and covers its posterior surface completely. Both sagittal and axial images are necessary to evaluate the patellar cartilage completely.

With the use of conventional spin-echo techniques, MRI has proved capable of detecting gross cartilaginous lesions, such as severe thinning in cases of osteoarthritis and chondromalacia patellae. Cartilaginous fractures overlying large osteochondral defects in cases of osteochondritis dissecans have also been described. However, moderate-size cartilage defects and subtle changes detectable by arthroscopy may not be visualized by standard MRI techniques.

Using cadaver knees, Gylys-Morin et al showed that high-field strength MRI was capable of detecting focal, surgically created full-thickness defects in the range of 2–3 mm in diameter.[18] These small defects were occult on T_1-weighted and balanced images. Defects of 3 mm, however, were visualized on T_2-weighted sequences. Intra-articular administration of gadolinium-DTPA revealed defects 2 mm in diameter on both T_1- and T_2-weighted sequences.

Recent reports have indicated that newer MR techniques such as chemical shift selective sequence (CHESS) and fast low-angle shot sequence (FLASH) may prove more sensitive than spin-echo techniques in the detection of early disease involving the articular cartilage. Konig et al, using both CHESS and FLASH techniques, were able to show a significant decrease in signal intensity from histologically proven degenerative cartilage compared to normal surrounding cartilage, whereas spin-echo images of some of the same lesions appeared normal.[19] Another potential advantage of both techniques is their ability to correct for chemical shift artifacts, which can obscure articular cartilage on T_2-weighted images.

One specific disorder of articular cartilage that MRI may be able to detect is chondromalacia patellae, a frequent source of pain in adolescents and young adults. It is thought to be caused by chronic trauma to the patellar cartilage, perhaps related to recurrent lateral subluxations of the patella. Arthroscopy is commonly employed to evaluate patients clinically suspected of having chondromalacia patellae. Computed arthrotomography has also been shown to be accurate in making the diagnosis.[20] In a recent study, Yulish et al showed excellent correlation between MRI findings and arthroscopy, using a standard spin-echo technique.[21] The MRI staging system developed by them includes:

Stage I   focal areas of swelling with areas of decreased signal intensity on T_1- and T_2-weighted images (corresponding to arthroscopy Grade I to II)

Stage II irregularity of the articular surface of the posterior patellar cartilage with focal thinning (corresponding to arthroscopy Grade III)

Stage III absence of cartilage with exposure of subchondral bone, and/or synovial fluid passing through an ulcer in the cartilage and extending to the bone (corresponding to arthroscopy Grade IV).

This last finding is optimally seen on T_2-weighted images, where synovial fluid produces a bright, high-intensity signal and is easily visible adjacent to the low-intensity cortical bone. Our own experience with the MR diagnosis of chondromalacia patellae has not been as encouraging as that of Yulish et al, but we hope to improve our accuracy through the use of high-resolution T_2-weighted images.

Osteochondritis dissecans (*Figures 8.40, 8.41, 8.43*)

The lesions of osteochondritis dissecans can be identified on T_1-weighted images as low-intensity foci in the subchondral portions of the lateral aspect of the medial femoral condyle, correlating precisely in location to the lesions identified on radiographs. The lesions frequently appear larger on the MR images than on corresponding radiographs. An advantage of MRI is its ability to evaluate the integrity of the higher signal articular cartilage overlying the bony subchondral defect. We recently identified two surgically confirmed cases of intact cartilage overlying large bony defects. The usual appearance is that of disruption of the cartilage, and the defect often correlates with the loose body found at surgery. The role of MRI in assessing treatment of osteochondritis dissecans has yet to be determined.

Avascular necrosis (*Figure 8.42*)

MRI has been shown to be sensitive in the detection of osteonecrosis, especially involving the femoral head.[22] We have also reported cases of osteonecrosis involving the distal femoral condyles.[6] Various patterns of abnormal signal intensity owing to osteonecrosis have been described. The most typical appearance is that of focal areas of homogeneous or heterogeneous low-signal intensity. Ring-like or band-like patterns of low-signal intensity surrounding central regions of higher signal intensity are also common. The areas of decreased signal intensity can be seen regardless of the pulse sequence used.

Medullary infarcts (*Figures 8.43–8.46*)

Infarcts in the distal femur and proximal tibia are readily seen on T_1-weighted spin-echo images as focal, discrete, inhomogeneous areas or rings of low-signal intensity surrounding central cores of higher signal intensity. The infarcts may be circular or have irregular margins.

Intra-articular osteochondral fragments (*Figures 8.37–8.39*)

Osteochondral loose bodies have varying signal intensity, depending on their composition. Fragments containing marrow have high-signal on T_1- and T_2-weighted images. The majority of fragments, however, do not contain marrow and are of low-signal intensity on all pulse sequences, thereby making their identification difficult. We have found that fragments greater than 5 mm–1 cm in diameter can usually be seen within the knee joint. Smaller fragments may be identified only if they are located adjacent to high-intensity structures such as the infrapatellar fat pad. If located within the suprapatellar bursa, the fragments may appear as filling defects surrounded by higher signal joint effusion, especially on T_2-weighted sequences.

References

1　GALLIMORE GW JR, HARMS SE, Knee injuries: high-resolution MR imaging, *Radiology* (1986) **160**:457–61.
2　BURK DL, KANAL E, BRUNBERG JA et al, 1.5-T surface-coil MRI of the knee, *AJR* (1986) **147**:293–300.
3　LI KC, HENKELMAN RM, POON PY et al, MR imaging of the normal knee, *J Comput Assist Tomogr* (1984) **8**:1147–54.
4　REICHER MA, HARTZMAN S, DUCKWILER GR et al, Meniscal injuries: detection using MR imaging, *Radiology* (1986) **159**:753–7.
5　REICHER MA, HARTZMAN S, BASSETT LW et al, MR imaging of the knee. Part I. Traumatic disorders, *Radiology* (1987) **162**:547–51.
6　HARTZMAN S, REICHER MA, BASSETT LW et al, MR imaging of the knee. Part II. Chronic disorders, *Radiology* (1987) **162**:553–7.
7　CRUES JV III, MINK J, LEVY TL et al, Meniscal tears of the knee: accuracy of MR imaging, *Radiology* (1987) **164**:445–8.
8　BELTRAN J, NOTO AM, MOSURE JC et al, Meniscal tears: MR demonstration of experimentally produced injuries, *Radiology* (1986) **158**:691–3.
9　STOLLER DW, MARTIN C, CRUES JV III et al, Meniscal tears: pathologic correlation with MR imaging, *Radiology* (1987) **163**:731–5.
10　MINK J, STOLLER DW, MARTIN C et al, MR of the knee. Pitfalls in interpretation. In: *Book of Abstracts* Vol 4 (Society of MR in Medicine: Berkeley, California, 1986) 1195.
11　REICHER MA, BASSETT LW, GOLD RH, High-resolution magnetic resonance imaging of the knee joint: pathologic correlations, *AJR* (1985) **145**:903–909.
12　REICHER MA, RAUSCHNING W, GOLD RH et al, High-resolution magnetic resonance imaging of the knee joint: normal anatomy, *AJR* (1985) **145**:895–902.
13　TOBLER TH, Makroskopische und histologische Befund am Kniegelenk Meniscus in Verschiedenen Lebensaitern Schweiz, *Med Wschr* (1926) **56**:1359.
14　BELTRAN J, NOTO AM, MOSURE JC et al, The knee: surface-coil MR imaging at 1.5 T1, *Radiology* (1986) **159**:747–51.
15　TURNER DA, PRODOMOS CC, PETASNICK JP et al, Acute injury of the ligaments of the knee: magnetic resonance evaluation, *Radiology* (1985) **154**:717–22.
16　LI DK, ADAMS ME, McCONKEY JP, Magnetic resonance imaging of the ligaments and menisci of the knee, *Radiol Clin North Am* (1986) **24**:209–27.
17　BELTRAN J, NOTO AM, HERMAN LJ et al, Joint effusions: MR imaging, *Radiology* (1986) **158**:133–7.
18　GLYLYS-MORIN VM, HAJEK PC, SARTORIS DJ et al, Articular cartilage defects: detectability in cadaver knees with MR, *AJR* (1987) **148**:1153–7.
19　KONIG H, SAUTER R, DEIMLING M et al, Cartilage disorders: comparison of spin-echo, CHESS, and FLASH sequence MR images, *Radiology* (1987) **164**:753–8.
20　BOVEN F, BELLEMANS MA, GEURTS J et al, The value of computed tomography scanning in chondromalacia patellae, *Skeletal Radiol* (1982) **8**:183–5.
21　YULISH BS, MONTANEZ J, GOODFELLOW DB et al, Chondromalacia patellae: Assessment with MR imaging, *Radiology* (1987) **164**:763–6.
22　TOTTY WG, MURPHY WA, GANZ WI et al, Magnetic resonance imaging in the early diagnosis of ischemic necrosis of the femoral head, *AJR* (1984) **143**:1273–80.

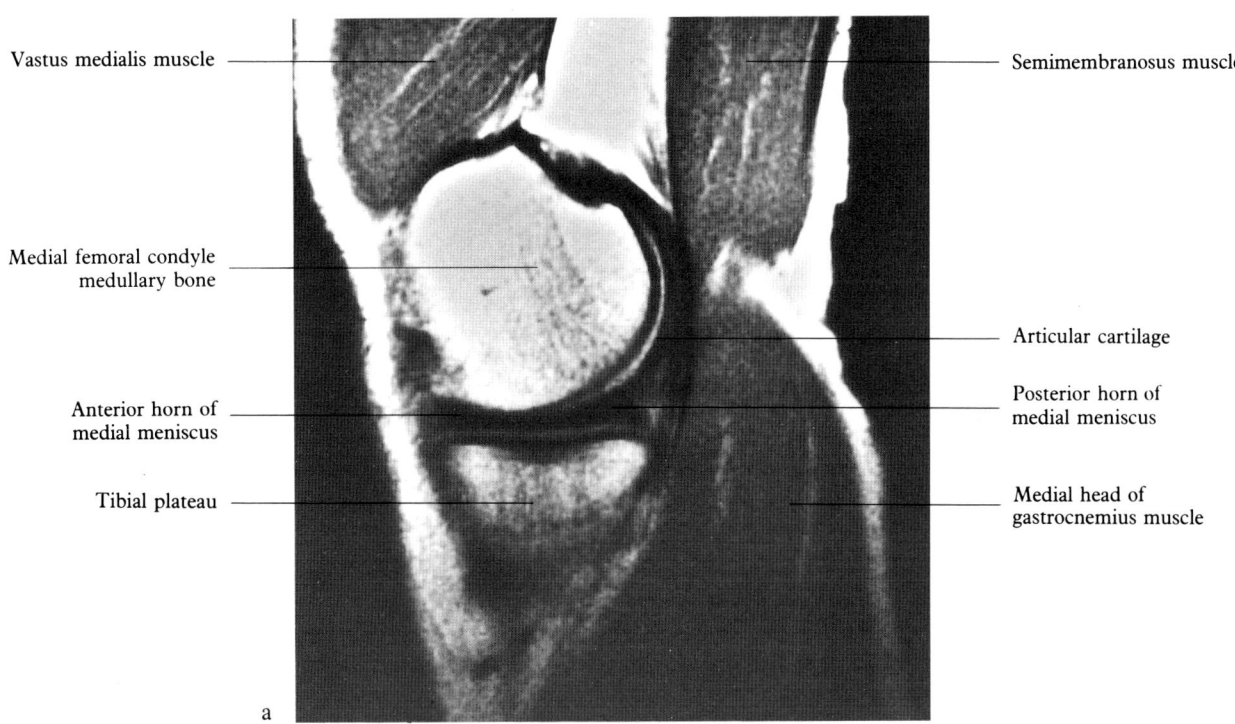

Figure 8.1

Normal sagittal anatomy (SE 800/28). (**a**) Sagittal image through the medial femoral condyle. Note the typical bow-tie appearance of the peripheral portion of the medial meniscus. (**b**) Sagittal image through the medial femoral condyle. (**c**) Intercondylar sagittal image at level of the posterior cruciate ligament. (**d**) Intercondylar sagittal image at the level of the anterior cruciate ligament. (**e**) Sagittal image through the lateral femoral condyle.

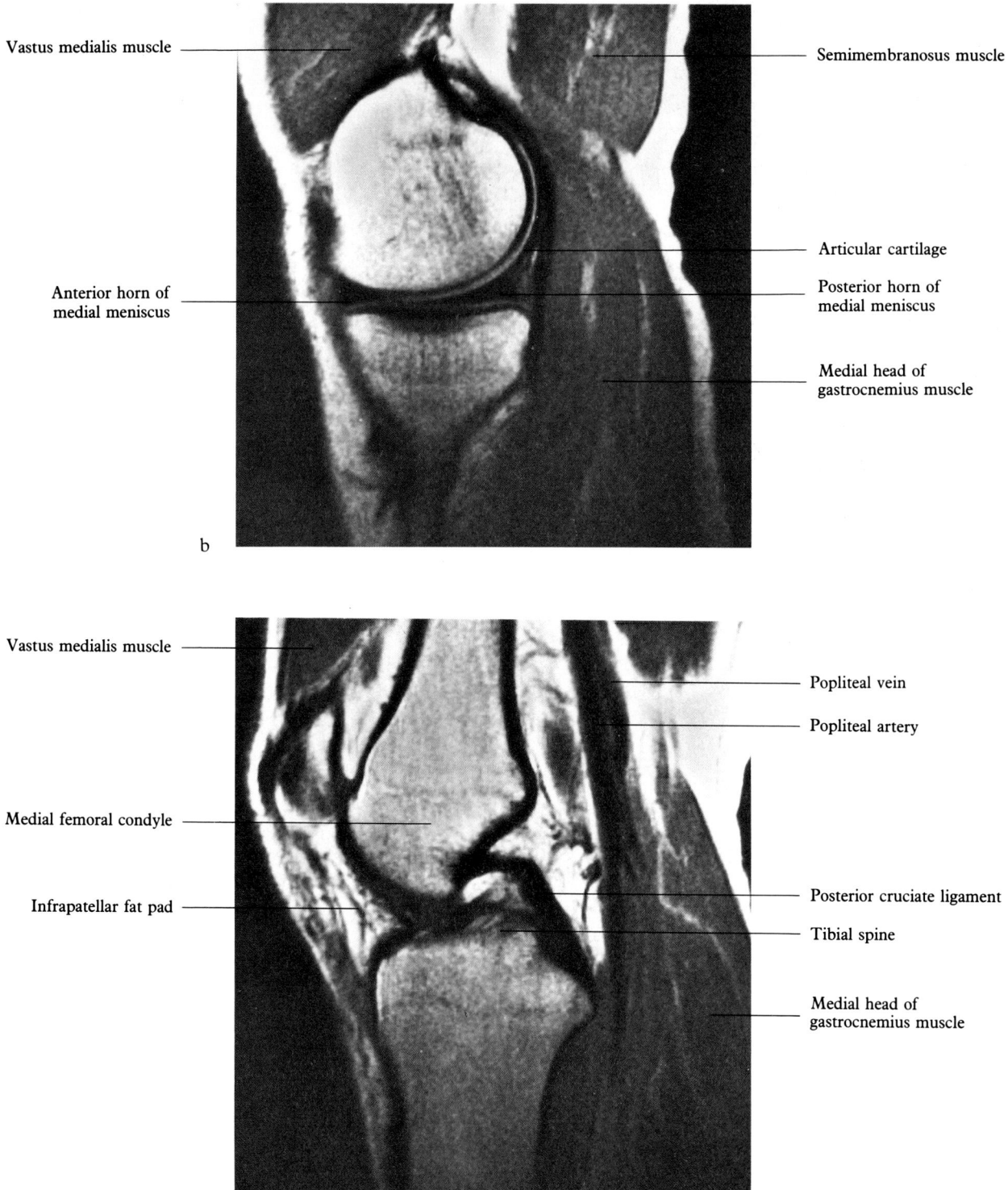

Figure 8.1 *continued*

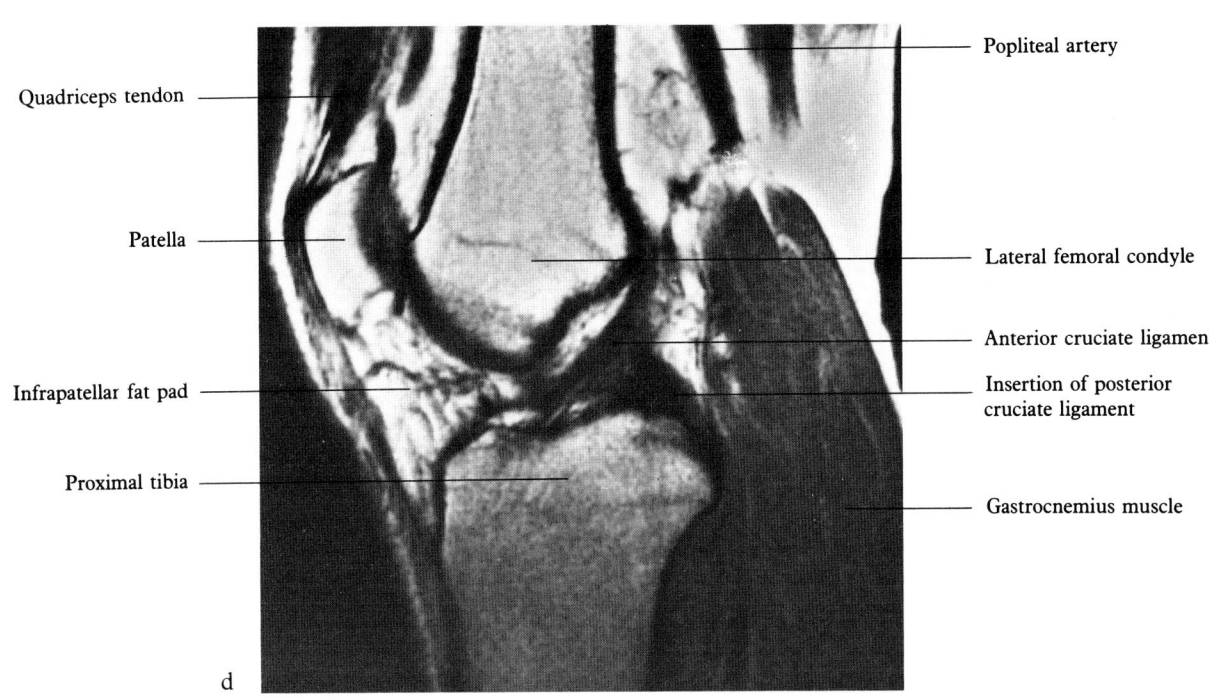

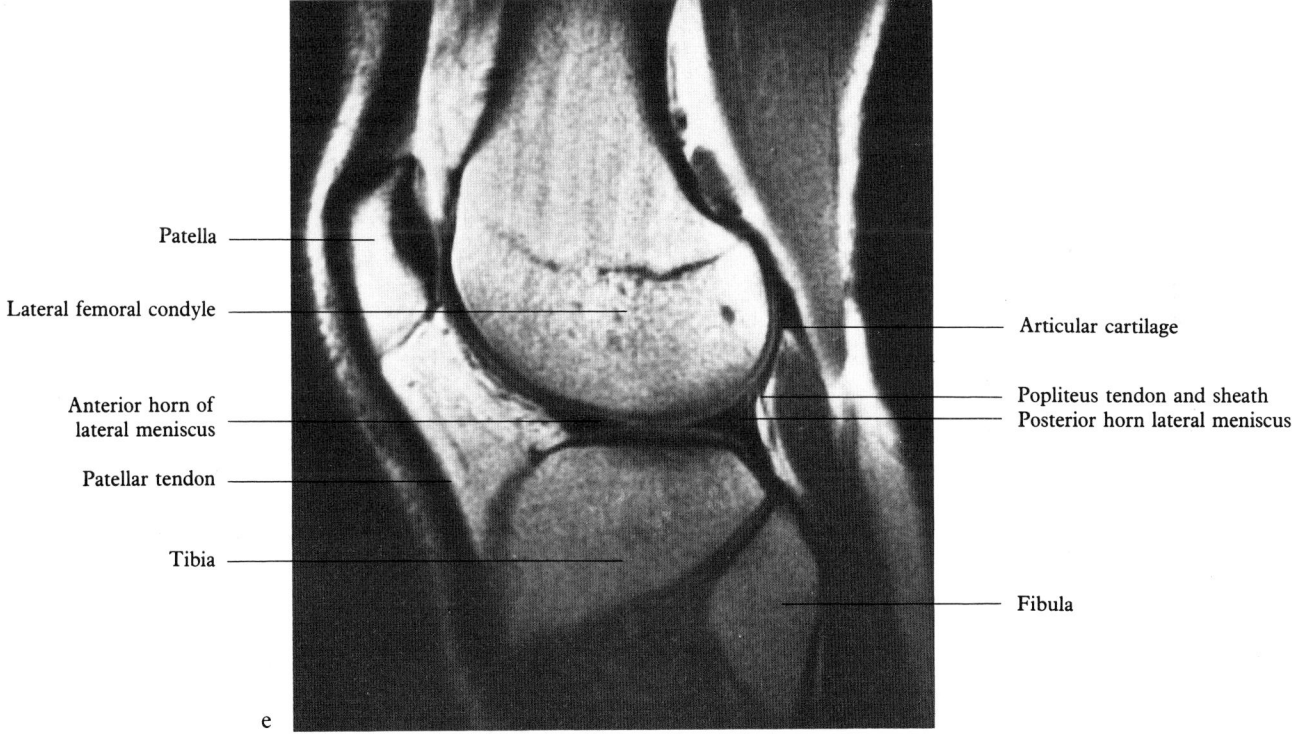

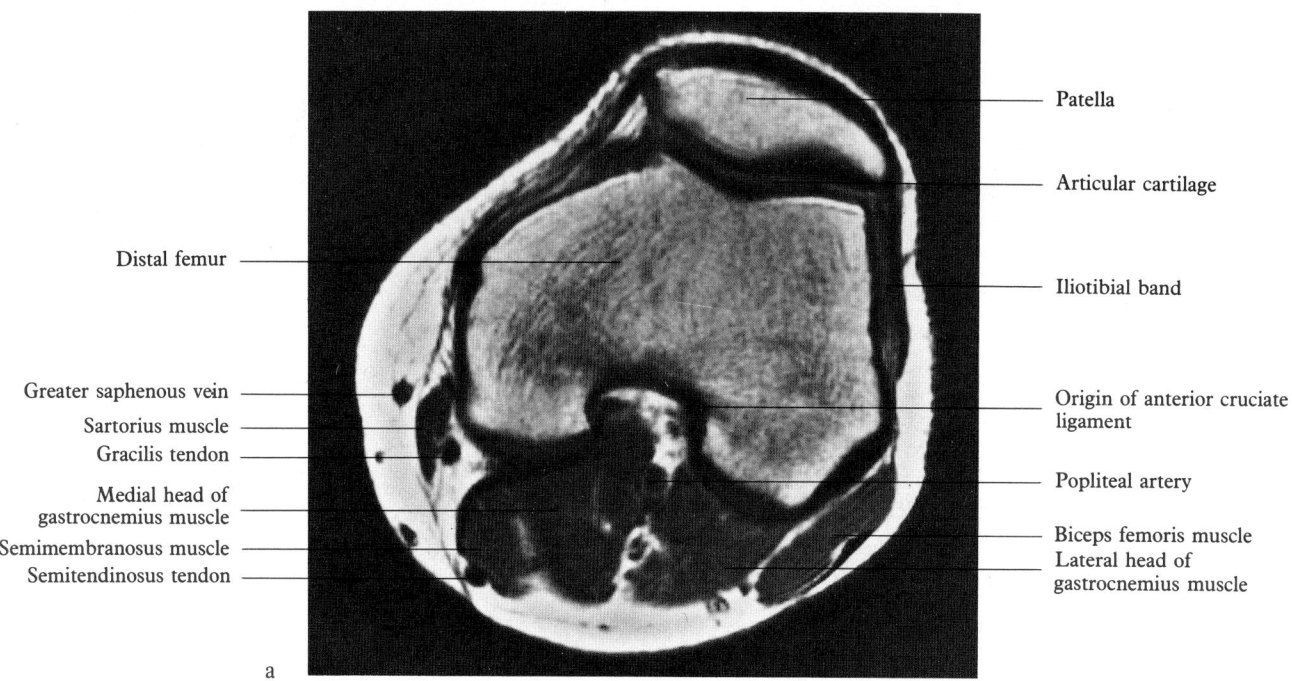

Figure 8.2

Normal axial anatomy (SE 800/28). (**a**) Axial image at the level of the mid patella and distal femur. (**b**) Axial image 1 cm, caudal to (**a**). (**c**) Axial image at the level of the femoral condyles. (**d**) Axial image at the level of the menisci. (**e**) Axial image at the level of the tibial plateau.

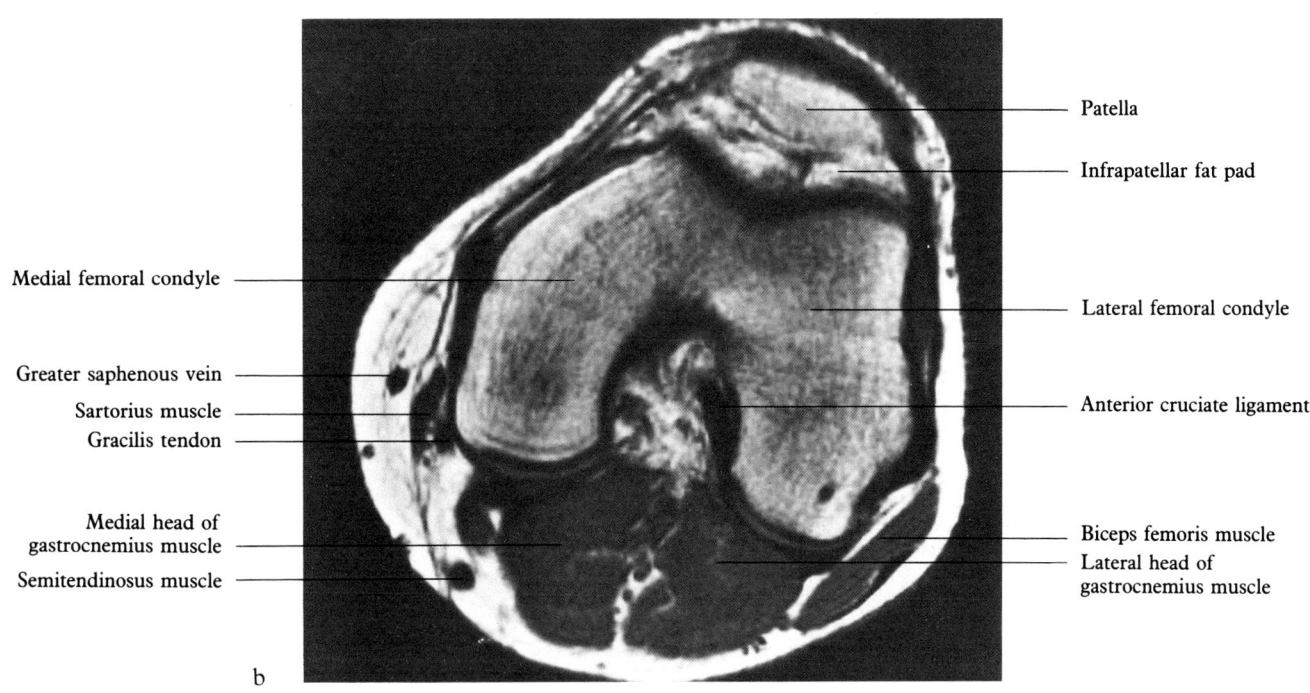

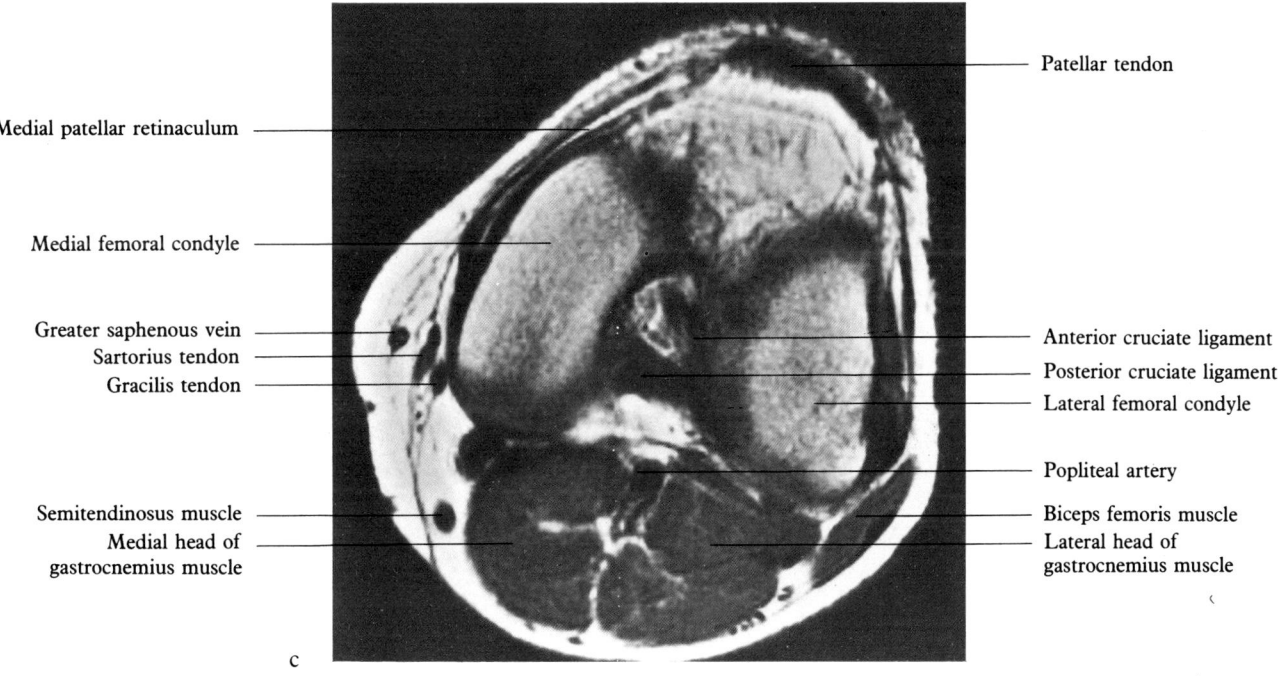

Figure 8.2 *continued*

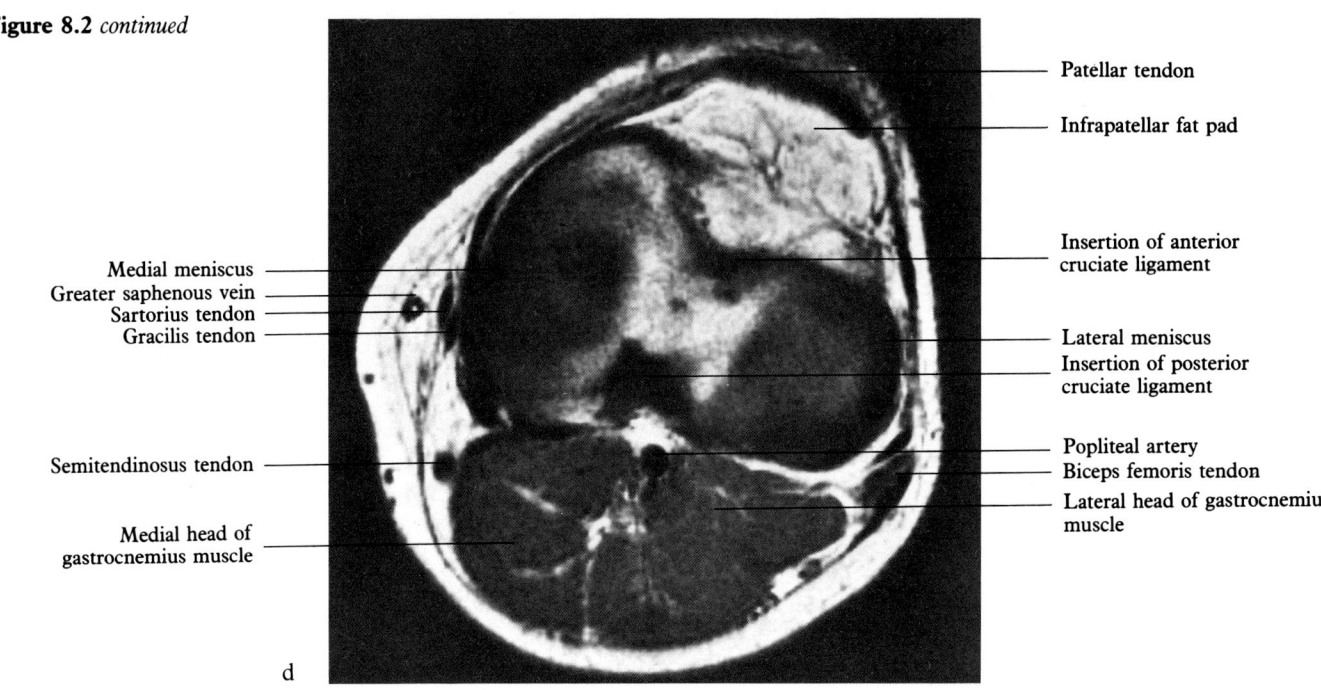

d

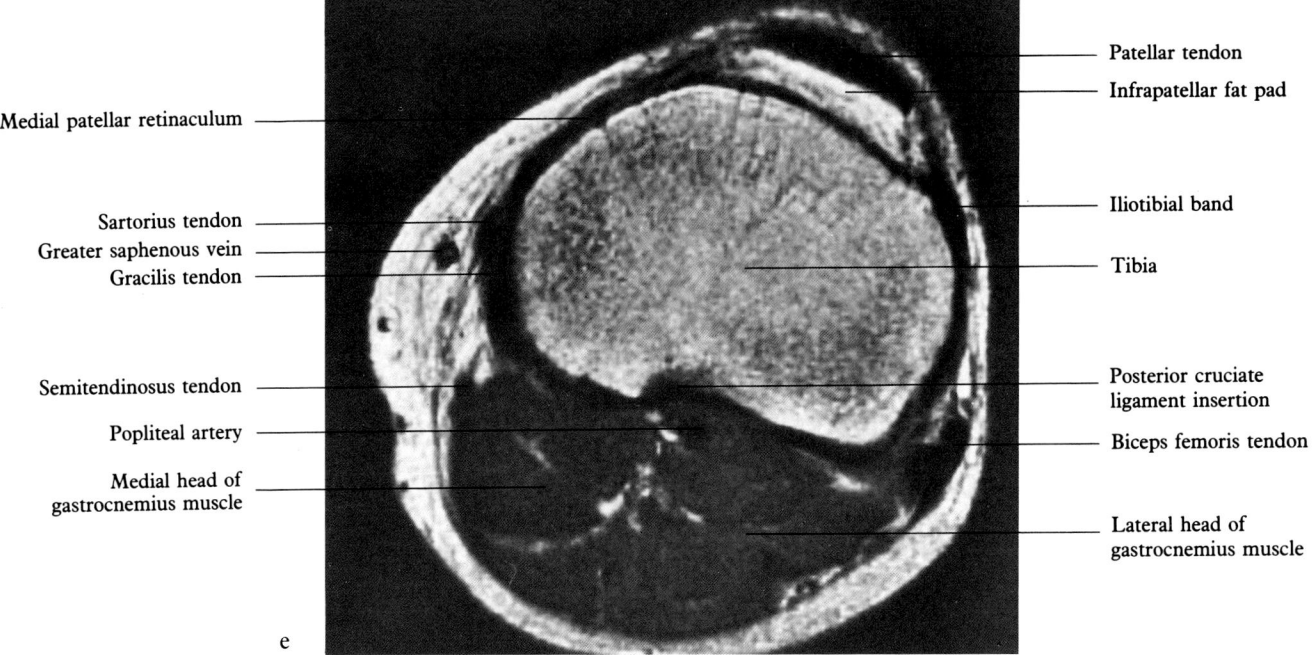

e

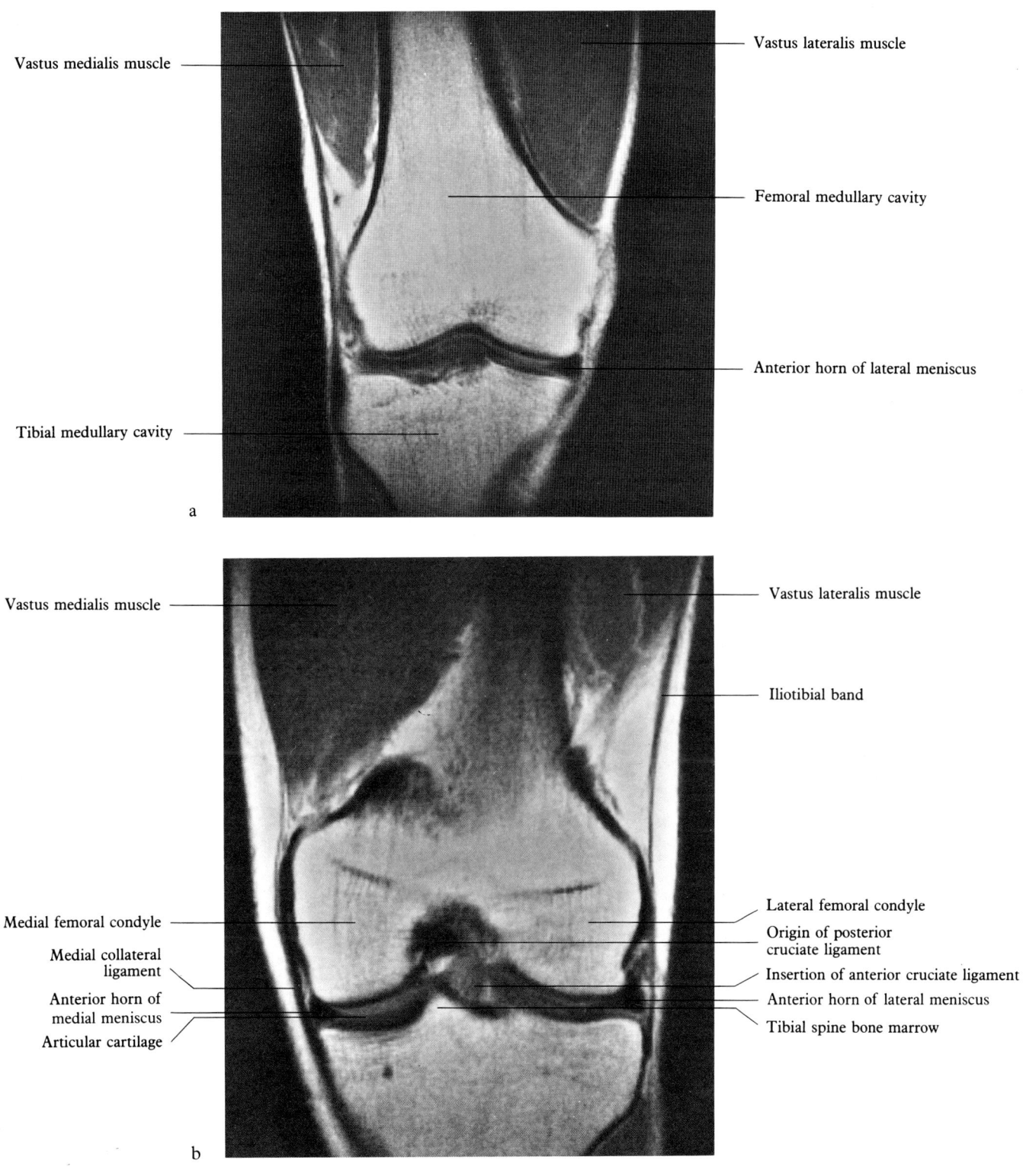

Figure 8.3

Normal coronal anatomy (SE 800/28). (**a**) Coronal image through the anterior horns of the menisci. (**b**) Coronal image at the level of insertion of the anterior cruciate ligament. (**c**) Coronal image through the intercondylar notch. (**d**) Coronal image at the level of the posterior cruciate ligament insertion.

Figure 8.3 *continued*

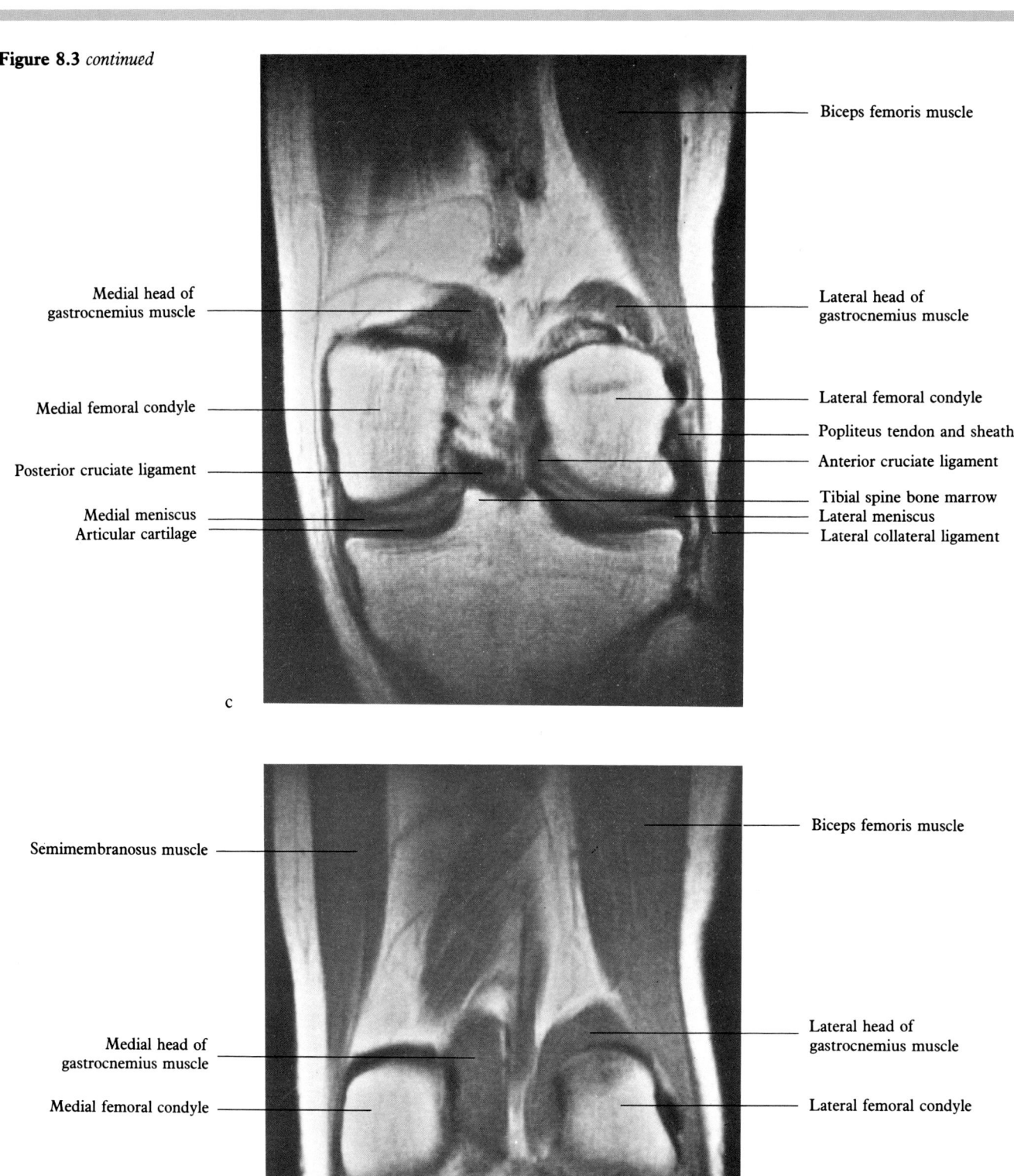

c

d

Grade 1 Normal

Grade 2 Tear unlikely

Figure 8.4
Grades of intrameniscal signal.

Grade 3 Intrameniscal tear, unlikely to be seen at arthroscopy

Grade 4 Definite tear

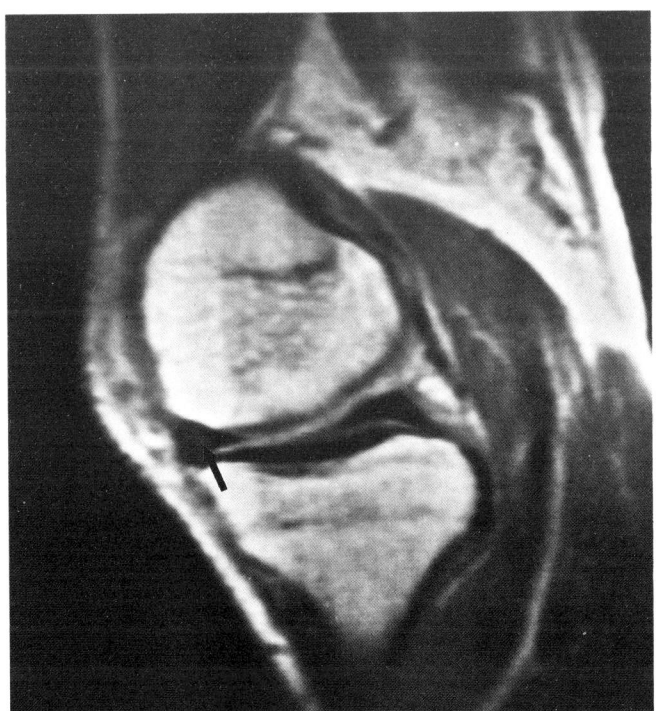

Figure 8.5
Grade II meniscal abnormality (tear unlikely) in a 58-year-old man with chronic medial knee pain. T_1-weighted sagittal MR image (SE 800/28) shows globular increased signal intensity in the anterior horn of the medial meniscus (arrow). No tear was seen at surgery.

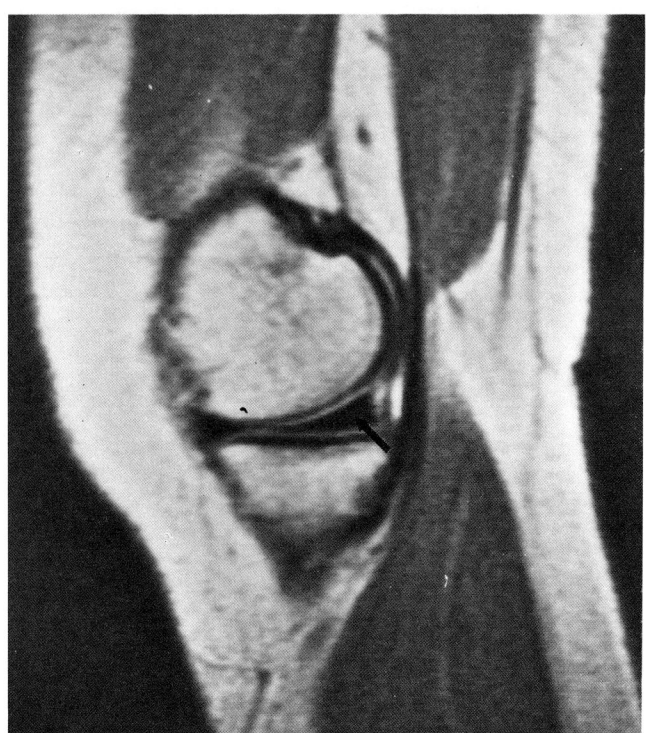

a

Figure 8.6

Grade III meniscal abnormality (possible tear) in a 24-year-old woman who sustained an athletic injury. (**a**) Initial sagittal MR image (SE 800/28) 1 week after injury shows a subtle, linear intersubstance signal, which does not extend to an articular margin (arrow). This scan was performed early in our experience and was falsely interpreted as a meniscal tear. Subsequent arthroscopy revealed a normal meniscus. (**b**) Repeat sagittal MR image (SE 800/28) 1 month later discloses similar meniscal changes (arrow). Because the patient remained symptomatic, a second arthroscopy was performed and again showed a normal meniscus. Symptoms decreased over the next few months and the patient was pain free at 6 months post arthroscopy.

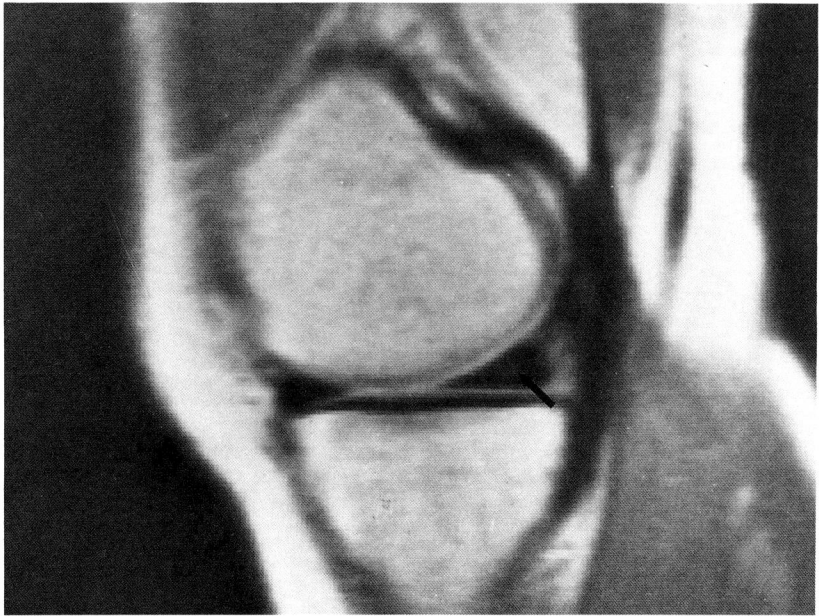

b

Figure 8.7

Grade III meniscal abnormality (possible tear) in a 17-year-old girl with acute knee pain. T_1-weighted sagittal MR image (SE 800/28) reveals a subtle linear signal in the posterior horn of the medial meniscus (arrow). Symptoms abated with conservative treatment and the patient was completely pain-free within 3 months.

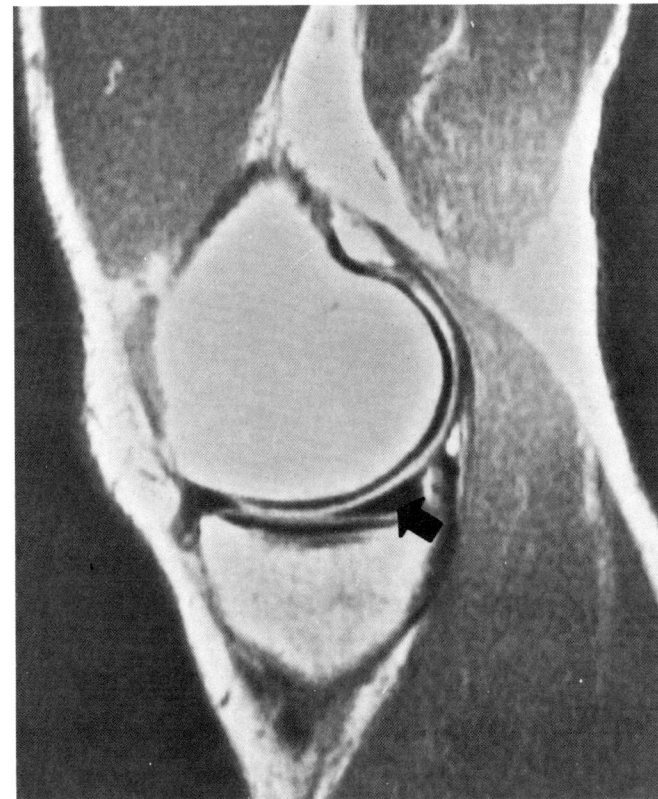

Figure 8.8

Grade IV meniscal abnormality (frank nonlinear tear) in a 38-year-old man with lateral knee pain. (**a**) T_1-weighted sagittal MR image (SE 800/28) shows globular increased signal intensity in the anterior horn of the lateral meniscus (arrow), extending to the superior and inferior articular margins. A frank tear was confirmed at arthroscopy. (**b**) T_1-weighted sagittal MR image (SE 800/28) of the anterior cruciate ligament (arrow) shows no evidence of disruption.

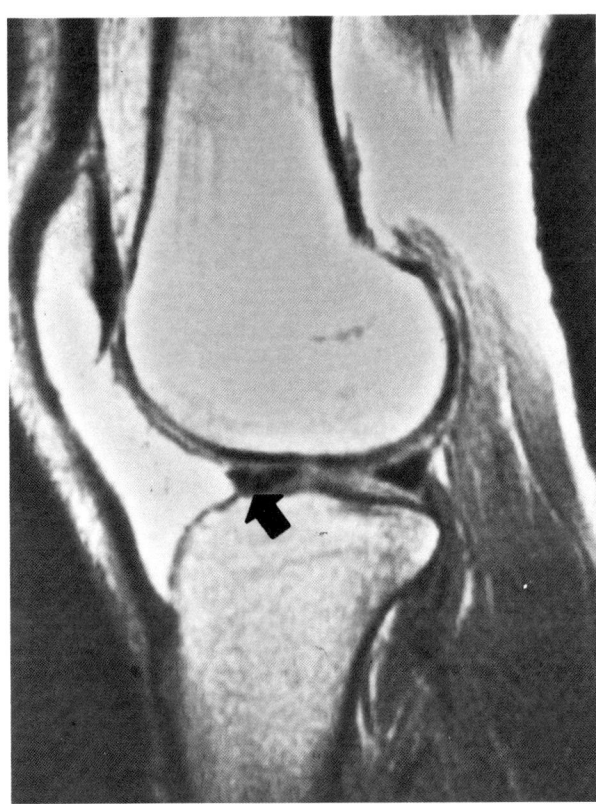

a

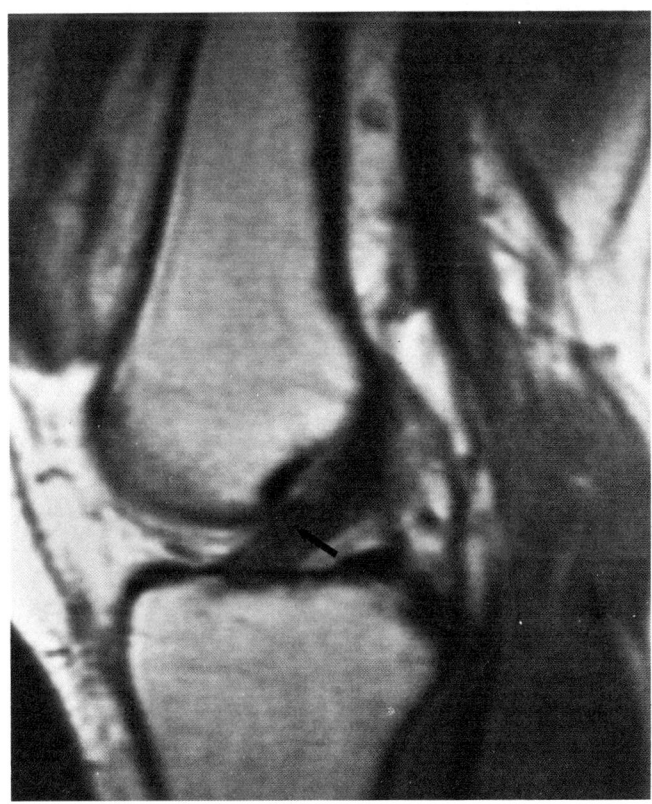

b

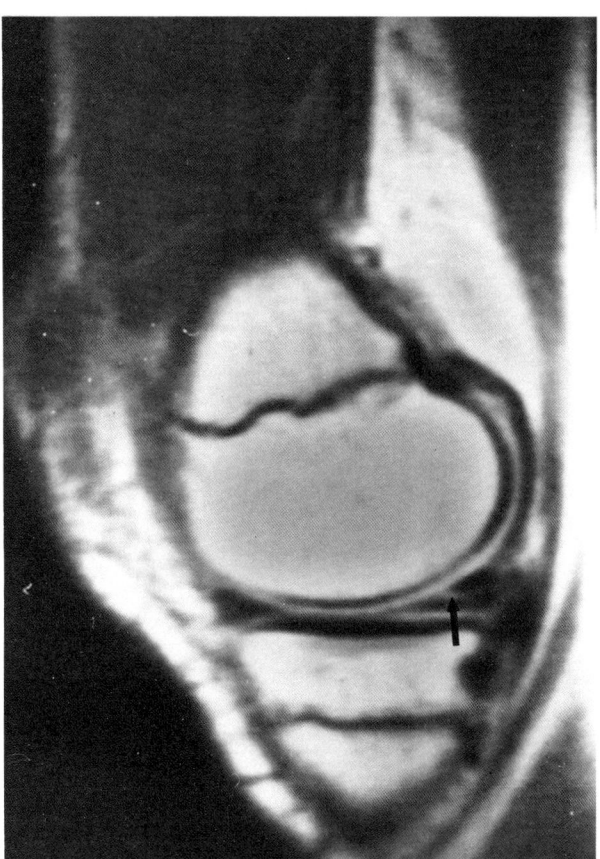

Figure 8.9

Grade IV meniscal abnormality (frank linear tear) in a 14-year-old girl with acute knee pain. T_1-weighted sagittal MR image (SE 800/28) reveals linear focus of increased signal intensity which extends to the superior articular margin of the posterior horn of the medial meniscus (arrow), confirmed surgically as a frank tear.

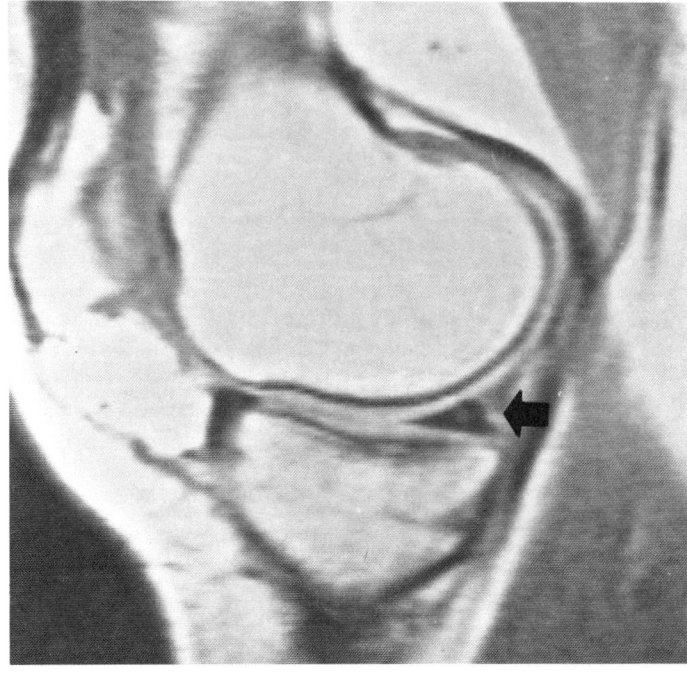

Figure 8.10

Grade IV meniscal abnormality (frank meniscal tear) in a 29-year-old woman with medial knee pain. T_1-weighted sagittal MR image (SE 800/28) shows truncation of the posterior horn of the medial meniscus, with an oblique tear posteriorly (arrow), surgically confirmed.

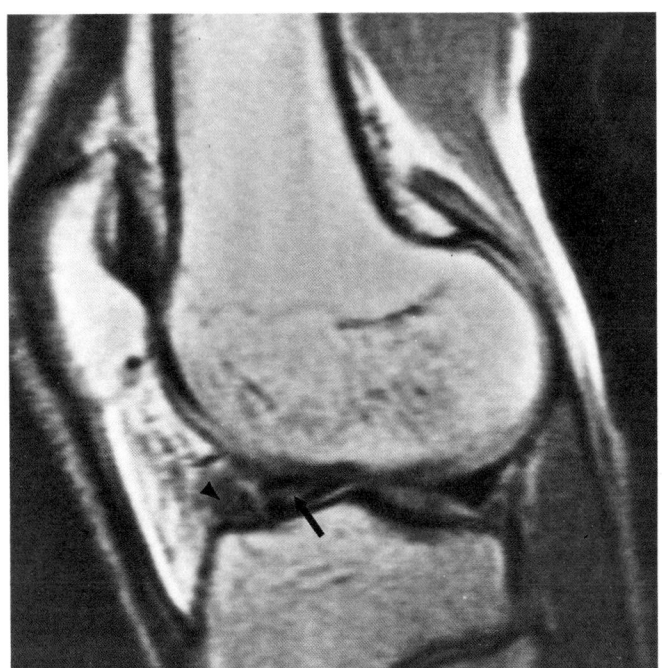

a

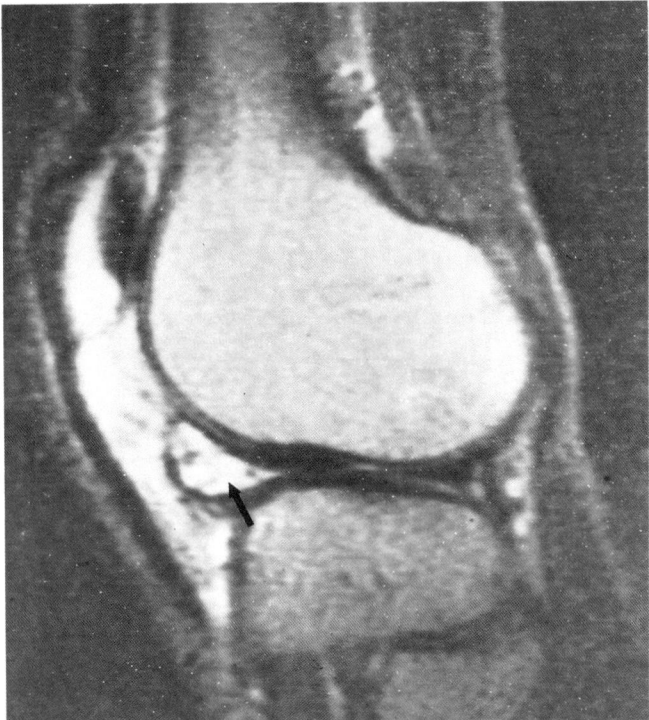

b

Figure 8.11
Grade IV meniscal abnormality (frank tear) with associated parameniscal cyst in a 26-year-old man with a 2 month history of knee pain secondary to athletic injury. (**a**) T_1-weighted sagittal MR image (SE 800/28) shows a frank linear tear in the anterior horn of the lateral meniscus (arrow) with associated focus of increased signal intensity (arrowhead) extending into the infrapatellar fat pad. (**b**) T_2-weighted sagittal MR image (SE 2000/56) confirms the presence of a parameniscal cyst which communicates with a horizontal tear of the anterior horn (arrow).

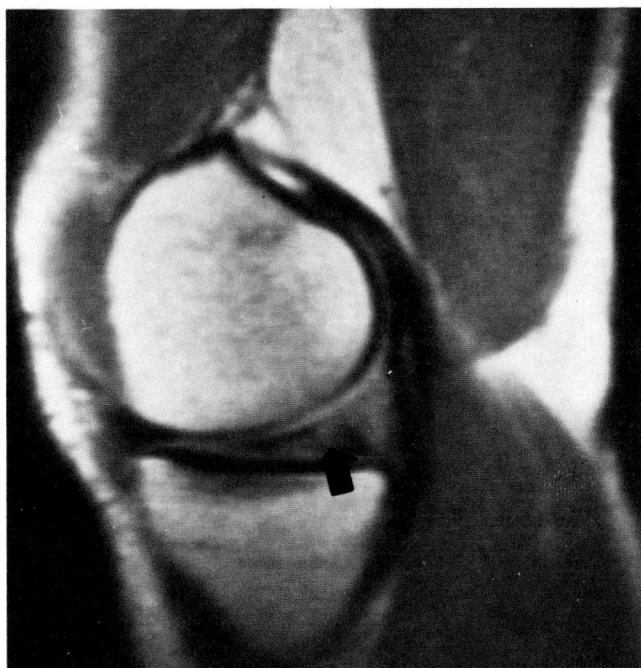

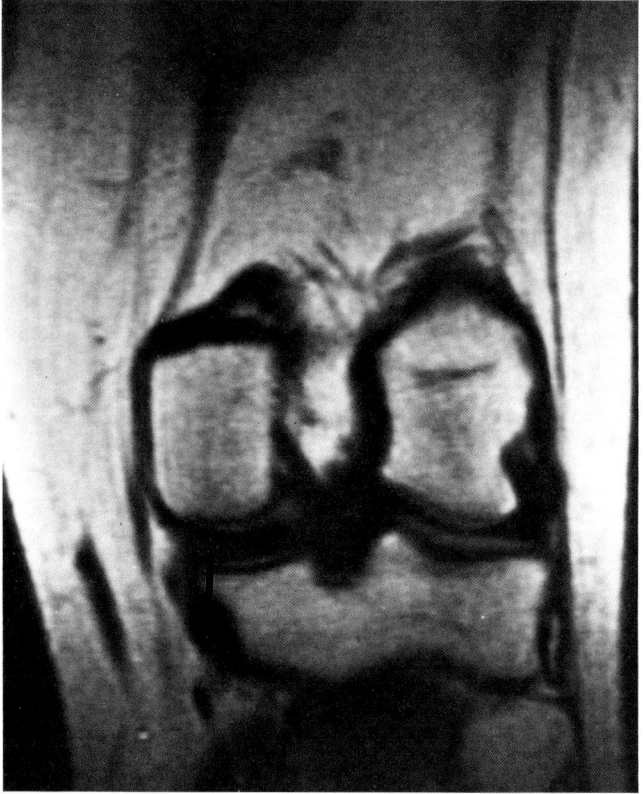

Figure 8.12
Frank meniscal tear (Grade IV intensity) in a 34-year-old woman with a clinically suspected tear of the medial meniscus. (**a**) T_1-weighted sagittal MR image (SE 800/28) discloses a horizontal linear focus of increased signal intensity extending to the para-articular margin (arrow). (**b**) T_1-weighted coronal MR image (SE 800/28) also clearly demonstrates the linear tear (arrow).

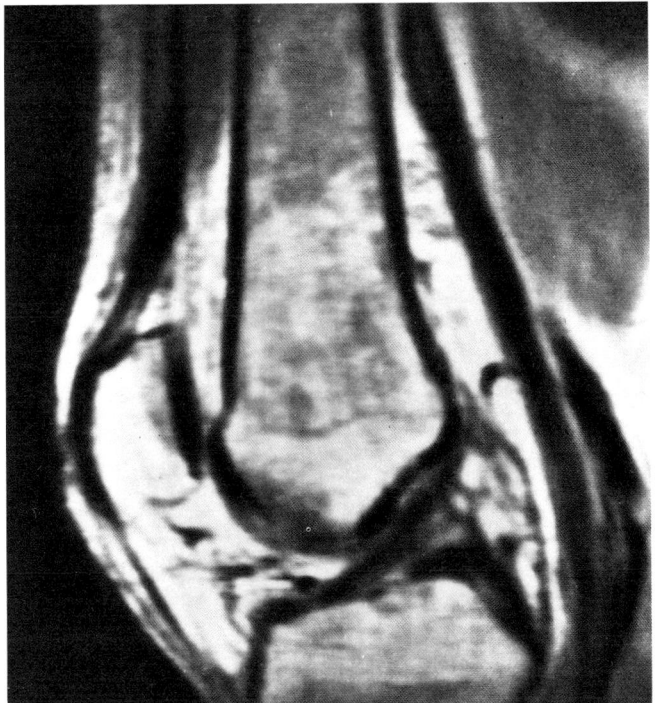

a

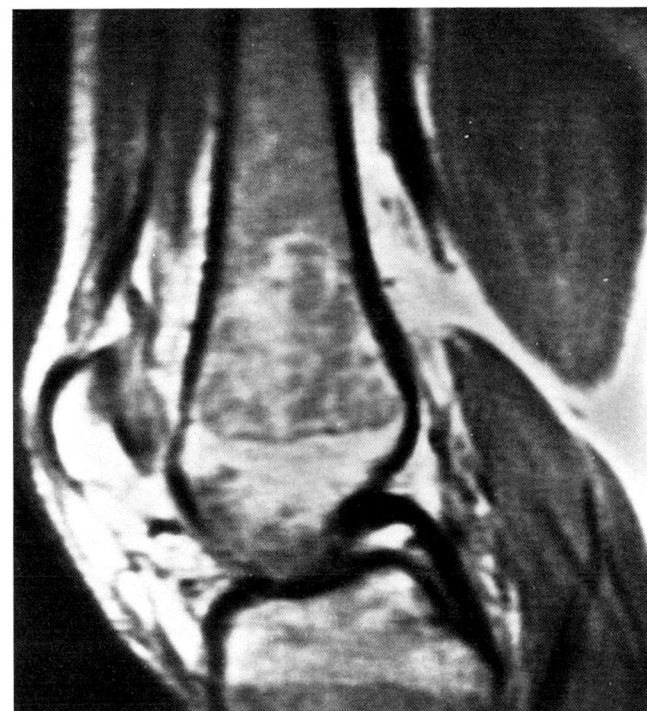

b

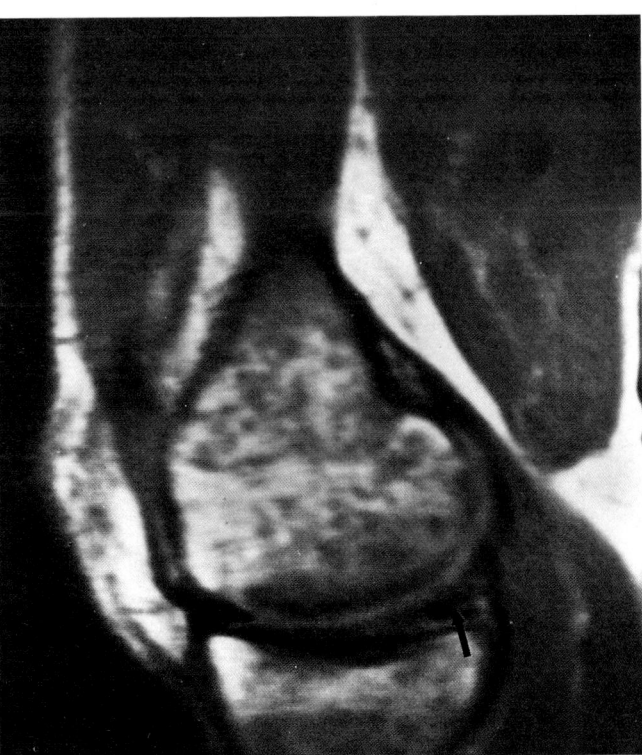

c

Figure 8.13

Bone marrow infiltration and tears of the medial meniscus (Grade IV abnormality) in a 42-year-old man with hairy-cell leukemia and long-standing medial knee pain. Radiographs were normal. (**a**) T_1-weighted sagittal MR images (SE 800/28) through the anterior cruciate ligament and (**b**) the posterior cruciate ligament disclose that both ligaments are normal. Inhomogeneous signal intensity emanating from the marrow is consistent with myelofibrosis, later confirmed by biopsy. (**c**) T_1-weighted sagittal MR image (SE 800/28) through the medial meniscus reveals a complex tear of the posterior horn (arrow).

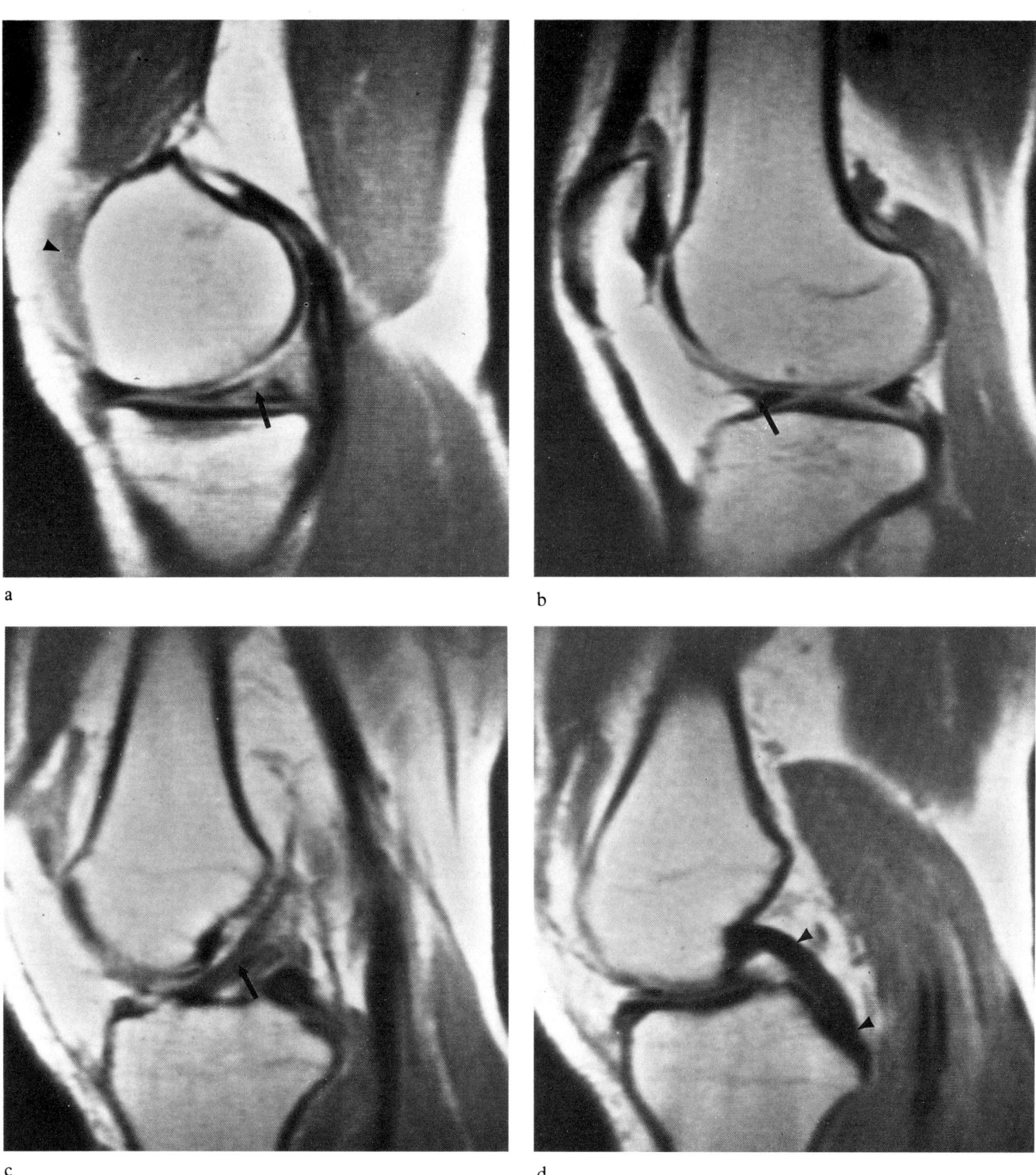

Figure 8.14

Frank meniscal tear (Grade IV abnormality) in a 54-year-old man with right knee pain and clinical suspicion of medial and lateral meniscal tears. (**a**) T_1-weighted sagittal MR image (SE 800/28) through the medial meniscus shows a Grade IV linear signal in the posterior horn (arrow), consistent with a frank tear. Joint effusion is also present (arrowhead). (**b**) T_1-weighted sagittal MR image (SE 800/28) through the lateral meniscus reveals no abnormality. Zone of intermediate signal between transverse meniscal ligament and anterior horn is well seen and should not be interpreted as an oblique tear (arrow). (**c**) The anterior cruciate ligament (arrow) and (**d**) posterior cruciate ligament (arrowhead) are well seen on sagittal images.

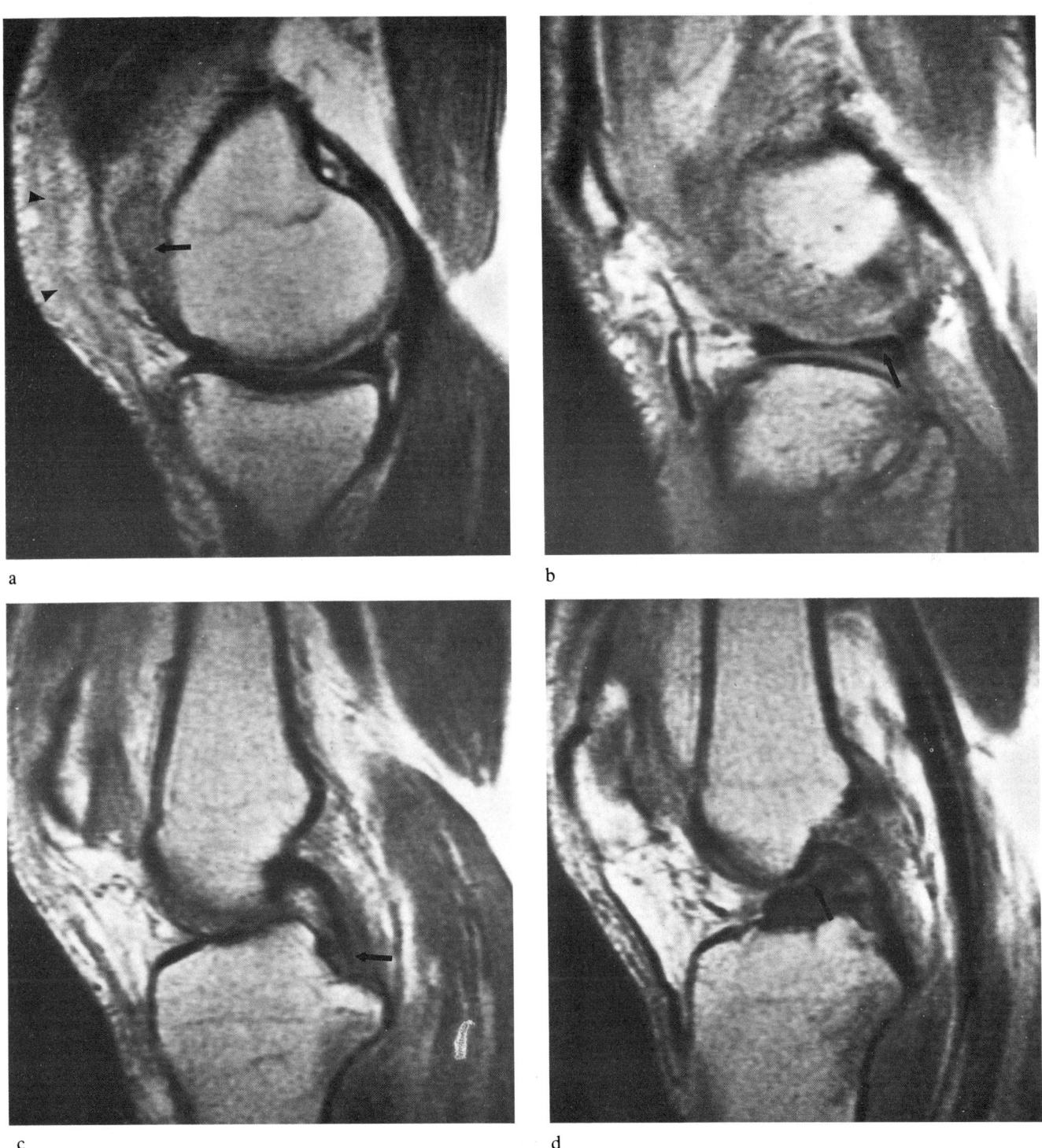

Figure 8.15

False-positive meniscal tear in a 49-year-old man involved in a motor vehicle accident. All images are T_1-weighted (SE 800/28). (**a**) Sagittal view of the medial meniscus shows no meniscal abnormality but only large effusion of intermediate signal intensity (arrow) and extensive prepatellar soft-tissue edema and hemorrhage of high intensity (arrowhead). (**b**) Sagittal image of the lateral meniscus reveals focus of increased signal intensity in the posterior horn (arrow), which appears to involve the inferior articular margin. This was interpreted as a frank tear (Grade IV abnormality). At surgery, mild degeneration of the meniscus was noted, but no obvious tear. (**c**) Sagittal image through the intercondylar notch discloses increased signal intensity emanating from the inferior portion of the posterior cruciate ligament (arrow). A partial tear was found at surgery. (**d**) Sagittal image slightly lateral to (**c**) shows gross abnormality of the anterior cruciate ligament, which appears to course more posteriorly than normal (arrow). Surgery revealed avulsion of the ligament from the femur.

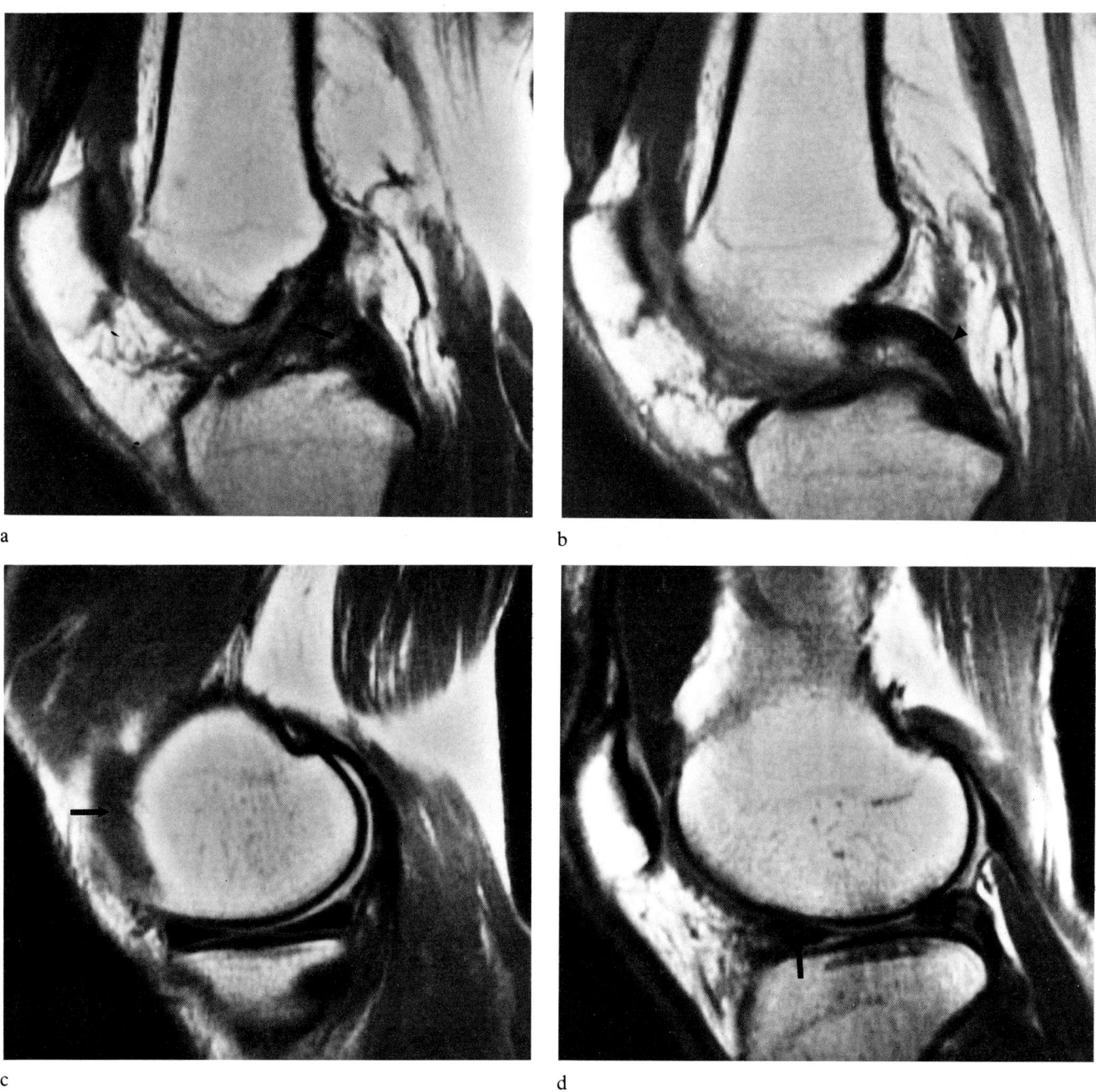

Figure 8.16

False-positive tear of the lateral meniscus in a 35-year-old man with chronic knee pain and recurrent joint effusions. All images are T_1-weighted (SE 800/28). (**a,b**) Sagittal images of (**a**) the anterior cruciate ligament (arrow) and (**b**) the posterior cruciate ligament (arrowhead) are normal. (**c**) Sagittal image of the medial meniscus shows no abnormality. Large joint effusion is present (arrowhead). (**d**) Sagittal image of the lateral meniscus reveals globular focus of increased signal intensity in the posterior horn (arrowhead), which appears to involve the superior articular surface. This was interpreted as a Grade IV meniscus abnormality consistent with a frank tear. Linear increased signal intensity not involving an articular margin is seen in the anterior horn (arrow). At arthroscopy, both menisci appeared normal.

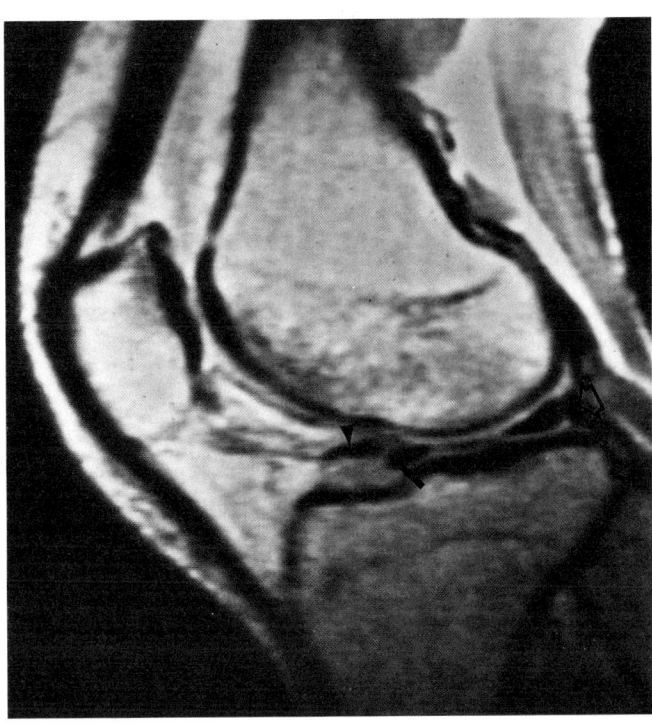

Figure 8.17
Transverse meniscal ligament appearing as a pseudotear. T_1-weighted sagittal MR image (SE 800/28) of the lateral meniscus of a 24-year-old man discloses intermediate signal intensity zone (arrow) between the anterior horn of the lateral meniscus and the transverse ligament (arrowhead). This should not be confused with a meniscal tear. Neither should the popliteus tendon coursing obliquely across the posterior horn of the normal lateral meniscus (open arrow) be mistaken for a meniscal tear.

Figure 8.18
Transverse meniscal ligament in a 26-year-old woman with lateral knee pain. T_1-weighted sagittal MR image demonstrates a normal lateral meniscus. The transverse ligament is well seen and may be misinterpreted as a tear of the anterior horn of the lateral meniscus (all images SE 800/28). (a) T_1-weighted sagittal MRI of the lateral meniscus. Zone of intermediate signal intensity (arrow) separates the transverse ligament from the anterior horn. (b) Slightly medial to (a): the low-intensity ligament is better defined (arrow). (c) Slightly medial to (b): the anterior horn of the lateral meniscus is no longer visible, but the transverse ligament is again seen (arrow). (d) Level of the anterior cruciate ligament. The transverse ligament (arrow) is seen anterior to a tibial attachment of ACL (arrowhead).

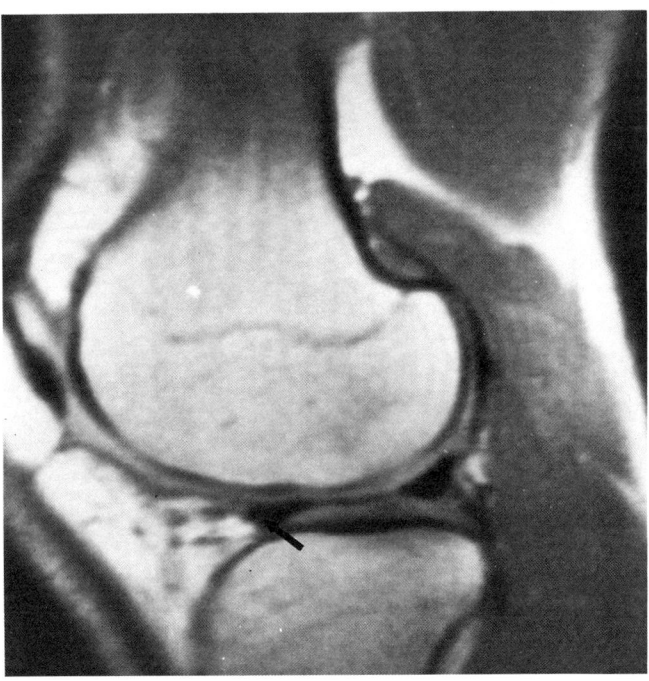

a

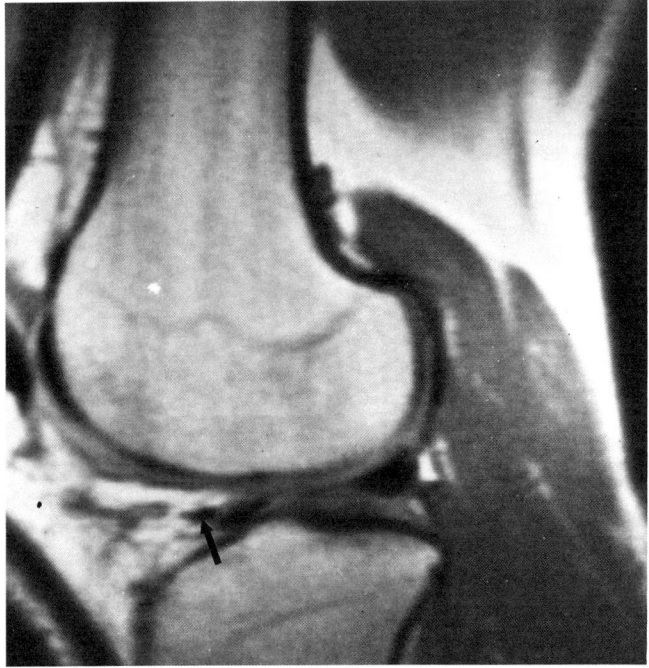

b

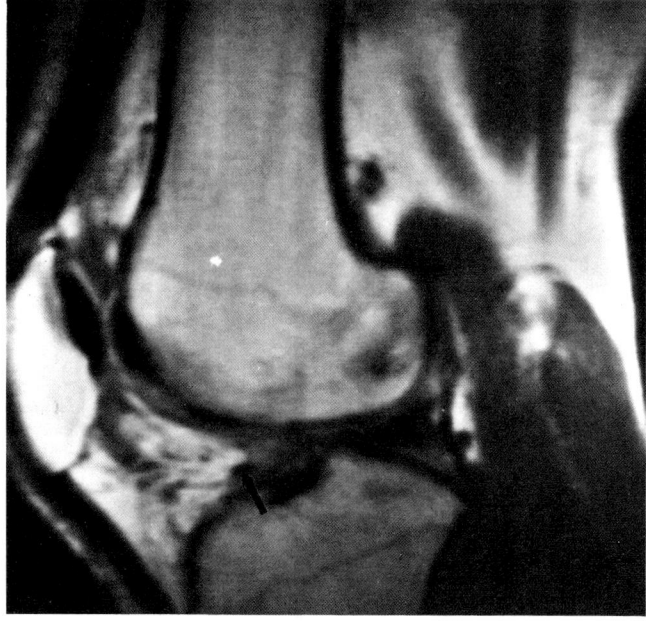

c

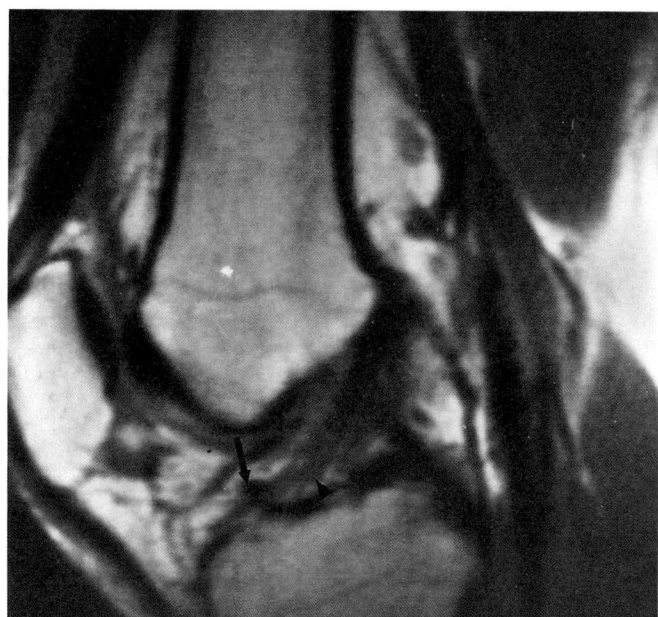

d

Figure 8.18 *continued*

Figure 8.19

A posterior cruciate ligament tear and false-positive anterior cruciate ligament tear in a 48-year-old man injured in a motor vehicle accident. All images are T_1-weighted (SE 800/28) and are slightly degraded by motion artifact. (**a**) Sagittal image through the PCL shows its complete disruption (arrow). (**b**) Sagittal image slightly lateral to (**a**) does not clearly identify the proximal portion of the ACL (arrow). Despite multiple attempts, the ACL could never be adequately seen and was therefore thought to be torn. Surgery revealed a complex tear of the PCL, but the ACL was normal. Perhaps stressed flexion views of the anterior cruciate ligament, as described by Turner et al,[15] would have shown a normal ACL.

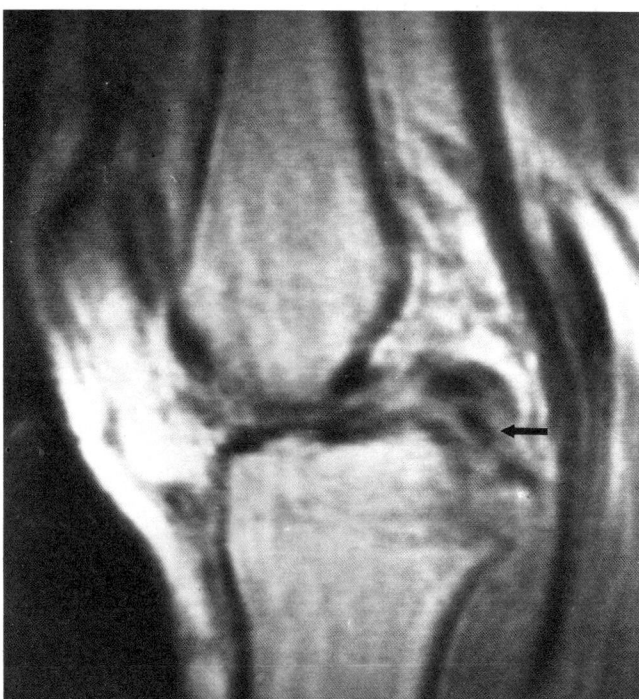

a

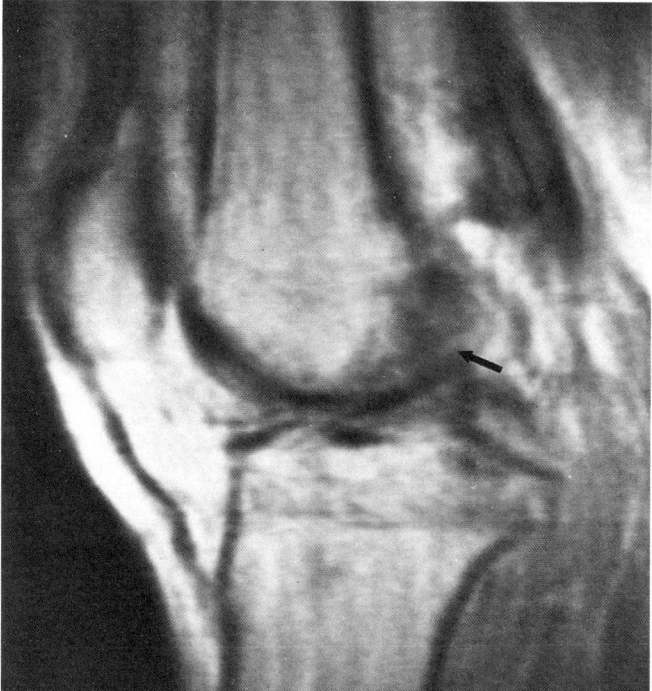

b

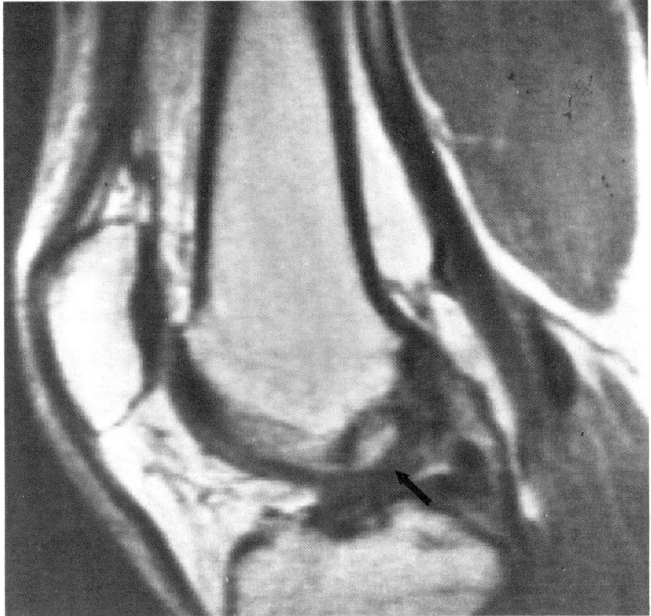
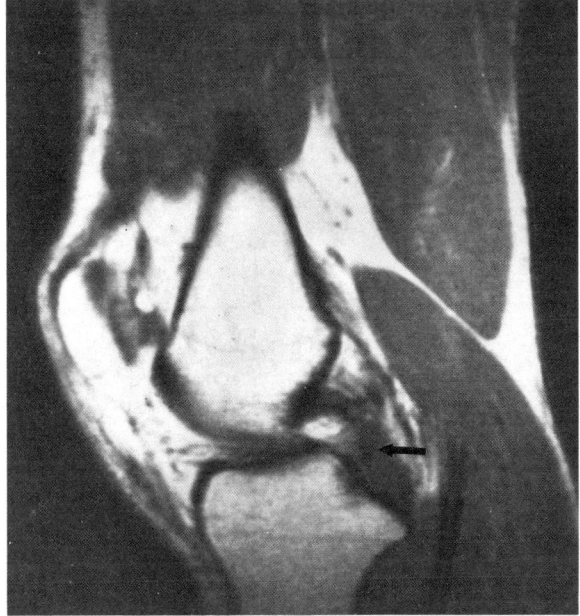

Figure 8.20

Partial tears of the anterior and posterior cruciate ligaments in a 34-year-old man injured in a skiing accident. (**a**) T_1-weighted sagittal MR image (SE 800/28) shows disruption of the ACL (arrow) consistent with a tear. (**b**) The PCL contains higher signal intensity than is normal, consistent with an intrasubstance hemorrhage (arrow). Surgery revealed partial tears of both cruciate ligaments.

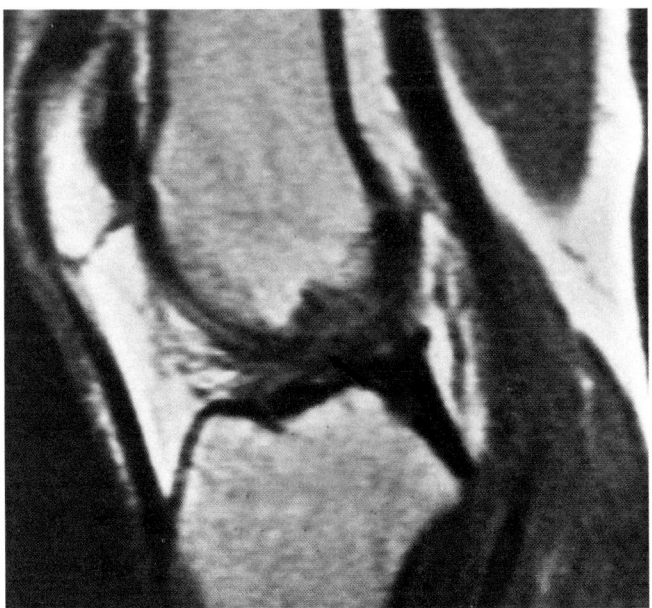
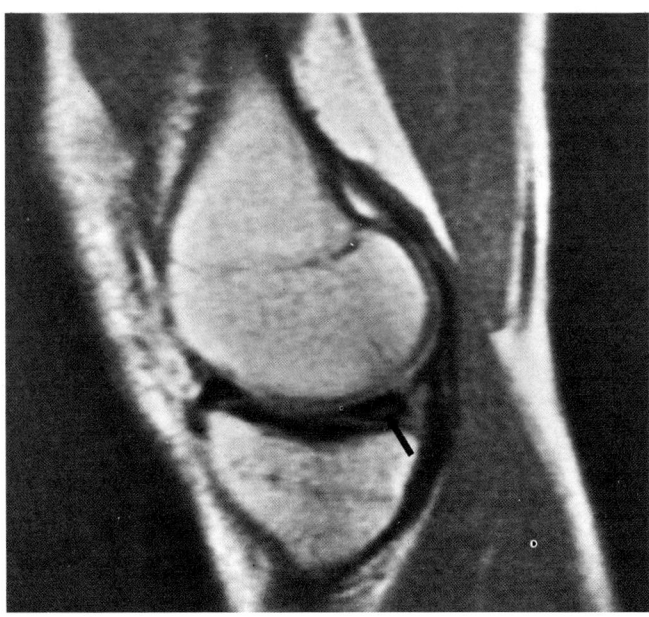

Figure 8.21

Anterior cruciate ligament tear and Grade III meniscal abnormality (possible tear) in a 35-year-old man who sustained an athletic injury. (**a**) T_1-weighted sagittal MR image (SE 800/28) at the level of the anterior cruciate ligament. There is complete disruption of the ACL (arrow). This was confirmed as a tear at surgery. (**b**) T_1-weighted sagittal MR image (SE 800/28) at the level of the medial meniscus. Linear focus of increased signal intensity (arrow) in the posterior horn extends close to but does not definitely involve the articular margin. At surgery, a frank tear was not seen.

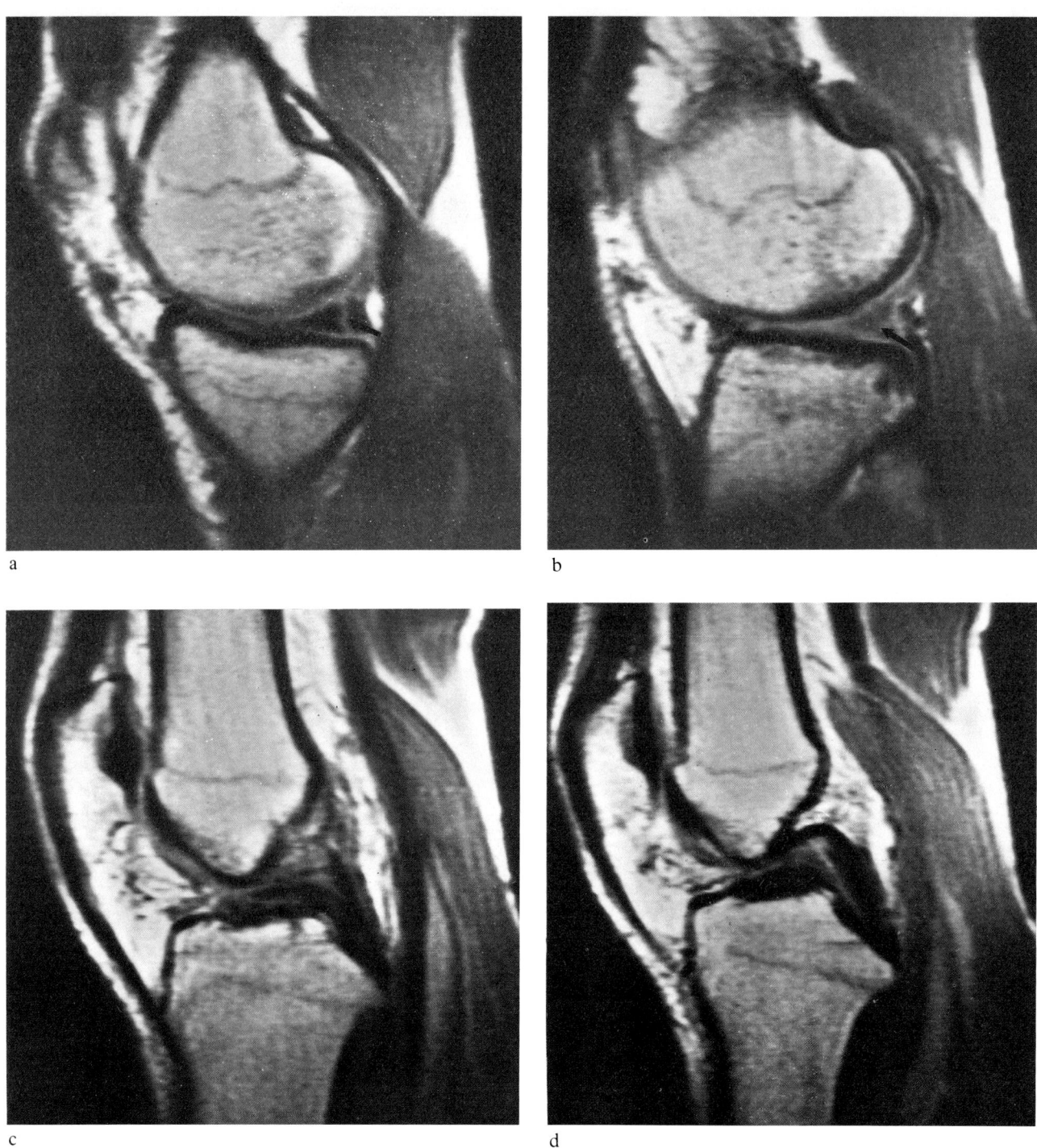

Figure 8.22

Anterior cruciate ligament tear and bilateral meniscal tears in a 29-year-old man involved in a skiing accident. (**a**) T_1-weighted sagittal MR image (SE 800/28) of the medial meniscus. A vertical bucket-handle tear involves the posterior horn (arrow). (**b**) T_1-weighted sagittal MR image (SE 800/28) of the lateral meniscus reveals diffuse increased signal intensity of the entire posterior horn (arrow), confirmed as a complex tear at surgery. (**c**) T_1-weighted sagittal MR image (SE 800/28) of the ACL discloses complete disruption with associated surrounding hemorrhage. (**d**) T_1-weighted sagittal MR image (SE 800/28) of the posterior cruciate ligament shows exaggerated bowing, a frequent finding in patients with ACL tears, thought to be due to anterior displacement of the tibia in relation to the femur. The PCL otherwise appears normal.

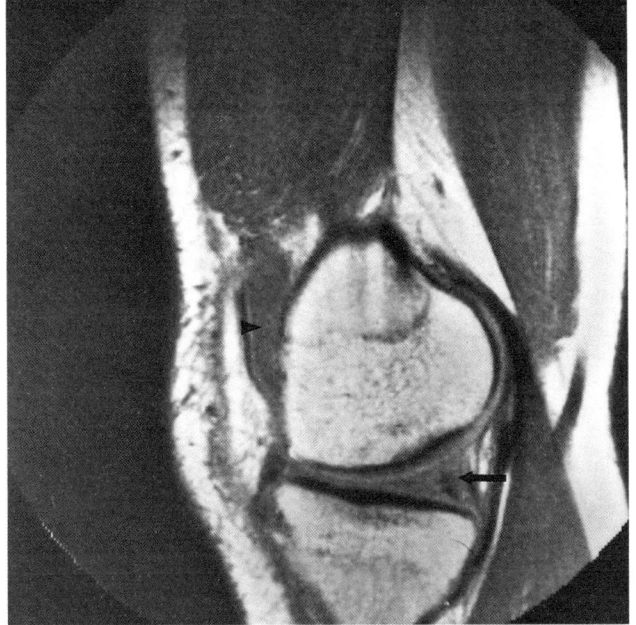
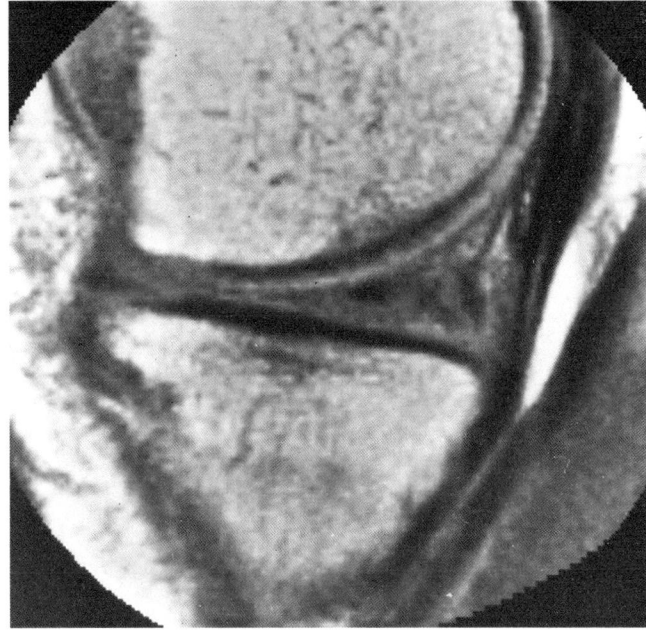
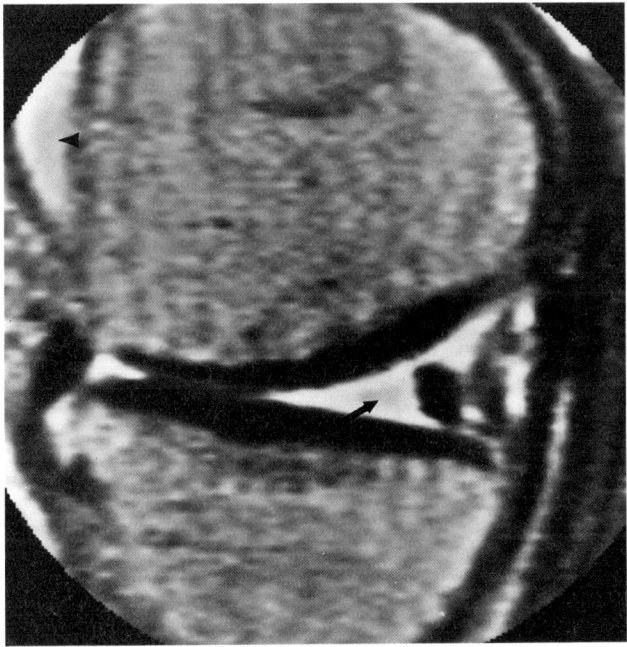

Figure 8.23

Complex meniscal and ligamentous injuries in a 24-year-old man involved in a motorcycle accident. (**a**) T_1-weighted sagittal MR image (SE 800/28) shows diffuse increased signal intensity in the posterior horn of the medial meniscus consistent with a complex tear (arrow). The anterior horn appeared too small on all images, and is avulsed from its normal attachments. An effusion is present in the suprapatellar bursa (arrowhead). (**b**) Closeup of lesion in (**a**). (**c** and **d**) T_2-weighted sagittal MR image (SE 2000/56) of medial meniscus show high signal intensity within the tear (arrow). Joint effusion is characterized by high-signal intensity (arrowhead). (**e**) T_1-weighted sagittal MR image (SE 800/28) at level of expected course of anterior cruciate ligament. The ligament is so severely torn that only the most distal portion is visible (arrow). (**f**) T_1-weighted sagittal MR image (SE 800/28) through the posterior cruciate ligament shows it to be intact. Subtle areas of increased signal intensity within the PCL possibly represent an intersubstance hemorrhage (white arrowhead). (**g**) T_1-weighted coronal MR image (SE 800/28) shows increased soft-tissue thickening in the expected location of the lateral collateral ligament (arrow). A tear was confirmed at surgery. Note the normal medial collateral ligament (arrowhead).

244 MRI atlas of the musculoskeletal system

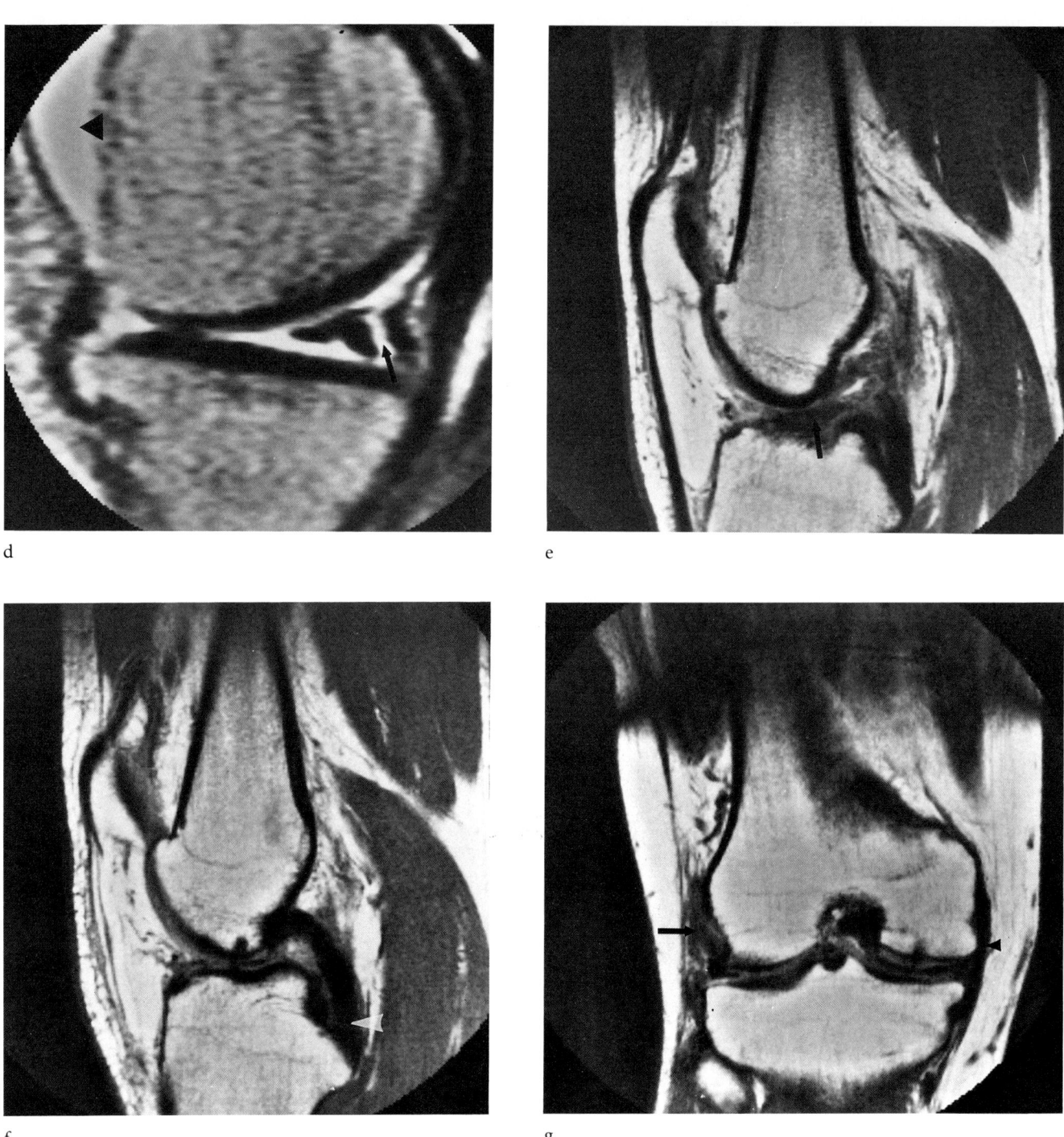

Figure 8.23 continued

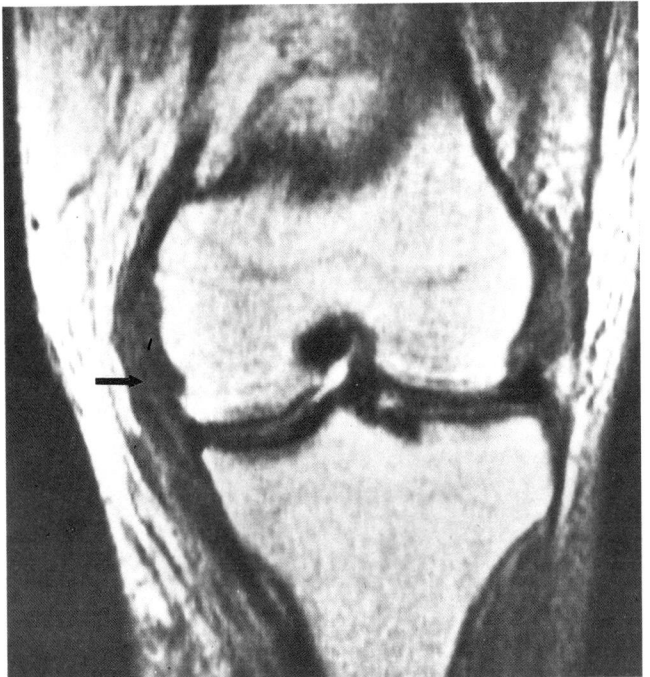
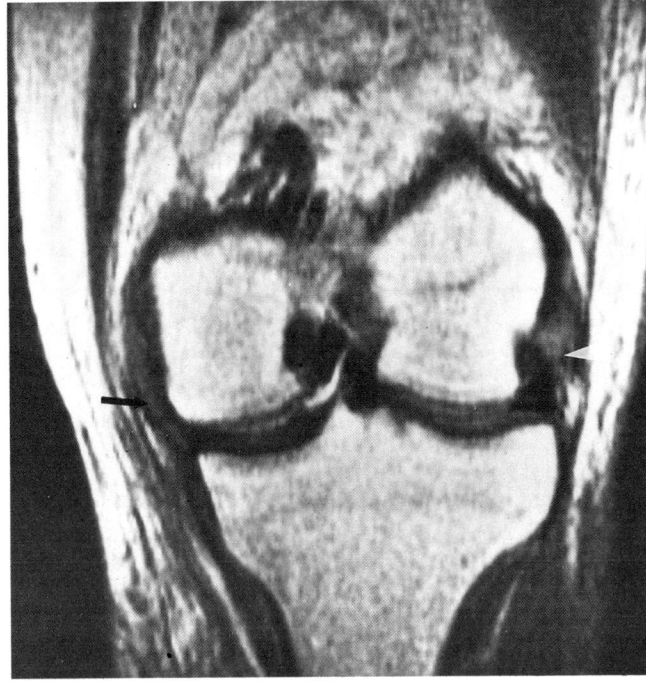
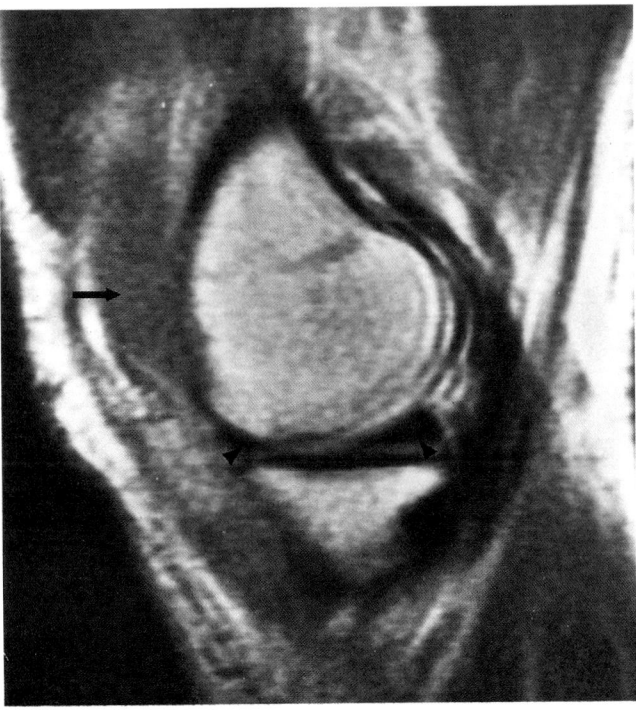

Figure 8.24

Multiple ligamentous injuries in a 67-year-old man involved in a motor vehicle accident. All images are T_1-weighted (SE 800/28). (**a**) Coronal image of the knee discloses complete disruption of the medial collateral ligament (arrow). Note the extensive hemorrhage separating the medial meniscus from subcutaneous fat. (**b**) Coronal image posterior to (**a**) again shows a tear of the medial collateral ligament (arrow). A hemorrhage separates the lateral collateral ligament (white arrowhead) from the adjacent bone, signifying ligamentous disruption. At surgery, a complete tear of the medial collateral ligament and partial tear of the lateral collateral ligament were identified. (**c**) Sagittal image at the level of the medial meniscus reveals effusion in the suprapatellar bursa (arrow). The anterior and posterior horns of the medial meniscus appear normal (arrowheads). (**d**) Sagittal image at the level of the lateral meniscus. Although subtle degenerative changes are present within the posterior horn, a frank tear is not identified. The infrapatellar ligament is well seen and reveals no evidence of a tear. Nevertheless, subtle focus of increased signal intensity within the ligament (arrowhead) is indicative of an intrasubstance hemorrhage. (**e**) Sagittal image at the level of the intercondylar notch shows disruption of the proximal portion of the anterior cruciate ligament (arrow). (**f**) Sagittal image slightly medial to (**e**) reveals a complete tear of the posterior cruciate ligament (arrow). All of the above findings were confirmed at surgery.

246 MRI atlas of the musculoskeletal system

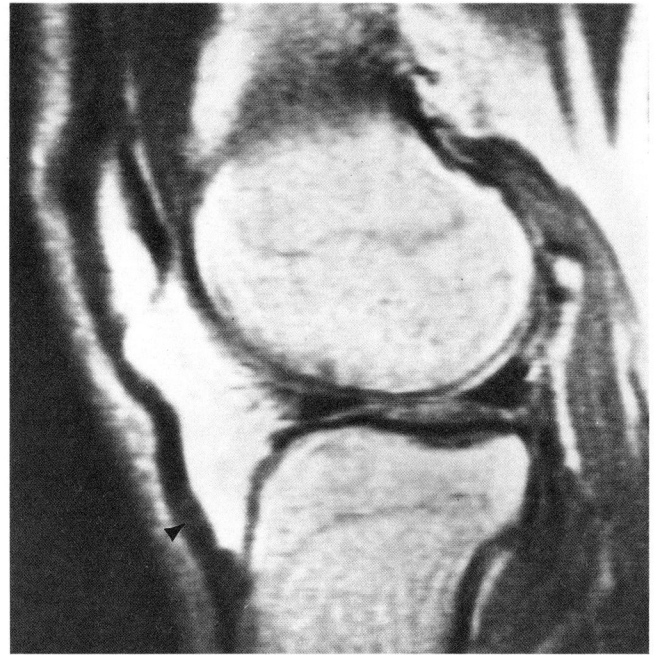

d

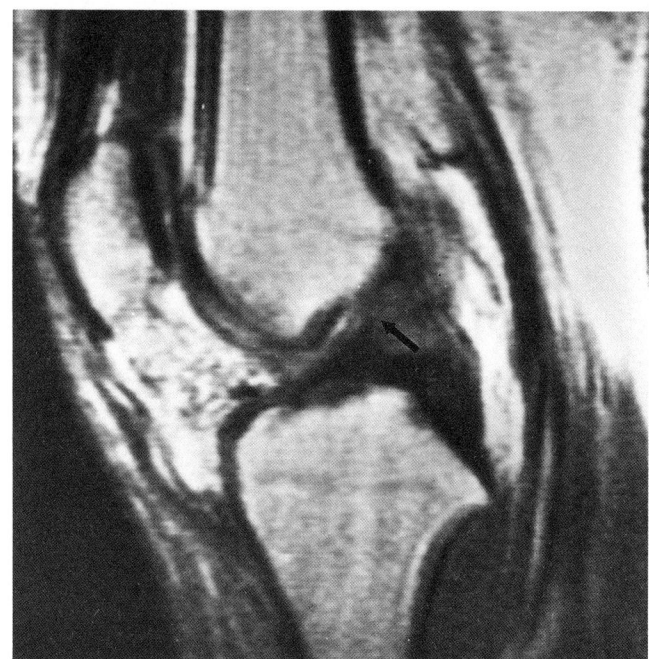

e

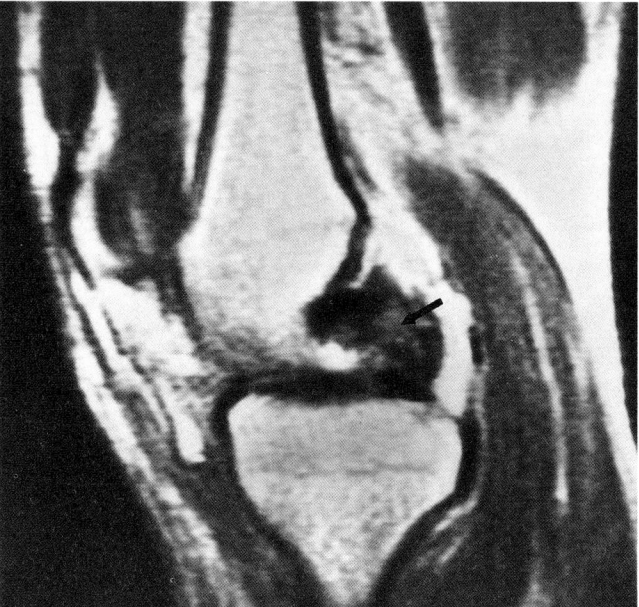

f

Figure 8.24 continued

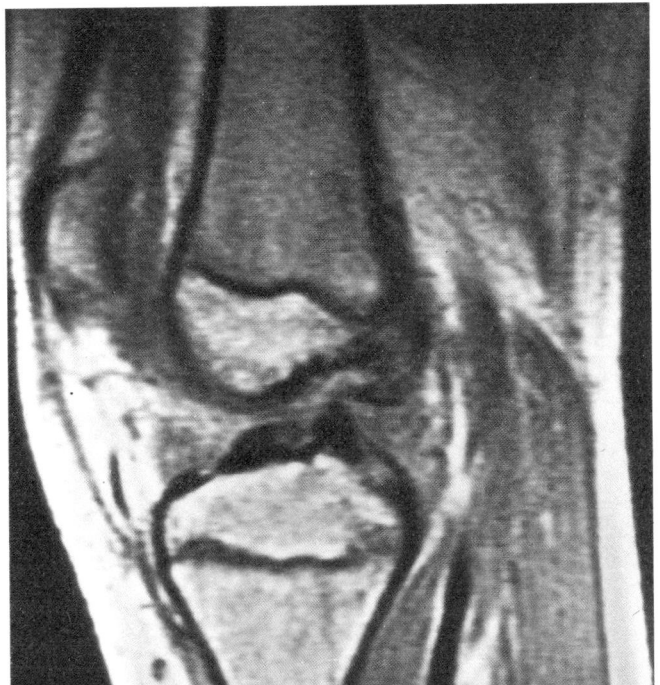

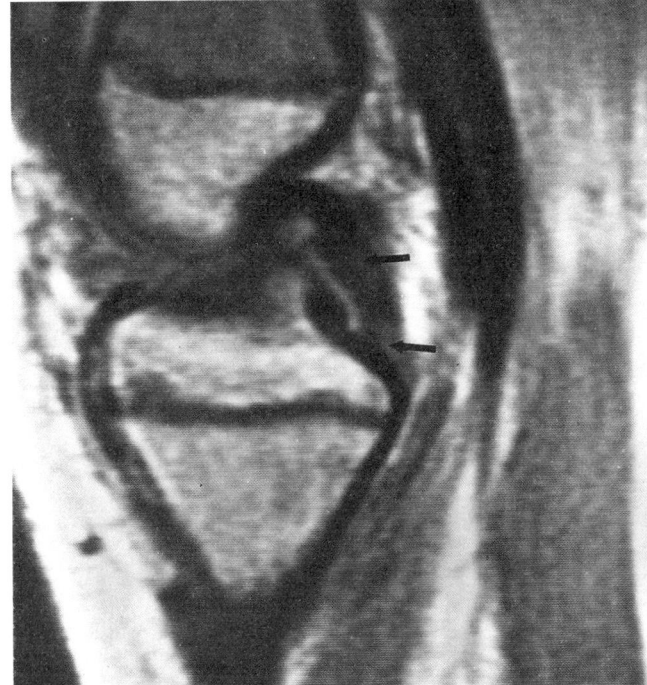

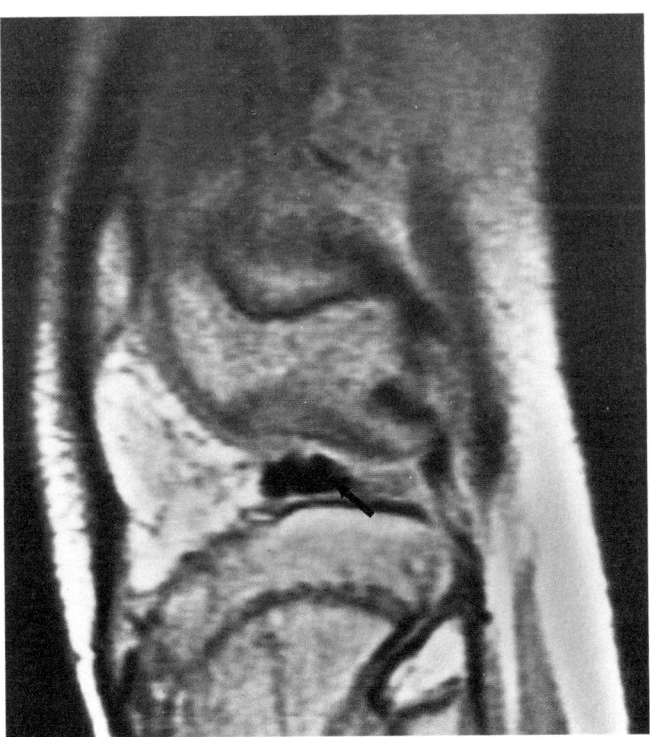

Figure 8.25

Cruciate and lateral meniscal tears in a 13-year-old boy involved in an automobile accident (all images SE 800/28). (**a**) T_1-weighted sagittal image through the intercondylar notch shows complete disruption of the anterior cruciate ligament. A frank tear was confirmed at surgery. (**b**) T_1-weighted sagittal image slightly medial to (**a**) reveals increased signal intensity within the otherwise intact posterior cruciate ligament, indicative of an intrasubstance hemorrhage (arrows). A frank tear was not identified at surgery. (**c**) T_1-weighted sagittal image of the lateral meniscus reveals gross distortion of the anterior and posterior horns, with anterior displacement of a large fragment of posterior horn (arrow). The medial meniscus (not shown) appeared normal. An extensive tear of the lateral meniscus was confirmed at surgery.

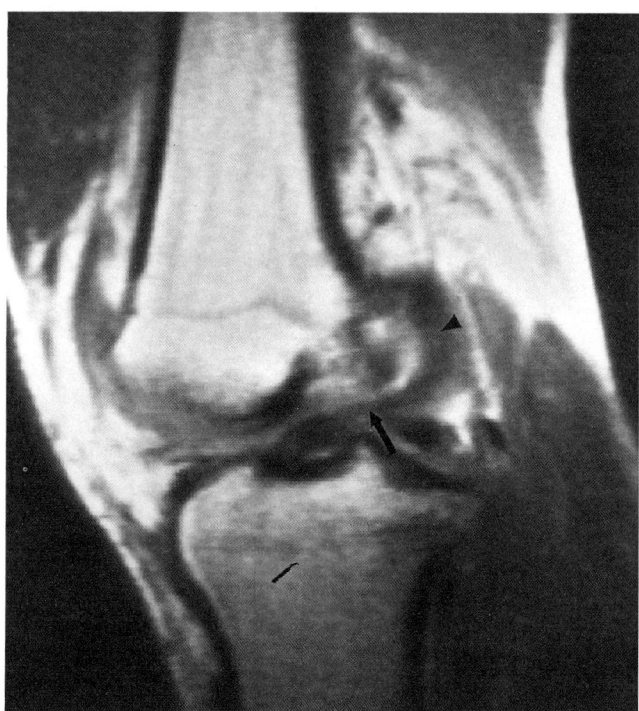

a

Figure 8.26

Anterior cruciate ligament tear in a 30-year-old man injured in a skiing accident. A positive drawer sign elicited on physical examination was consistent with a tear of the ACL (**a**) T_1-weighted sagittal MR image (SE 800/28) shows posterior displacement of the ACL (arrow) and avulsion of its proximal attachment from the femur (arrowhead). (**b**) Sagittal image slightly medial to (**a**) shows normal continuity of the posterior cruciate ligament (arrow). Surgery disclosed avulsion of the ACL from the femur.

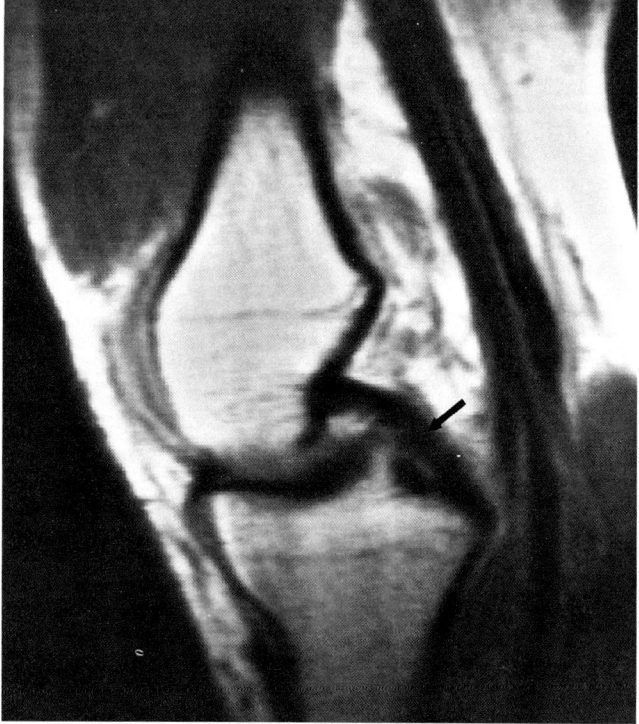

b

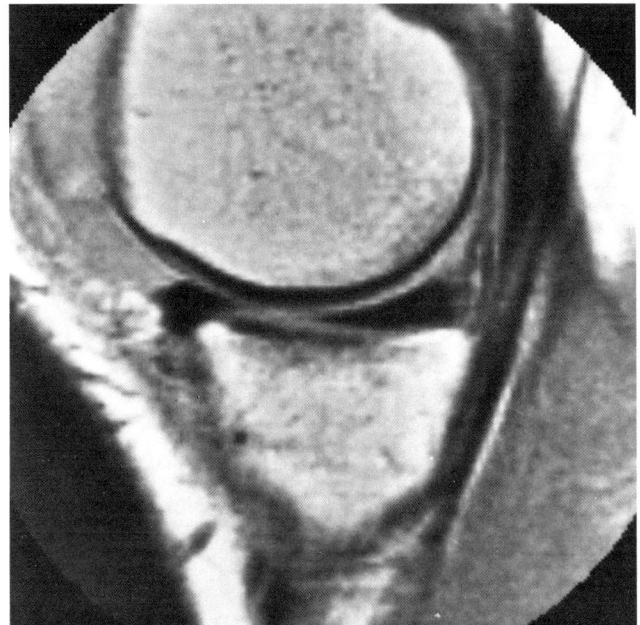

a

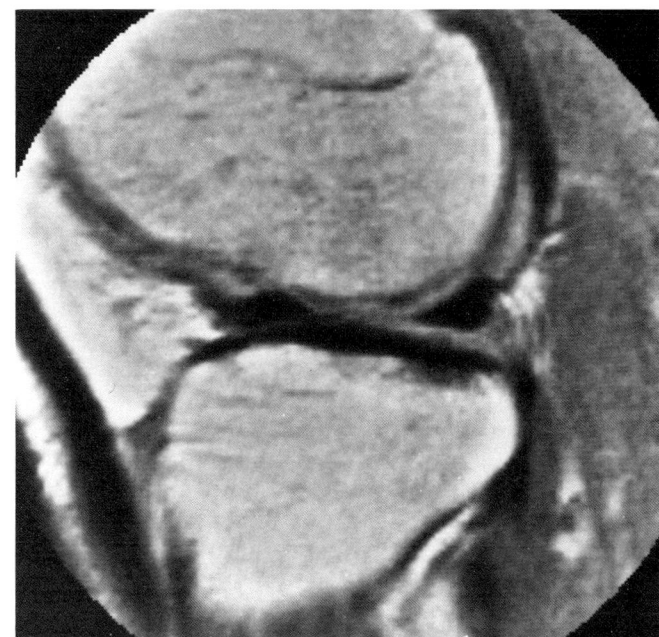

b

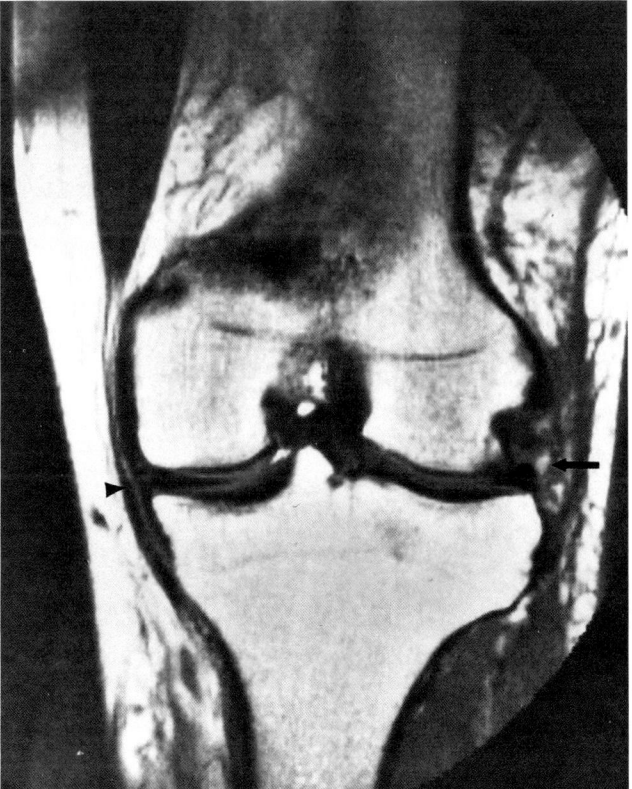

c

Figure 8.27

Lateral collateral ligament tear in a 40-year-old man with suspected meniscal and lateral collateral ligament tears. (**a**) Medial and (**b**) lateral menisci are entirely normal on T_1-weighted sagittal MR images (SE 800/28). (**c**) T_1-weighted coronal MR image (SE 800/28) shows disruption of the lateral collateral ligament (arrow), with surrounding intermediate signal intensity implying hemorrhage and edema. The medial collateral ligament (arrowhead) is normal. These findings here confirmed at surgery.

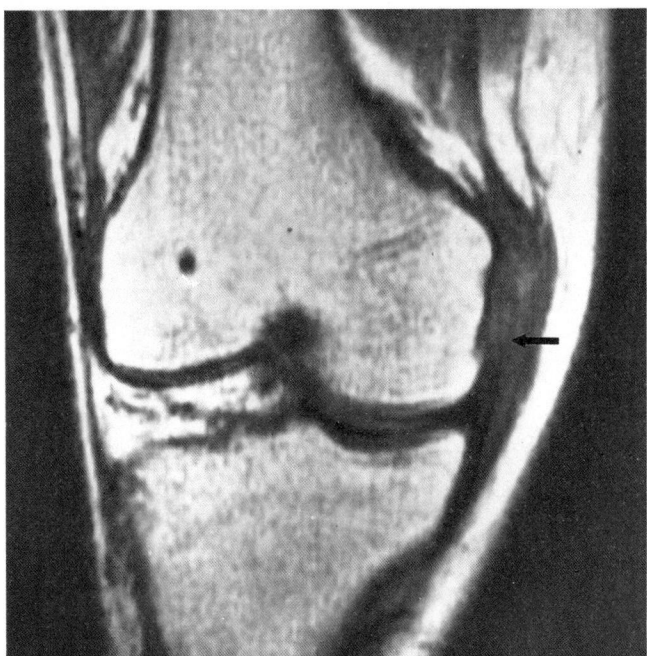

Figure 8.28

Pellegrini–Stieda disease. A 42-year-old man with chronic medial knee pain. T_1-weighted coronal MR image (SE 800/28) shows marked thickening of the medial collateral ligament (arrow) consistent with chronic inflammation.

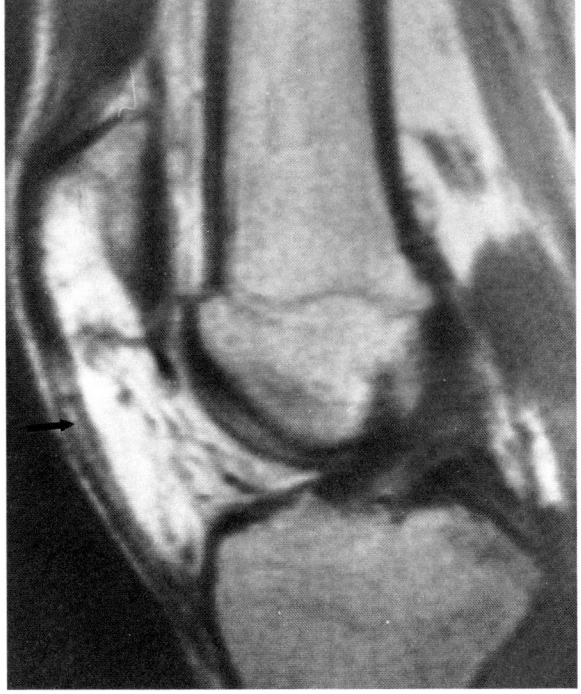

a

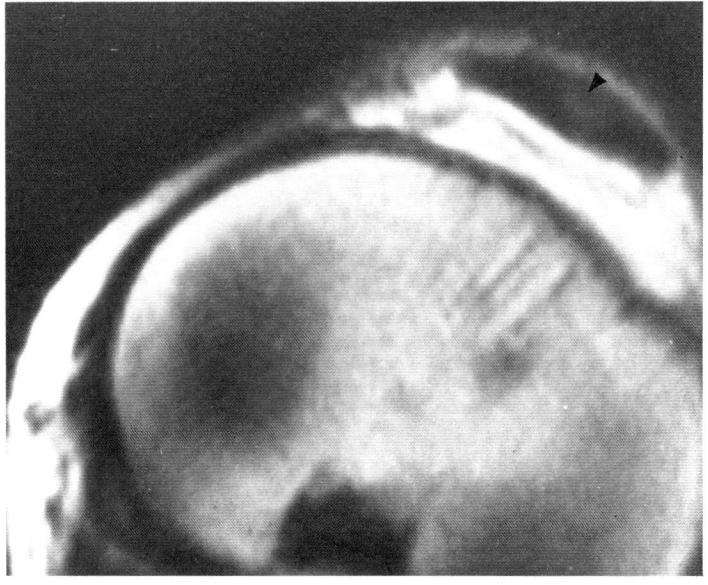

b

Figure 8.29

Patellar tendinitis in a 26-year-old male athlete with persistent tenderness inferior to the patella, despite normal radiographic examination. All images are T_1-weighted (SE 800/28). (**a**) Sagittal image through the patellar tendon shows a linear band of increased signal intensity within the patellar tendon, consistent with inflammation (arrow). (**b**) Axial image shows round focus of intermediate signal intensity (arrowhead) within normal low-intensity tendon. The symptoms resolved with conservative treatment.

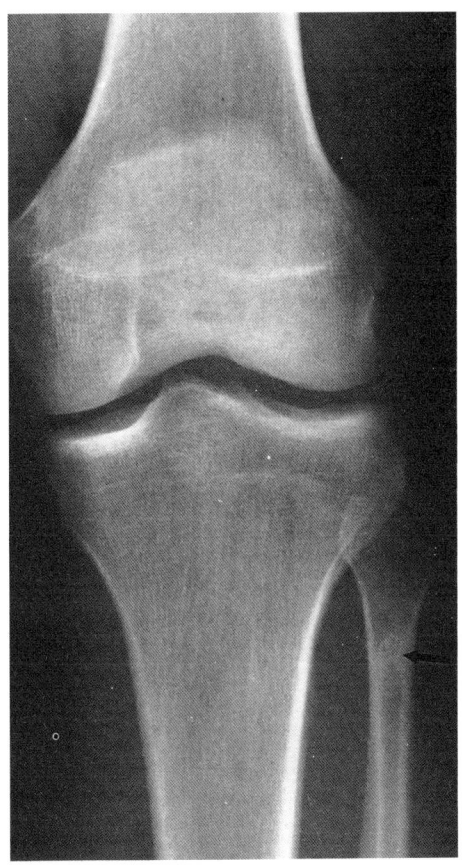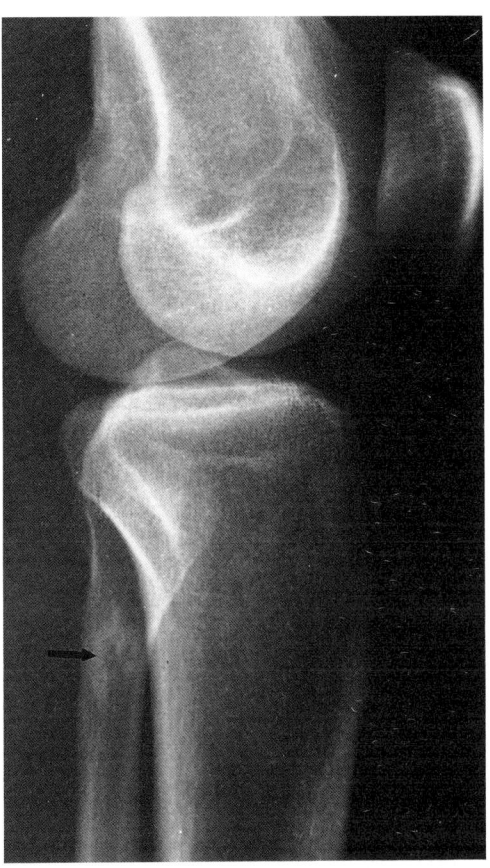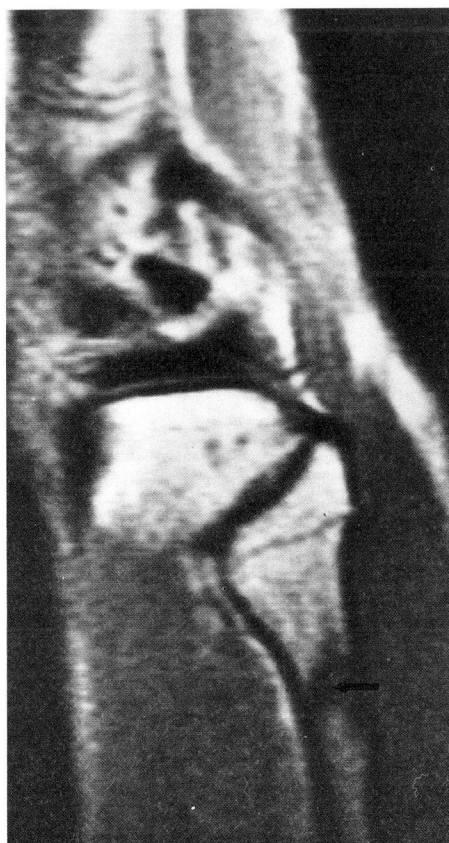

Figure 8.30

Stress fracture of the proximal part of the fibula in a 36-year-old man. (**a**) Anteroposterior and (**b**) lateral radiographs of the fibula show callus (arrow). (**c**) T_1-weighted sagittal MR image (SE 800/28) of the fibula reveals a corresponding region of decreased signal intensity (arrow).

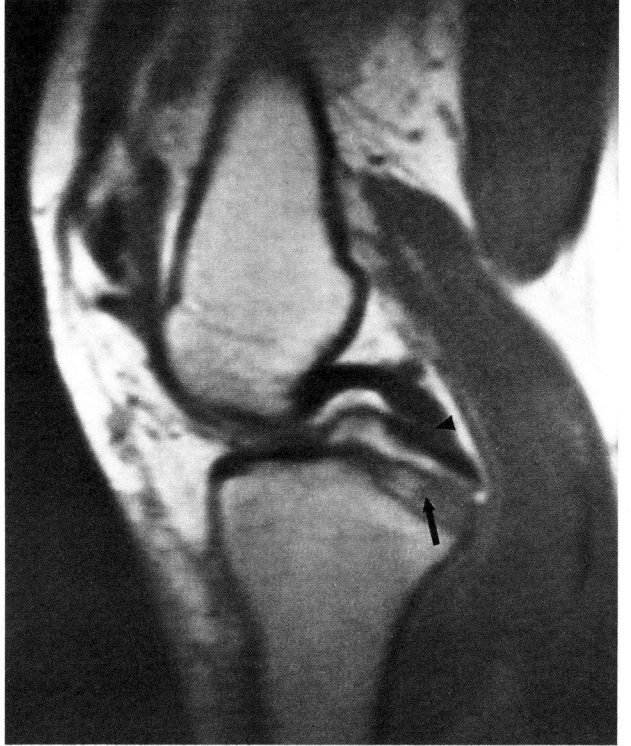

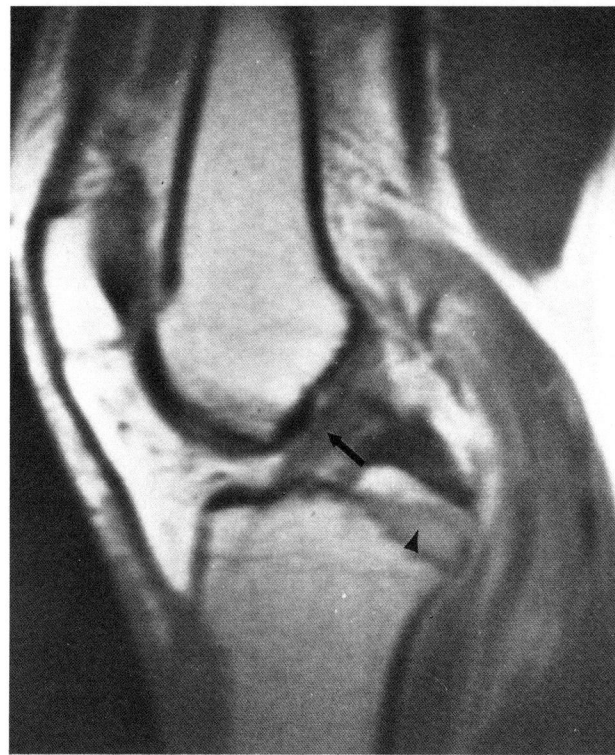

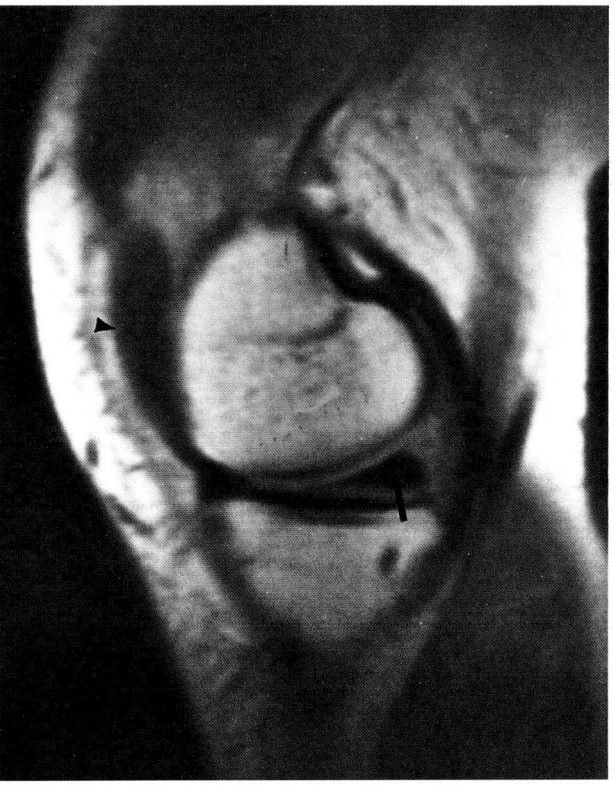

Figure 8.31

Tibial plateau fracture and associated posterior cruciate ligament avulsion in a 39-year-old woman injured in a motor vehicle accident. Radiographs revealed fracture of the posterior aspect of the tibial plateau. (**a**) T_1-weighted sagittal MR image (SE 800/28) clearly shows the fracture (arrow) and avulsion of bone at the tibial attachment of the PCL. Increased signal intensity within the PCL (arrowhead) represents an intraligamentous hemorrhage. (**b**) T_1-weighted sagittal MR image (SE 800/28) slightly lateral to (**a**) shows the anterior cruciate ligament intact (arrow). The fracture is again seen posteriorly (arrowhead). (**c**) T_1-weighted sagittal MR image (SE 800/28) through the medial meniscus discloses subtle linear increased signal intensity in the posterior horn (arrow), but no evidence of a definite tear. Joint effusion is characterized by low-intensity signal (arrowhead).

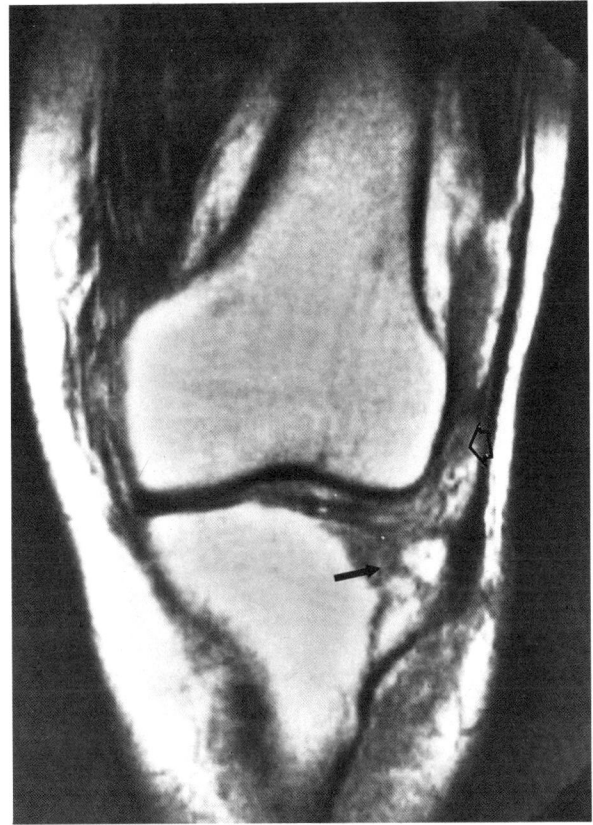
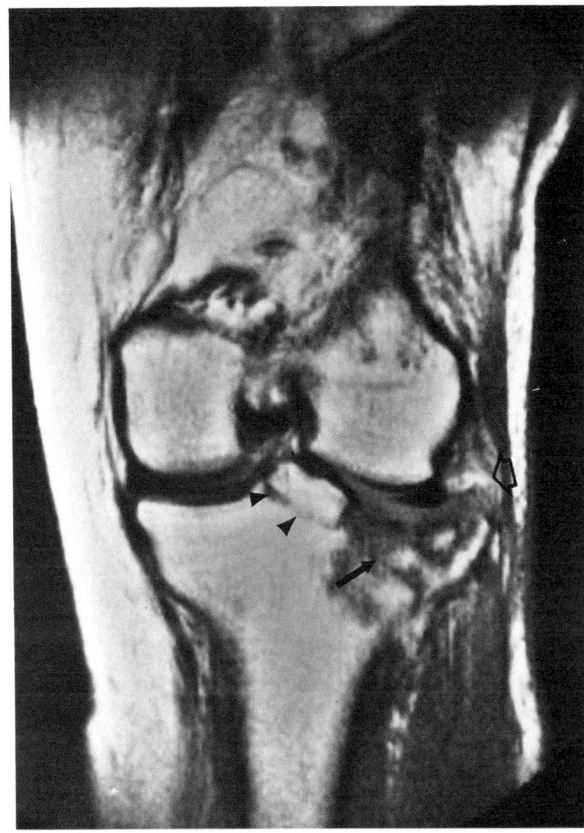
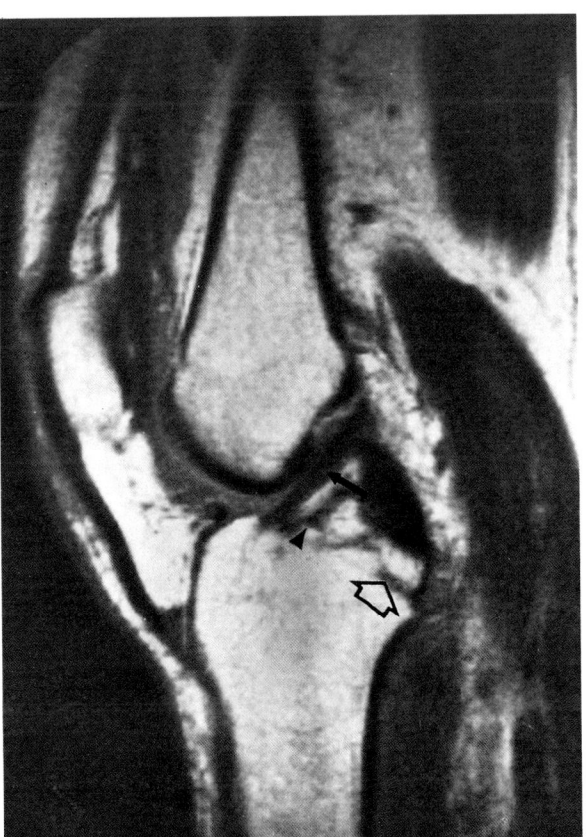

Figure 8.32

Tibial plateau fracture. A 24-year-old man with extensive right lower extremity injuries sustained in a motorcycle accident. All images are relatively T_1-weighted (SE 1000/28). (**a,b**) Coronal images demonstrate a comminuted fracture of the lateral tibial plateau (arrow). The fracture line appears to extend to the tibial tuberosity on these views (arrowheads). The lateral collateral ligament is also extensively torn (open arrow). (**c,d**) Sagittal images at the level of the anterior cruciate ligament show the ACL to be intact (arrow). The fracture seems to involve the tibial plateau posterior to the ACL attachment (arrowhead). The fracture does involve the tibial attachment of the posterior cruciate ligament (open arrow). (**e**) Sagittal image slightly lateral to (**c**) again shows avulsion of the PCL from the tibia (arrow). An intrasubstance ligament tear is not seen. (**f**) Sagittal image at the level of the lateral meniscus shows the posterior horn to be deformed and attenuated, especially along the inferior articular margin (arrow). Low-signal regions in the proximal tibia represent bone hemorrhage and edema. Note the high-signal fat within the patellofemoral joint and suprapatellar bursa (arrowhead). At surgery, all the MR findings were confirmed.

254   *MRI atlas of the musculoskeletal system*

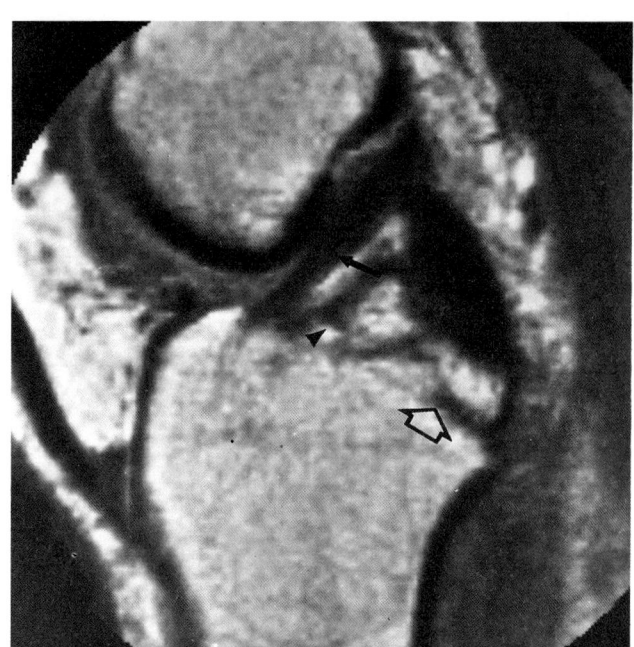

d

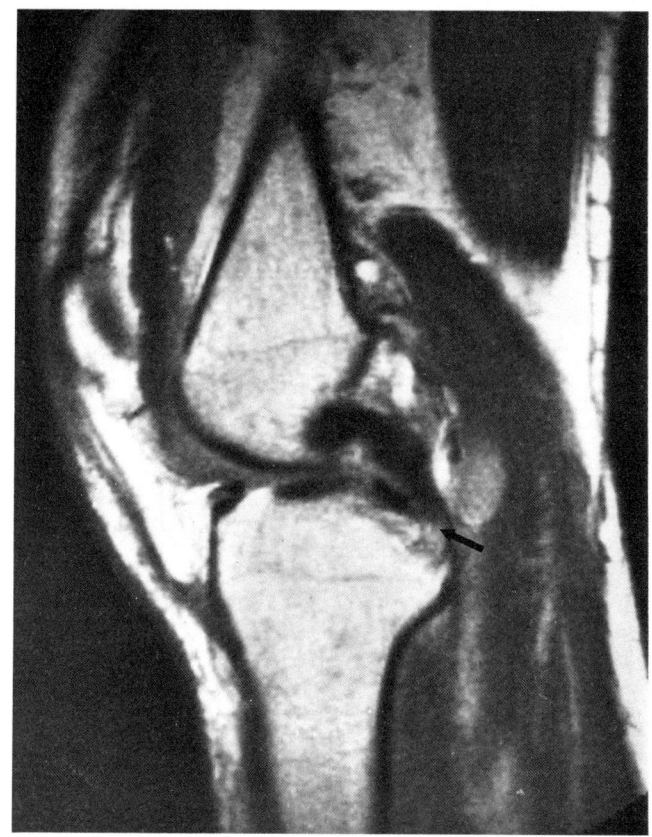

e

Figure 8.32 *continued*

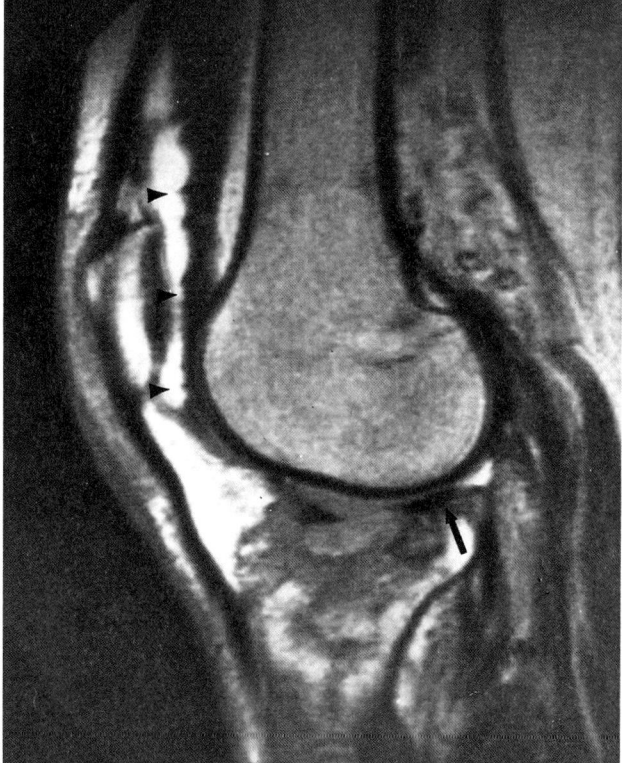

f

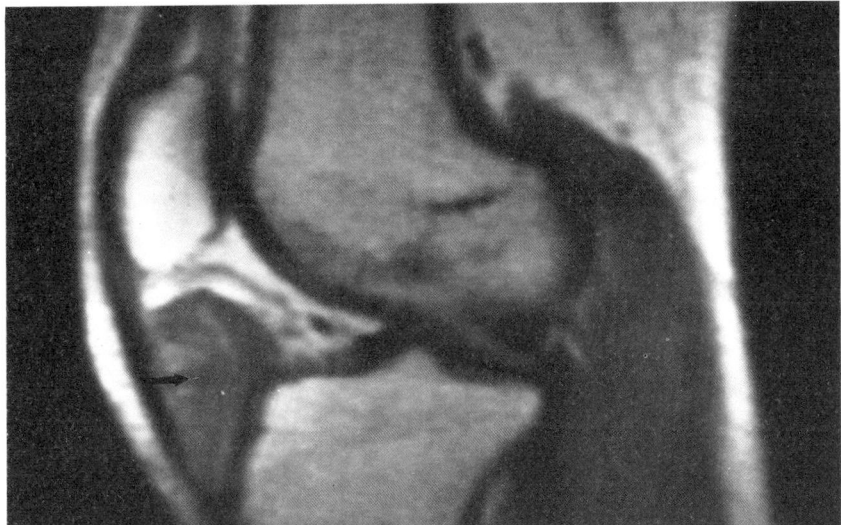

a

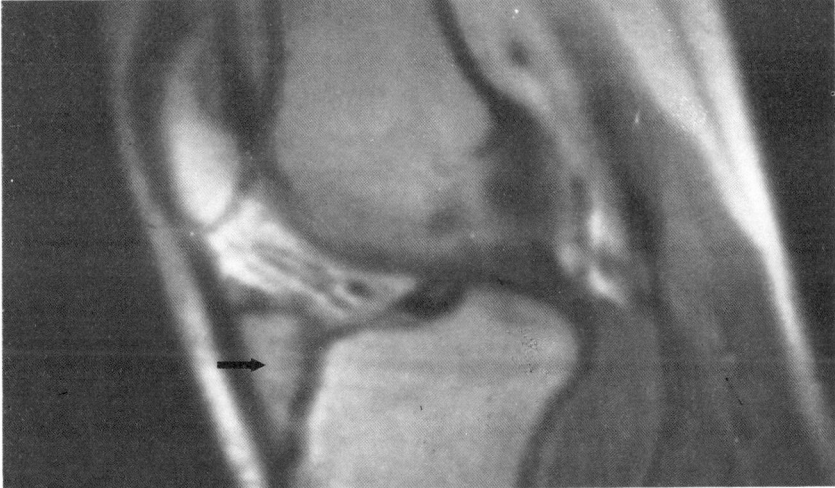

b

Figure 8.33

Infrapatellar bursa effusion. A 30-year-old woman who developed acute pain and swelling immediately below the patella after stepping off a curb. (**a**) Initial sagittal MR image (SE 800/28) showed an intermediate signal intensity fluid collection in the intrapatellar bursa (arrow). The patient declined treatment and returned for a repeat study. (**b**) Sagittal image 1 month later showed the collection to be markedly decreased in size (arrow). Symptoms subsequently resolved.

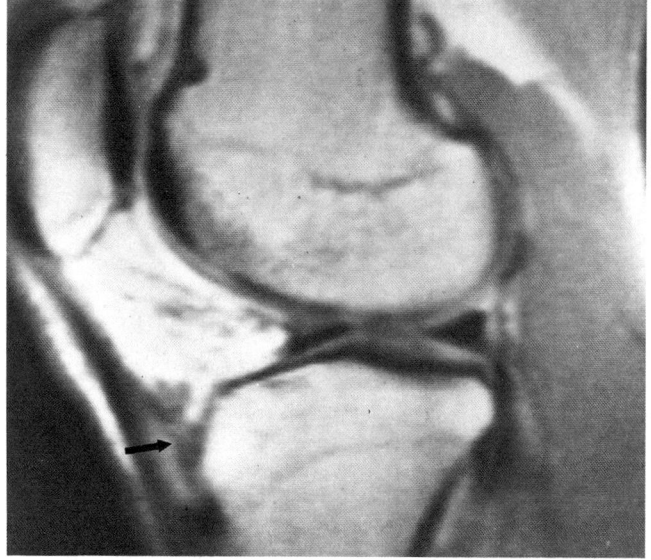
a

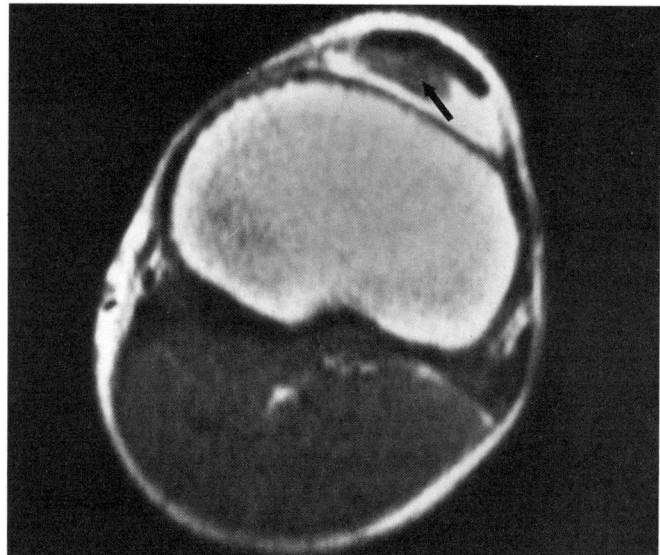
b

Figure 8.34

Effusion of an infrapatellar bursa in a 20-year-old man with recurrent infrapatellar swelling and pain. (a) T_1-weighted sagittal and (b) axial MR images (SE 800/28) show fluid collection in the infrapatellar bursa (arrow), characterized by intermediate signal intensity posterior to the infrapatellar ligament and anterior to the infrapatellar fat pad. The bursa was aspirated and injected with corticosteroid medication. A scan 3 months later (not shown) revealed complete disappearance of the bursa.

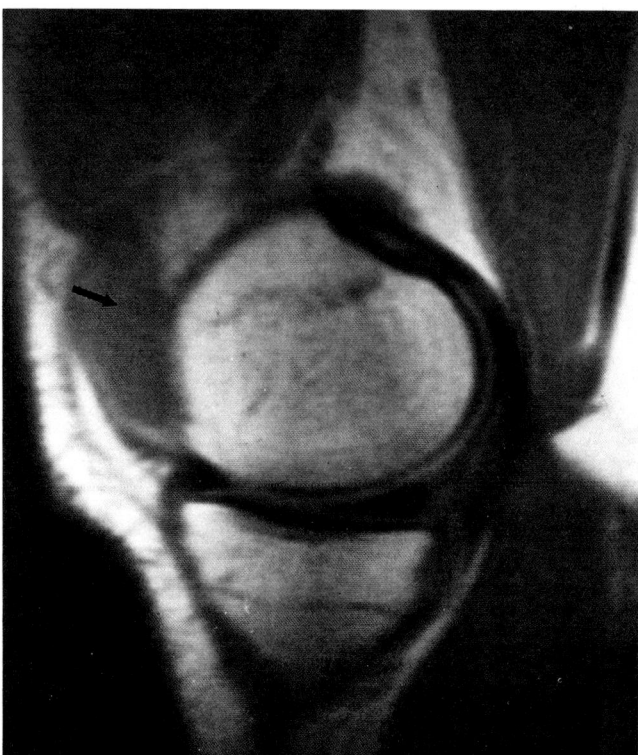
a

Figure 8.35

Effusion and normal (Grade I) meniscus in a 50-year-old man with strong clinical suspicion of medical meniscal tear and suspicion of possible tear of anterior cruciate ligament (all images SE 800/28). (a) T_1-weighted sagittal image shows large effusion in the suprapatellar bursa (arrow). Anterior and posterior horns of the medial meniscus are normal. (b) T_1-weighted sagittal image at the level of the ACL discloses the intact ligament (arrows). (c) T_1-weighted sagittal image reveals the normal posterior cruciate ligament. (d) Joint effusion is also well seen on T_1-weighted axial image (arrow). Despite the absence of significant MR findings, the patient was taken to surgery, where no abnormalities were identified.

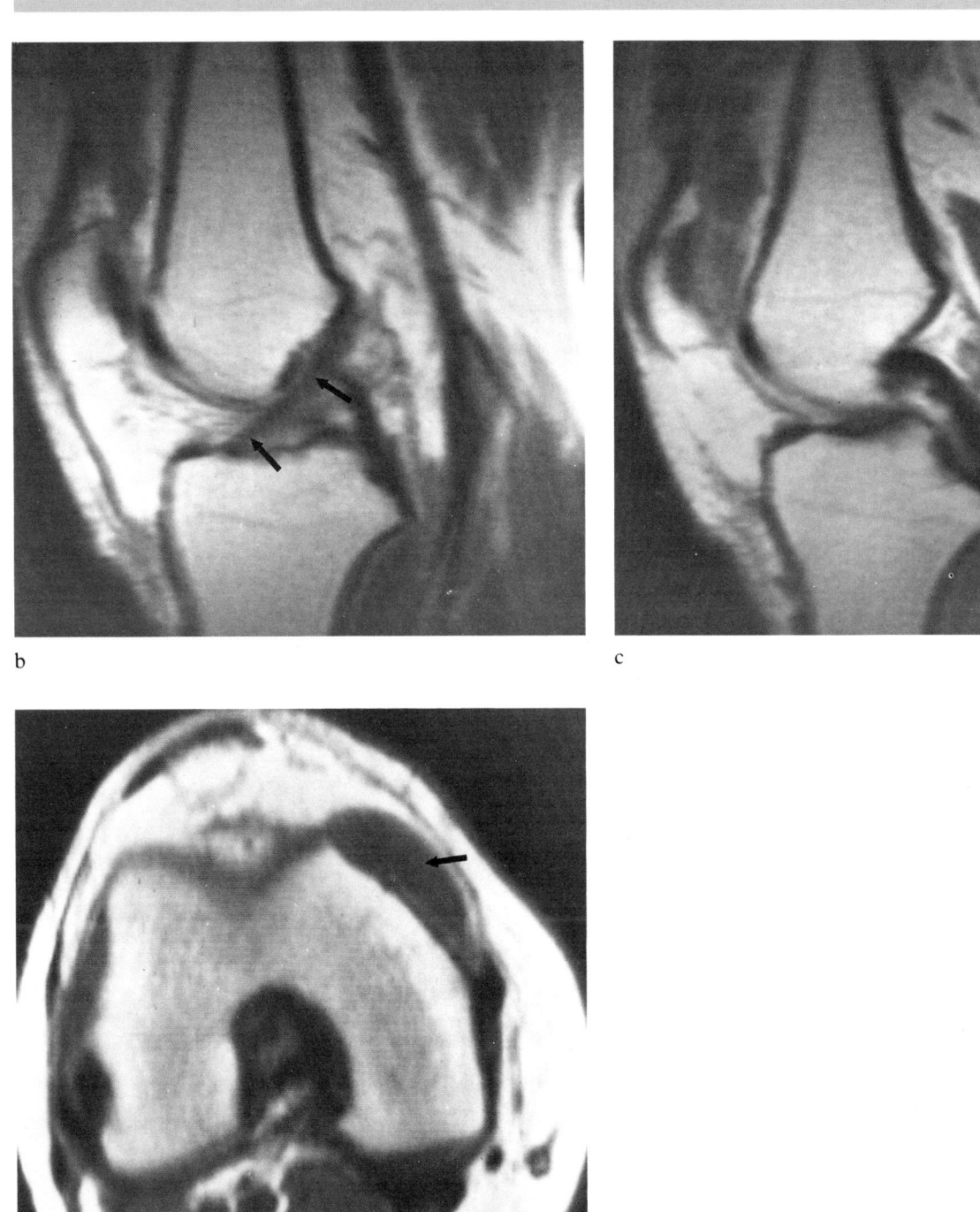

b

c

d

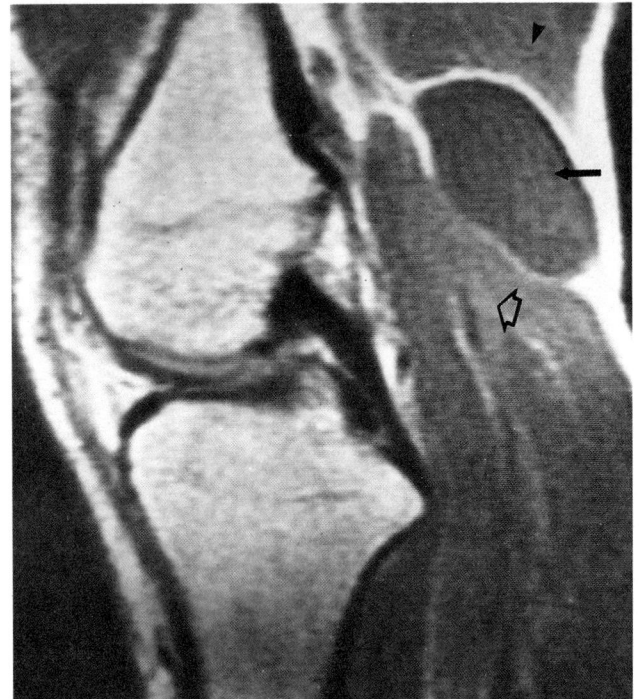

a

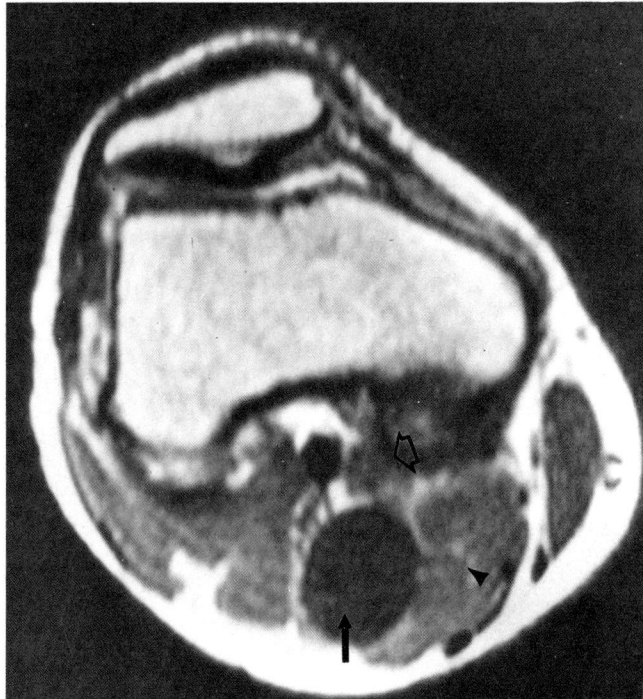

b

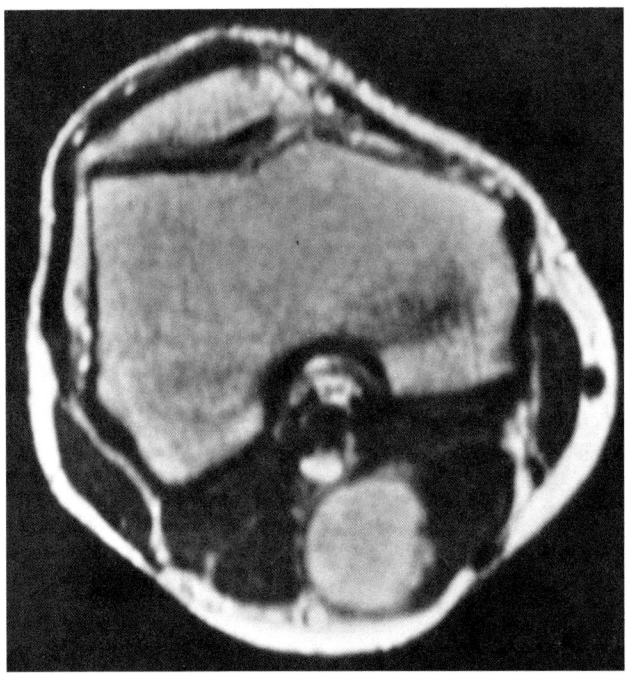

c

Figure 8.36

Popliteal cyst. A 50-year-old man with a palpable mass in the popliteal fossa. (**a**) T_1-weighted sagittal and (**b**) axial MR images (SE 800/28) reveal a well circumscribed mass (arrow) between the semimembranosus (arrowhead) and medial gastrocnemius muscles (open arrow). Signal intensity of mass is slightly less than intensity of the surrounding musculature. (**c**) T_2-weighted axial MR image (SE 2000/56) reveals the mass has increased signal intensity, characteristic of fluid within the cyst-like structure.

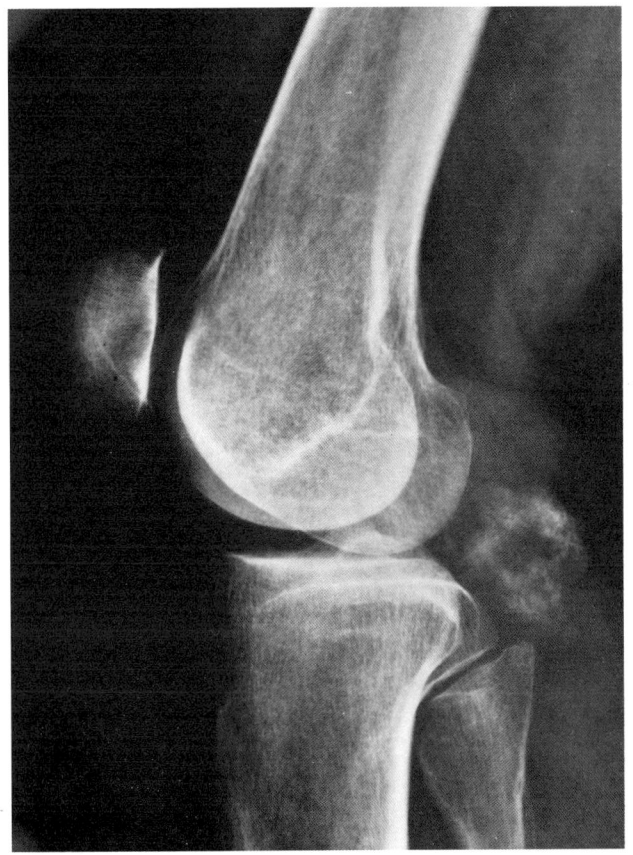
a

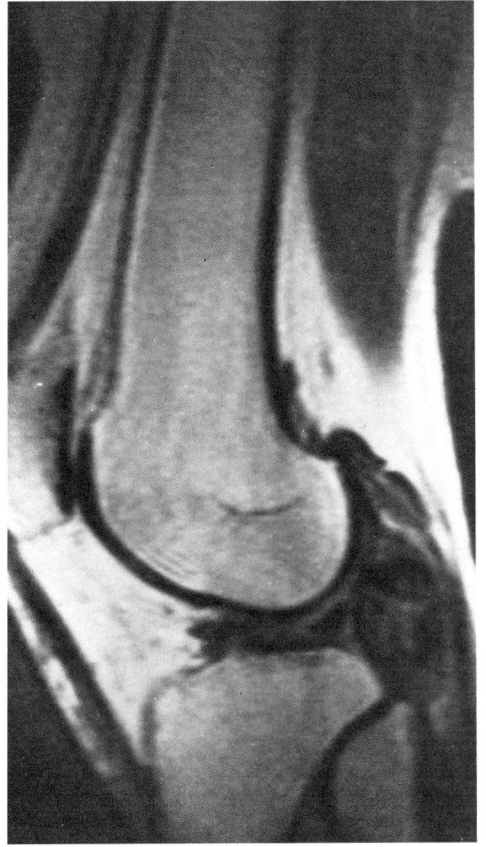
b

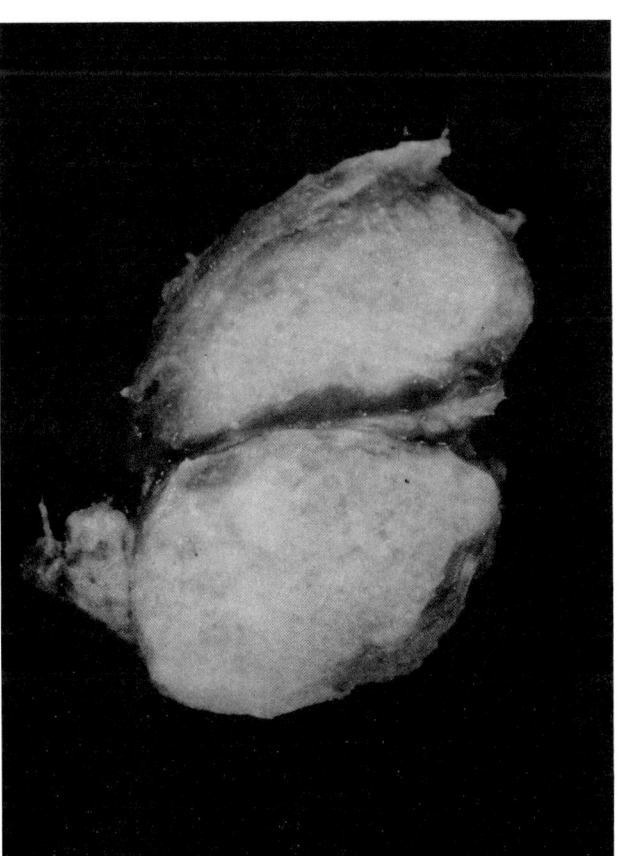
c

Figure 8.37

Calcified loose body. A 55-year-old man with a hard palpable mass in the posterior knee. (**a**) Radiograph shows a bilateral calcified mass in the popliteal fossa. (**b**) T_1-weighted sagittal MR image (SE 800/28) reveals a well circumscribed mass of mixed signal intensity. The low-signal regions represent calcification seen on the X-ray. (**c**) The surgical specimen shows the mass to be a large calcified osteochondral loose body.

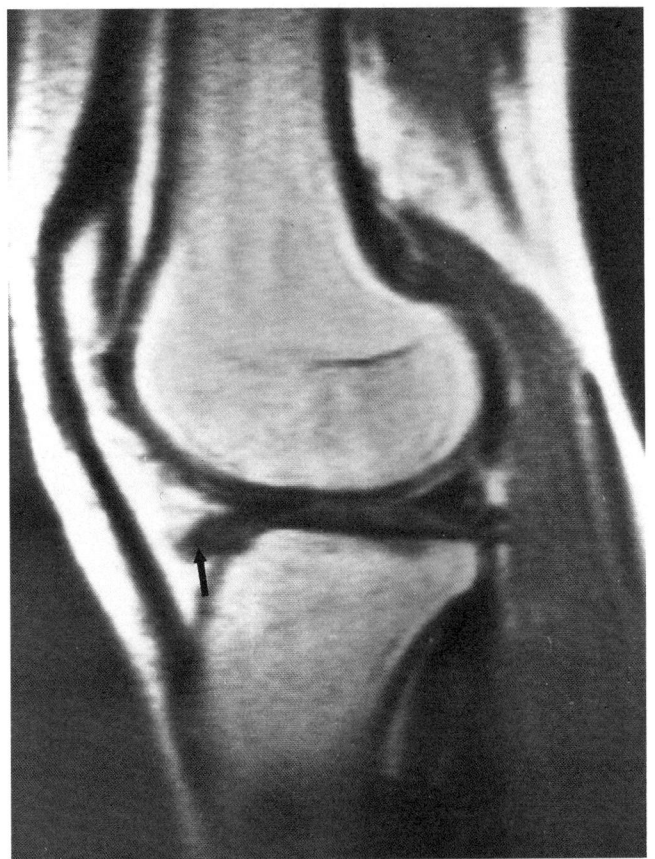
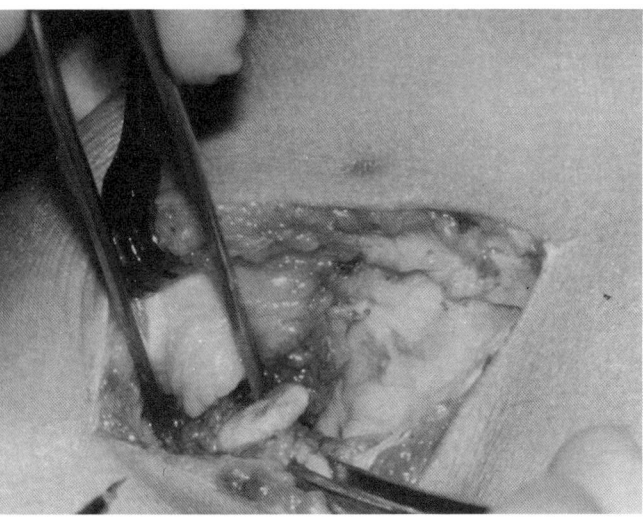

Figure 8.38

Calcified loose body. A 31-year-old woman with anterior knee pain. Radiograph showed a small calcified loose body in the anterior knee compartment. At arthroscopy the loose body could not be located. (**a**) MR (SE 800/28) was performed, localizing a 2 cm loose body (arrow) anterior to the anterior horn of the lateral meniscus, embedded in the infrapatellar fat pad. (**b**) Surgical exploration revealed a calcified fragment in the location depicted on the MR image.

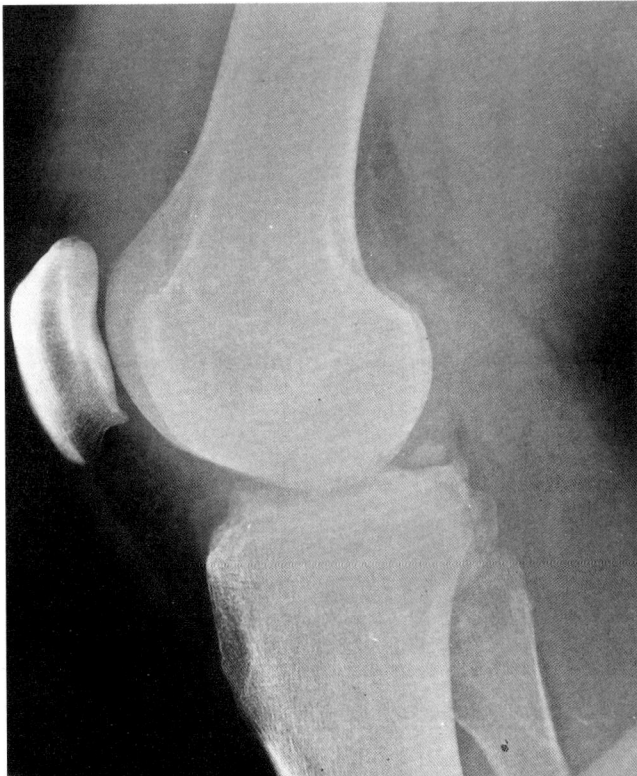
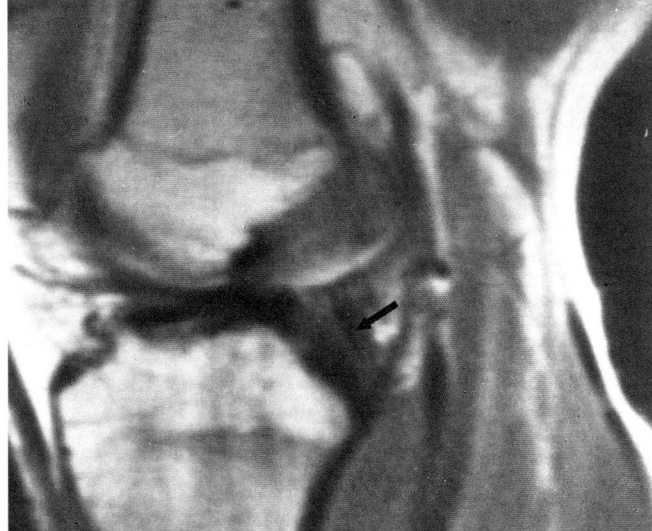

Figure 8.39

Loose bodies. A 21-year-old man with recurrent knee pain. (**a**) Lateral radiograph of the knee shows multiple osteochondral loose bodies in the posterior knee compartment. (**b**) Loose bodies are very difficult to identify on the T_1-weighted sagittal images (SE 800/28). In retrospect, low signal structure (arrow) is seen posteriorly; this probably represents one of the larger fragments. Fragments less than 5 mm are routinely difficult to identify by MRI.

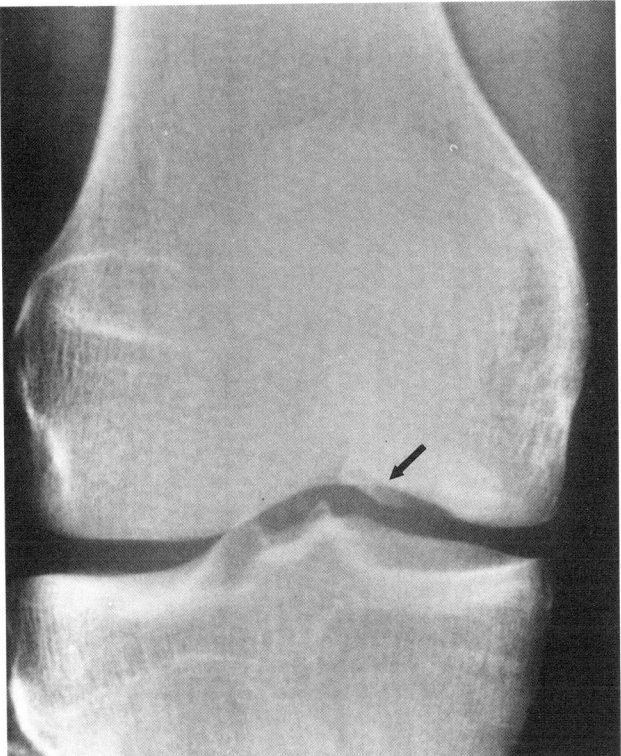

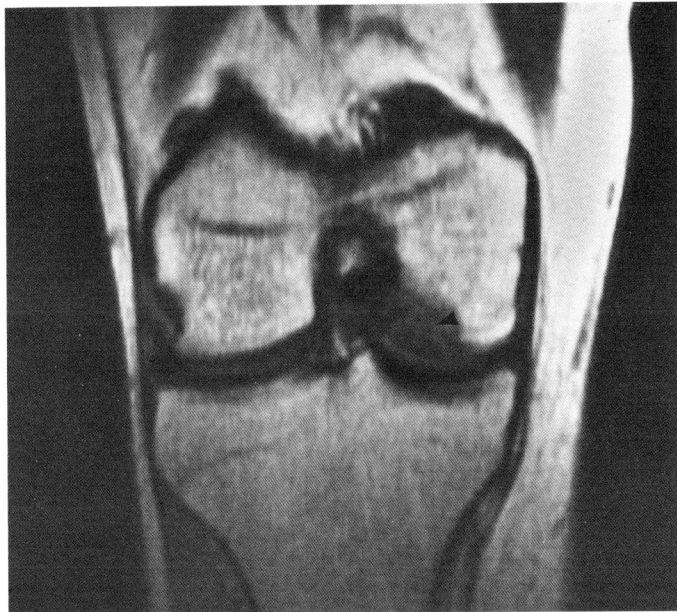

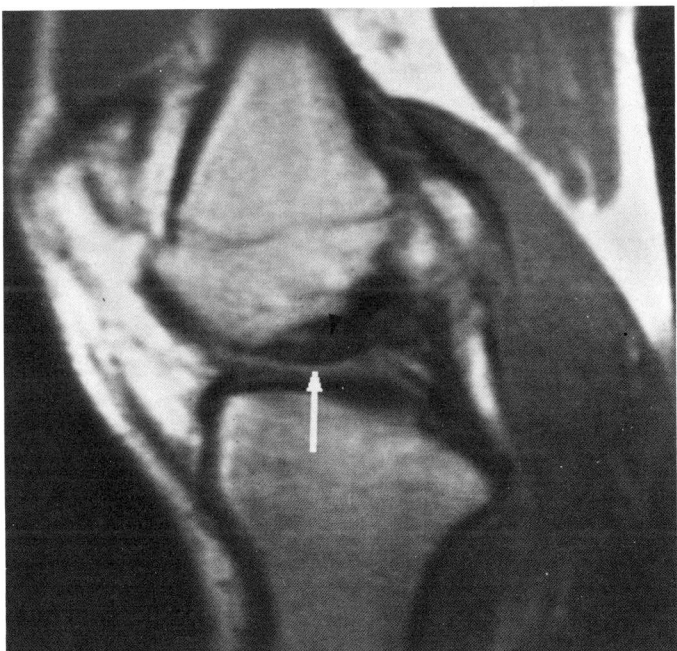

Figure 8.40

Osteochondritis dissecans in a 19-year-old man. (**a**) Anteroposterior radiograph reveals a loose fragment overlying the medial femoral condyle (arrow). (**b**) T_1-weighted coronal and (**c**) sagittal MR images (SE 800/28) show a low-intensity osteochondral focus (arrowhead) and disruption of the articular cartilage (arrow) consistent with the defect of osteochondritis dissecans. These findings were confirmed at surgery.

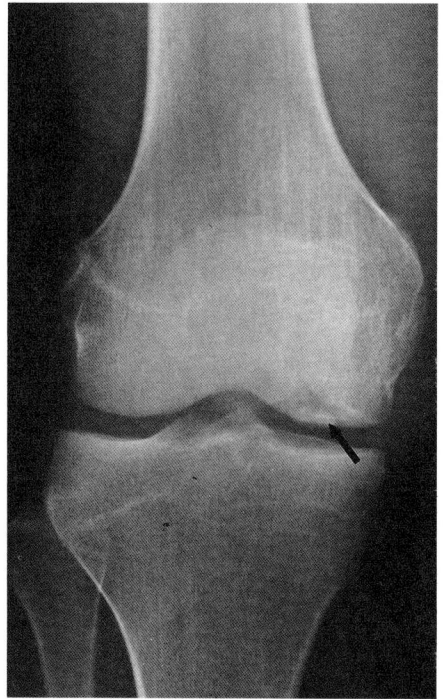

a

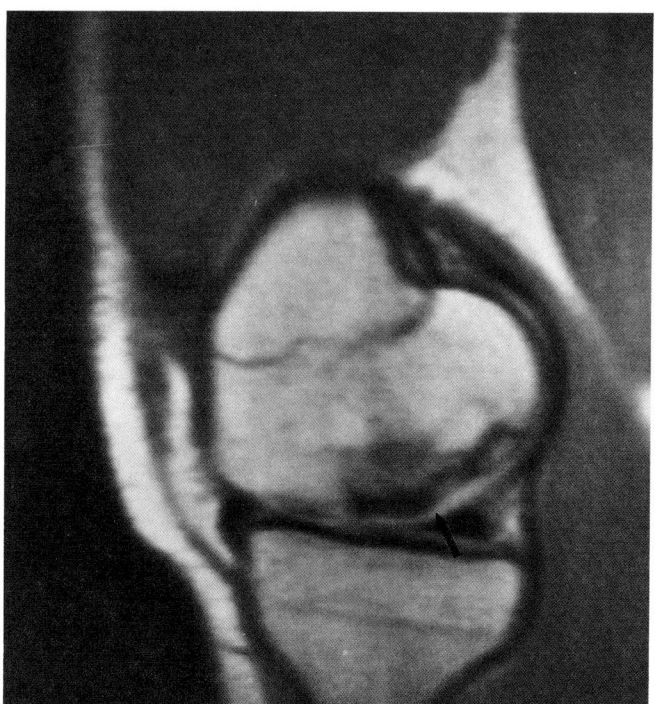

b

Figure 8.41

Osteochondritis dissecans in a 17-year-old girl with severe right knee pain.
(**a**) Anteroposterior radiograph shows an osteochondral defect in (arrow) the medial femoral condyle, consistent with osteochondritis dissecans.
(**b**) T_1-weighted sagittal and (**c**) coronal MR images (SE 800/28) reveal low-signal intensity in the subchondral bone. Overlying articular cartilage (arrow) is intact, with no evidence of loose fragment. At surgery, the articular cartilage overlying the bony defect was found not to be fractured. Holes were drilled through the articular surface to promote healing. Six months later, symptoms had greatly abated.

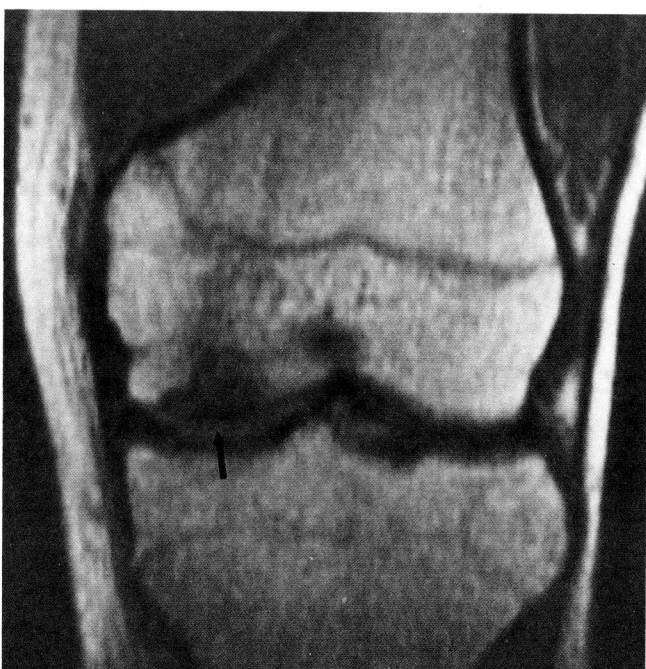

c

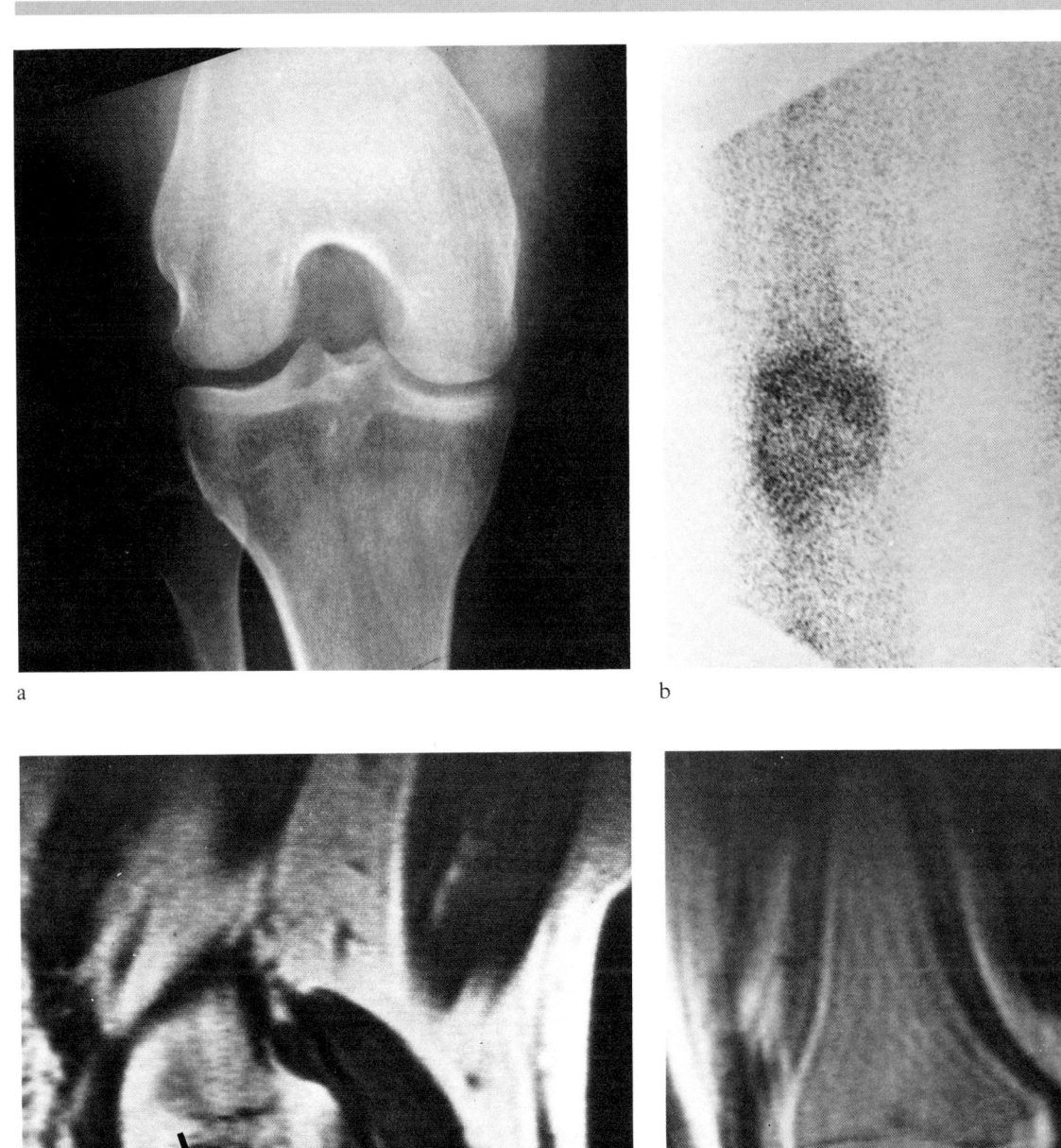

Figure 8.42

Osteonecrosis of the distal femur in a 25-year-old woman with systemic lupus erythematosus and right knee pain. (**a**) Anteroposterior radiograph and (**b**) ^{99m}technetium-MDP bone scan are normal. (**c**) T_1-weighted sagittal and (**d**) coronal MR images (SE 800/28) contain low-intensity regions in the femoral condyles (arrows), consistent with osteonecrosis. These findings were confirmed by core biopsy.

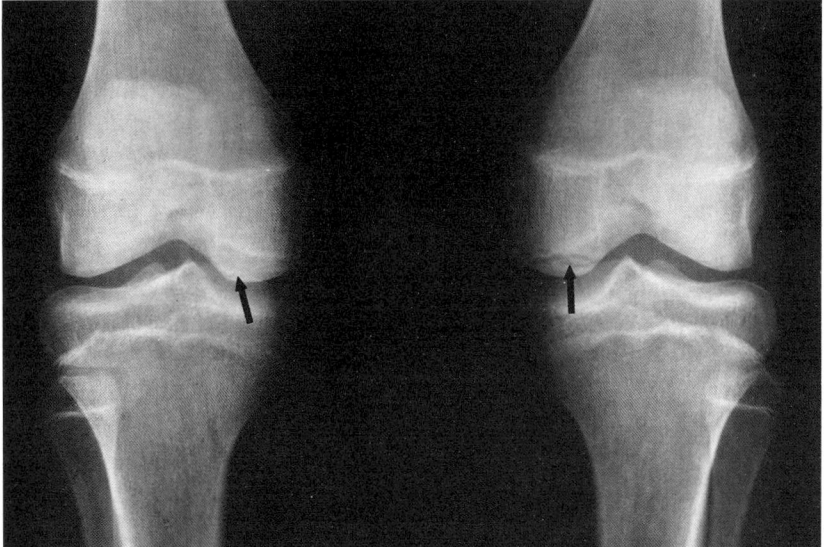

Figure 8.43

A 15-year-old girl with systemic lupus erythematosus on steroid therapy who presented with bilateral severe knee pain. (a) Anteroposterior radiograph shows bilateral subchondral femoral defects consistent with bilateral osteochondritis dissecans (arrow). (b) T_1-weighted MR image (SE 800/28) of the knee performed without a surface coil shows extensive bone infarcts not identified on the radiographs.

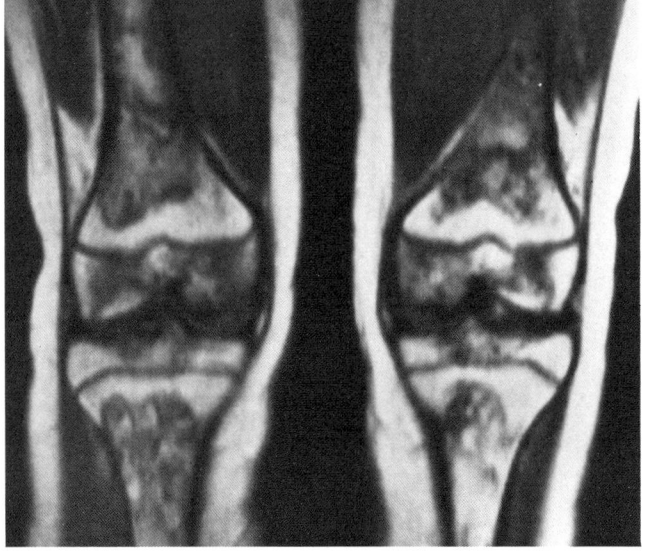

Figure 8.44

Medullary bone infarcts in the distal femora of a 51-year-old woman receiving long-term corticosteroid therapy for systemic lupus erythematosus. Radiographs were within normal limits. T_1-weighted coronal MR image (SE 800/28) of the knees, using body coil, shows multiple low-intensity infarcts.

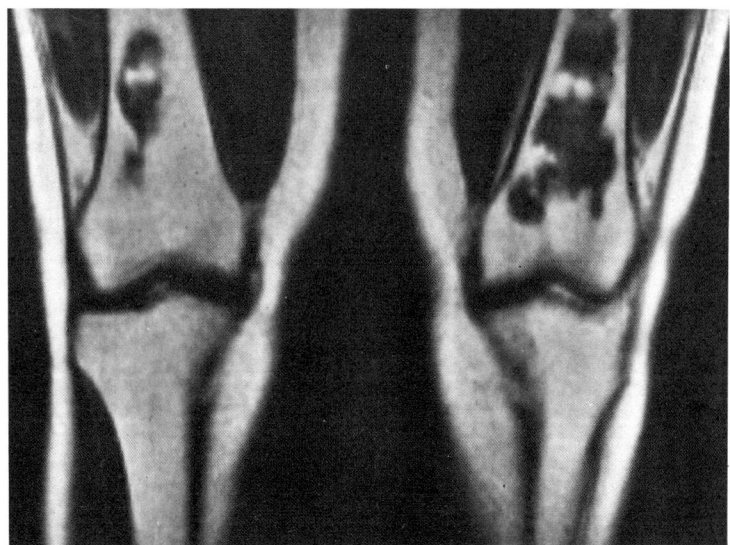

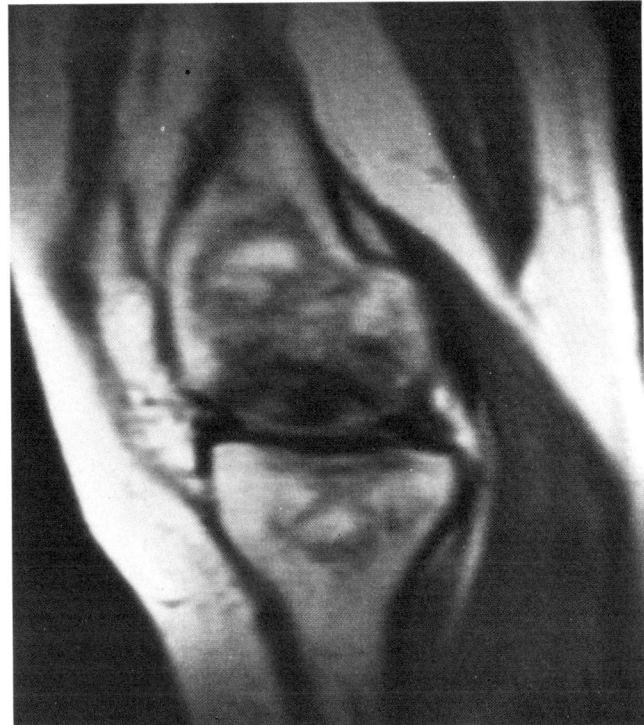

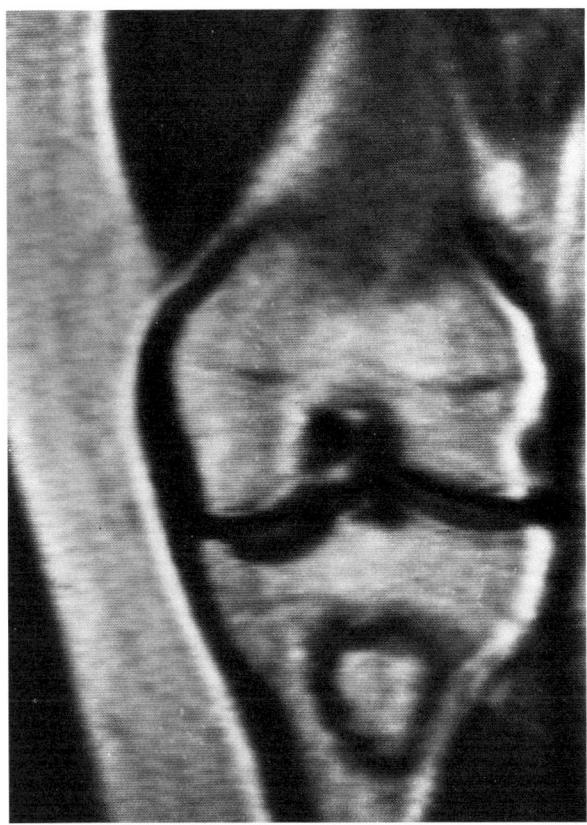

Figure 8.46

Medullary bone infarct in a 29-year-old woman with pancreatitis. Radiographs were normal. T_1-weighted coronal MR image (SE 800/28) shows bone infarct in the proximal tibia. High-signal central core surrounded by a rim of low-signal intensity is one of many MR patterns characteristic of infarct.

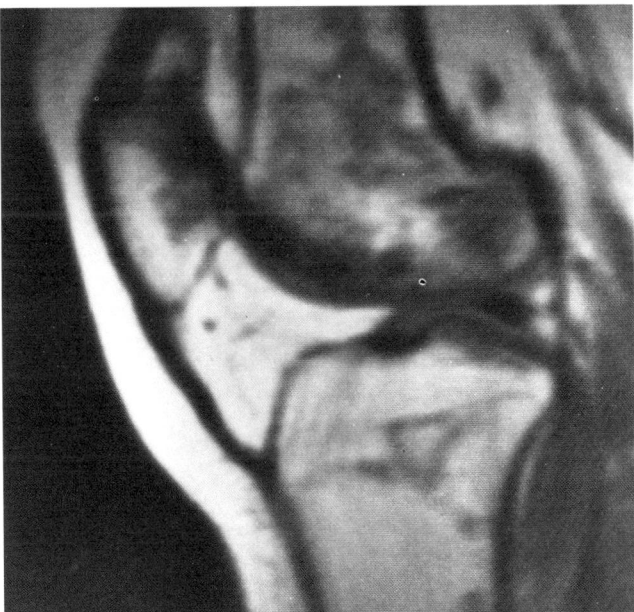

Figure 8.45

Medullary bone infarcts in the tibia, femur and patella of a 40-year-old woman with systemic lupus erythematosus. (**a,b**) T_1-weighted sagittal MR images (SE 800/28) of the (**a**) medial and (**b**) lateral aspects of the knee show multiple infarcts of low-signal intensity, and several with high-intensity cores.

9

The ankle and foot

Julia R Crim and Lawrence W Bassett

The anatomy shown in this chapter was drawn from several anatomy texts,[1-5] and correlated with cadaveric sections. T_1-weighted images were obtained in the axial plane from superior to inferior (Figure 9.1), the coronal plane from posterior to anterior (Figure 9.2), and the sagittal plane from medial to lateral (Figure 9.3). The foot was not braced in a neutral position as in CT scanning because of patients' discomfort and motion in that position. Instead, the foot was held in a comfortable degree of plantar flexion, and the scan plane for true axial or coronal images of the foot was determined from a sagittal scout image. In clinical situations, an oblique axial-coronal plane was sometimes used (Figure 9.6).

MR of the ankle and foot is still in the early stages of clinical investigation, although several studies indicate its promise.[6-7] MR is clearly able to delineate soft tissue structures of clinical interest, some of which cannot be imaged by other means. Patients with metal implants can be studied safely, and the surrounding bone marrow can in many cases be evaluated (Figure 9.4).

Ligament and tendon injuries have been identified (Figures 9.5–9.7). Integrity of the peroneal retinacula (Figure 1a–c) can be documented by MRI in individuals with suspected peroneal tendon subluxation.

The calcaneonavicular (Figures 9.1d, 9.2e) and long plantar (Figure 9.3c) ligaments, important structures in the maintenance of the arches of the foot, can now be imaged for the first time. Direct multiplanar imaging is useful in the evaluation of the complex structures of the hindfoot, such as the subtalar joint.

References

1 HOLLINSHEAD WH, *Textbook of Anatomy*, 3rd edn (Harper & Row: Philadelphia, 1974).

2 SOLOMON MA, GILULA LA, OLOFF LM et al, CT scanning of the foot and ankle: 1. Normal anatomy, *AJR* (1986) **146**:1192–1203.

3 McMINN RMH, HUTCHINGS RT, LOGAN BM, *Color Atlas of Foot and Ankle Anatomy* (Appleton Century Crofts: East Norwalk, CT, 1982) 72–83.

4 CLEMENTE C, *Anatomy: A Regional Atlas of the Human Body*, 2nd edn (Urban & Schwarzenberg: Baltimore, 1981).

5 WARWICK R, WILLIAMS PL, *Gray's Anatomy: 35th British Edition* (Saunders: Philadelphia, 1973).

6 BELTRAN J, NOTO AM, MOSURE JC et al, Ankle: surface coil MR imaging at 1.5T. *Radiology* (1986) **161**:203–209.

7 SARTORIS DJ, RESNICK D, Pictorial review: Cross-sectional imaging of the foot and ankle, *Foot Ankle* (1987) **8**(2):59–80.

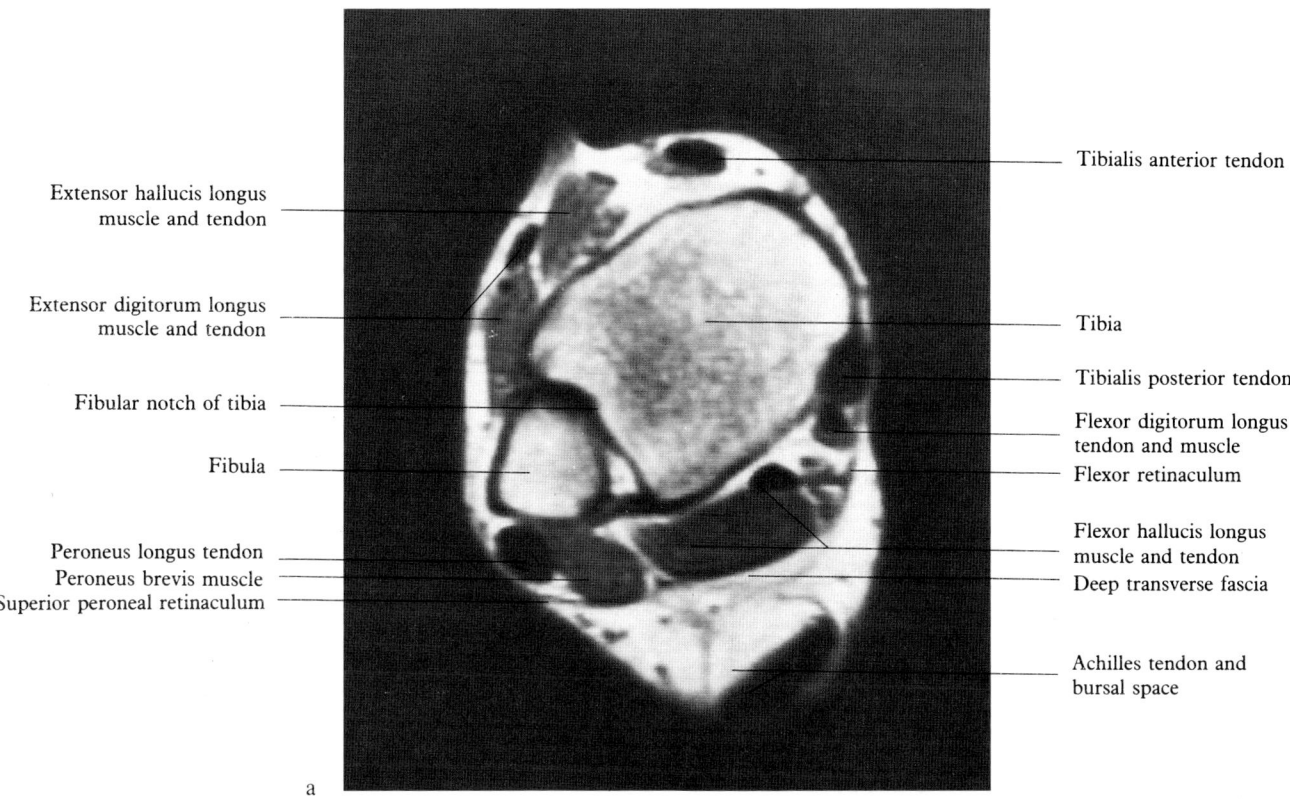

Figure 9.1

Normal anatomy, axial. (**a**) Distal tibia (SE 800/30). The muscles of the leg are divided into anterior, lateral and posterior groups. The anterior muscles are the tibialis anterior, extensor hallucis longus, and extensor digitorum longus. The peroneus longus and brevis muscles make up the lateral group. Posteriorly, the Achilles tendon is formed from the gastrocnemius and soleus muscles. It is separated from the deep muscles of the posterior compartment, the tibialis posterior, flexor digitorum longus and flexor hallucis longus, by a fat-filled bursal space and deep transverse fascia. At this level, the deep transverse fascia merges medially into the flexor retinaculum and laterally into the superior peroneal retinaculum. (**b**) Tibiotalar joint (SE 800/30). The anterior tibial artery and vein, together with the deep peroneal nerve, are located deep to the extensor hallucis longus muscle. The tibialis posterior vessels and tibial nerve are between the flexor hallucis longus and flexor digitorum longus. The sural nerve and lesser saphenous vein are posterior to the peroneus brevis. (**c**) Lateral malleolus (SE 800/30). The foot is plantar flexed, so the tuberosity of the calcaneus and the insertion of the Achilles tendon are seen. Several components of the deltoid ligament can be distinguished: deep fibers at their insertion on the body of the talus, superficial fibers extending towards the calcaneus, and anterior fibers coursing towards the talar neck and navicular. The anterior and posterior talofibular ligaments are components of the lateral collateral ligament. (**d**) Spring ligament (SE 800/30). The calcaneonavicular or spring ligament extends from the sustentaculum tali and anteromedial calcaneus to the medial aspect of the navicular. The tibialis posterior tendon inserts on the navicular inferior to the spring ligament, and these structures may be difficult to separate on axial images. (**e**) Calcaneus (SE 800/30). A normal dorsolateral depression in the cortex of the calcaneus is seen as a wedge-shaped defect at this level. The medial plantar nerve and vessels course anteriorly deep to the abductor hallucis muscle.

Figure 9.1 *continued*

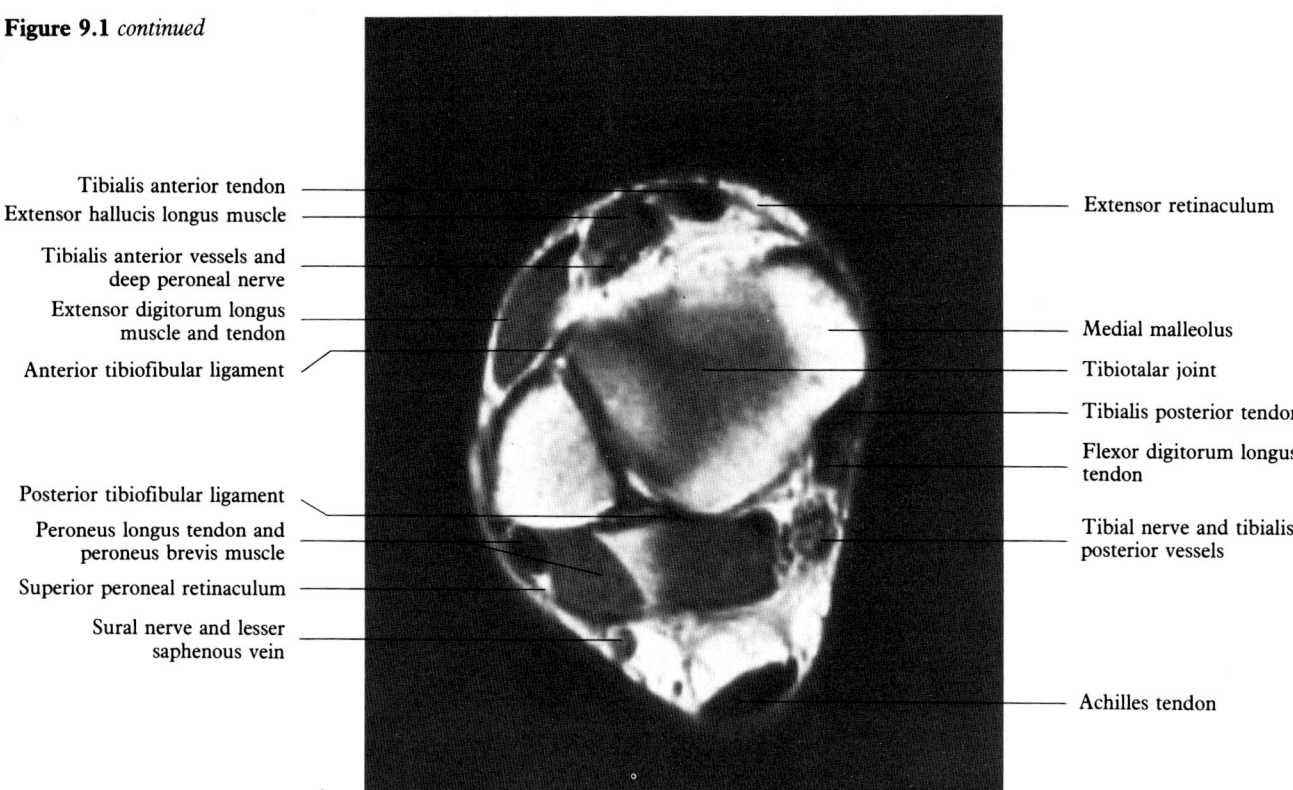

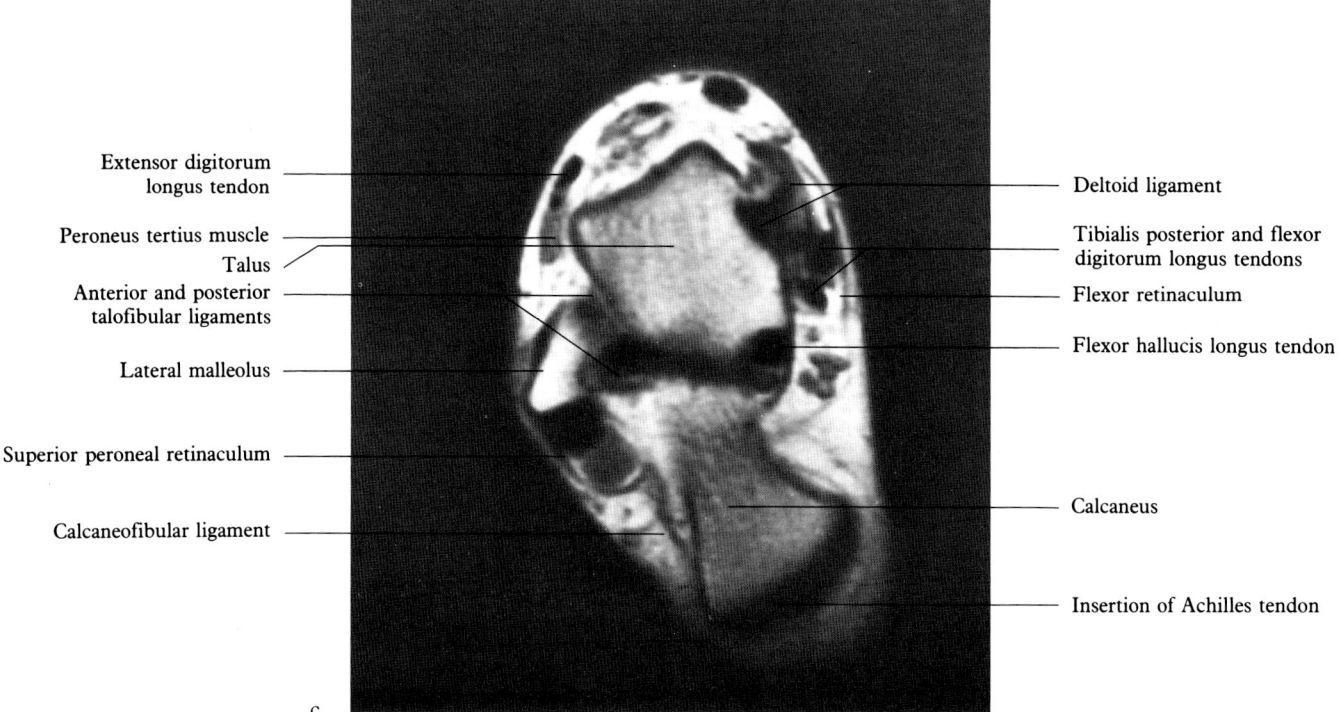

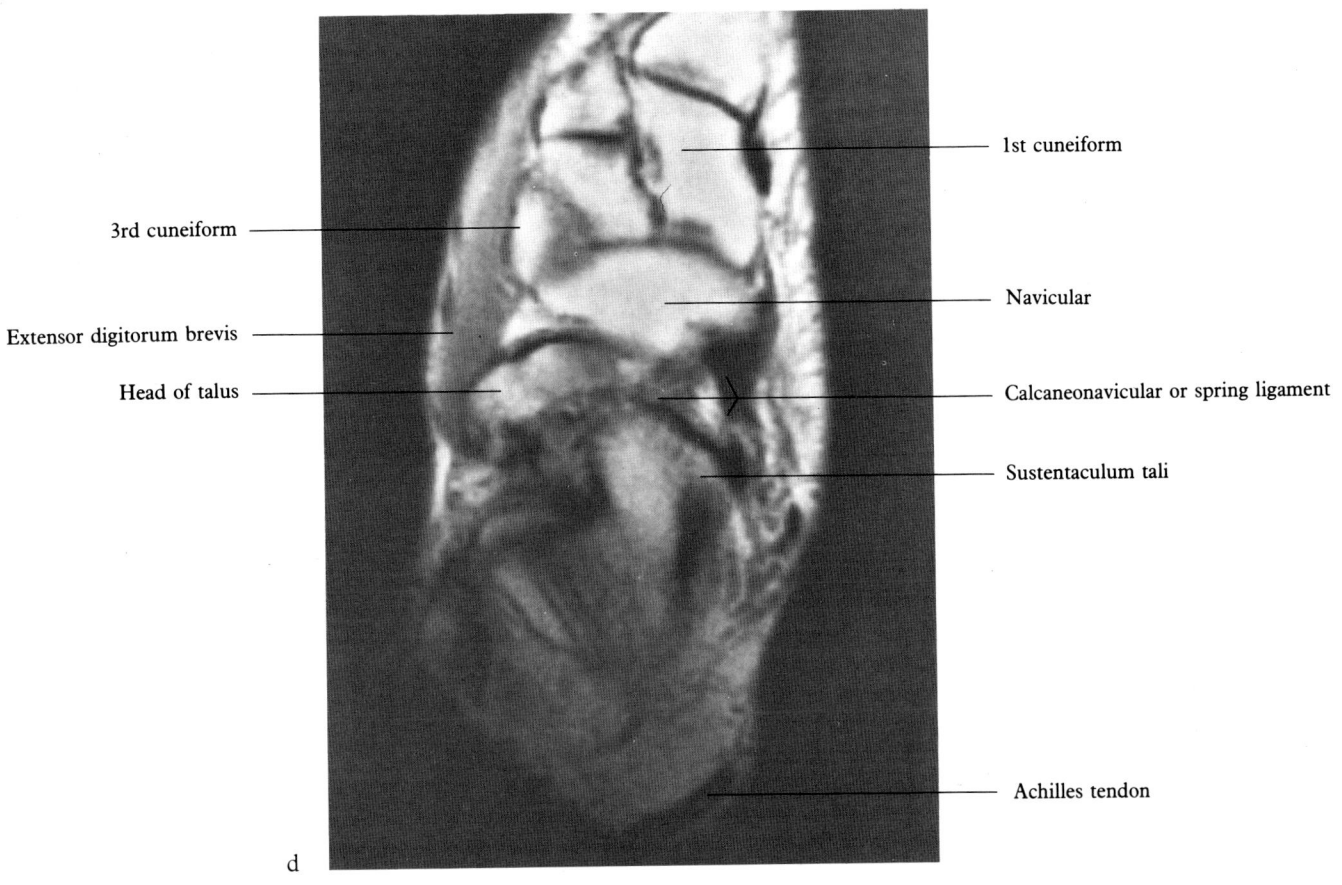

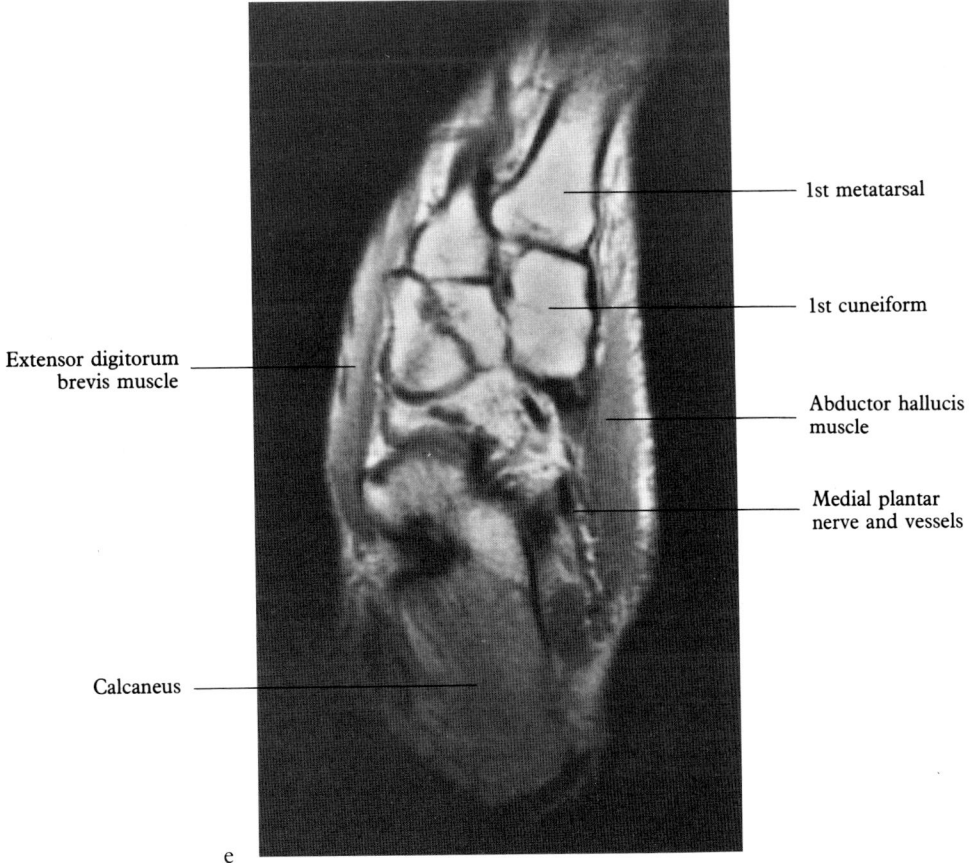

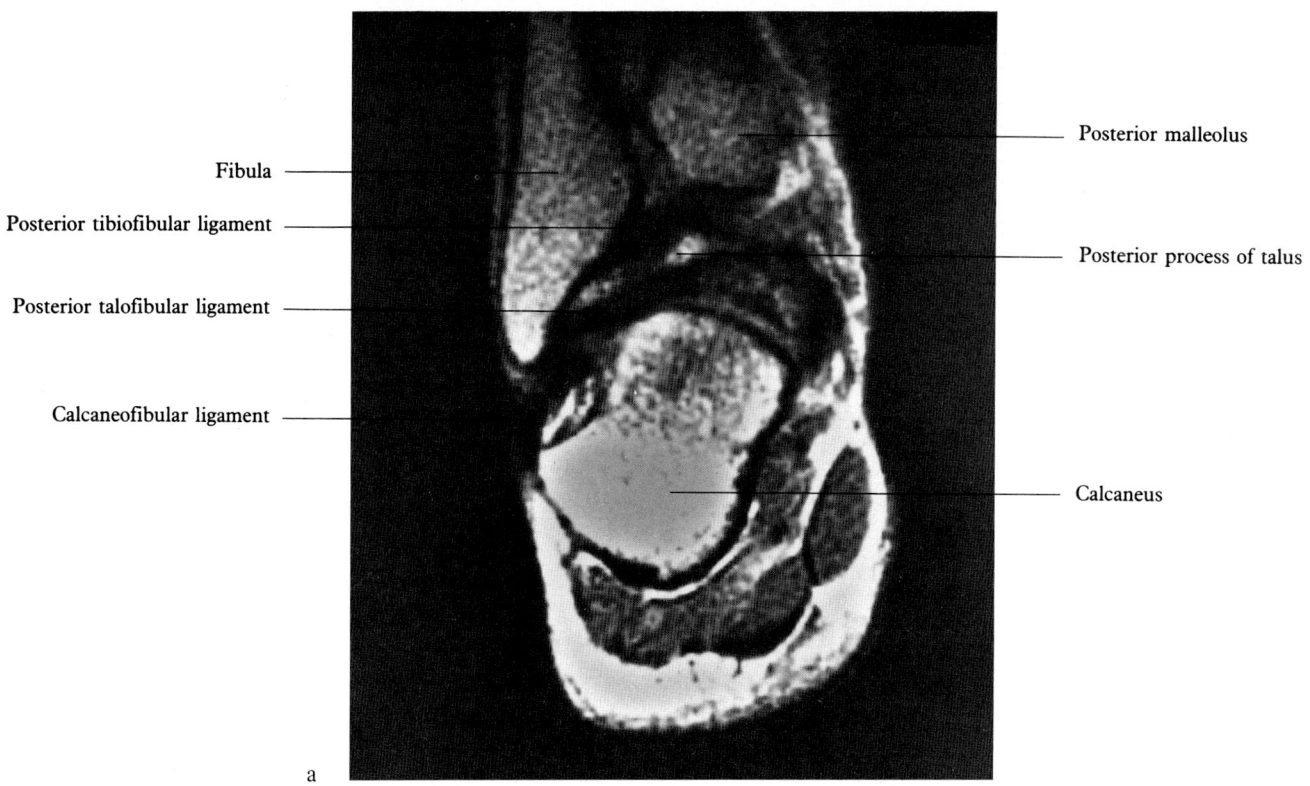

Figure 9.2

Normal anatomy, coronal. (**a**) Posterior malleolus (SE 800/30). The foot is plantar-flexed, resulting in an oblique plane through the calcaneus. Note the posterior tibiofibular and talofibular ligaments laterally. (**b**) Lateral malleolus (SE 800/30). The foot is plantar-flexed. The calcaneofibular ligament is seen laterally, and the anterior tibiotalar and tibiocalcaneal ligaments medially. (**c**) Posterior facet, talocalcaneal (subtalar) joint (SE 800/30). Note that the posterior portion of the sustentaculum tali normally does not articulate with the talus. The intrinsic plantar muscles include, superficially, the abductor hallucis, flexor digitorum brevis and abductor digiti minimi. Deep to these is the quadratus plantae. (**d**) Middle facet, talocalcaneal joint (SE 800/30). The interosseous ligament is seen within the sinus tarsi. Peroneus brevis and peroneus longus tendons are held in position by the inferior peroneal retinaculum laterally and the calcaneus medially. (**e**) Anterior facet, talocalcaneal joint (SE 800/30). Low-signal areas within the talus and calcaneus are due to the partial volume effect from the calcaneocuboid and talonavicular joints, located immediately anterior to this image. The plantar calcaneonavicular ('spring') ligament extends from the anteromedial calcaneus, and will insert on the medial aspect of the navicular. The peroneus longus tendon is deep to the long plantar ligament. (**f**) Cuneiforms (SE 800/30). The tuberosity of the base of the fifth metatarsal lies lateral to the cuboid. The tendon of the peroneus longus is coursing medially toward its insertion on the first metatarsal and first cuneiform.

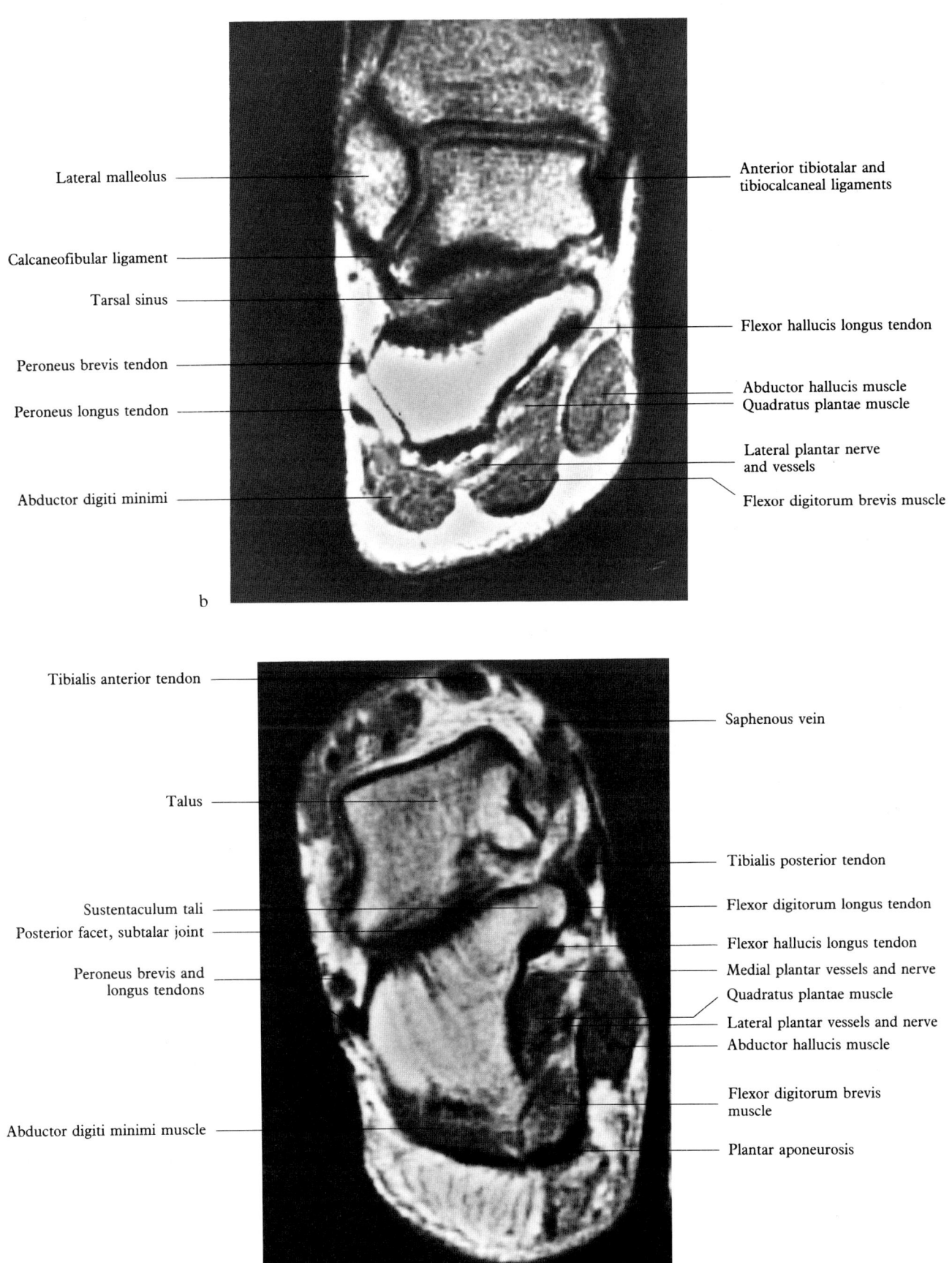

Figure 9.2 continued

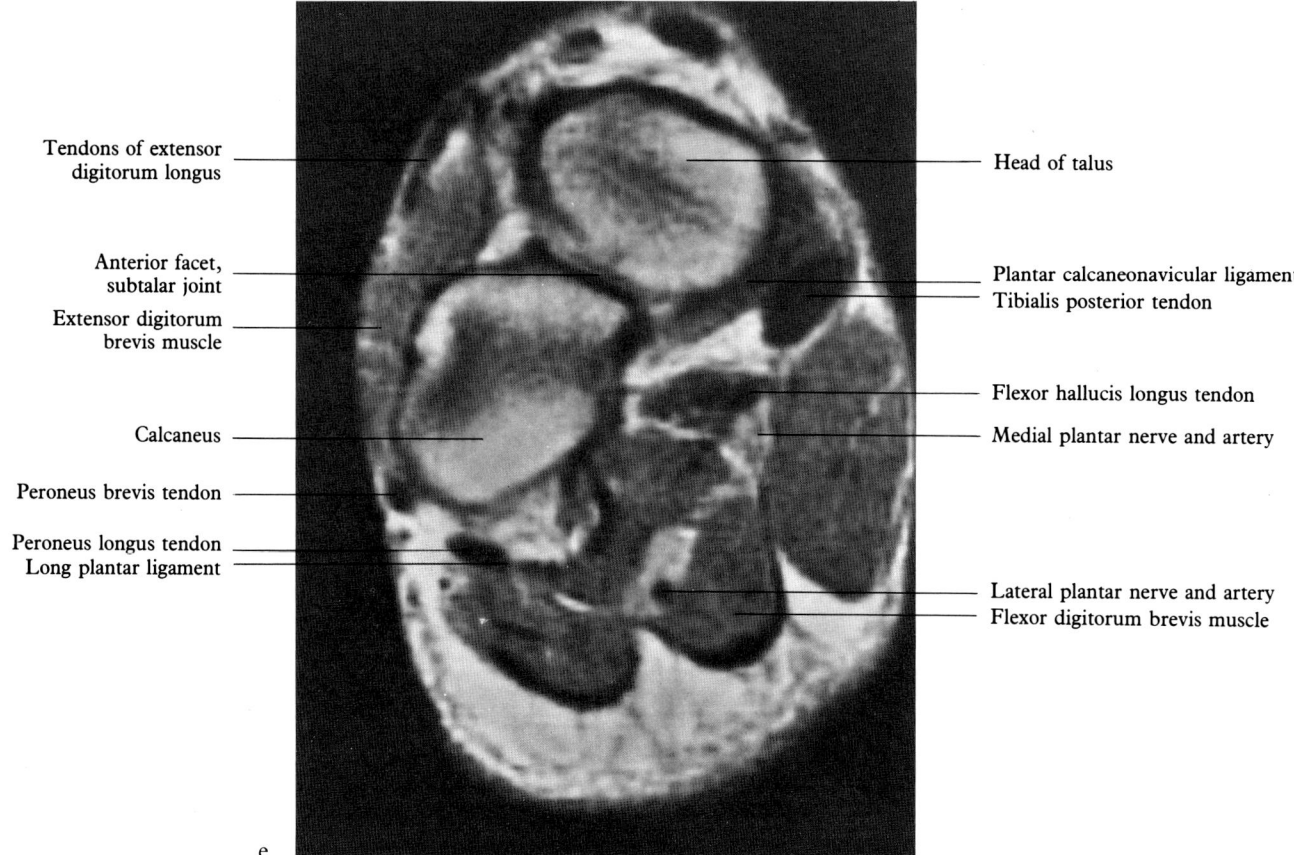

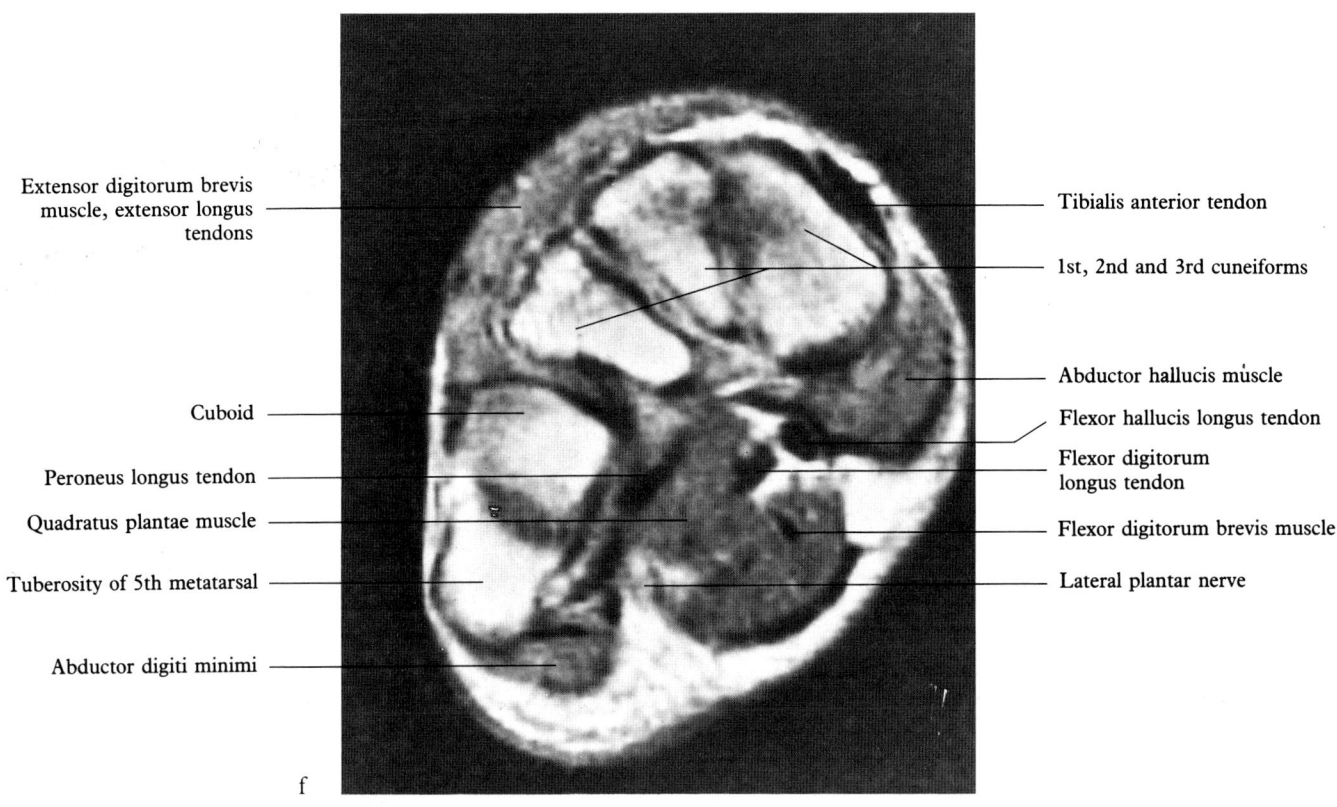

274 MRI atlas of the musculoskeletal system

Figure 9.3

Normal anatomy, sagittal. (a) Medial malleolus (SE 800/30). The tibialis posterior tendon is seen extending anteriorly to its navicular insertion. (b) Sustentaculum tali (SE 800/30). The insertion of deep fibers of the anterior tibiotalar ligament is seen. The flexor hallucis longus tendon runs in a groove below the sustentaculum tali, which acts as a pulley during muscle contraction. The plantar aponeurosis is normally prominent and devoid of signal. The posterior tibial artery and tibial nerve divide as they reach the plantar aspect of the foot into medial and lateral plantar branches. (c) Sinus tarsi (SE 800/30). The anterior and posterior facets of the talocalcaneal joint are both seen, separated by the sinus tarsi. The long plantar ligament extends from the posterior portion of the calcaneus to the cuboid and metatarsals; the band deep to it may represent the short plantar ligament. (d) Posterior subtalar facet (SE 800/30). The posterior subtalar facet is longer laterally than medially, and ends anteriorly at the angle of Gissane. The Achilles tendon is inserted along the posterior surface of the calcaneus.

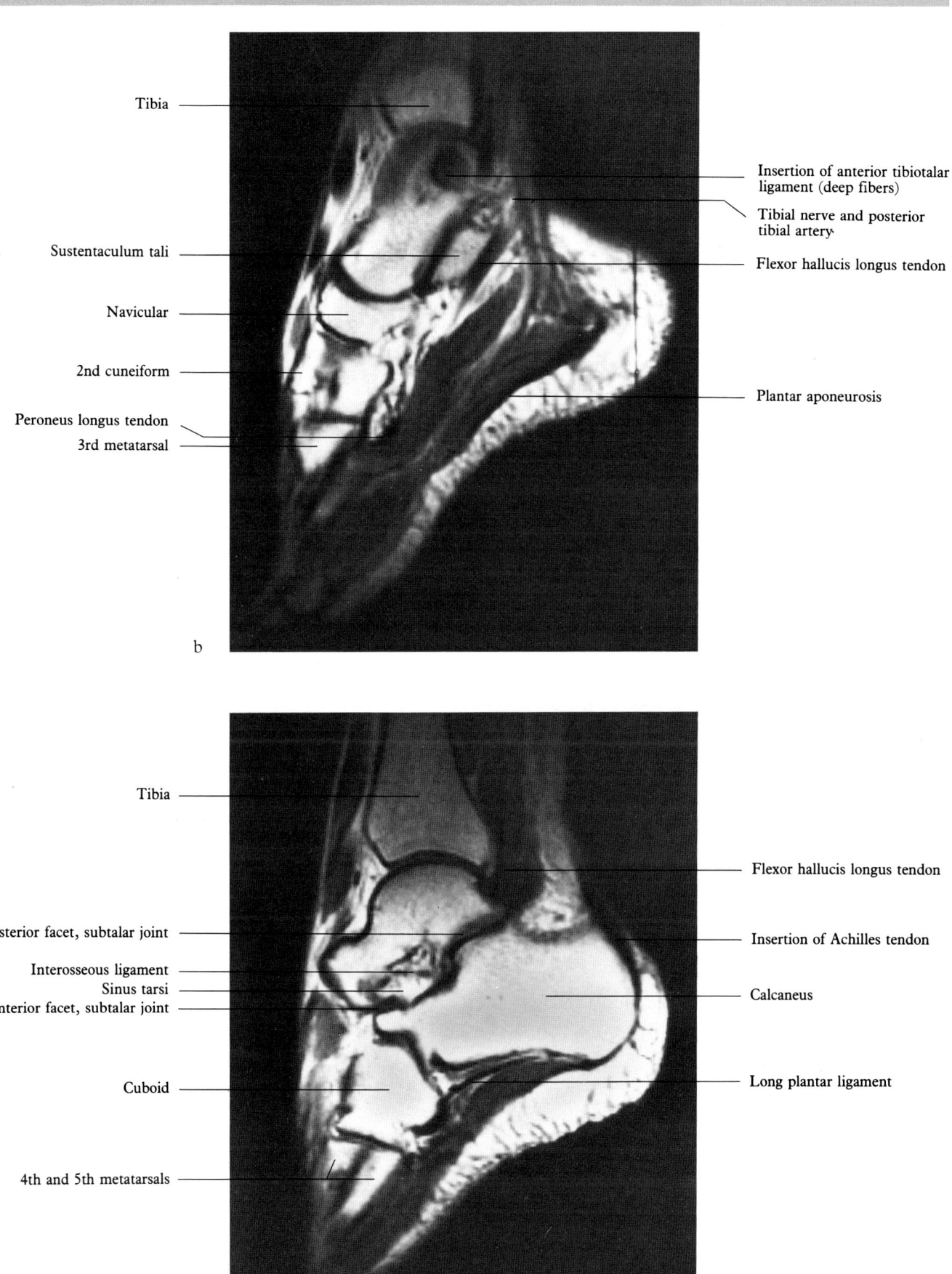

Figure 9.3 continued

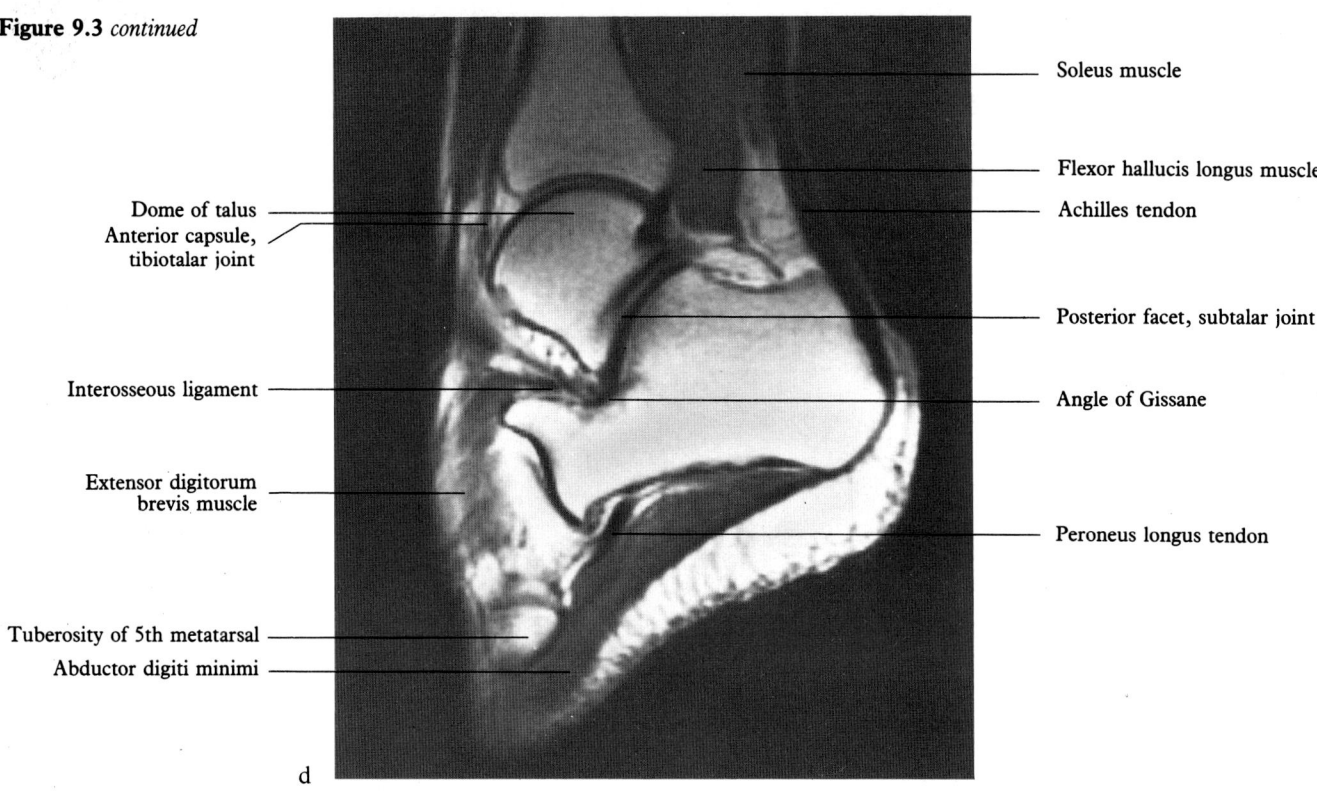

- Soleus muscle
- Flexor hallucis longus muscle
- Achilles tendon
- Dome of talus
- Anterior capsule, tibiotalar joint
- Posterior facet, subtalar joint
- Interosseous ligament
- Angle of Gissane
- Extensor digitorum brevis muscle
- Peroneus longus tendon
- Tuberosity of 5th metatarsal
- Abductor digiti minimi

d

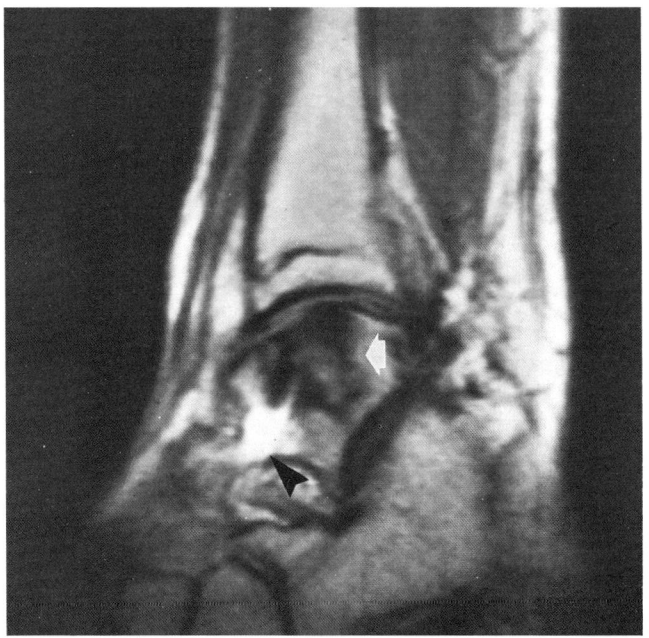

Figure 9.4

Osteonecrosis (SE 800/30). Despite the metal artifact (arrowhead) from a surgical screw, the talar dome osteonecrosis (arrow), with flattening of the articular surface, is well seen.

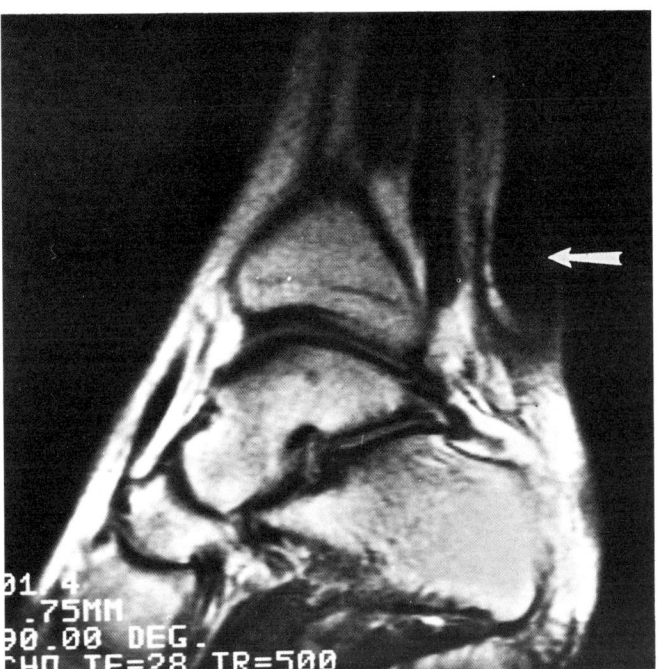

Figure 9.5

Complete tendon rupture (SE 500/28). The Achilles tendon (arrow) is thickened and retracted superiorly in this complete rupture.

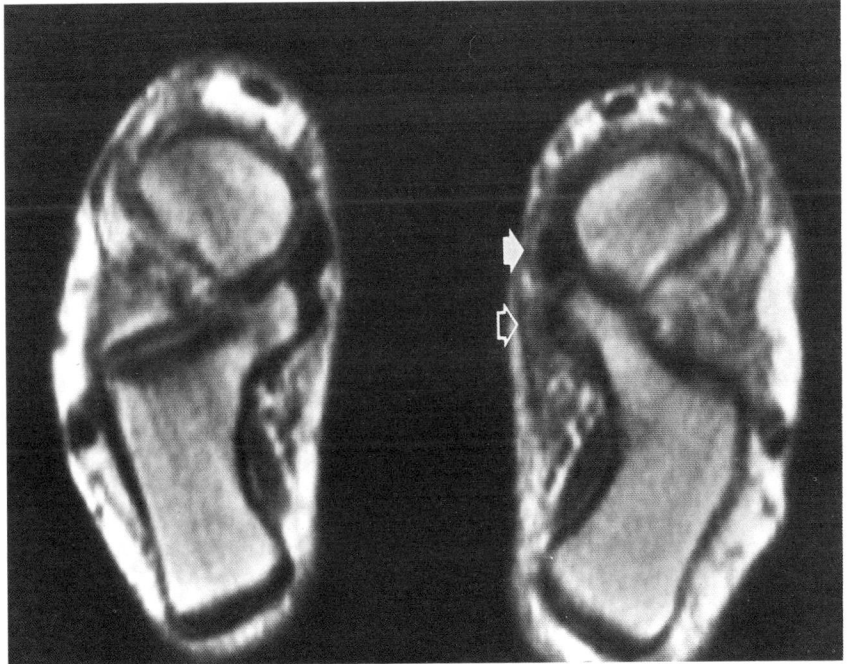

Figure 9.6

Partial tendon rupture (SE 600/30). The patient presented clinically with signs of left posterior tibial tendon tear. The tendon sheath (solid arrow) is enlarged and contains an area of abnormally high signal intensity, but the inner signal void region represents intact tendon fibers, indicating this is a partial tear. The flexor digitorum longus tendon (open arrow) is also abnormal. The normal right side was included for comparison.

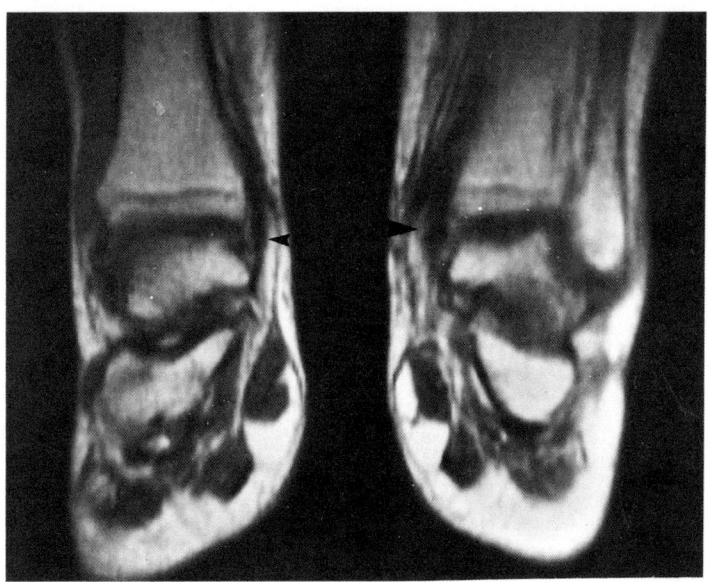

Figure 9.7
Ligament injury (SE 500/28). Compared to the normal right side (small arrowhead), the left deltoid ligament (larger arrowhead) is thickened. The increased signal intensity of the ligament represents edema.

10
Pediatrics

Theodore R Hall and
Hooshang Kangarloo

Magnetic resonance imaging (MRI) provides a noninvasive and reliable methodology for the evaluation of the musculoskeletal system in children. In this chapter, we have focused on developmental anatomy as depicted by MRI as well as on pathological conditions which are unique to children. Obviously, there is overlap between some of the pathological abnormalities in children and adults.

Imaging techniques

We have routinely used spin-echo pulse sequences for the evaluation of the musculoskeletal system in children. T_1-weighted short TR, short TE images are initially obtained and if necessary are supplemented by long TR, long TE (T_2-weighted) pulse sequences. Solonoid band or planar surface coils increase the signal-to-noise ratio and are utilized for detailed evaluation of the joints and the spine. In instances where simultaneous comparison with contralateral side is desirable, examination is performed in the head or body coil. Evaluation of the spine, extremities, and joints are performed by utilizing appropriate size 'pediatric' surface coils. Coronal and sagittal images are common initial planes obtained for the evaluation of the spine or extremities. Axial planes may be a useful adjunct for the initial imaging of the pelvis and shoulder. In general, it is best to tailor each pulse sequence and imaging plane for a specific patient and suspected disease entity. With the most commonly utilized pulse sequences in the pediatric age group (short TR, short TE spin-echo) the MR signal intensity may be high (marrow), low (cortex), or intermediate (cartilage and muscle). The intensity of the MR signal with this pulse sequence is the result of added signals from lipid and water protons. In contrast, the MR signal intensity in the chemical shift imaging technique results from subtraction of the above mentioned signals. Thus, chemical shift imaging, although uncommonly used, may be useful in some instances, particularly in differentiating fatty and non-fatty marrow.

The anatomical details of joints in children as depicted by MRI are similar to those of adults with the following exceptions. The growth plate (physis) produces a zone of low-signal intensity and is wider in younger children than in older children; in adolescents, even after radiographic evidence of closure, a line of low-signal intensity can be identified at the site of the former growth plate. The ossified portion of the epiphysis contains marrow, and therefore has a higher signal intensity than the surrounding cartilaginous portion.

Sedation

Since $\bar{M}R$ signal acquisition is carried out during the entire examination time, child motion, even if it is sporadic, will significantly distort the image. In children under 6 to 7 years of age, we have routinely used chloral hydrate, 50 to 75 mg/kg orally, for sedation approximately 30 minutes prior to scanning. An apnea monitor is used for sedated patients.

Bone marrow (*Figures 10.1–10.4*)

MR is the modality of choice for evaluation of bone marrow in children. The MR signal intensity of bone marrow is related to the ratio of fat to other cells. In normal children the fat-to-cell ratio is approximately 50 per cent, and the signal is strong and uniform, particularly in the metaphyses and diaphyses. When signal intensity differences of the marrow are compared between two patients or between two examinations on the same patient, the results may not be reliable. Thus comparison with the signal intensity of an adjacent structure such as muscle provides more accurate information regarding the marrow. The signal intensity of marrow has been correlated with its cellularity in normal and abnormal children. The signal intensity of the epiphysis and metaphysis should be analyzed for uniformity and relative brightness, and compared to that of muscle. The ratios of the signal intensity of epiphyseal and metaphyseal marrow compared to muscle are given in Table 10.1.

Table 10.1 Ratios of signal intensity in marrow of metaphyses and epiphyses compared to muscle.

Type of marrow	*Metaphysis: muscle*	*Epiphysis: muscle*
Normal marrow	3.25	3.03
Aplastic anemia, untreated	3.21	3.14
Sickle cell anemia	1.57	3.02
Aplastic anemia, treated	1.81	3.30

Source: Kangarloo, Dietrich, Taira et al.[1]

The ratios in Table 10.1 represent signal intensities taken from the same image and are independent of mechanical parameters such as amplifier gain and coil tuning. The ratios more accurately characterize the physiologic state of the marrow than the intensities alone. When evaluating bone marrow, direct observation produces subjective analysis, digital values produce a more accurate objective analysis. The latter eliminates the substantial error present in subjective analyses, even when optical density measurements are made from film images. Film images may suffer from non-linear film characteristics and field non-uniformities that may be introduced by multi-format cameras.

Normal marrow

In the normal marrow of children, the signal intensity of the epiphyses, metaphyses, and diaphyses is uniform, and about three times greater than the intensity of adjacent muscle. One should remember that in the bones of growing children, the signal intensity of the epiphyses may be slightly greater than that of the metaphyses.

Hyperplastic marrow (*Figure 10.10*)

Hyperplastic marrow emits a signal that is less intense than that of normal marrow. This may be observed in the bones of patients with sickle-cell anemia, hemophilia, and aplastic anemia. In patients with anemia, the muscles and metaphyses may have identical intermediate signal intensity, while the signal intensity of the epiphyses remains close to normal. (The ratios of the signal intensities of the epiphyses and metaphyses compared to muscle are given in Table 10.1.) Hence, the signal intensity in anemic patients is significantly lower than normal in the metaphyses and diaphyses, but approaches normal intensity in the epiphyses. When marrow cellularity increases significantly, such as with erythroid hyperplasia resulting from sickle cell anemia and with treated aplastic anemia, the volume of marrow fat decreases, and the signal intensity of the metaphyses appears less than normal. However, the signal intensity from the epiphyses may be unchanged from the normal. In a series of patients with sickle-cell anemia compared to a control population of children, the ratio of signal intensities of the metaphyses to muscle was one-half the value for normal, but the ratio of the signal intensity of the epiphyses to muscle was similar to normal. This comparison indicates that most of the hematopoiesis occurred in the metaphyses rather than the epiphyses. A peculiar decreased signal intensity at the periphery of the epiphyses in these patients may imply more hematopoiesis in this region. Partial volume averaging from the adjacent cartilage, however, could conceivably produce a similar finding. When the marrow is hyperactive in treated aplastic anemia, the signal intensity of the metaphyses and diaphyses decreases relative to their intensity prior to treatment. When marrow is hyperplastic with a significantly greater volume of hematopoietic cells than in fat, the signal intensity decreases compared to normal.

Aplastic marrow (*Figure 10.1*)

In patients with untreated aplastic anemia, the signal of the metaphyses and epiphyses is normal high-intensity fat. The ratios of metaphyses and epiphyses signal intensities are compared to muscle in Table 10.1. When the marrow is hypocellular such as in untreated aplastic anemia, its signal intensity is high because of overabundant fat. In one patient with untreated aplastic anemia the ratio of signal intensities between metaphyses and epiphyses to muscle was virtually normal. In another patient with aplastic anemia who was in the recovery phase, when a bone marrow biopsy showed hypercellularity the ratio of signal intensities between the metaphyses and muscle was less than normal, but the ratio between the epiphyses and muscle approximated normal. These findings again indicate more active hematopoiesis in the metaphyses than in the epiphyses. Marrow infiltrated by leukemic cells exhibits a signal intensity that is decreased in a patchy, non-uniform fashion. However, we were unable to differentiate hypocellular marrow that contained a small amount of leukemic cells from hypocellular marrow without leukemic cells.

Leukemia (*Figure 10.2*)

The marrow of untreated leukemic patients shows non-uniform patchy signal intensity. Hypointense areas on T_1-weighted images may become hyperintense in T_2-weighted images, reflecting increased water content in the tumor-infiltrated areas.

Primary bone neoplasms (*Figures 10.5, 10.6, 10.15–10.18*)

By far the most frequent primary malignant neoplasm of bone occurring in the pediatric age group is osteosarcoma, with Ewing's sarcoma a distant second.

Osteosarcoma (*Figure 10.15*)

Histologically, the stroma of osteosarcoma is comprised of proliferating spindle cells that produce osteoid. The tumor most commonly arises in the second decade. Boys are more commonly affected, with a male : female ratio of 1.5 : 1. Pain, swelling, and limitation of motion are the most common complaints. The metaphyses of the long tabular bones are the most commonly affected sites (see Table 10.2).

Osteosarcoma may occur in multiple sites either as synchronous (concurrent) or metachronous (differing time of onset) lesions. Multiple, simultaneously occurring osteosarcomas, referred to as multi-centric sclerosing osteosarcoma, usually appear in the first decade of life and may be bilaterally symmetric. Osteosarcoma is typically eccentric with a metaphyseal location and cortical disruption. In images with short TR and TE times (T_1-weighted images), the tumor often has a mixed signal intensity with prominent areas of low-signal intensity representing mineralized tumor osteoid. With longer pulsing sequences, the primary tumor shows increased signal intensity due to T_2-shortening of the neoplastic tissue. Involvement of the adjacent soft tissue is better appreciated in T_2-weighted images.[3-5] Coronal images provide an excellent delination of the superior/inferior extent of the lesion, while axial images are very useful for evaluating the involvement of adjacent tissues.

Periosteal osteosarcoma

Occurring most frequently in the second decade of life, periosteal osteosarcoma is a rare tumor that affects the diaphyseal or diametaphyseal regions of the long bones. The tibia is the bone most frequently involved. The lesion begins in the periphery of the cortex and displaces the overlying soft tissues; the medullary canal and endosteal margin of the cortex are usually preserved. The lesion has a broad base and is characterized by tumor mineralization, cortical thickening, radiating spicules of periosteal new bone, and Codman triangles of new bone. Its signal characteristics are similar to those of classical osteosarcoma in both T_1- and T_2-weighted images.

Ewing's sarcoma

Ewing's sarcoma is the most common primary malignant skeletal neoplasm in children under 10 years of age. The tumor is composed of small, darkly-staining cells arranged in broad sheets. The peak incidence is between the first and third decades. Boys are more often affected than girls by a ratio of 3:2. Characterized clinically by localized pain, swelling, fever, weight loss, anemia, and leukocytosis, the most common sites of involvement include the long tubular bones (femur, tibia, humerus, and fibula, according to their decreasing incidence), pelvic girdle, vertebrae, ribs, and

Table 10.2 Relative frequency of sites of osteosarcoma.

Location	Frequency (per cent)
Femur	40
Tibia	16
Humerus	15

Source: Greenfield.[2]

scapula. Flat bone involvement is reported to be more frequent in patients over 9 years of age.[6]

A diaphyseal predilection, with occasional involvement of the metaphyseal region, is characteristic. The lesion is highly destructive with permeative osteolysis, poorly defined margins, cortical disruption and longitudinal, multilaminar or perpendicular spiculated periosteal reactions classically described as 'onionskin' and 'hair-on-end' type, respectively. In the axial skeleton, Ewing's sarcoma often manifests osteolysis and vertebral collapse. Occasionally, it leads to a uniform sclerosis of an affected vertebral body. A paravertebral mass is a commonly associated feature. Metastases to skeletal and extraskeletal sites are frequent.

The MR signal characteristics of Ewing's sarcoma are similar to those of most marrow-replacing lesions; the signal intensity is less than or equal to that of muscle on short TR, short TE pulse sequences (T_1-weighted images). Often associated with the skeletal lesion, the soft-tissue tumor mass is of a similar intermediate-signal intensity. T_2-weighted pulse sequences better demonstrate adjacent soft-tissue extension, with a slight increase in signal intensity of both the marrow and extra-osseous tumor components.[3,4,5,7]

Primary soft-tissue neoplasms (*Figure 10.18*)

For the evaluation of soft-tissue neoplasms, MRI is equal and in some cases superior to CT.[8-10] Both modalities depict the size of the tumor, the extent of its involvement of adjacent anatomic compartments, individual muscles, neurovascular structures, and joints. Recently published data suggest that evaluation of anatomic compartments and individual muscles is more accurately accomplished with MRI than with CT.[10]

Soft-tissue lesions that we have commonly evaluated using MRI are hematomas, lipomas, hemangiomas, lymphangiomas, and post-traumatic conditions such as myocitis and soft-tissue scar. The signal intensities of hematomas vary with age; sub-acute hematomas have a high-signal intensity on T_1-weighted images, acute and chronic hematomas a low-signal intensity. Lipomas have characteristic high-signal intensity on T_1-weighted sequences.[11] Although immature fatty tumors such as lipoblastomatosis contain areas of high-signal intensity on T_1-weighted images, the characteristic high-signal intensity of the lipomas does not appear. On T_1-weighted images hemangiomas and lymphangiomas are indistinguishable and both have low-signal intensity. On T_2-weighted images both hemangiomas and lymphangiomas show a significant increase in signal intensity (higher than fat) which is suggestive of these two lesions. Scars and myocitis have low-signal intensity both on short T_1-weighted and long T_2-weighted sequences.[12]

Pediatric spine (*Figures 10.19–10.38*)

The use of MRI for the evaluation of pediatric spinal pathology is well established in clinical practice. Accurate image evaluation demands a thorough knowledge of normal anatomy and of age-related changes in spinal growth and development. As in all other regions of the body of a child, spinal and extraspinal pathology differs from that of adults. Imaging sequences and planes should be tailored to the particular disease process and the involved anatomical region.

Normal anatomy (*Figures 10.19–10.21*)

The fusion of the upper and lower halves of two successive sclerotomes with the intersegmental tissue forms the fetal precartilaginous vertebral body. The intervertebral disc is formed by cells originating in the cephalic part of the underlying sclerotome. While the notochord in the vertebral body completely regresses, it persists between the vertebrae, and expands to form a gelatinous center of the intervertebral disc called the nucleus pulposus, which is later surrounded by the circular fibers of the annulus fibrosus.[13]

Ossification of the vertebra begins in the embryonic period and may continue into the twenty-fifth year. In the prenatal period, three primary ossification centers appear: one in the centrum and one in each half of the vertebral arch. Fusion of the halves of the vertebral arch is usually complete by one year of age. A neurocentral joint forms where the vertebral arch and body articulate, and this joint fuses with the centrum between the third and sixth years of age.[13]

Five secondary ossification centers appear just after puberty: one for the tip of the spinous process, one for each transverse process, and one for each ring apophysis. Fusion of the secondary centers is completed by age twenty-five.[13]

Congenital anomalies (*Figures 10.22–10.26*)

Congenital anomalies of the spine are well suited for evaluation by MRI. Transaxial images usually supplement sagittal images and, in some cases, coronal images, to detect the level of the conus medullaris and the nerve roots of the cauda equina.[14] Recent data suggest that there is no significant difference in the level of the conus medullaris location among various age groups of normal children, indicating the normal conus medullaris does not ascend throughout childhood.[15] Because of technical limitations in MRI, conventional myelography should supplement evaluation of the tethered cord to demonstrate dural adhesions.[16]

Tumor (Figures 10.27–10.32)

For the evaluation of neoplasia affecting the spine and adjacent soft-tissue elements, both short TR, TE and long TR, TE sequences are utilized.[11,12] In our experience, tumor differentiation from marrow fat is best determined on the T_1-weighted (short TR, TE) pulse sequences. These sequences were also best for evaluating extradural components of a lesion. The long TR, TE T_2-weighted sequences provided the best evaluation of adjacent paraspinal soft-tissue pathology. These observations correlate with those made by other authors.[3,8,10,17,18] Of note is the limited ability to detect bone marrow abnormality when the involvement of the spine is diffuse. This is contrary to the reported experience of some authors who suggest that MRI can readily diagnose leukemic infiltrates.[19]

Trauma (Figures 10.36–10.38)

MRI of the craniocervical junction is best performed in the sagittal projection. Although computed tomography, conventional radiography, and polytomography are better for bone detail,[20,21] MRI is superior for soft-tissue evaluation. Spinal trauma and degenerative disease are also well-suited for magnetic resonance evaluation as reported by others.[17,22] T_1-weighted sequences are best for anatomic detail of the bony portion of the spine and of disc problems.[17,22]

References

1. KANGARLOO H, DIETRICH RB, TAIRA RT et al, MR imaging of bone marrow in children, *J Comput Assist Tomogr* (1986) **10**(2):205–9.
2. GREENFIELD GB, *Radiology of bone diseases*. JP Lippincott: Philadelphia, 4 edn, 1986) 562.
3. BRADY TJ, ROZEN BR, PYKERT IL et al, NMR imaging of leg tumors, *Radiology* (1983) **149**:181–7.
4. MOON KL, GENANTG HK, HELMS CA et al, Musculoskeletal applications of nuclear magnetic resonance, *Radiology* (1983) **147**:161–71.
5. ZIMMER WD, BERQUIST TH, McLEOD RA et al, Bone tumors: magnetic resonance imaging versus computed tomography, *Radiology* (1985) **155**:709–18.
6. MURRAY RO, JACOBSON HC, The radiology of skeletal disorders. Exercises in diagnosis. (Williams and Wilkins: Baltimore 1985.)
7. TOTTY WG, MURPHY WA, LEE JKT, Soft tissue tumors: MR imaging, *Radiology* (1986) **160**:135–41.
8. AISEN AM, MARTEL W, BRAUNSTEIN EM et al, MRI and CT evaluation of primary bone and soft tissue tumors, *AJR* (1986) **146**:749–56.
9. COHEN MD, DE ROSA GP, KLEIMAN M et al, Magnetic resonance evaluation of disease of the soft tissues in children, *Pediatrics* (1987) **79**(5):696–701.
10. DEMAS BE, HEELAN RT, LANE J et al, Soft tissue sarcomas of the extremities: comparison of MR and CT in determining the extent of disease, *AJR* (1988) **150**:615–20.
11. SUNDARAM M, McGUIRE MH, HERVOLD DR et al, High signal intensity soft tissue masses on T_1-weighted pulsing sequences, *Skeletal Radiol* (1987) **16**:30–36.
12. SUNDARAM M, McGUIRE MH, SCHAJOWICZ F, Soft tissue masses: histologic basis for decreased signal (short T_2) on T_2-weighted MR images, *AJR* (1987) **148**:1247–50.
13. LANGMAN J, *Medical embryology*. Williams and Wilkins: Baltimore 1975) 85.
14. BARNES PD, LESTER PD, YAMANASHI WS et al, MRI in infants and children with spinal dysraphism, *AJR* (1986) **147**: 339–46.
15. WILSON DA, PRINCE JR, MR imaging determination of the conus medullaris in normal children and in children with tethered cord, presented at The Society of Pediatric Radiology, 31st Annual Meeting, Coronada, California 1988 (in press).
16. HAUS JS, BENSON JE, KAUFMAN B et al, Demonstration of a diastematomyelia and associated abnormalities with MR imaging, *AJNR* (1985) **6**:215–19.
17. MARAVILLA KR, LESH P, WEINREB JC et al, Magnetic resonance imaging of the lumbar spine with CT correlation, *AJNR* (1985) **6**:237–45.
18. PETTERSSON H, GILLESPY T, HAMLIN DJ et al, Primary musculoskeletal tumors: examination with MR imaging compared with conventional modalities, *Radiology* (1987) **164**:237–41.
19. MOORE SG, GOODING CA, BRASCH RC et al, Bone marrow in children with acute lymphocytic leukemia: MR relaxation times, *Radiology* (1986) **160**:237–9.
20. HREIDARSSON S, MAGRAM G, SINGER H, Symptomatic atlantoaxial subluxation in Down's Syndrome, *Pediatrics* (1982) **69**:568–71.
21. LEE BCP, DECK MDF, KNEELAND JB et al, MR imaging of the craniocervical junction, *AJNR* (1985), **144** : 1123–9.
22. EDELMAN RR, SHOUKIMAS CM, STARK DD et al, High-resolution surface coil imaging of lumbar disk disease, *AJR* (1985) **144**:1123–9.

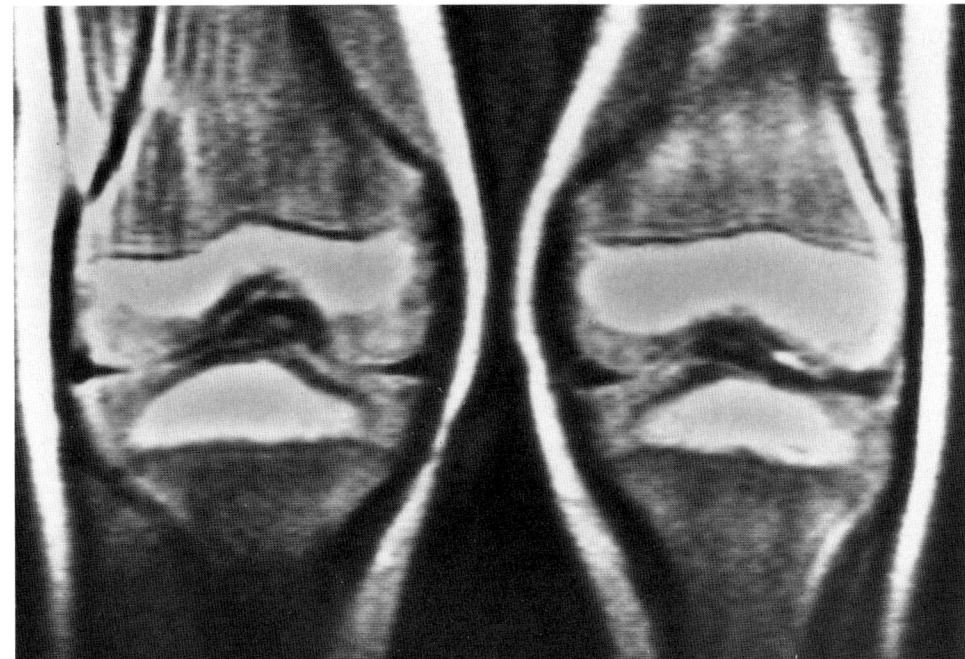

Figure 10.1

Aplastic anemia. Signal intensity of the marrow is returning to normal on a 14-year-old girl with treated aplastic anemia.

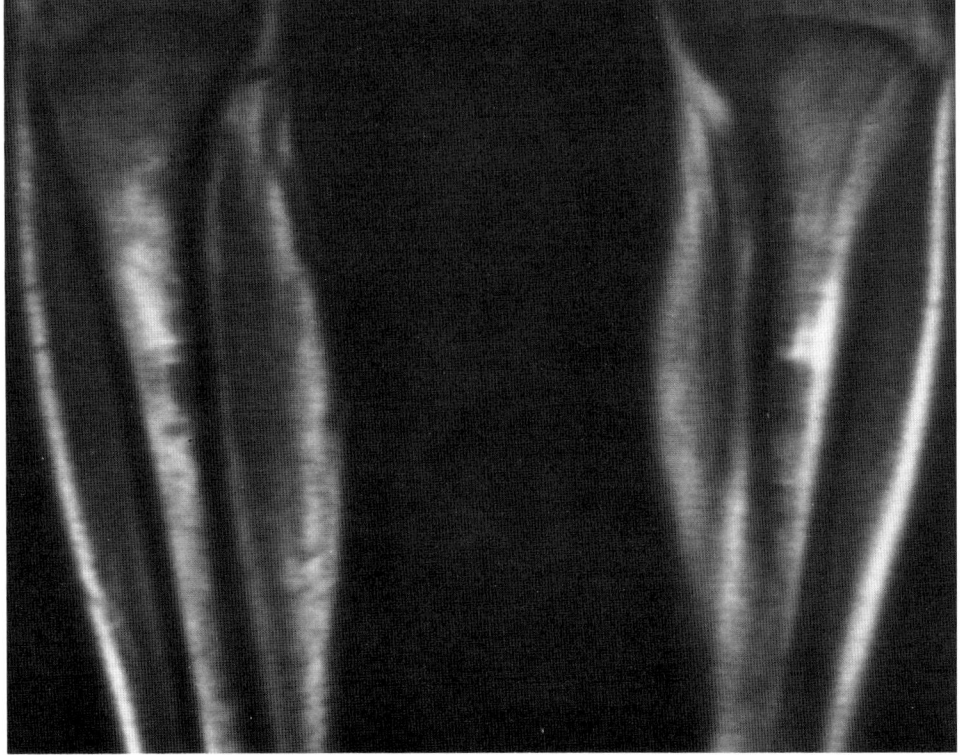

Figure 10.2

Leukemia. An 8-year-old girl with leg pain. (**a**) Coronal MR image (SE 300/18) shows patchy irregular signal intensity in the marrow space of the tibia. (**b**) Leukemia. Coronal MR image (SE 1500/56) of the tibia shows inhomogeneous signal intensity of the marrow.

a

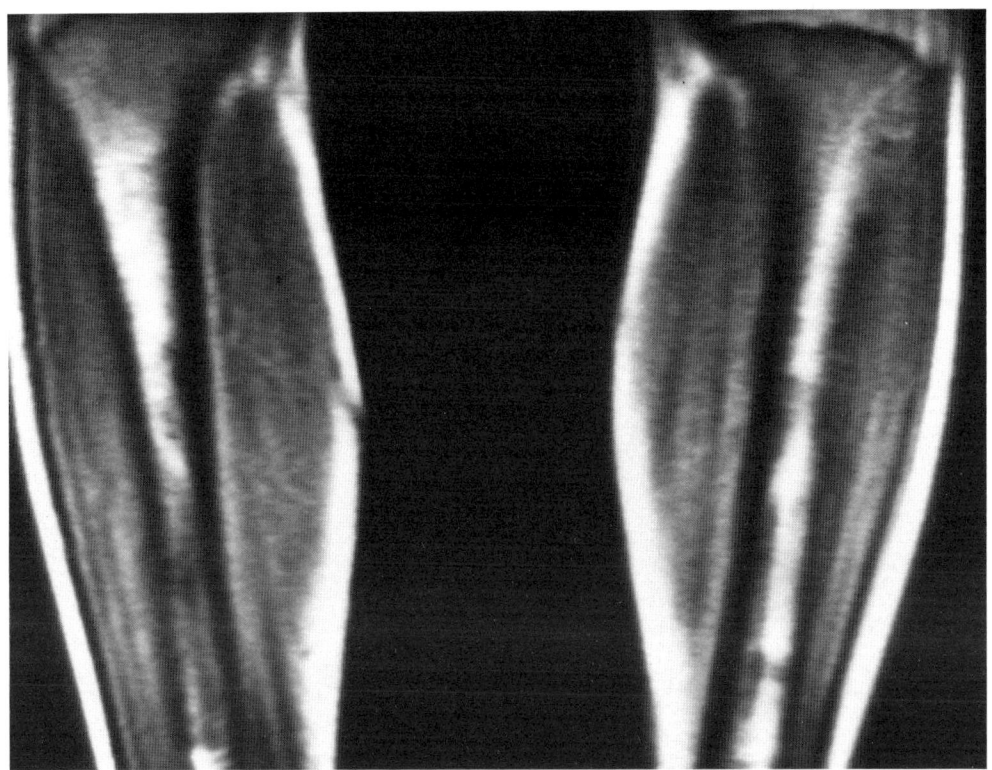

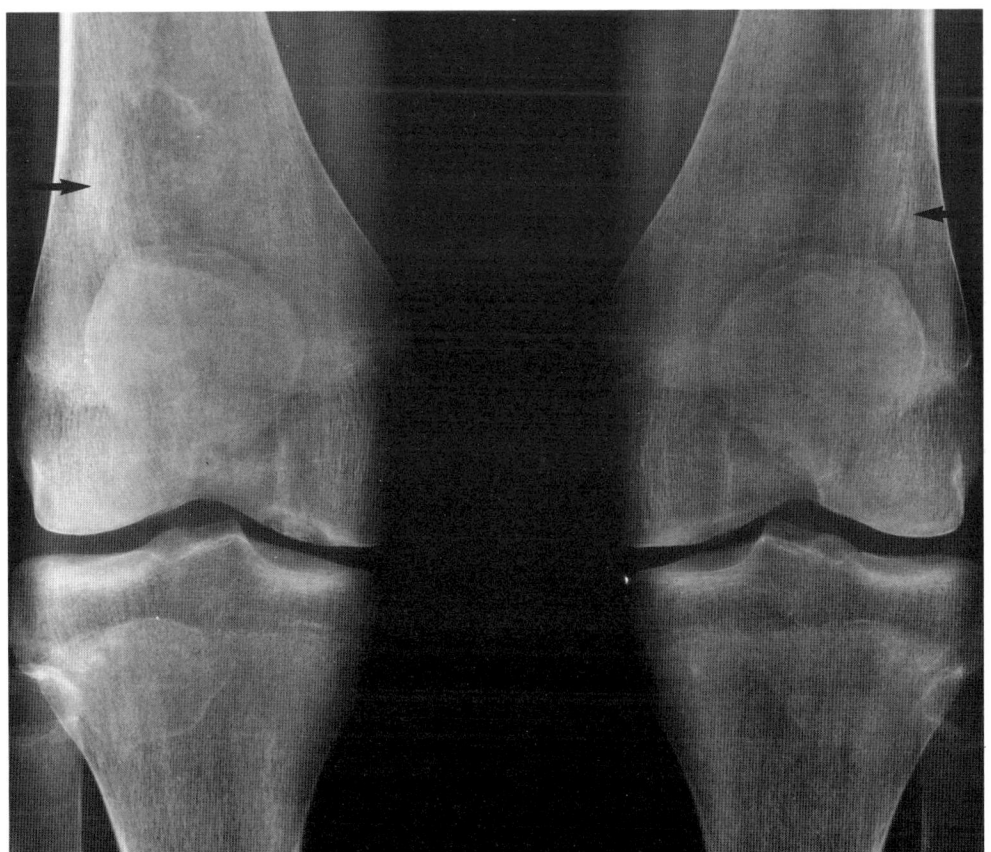

Figure 10.3

Infarct. (**a**) AP radiograph of knees was obtained for pain. Bilateral radiolucent defects are noted at the distal ends of the femurs (arrows). (**b**) Radionuclide bone scan demonstrates non-specific increase in activity at the distal diaphyseal end of the right femur. (**c**) T_1 coronal MR image (SE 500/28) demonstrates a large and smaller area of signal void in the diaphysis of the right and left femurs. (**d**) Coronal image (SE 1500/56) shows the heterogeneous signal intensity of the bone infarct with decreased signal in areas of calcification. The membrane shows increased signal intensity.

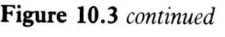

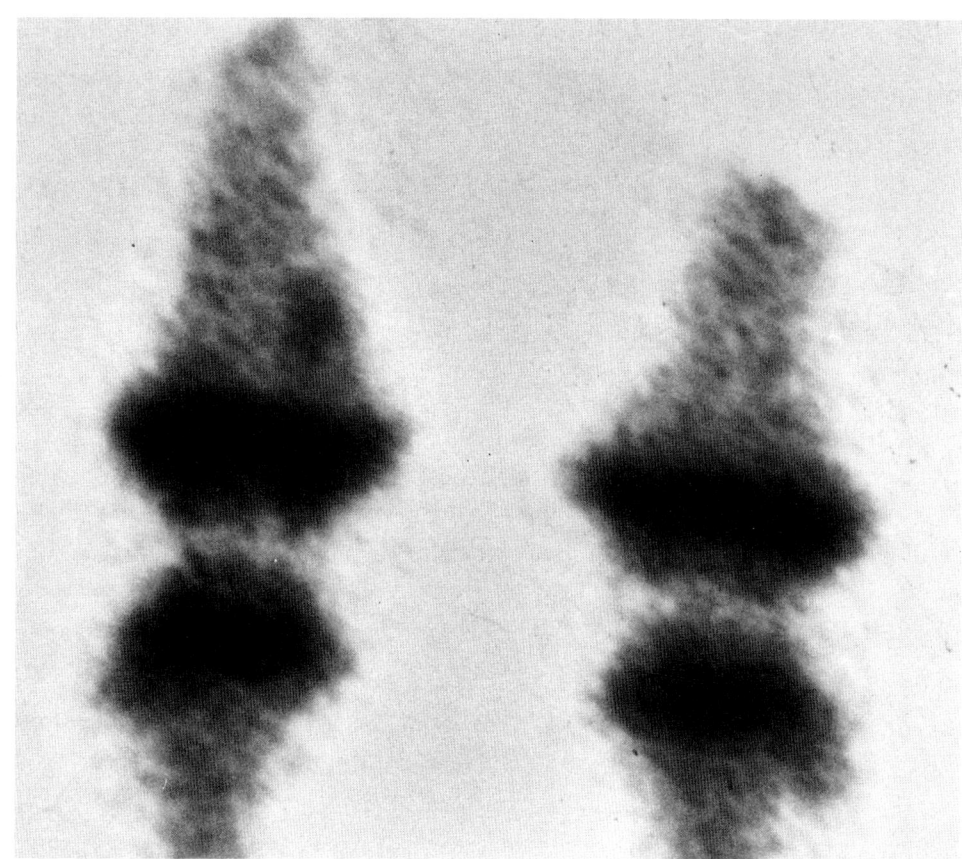

b

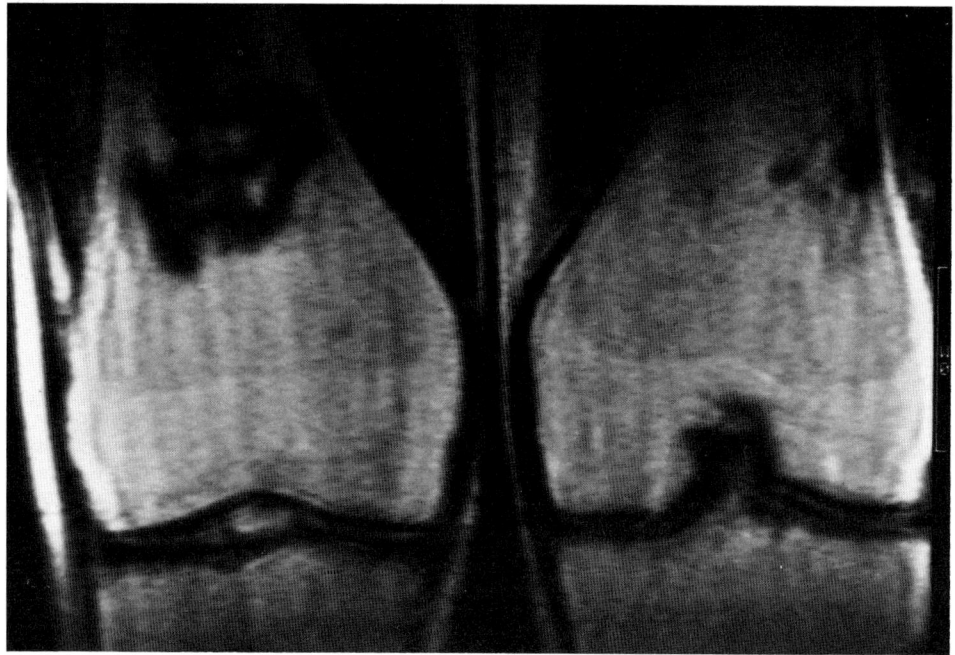

c

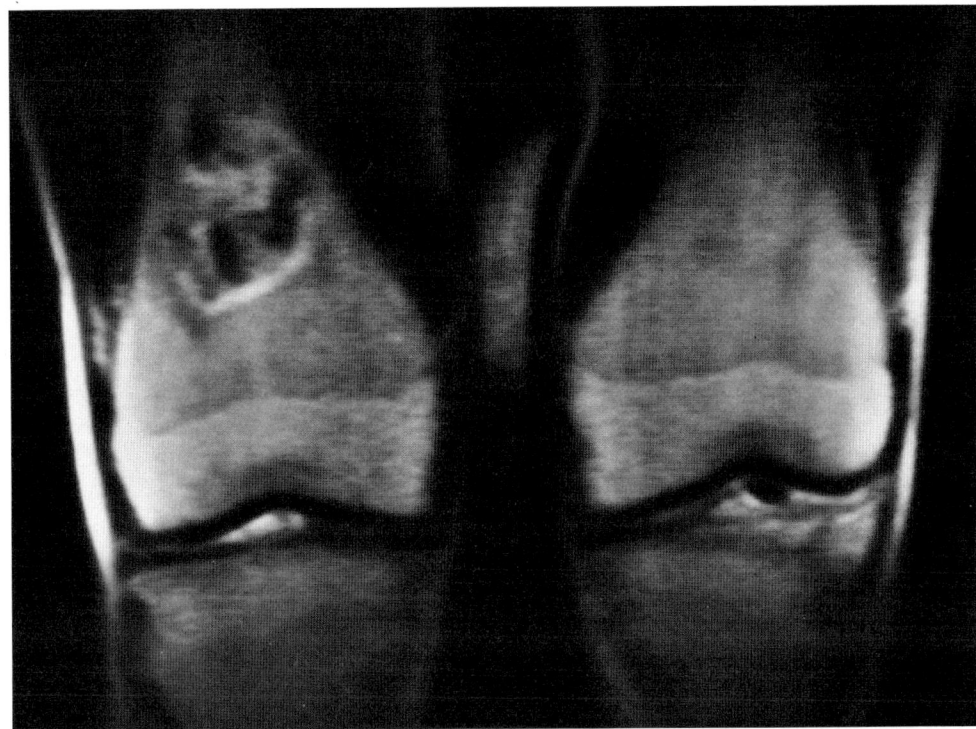

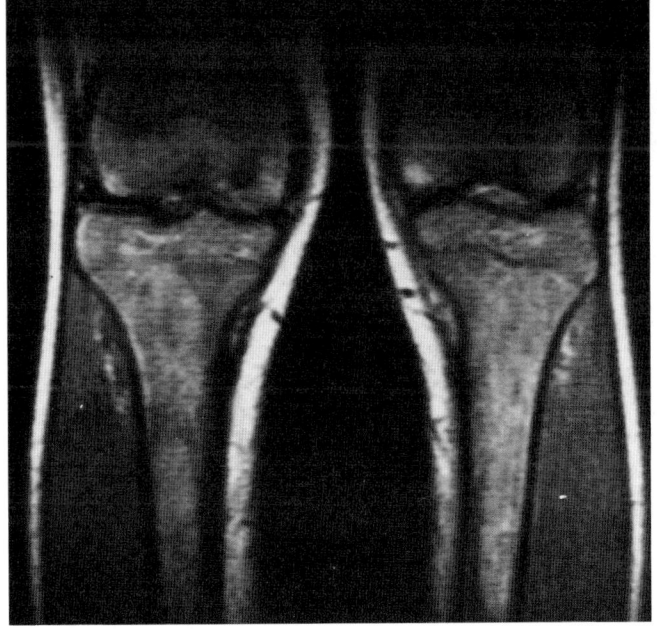

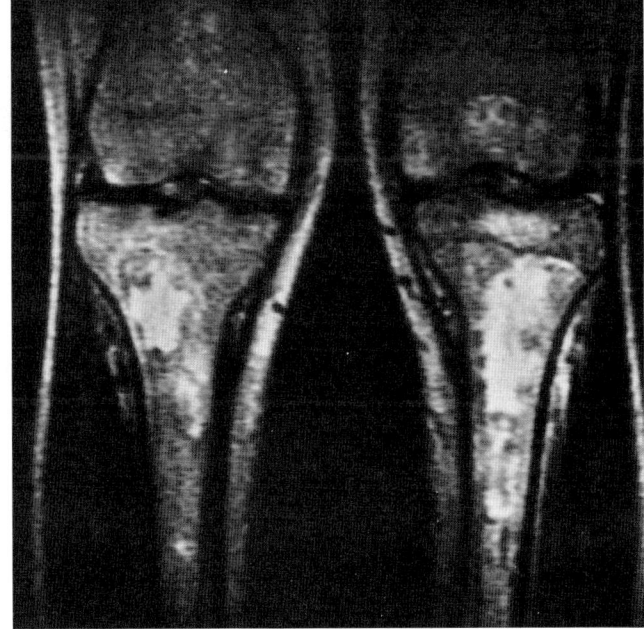

Figure 10.4

Sickle-cell anemia. A 14-year-old boy presented in 'crisis' with lower limb pain and low grade fever. (**a**) Coronal MR image (SE 500/28) clearly demonstrates decreased marrow signal intensity with mottled changes diffusely shown throughout both lower limbs. (**b**) Coronal MR image (SE 2000/84) more clearly depicts the irregular pattern of the marrow signal. A well defined bone infarct is present at the metaphyseal end of the right tibia.

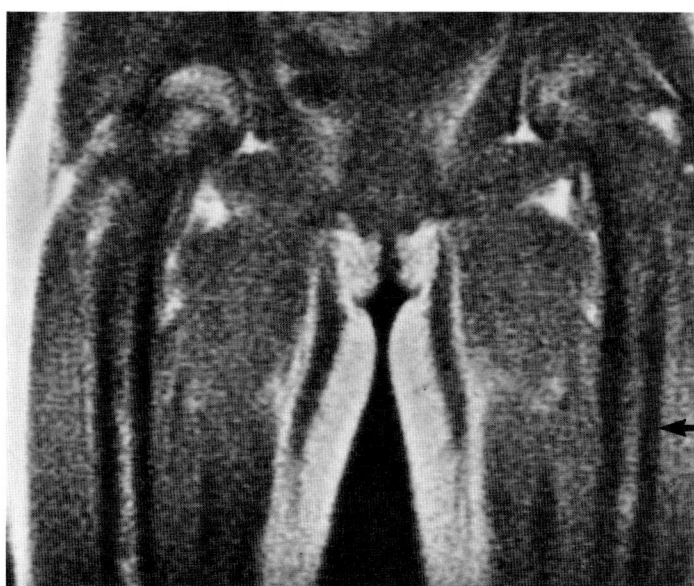

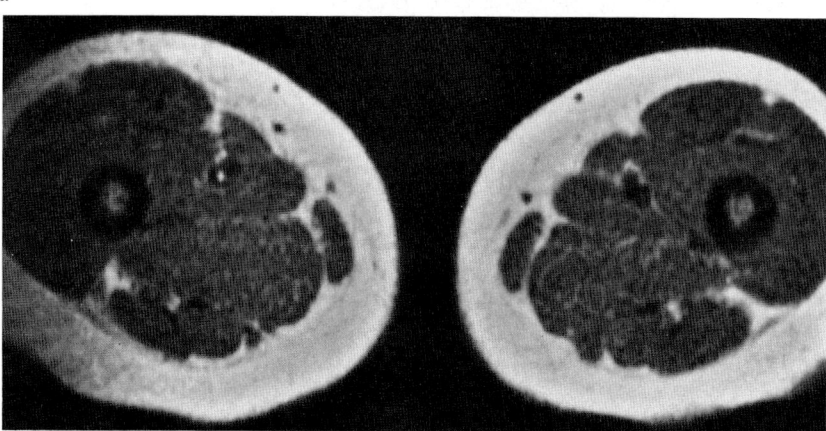

Figure 10.5

Neuroepithelial tumor. A 17-year-old girl with a marrow biopsy showing infiltration by small round cells. (**a**) Coronal MR image (SE 500/18) shows decreased signal intensity with intact cortex (arrow). (**b**) Axial scans (SE 500/18) demonstrate marrow signal loss.

a

b

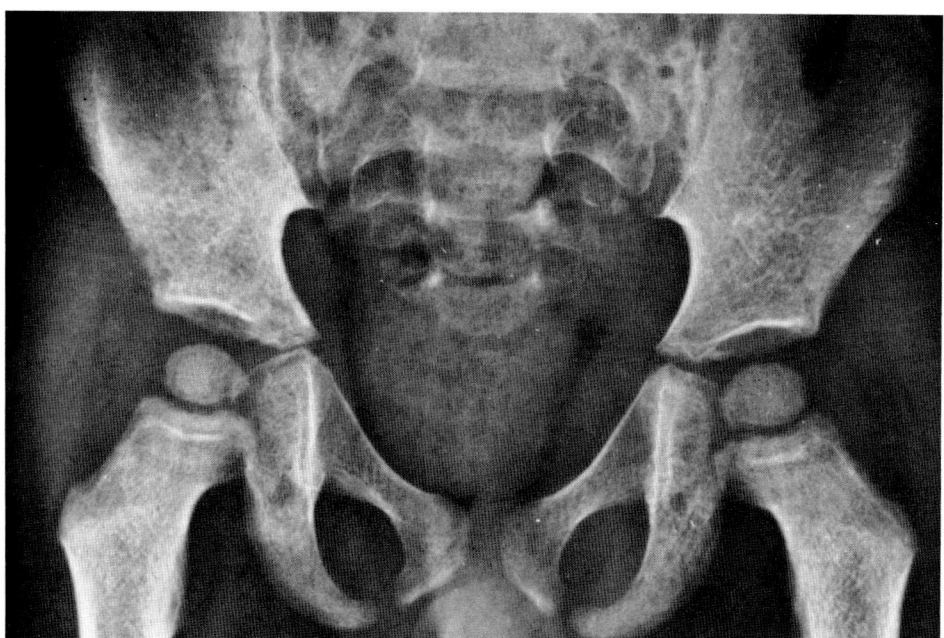

Figure 10.6

Neuroblastoma. (**a**) AP radiograph of the pelvis shows mottled trabecula pattern around the acetabula. There is subtle osteopenia of both femoral necks. (**b**) Radionuclide bone scan using ^{99m}Tc methylene diphosphonate shows diffuse increase of activity of innominate bones and proximal femurs as well as the right iliac bone. (**c**) Axial MR image (SE 500/28) of a 4-year-old boy with marrow metastases in both femurs and acetabula. Arrow delineates prominent soft-tissue component.

a

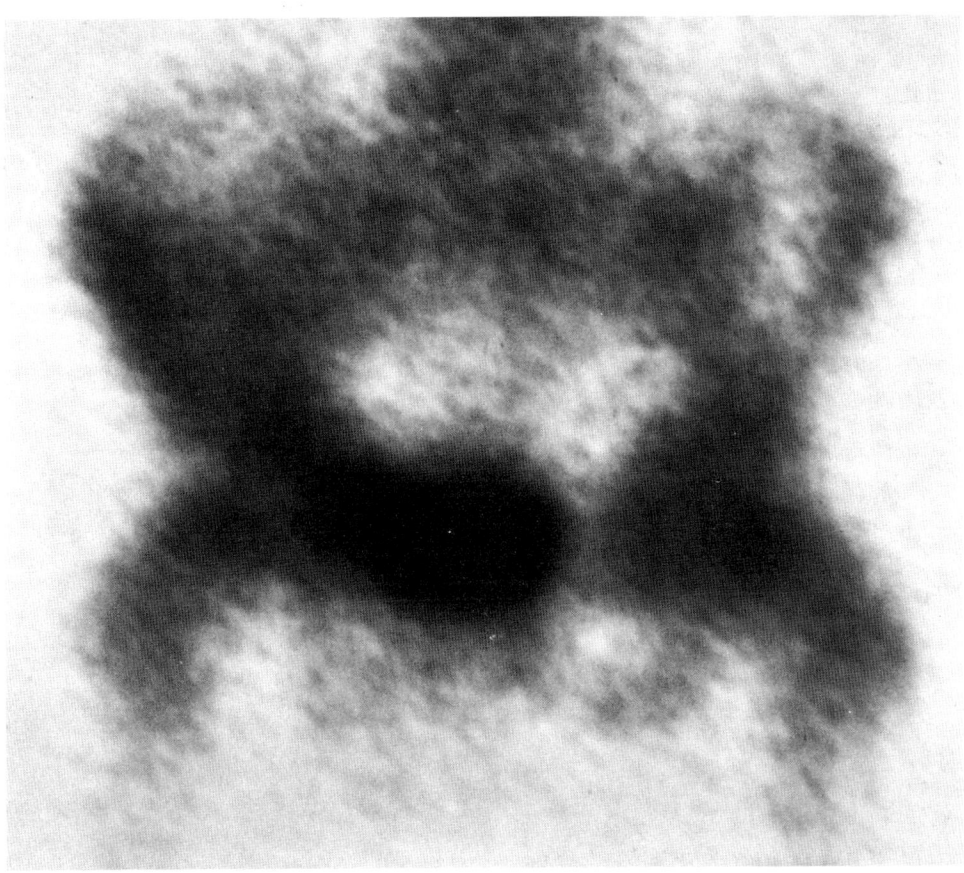
b
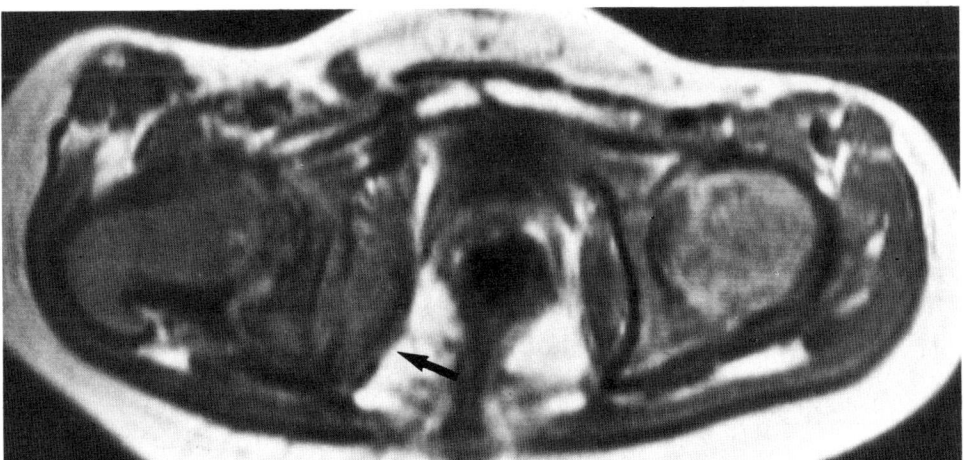
c

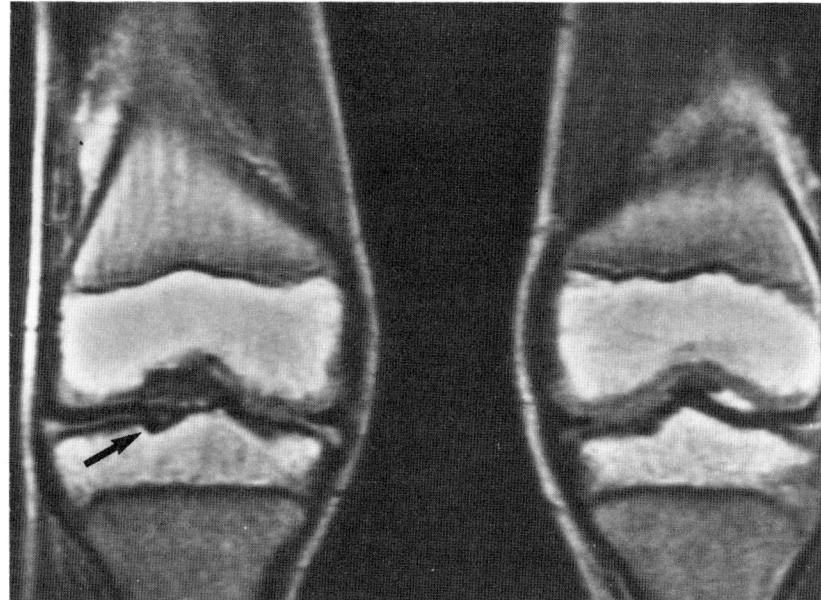

Figure 10.7

Hemophiliac arthropathy. An 11-year-old boy with chronic knee pain. Coronal MR image (SE 500/28) demonstrates medial femoral condyle enlargement. In addition, a focal subchondral erosion is noted at the tibial plateau (arrow).

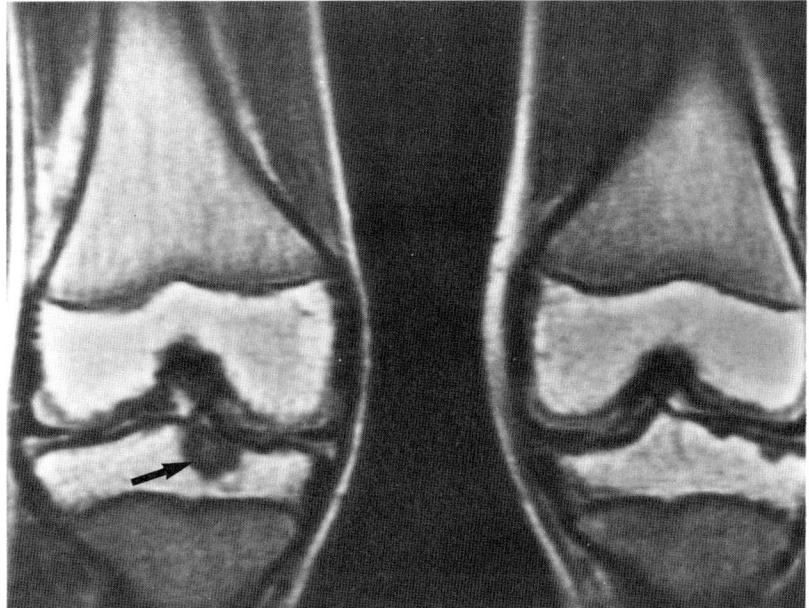
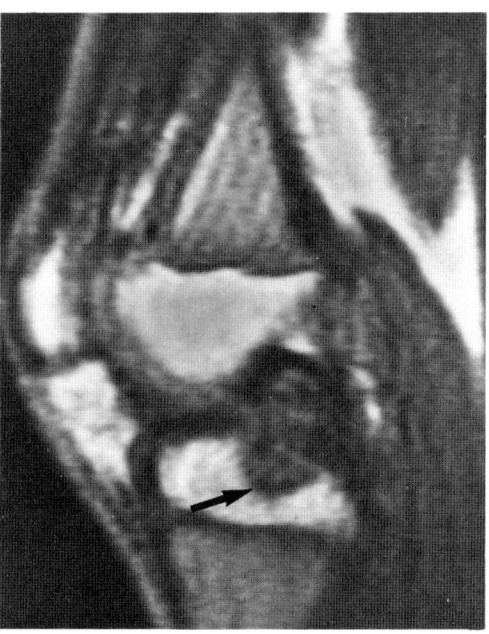

a b

Figure 10.8

Hemophiliac arthropathy. A 9-year-old boy with recurrent joint effusions. (**a**) Coronal and (**b**) sagittal MR images (SE 500/28) demonstrate a large subchondral cyst secondary to intra-osseous hemorrhage (arrow).

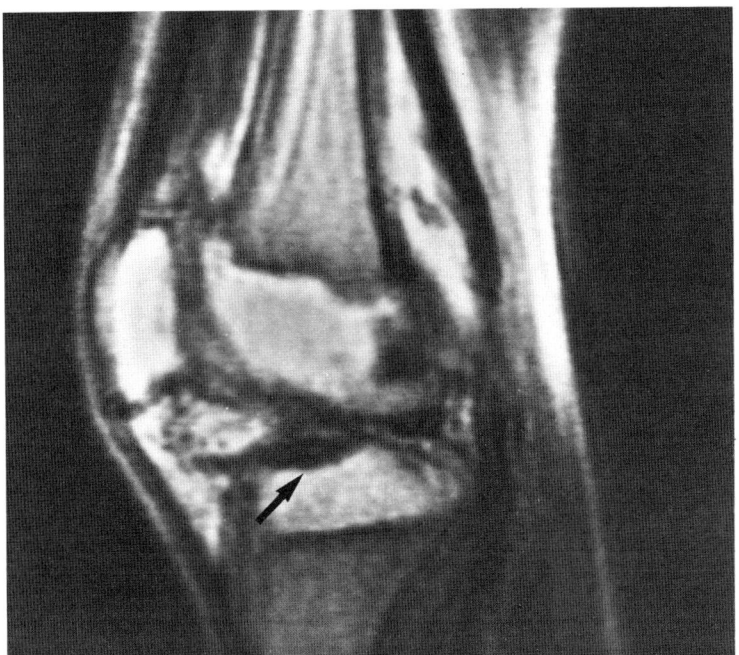

Figure 10.9

Hemophiliac arthropathy. Sagittal MR image (SE 500/28) in this 16-year-old asymptomatic boy shows a large subchondral erosion (arrow) extending along the anterior portion of the tibial plateau.

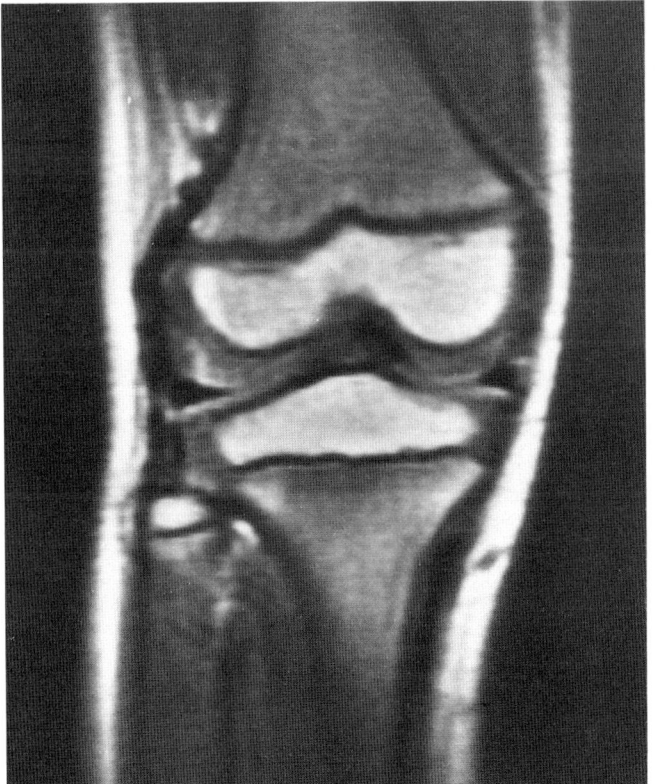

Figure 10.10

Hemophilia. Coronal image (SE 500/28) demonstrates decreased marrow signal secondary to anemia with preserved signal in the epiphyseal centers.

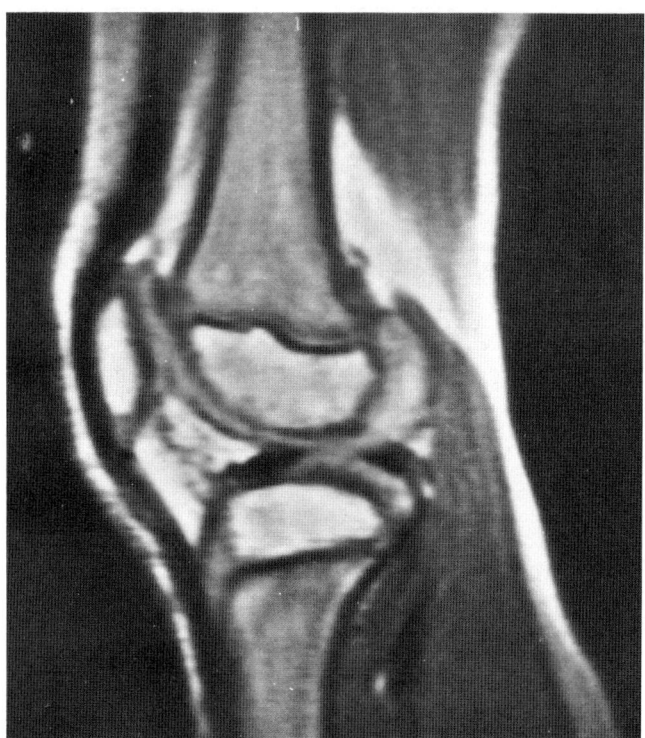

Figure 10.11

Hemophilia. A 5-year-old boy with joint pain. Asymmetric cartilage hypertrophy is demonstrated on this sagittal image (SE 500/28).

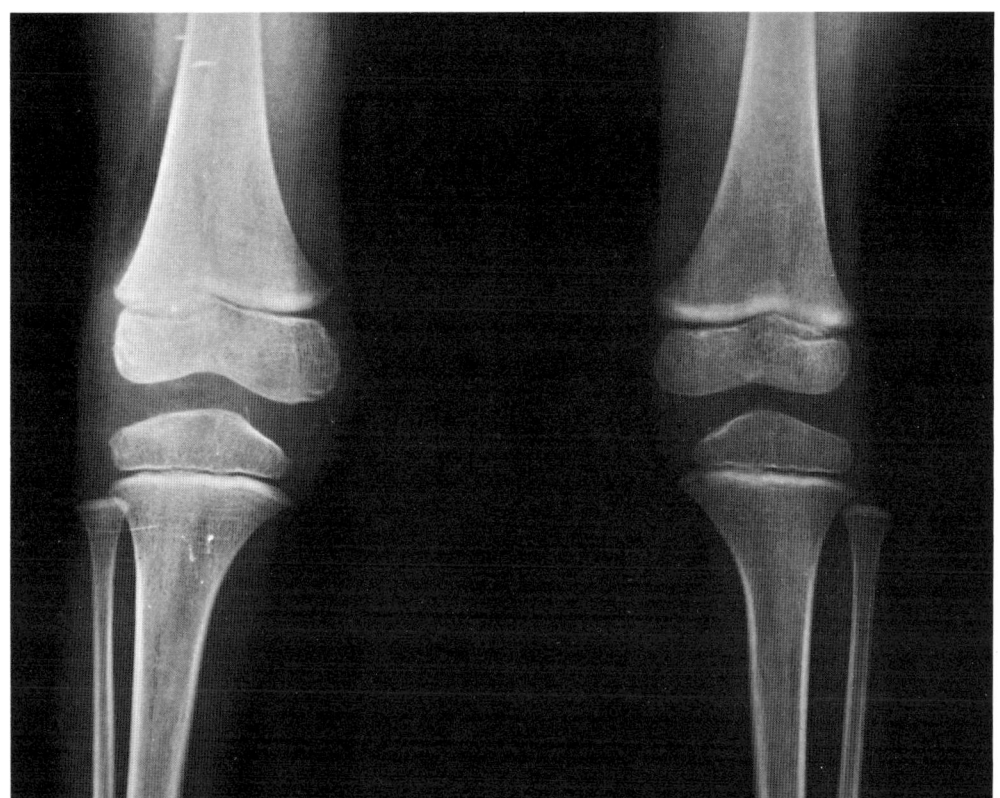
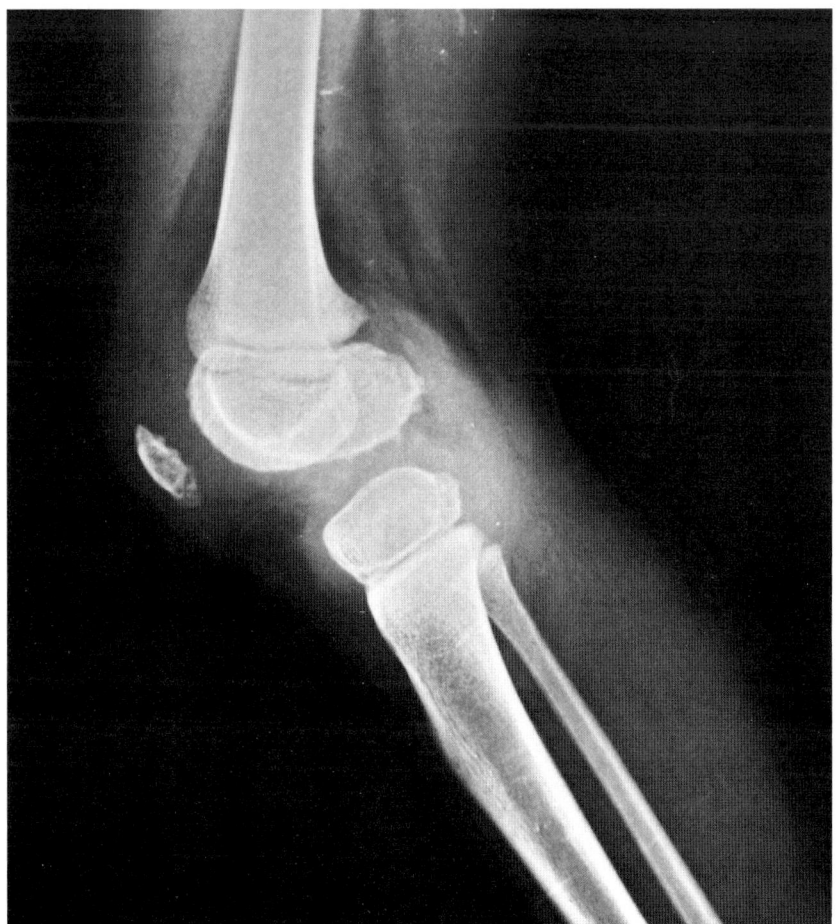

Figure 10.12
Benign hypertrophy. This 4-year-old girl presented with focal enlargement of the right knee. Past history was negative for previous trauma or infection. (**a**) AP and (**b**) lateral radiographs demonstrate focal enlargement of distal end of the femur and proximal end of the tibia on the right side. No joint effusion is noted at this time. (**c**) Coronal and (**d**) axial MR images (SE 500/28) obtained at the time of the radiographs show benign cartilage hypertrophy and osseous enlargement of the right knee. These findings are most likely related to a previous episode of joint hyperemia with subsequent overgrowth.

Figure 10.12 continued

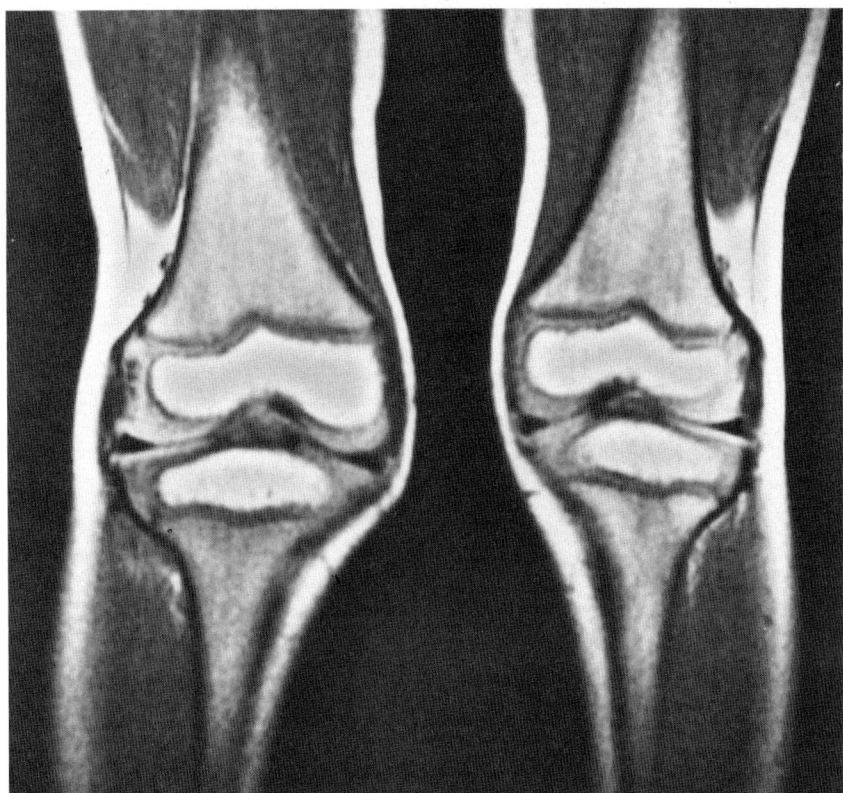

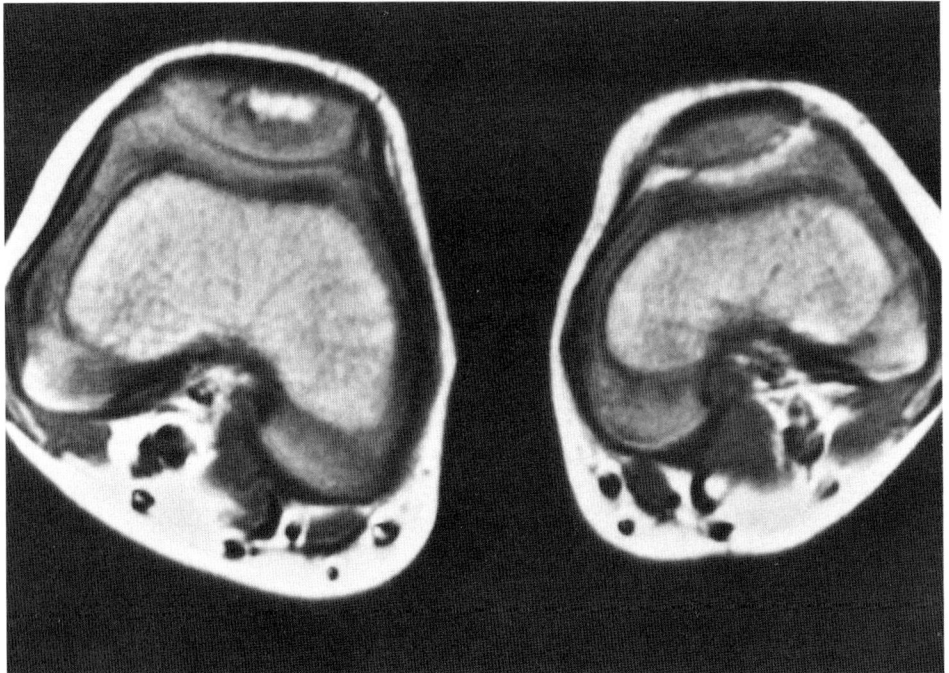

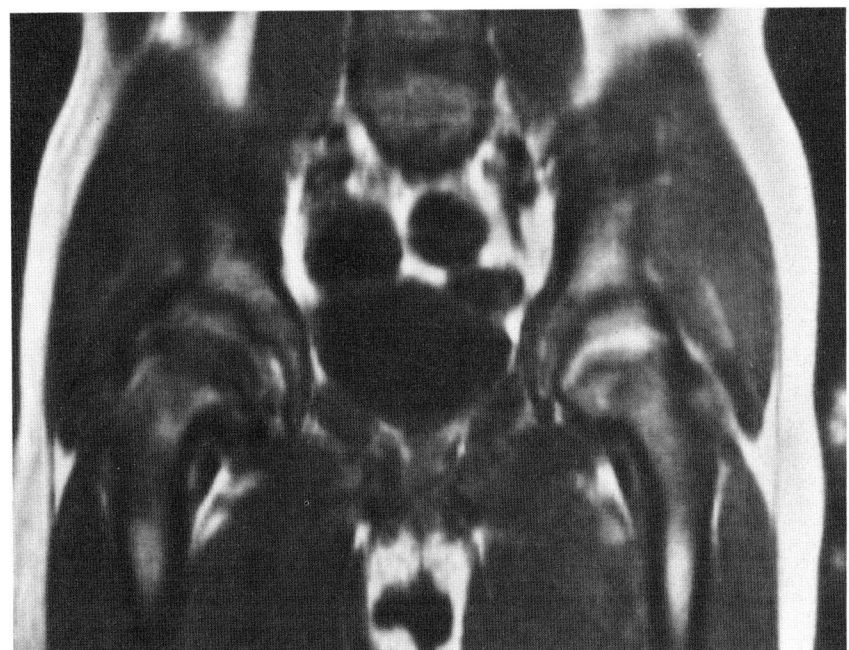

Figure 10.13

Legg-Calve-Perthes disease. A 5-year-old with painful gait. Coronal MR image (SE 500/28) shows marked decrease in signal in the right femoral head with flattening of the left femoral head. Bilateral asymmetric involvement of the femoral heads is common in LCP.

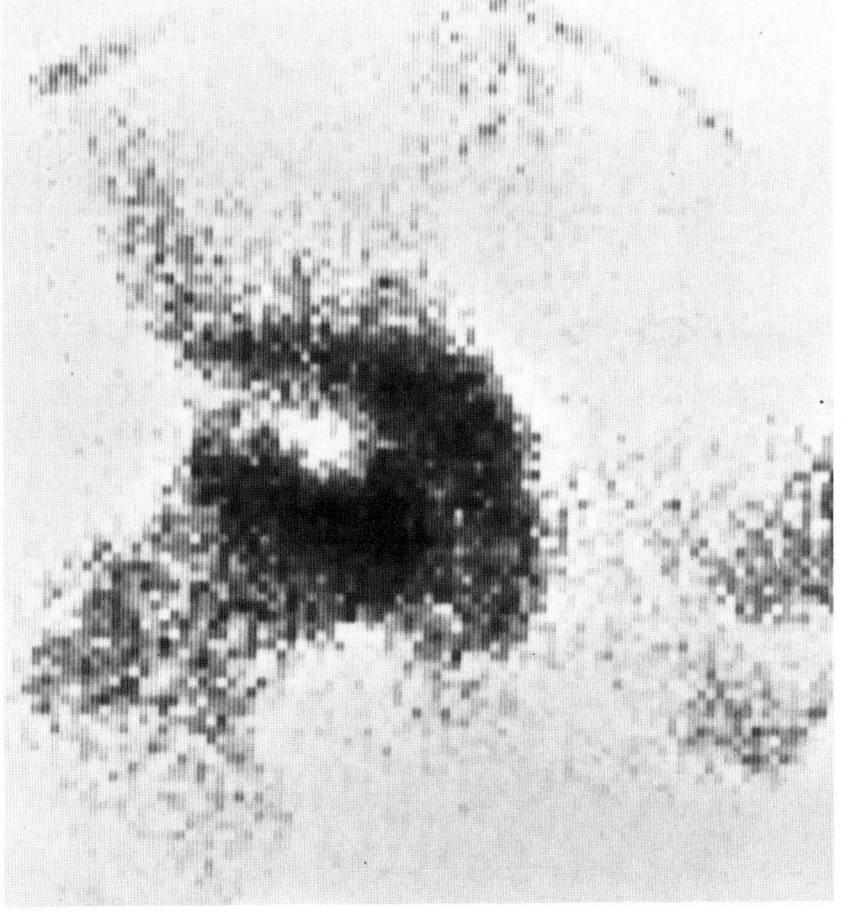

Figure 10.14

Osteonecrosis. This older child on steroid therapy complained of right knee pain. Initial imaging evaluation of the knee was negative. (**a**) Radionuclide bone scan depicts a focal area of decreased uptake in the right femoral head. (**b**) Coronal MR image (SE 500/28) clearly shows a corresponding area of decreased signal intensity in the right femoral head.

Figure 10.14 continued

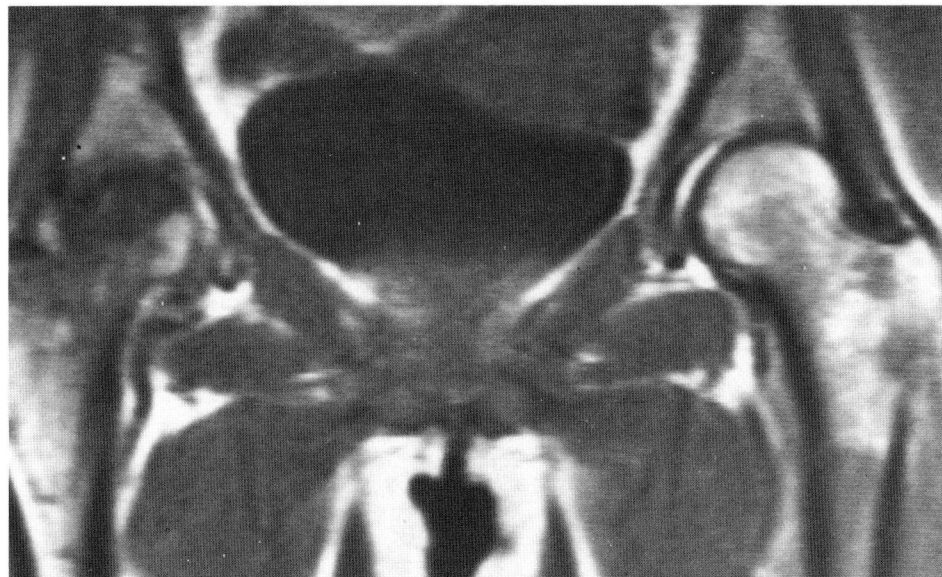

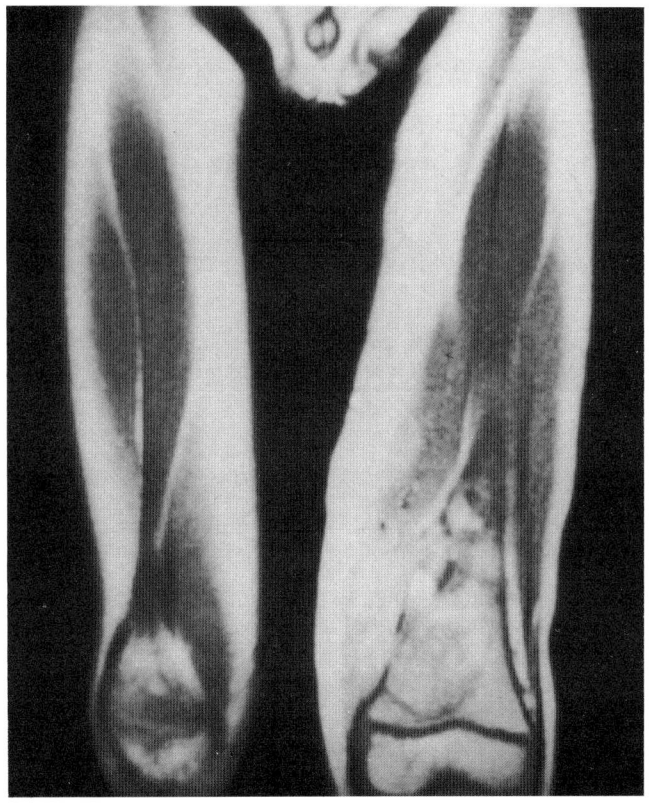

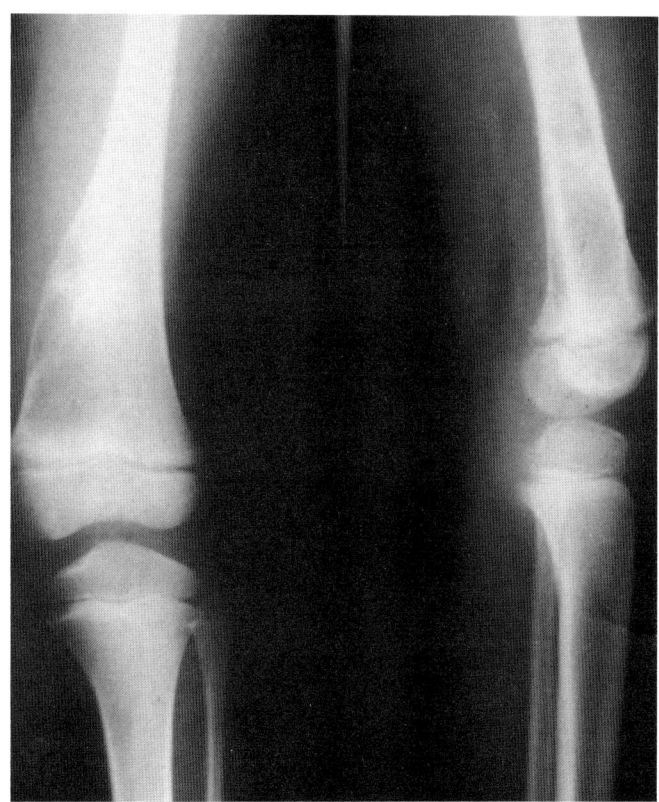

Figure 10.15

Telangiectatic osteosarcoma. A 4½-year-old girl presented with left knee pain and swelling. (**a**) AP and lateral radiographs depict an eccentric lytic lesion in the metaphyseal portion of the femur. (**b**) Coronal MR image (SE 1500/56) clearly demonstrates the osseous as well as soft-tissue extent of the lesion. High-signal areas at the proximal end of the lesion may be related to hemorrhage.

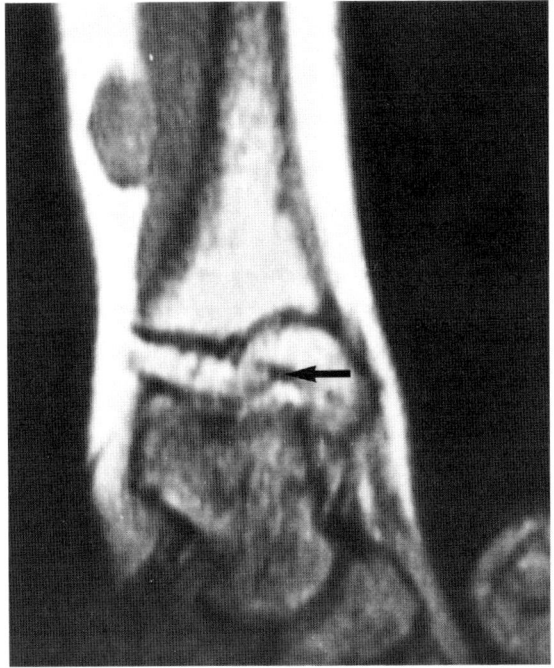
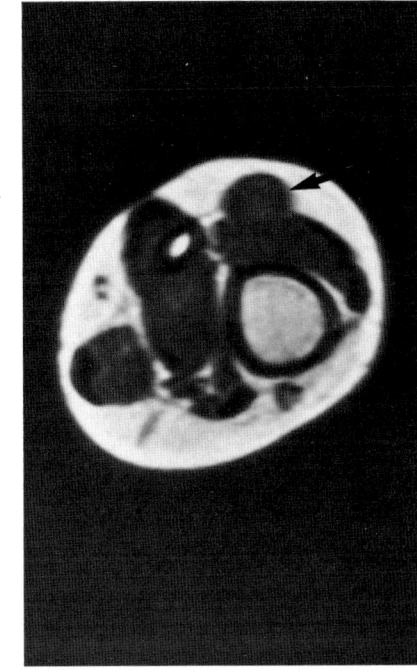

Figure 10.16

Neurofibromatosis. A 14½-year-old girl with a palpable, non-tender lump just above the ankle. (**a**) Coronal image (SE 500/28) shows intermediate-signal intensity of biopsy-proven neurofibroma. Arrow shows the pseudoarthrosis from a non-healed fracture. (**b**) Axial image (SE 300/18) shows the well-defined mass with a low-signal 'pseudocapsule' separating the lesion from muscle and adjacent subcutaneous fat (arrow).

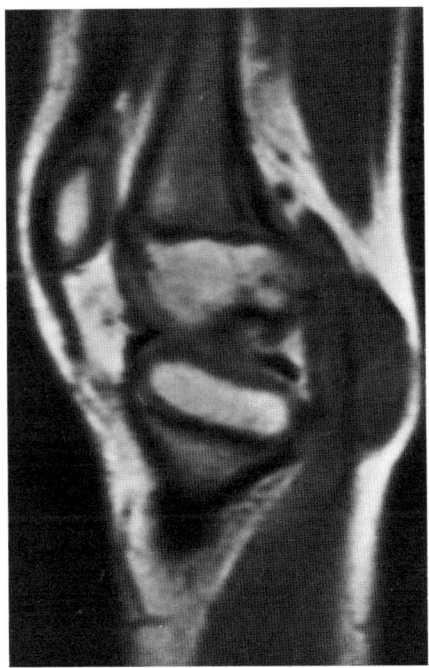
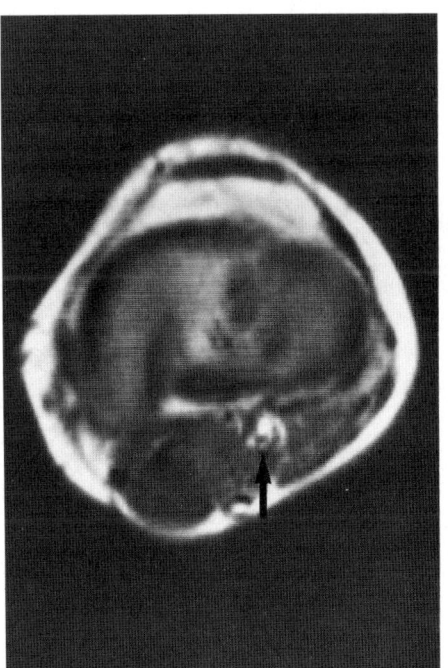
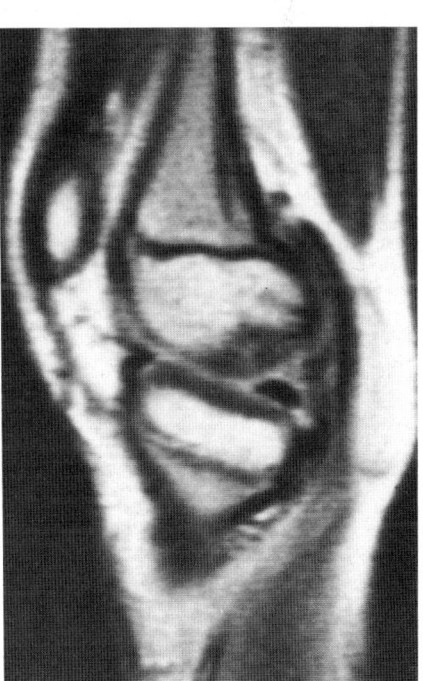

Figure 10.17

Baker's cyst. A 6-year-old girl with popliteal mass. (**a**) Sagittal image (SE 500/28) shows intermediate-signal mass in the subcutaneous fat. (**b**) Axial image (SE 400/28) defines the relationship of the cyst to the neurovascular bundle (arrow). (**c**) Sagittal image (SE 1500/56) shows marked T_2-shortening with high-signal intensity.

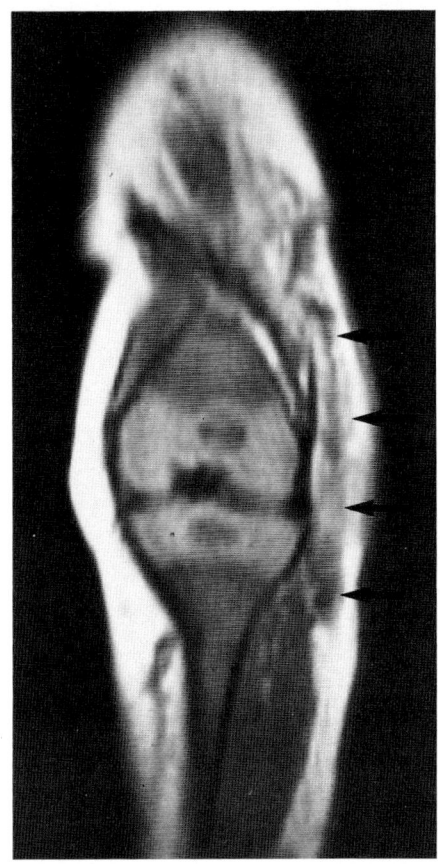
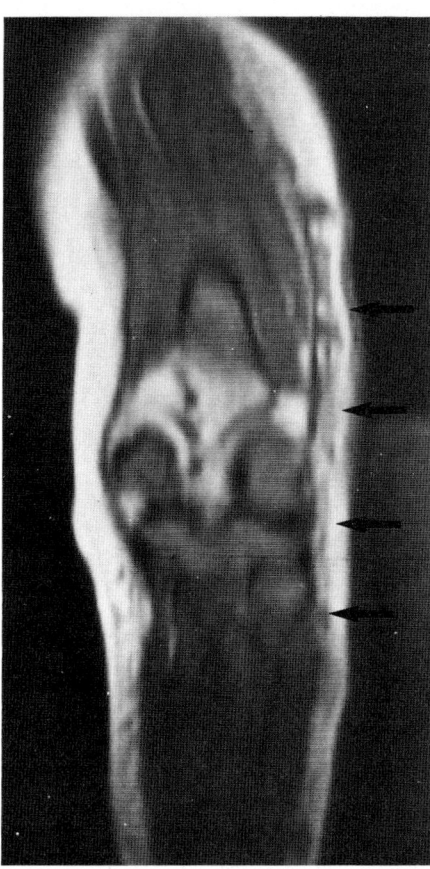
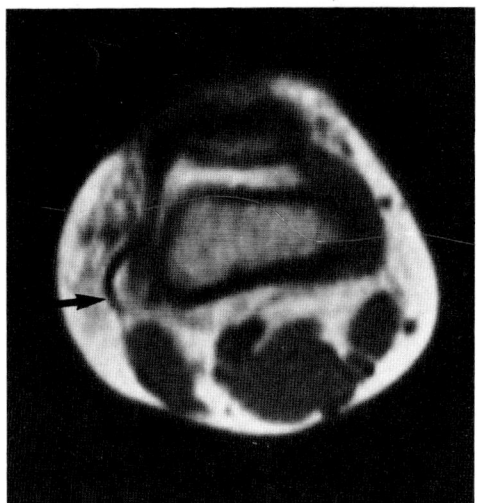
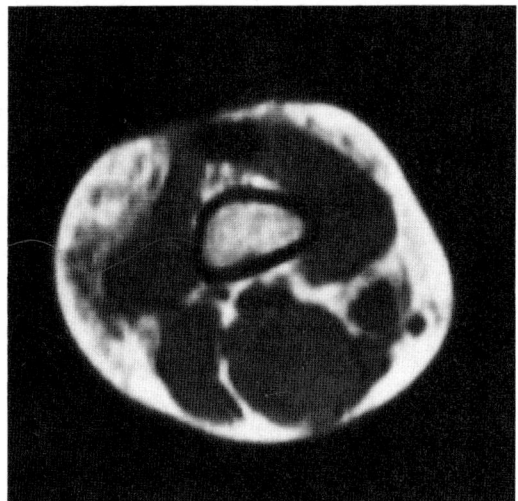

Figure 10.18

Hemangioma. A 10-year-old boy with a cutaneous hemangioma. (a) and (b) coronal images (SE 500/28) show the longitudinal extent of the lesion (arrows), (c) and (d) axial scans (SE 500/18) demonstrate mixed signal intensity lesion with low-signal, serpinginous vascular channel (arrows).

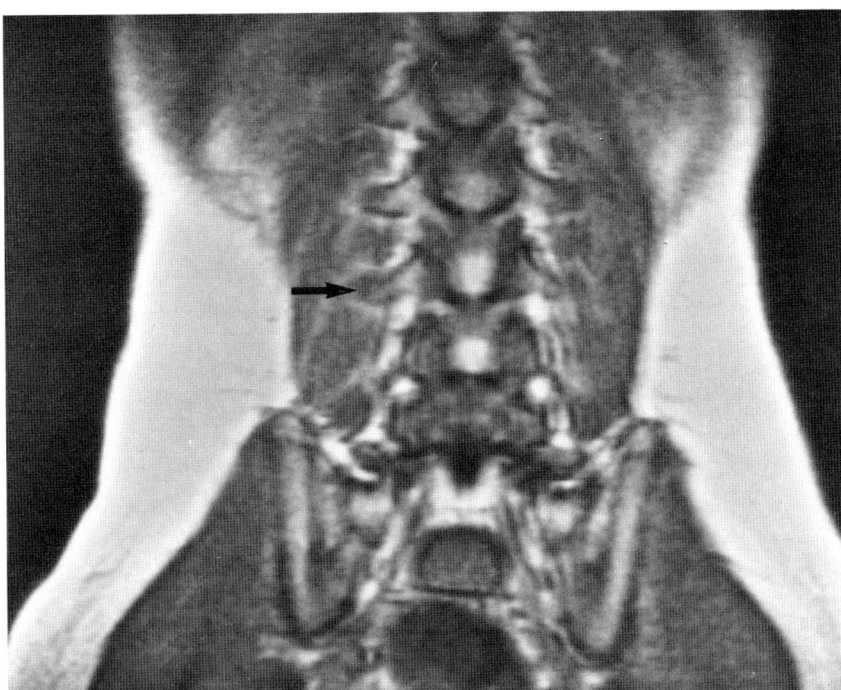

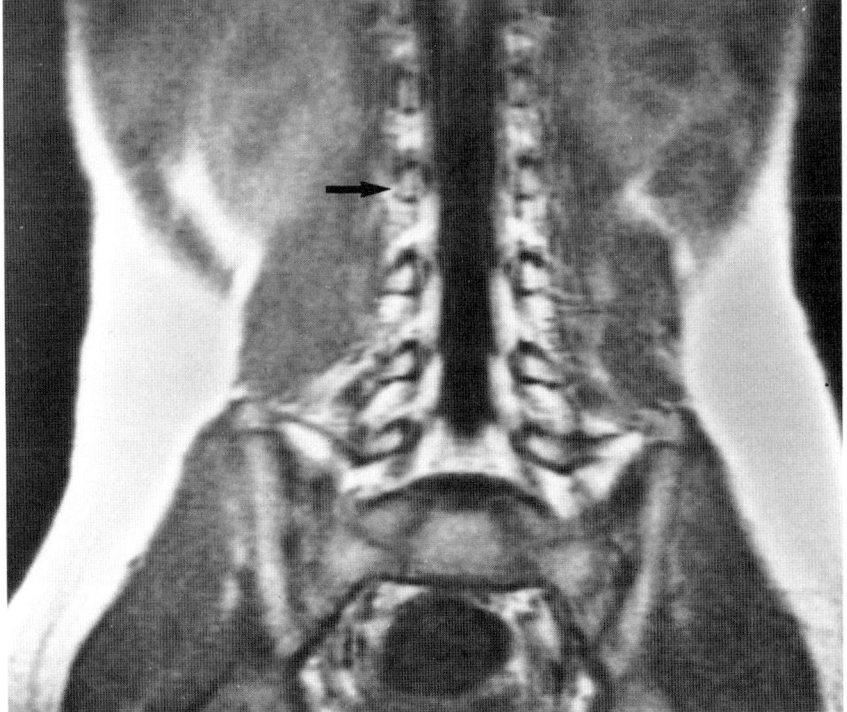

Figure 10.19

Normal coronal anatomy. Coronal MR images (SE 500/28) of a 2-year-old girl. In this plane, structures such as the (**a**) transverse process (arrow), (**b**) pedicles (arrow), (**c**) basivertebral vein (arrow), and (**d**) intervertebral disks (arrow) are clearly shown.

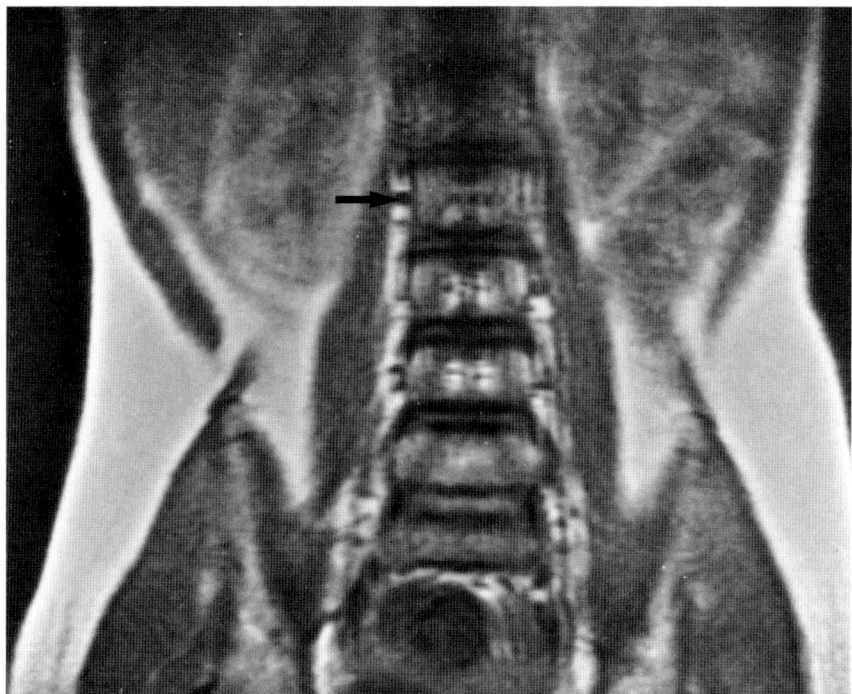

c

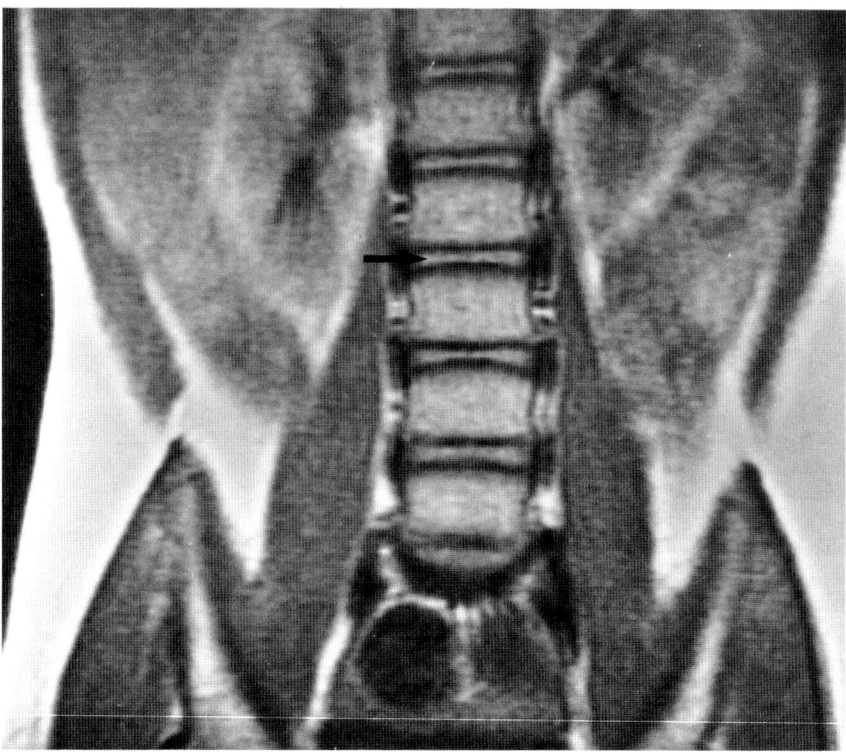

d

Figure 10.19 continued

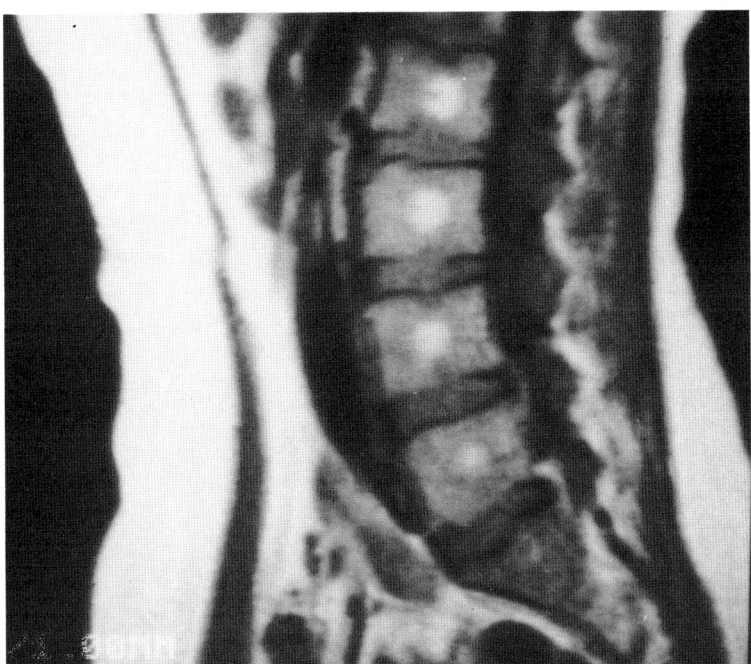

Figure 10.20

Normal variant. Sagittal MR image (SE 500/28) of a normal 15-year-old girl. Rounded areas of high-signal intensity in the center of the vertebral bodies may represent areas of decreased cellularity and increased fat.

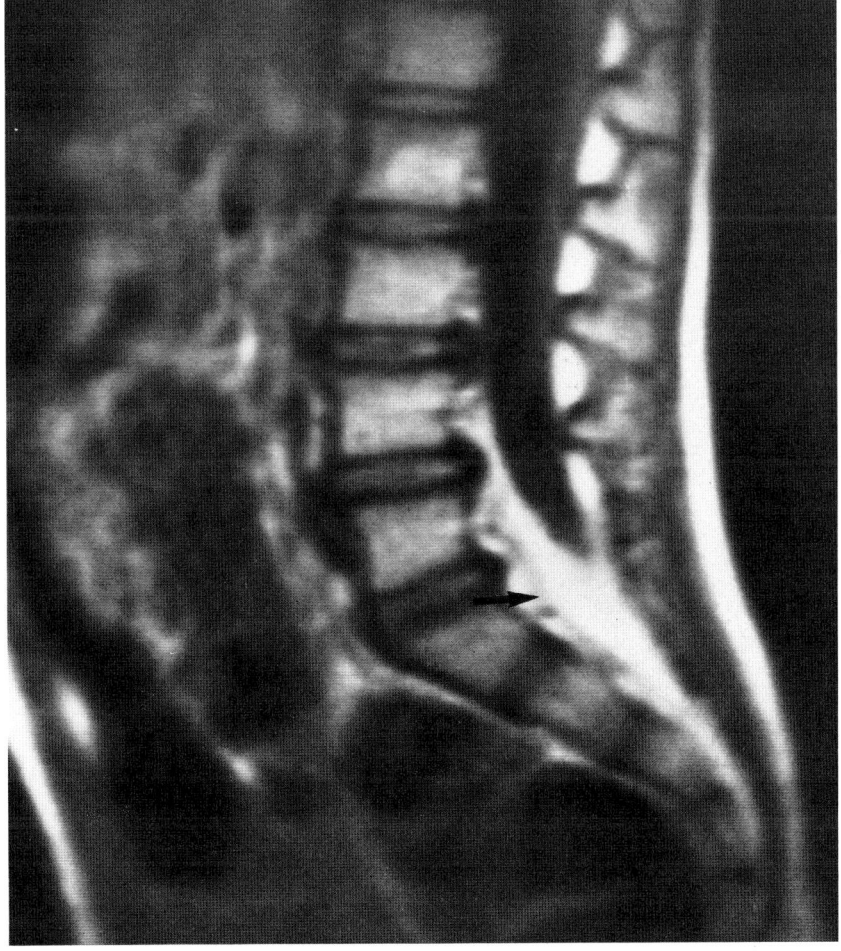

Figure 10.21

Normal variant. Sagittal MR image (SE 500/28) of a 5-year-old boy shows the prominent epidural fat in the anterior sacral space (arrow).

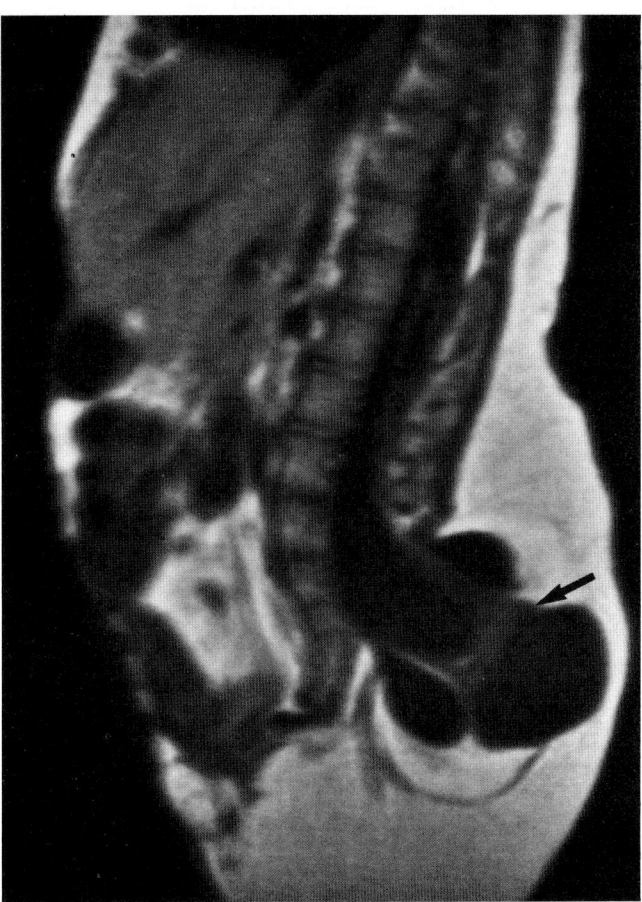

Figure 10.22

Meningocele with myelocystocele. Multiple congenital anomalies and a lumbar mass were noted in this 2-year-old girl. (**a**) Sagittal and (**b**) axial MR images (SE 500/28) demonstrate the large septated meningocele (arrow) and the tethered cord with the widened central canal. Existrophy of the bladder is also present.

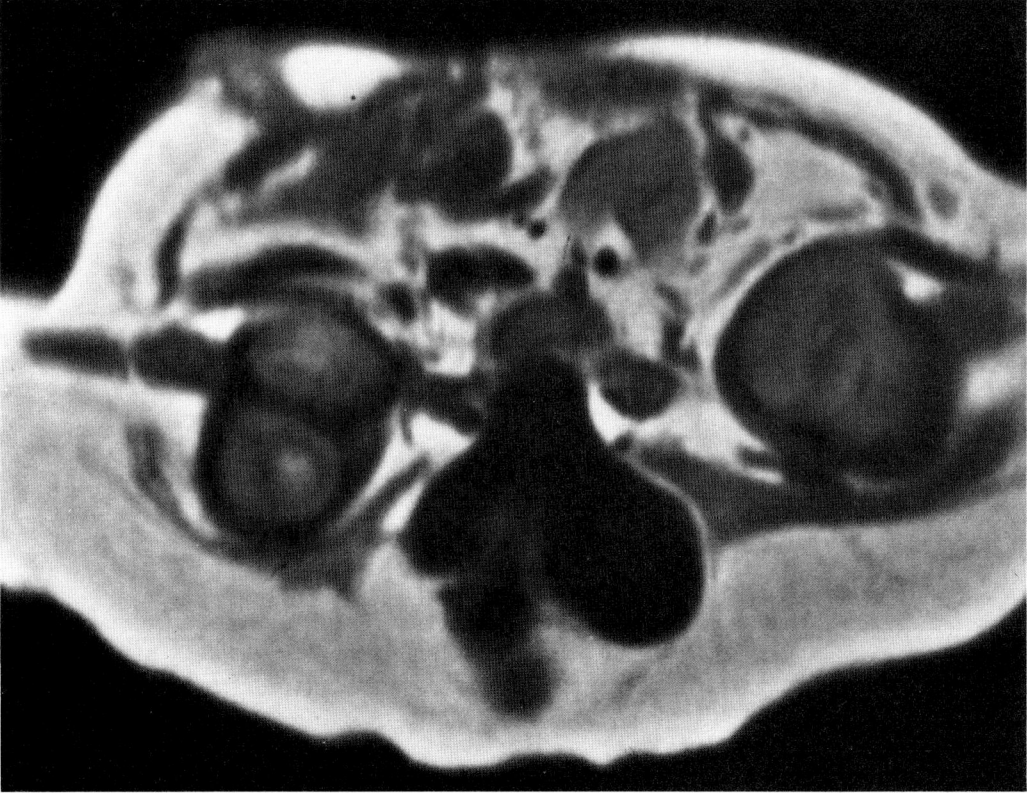

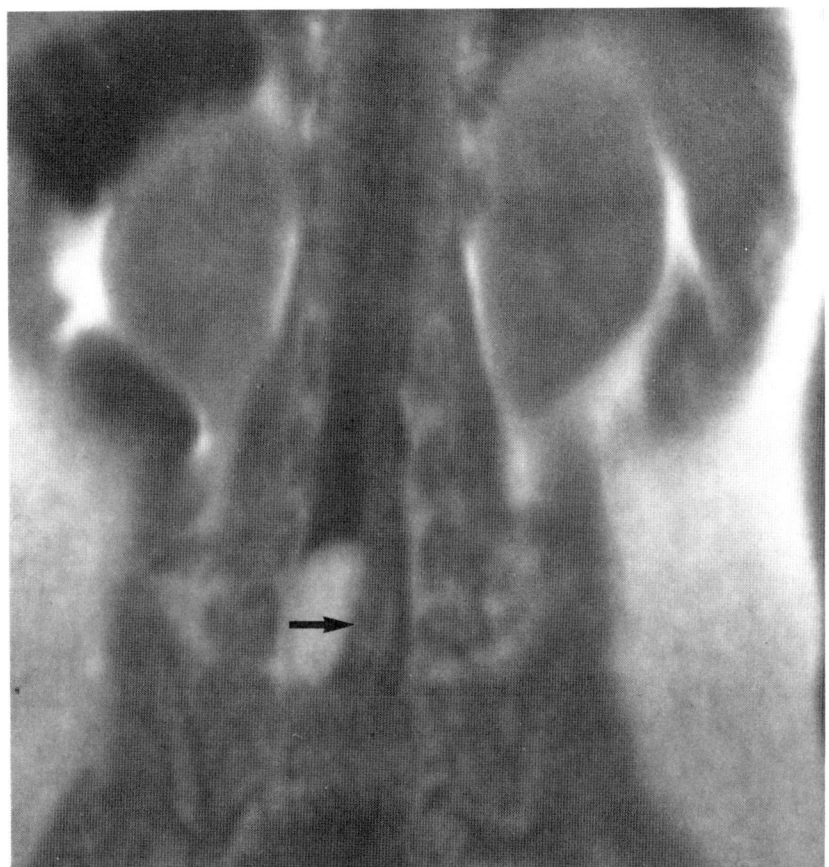

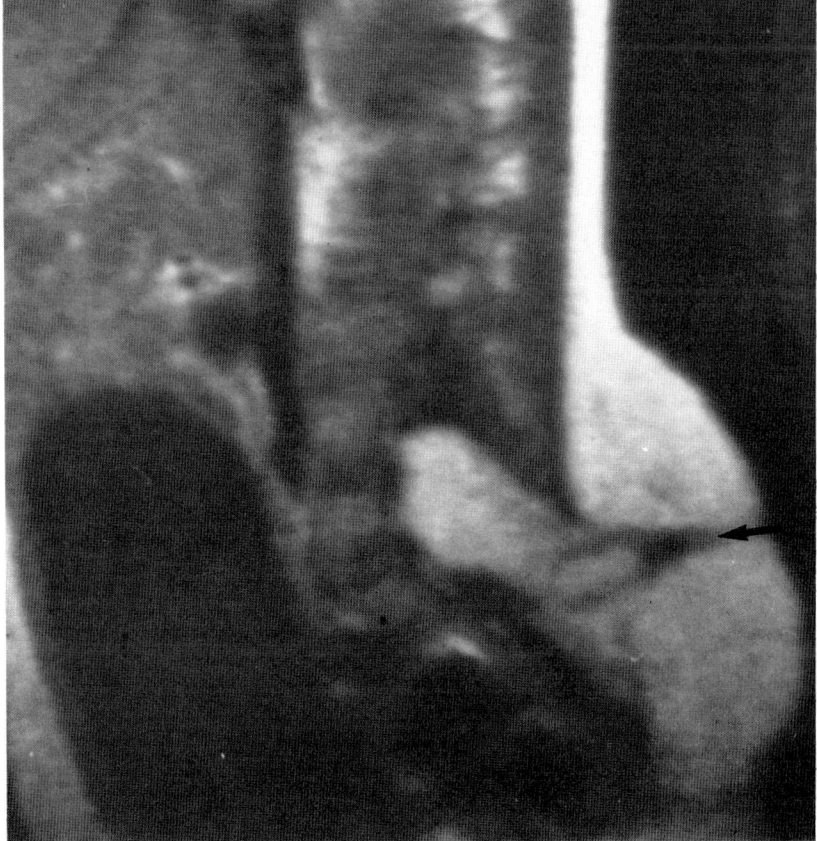

Figure 10.23
Tethered cord and lipoma. A 7-month-old girl with a sacral mass. (**a**) Coronal and (**b**) sagittal MR images (SE 500/28) show the high-signal-intensity lipoma displacing the cord (arrow) with cutaneous sinus tracts (arrow).

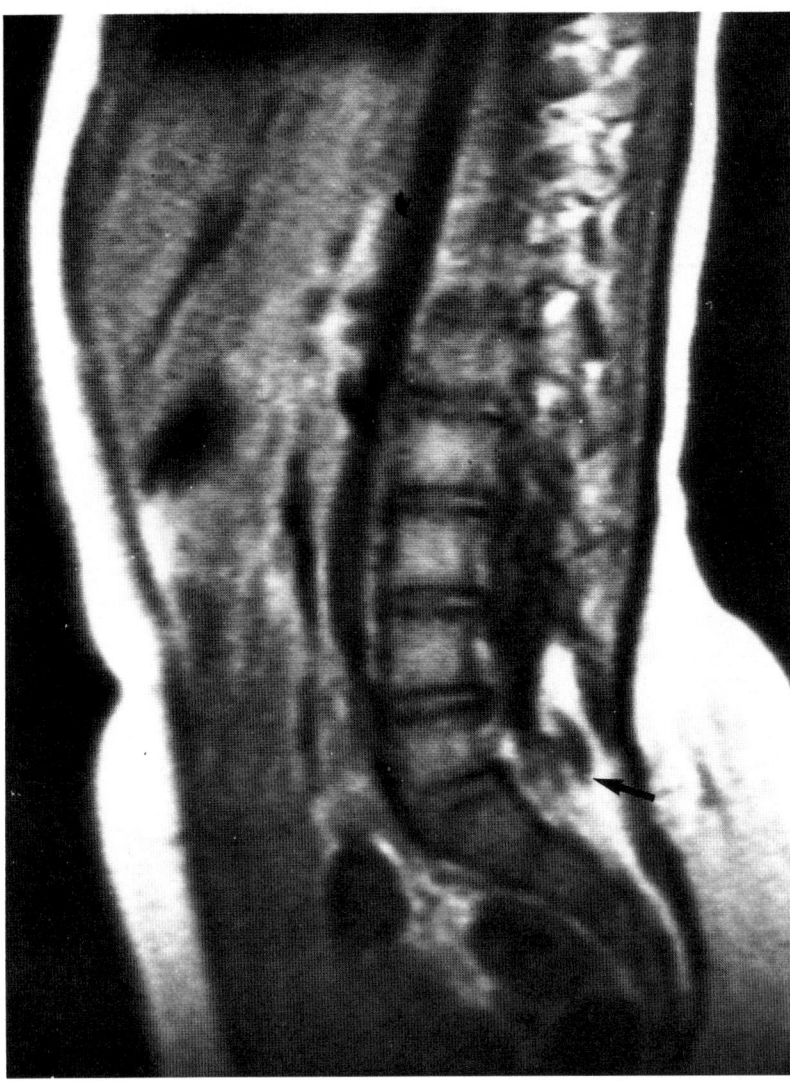

Figure 10.24
Tethered cord. Sagittal MR image (SE 500/28) demonstrates tethered nerve roots with surrounding lipoma (arrow).

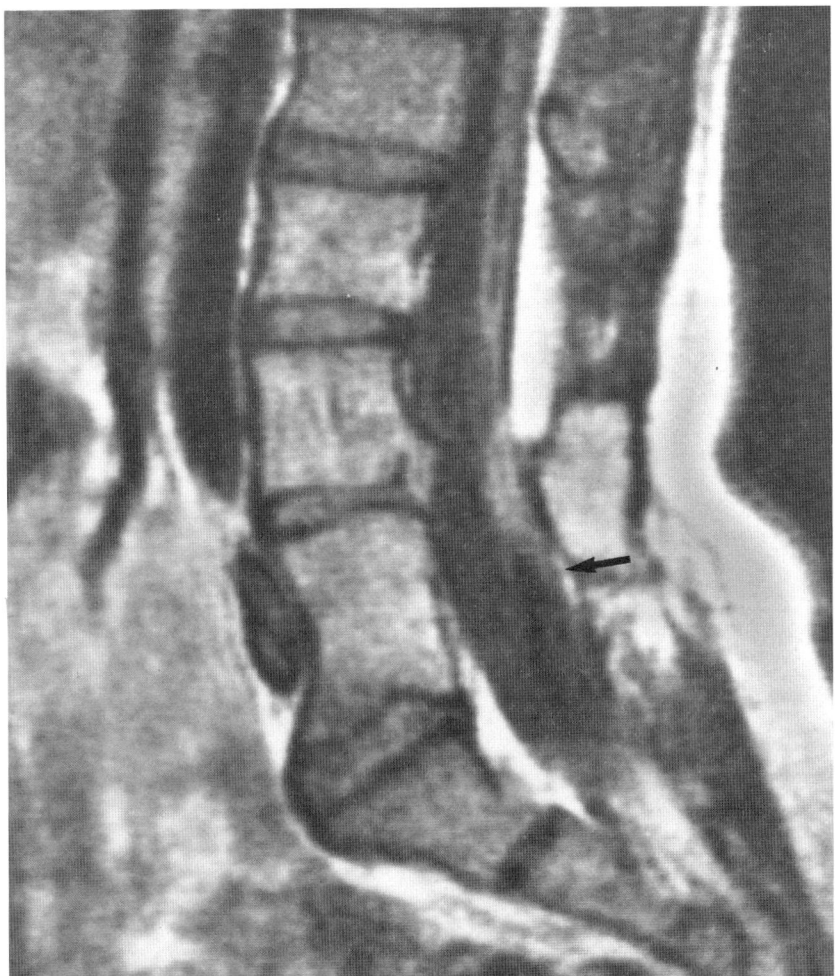

a

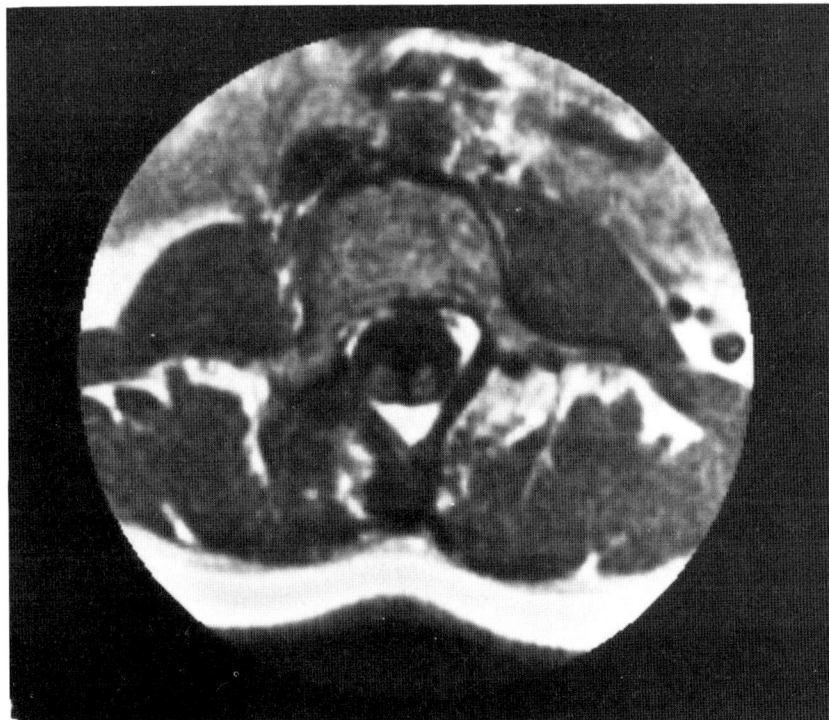

b

Figure 10.25

Diastematomyelia. A 17-year-old girl with symptoms of a tethered cord. (**a**) Sagittal MR image (SE 500/28) depicts a block vertebra, a lipoma, a dilated central canal, and a tethered cord (arrow). (**b**) The transaxial T_1-weighted image (SE 500/28) best depicts the division of the cord.

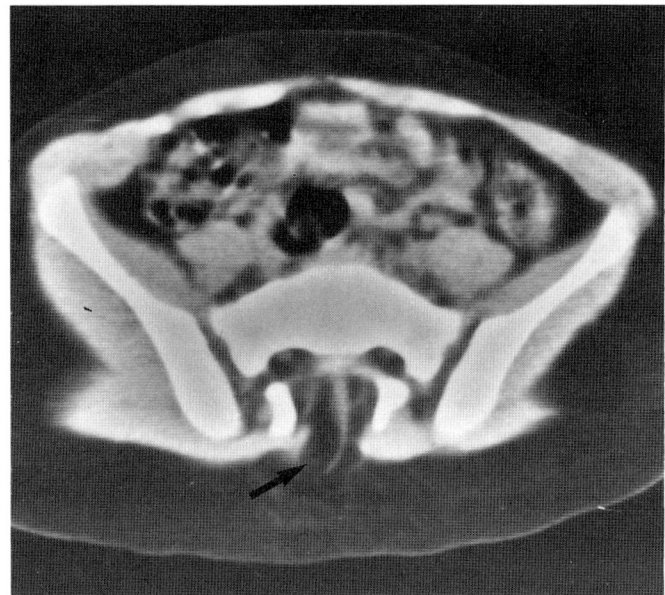

Figure 10.26

Lipomeningocele. (**a**) CT scan. Axial scan through S1 level demonstrates a lucent mass extending from the canal through the posterior defect of the sacrum (arrow). (**b**) Axial MR image (SE 500/28) shows the defect in posterior elements of S1 with the lipomatous component of this lipomeningocele. (**c**) Axial image (SE 500/28) shows large lipoma with meningocele. (**d**) Interleaved sagittal image (SE 500/28) shows the craniocaudal extent of the lipomatous component as well as the meningocele. (**e**) Interleaved sagittal image (SE 500/28) demonstrates the extent of the soft-tissue pathology outside the sacral canal (arrow).

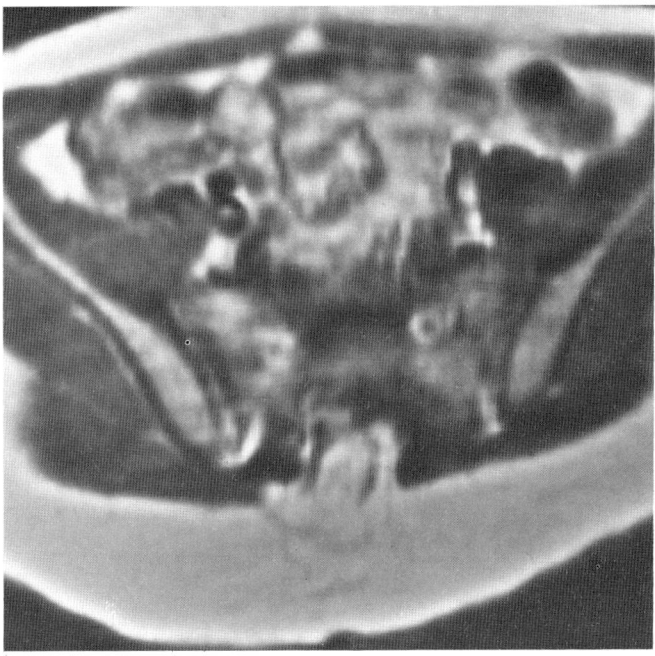

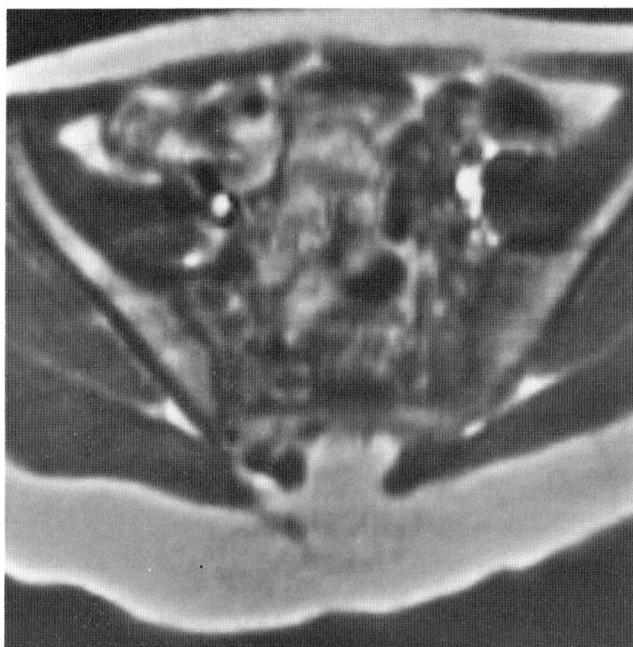

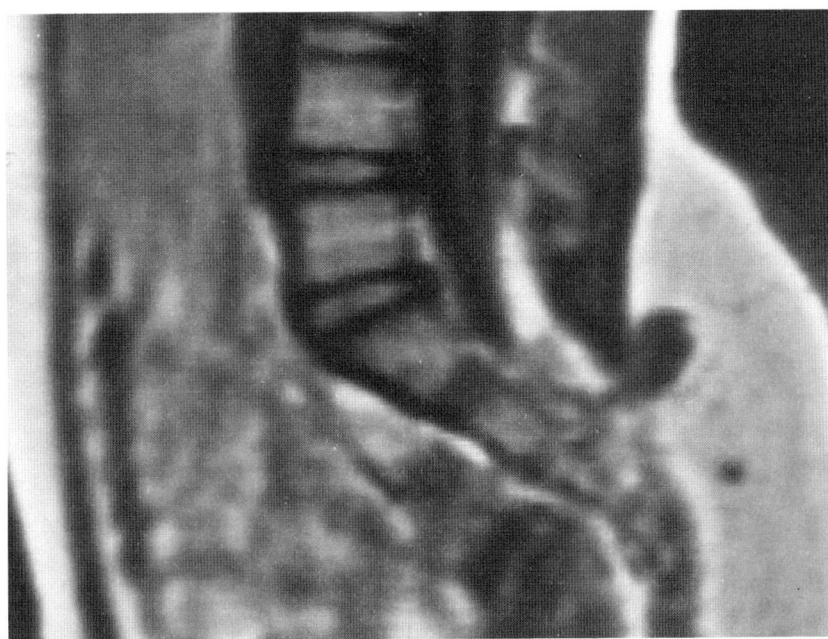

d

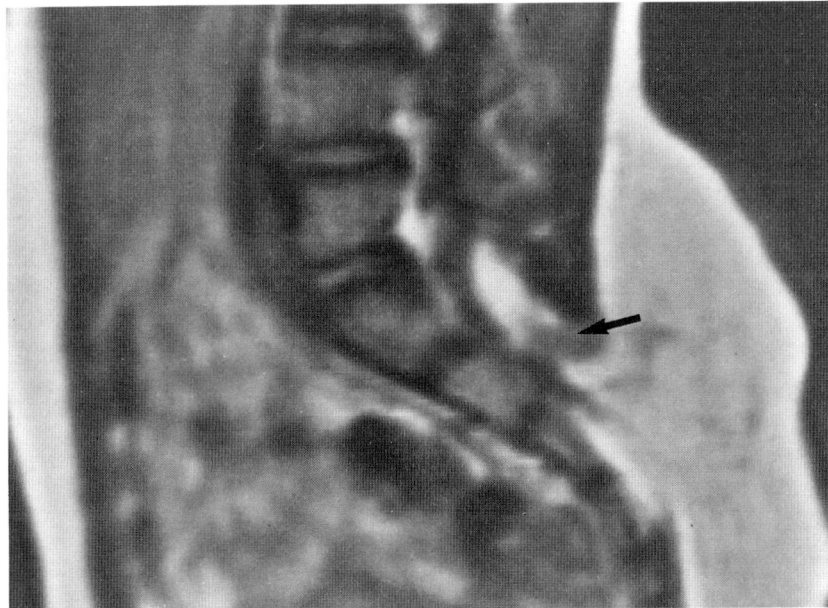

e

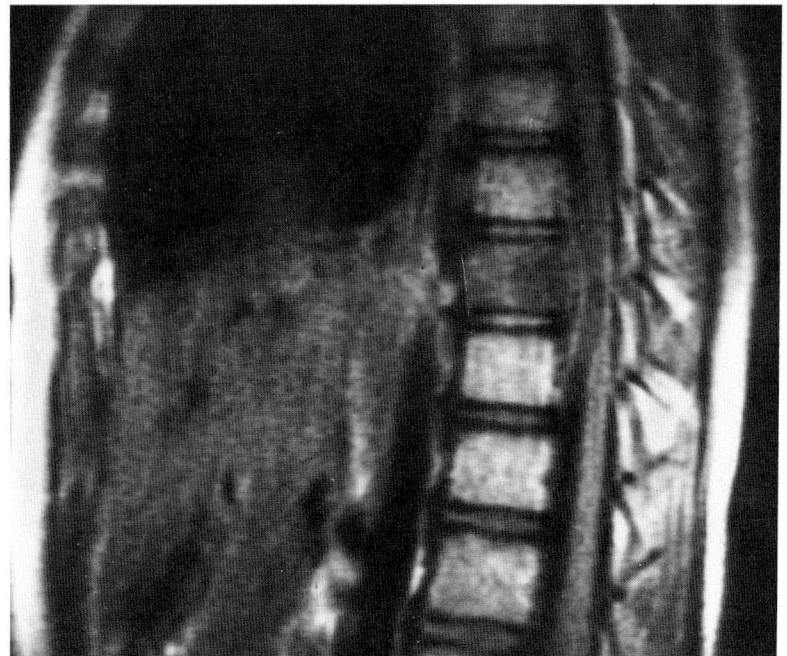

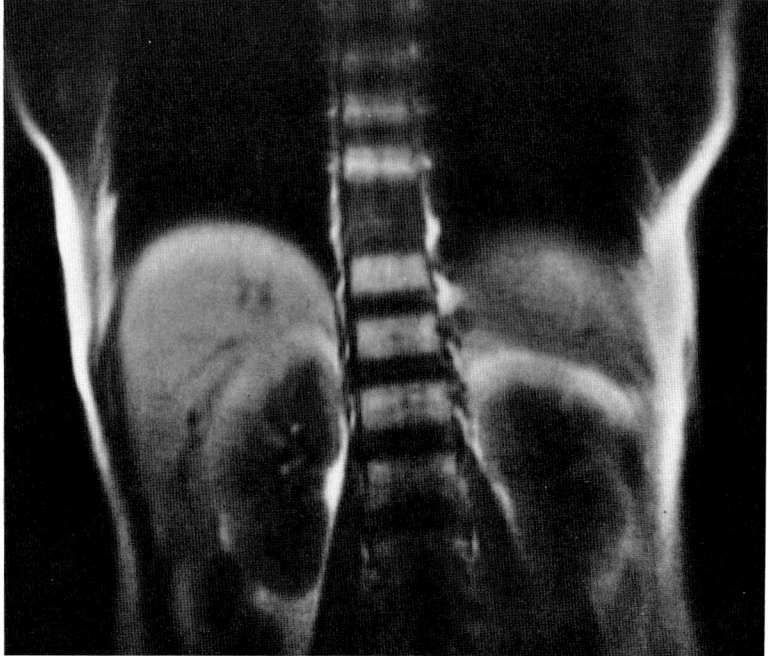

Figure 10.27

Lymphoma. This 8-year-old girl presented with back pain. (a) Sagittal MR image (SE 500/28). The signal intensity of the affected vertebral body is less than the other vertebrae. Extradural extension is also demonstrated with mass effect on the cord. (b) Coronal MR image (SE 500/28) shows no evidence of lateral extension, and the intervertebral disk is preserved without evidence of paraspinal mass.

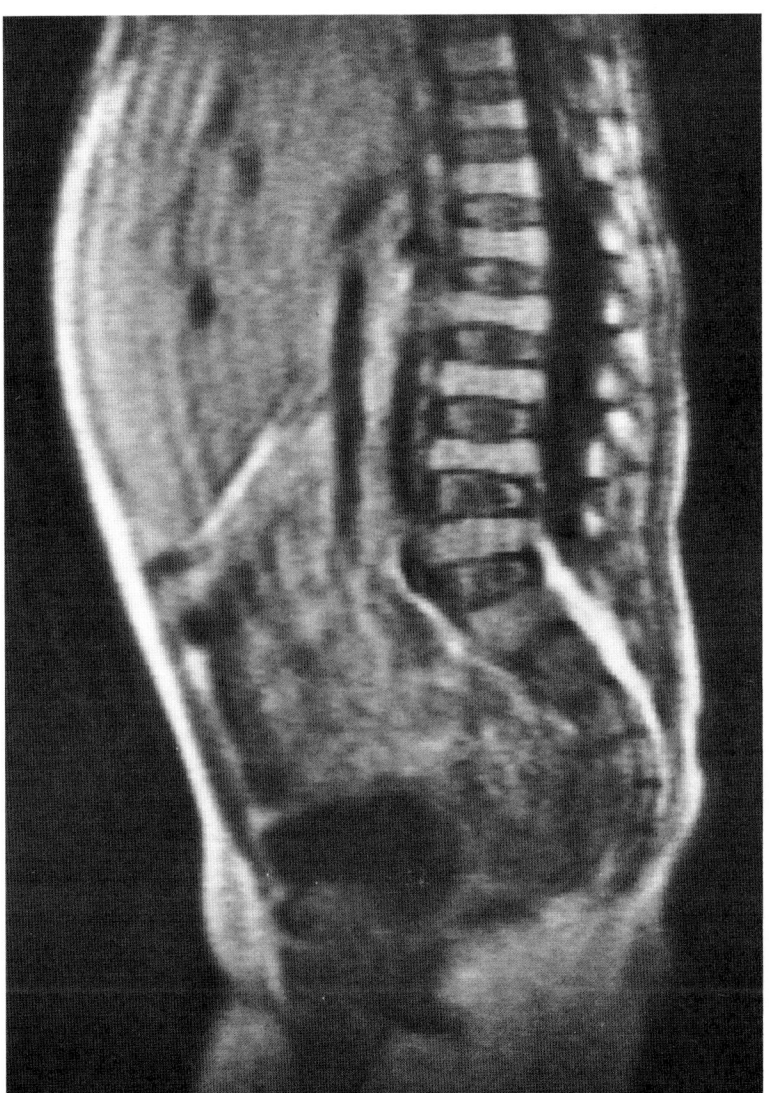

Figure 10.28

Acute lymphoblastic leukemia. This 5-year-old girl presented with fever, increasing abdominal girth and back pain. Sagittal MR image (SE 500/28) demonstrates universal vertebral plana, with homogeneous marrow signal. Diffuse replacement of bone marrow by tumor is not easily seen when all bodies are involved.

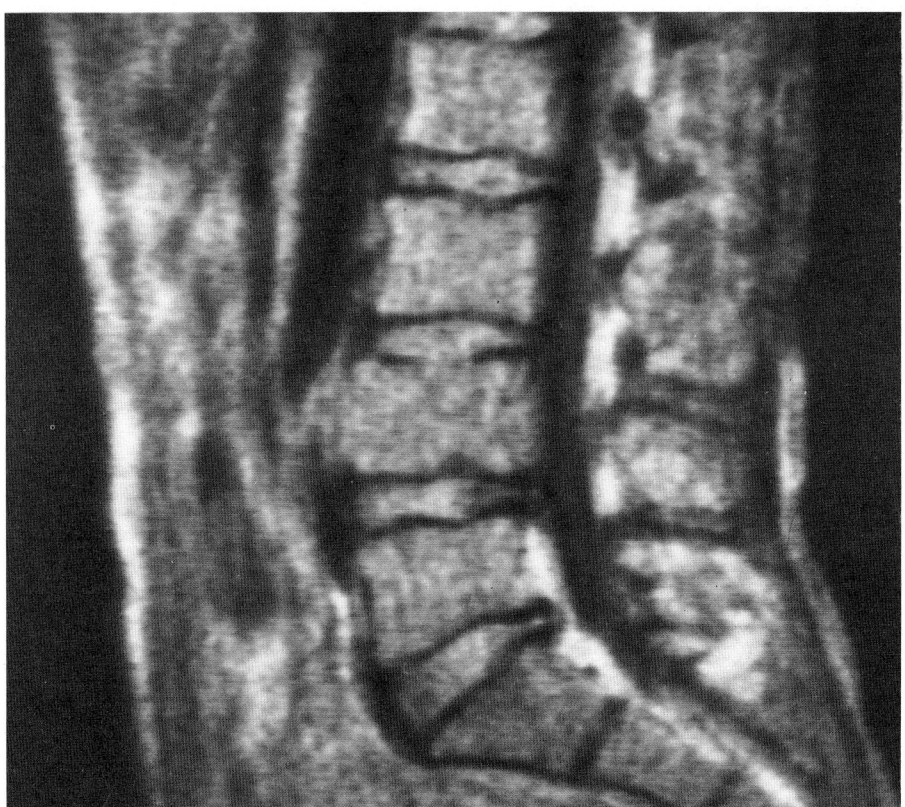

a

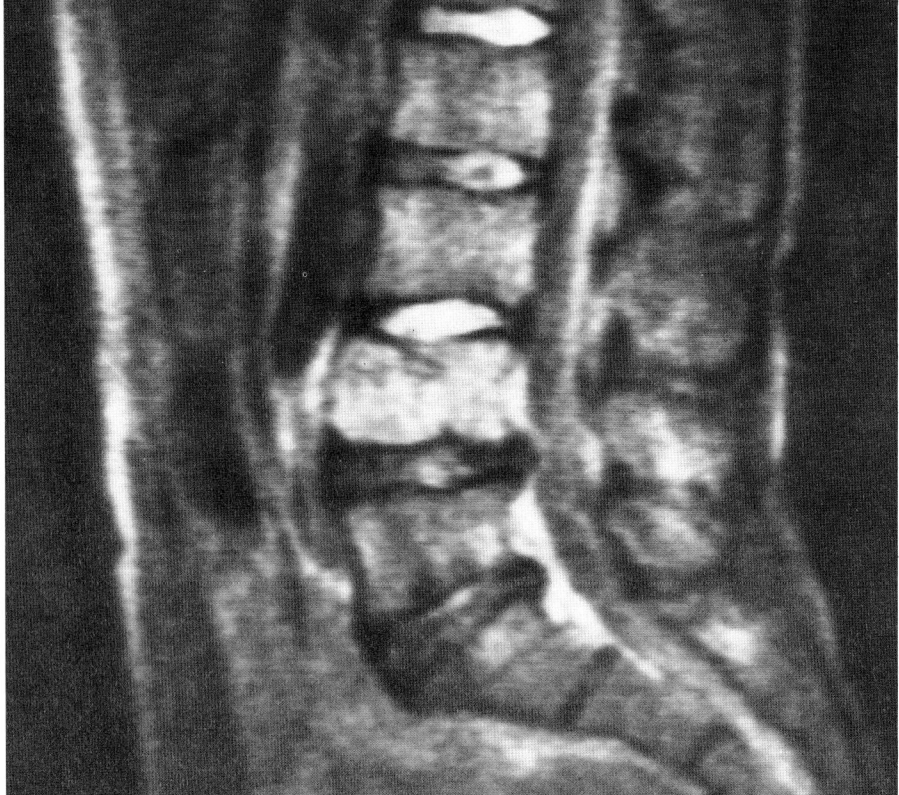

b

Figure 10.29

Metastatic Ewing's sarcoma. Ewing's sarcoma of the femur was diagnosed in this 20-year-old man with back pain.
(**a**) Sagittal MR image (SE 500/28) shows a distortion of the shape of the involved vertebral body. The signal intensity is similar to that of bone marrow in the adjacent bodies.
(**b**) Sagittal MR image (SE 2000/56) shows T_2-shortening with high-signal intensity of the involved vertebral body.

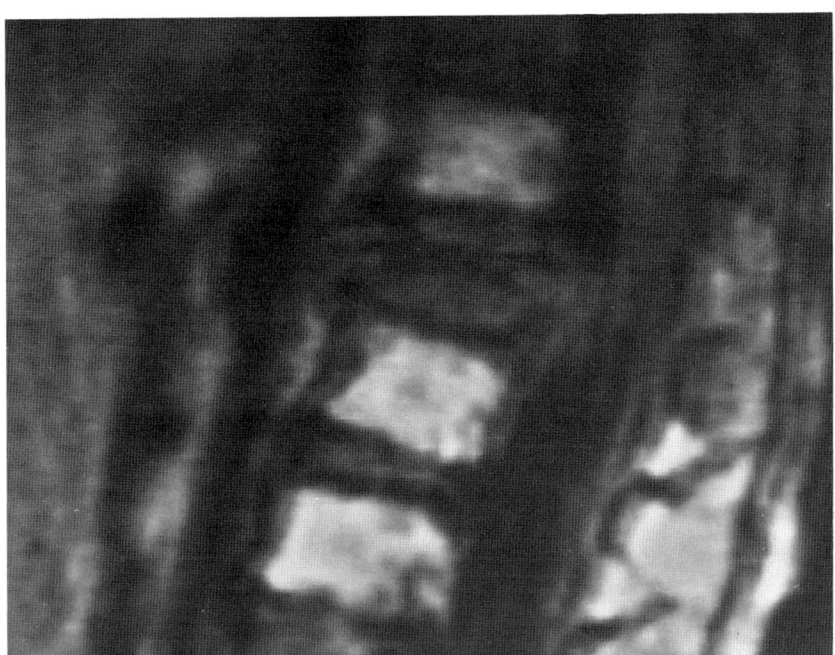

Figure 10.30

Ewing's sarcoma post-irradiation. Sagittal MR image (SE 500/28) shows the partially collapsed L1 vertebral body as well as increased signal intensity of contiguous bodies within the radiation portal.

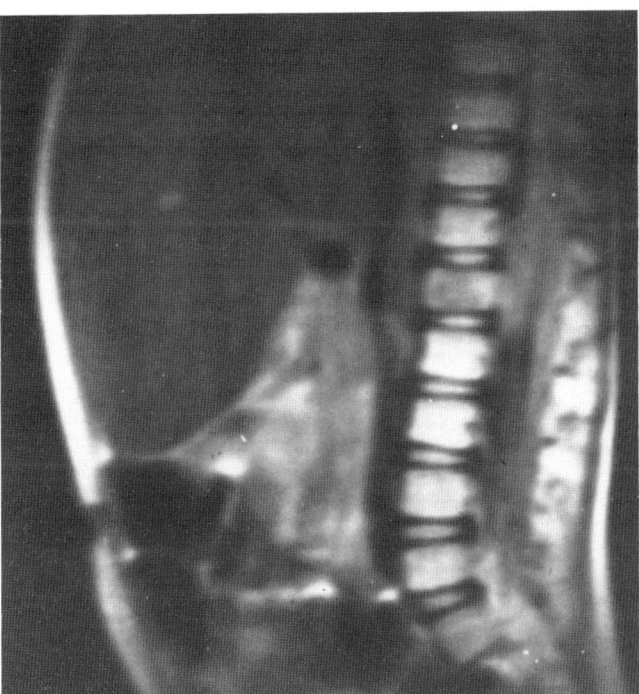

Figure 10.31

Primary Ewing's sarcoma. (a) This 3-year-old girl presented with back pain. Sagittal MR image (SE 500/28) showed involvement of L1 with posterior extension of the tumor into the spinal canal. (b) Sagittal MR image with T_2-weighting (SE 2000/84) shows the anterior extension of soft-tissue mass over contiguous segments. (c) Sagittal MR image with less T_2-weighting (SE 1500/56) shows less signal intensity in the spinal canal than the heavily weighted T_2-image. The soft-tissue mass anterior to the contiguous vertebrae is again shown here.

a

312 MRI atlas of the musculoskeletal system

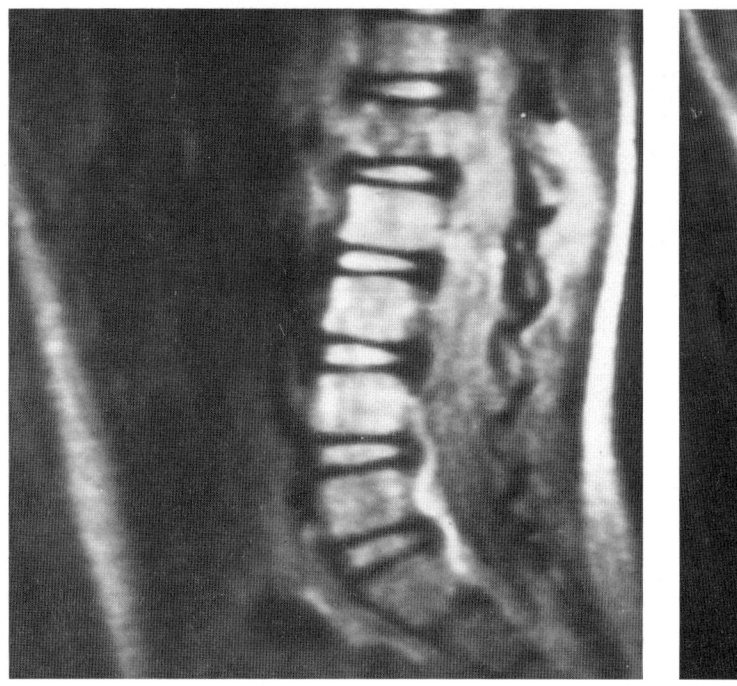
b

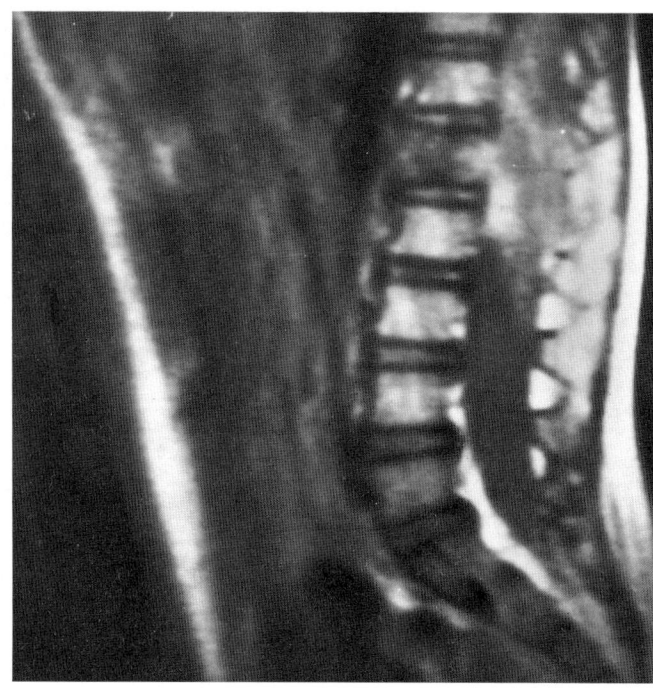
c

Figure 10.31 continued

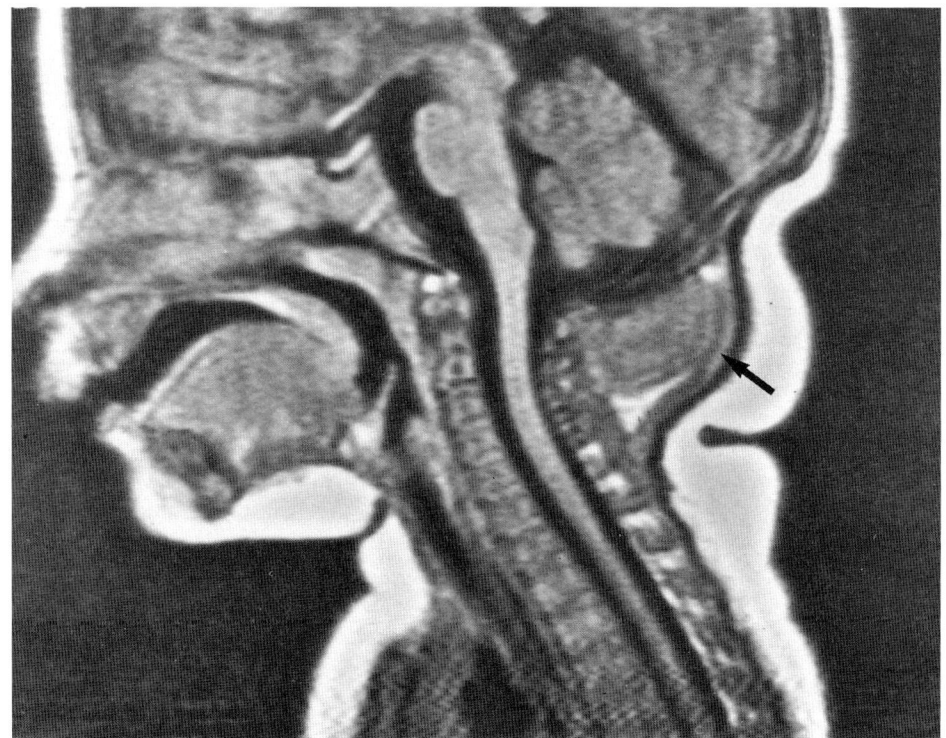

Figure 10.32

Extraspinal tumor: lipoblastoma. An obvious neck mass was palpated in this 6-month-old infant. Sagittal MR image (SE 500/28) clearly demonstrates extracranial location and lack of spinal cord involvement (arrow). The intermediate signal intensity may be due to fibrous stroma within the tumor.

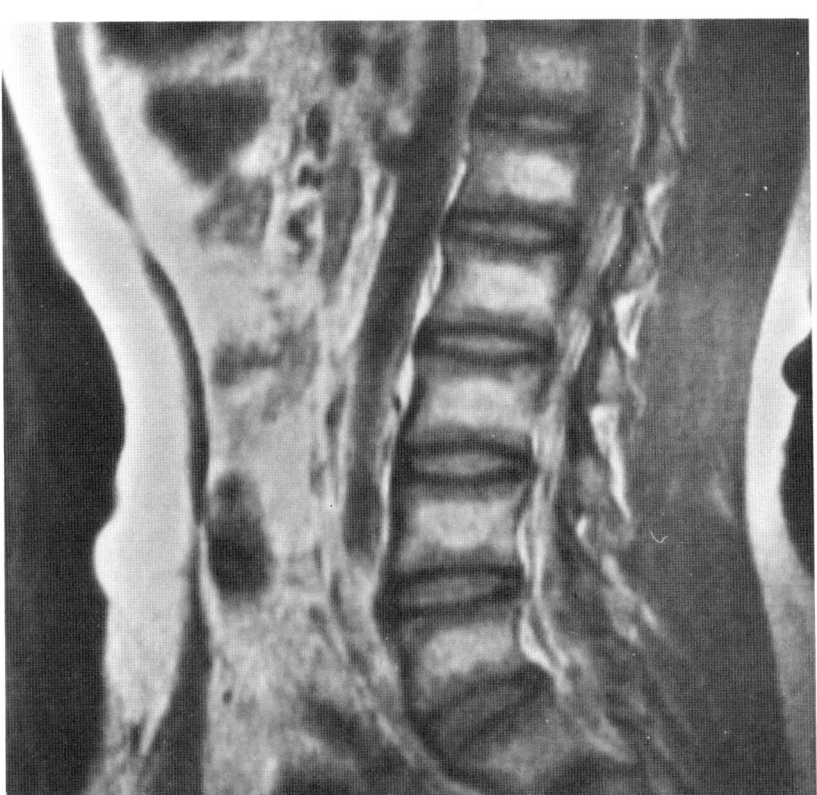

Figure 10.33

Metabolic bone disease. A young man with known renal osteodystrophy. Sagittal MR image (SE 500/28) demonstrates endplate sclerosis with decreased signal intensity, giving a reverse 'rugger jersey' appearance.

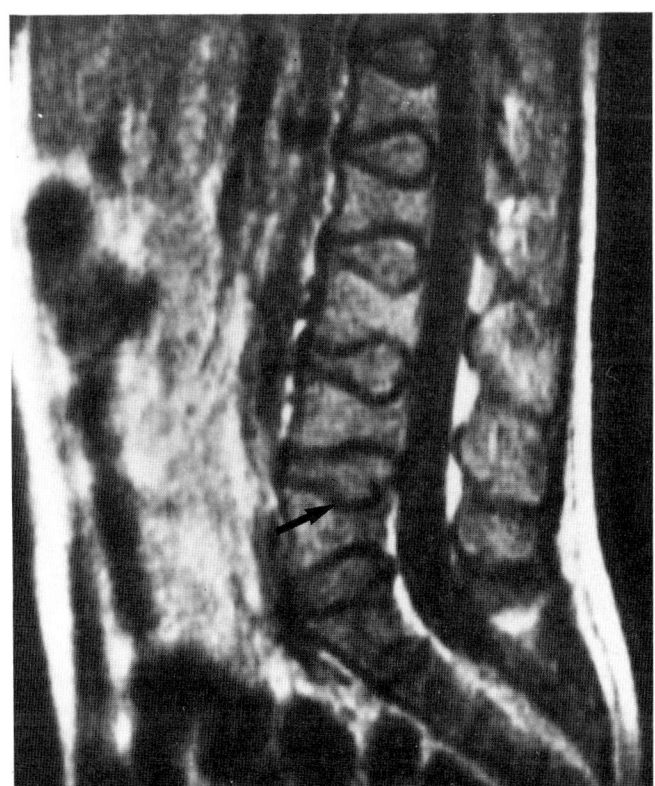

Figure 10.34

Sickle-cell anemia. This 14-year-old boy presented in 'sickle crisis'. Sagittal MR image (SE 500/28) clearly shows vertebral body endplate depressions from severe osteopenia with a characteristic Reynold's sign (arrow) of endplate ischemia at L5.

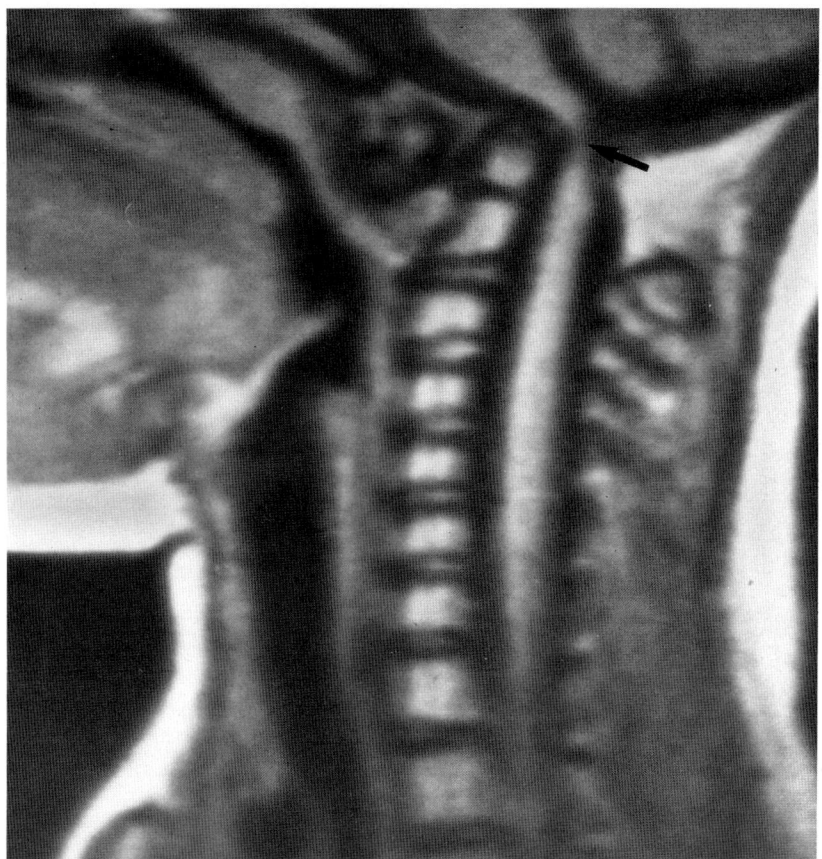

a

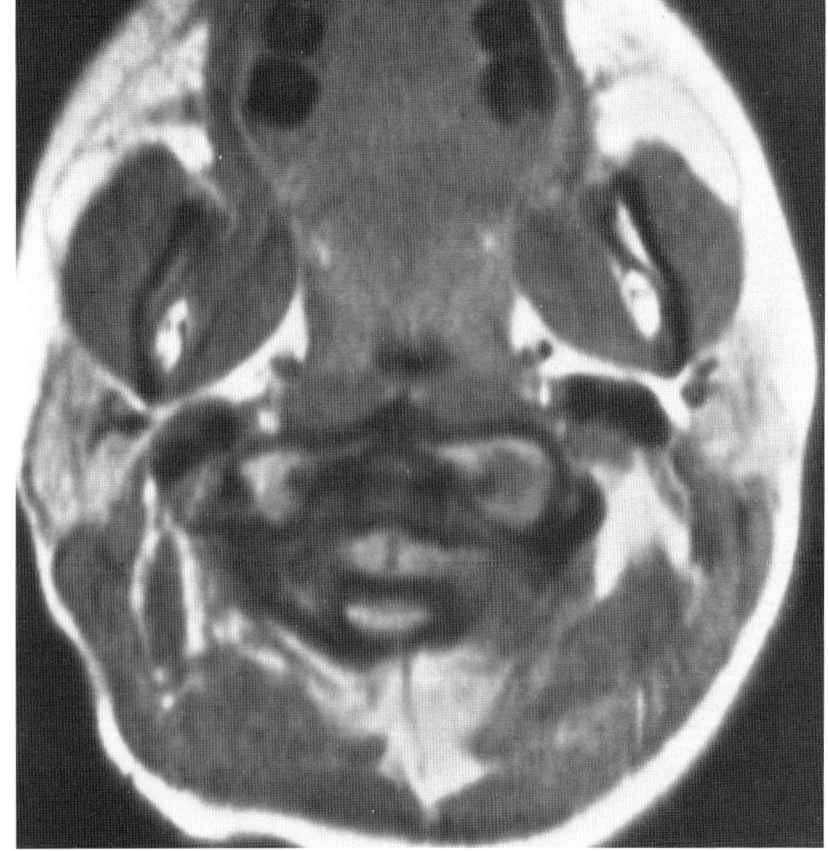

b

Figure 10.35
Down's syndrome. An 8-year-old girl with symptoms of cord compression. (**a**) Sagittal and (**b**) transaxial MR images (SE 500/28) show atlantoaxial subluxation and compression of the cord (arrow). This has been described in 12 to 20 per cent of patients with Down's Syndrome.[19]

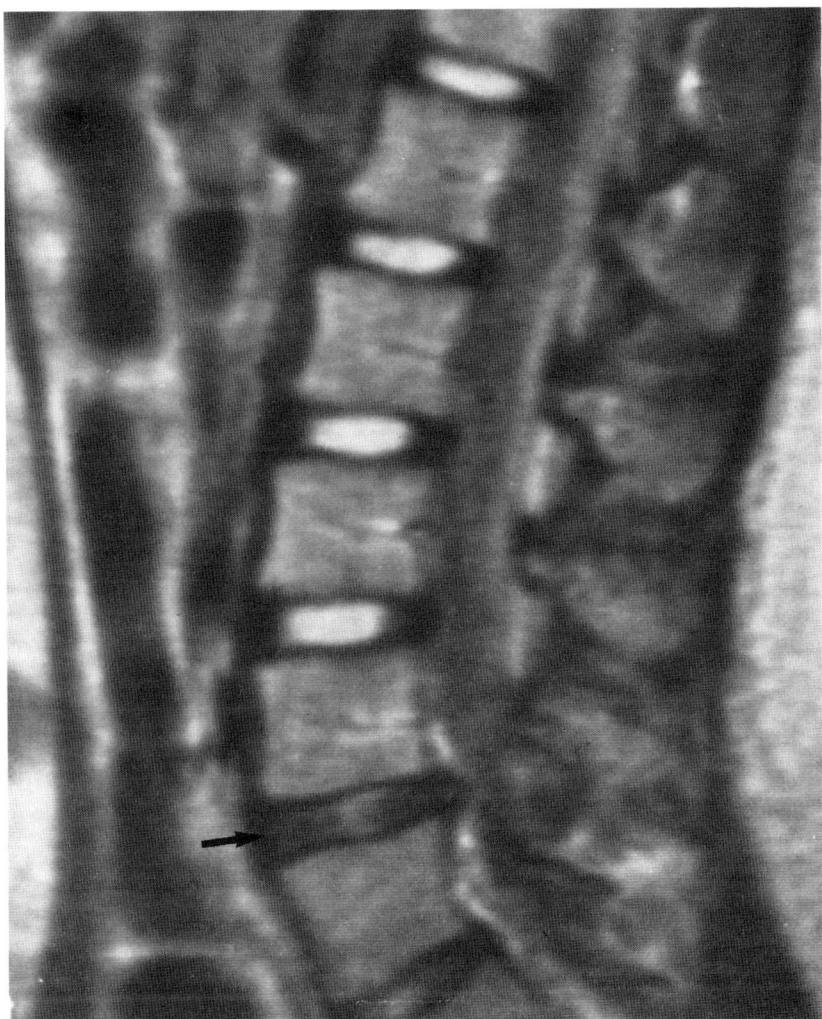

Figure 10.36

Degenerative disk disease. A 15-year-old gymnast with back pain. Sagittal MR image (SE 500/28) shows a degenerative disk (arrow) with loss of disk height, and decreased signal intensity from the nucleus pulposus. A posterior disk bulge is also evident.

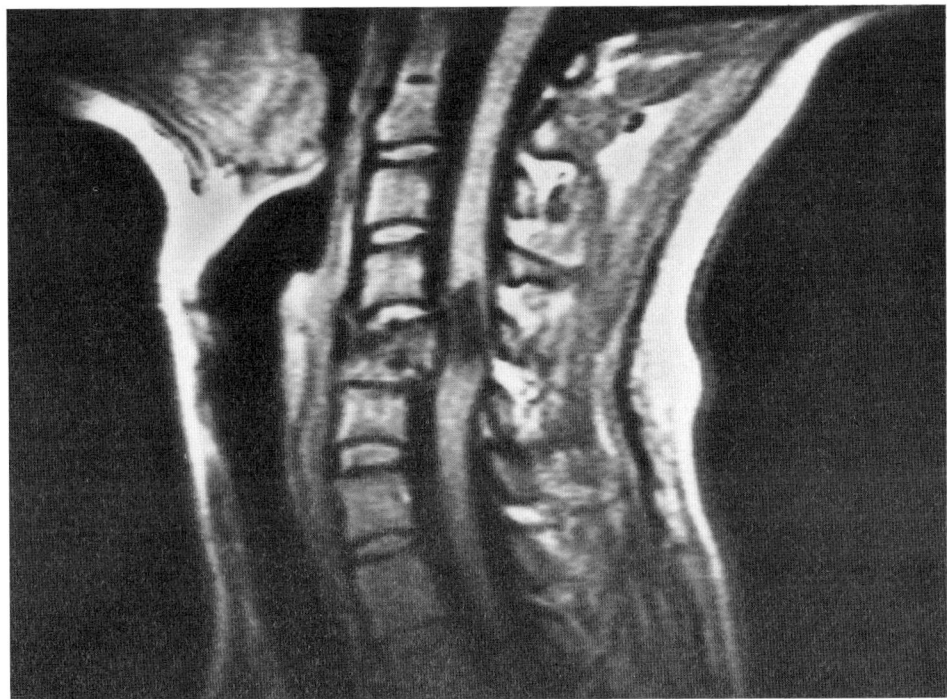

Figure 10.37

Cervical vertebral fracture. Sagittal MR image (SE 500/28). A 21-year-old woman was involved in a motor vehicle accident. Re-evaluation several months after the accident showed a crush fracture of the C5 vertebral body and cystic degenerative changes in the cord.

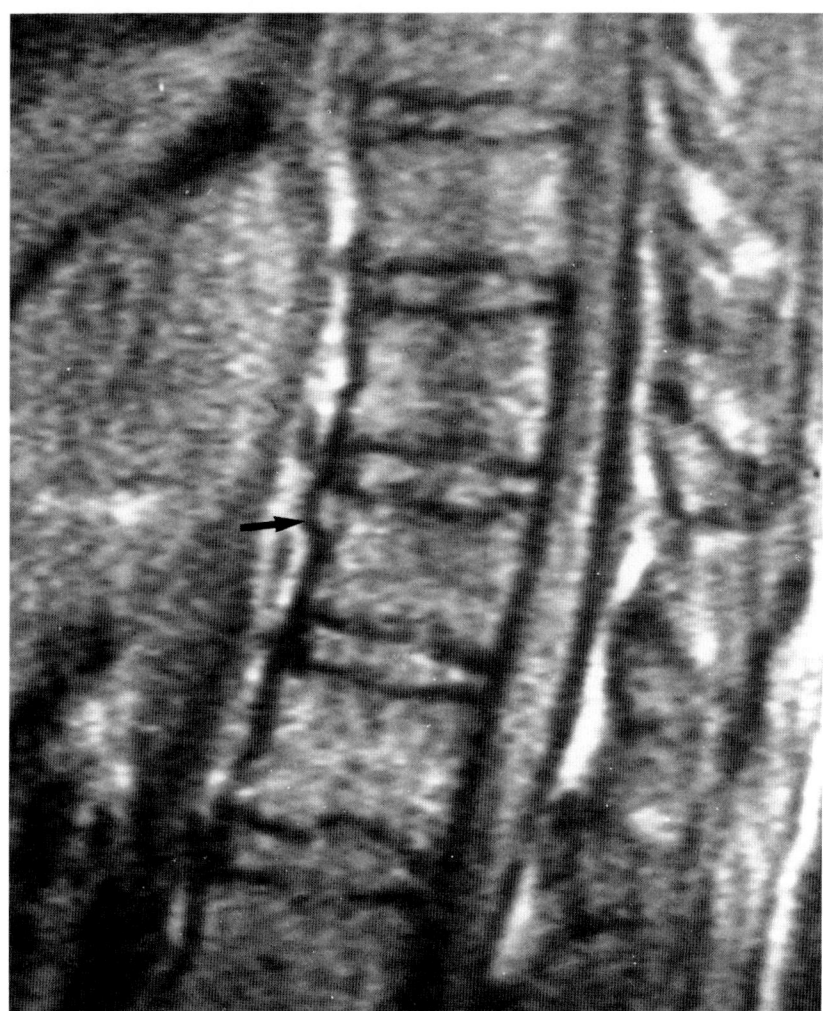

Figure 10.38

Corner fracture. A 17-year-old gymnast with a hyperflexion injury of her back. Sagittal MR image (SE 500/28) demonstrates the corner fracture (arrow) with slight displacement.

11
Tumors and tumor-like conditions

Leanne L Seeger and
Lawrence W Bassett

The features of magnetic resonance imaging (MRI) that are advantageous for the evaluation of musculoskeletal tumors are its excellent depiction of differences in soft-tissue contrast, the high-signal intensity produced by normal bone marrow, and the capability of imaging directly in any plane. Because MRI is usually not reliable for making specific pathologic diagnoses or in differentiating benign from malignant lesions, correlative plain radiographs are essential. The major value of MRI is in defining the intra- and extra-osseous extent of tumors and their relationship to surrounding structures.

Both T_1- and T_2-weighted images are required for the full evaluation of musculoskeletal tumors. In general, tumors have intermediate- to low-signal intensity in T_1-weighted images and high-signal intensity in T_2-weighted images. Exceptions include lipomas, which have high-signal intensity in both T_1- and T_2-weighted images,[1] and relatively acellular tumors containing abundant collagen, that may retain a low-signal intensity in both T_1- and T_2-weighted images.[2]

Analysis of numerical T_1- and T_2-values has not proved useful in determining tumor type. Not only is there unacceptable overlap between the T_1- and T_2-values of benign and malignant lesions,[3] but since tissue relaxation times are field-strength dependent,[4] the differentiation of lesions by their differing relaxation times is fraught with hazard.

Imaging of the extremities may be done with a specialized surface coil or with the body coil, and the decision as to which to use depends on the individual case. Surface coils provide a higher signal-to-noise ratio and hence better resolution than is possible with body coils.[5-8] There are situations, however, when the body coil is useful in order to image simultaneously the contralateral normal extremity for comparison. In addition, it is often impossible to obtain high-resolution images of the entire bulk of a large tumor due to signal fall-off outside the area of the coil.

This chapter presents our experience with MRI in the evaluation of primary tumors of bone, tumor metastatic to bone, marrow-infiltrative disorders, soft-tissue tumors, and conditions which may be confused with tumors.

Primary tumors of bone

Malignant bone tumors

The extent of bone marrow involvement by a tumor is best evaluated in T_1-weighted images, and is characterized by replacement of the high-signal intensity of normal marrow by the low- to intermediate-signal intensity of the tumor. In T_2-weighted images, the signal intensity of the tumor is increased compared to T_1-weighted images, and may be even higher than that of normal marrow, depending on the amount of T_2-weighting and tumor type. T_2-weighted images provide higher contrast between normal and abnormal soft tissues, and therefore are preferred for evaluating the soft-tissue component of musculoskeletal tumors. In both T_1- and T_2-weighted images, osteosclerotic tumors may show foci of low-signal intensity in regions of tumor mineralization.

In the extremities, MRI in the coronal plane is well suited for evaluating the extent of tumor infiltration. Imaging in the axial plane is superior for evaluating the relationship of the tumor to the surrounding normal anatomy, including the vital neurovascular structures. MRI compares favorably with CT for determining the extent of tumor involvement in the extremities,[9-13] and is also useful in monitoring tumor recurrence. However, MRI lacks specificity, and is poor at depicting cortical bone detail and tumor matrix mineralization.[13-14]

Osteosarcoma, the most common primary tumor of bone after myeloma, is frequently located in a lower extremity (Figures 11.1–11.5). Possible treatment includes a limb salvage procedure, which requires an exact assessment of the extent of the lesion.[12,15] MRI in the coronal or sagittal plane is particularly effective in defining the extent of the tumor and in demonstrating intramedullary 'skip' lesions which may alter the scope of surgical resection.

Ewing's sarcoma frequently has a large soft-tissue component, and MRI is an excellent method to depict its extent (Figures 11.6–11.7). The degree of bone marrow involvement and soft-tissue extension of chondrosarcoma (Figures 11.8–11.9) and fibrosarcoma (Figure 11.10) are also well shown with MRI.

Benign bone tumors

MRI is useful for the preoperative evaluation of aggressive benign lesions which require surgical intervention, such as giant cell tumor (Figure 11.11). Monitoring for tumor recurrence can also be achieved with MRI (Figure 11.12).

Although MRI is very sensitive for the detection of bone tumors, it is non-specific. MRI is seldom useful for classifying tumors pathologically, and conventional radiography is far more diagnostic. However, MRI may reveal some specific features of certain benign tumors. For example, quiescent non-ossifying fibromas manifest a homogeneous signal void characteristic of fibrous tissue (Figure 11.13). Like their malignant counterparts, the majority of actively growing benign tumors show intermediate- to low-signal intensity on T_1-weighted images and high-signal intensity on T_2-weighted images. Enchondromas tend to be well circumscribed (Figure 11.14) and in both T_1- and T_2-weighted images may show punctate areas of signal void representing mineralized tumor cartilage (Figure 11.15). Regions of early malignant transformation to chondrosarcoma are as difficult to distinguish from the parent enchondroma in MR images as they are in plain radiographs. Osteochondromas exhibit high-signal-intensity marrow continuous with that of the parent bone (Figure 11.16). Traumatic exostoses may manifest an inhomogeneous internal signal intensity and irregular external margins (Figure 11.17). Enostoses (bone islands),[16] which may be incidentally encountered in MR images, have low- to intermediate-signal intensity with all pulse sequences (Figure 11.18). Osteoid osteoma is characterized by a nidus of intermediate-signal intensity surrounded by reactive bone of very low signal intensity in both T_1- and T_2-weighted images (Figure 11.19). Non-calcified periosteal tissue shows intermediate-signal intensity in T_1-weighted images, which increases in T_2-weighted images (Figure 11.20d and e).

Tumor metastatic to bone

Conventional radiography, radionuclide bone scanning and CT all rely heavily on destruction and turnover of bone for the depiction of metastatic disease. Since the majority of skeletal metastases seed the marrow through hematogenous spread, MRI is more sensitive than these other modalities for the early detection of bony metastases (Figures 11.21–11.22). In T_1-weighted pulse sequences, they typically appear as areas of low-signal intensity which are well circumscribed from the surrounding high-signal intensity of normal marrow. Metastases generally are not as easily identified in T_2-weighted images, where their signal intensity increases to become equal to or higher than that of the surrounding marrow (Figures 11.23–11.24).

Although whole-body radionuclide bone scanning remains the most appropriate screening tool for detecting skeletal metastases, there are four circumstances in which MRI may have a significant impact on the management of the patient with osseous metastatic disease: (1) in the detection of metastases in symptomatic patients in whom conventional studies (radiographs and radionuclide bone scans) are equivocal or negative (Figure 11.25); (2) in the asymptomatic patient for whom there exists a high suspicion of metastatic disease and an equivocal radiographic or scintigraphic abnormality; (3) in the detection of metastases at sites which are difficult to evaluate by conventional radiography or scintigraphy, such as the symphysis pubis, sacroiliac regions, and the sternum (Figures 11.26–11.27); and (4) in the determination of the extent of metastatic disease when planning palliative surgery or radiotherapy (Figures 11.28–11.29). MRI has also been useful to determine the etiology of vertebral collapse in osteopenic patients with a known primary malignancy (Figures 11.30–11.31).

Marrow infiltrative disorders

When evaluating the bone marrow, it is important to be aware of variations that normally occur with aging.[17-18] In childhood, prior to the conversion of hematopoietic red marrow to fatty yellow marrow, the bone marrow signal in the extremities is generally of intermediate to low intensity. The conversion from red to yellow marrow is a function of both the age of the patient and the site in the skeleton. In addition, hematopoietic disorders may result in the reconversion of fatty marrow to hematopoietic marrow, thus changing the marrow signal intensity.

MRI is more sensitive than radionuclide bone scanning in identifying the marrow infiltrates of myeloma (Figure 11.32).[19] As with other malignancies that have invaded the bone marrow, the signal intensity in T_1-weighted images is abnormally low, but increases with T_2-weighting (Figure 11.32). Bone marrow involvement by leukemia (Figures 11.33–11.34) and lymphoma (Figure 11.35) may be evident in MR images, even in the presence of a normal radionuclide bone scan. MRI may be useful in assessing the effect of treatment in patients with hematologic disorders.[20]

Soft-tissue tumors

Because of its inherent soft-tissue contrast, MRI is ideally suited to the evaluation of malignant and benign tumors arising in fat, muscle, nerves, and synovium.[21] However, it has not yet proved useful in differentiating benign from malignant lesions. The selection of imaging planes should be determined according to the location of the tumor. For tumors in the extremities, coronal and sagittal planes provide the optimal assessment of lesional extent. The axial plane is useful in defining the relationship of the tumor to adjacent bones, muscles, tendons, and neurovascular structures.

Malignant soft-tissue tumors

T_1-weighted images are particularly useful to evaluate the invasion by the tumor of adjacent bones (Figure 11.36), while T_2-weighted images are more effective in defining the extent of the tumor and its relationship to neighboring neurovascular structures (Figure 11.37). Adequate T_2-weighting is required for differentiation of the tumor from surrounding edema (Figure 11.38).[22] MRI is also useful to exclude or monitor soft-tissue tumor recurrence (Figure 11.39).

Benign soft-tissue tumors

The same MR imaging principles used for evaluating malignant soft-tissue tumors can be applied to benign soft-tissue tumors. This includes the use of both T_1- and T_2-weighted pulse sequences, and multiplanar imaging to determine the full extent of the tumor.

Neurofibroma, one of the most common benign neural tumors, may be observed in the spine or extremities. In both locations, MRI can reveal the size of the neurofibroma, its displacement of neighboring soft-tissue structures, and any accompanying bone destruction (Figures 11.40–11.41).

A variety of soft-tissue tumors are found in the vicinity of joints. MRI is capable of providing information about these lesions that previously was available only through invasive methods such as arthrography, arthroscopy, or surgery.[23]

Pigmented villonodular synovitis occurs most frequently in the knee, and may present clinically as a non-specific joint effusion. These tumors are associated with intra-articular hemorrhage and hemosiderin deposition. The findings in the MR image will therefore vary with the timing of the scan in relation to bleeding episodes, and with the pulse sequences that are used (Figure 11.42). T_2-weighted images may disclose a mixture of low-intensity signal, due to foci of hemosiderin, and high-intensity signal, due to foci of reactive synovitis and intra-articular fluid.[23]

Ganglion cysts are extra-articular, fluid-filled, synovial-lined masses that probably arise as diverticulae from the joint or adjacent tendon sheaths (Figures 11.43–11.44). Their signal intensity on T_1-weighted images is low to intermediate. Due to their fluid content, ganglion cysts appear homogeneous, and become very bright on T_2-weighted images.

Meniscal cysts are fluid collections that extend from meniscal tears, and are thought to result from repeated trauma (Figure 11.45). The diagnosis depends on the identification of the fluid within the cyst communicating with a meniscal tear,[24] and is therefore best made with T_2-weighted images.

Lipoma, one of the most common benign soft-tissue tumors, may present as a palpable mass or may be an unsuspected incidental finding in MR images. Lipomas may be subcutaneous (Figure 11.46) or intramuscular (Figure 11.47). As they are composed of fat, their signal characteristics are identical to subcutaneous fat in both T1- and T_2-weighted images.[1] A surrounding thin, fibrous capsule of low-signal intensity can often be identified. An abnormal collection of fat such as the 'buffalo hump' associated with Cushing's disease may mimic a lipoma (Figure 11.48), but can usually be distinguished from it by virtue of its characteristic location and lack of well defined borders.

Tumor-like conditions

Numerous benign conditions that mimic tumors may be encountered on MRI examinations. In order to avoid these diagnostic pitfalls, it is important to correlate relevant clinical information and radiographic changes with the MRI findings.

Paget's disease may be confused with benign or malignant tumors of bone (Figures 11.49–11.50). Since MRI lacks the resolution of bone detail that is a key feature of conventional radiography and CT, even moderately advanced cortical and trabecular thickening that is obvious in those examinations is often obscured in MR images.

Many non-tumorous conditions may be discriminated from malignancy because they show a homogeneous low-signal intensity on both T_1- and T_2-weighted images. For example, hemosiderin deposits from transfusions appear as foci of signal void on both T_1- and T_2-weighted images (Figure 11.51).[25] Bone islands are frequently responsible for small, well-circumscribed foci of low-intensity signal within the bone marrow on both T_1- and T_2-weighted images (Figures 11.52–11.53).

Aneurysms may be mistaken for soft-tissue tumors. Although correlative clinical history and physical examination may be required to make the final diagnosis, MRI provides valuable information regarding the origin and extent of the aneurysm (Figures 11.54–11.55). In cases of arteriovenous malformations and hemangiomas, MRI is useful in defining the extent of the lesions prior to treatment (Figure 11.56) and in seeking possible recurrence following therapy.[26]

References

1 DOOMS GC, HRICAK H, SOLLITTO RA et al, Lipomatous tumors and tumors with fatty component: MR imaging potential and comparison of MR and CT results, *Radiology* (1985) **157**: 479–83.

2 SUNDARAM M, McGUIRE MH, SCHAJOWICZ F, Soft tissue masses: histologic basis for decreased signal (short T_2) on T_2-weighted MR images, *AJR* (1987) **148**:1247–50.

3 ZIMMER WD, BERQUIST TH, McLEOD RA et al, Bone tumors: magnetic resonance imaging versus computed tomography, *Radiology* (1985) **155**:709–18.

4 KOENIG SH, BROWN III RD, ADAMS D et al, Magnetic field dependence of $1/T_1$ of protons in tissue, *Invest Radiol* (1984) **19**:76–81.

5 SCHENCK JF, FOSTER TH, HENKES JL et al, High-field surface-coil MR imaging of localized anatomy, *AJNR* (1985) **6**:181–6.

6 FISHER MR, BARKER B, AMPARO EG et al, MR imaging using specialized coils, *Radiology* (1985) **157**:443–7.

7 LUFKIN RB, VOTRUBA J, REICHER M et al, Solenoid surface coils in magnetic resonance imaging, *AJR* (1986) **146**:409–12.

8 KULKAMI MV, PATTON JA, PRICE RR, Technical considerations for the use of surface coils in MRI, *AJR* (1986) **147**:373–8.

9 HUDSON TM, HAMLIN DJ, ENNEKING WF et al, Magnetic resonance imaging of bone and soft tissue tumors: early experience in 31 patients compared with computed tomography, *Skeletal Radiol* (1985) **13**:134–46.

10 BLOEM JL, FALKE THM, TAMINIAU AHM et al, Magnetic resonance imaging of primary malignant bone tumors, *RadioGraphics* (1985) **5**(6):853–86.

11 AISEN AM, MARTEL W, BRAUNSTEIN EM et al, MRI and CT evaluation of primary bone and soft-tissue tumors, *AJR* (1986) **146**:749–56.

12 SUNDARAM M, McGUIRE MH, HERBOLD DR et al, Magnetic resonance imaging in planning limb-salvage surgery for primary malignant tumors of bone, *J Bone Joint Surg (1986)* **68A**(6):809–19.

13 BOYKO OB, CORY DA, PROVISOR A et al, MR imaging of osteogenic and Ewing's sarcoma, *AJR* (1987) **148**:317–22.

14 BOHNDORF K, REISER M, LOCHNER B et al, *Skeletal Radiol* (1986) **15**:511–17.

15 STEINBACH LJ, MRI contrast gives assist in limb salvaging efforts, *Diagn Imaging* (October 1986) 104–9.

16 RESNICK D, NEMCEK AA, HAGHIGHI P, Spinal enostoses (bone islands), *Radiology* (1983) **147**:373–6.

17 KRICUN ME, Red-yellow marrow conversion: its effect on the location of some solitary bone lesions, *Skeletal Radiol* (1985) **14**:10–19.

18 DOOMS GC, FISHER MR, HRICAK H et al, Bone marrow imaging: magnetic resonance studies related to age and sex, *Radiology* (1985) **155**:429–32.

19 DAFFNER RH, LUPETIN AR, DASH N et al, MRI in the detection of malignant infiltration of bone marrow, *AJR* (1986) **146**:353–8.

20 McKINSTRY CS, STEINER RE, YOUNG AT et al, Bone marrow in leukemia and aplastic anemia: MR imaging before, during, and after treatment, *Radiology* (1987) **162**:701–7.

21 TOTTY WG, MURPHY WA, LEE JKT, Soft-tissue tumors: MR imaging, *Radiology* (1986) **160**:135–41.

22 BELTRAN J, SIMON DC, KATZ W et al, Increased MR signal intensity in skeletal muscle adjacent to malignant tumors: pathologic correlation and clinical relevance, *Radiology* (1987) **162**:251–5.

23 SUNDARAM M, McGUIRE MH, FLETCHER J et al, Magnetic resonance imaging of lesions of synovial origin, *Skeletal Radiol* (1986) **15**:110–16.

24 BURK DL, DALINKA MK, KANAL E et al, Meniscal and ganglion cysts of the knee: MR evaluation, *AJR* (1988) **150**:331–6.

25 BRASCH RC, WESBEY GE, GOODING CA et al, Magnetic resonance imaging of transfusional hemosiderosis complicating thalassemia major, *Radiology* (1984) **150**:767–71.

26 KAPLAN PA, WILLIAMS SM, Mucocutaneous and peripheral soft-tissue hemangiomas: MR imaging, *Radiology* (1987) **163**: 163–6.

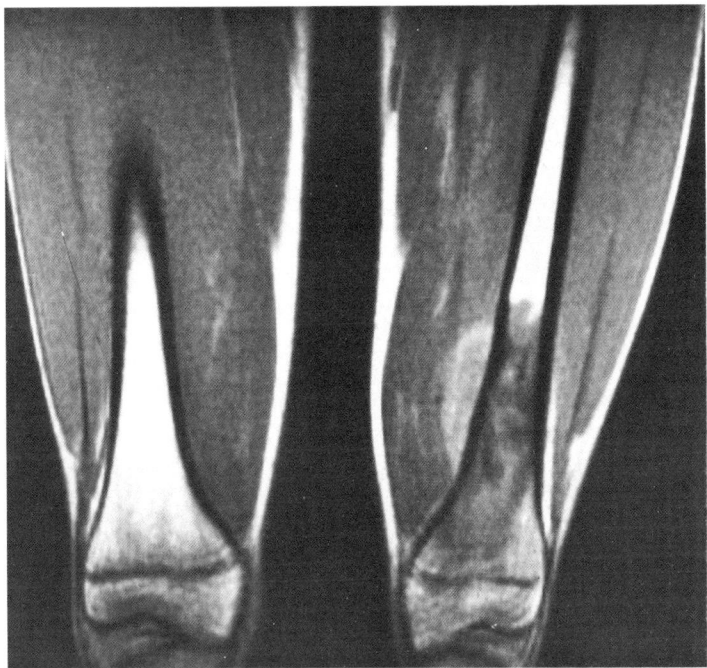

a

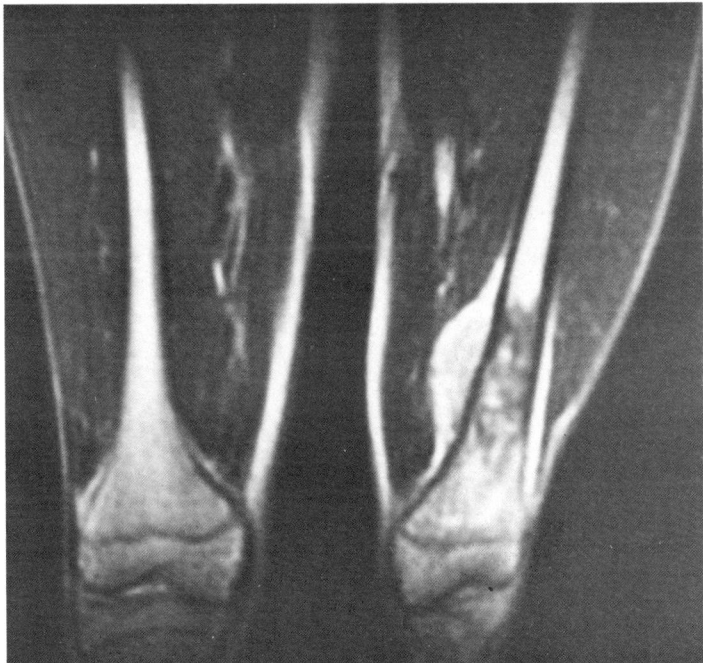

b

Figure 11.1

Osteosarcoma. A 13-year-old boy with mass in left inner-thigh. (**a**) T_1-weighted coronal scan (SE 500/30) shows extent of bone marrow involvement in distal femur, characterized by low-intensity signal within narrow space. To aid manufacture of limb salvage prosthesis, measurement is made from lower femoral condyle to upper extent of abnormal marrow. (**b**) T_2-weighted image (SE 2000/84) more clearly shows extraosseous component of tumor. Extent of medullary involvement is more difficult to determine because much of tumor has signal intensity equal to that of marrow.

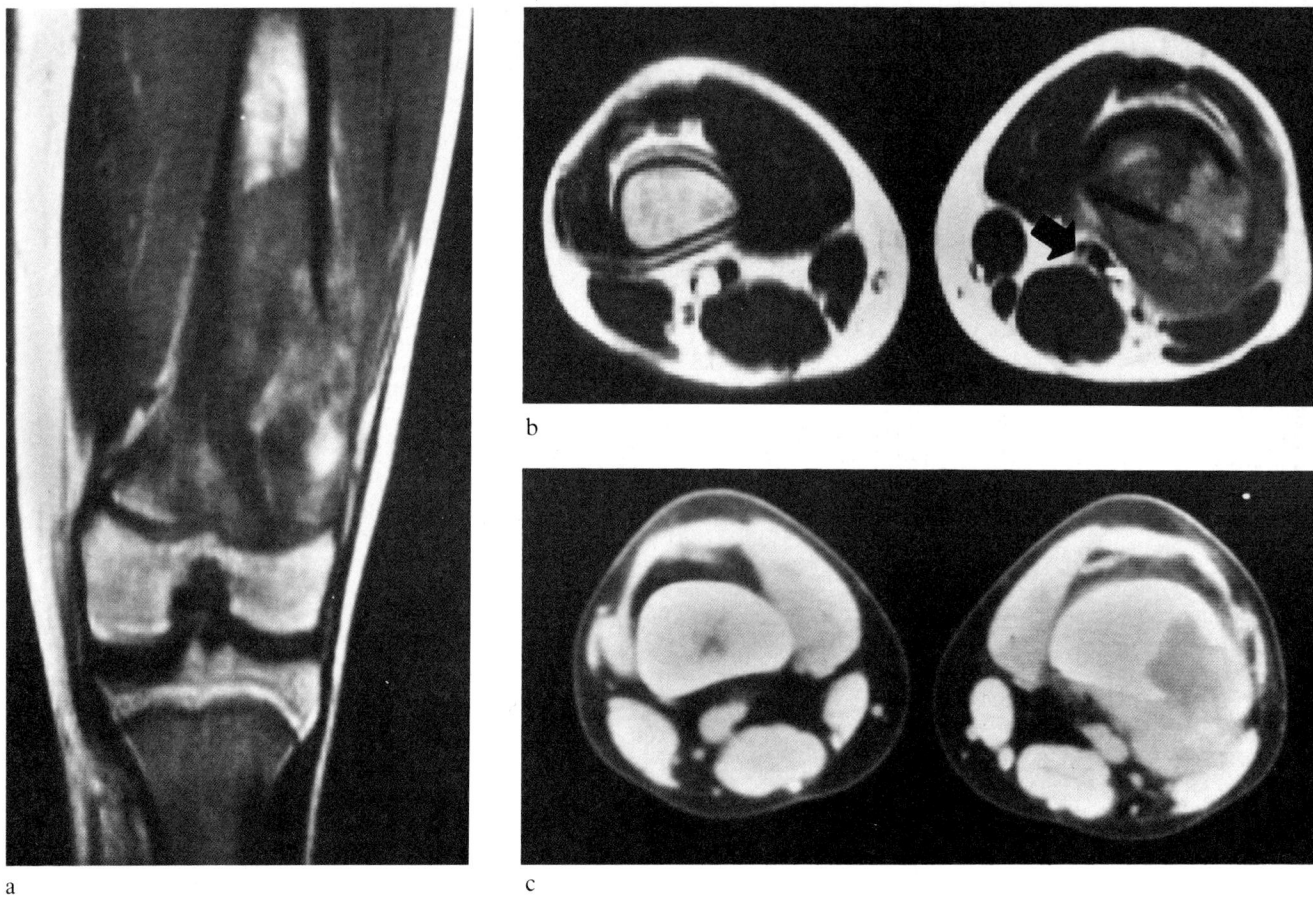

Figure 11.2

Osteosarcoma. An 11-year-old boy with mass in left distal thigh. (**a**) Coronal T_1-weighted image (SE 500/28) shows extent of marrow involvement. Foci of high-signal intensity within tumor probably represent hemorrhage. (**b**) Axial image (SE 500/28) more clearly illustrates relationship of tumor to surrounding muscles and neurovascular bundle. Although tumor has compressed biceps femoris muscle, muscle signal is normal, and fat plane between muscle and tumor is intact. Neurovascular bundle (arrow) is free of tumor. (**c**) Contrast-enhanced CT for comparison.

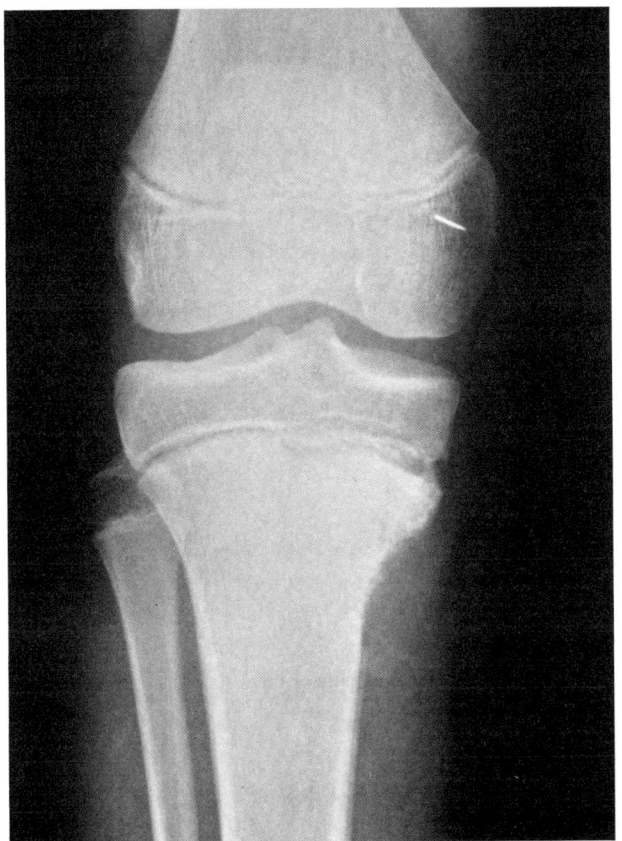
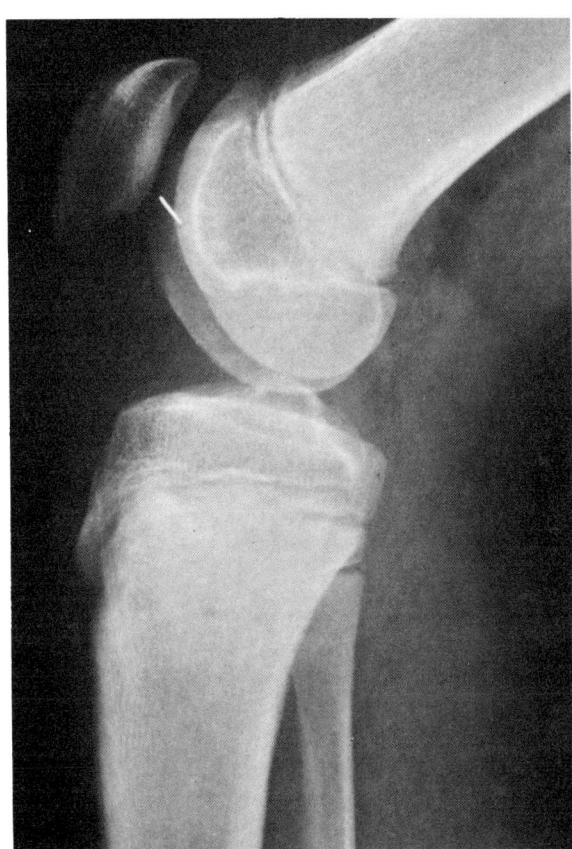

a b

Figure 11.3

Osteosarcoma. A 13-year-old boy with osteosarcoma of tibia. (**a,b**) Anteroposterior and lateral radiographs show sclerotic lesion of proximal tibial metaphysis, with destruction of medial cortex. (Needle tip in soft tissues has been present for many years.) (**c,d**) Coronal and sagittal images (SE 800/20). Signal intensity of tumor is not markedly different from that of normal hematopoietic marrow of proximal femur, but soft-tissue mass and adjacent cortical destruction are obvious. (Note prominent artifact from needle.) (**e,f**) Sagittal double-echo images (SE 1500/20–90). (*Courtesy of Richard Ofstein, Long Beach, CA*)

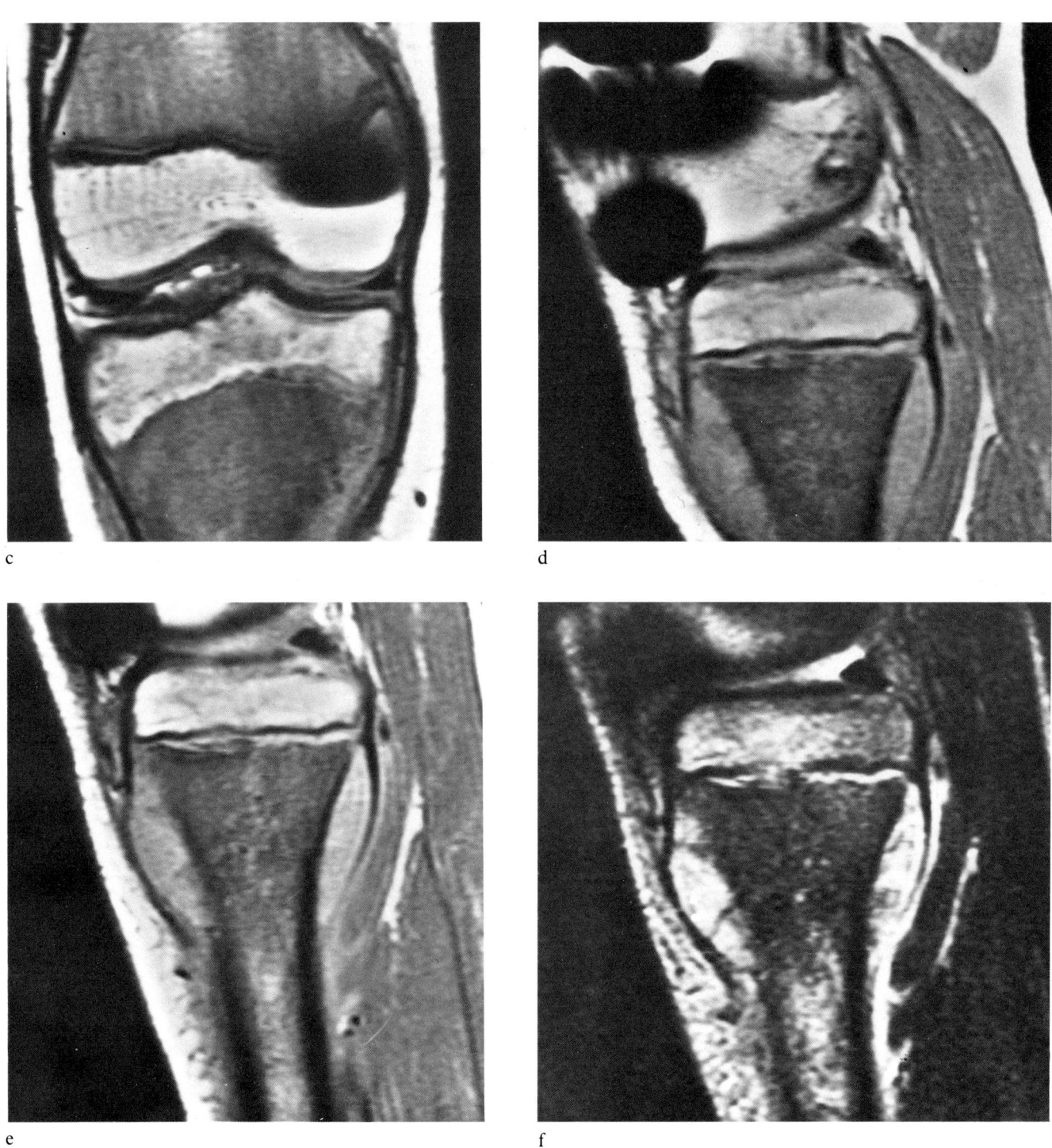

Figure 11.3 continued

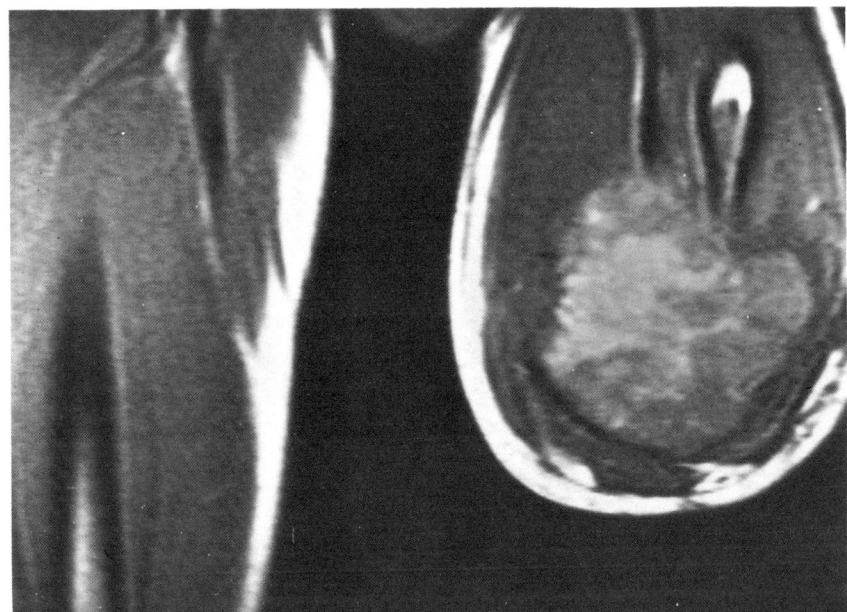

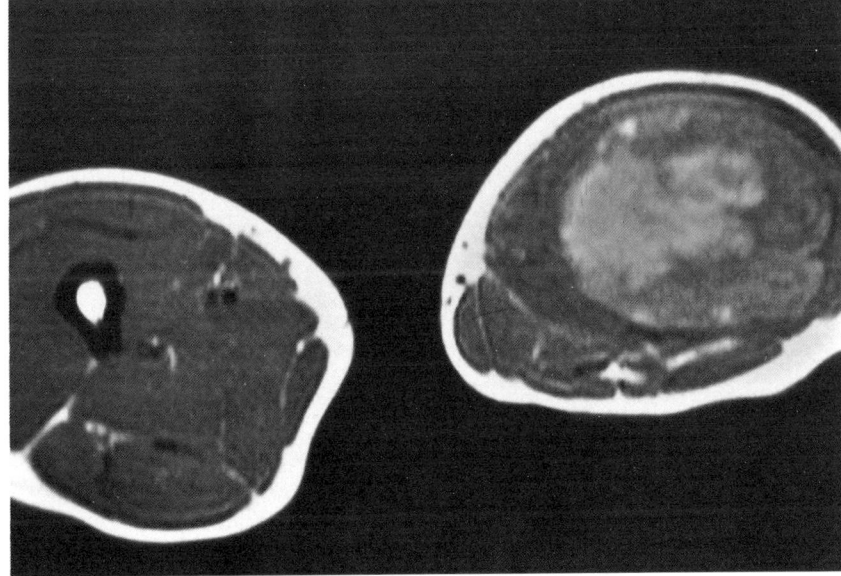

Figure 11.4

Recurrent osteosarcoma. A 19-year-old man 1 year following above-the-knee amputation. T_1-weighted coronal (**a**) and axial (**b**) images (SE 500/28) disclose large soft-tissue mass at amputation site. High-signal intensity within recurrent tumor is secondary to hemorrhage.

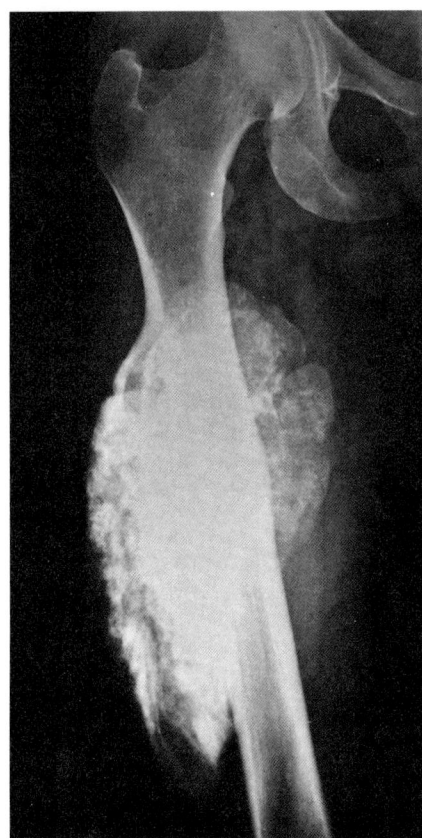
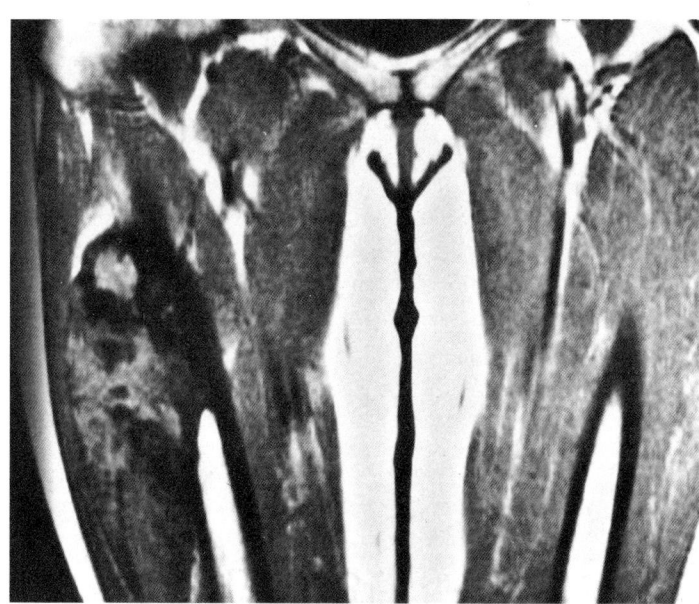

Figure 11.5

Parosteal osteosarcoma. A 44-year-old woman with parosteal osteosarcoma of right femur.
(**a**) Anteroposterior radiograph.
(**b**) Coronal image (SE 500/28) depicts soft-tissue mass extending distal to medullary involvement.

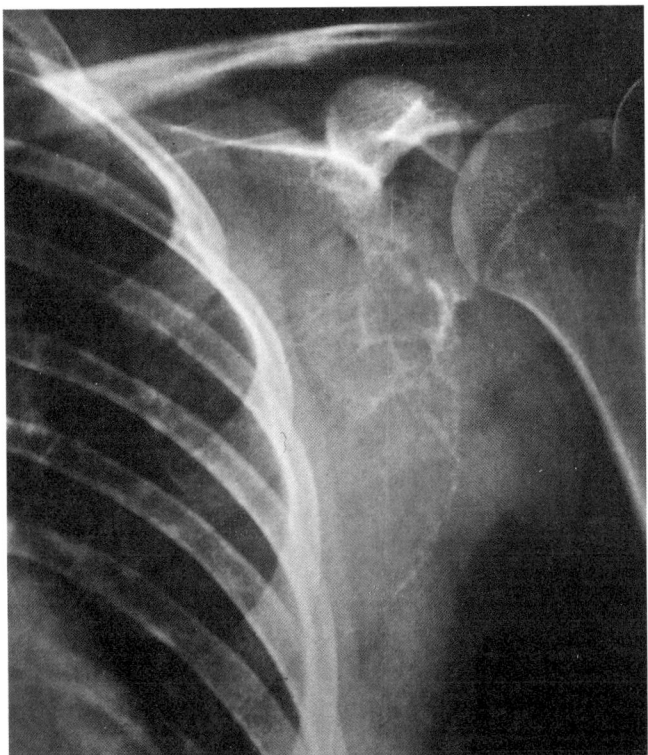
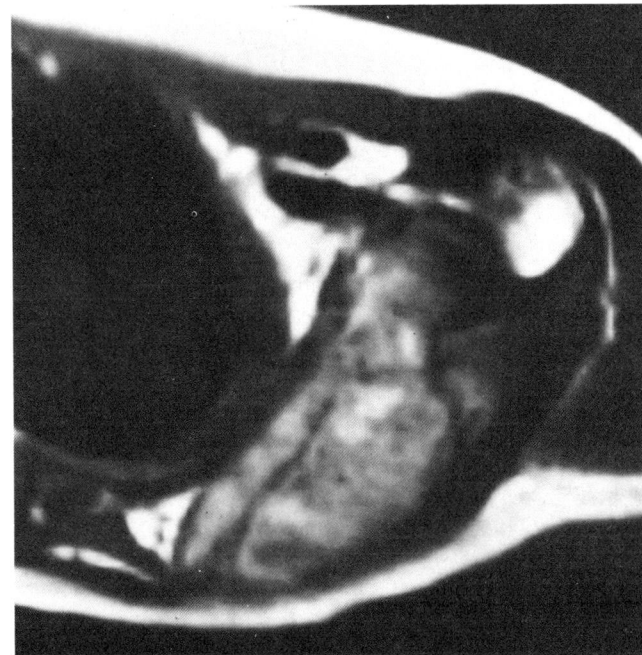
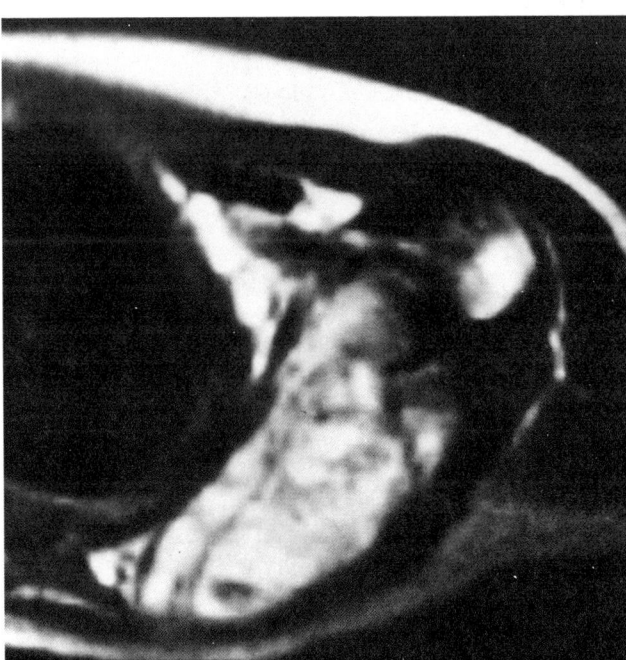

Figure 11.6

Ewing's sarcoma. A 15-year-old girl with left shoulder mass. (a) Anteroposterior radiograph shows bubbly, expansile lesion of scapula. (b,c) Axial double-echo MR images (SE 2000/35–70) through mid-scapula show inhomogeneous mass on first echo (b) with increased signal intensity on second echo (c). (d,e) Axial images (SE 2000/35–70) at lower aspect of scapula; lesion is more homogeneous in signal intensity on both first (d) and second (e) echoes. (f,g) Full extent of sarcoma is seen in sagittal images (SE 2000/35–70). There is little difference in signal intensity between first (f) and second (g) echoes. (Courtesy of Jeffrey Eckardt, Los Angeles, CA)

328 MRI atlas of the musculoskeletal system

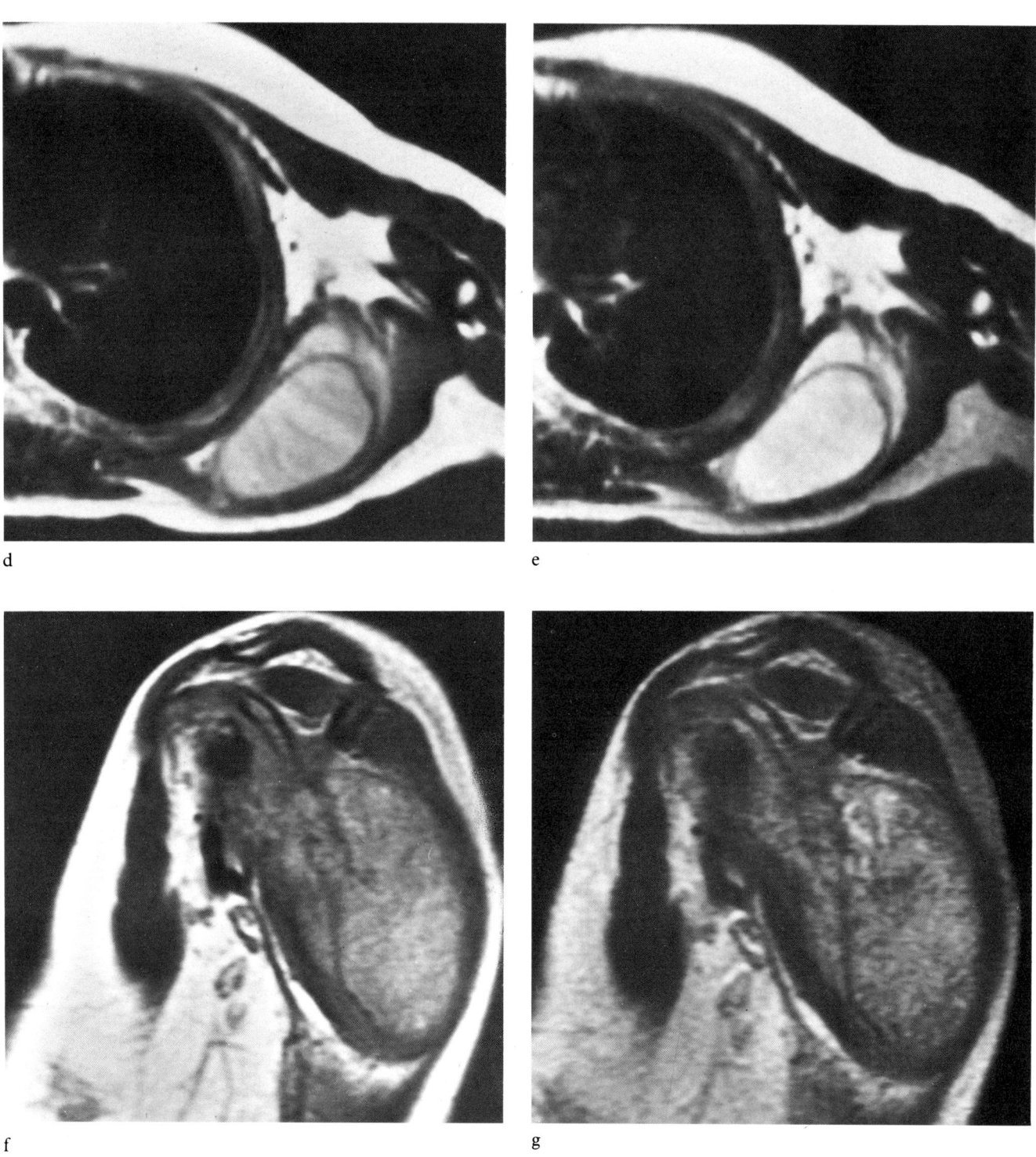

Figure 11.6 continued

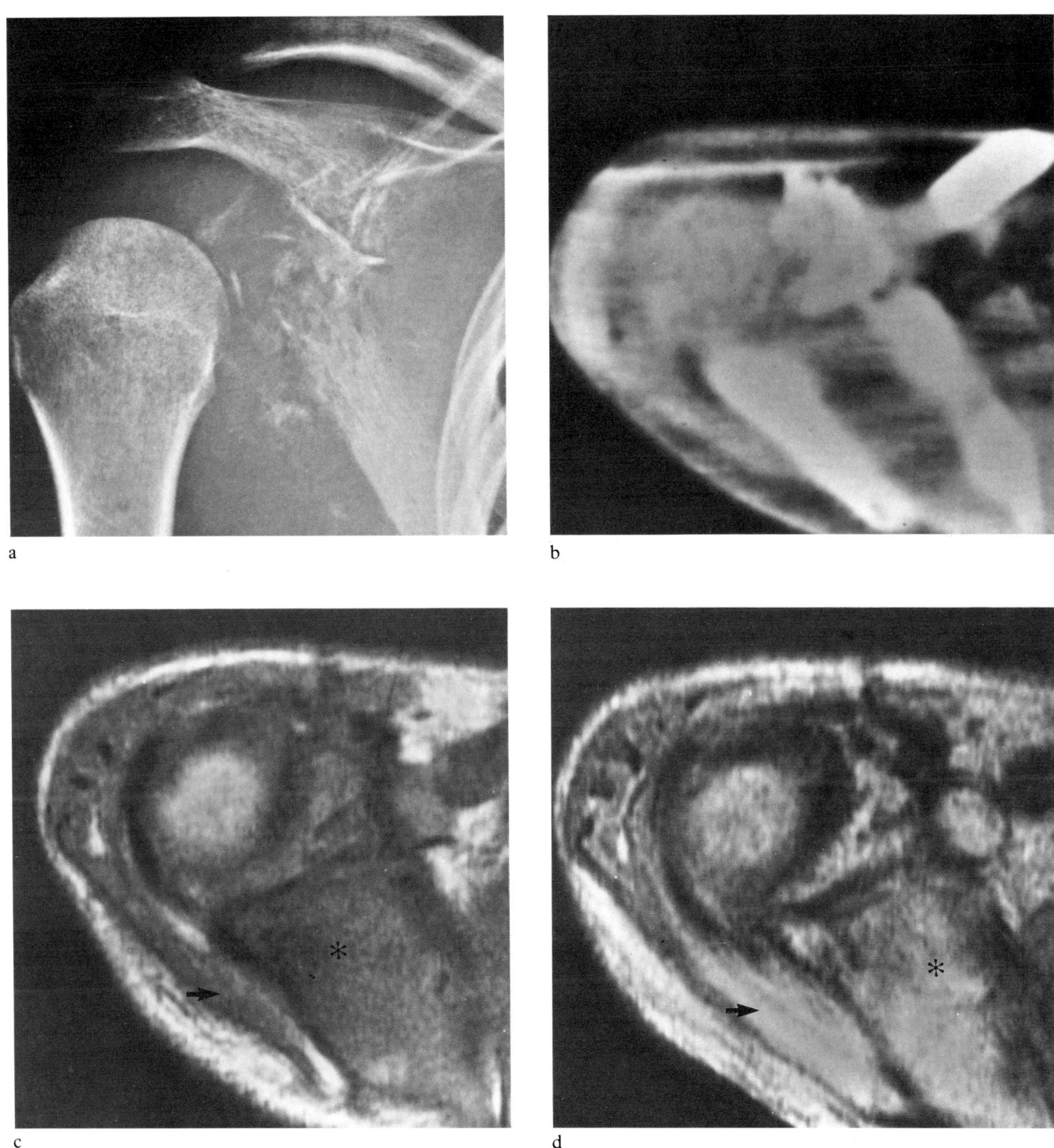

Figure 11.7

Ewing's sarcoma. A 22-year-old man with right shoulder mass. (a) Anteroposterior radiograph shows destruction of scapula. (b) CT image is compromised by beam-hardening from clavicles. (c,d) T_1-weighted (SE 500/28) (c) and T_2-weighted (SE 2000/56) (d) axial images at level of humeral head. Scapula is replaced by a large soft-tissue mass (★) which shows intermediate-signal intensity in T_1-weighted image and high-signal intensity in T_2-weighted image. Tumor has invaded infraspinatus muscle (arrow). Edema in superficial soft tissues may be due to preoperative radiotherapy.

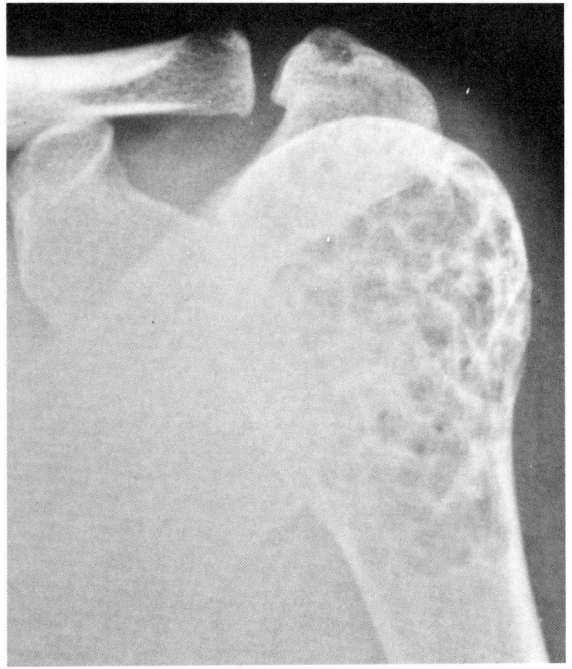
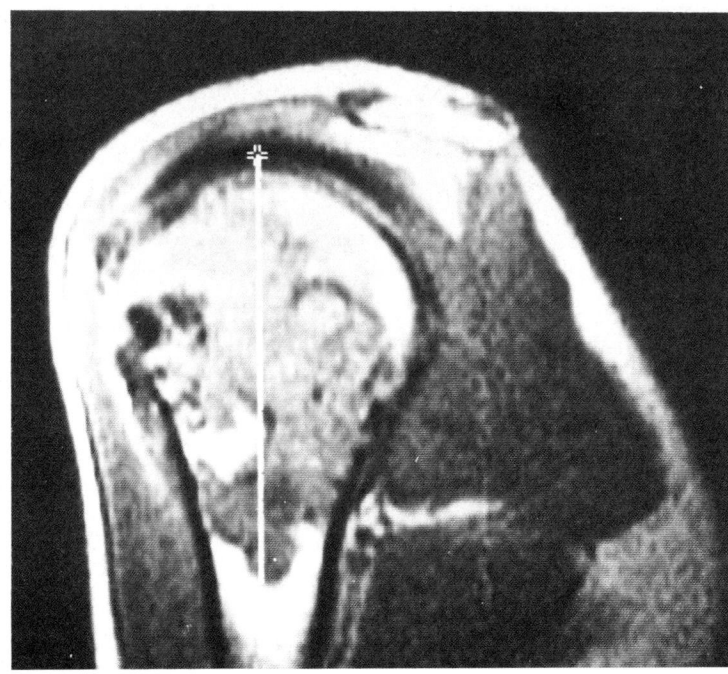
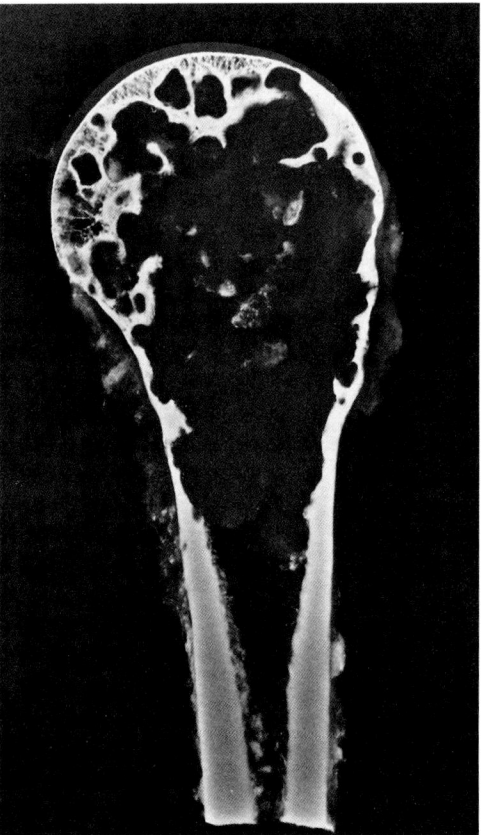

Figure 11.8

Chondrosarcoma. A 36-year-old woman with pain in left shoulder. (a) Anteroposterior radiograph shows bubbly lesion of proximal humerus.
(b) Sagittal scan (SE 500/28) discloses low-signal intensity tumor. Cursor was used to calculate exact length of tumor for limb-salvage procedure.
(c) Specimen radiograph.
(*Courtesy of Joseph M Mirra, Los Angeles, CA*)

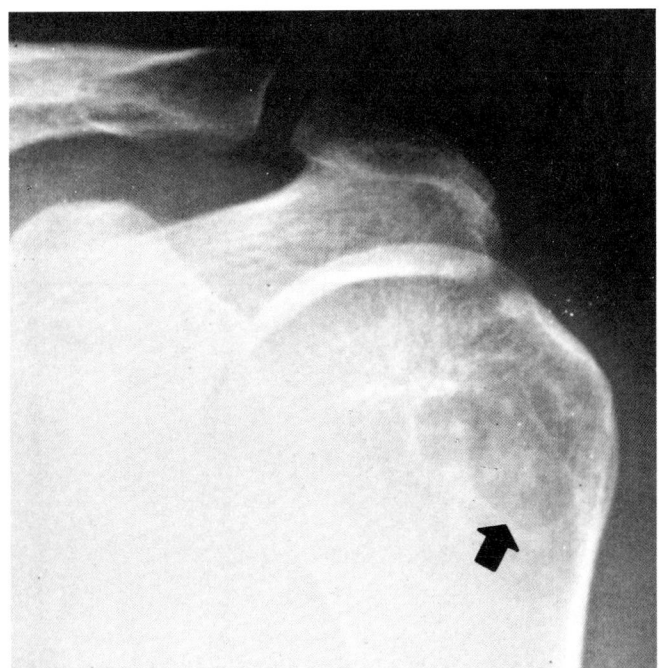
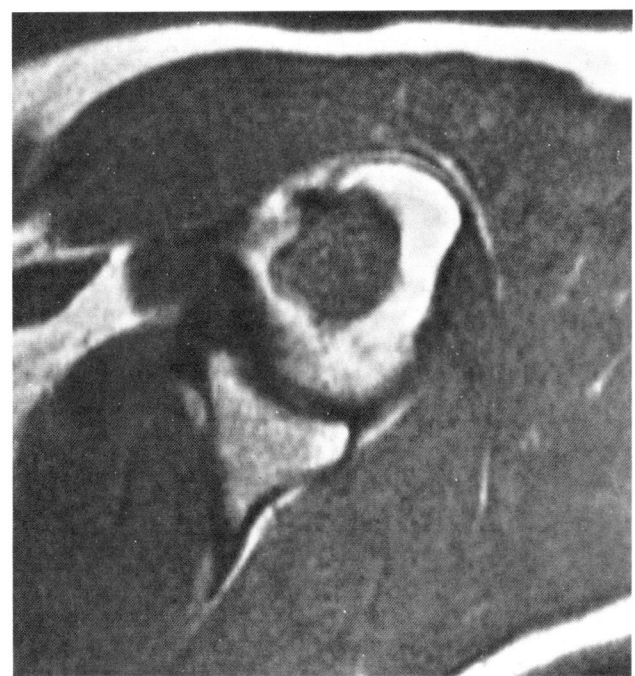
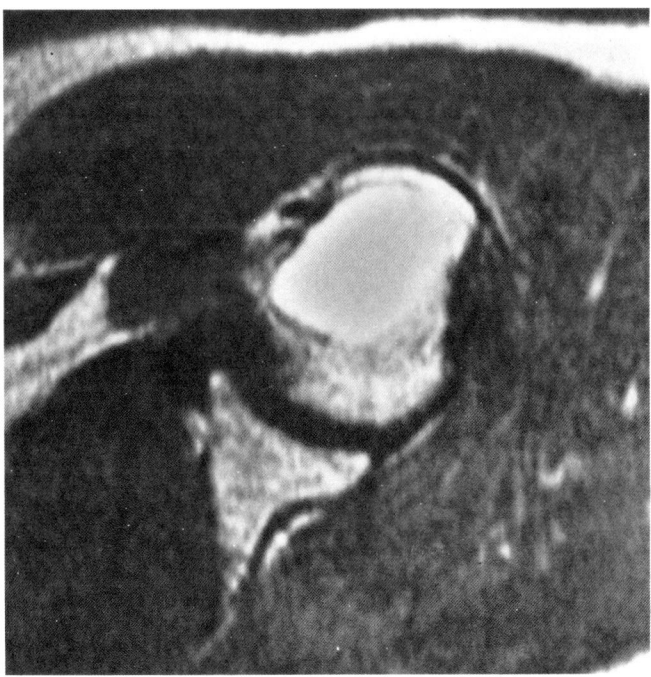

Figure 11.9

Chondrosarcoma. A 33-year-old man with painful left shoulder. (a) Anteroposterior radiograph reveals well-circumscribed lytic lesion in proximal humerus (arrow). (b) T_1-weighted axial image (SE 500/28) shows well-defined region of low-signal intensity at site of lytic lesion. (c) T_2-weighted axial image (2000/56) reveals that lesion has high-signal intensity.

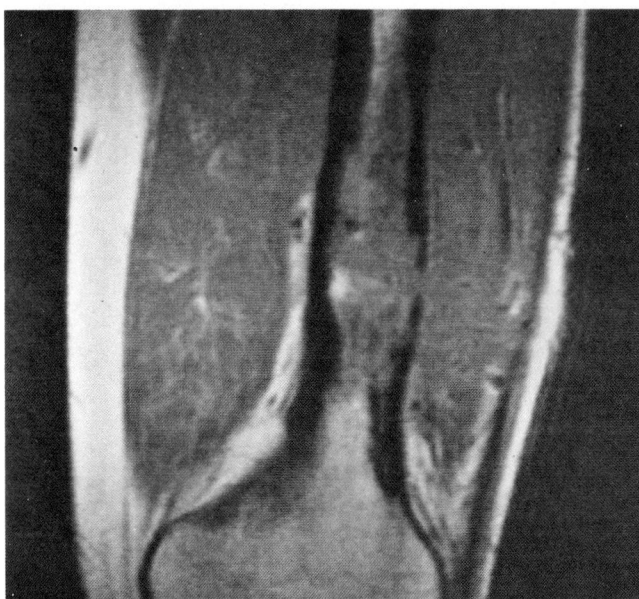

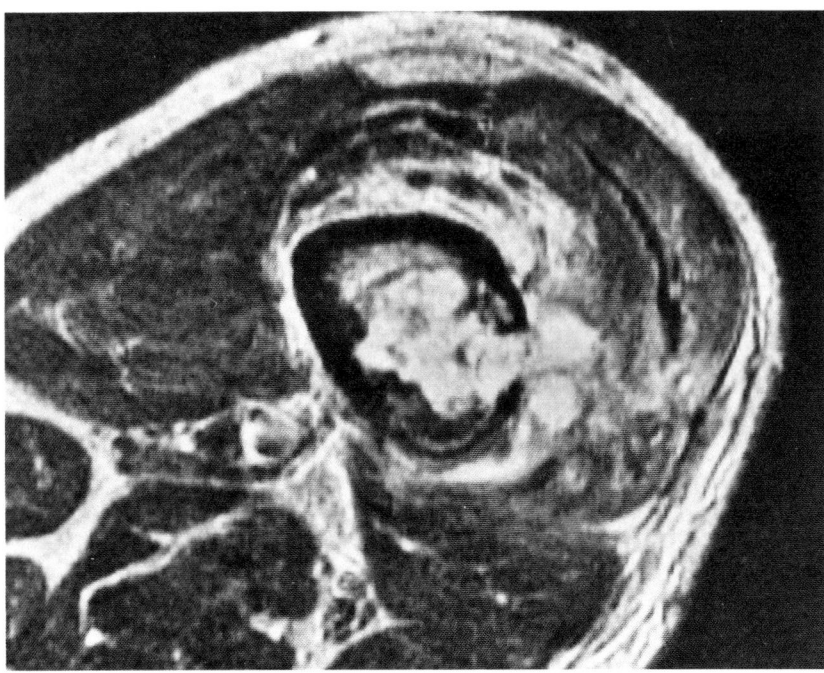

Figure 11.10

Malignant fibrous histiocytoma. A 41-year-old man with thigh pain. (**a**) Coronal T_1-weighted image (SE 500/30) depicts low-signal-intensity tumor in distal diaphysis of femur, and destruction of cortex. (**b**) Axial T_2-weighted image (SE 2500/85) reveals extent of soft-tissue component of tumor.

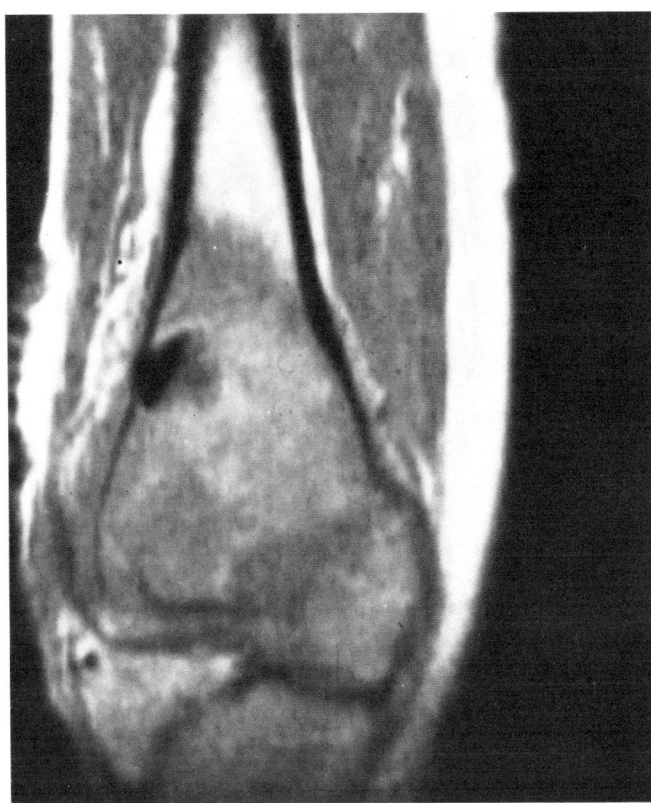

a

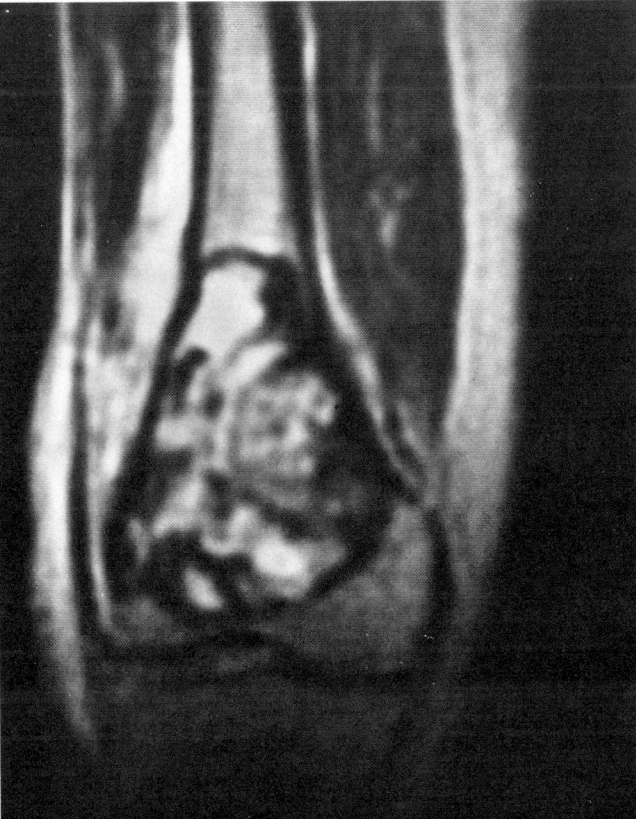

b

Figure 11.11

Giant cell tumor. A 30-year-old man who was referred for preoperative evaluation of biopsy-proven right femoral giant cell tumor. (**a**) T_1-weighted coronal image (SE 500/28) shows extensive marrow replacement by tumor. High-signal-intensity blood in adjacent soft tissues is result of recent biopsy. Signal void is due to plug of methyl methacrylate introduced at time of biopsy. (**b**) T_2-weighted coronal image (SE 2000/84) demonstrates well-defined margins and internal septations between lobules of high-signal-intensity tumor.

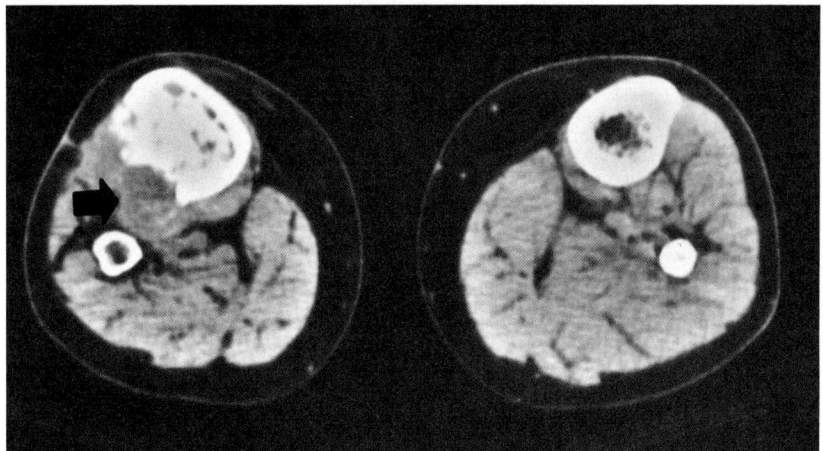

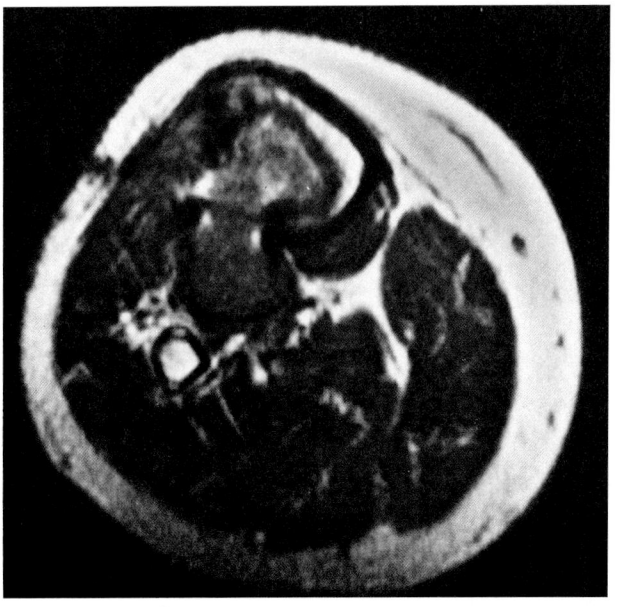

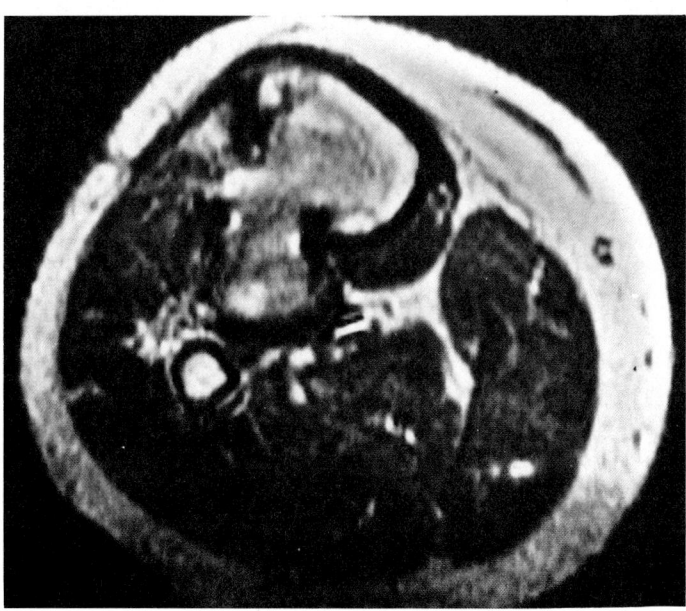

Figure 11.12

Recurrent giant cell tumor. A 29-year-old man 1 year following curettage and bone grafting of a giant cell tumor in right distal femur. (**a**) CT scan (non-contrast-enhanced) discloses soft-tissue mass (arrow) and scalloping of adjacent bone, representing tumor recurrence. (**b**) Axial T_1-weighted image (SE 500/28) reveals that soft-tissue mass is contiguous with medullary cavity. (**c**) Axial T_2-weighted image (SE 2000/56) shows that intraosseous and extraosseous portions of tumor both manifest high-signal intensity.

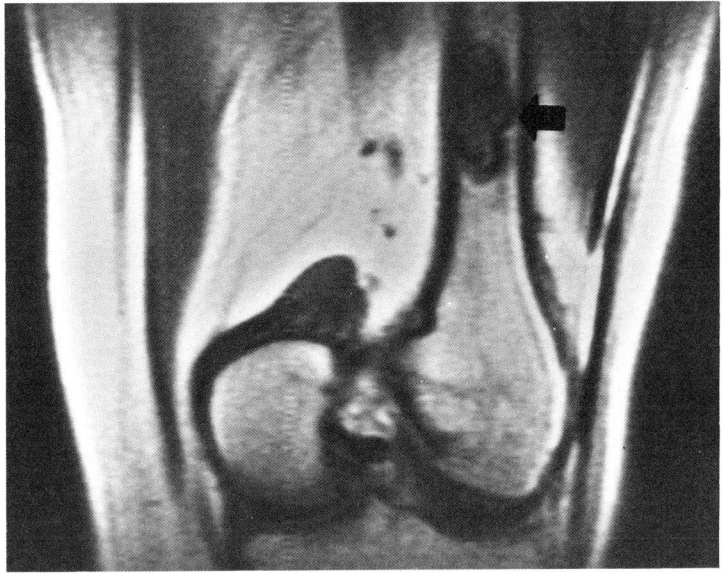
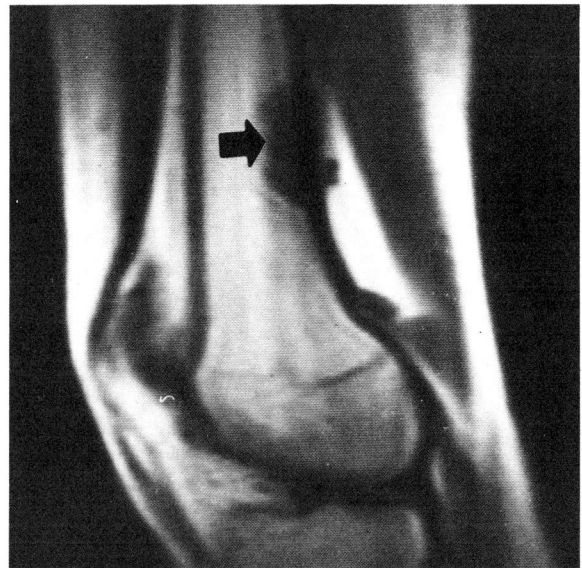

Figure 11.13

Non-ossifying fibroma. An 18-year-old woman with knee pain, referred for evaluation of menisci. Femoral non-ossifying fibroma was incidentally noted in radiographs. (**a,b**) T_1-weighted coronal (**a**) and sagittal (**b**) images (SE 500/28) show typical lobulation and sharp borders of non-ossifying fibroma (arrow).

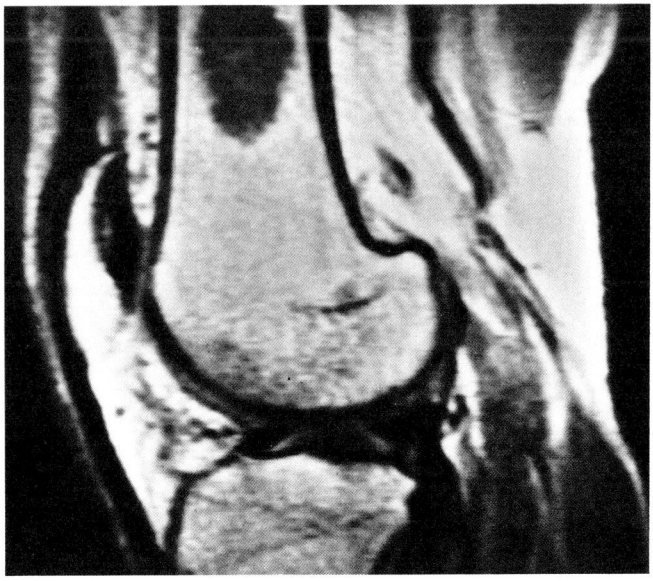
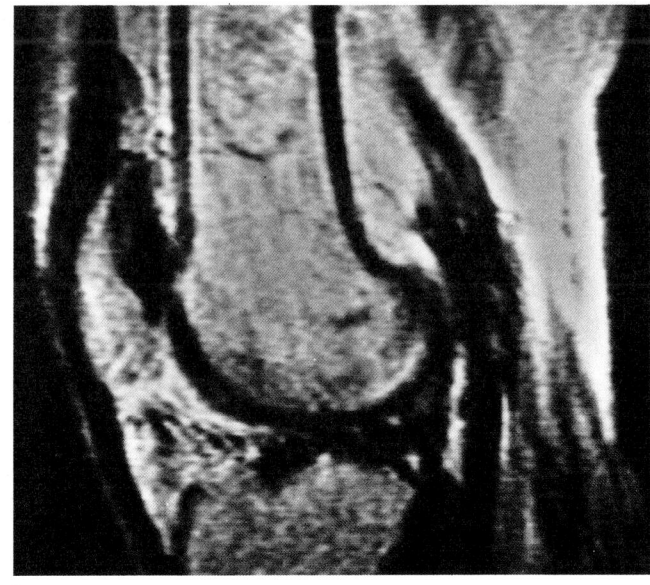

Figure 11.14

Enchondroma. A 40-year-old man referred for evaluation of possible meniscal tear. (**a**) Unexpected lesion with low-signal intensity is identified in femur in T_1-weighted sagittal image (SE 800/28). (**b**) In T_2-weighted sagittal image (SE 2000/56), lesion has high-signal intensity, but is distinguished from surrounding marrow by well-defined low-signal borders and inhomogeneous interior. Subsequent radiographs revealed typical calcifications of an enchondroma.

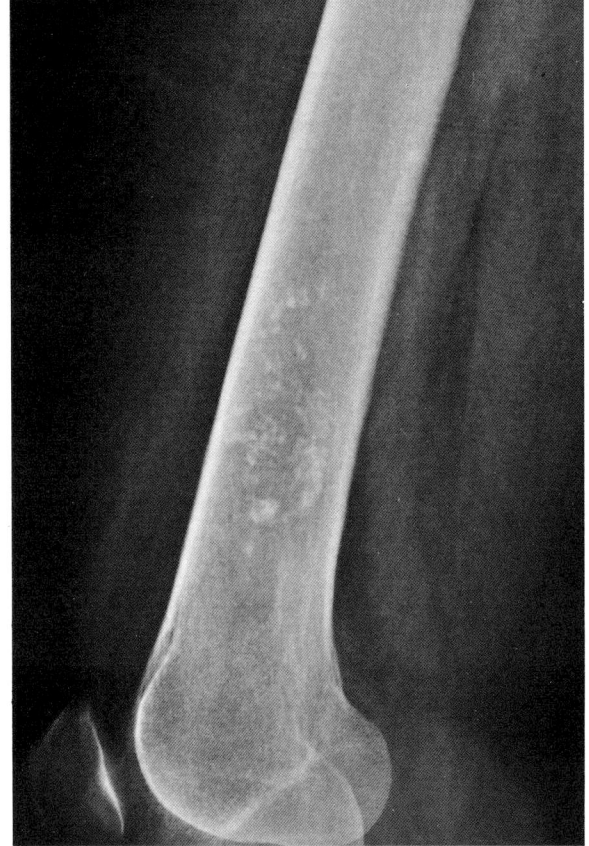
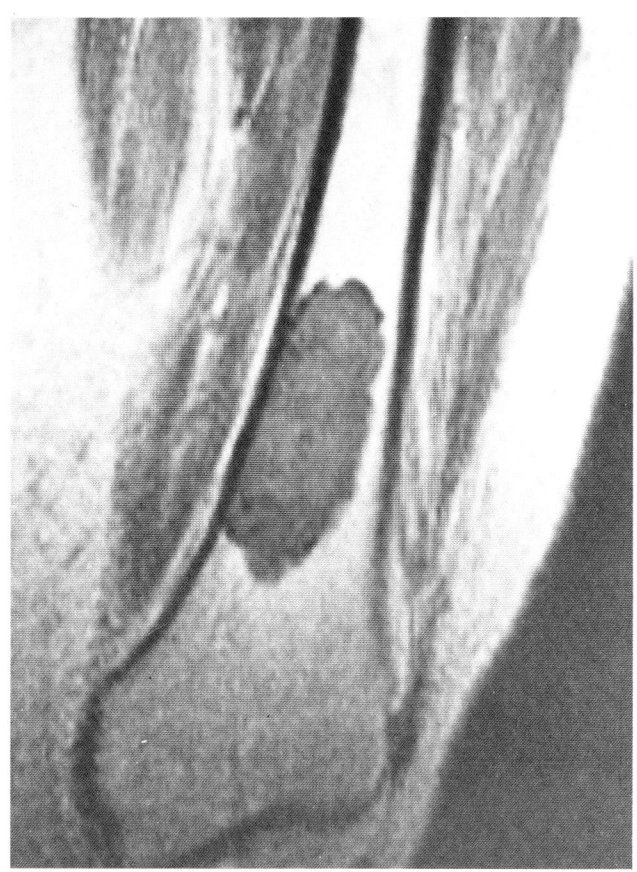
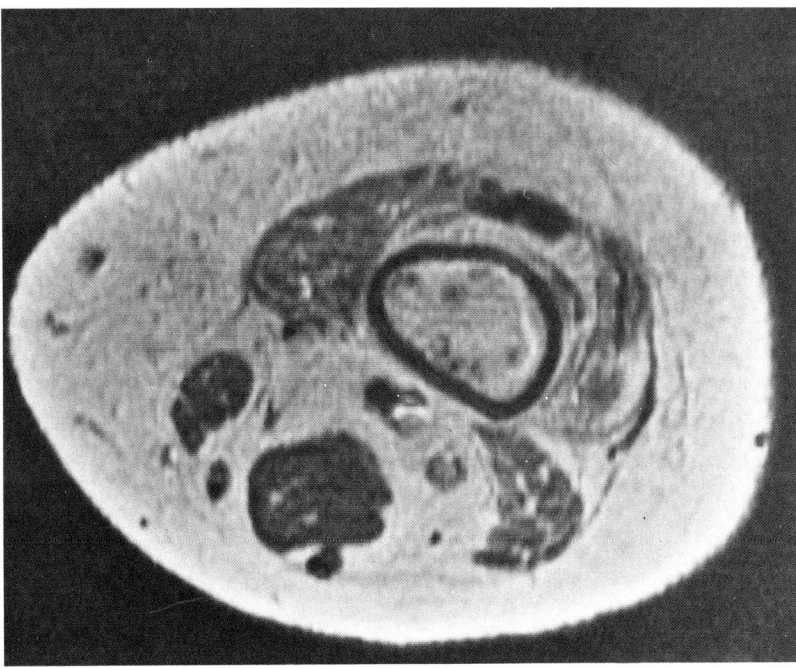

Figure 11.15

Enchondroma. A 56-year-old woman with abnormal radiograph of left femur. (**a**) Lateral radiograph shows well-demarcated lesion characterized by calcified cartilage matrix. (**b**) T_1-weighted coronal image (SE 500/28) demonstrates lesion with low-signal intensity and sharp margins. (**c**) Axial T_2-weighted image (SE 1500/56) discloses that, except for the punctate calcifications in its matrix, tumor has high-signal intensity. (**d**) Axial T_2-weighted image (SE 1500/56) near superior margin of lesion reveals that tumor has signal intensity nearly identical to that of surrounding marrow. (**e**) In SE 2000/84 image, tumor (arrow) shows higher signal intensity than surrounding normal marrow.

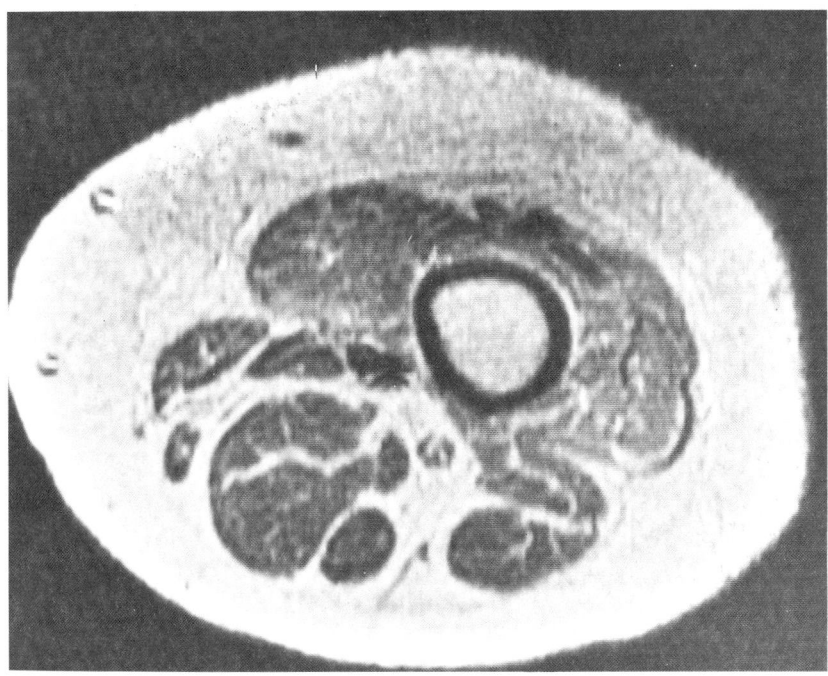

d

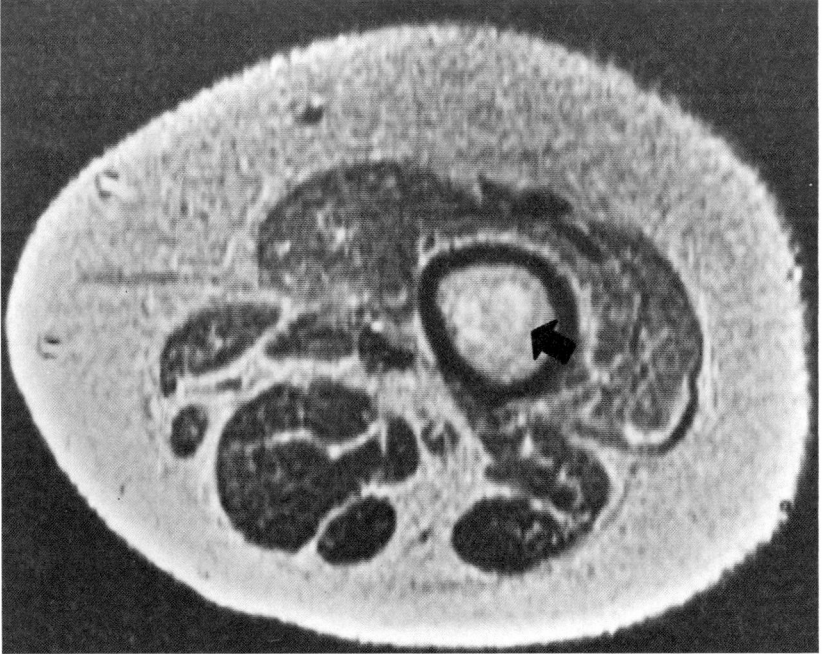

e

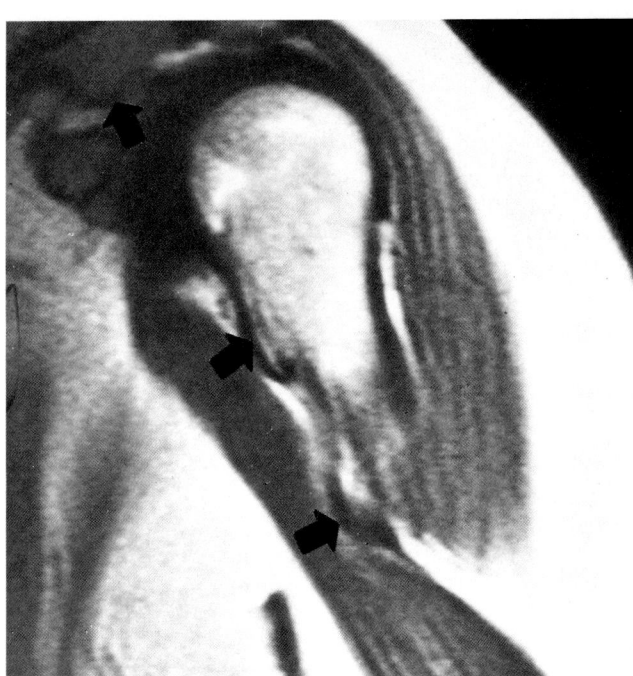

Figure 11.16

Multiple hereditary exostoses (osteochondromas). A 38-year-old woman with multiple hereditary exostoses. T_1-weighted coronal image (SE 500/28) shows exostoses (arrows) of medial humerus and distal clavicle. Characteristic features include continuity of marrow and cortex from parent bones to lesions, and modeling deformity of humeral metaphysis. Decreased signal intensity in upper aspect of image is due to position of surface coil.

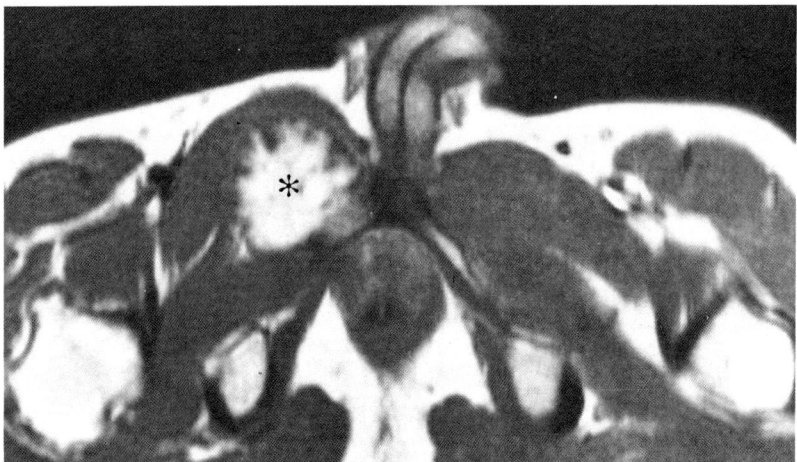

a

Figure 11.17

Osteochondroma. A 23-year-old man with large osteochondroma of ischium. Axial (**a**) and coronal (**b**) T_1-weighted images (SE 500/28) show high-signal-intensity marrow extending from ischium into osteochondroma (★). Benign features include well-defined signal void indicating intact cortical margins around lesion, and absence of soft-tissue mass. Surgical excision revealed benign osteochondroma.

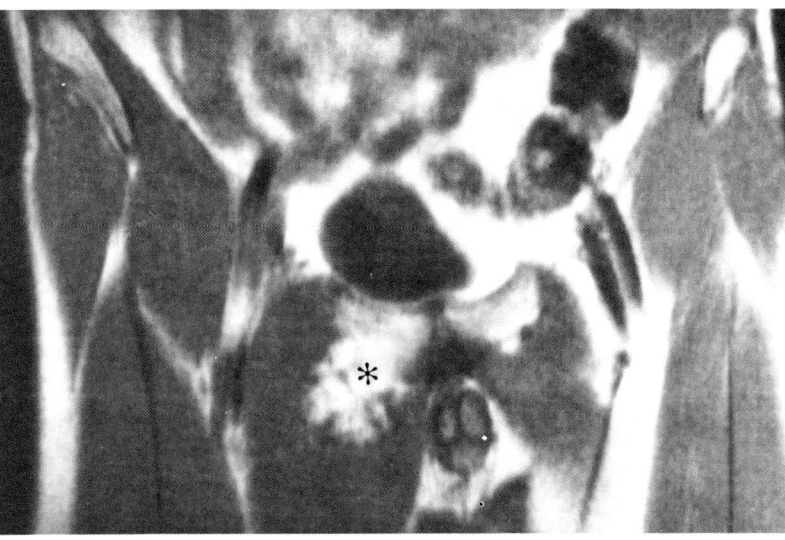

b

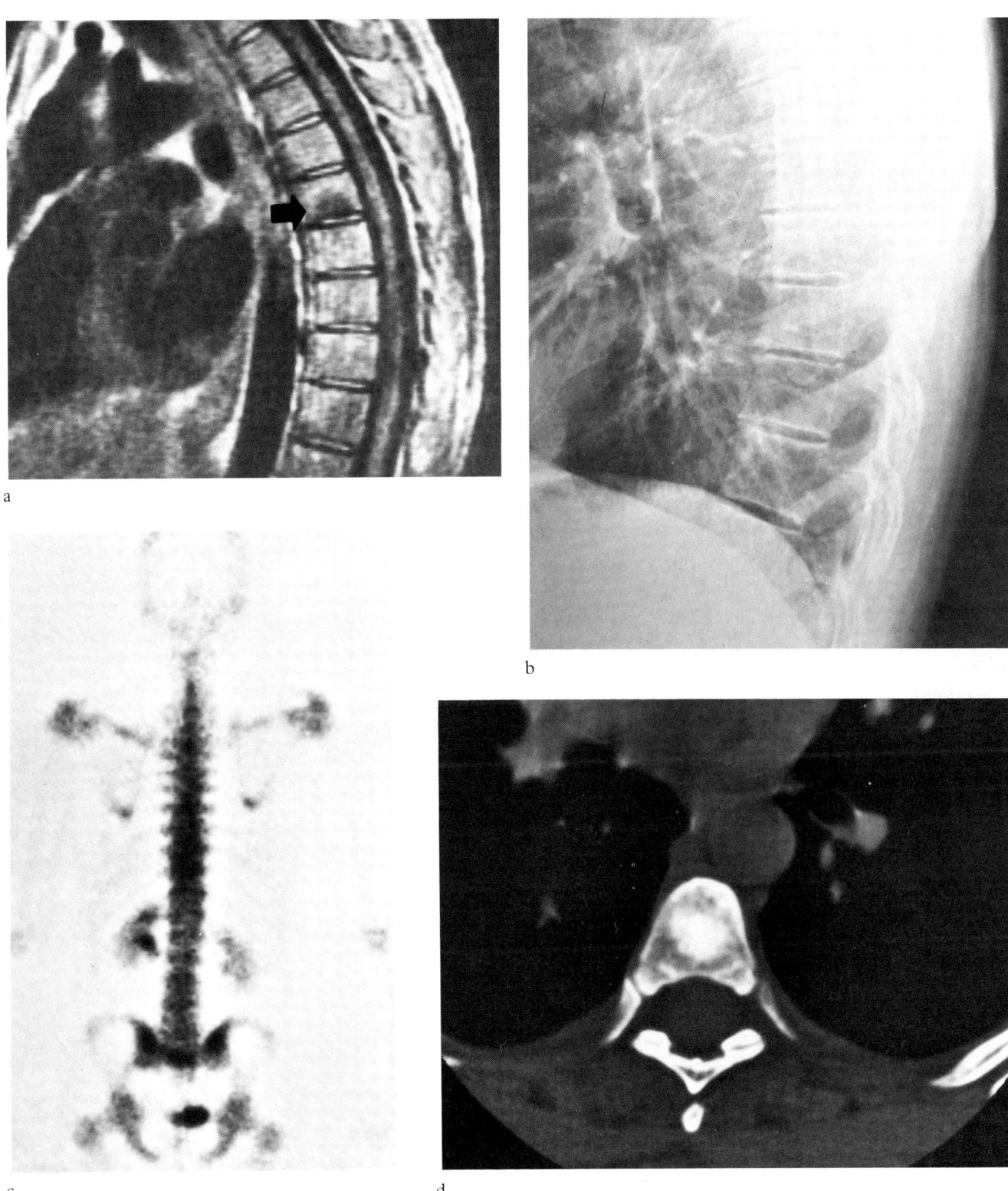

Figure 11.18

Vertebral enostosis (bone island). A 43-year-old man with coincidental leg numbness. (**a**) Sagittal T_1-weighted image (SE 500/28) shows focus of low-signal intensity (arrow) adjacent to inferior endplate of seventh thoracic vertebra. (**b**) In lateral radiograph, lesion is obscured by overlying scapulae. (**c**) Lesion is not evident in radionuclide bone scan. (**d**) CT scan reveals typical enostosis.

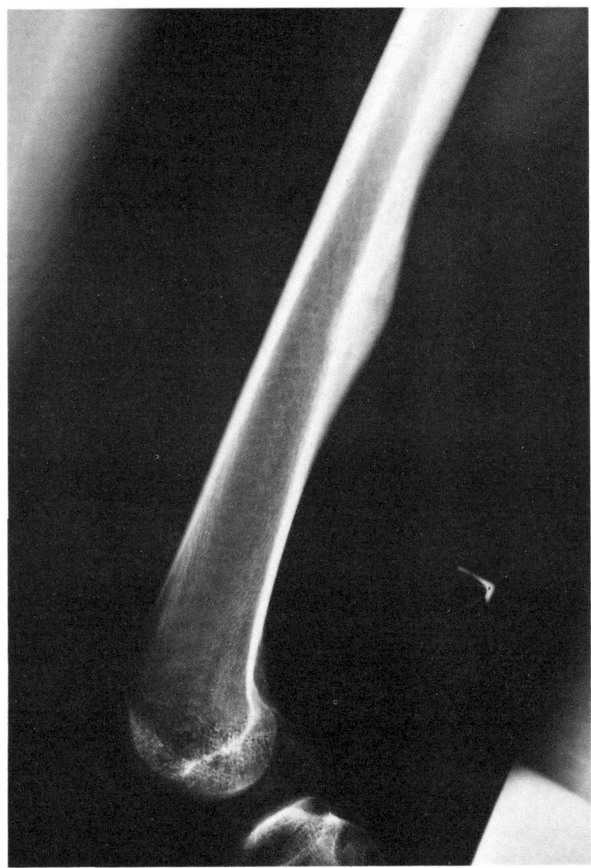

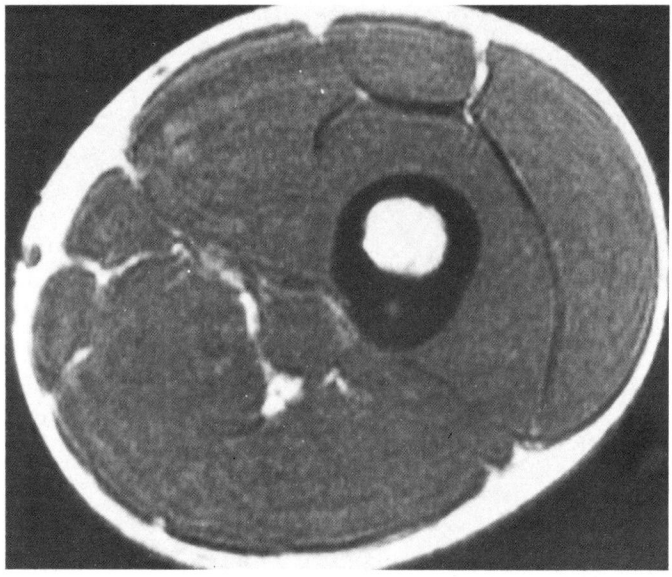

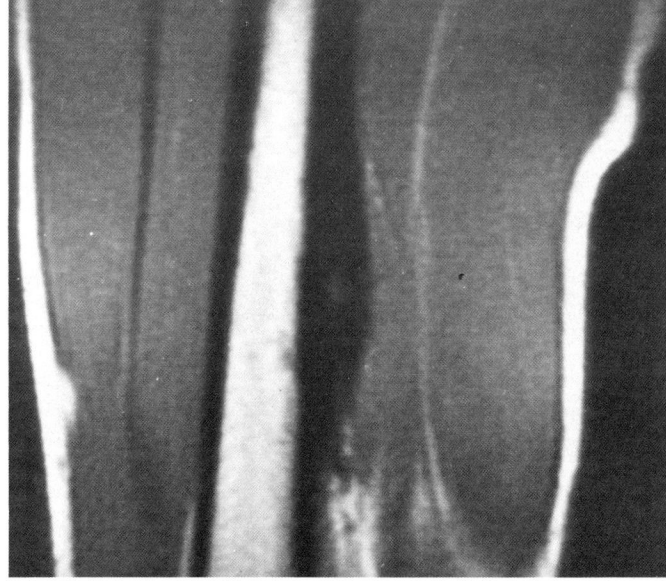

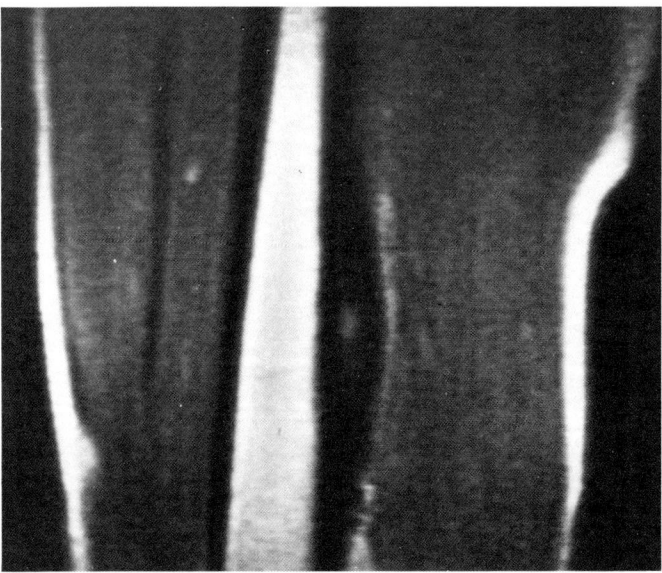

Figure 11.19

Osteoid osteoma. An 18-year-old man with left leg pain relieved by aspirin. (**a**) Lateral radiograph shows exuberant cortical thickening posteriorly, without a definite nidus of radiolucency. Axial (**b**) and sagittal (**c**) T_1-weighted images (SE 500/28) show thickened cortex represented by signal void. Nidus is identified as focus of intermediate-signal intensity in center of thickened cortex. (**d**) Sagittal T_2-weighted image (SE 2000/56). Signal intensity of nidus is slightly increased, and surrounding cortical bone remains devoid of signal.

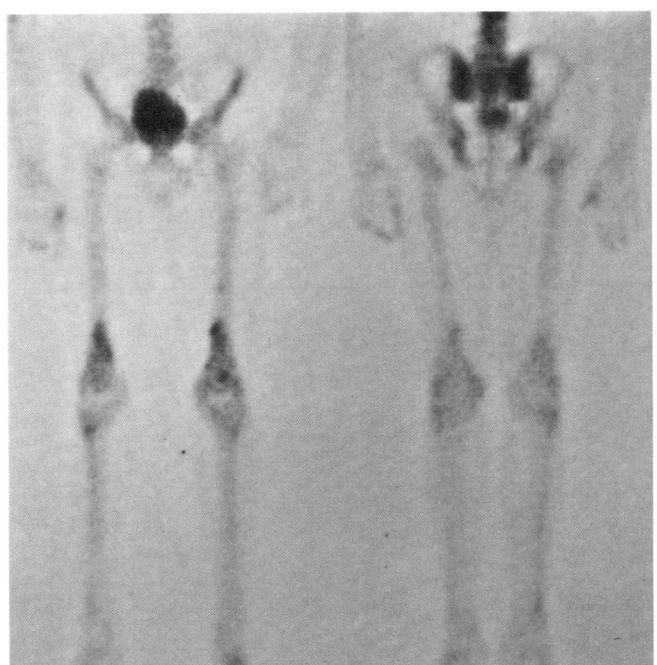

a

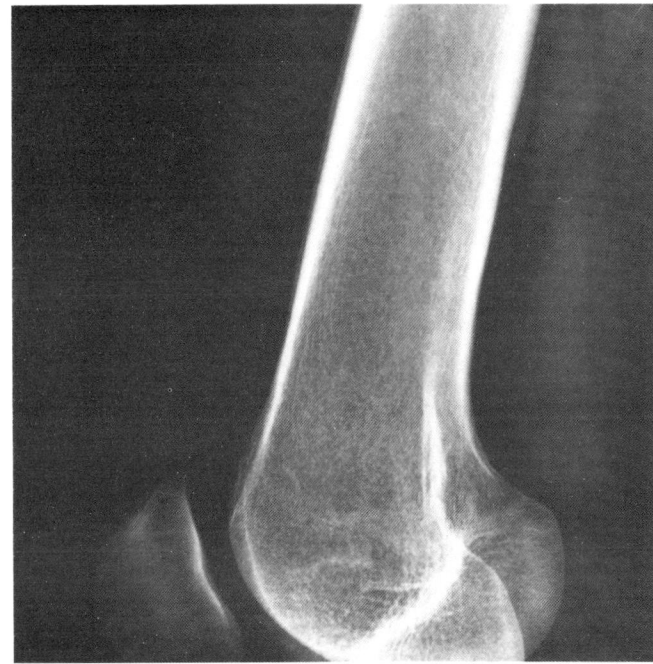

b

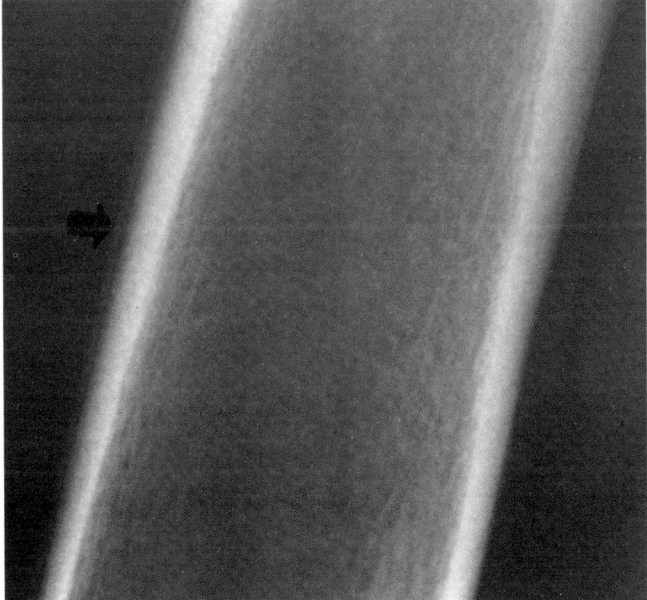

c

Figure 11.20

Hypertrophic pulmonary osteoarthropathy. A 58-year-old man with lung carcinoma. (**a**) Radionuclide bone scan reveals moderately increased isotope accumulation in distal femura. (**b**) Lateral radiograph of right distal femur. (**c**) Close-up of (**b**) shows single, uniform layer of periosteal new bone (arrow). (**d**) Sagittal T_1-weighted image (SE 500/28) discloses intermediate-signal intensity of thickened periosteum (arrows). An adjacent stripe of high-intensity signal represents the prefemoral fat. The periosteal new bone is too thin and too close to the cortex to image. (**e**) Sagittal T_2-weighted image (SE 1500/56) reveals that periosteum has high-signal intensity and cannot be distinguished from adjacent fat.

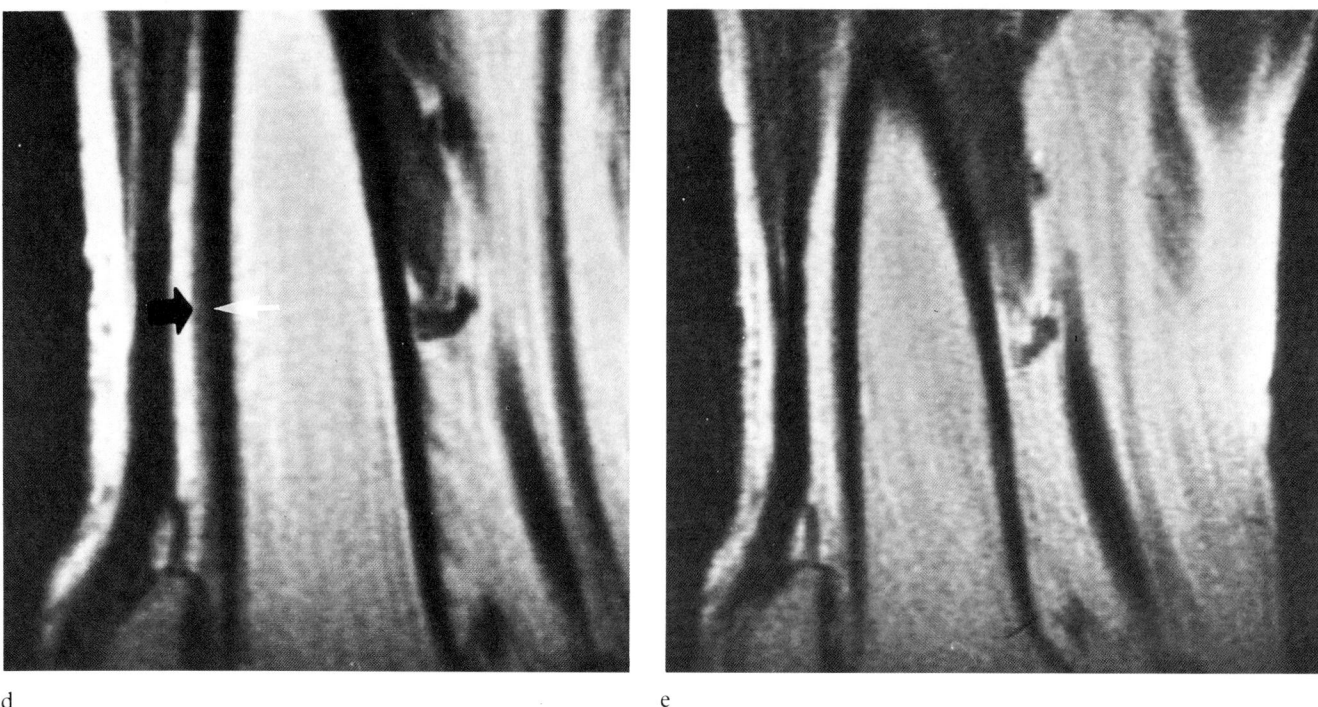

d   e

Figure 11.20 *continued*

Figure 11.21

Metastasis. A 68-year-old man who was referred for preoperative staging of prostatic carcinoma. (**a**) T_1-weighted axial image (SE 500/28) reveals focus of low-signal intensity in right ischium (arrow). (**b**) CT, performed 2 weeks after MRI examination, does not reveal lesion. (**c**) Radionuclide bone scan (posterior view) shows normal isotope accumulation at site of lesion. (**d**) CT scan 6 months later shows blastic metastasis (arrow).

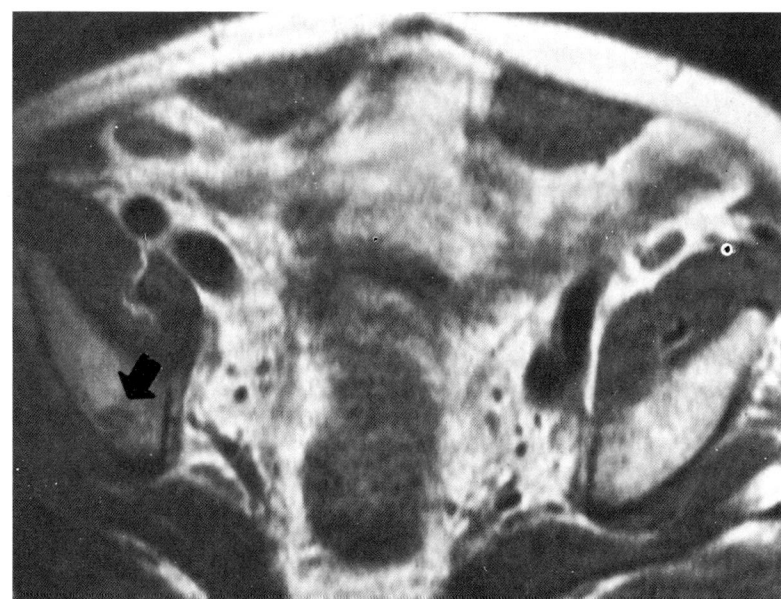

a

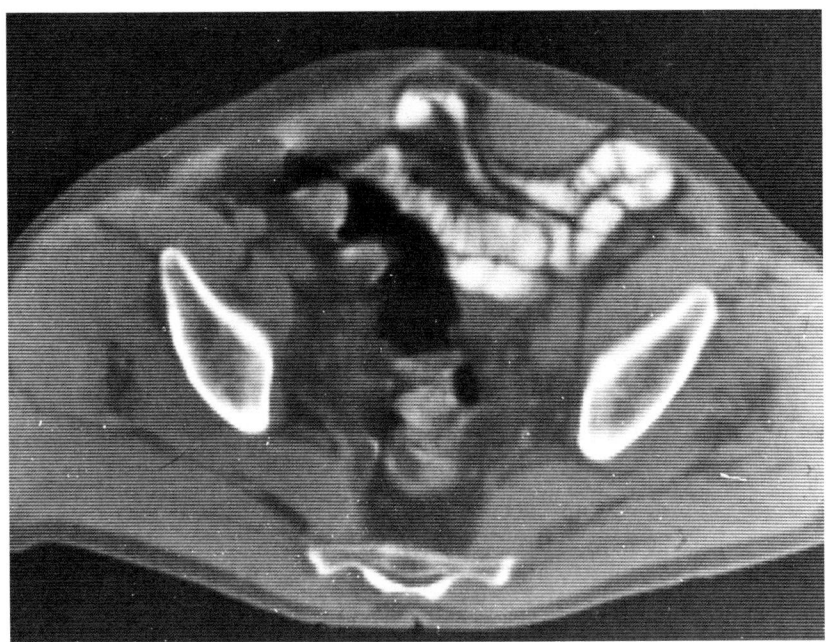

b

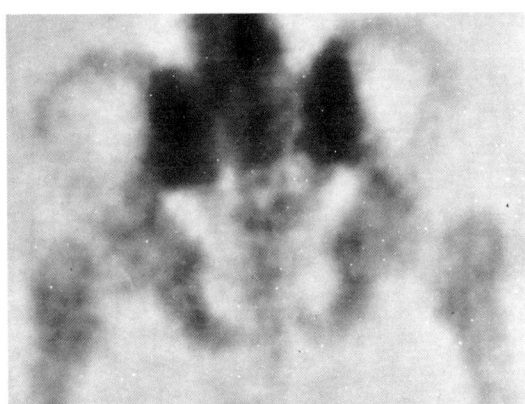

c

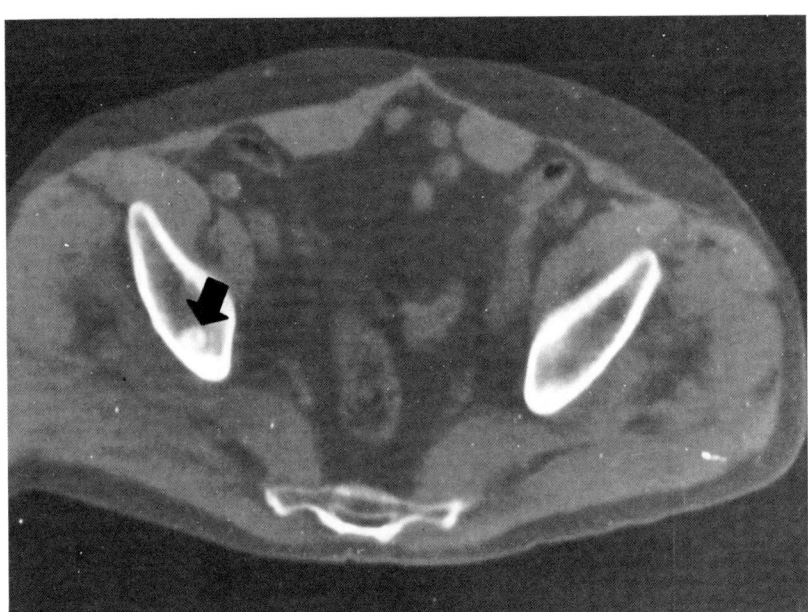

d

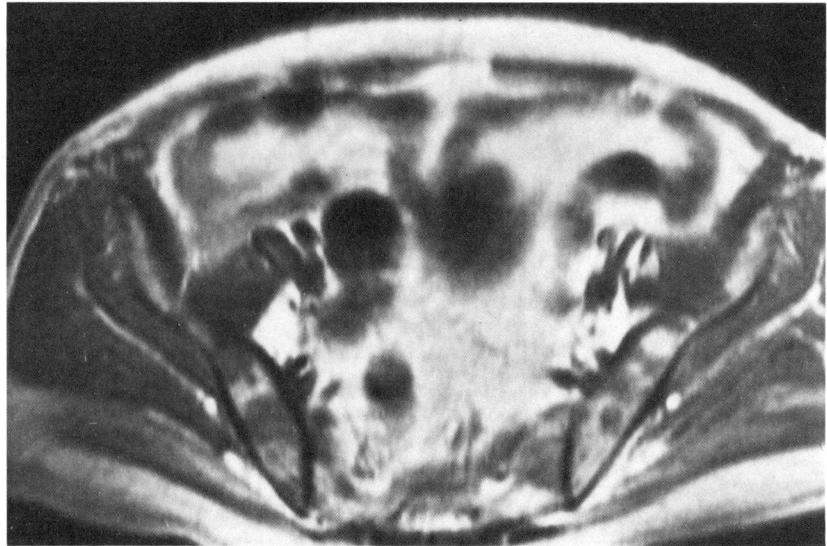

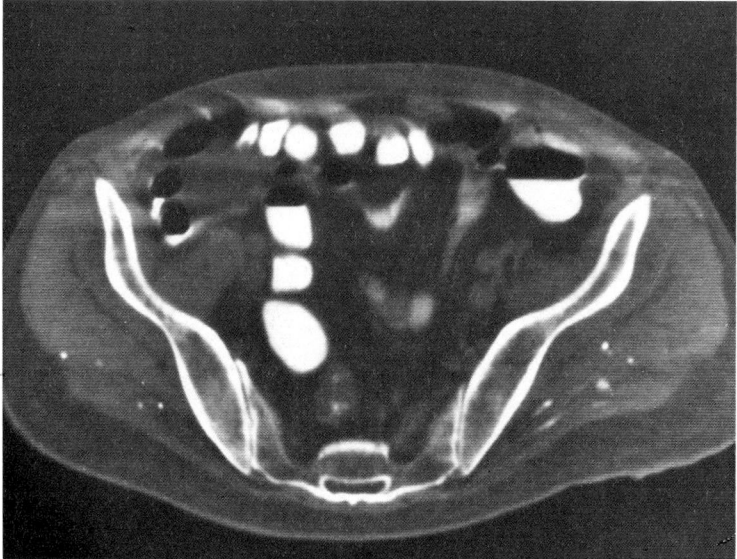

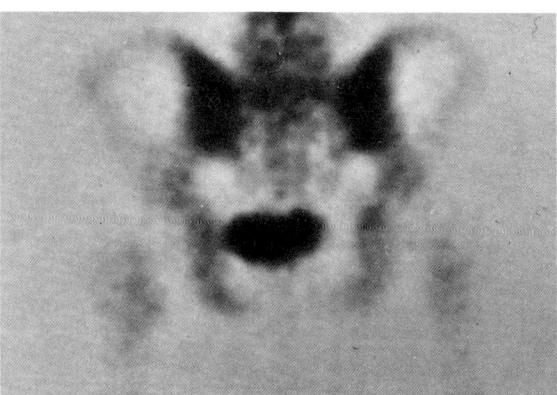

Figure 11.22

Metastasis. A 61-year-old man with small cell carcinoma of the lung, who complained of rectal spasm and urinary incontinence. (**a**) Axial T_1-weighted image (SE 500/28), performed to rule out sacral mass, unexpectedly reveals multiple low-signal-intensity metastases within marrow of iliac wings. (**b**) CT, also performed to rule out sacral mass, fails to show iliac lesions. (**c**) Radionuclide bone scan, performed 1 week before MRI is normal.

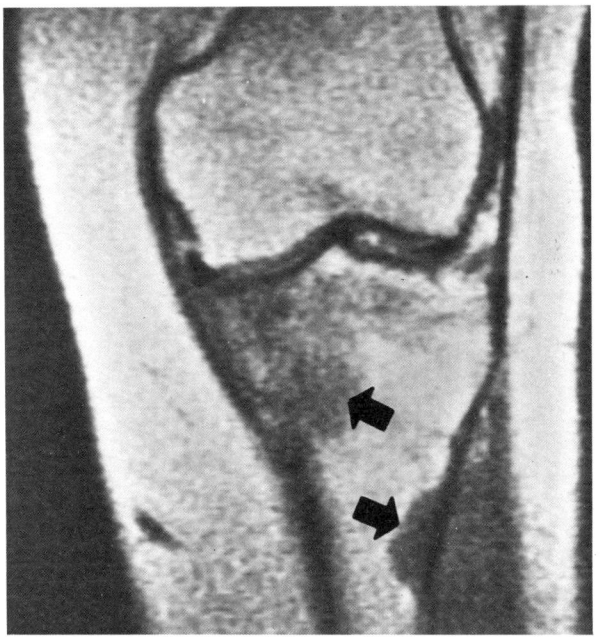
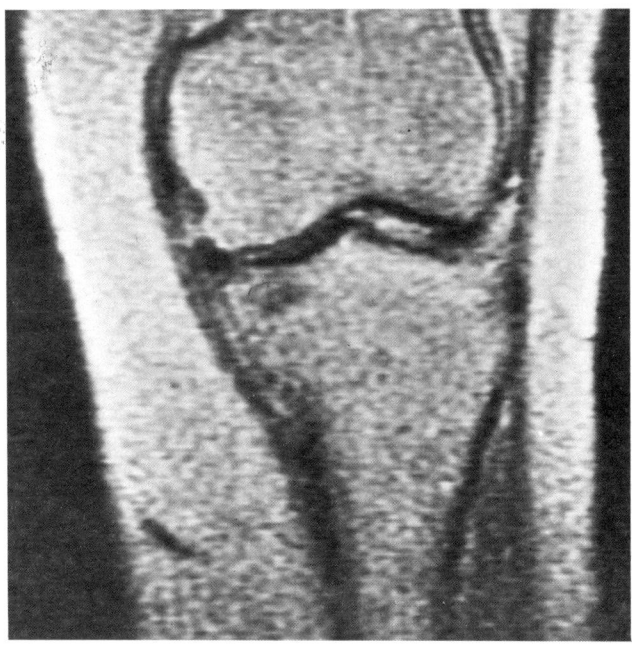

Figure 11.23

Metastasis. A 58-year-old woman with history of breast carcinoma and recent onset of right knee pain. (**a**) T_1-weighted coronal image (SE 500/28) shows 2 low-signal metastases (arrows) in proximal tibia. (**b**) In T_2-weighted coronal image (SE 2000/56), metastases have high-signal intensity and cannot be distinguished from normal bone marrow.

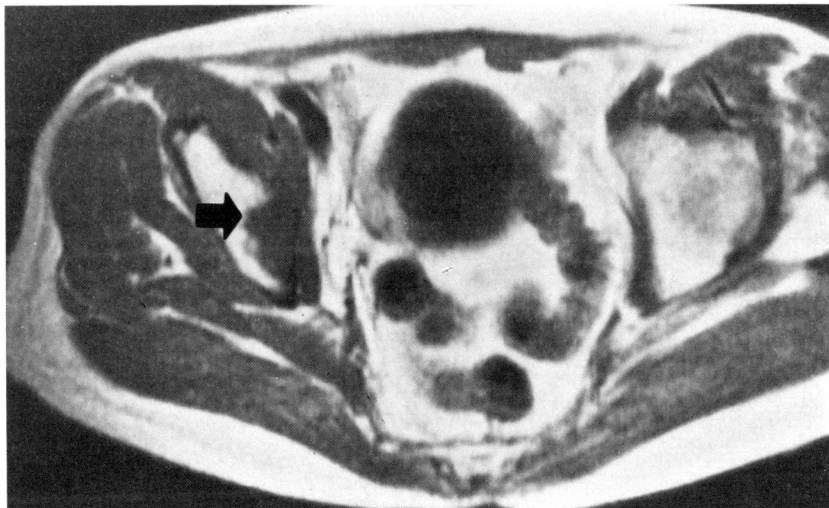
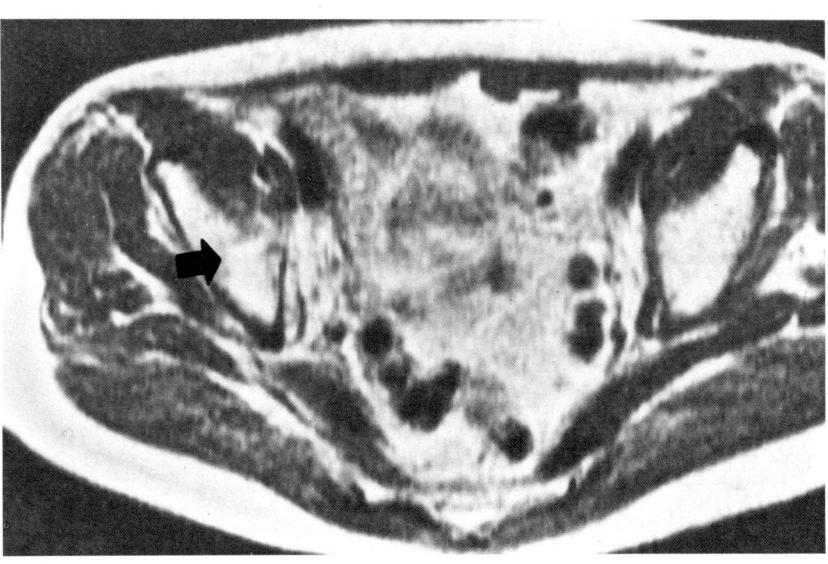

Figure 11.24

Metastasis. A 61-year-old man with prostatic carcinoma and right hip pain. (**a**) Axial T_1-weighted image (SE 300/18) shows low-signal-intensity metastasis in right ilium (arrow). (**b**) In axial T_2-weighted image (SE 2000/84), signal intensity of lesion (arrow) is higher than that of normal bone marrow.

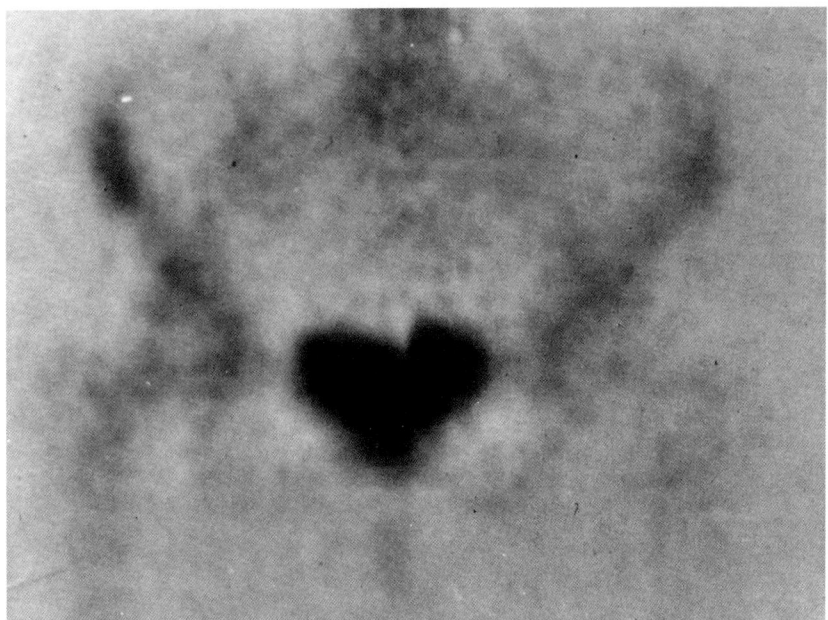

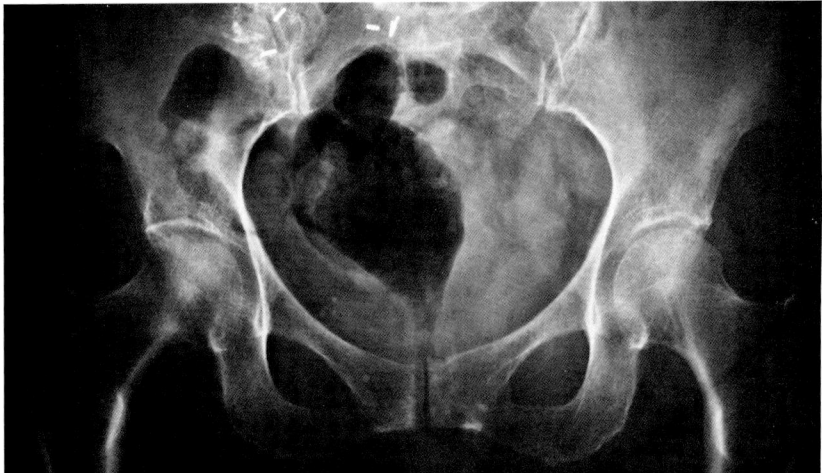

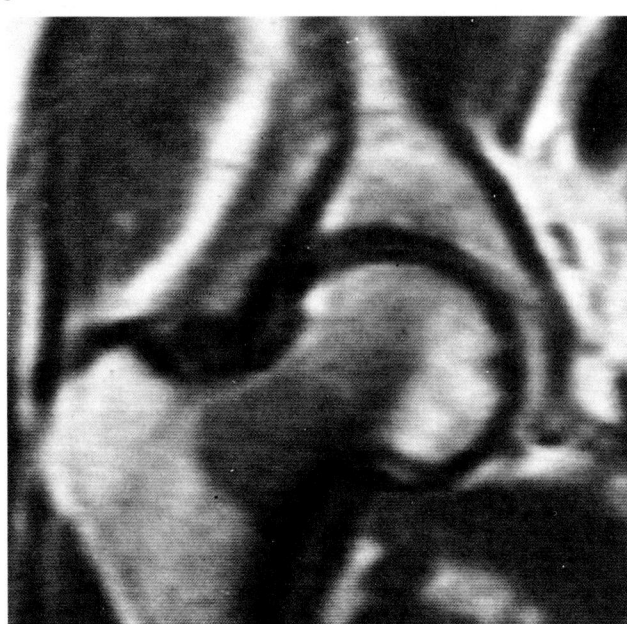

Figure 11.25

Metastasis. A 73-year-old woman with colon carcinoma and hip pain. (**a**) Radionuclide bone scan is normal. (**b**) Anteroposterior radiograph of pelvis interpreted as normal. (**c**) Coronal image (SE 500/28) shows low-signal-intensity metastasis in right femoral neck. (**d**) Radionuclide bone scan 3 months after MRI shows increased accumulation of isotope at site of lesion.

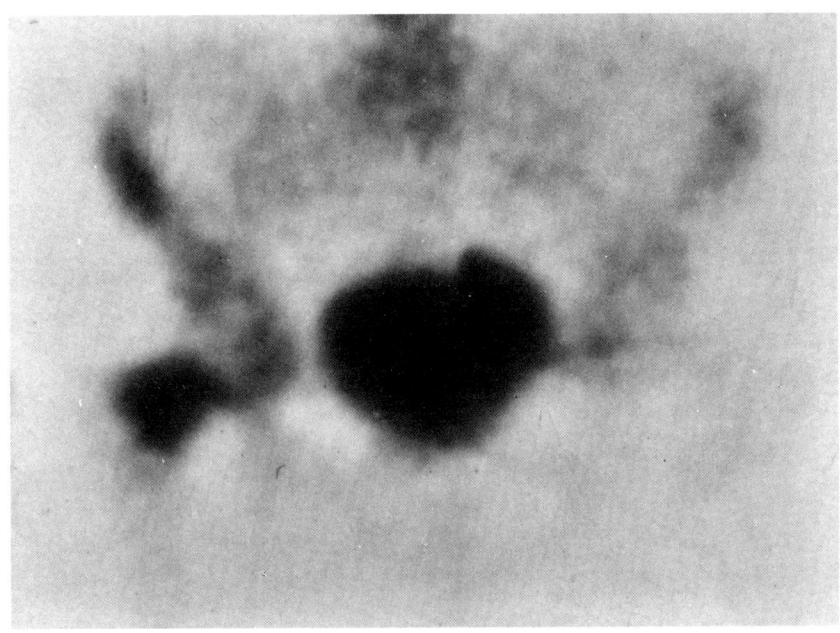
d

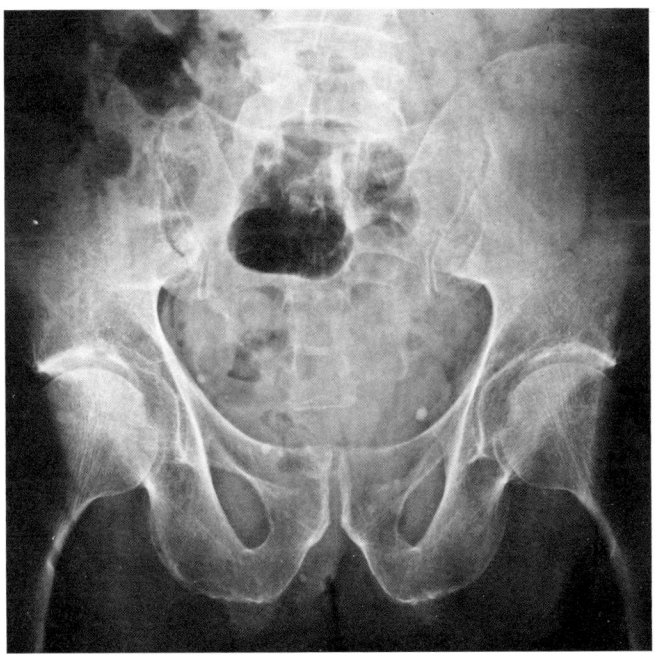
a

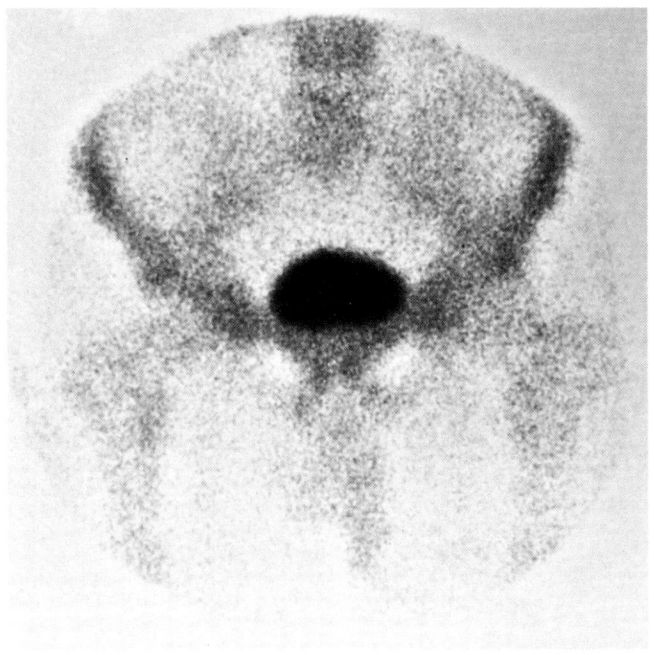
b

Figure 11.26

Metastasis. A 66-year-old man with adenocarcinoma of lung and pain in symphysis pubis. (a) Radiograph is normal. (b) Radionuclide bone scan is normal. T_1-weighted coronal (c) and axial (d) images (SE 200/18) reveal low-signal-intensity metastasis (★) in left pubis.

348 MRI atlas of the musculoskeletal system

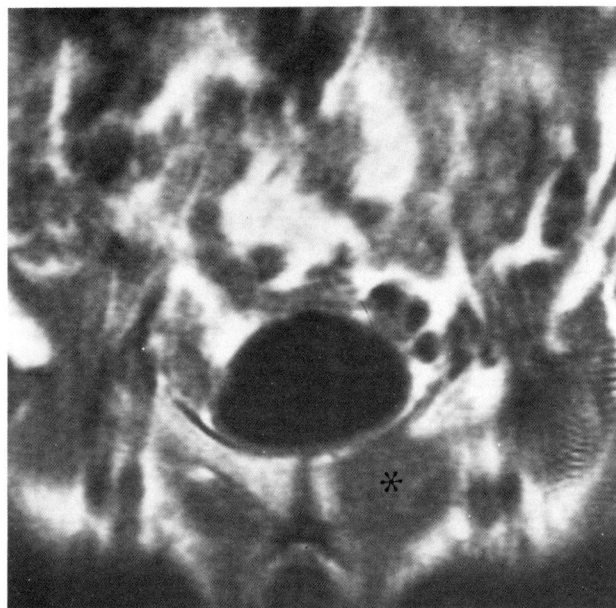

c

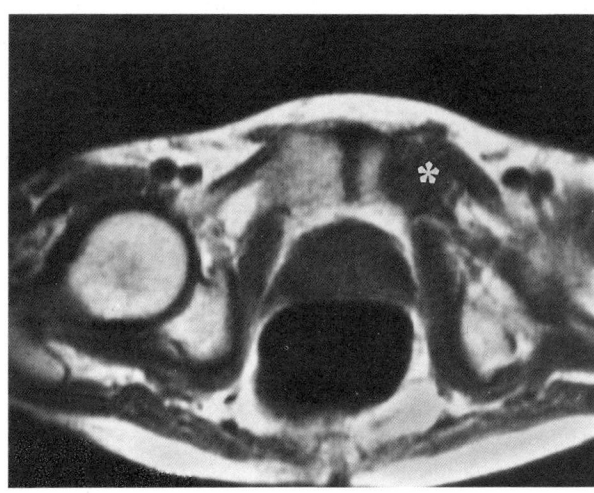

d

Figure 11.26 continued

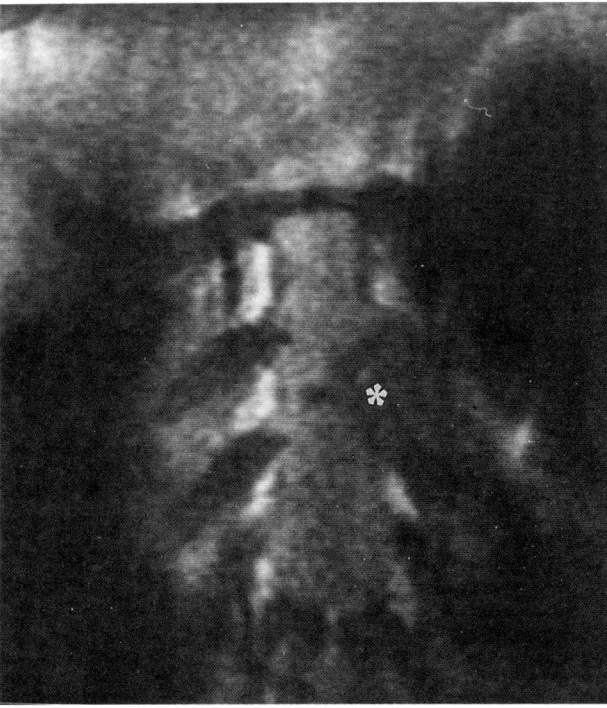

Figure 11.27

Metastasis. A 38-year-old woman with breast carcinoma and sternal mass. T_1-weighted coronal MRI (SE 500/28) shows destruction of left side of sternum by metastatic lesion (★).

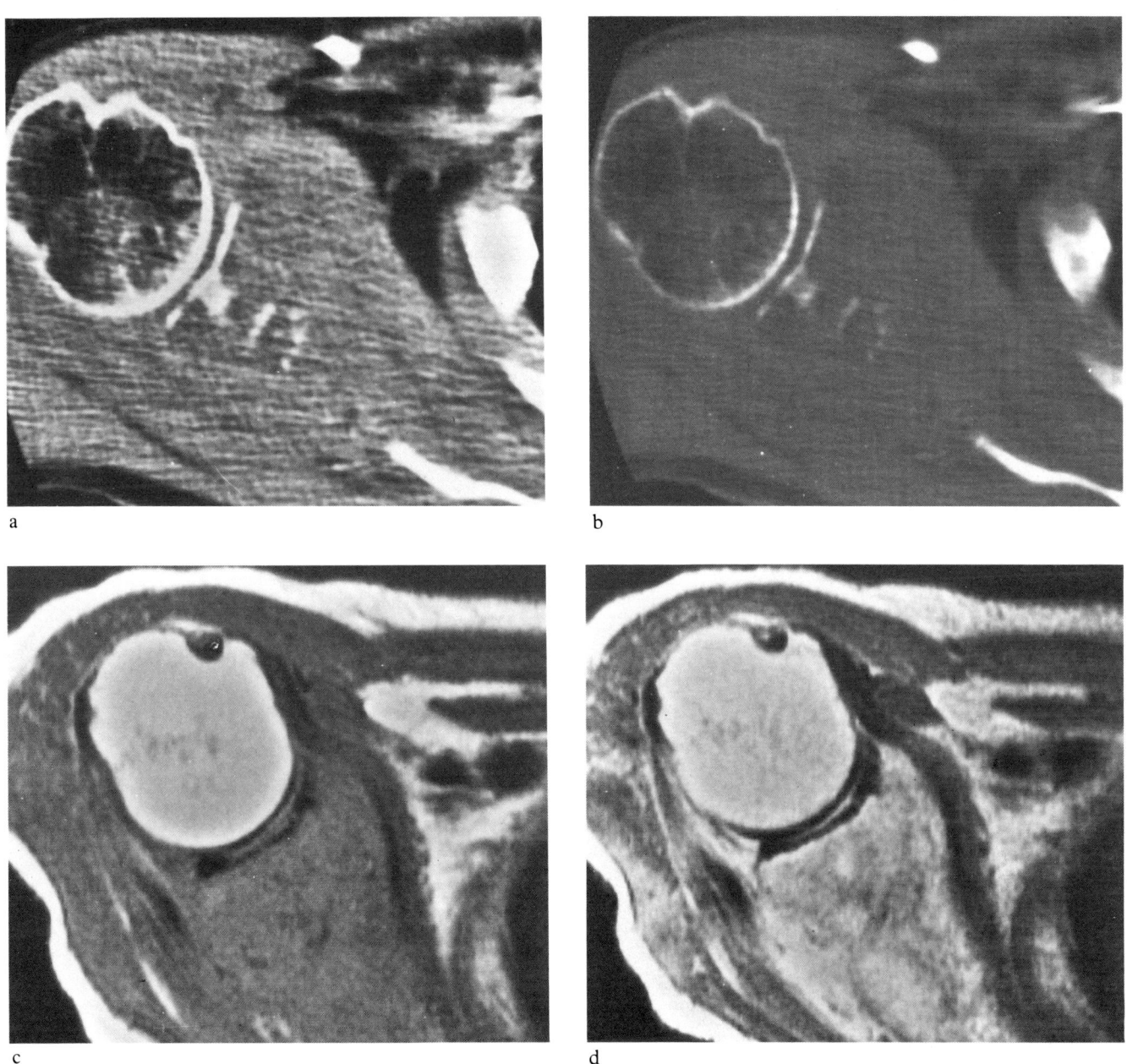

Figure 11.28

Metastasis. A 58-year-old man with laryngeal carcinoma and right shoulder pain. CT scans with soft-tissue (**a**) and bone (**b**) windows show extensive destruction of scapula. (**c**) Axial T_1-weighted image (SE 500/28) shows soft-tissue component of tumor better than does CT. (**d**) T_2-weighted image (SE 1500/56) more accurately demonstrates extent of soft-tissue mass. Regions of high-signal intensity in deltoid and infraspinatus muscles are attributable to edema.

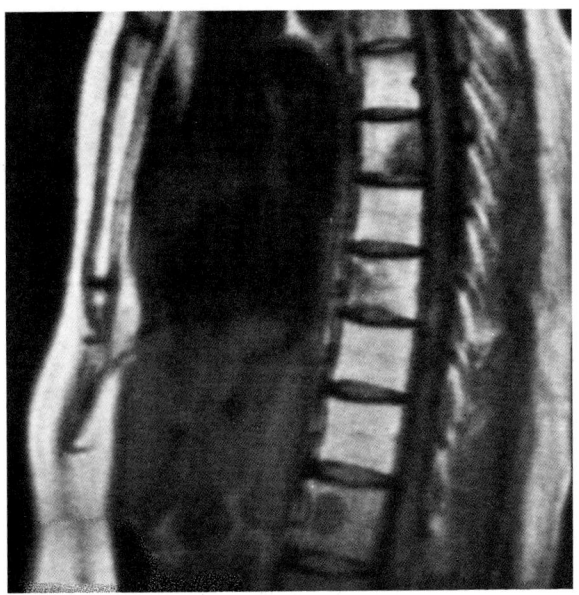

Figure 11.29

Metastasis. A 27-year-old man with history of melanoma and back pain. T_1-weighted sagittal image (SE 500/28) shows typical findings of metastasis: multiple low-signal-intensity lesions within normally high-signal-intensity bone marrow. Spinal cord is well visualized, and there is no evidence of tumor extending into spinal canal.

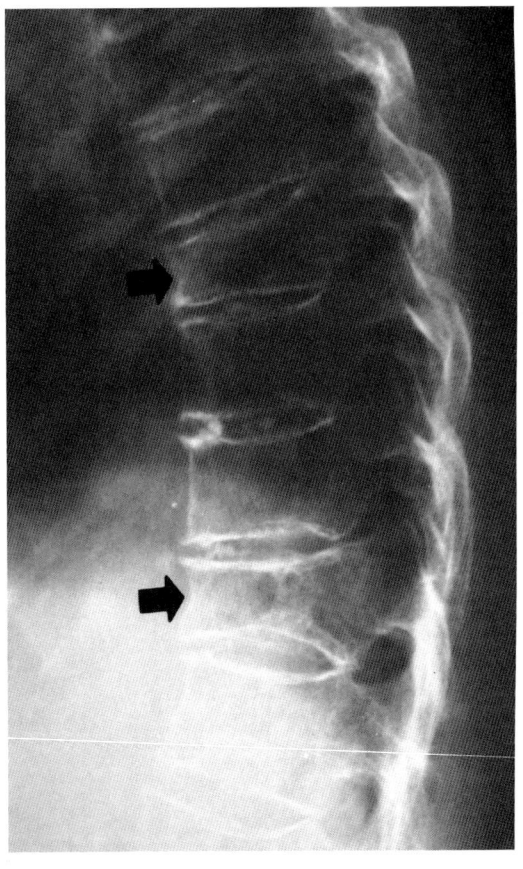

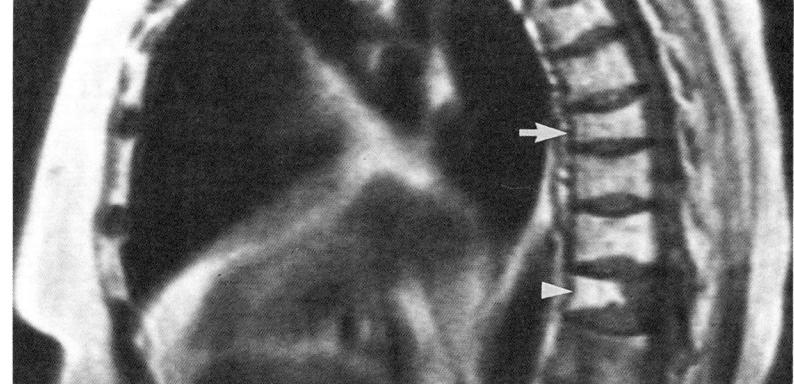

Figure 11.30

Metastasis. A 67-year-old woman with history of breast carcinoma and bilateral lower extremity weakness. (**a**) Lateral radiograph of thoracic spine shows diffuse osteopenia, and partial collapse of 2 vertebrae (arrows). (**b**) Sagittal T_1-weighted image (SE 500/28) discloses normal bone marrow signal within *upper* compressed vertebra (arrow) indicating that fracture was due to osteoporosis. However, posterior aspect of *lower* vertebra manifests replacement of bone marrow by metastatic tumor (arrowhead).

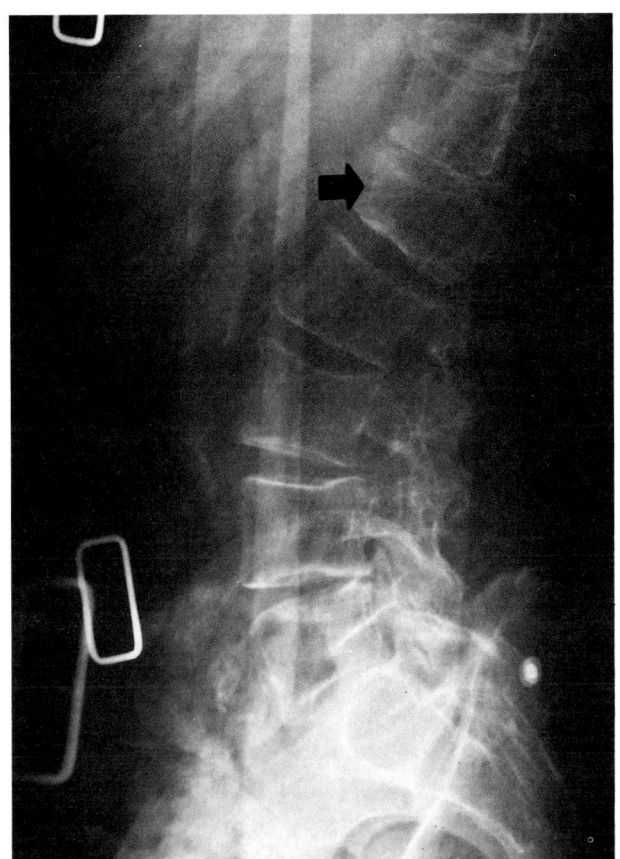
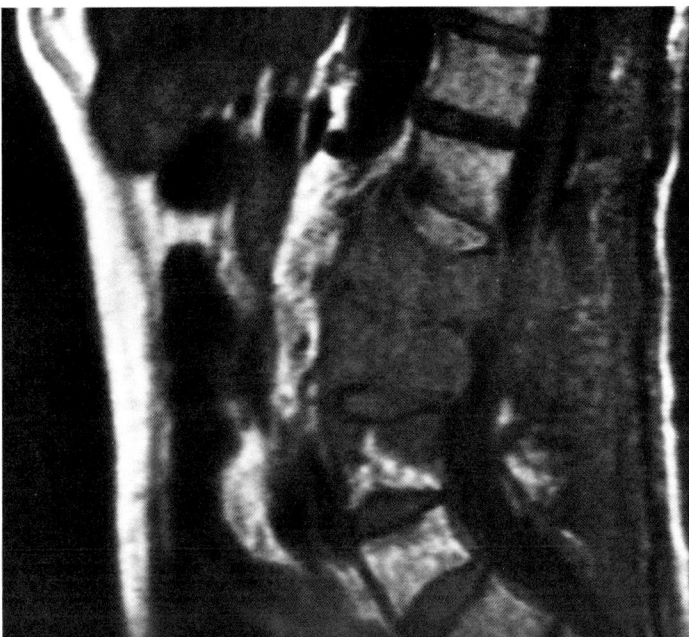

Figure 11.31

Metastasis. An 80-year-old man with lung carcinoma, increasing back pain, and lower extremity weakness. (**a**) Lateral radiograph reveals osteopenia and compression of L1 (arrow). (**b**) Sagittal image (SE 500/28) shows low-signal-intensity metastases replacing marrow of L2–L3 and superior aspect of L4. Tumor extends posteriorly into spinal canal. However, L1 vertebra manifests no evidence of metastasis; bone marrow signal intensity is normal except at inferior endplate where diminished signal is consistent with compression fracture. Inhomogeneous signal intensity of tissues posterior to canal is due to previous surgery.

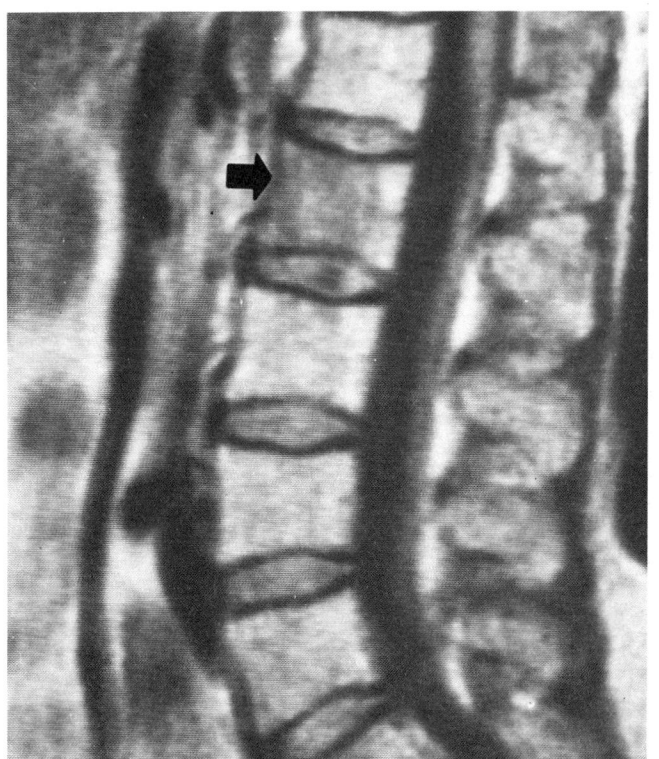 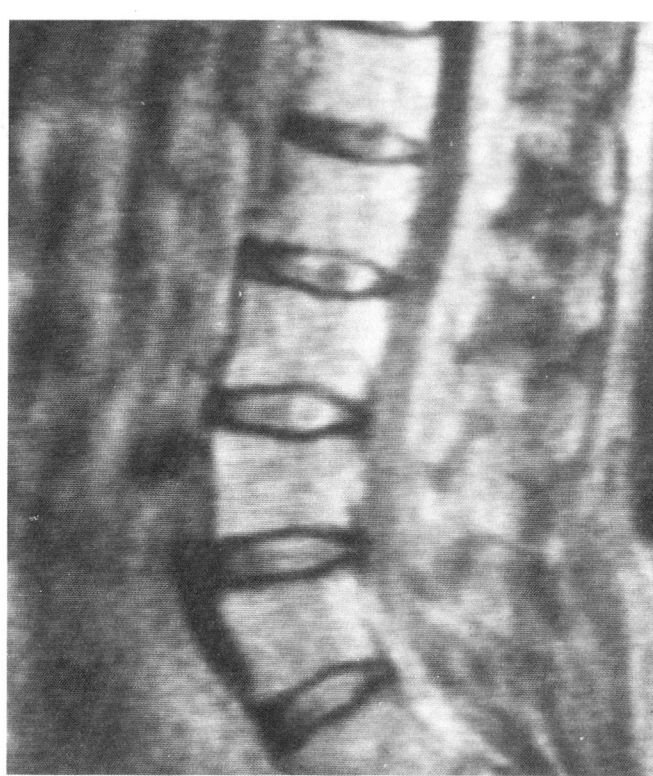

Figure 11.32

Multiple myeloma. A 63-year-old man with back pain. (a) T_1-weighted sagittal image (SE 300/28) shows low-signal-intensity region in L2 vertebral body (arrow) representing myeloma infiltrate. (b) In T_2-weighted sagittal image (SE 2000/56), myeloma infiltrate cannot be distinguished from normal bone marrow.

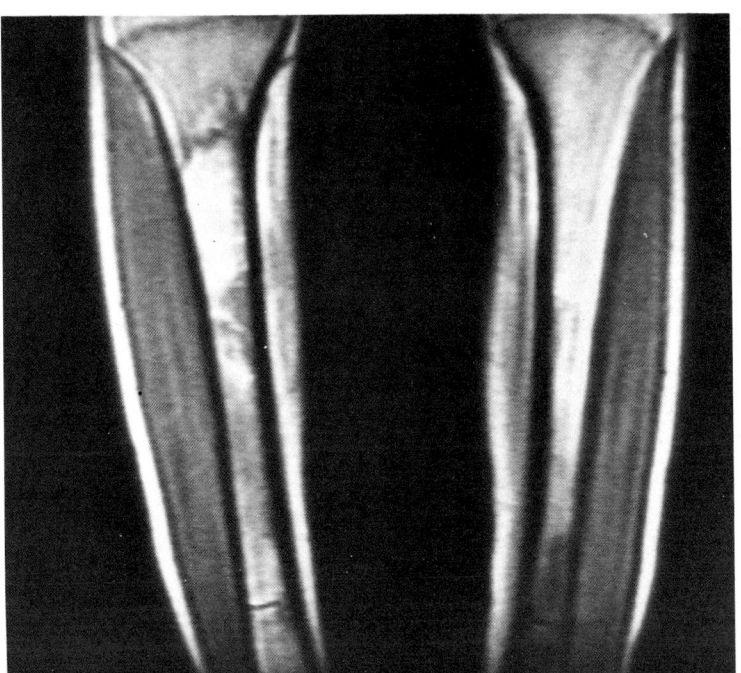

Figure 11.33

Acute lymphocytic leukemia. A 15-year-old boy with pain in both legs. Coronal T_2-weighted image (SE 500/28) reveals multiple regions of low-signal intensity representing intramedullary leukemic infiltrates.

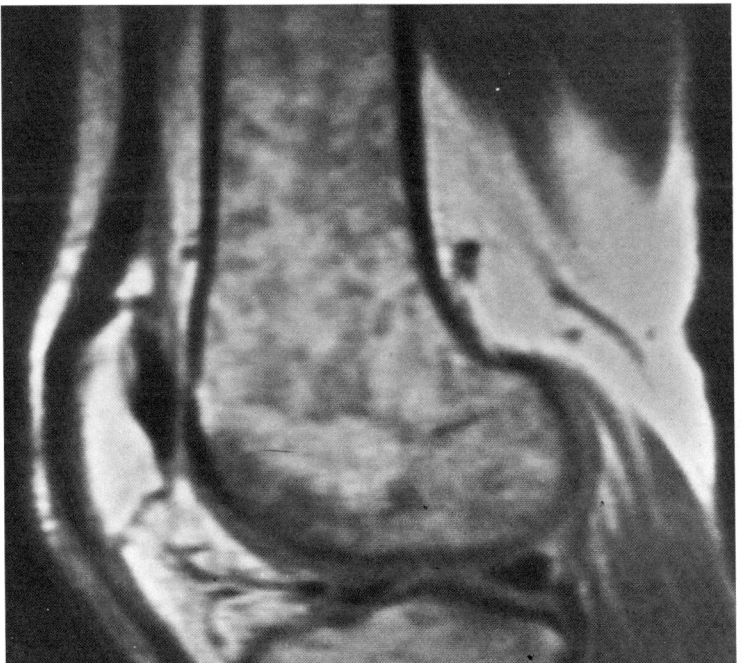

Figure 11.34

Hairy cell leukemia. A 44-year-old man with hairy cell leukemia treated by splenectomy, and referred for suspected meniscal tear of knee. Sagittal T_1-weighted image (SE 800/28) reveals diffusely mottled signal intensity of bone marrow. With only T_1-weighted imaging, the low-signal intensity could represent either leukemic infiltrates or myelofibrosis.

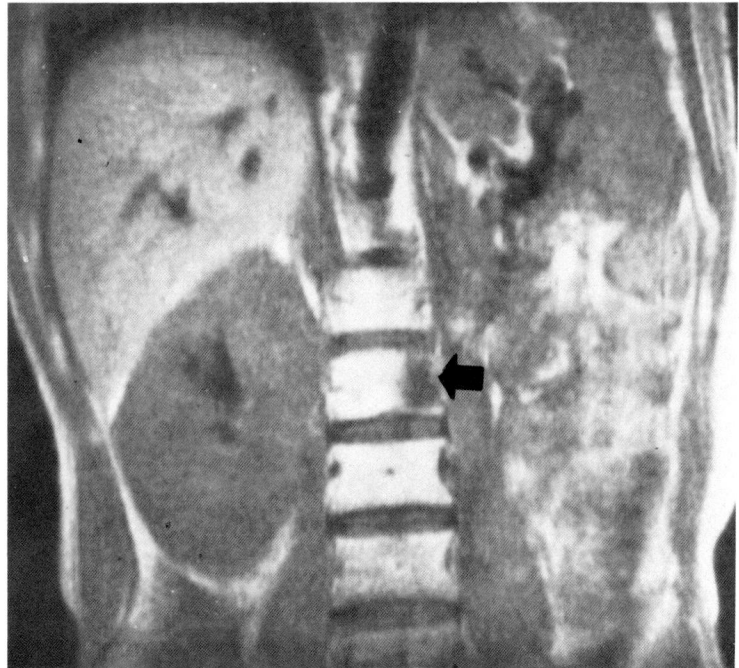

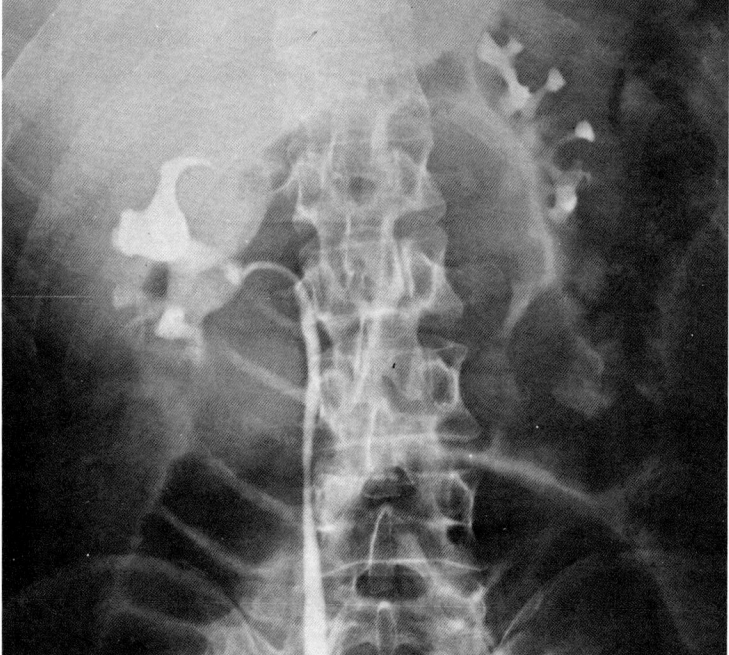

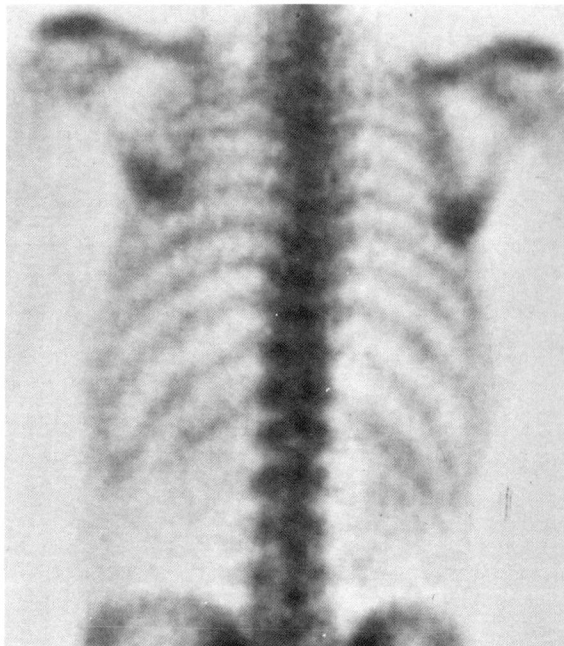

Figure 11.35

Lymphoma. A 38-year-old man with renal lymphoma. (**a**) Coronal image (SE 300/18) shows enlarged right kidney and unexpected involvement of L2 body (arrow). (**b**) Radiograph from retrograde urogram. L2 body appears normal. (**c**) Radionuclide bone scan is normal.

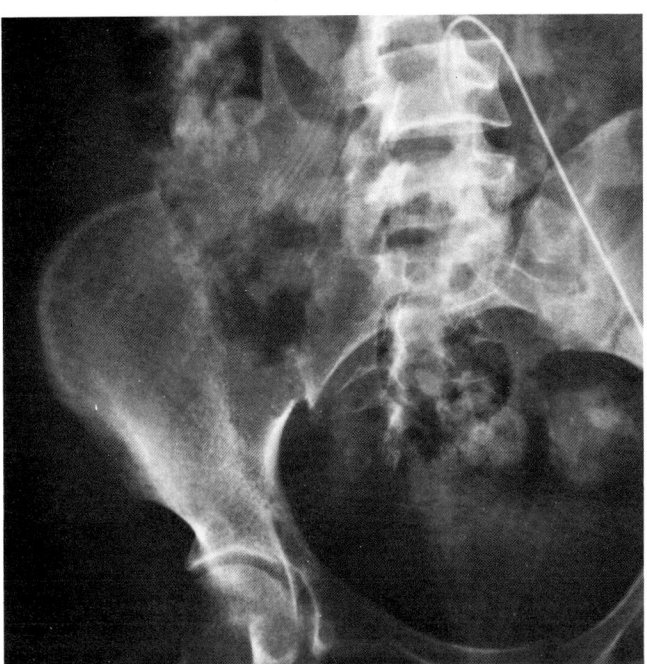

Figure 11.36

Soft-tissue leiomyosarcoma. A 23-year-old woman with biopsy-proven leiomyosarcoma of right buttock. (**a**) Scout radiograph of angiogram shows destruction of right sacrum and ilium. (**b–d**) Coronal images (SE 500/28) from posterior to anterior show extent of soft-tissue tumor and its relationship to adjacent bony structures and spinal canal. Tumor has destroyed portion of sacrum and ilium (**b**). Subcutaneous low-signal-intensity region (arrow) is due to previous biopsy. There is no evidence of involvement of spinal canal (**c**). An unexpected metastasis is identified in L2 vertebra (arrow) (**d**). Low-signal-intensity foci within tumor probably represent tumor necrosis. (**e**) Sagittal image (SE 500/28) reveals invasion of iliopsoas muscle (arrow). (**f**) Axial image (SE 500/28) discloses another unexpected metastasis in left ileum (arrow). (**g**) CT for correlation.

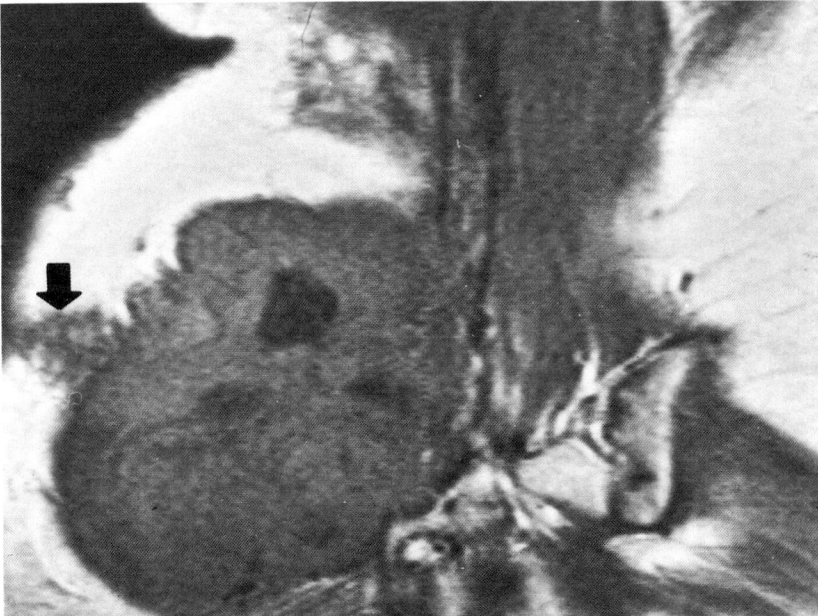

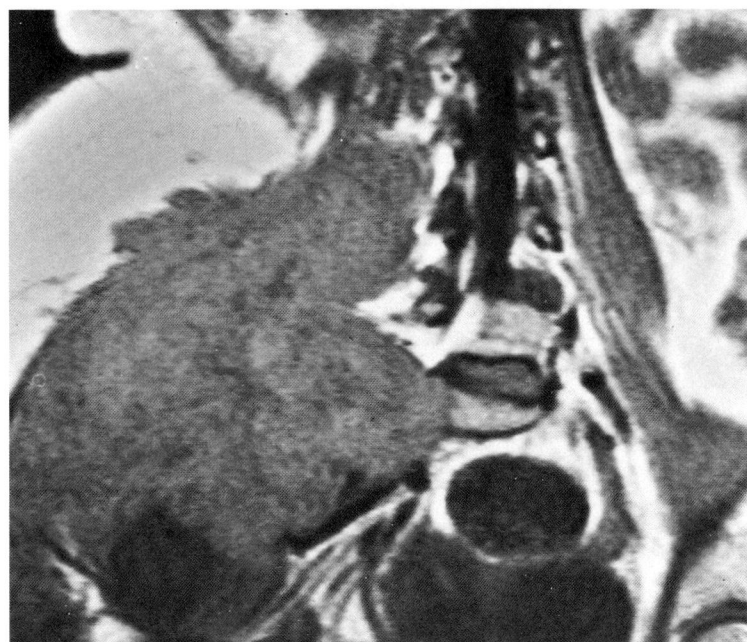

c

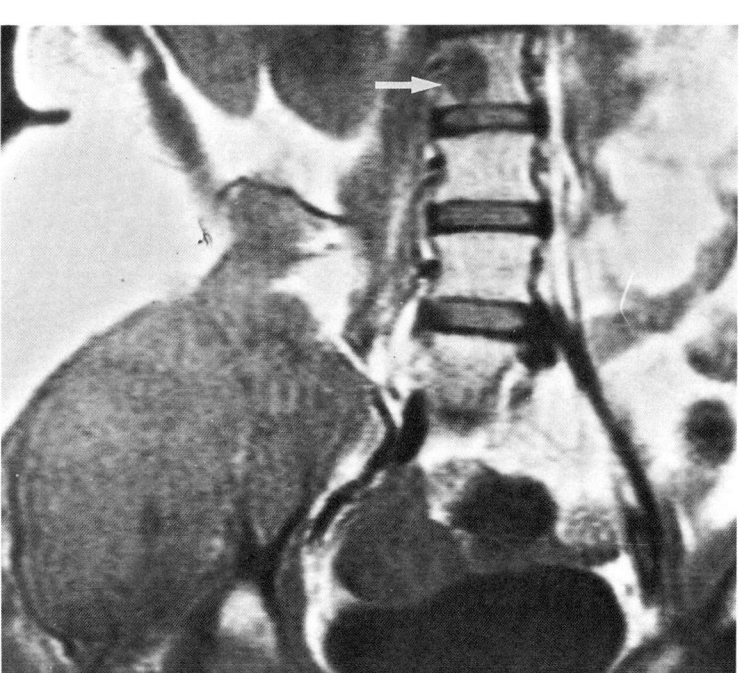

d

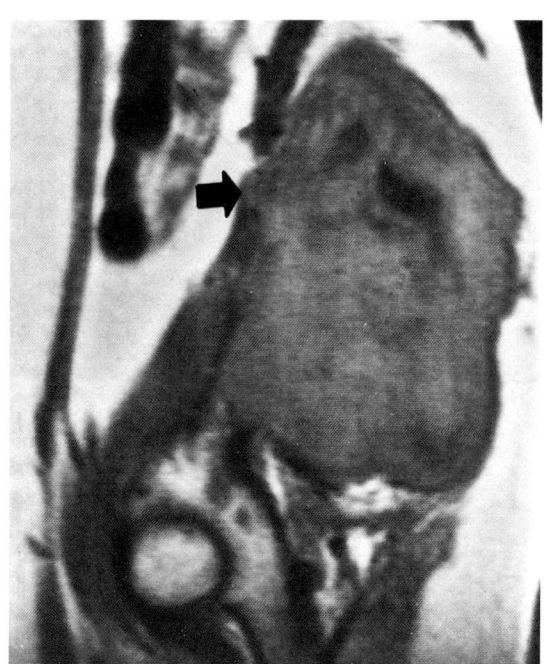

e

Figure 11.36 *continued*

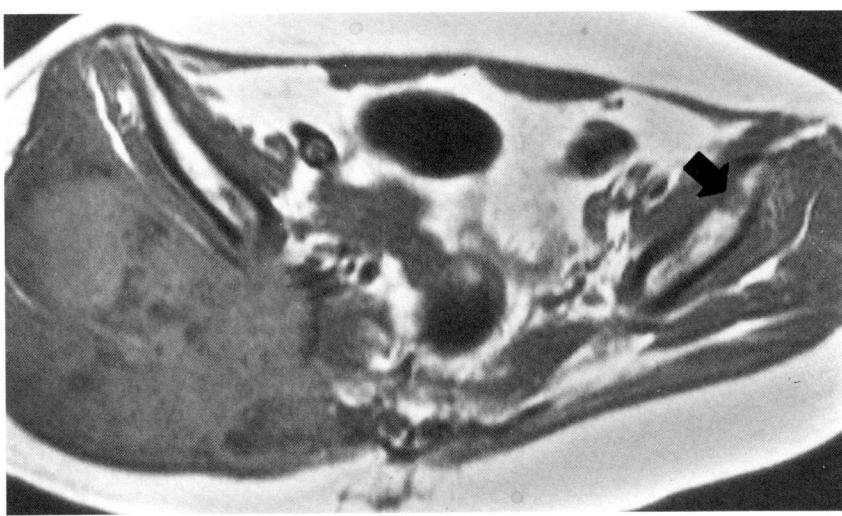

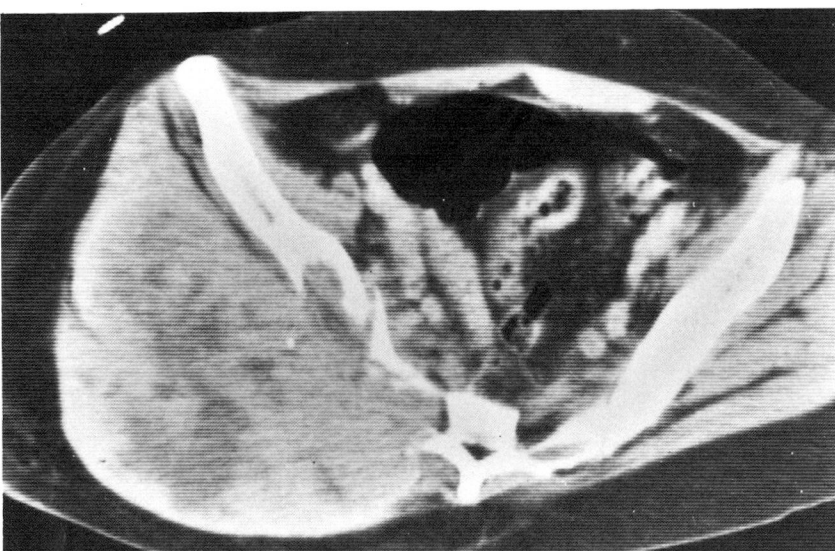

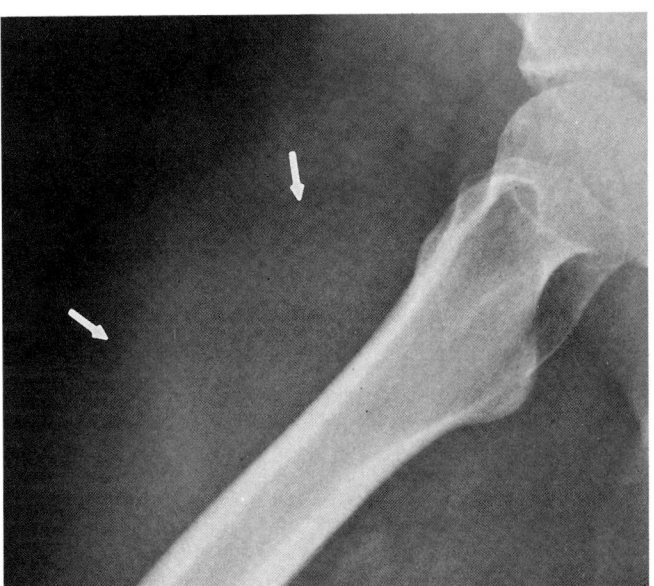

Figure 11.37

Liposarcoma. A 45-year-old man with large soft-tissue mass of right thigh. (**a**) Frog-leg lateral radiograph shows soft-tissue mass (arrows). (**b**) Contrast-enhanced CT scan discloses that tumor (★) within vastus intermedius muscle is well-defined and contains several areas of low density. (**c,d**) Axial MR images. Well-circumscribed soft-tissue mass (★) has signal intensity close to that of surrounding muscles on T_1-weighted image (SE 600/20), and has high-signal intensity on T_2-weighted image (SE 2000/85). Skin defect is from biopsy. (**e**) Sagittal T_2-weighted image (SE 2000/85).

358 MRI atlas of the musculoskeletal system

Figure 11.37 continued

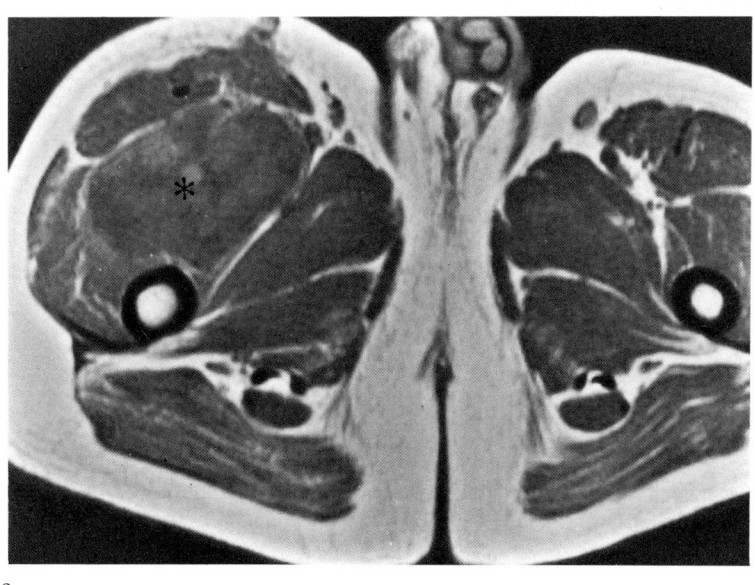

b

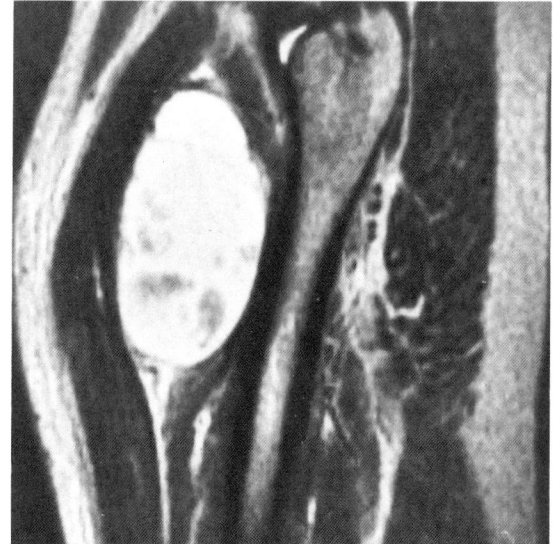

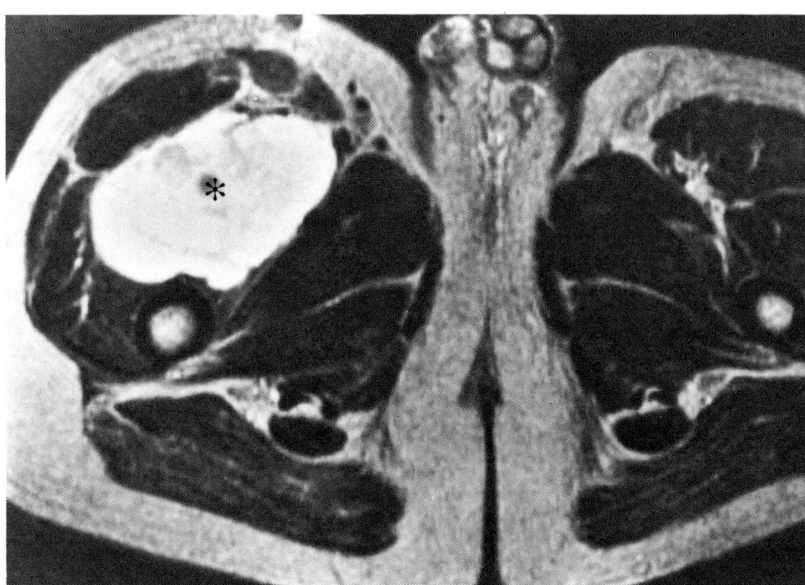

c

e

d

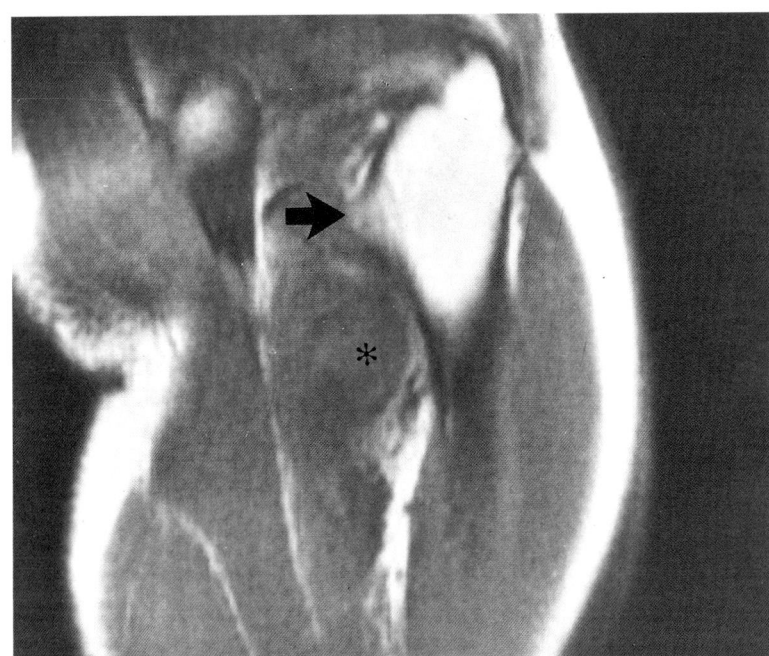

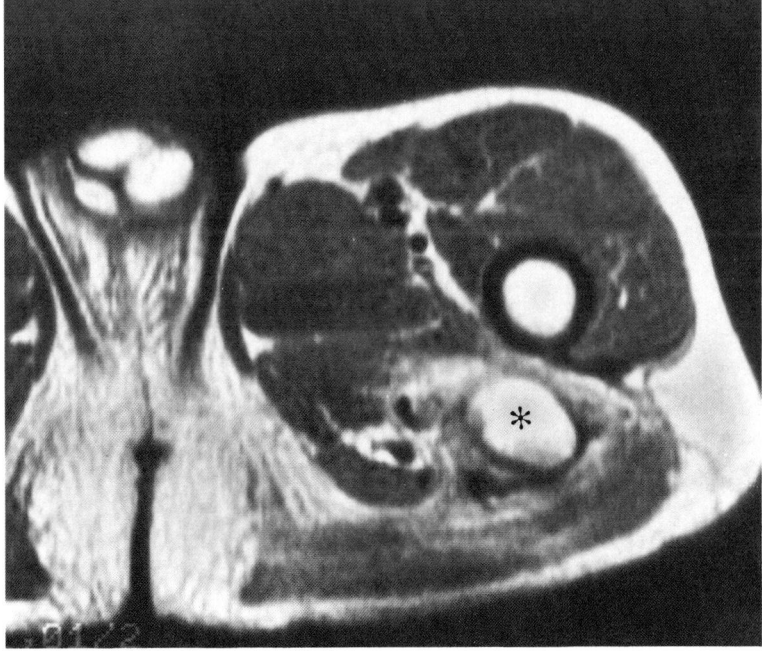

Figure 11.38

High-grade soft-tissue sarcoma. A 63-year-old man with mass in left thigh. (**a**) Coronal T_1-weighted image (SE 500/18) shows mass (★) medial to femur, below lesser trochanter (arrow). Adjacent cortex is intact. (**b**) Axial T_2-weighted image (SE 1500/56) discloses well-defined mass (★) with high signal intensity. At surgery, high-intensity signal of muscles surrounding tumor was found to represent edema rather than tumor invasion.

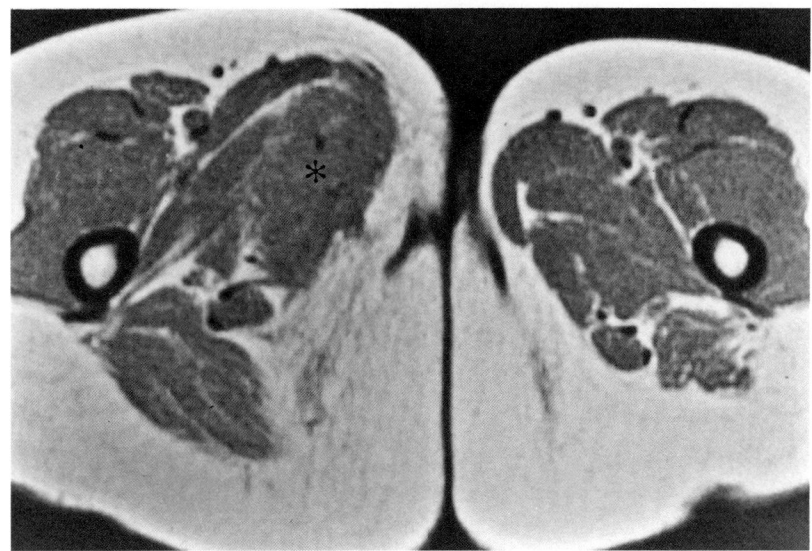

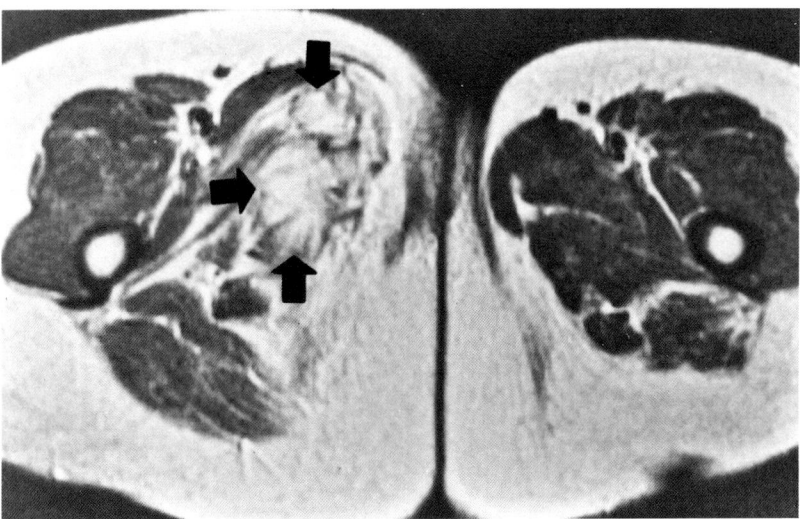

Figure 11.39
Recurrent soft-tissue desmoid tumor. A 55-year-old woman who had undergone resection of desmoid tumor of right thigh 6 years previously, and now complains of hard mass at site of surgery. (**a**) Axial T_1-weighted image (SE 600/20) shows mass in region of adductor magnus muscle (★). (**b**) Axial T_2-weighted image (SE 2000/85) more clearly separates borders of lobulated high-signal-intensity mass (arrows) from surrounding tissues. Adjacent muscles show abnormally high-signal intensity, representing edema.

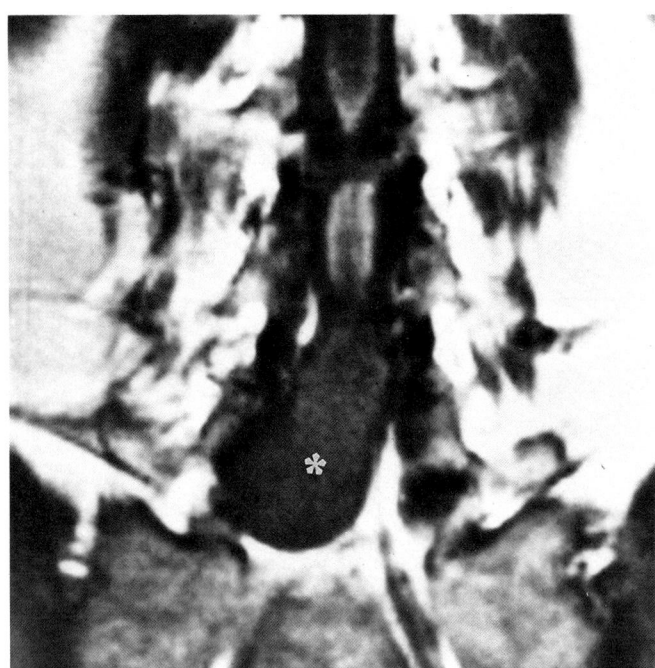

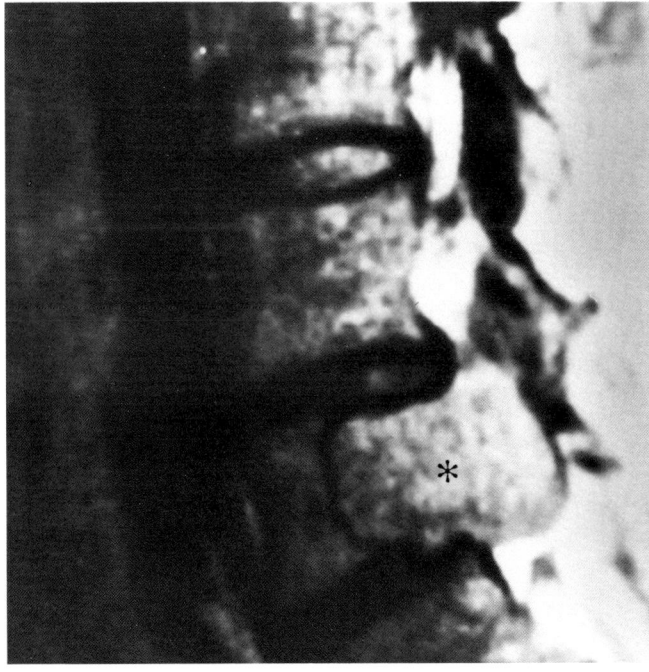

Figure 11.40

Spinal neurofibroma. A 41-year-old man with neurofibromatosis. (**a**) T_1-weighted coronal image (SE 300/30) at level of L5 shows well-defined tumor mass (★) extending into spinal canal and eroding surrounding bone. (**b**) In T_2-weighted sagittal image (SE 1500/35), tumor mass (★) has high-signal intensity and erodes adjacent L5 vertebral body. (*Courtesy of Paul E Berger, Long Beach, CA*)

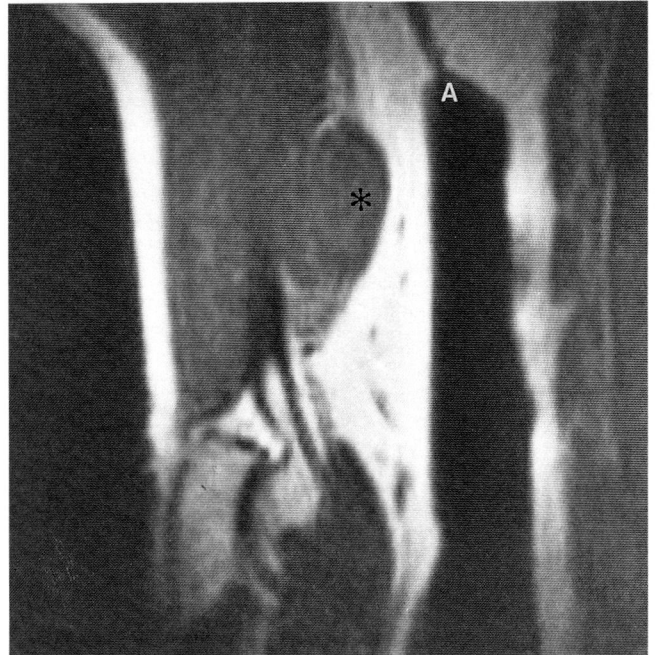
a

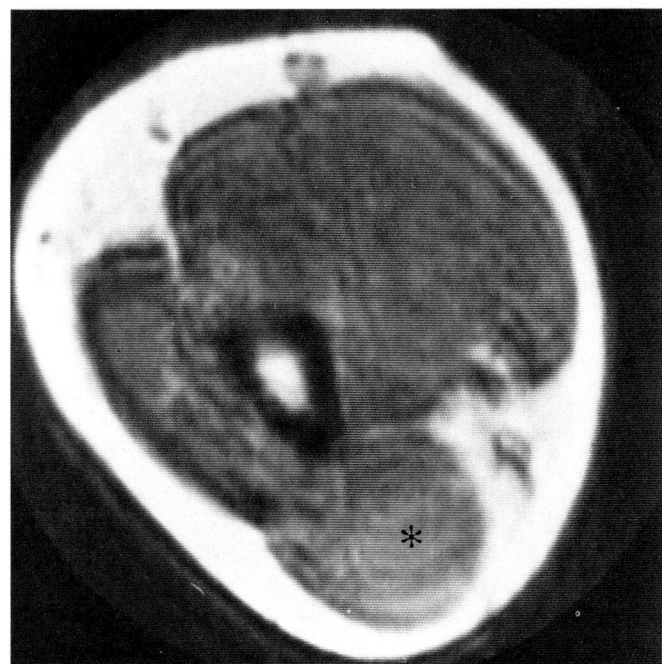
b

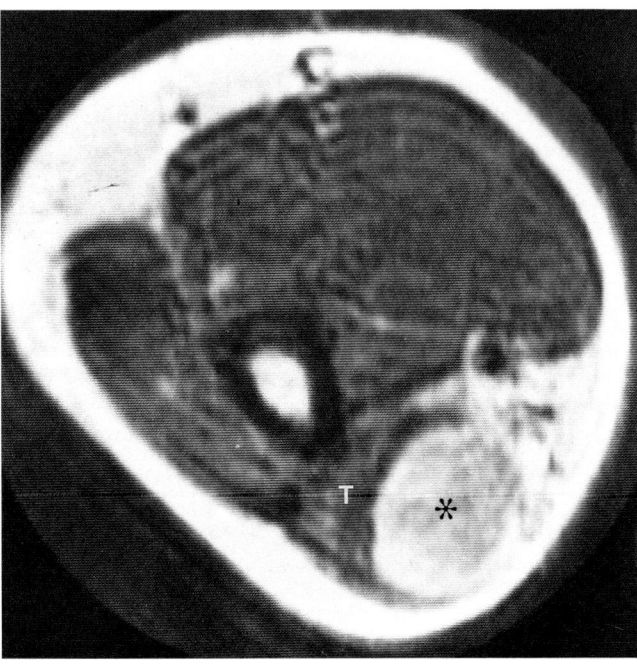
c

Figure 11.41

Ulnar nerve neurofibroma. A 64-year-old man with neurofibromatosis and mass above right elbow. (**a**) Coronal T_1-weighted image (SE 300/28) reveals intermediate-signal intensity mass (★) above elbow. A = axilla. (**b**) In axial T_1-weighted image (SE 600/28), mass is well circumscribed. (**c**) In axial T_2-weighted image (SE 1500/56), mass (★) has high-signal intensity and can be differentiated from the intermediate-signal intensity of triceps muscle (T) which is displaced but not invaded.

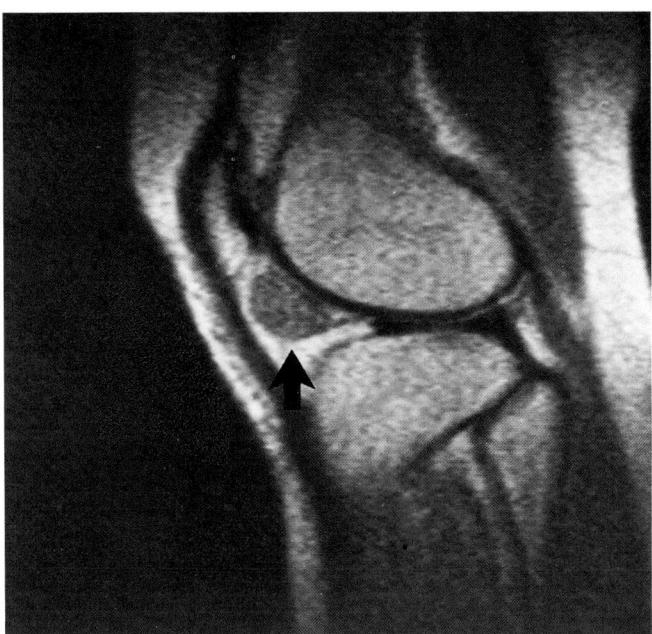

Figure 11.42

Pigmented villonodular synovitis. A 21-year-old female gymnast who complained of knee pain. MRI was performed to rule out torn meniscus. Sagittal image (SE 1000/28) shows intermediate-signal-intensity mass (arrow) displacing infrapatellar fat. Surgery revealed localized pigmented villonodular synovitis.

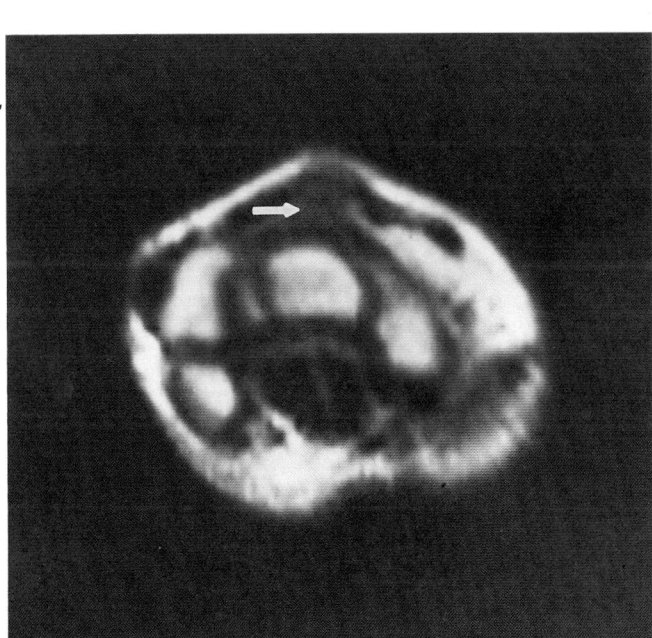
a

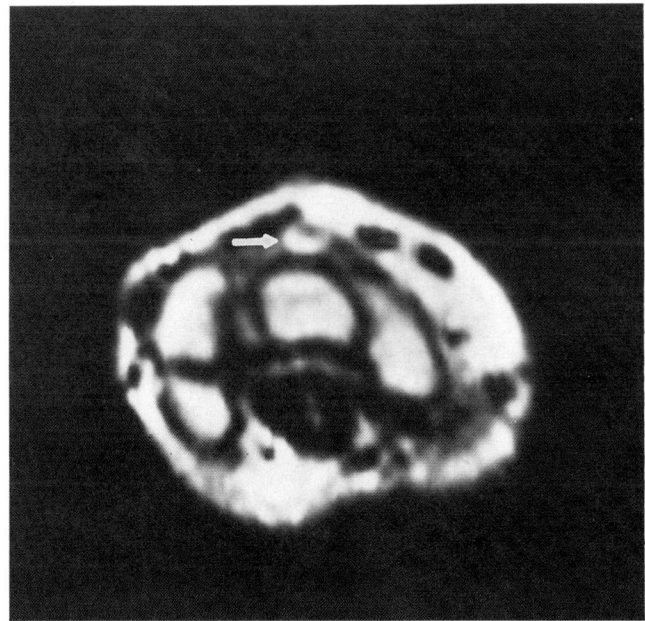
b

Figure 11.43

Ganglion cyst of wrist. A 28-year-old woman with painful lump on dorsal surface of wrist. (**a**) Axial T_1-weighted image (SE 600/28) shows homogeneous intermediate-signal-intensity mass (arrow). (**b**) In axial T_2-weighted image (SE 2000/56), cyst (arrow) has high-signal intensity throughout.

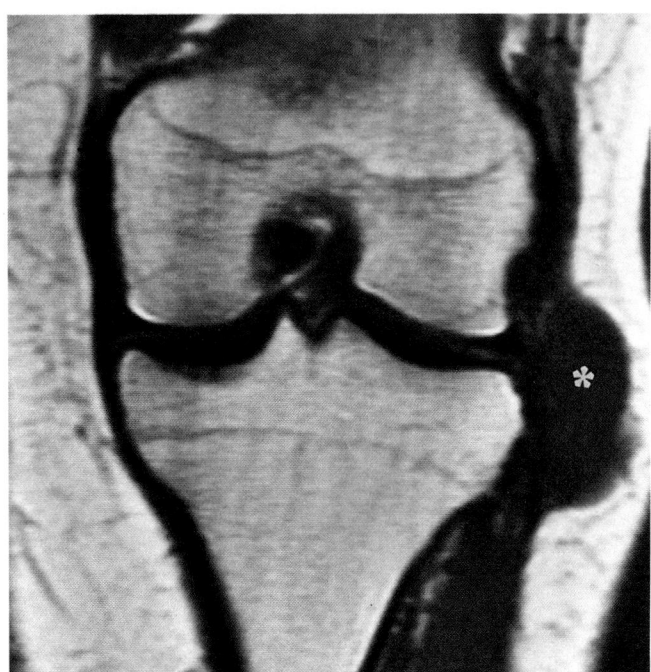

Figure 11.44

Ganglion cyst of knee. A 44-year-old man with mass on lateral aspect of knee. (**a**) Coronal image (SE 800/20) discloses homogeneous low-signal-intensity mass (★). Menisci were normal. (**b,c**) Axial double-echo images (SE 1500/20–90) at level of tibial plateau. Mass (★) has intermediate-signal intensity on first echo (**b**), and high-signal intensity on second echo (**c**). Note septation (arrow). (**d,e**) Axial double-echo images (SE 1500/20–90) 2 cm below (**b**). Second echo image (**e**) shows high-signal-intensity extension (pseudopod) of cyst between tibia and fibula (arrow). Location, septation and pseudopod are characteristic of tibiofibular ganglion cyst of knee. (*Courtesy of Richard Ofstein, Long Beach, CA*)

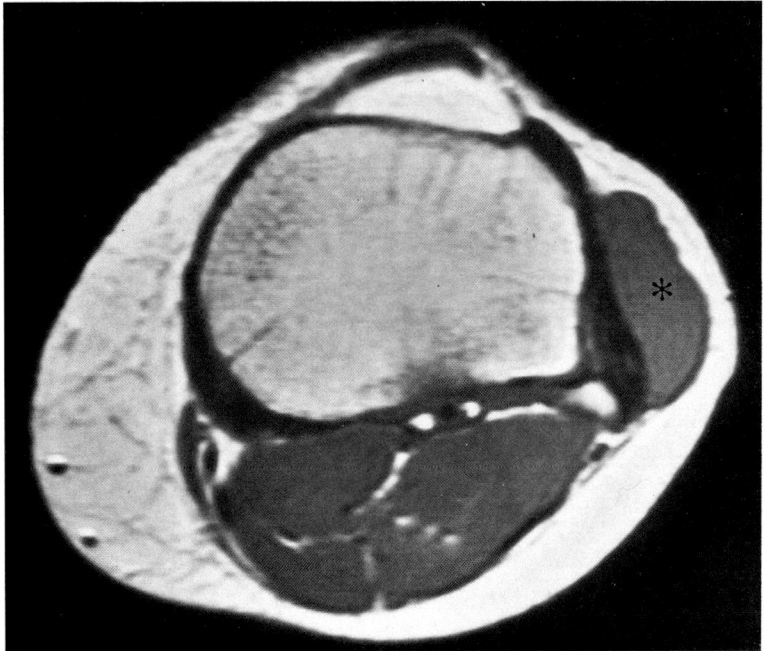

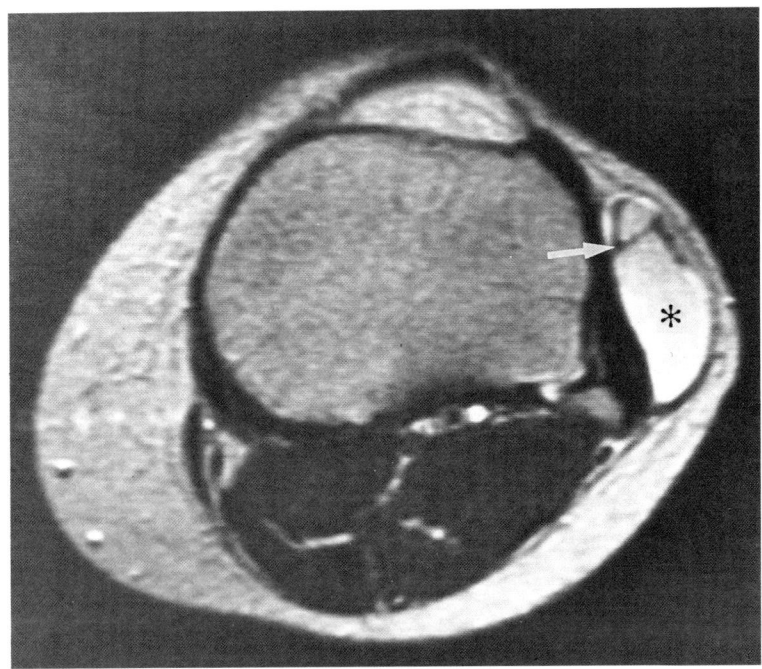

c

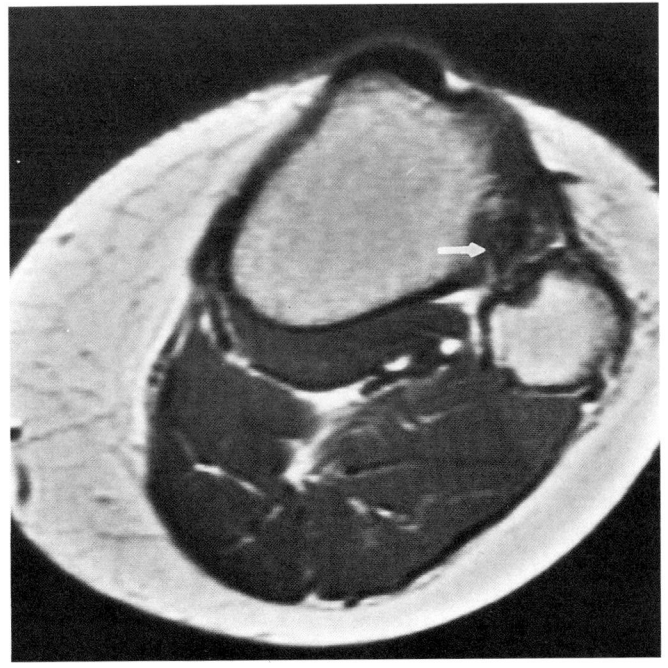

d

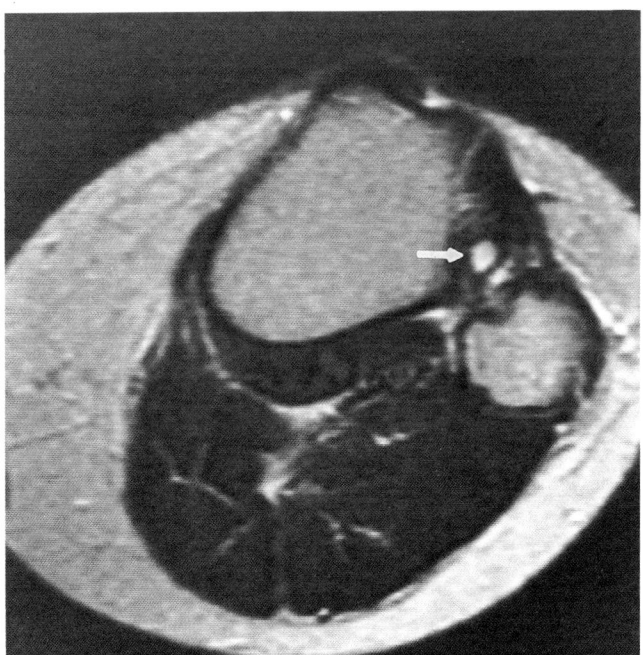

e

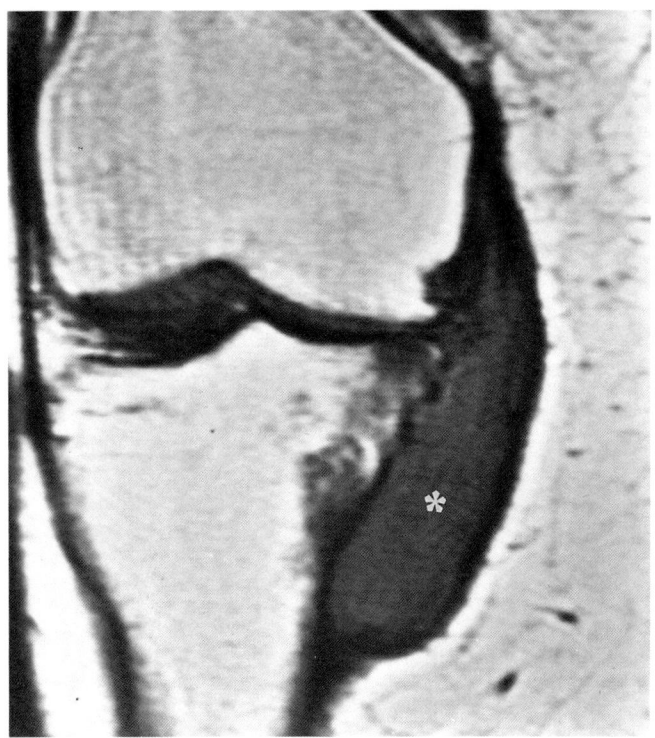

a

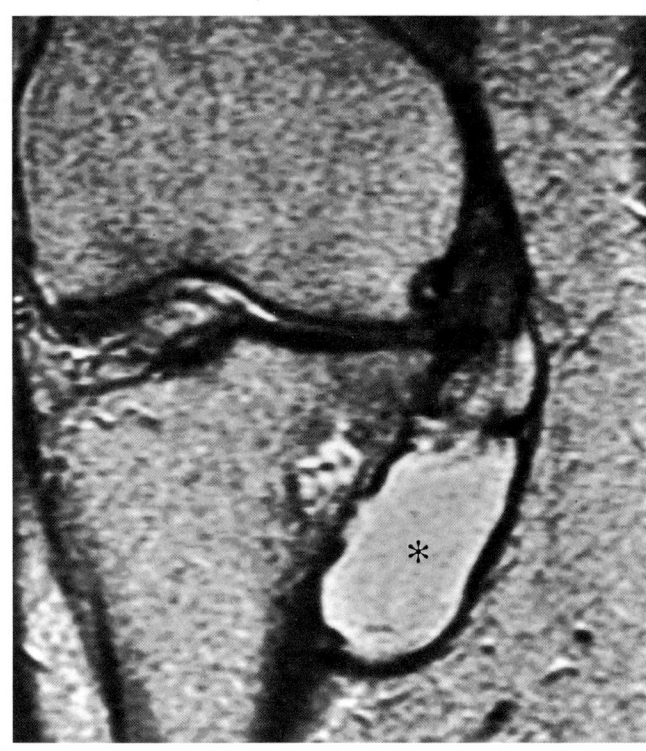

b

Figure 11.45

Meniscal cyst. A 75-year-old woman with knee pain. (**a**) Coronal T_1-weighted image (SE 800/20) discloses cyst (★), of intermediate-signal intensity, eroding tibia. (**b**) In coronal T_2-weighted image (SE 1500/90), cyst (★) has high-signal intensity. Communication with joint is not well demonstrated in this image. (*Courtesy of Richard Ofstein, Long Beach, CA*)

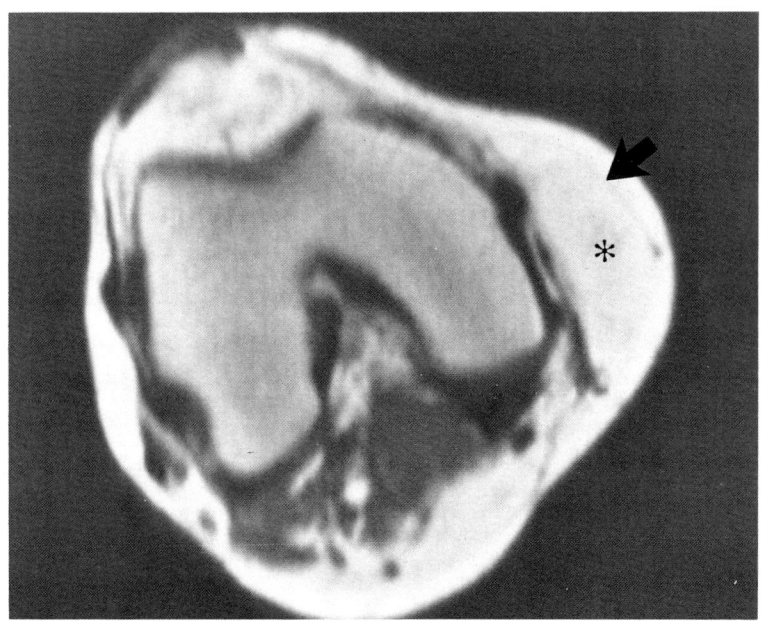

Figure 11.46

Subcutaneous lipoma. A 53-year-old woman with soft mass on medial aspect of knee. Axial T_1-weighted image (SE 500/28) shows lipoma (★) with signal intensity equal to that of surrounding subcutaneous fat. Note thin, intermediate-signal-intensity capsule (arrow).

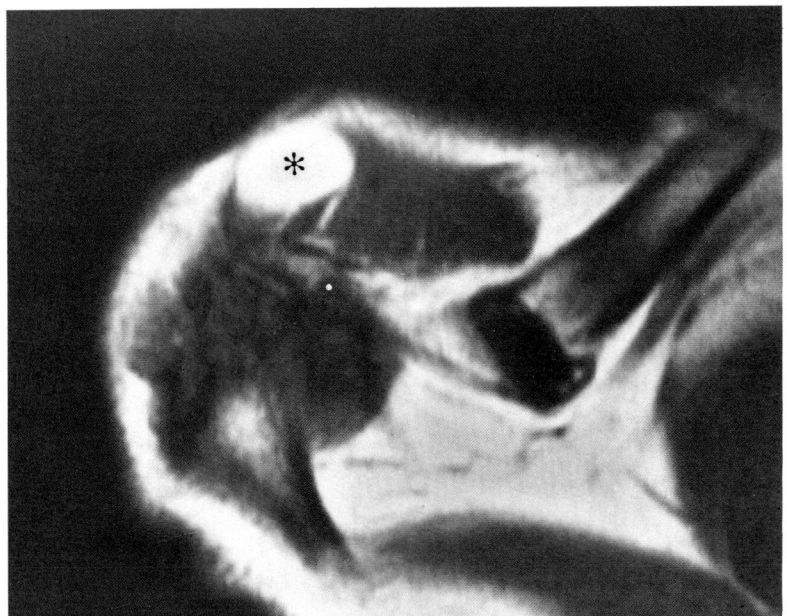

Figure 11.47

Intramuscular lipoma. A 58-year-old man who was referred for evaluation of shoulder pain. Axial T_1-weighted image (SE 500/28). Lipoma (★) was incidentally noted within deltoid muscle.

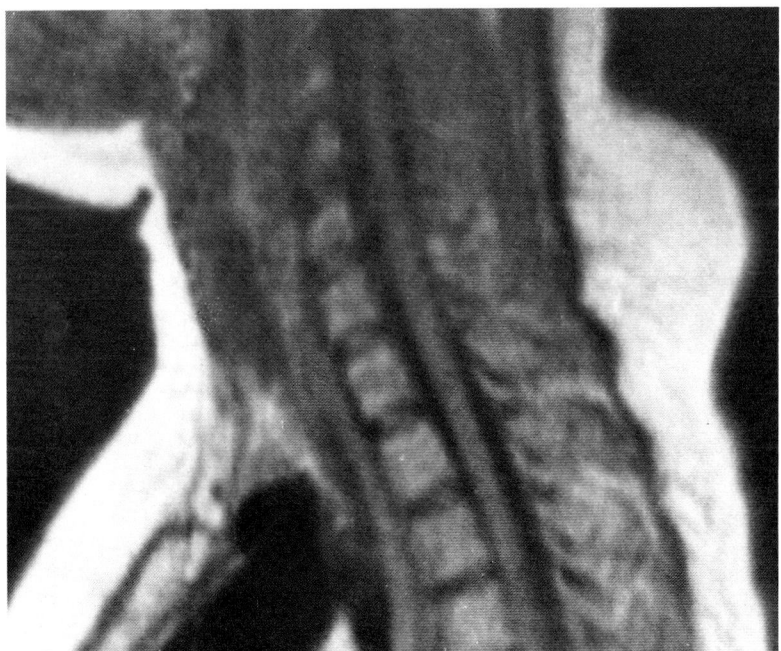

Figure 11.48

'Buffalo hump.' A 20-year-old woman who complained of fullness in posterior soft-tissue of neck. Sagittal T_1-weighted image (SE 500/28) reveals thickened tissue with signal intensity identical to that of surrounding fat.

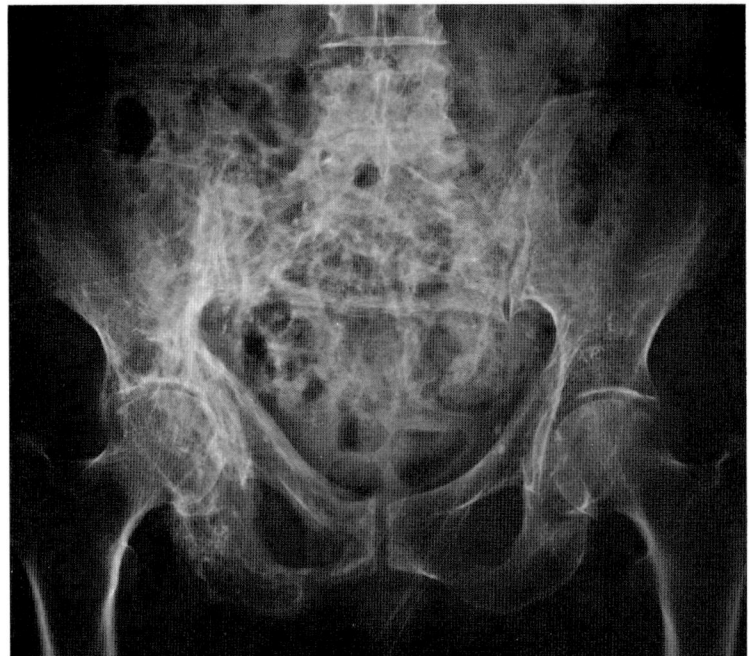
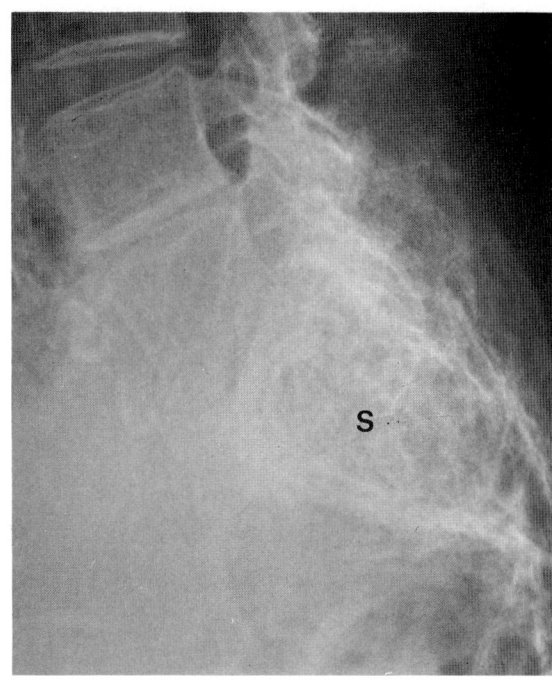

Figure 11.49

Paget's disease. An 88-year-old woman with history of breast carcinoma and new onset of bone pain. (**a**) Anteroposterior radiograph of pelvis and (**b**) lateral radiograph of sacrum (S) reveal extensive pagetic changes. (**c**) Axial CT scan at level of S1 shows osteolysis throughout sacrum and right ilium. (**d**) Axial T_1-weighted image (SE 500/28) discloses inhomogeneous high-signal intensity throughout expanded sacrum and right ilium. Absence of low-signal intensity at sites of osteolysis in CT scan excludes metastatic disease.

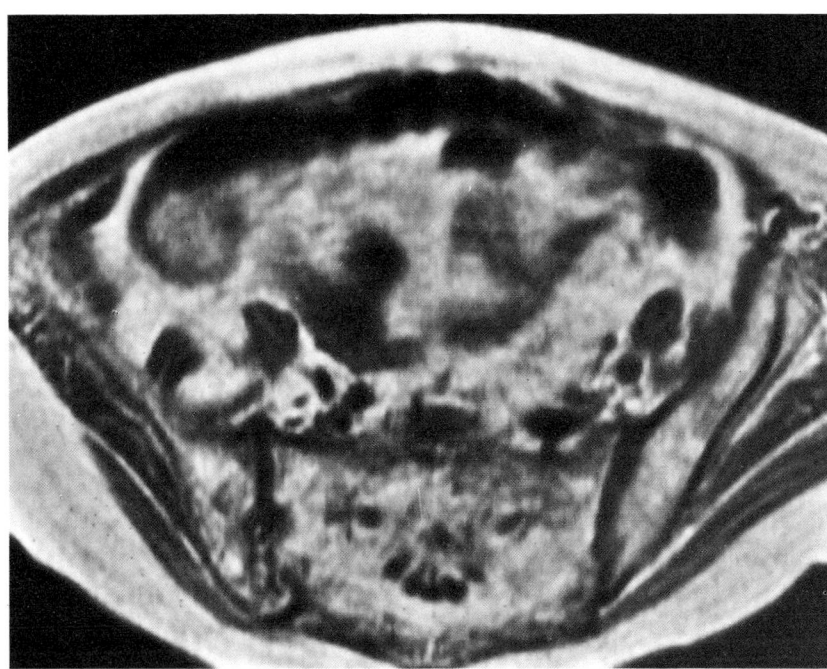

d

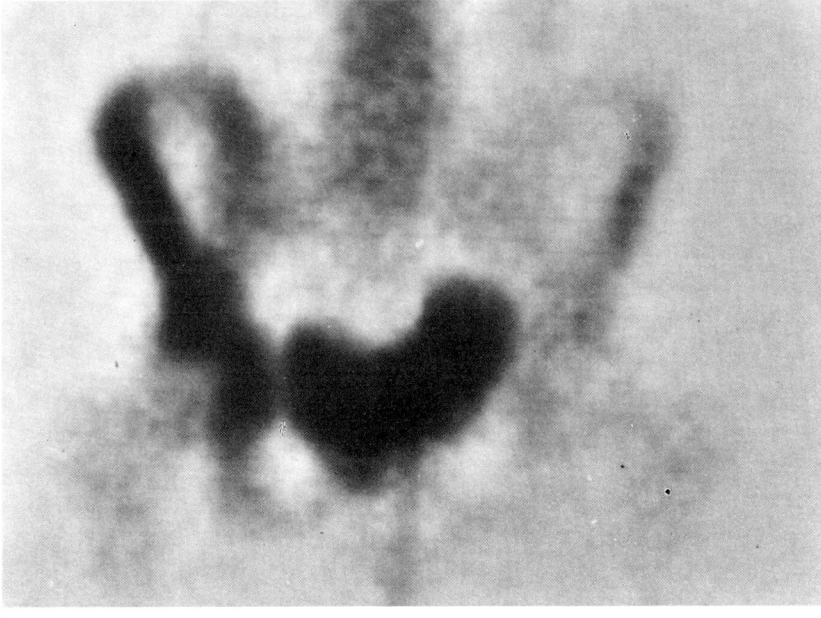

a

Figure 11.50

Paget's disease. A 66-year-old woman with history of breast carcinoma and right hip pain. (**a**) Radionuclide bone scan shows abnormally increased and contiguous radionuclide uptake throughout right ilium and ischium, typical of Paget's disease. (**b**) Anteroposterior radiograph discloses increased density of right ilium and ischium, with thickening of iliopectineal line (arrow). (**c**) Coronal T_1-weighted image (SE 500/30) reveals thickened cortex, manifested by signal void (arrow), and diffuse mottling of signal intensity of marrow of right ilium and ischium. (**d**) Axial T_1-weighted image (SE 500/30) discloses uniformly decreased signal intensity of marrow in right acetabulum. (**e**) In axial T_2-weighted image (SE 2000/85), signal intensity of pagetic bone marrow has increased to that of normal contralateral acetabular marrow.

370 MRI atlas of the musculoskeletal system

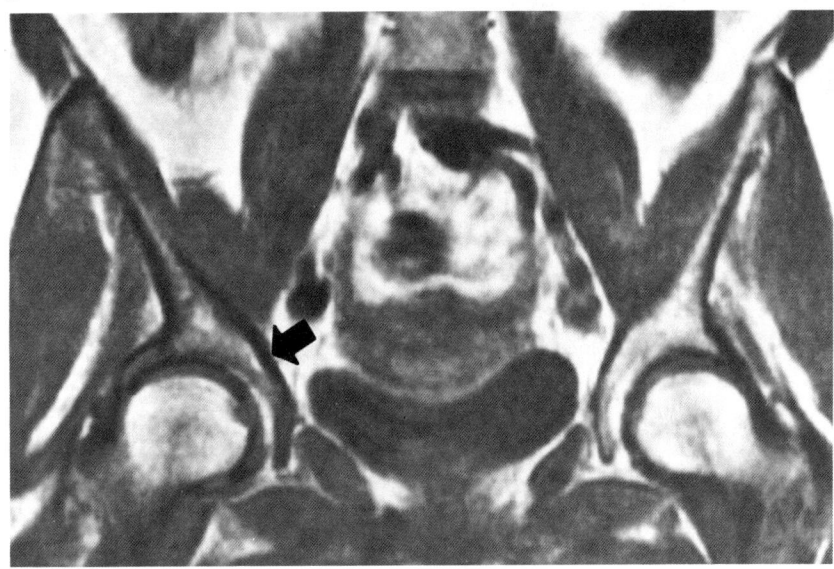

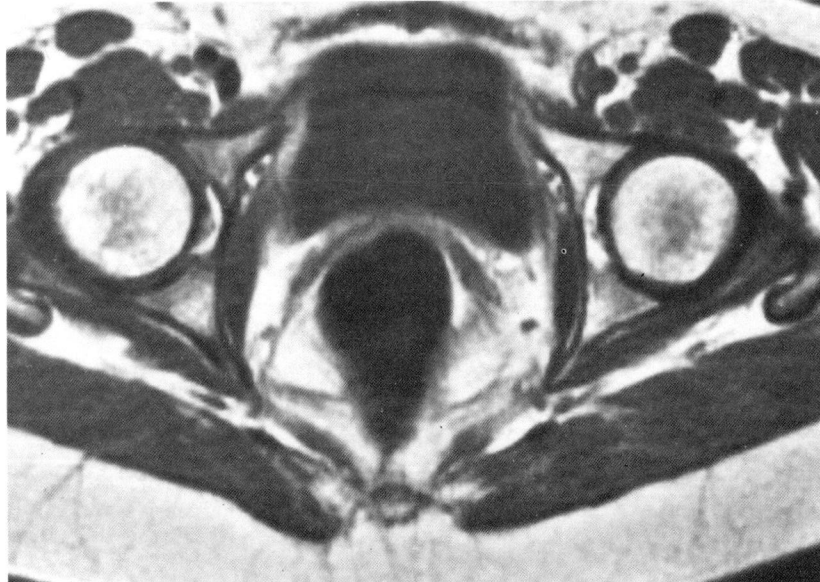

Figure 11.50 continued

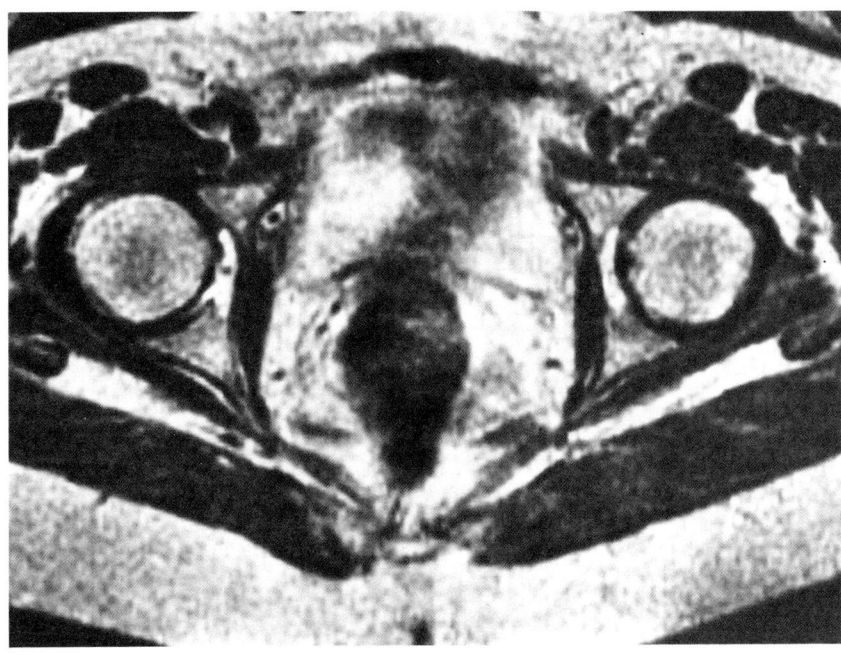

e

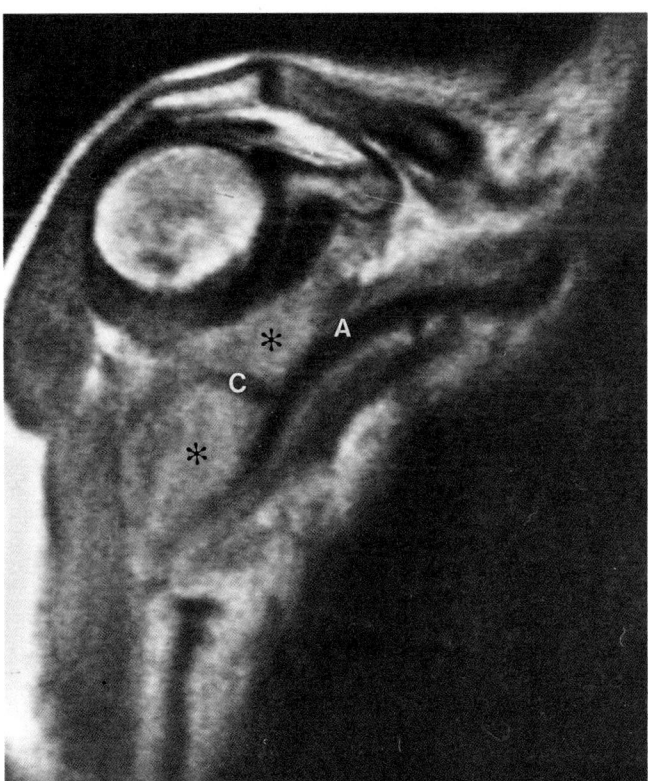

a

Figure 11.51

Soft-tissue hematoma and hemosiderin deposits in bone marrow. A 29-year-old woman with aplastic anemia and severe thrombocytopenia, who was referred for evaluation of rapidly enlarging mass in right axilla. (a) Coronal T_1-weighted image (SE 800/28). Intermediate-signal-intensity hematoma (★) is present lateral to axillary artery (A) and surrounds posterior circumflex humeral artery (C). (b) Axial T_1-weighted image (SE 500/28). At this level, low-signal-intensity hematoma is well circumscribed, and lies within coracobrachialis and short head of biceps muscles. (c) Axial T_2-weighted image (SE 2000/56), slightly inferior to (b). Hematoma has increased signal intensity. Coronal T_1-weighted image (SE 800/28) (d) through humeral shaft shows large area of signal void in marrow, which persists in T_2-weighted image (SE 2000/56) (e). Signal void was felt to represent hemosiderin deposits due to multiple transfusions.

372 MRI atlas of the musculoskeletal system

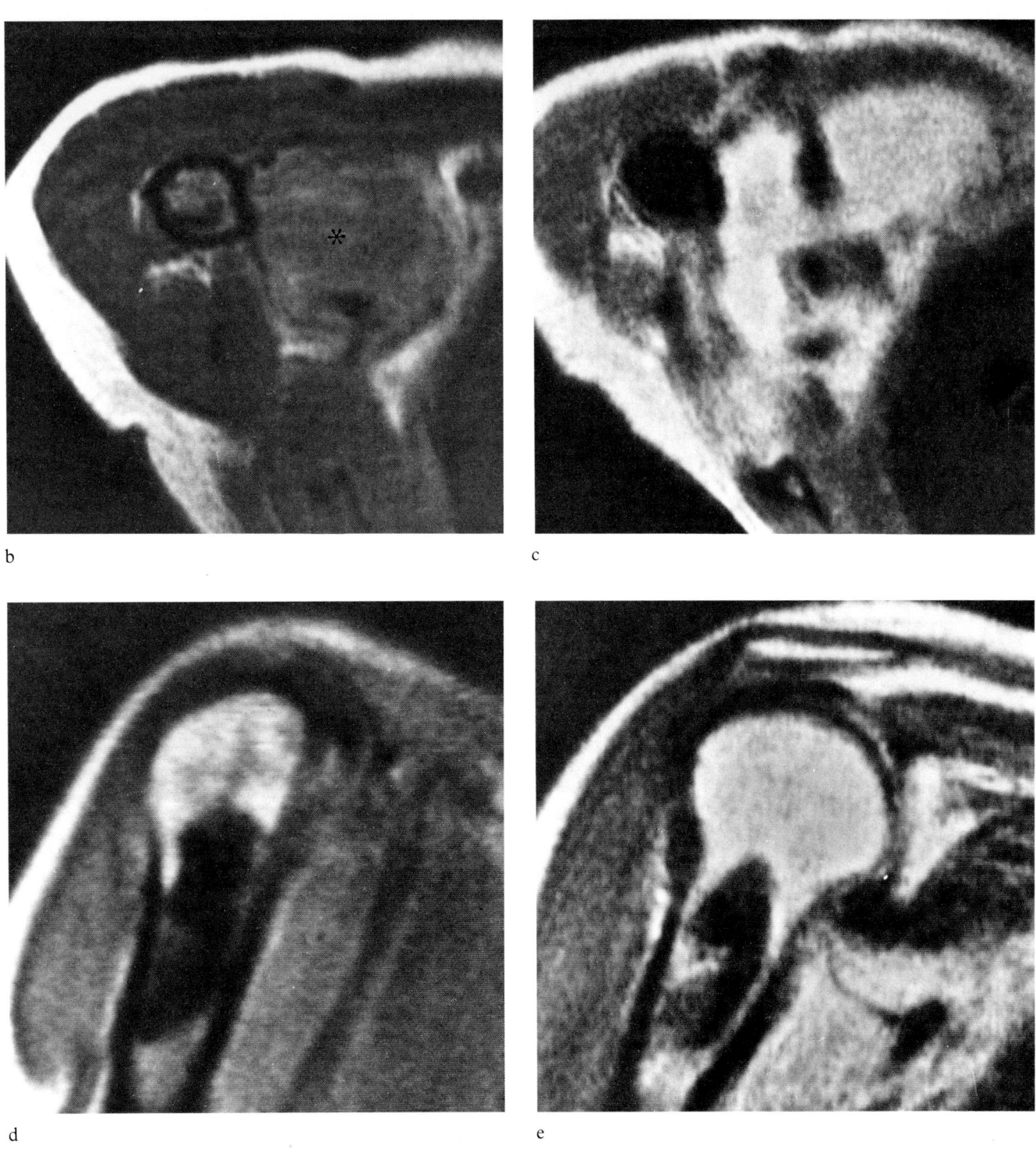

Figure 11.51 *continued*

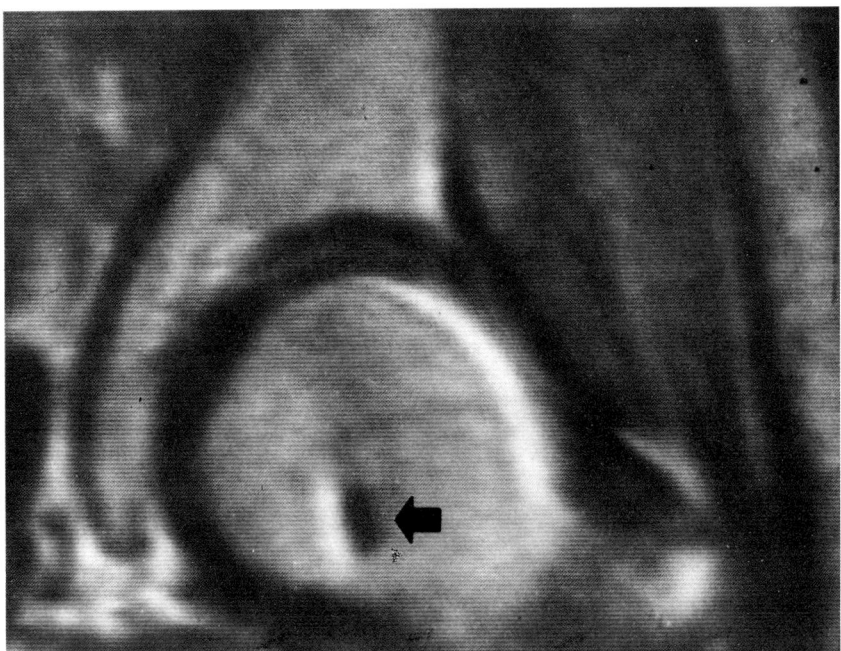

a

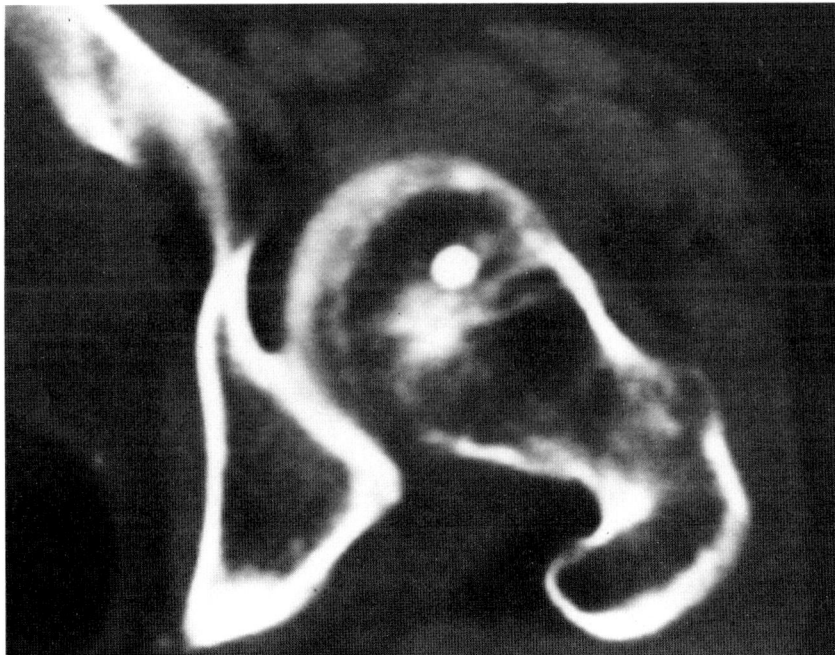

b

Figure 11.52

Bone island (enostosis). A 49-year-old man with metastatic adenocarcinoma of unknown primary. (**a**) Coronal image (SE 500/28) reveals focus of signal void (arrow) within left femoral head. Metastasis could not be excluded on T_1-weighted image alone. (**b**) CT scan shows area of MR signal void to be well circumscribed, and of same density as cortical bone. (**c**) Radionuclide bone scan is normal. (**d**) Anteroposterior radiograph of femoral head and neck shows typical bone island.

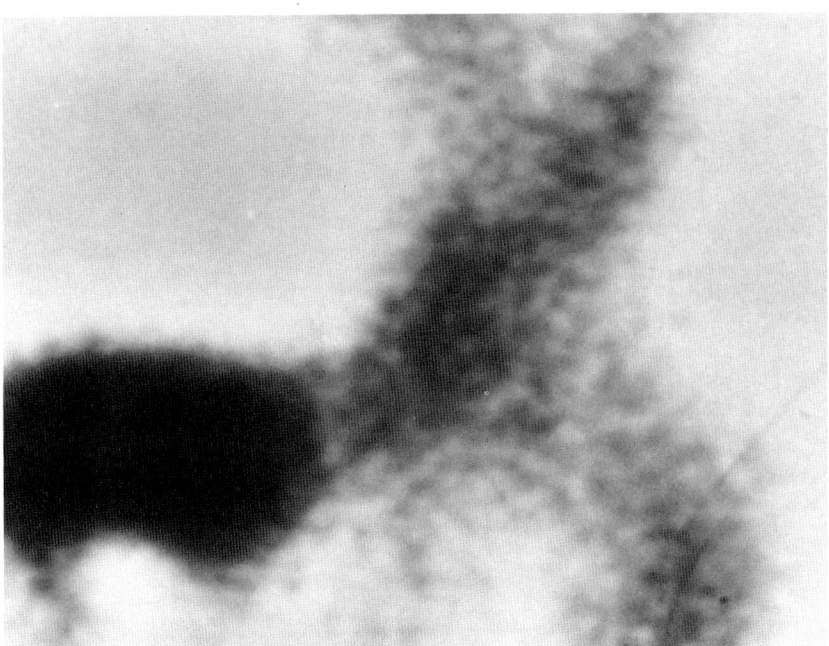

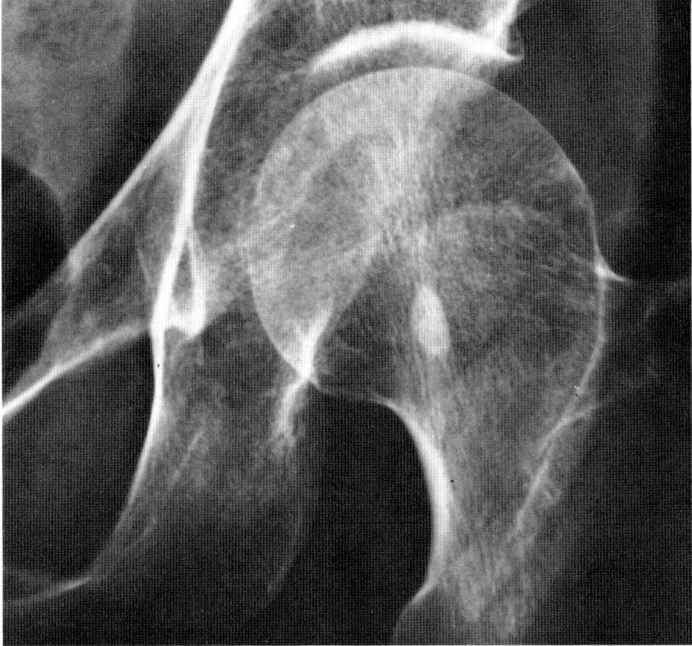

c

d

Figure 11.52 continued

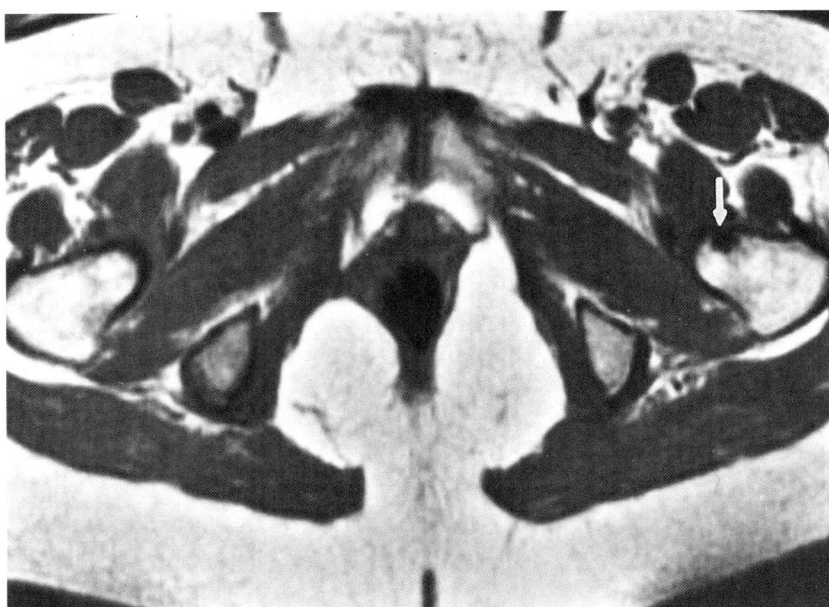

a

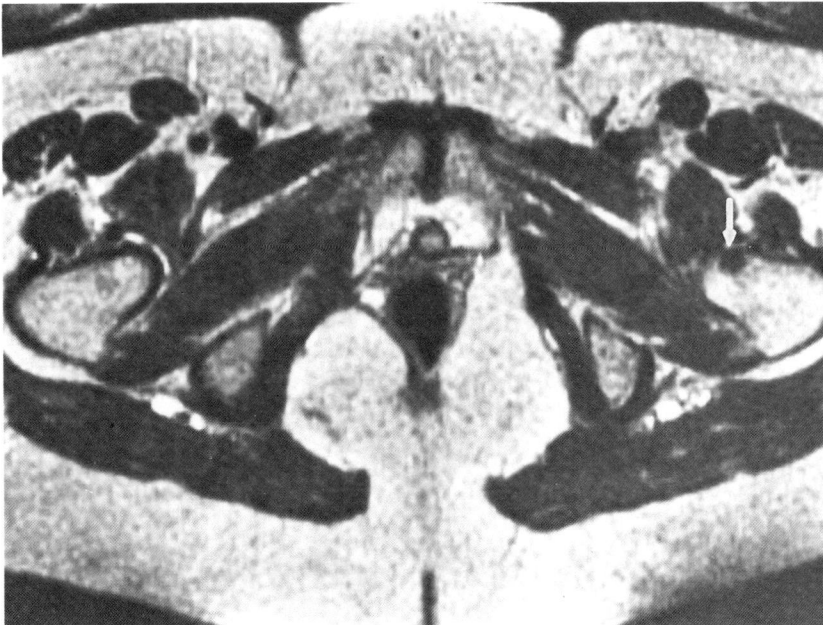

b

Figure 11.53

Bone island (enostosis). A 66-year-old woman with history of breast carcinoma. (**a**) Axial T_1-weighted image (SE 500/30) depicts bone island as a signal void (arrow) in left femoral neck. (**b**) T_2-weighted image (SE 2000/85). Unlike metastases, bone islands have low-signal intensity on T_2-weighted images.

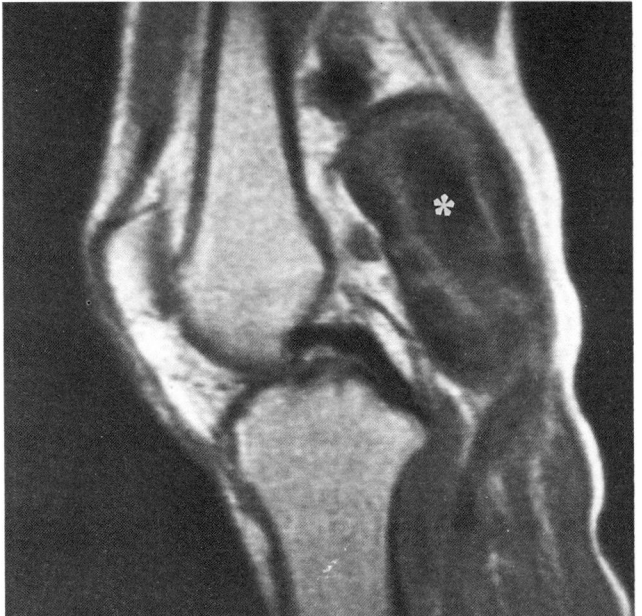

a

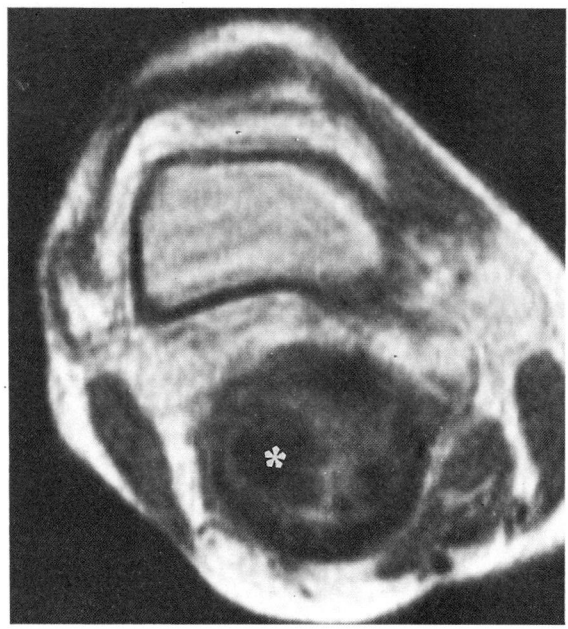

b

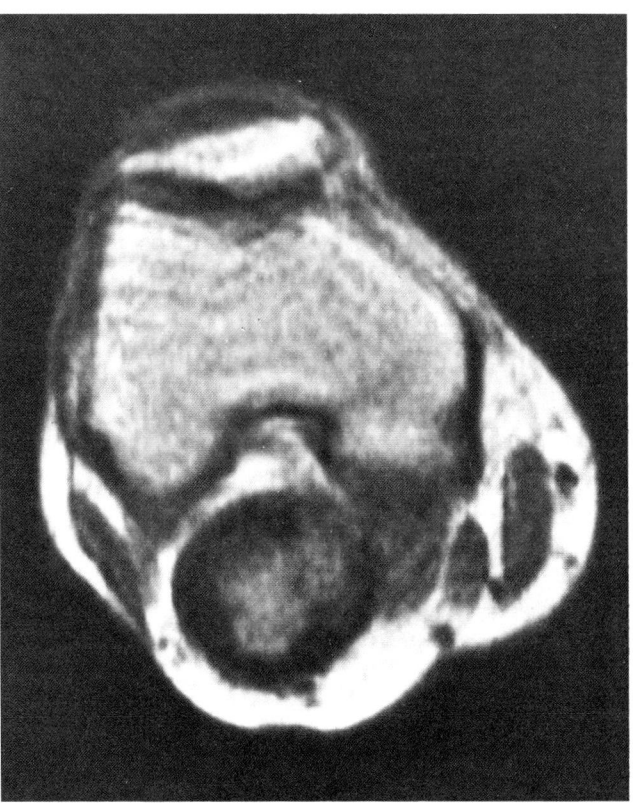

c

Figure 11.54

Popliteal aneurysm. A 69-year-old man with pulsatile mass in popliteal fossa. (**a**) Sagittal image (SE 800/28) discloses that superior aspect of aneurysm shows intermediate-signal-intensity thrombus surrounding central signal void of flowing blood (★). (**b**) Axial image (SE 800/28) of superior aspect of aneurysm. (**c**) Axial image (SE 800/28) of inferior aspect of aneurysm depicts signal void of flowing blood around central intermediate-signal intensity of thrombus.

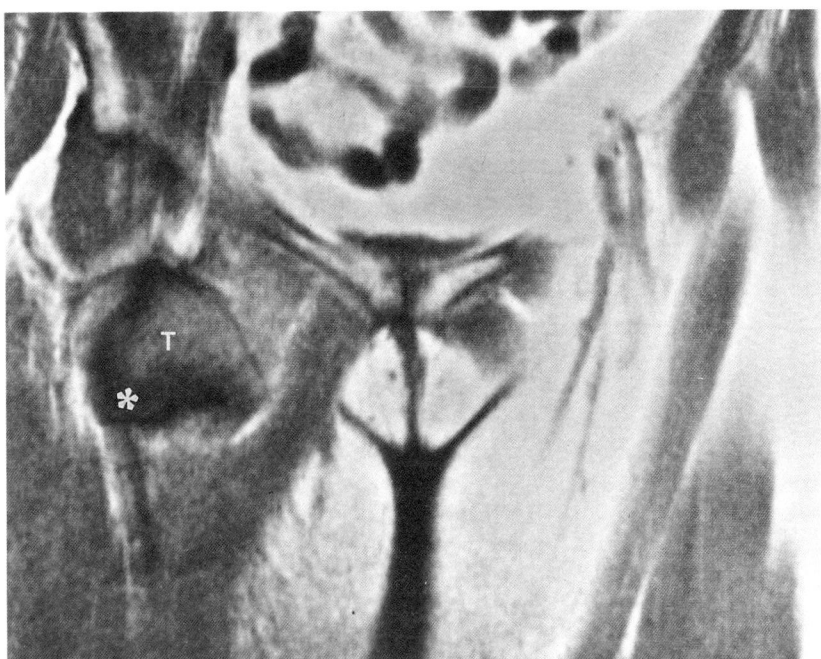

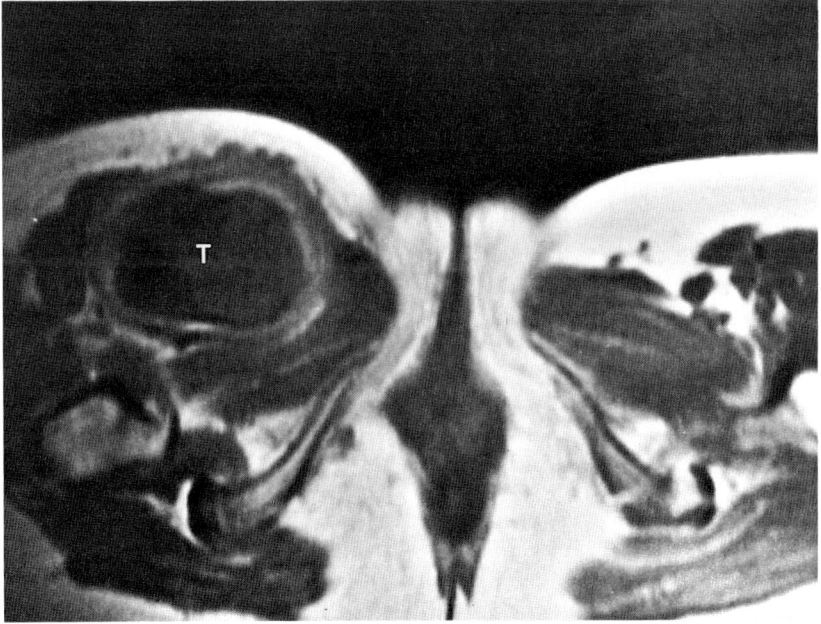

Figure 11.55

Femoral artery aneurysm. A 47-year-old woman who noted pulsatile mass in right groin following femoral artery puncture for arteriography. (**a**) Coronal T_1-weighted image (SE 400/28) depicts aneurysm as combination of signal void of flowing blood (★) and intermediate signal intensity thrombus (T). (**b**) Axial T_1-weighted image (SE 500/28) discloses that aneurysm is largely occupied by intermediate-signal-intensity thrombus. Signal void of flowing blood is located in lateral aspect of aneurysm.

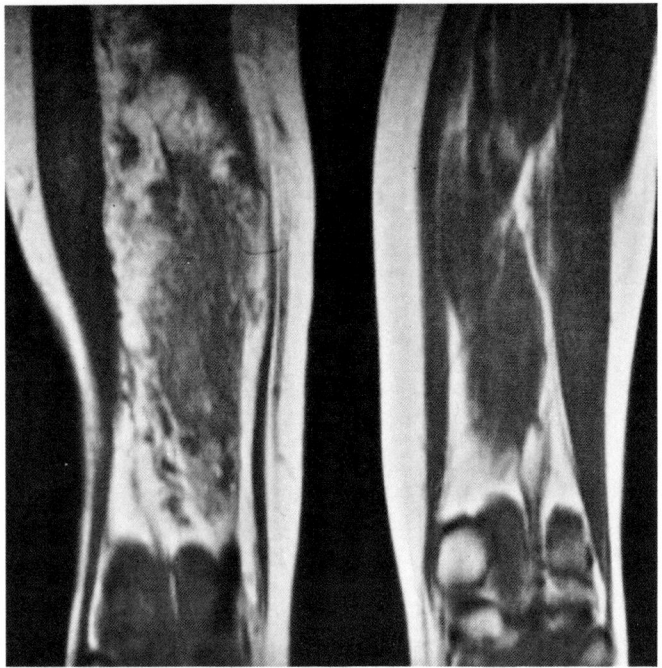

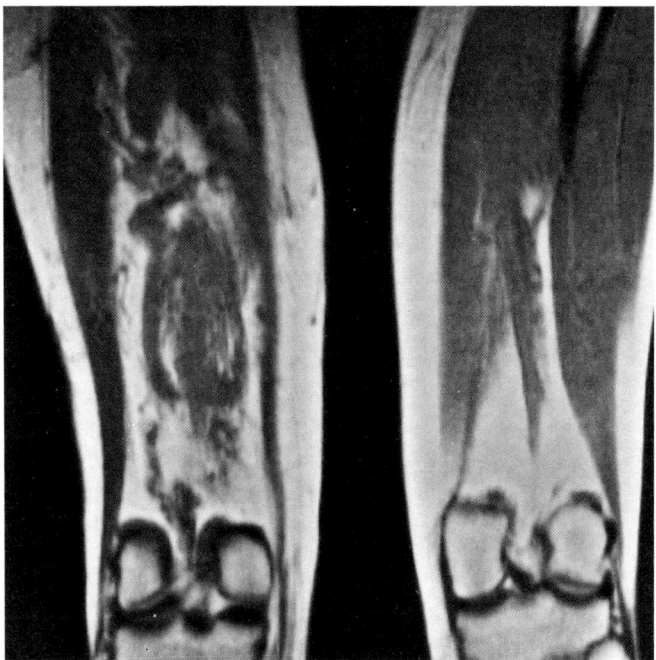

Figure 11.56

Arteriovenous malformation. A 32-year-old woman with arteriovenous malformation of right lower extremity. (**a**) Coronal image (SE 500/28) shows full extent of tortuous, serpiginous vessels in right thigh compared to normal left thigh. (**b**) Coronal image (SE 500/28), 1 cm anterior to (**a**).

12
Musculoskeletal infection

John SH Tang and Richard H Gold

Although antibiotics and sophisticated surgical therapy have led to a progressive decrease in its mortality and morbidity, osteomyelitis remains a diagnostic problem that is rendered even more difficult when underlying bone abnormalities, such as deformities from previous trauma or surgery, are present. In addition, it is often difficult to distinguish clinically between soft-tissue, bone or joint infection.[1-3]

Osteomyelitis can be divided into three groups: hematogenous osteomyelitis (19 per cent); osteomyelitis secondary to contiguous soft-tissue infection (47 per cent); and osteomyelitis associated with peripheral vascular disease (34 per cent).[1] The medullary canal of the metaphysis is the most common site of hematogenous osteomyelitis. In neonates, because transphyseal blood vessels course between the metaphysis and the epiphysis, osteomyelitis can spread rapidly from an infected metaphyseal focus to the epiphysis and subjacent joint, possibly resulting in premature epiphyseal fusion and secondary growth aberration. In childhood, although the growth plate acts as a barrier to infection, pyoarthrosis secondary to osteomyelitis commonly occurs in the shoulder and hip, where the synovial-lined capsule extends beyond the epiphyseal plate, and infection in the metaphysis may spread to the joint simply by breaching the cortex. In adults, septic arthritis is rarely seen as a complication of osteomyelitis, despite the presence of anastomoses between the metaphyseal and epiphyseal blood vessels.

Osteomyelitis is most frequently encountered in the long bones, particularly those in the lower limbs, and, in adults, the spine as well. In neonates, multifocal infection is not uncommon. Because extensive and often permanent damage to the bone or joint is a frequent sequela of delayed diagnosis or inadequate therapy, sensitive and accurate imaging examinations are essential whenever a diagnosis of infection involving the musculoskeletal system is considered clinically.

Imaging modalities and musculoskeletal infection

In the earliest stage of osteomyelitis or septic arthritis, conventional radiographs may be normal. Although the ^{99m}technetium-labeled diphosphonate bone scan has until recently been recognized as the most sensitive imaging modality for detecting osteomyelitis, limited spatial resolution may prevent even three-phase scans from distinguishing involvement of bone from that of joints and soft tissue. Moreover, the ^{99m}technetium bone scan has poor specificity for active infection.[4-7] The ^{111}indium-labeled leukocyte scan is more specific for infection, but less accurate in disclosing abscesses of bone than of soft tissue, and is often unsuccessful in depicting foci of chronic osteomyelitis and vertebral osteomyelitis.[7-10] The gallium-67 citrate scan is too non-specific for detecting bone infection, having a high incidence of false-positive results in healing fractures, heterotopic bone formation, neoplasms, and noninfected arthroplasties; false-negative results are common in cases of active osteomyelitis under treatment with antibiotics.[8-10]

Computed tomography (CT) has proved a useful adjunct to conventional radiographs and radionuclide studies in the evaluation of selected patients who may have bone or joint infections.[11-13] CT is capable of providing reliable information regarding the extent of disease, even when plain radiographs are normal. Intraosseous or soft-tissue gas, sequestra and foreign bodies can also be demonstrated with CT. However, CT is usually performed only when the diagnosis or extent of disease remains in doubt after conventional imaging, and is less useful for patients with metallic joint prostheses or internal fixation devices.

Magnetic resonance imaging (MRI) has been recognized as useful in evaluating musculoskeletal disease because of its capability of demonstrating anatomic details and its high sensitivity in detecting pathologic changes in the bone marrow and soft tissue. Although the number of patients reported has thus far been small, preliminary evaluation has shown MRI to be a promising modality for detecting the presence and local extent of musculoskeletal infection.[14-24] In this chapter we illustrate the role and limitations of MRI in the diagnosis of musculoskeletal infection.

Technical consideration of MR imaging

Although T_1-weighted MR images are sensitive in depicting pathologic lesions of medullary bone and adipose tissue, they exhibit no pathognomonic changes of active infection. Only when the abnormally low-intensity signal on T_1-weighted images changes to a high-intensity signal on T_2-weighted images, even brighter than that of normal marrow or subcutaneous fat, is active infection strongly suggested. Therefore, although T_1-weighted images can be used as a nonspecific screen for an abnormality in the marrow or soft tissue, both T_1- and T_2-weighted sequences are needed to distinguish musculoskeletal infection from other abnormalities. The pulse sequences commonly used are spin-echo T_1-weighted (TE = 20 to 30 msec; TR = 300 to 500 msec), and T_2-weighted (TE = 56 to 120 msec; TR = 1500 to 3000 msec) images. In order to improve the signal-to-noise ratio and image quality, surface coils should be used whenever possible.

One of the advantages of MRI over other imaging modalities is its capability of providing multiplanar images. In imaging the extremities and flat bones, both coronal and sagittal images suffer from the effect of partial volume averaging, and are therefore usually employed only to screen for or localize the focus of disease. Axial images are then obtained, and are usually more effective in demonstrating the intra- and extraosseous extent of the disease. For studying joints, coronal and sagittal images are the most suitable. For evaluating the spine, sagittal images are the most effective in disclosing changes in the intervertebral disks and the relationship of the disease process to the thecal sac and neural structures, while coronal and axial images are more useful for depicting paraspinal extension of disease.

MRI features of musculoskeletal infection

The signal intensity of tissue in spin-echo MR images is dependent on several tissue parameters (hydrogen density, T_1 and T_2 relaxation times, chemical shift, etc), and instrument parameters (TE and TR, field strength, flip angle, tuning of center frequency, etc). Although the signal intensity in the tissue varies with changes in TE and TR, the relative intensities of most normal musculoskeletal structures remain constant throughout most combinations of TE and TR: cortical bone, fibrocartilage, tendons, and ligaments have low-signal intensity, while bone marrow and subcutaneous fat have high-signal intensity. Fluid-filled structures, inflammatory tissue, edematous tissue, most tumors, and acute hemorrhage are characterized by prolonged T_1 and T_2 values, and therefore are of low-signal intensity on T_1-weighted images and of high intensity on T_2-weighted images.[25-27]

Based on current concepts of the pathophysiology of osteomyelitis,[1-2] the abnormal patterns of MR signal intensity can be explained as follows:

Acute osteomyelitis (*Figures 12.1–12.4*)

Because of an increase in intramedullary water caused by edema, exudate, hyperemia and ischemia, the infected marrow manifests a low-signal intensity on T_1-weighted images, which becomes high on T_2-weighted images. During this early stage, the lesions are usually homogeneous with an ill-defined margin. With further localized bone destruction the margins tend to become more distinct and, on T_2-weighted images, are characterized by a low-intensity rim surrounding a bright central core. The histologic changes of resolved or resolving infections reflect the resolution of both the inflammatory exudate and the secondary ischemia. Unless there is complete restitution and remodeling of the bone, as may occur in infants and children, healing usually results in fibrosis and irregular bony overgrowth, leading to a lower proton density and shorter T_2 values. Fracture or surgery, with reactive new bone formation and fibrosis within the medullary canal, lead to replacement of the normal marrow in an uneven pattern, appearing on

MR images as areas of mixed low, intermediate and high-intensity signals that are relatively constant on T_1- and T_2-weighted sequences.

Subacute osteomyelitis and Brodie's abscess (*Figures 12.8–12.11*)

The abscess is often surrounded by a dense, fibrous capsule and sometimes by a rim of sclerotic bone. The abscess may rarely breach the cortex, and extend extraosseously. Associated sequestra are uncommon. The signal intensity of a Brodie's abscess is low-to-intermediate on T_1-weighted images, and high on T_2-weighted images. The signal is homogeneous and on both T_1- and T_2-weighted images is sharply outlined by a low-intensity rim representing a dense fibrous capsule or sclerotic reactive bone.

Chronic osteomyelitis (*Figures 12.12–12.14*)

Irregular, thick, new bone formation and fibrosis produce an inhomogeneous, irregular focus of low-signal intensity that is relatively constant on T_1- and T_2-weighted images. Sequestra, isolated fragments of devitalized bone, usually cortical, surrounded by exudate, are common. The signal intensity of a sequestrum is similar to that of its site of origin, i.e., low-intensity if derived from cortical bone, and higher intensity if derived from cancellous bone. If active infection is present in association with chronic osteomyelitis, the lesion tends to manifest low-signal intensity on T_1-weighted images, with or without a centrally placed low- or high-intensity sequestrum, and tends to increase in intensity on T_2-weighted images. Cloacae and sinus tracts appear as low-to-intermediate signal intensity on T_1-weighted images and may not be altered significantly on T_2-weighted images unless they contain exudate, which results in a high-signal intensity.

Septic arthritis (*Figures 12.5–12.6*)

Intra-articular effusion or exudate and destruction of articular cartilage, with or without subchondral bone involvement, are the major pathologic changes. MRI is very sensitive in the depiction of increased joint fluid[16] and medullary bone destruction. Destruction of articular cartilage results in loss of its normal intermediate intensity signal. Because the destroyed articular cartilage leaves the joint bounded only by subchondral cortical bone, the joint space on T_2-weighted images appears lower in signal intensity and more distinctly outlined than normally. In the absence of medullary involvement, the pattern is similar to that of any noninfectious inflammatory, traumatic or degenerative arthritis.

Vertebral osteomyelitis (*Figures 12.19–12.20*)

The unique anatomy of the spine is responsible for a typical morphologic pattern of involvement by osteomyelitis. Although the intervertebral disks are almost invariably affected, with or without involvement of the adjacent vertebral bodies, the posterior elements are rarely affected. Paravertebral soft-tissue swelling is often present. The resultant changes in MR signal intensity are characteristic and summarized as follows:

1. T_1-weighted images reveal decreased signal intensity of the intervertebral disk and adjacent vertebral bodies, with loss of the normal margins between them.

2. On T_2-weighted images, foci of active infection in the vertebral bodies and disks yield increased signal intensity.

3. There is loss of the configuration of the intranuclear cleft of the disk (the central portion of disk which in adults normally has a streaky, linear, low-signal intensity).

4. Tuberculous spondylitis is characterized by prominent bilateral paraspinal abscesses and anterior destruction of the vertebral bodies. Calcifications, a typical radiographic feature of long-standing disease, may not be depicted on MR images.

The origin of the changes in signal intensity of vertebral osteomyelitis, active or healed, is similar to that for other bones. In contradistinction, although the signal intensities of the disk and adjacent vertebral bodies in spondylosis are usually decreased on T_1-weighted images, they are not significantly altered on T_2-weighted images.

Healed osteomyelitis (*Figures 12.21–12.22*)

Unless there is complete restitution and remodeling of bone, events that are most likely to occur in infants and children, healed osteomyelitis usually features irregular new bone formation and fibrosis. In the absence of active infection, the appearance on MR images is therefore characterized by inhomogeneous and irregular low-signal intensity, which is relatively constant from T_1- to T_2-weighted sequences.

Soft-tissue infection (*Figures 12.7, 12.15–12.18*)

Because the inflammatory soft-tissue changes associated with active osteomyelitis have histologic and morphologic features identical to those of the affected underlying bone, their signal intensities are similar (low-intensity on T_1-weighted images and high-intensity on T_2-weighted images). On T_1-weighted images, the normally high-intensity signal of subcutaneous fat and intermuscular fatty septae is replaced by the intermediate signal intensity of the inflammatory infiltrate. In the absence of an abscess, the margins of the infiltrate are ill-defined. Scarring of the soft tissue after trauma or surgery, and healed inflammation, appear as areas of low-to-intermediate signal intensity on both T_1- and T_2-weighted images. Osteomyelitis is easily distinguished from regional soft-tissue infection alone because the signal intensity of the uninvolved medullary canal is normal.

MR differential diagnosis

The rather specific MR pattern of vertebral osteomyelitis, as evidenced by unique changes of signal intensities and characteristic anatomical involvement, is easily differentiated from primary and metastatic bone tumors and spondylosis.

Based upon our evaluation of a rather limited number of cases of osteomyelitis, it appears that in the bones of the extremities and in flat bones, although the signal intensity of most bone tumors is similar to that of osteomyelitis, the two lesions can be differentiated by their different morphologic MR patterns. Unlike the adjacent homogeneous signal of acute or subacute osteomyelitis of the extremities, and the infiltrative morphologic pattern of adjacent involved soft tissue, malignant bone tumors tend to manifest an inhomogeneous signal in the medullary cavity and usually a well demarcated extraosseous soft-tissue component.[28-29]

Although on both T_1- and T_2-weighted images the margins of the lesions in early acute osteomyelitis are usually ill-defined, with further localized medullary destruction the margins tend to become better demarcated and circumscribed by a low-intensity rim representing a fibrous capsule or reactive bone formation. In contradistinction, the margins of malignant bone tumors usually appear sharply defined on T_1-weighted images, and ill-defined on T_2-weighted images. A homogeneous, round-to-oval focus of Brodie's abscess can be differentiated from most benign tumors, which are usually inhomogeneous, lobulated, expansile, and, in the case of aneurysmal bone cyst, may contain a fluid level.

Although the infiltrative, homogeneous morphologic pattern of soft tissue infection is easily differentiated from the well circumscribed, inhomogeneous pattern of most soft-tissue neoplasms, soft-tissue infection cannot be differentiated from the changes of noninfectious inflammation, trauma or surgery.

Advantages and limitations of MRI for evaluation of musculoskeletal infection

Advantages

1. MRI is a nonradiation-producing, noninvasive imaging modality capable of providing multiplanar images.

2. MRI is useful in the evaluation of possible osteomyelitis of the spine and extremities. Based on a limited number of cases, it appears to be more sensitive than most other imaging modalities and as sensitive as radionuclide studies.

3. Because of its excellent capability in distinguishing involvement of bone from that of joints and soft tissue, MRI has greater accuracy than radionuclide studies or CT scans in delineating the extent of disease.

4. MRI is probably the ideal imaging modality for detecting and localizing possible foci of active infection in regions of chronic osteomyelitis, even when complicated by previous fracture or surgery.

5. MRI seems ideal for guiding biopsy and surgery. Further evaluation is needed to confirm that it is a reliable monitor of effective therapy, as indicated by a return from an abnormally high-signal intensity to normal intensity.

Limitations

1. MRI is contraindicated for patients with neurovascular surgical clips, cardiac pacemakers or other electronic implants such as cochlear implants and spinal cord stimulators.

2. MRI is of limited value in the investigation of infection immediately following fracture or surgery.

3. Artifacts induced by metallic implants may hamper the interpretation of MR images.

4. Because at the current state of the science the spatial resolution of MRI is inferior to that of CT, calcifications, small sequestra, foreign bodies and

intra- or extraosseous gas could be obscured on MR images.

5. Because of its high cost, MRI should be employed only when the results of conventional radiographs or radionuclide studies are equivocal or do not correlate with clinical findings.

References

1. WALDVOGEL FA, MEDOFF G, SWARTZ MN, Osteomyelitis: A review of clinical features, therapeutic considerations and unusual aspects (First Part), *N Engl J Med* (1970) **282**: 198–205.
2. BONAKDAR-POUR A, GAINES VD, The radiology of osteomyelitis, *Orthop Clin North Am* (1983) **14**(1):21–37.
3. PAUS B, Tumor, tuberculosis and osteomyelitis of the spine: Differentiated diagnostic aspects, *Acta Orthop Scand* (1973) **44**:372.
4. SULLIVAN DC, ROSENFIELD NS, OGDEN J et al, Problems in the scintigraphic detection of osteomyelitis in children, *Radiology* (1980) **135**:731–6.
5. LEWIN JS, ROSENFIELD NS, HOFFER PB et al, Acute osteomyelitis in children: Combined Tc-99m and Ga-67 imaging, *Radiology* (1986) **158**:795–804.
6. TUMEH SS, ALIABADI PA, WEISSMAN BN et al, Chronic osteomyelitis: bone and Gallium scan patterns associated with active disease, *Radiology* (1986) **158**:685–8.
7. MAURER AH, MILLMOND SSH, KNIGHT LC et al, Infection in diabetic osteoarthropathy: Use of Indium-labeled leukocytes for diagnosis, *Radiology* (1986) **161**:221–5.
8. McAFEE JG, SAMIN A, In-111 labeled leukocytes: a review of problems in image interpretation, *Radiology* (1985) **155**: 221–9.
9. GEORGI P, KAPS HP, SINN HJ, Leukocyte scanning of inflammatory processes in the spinal column, *Radiology* (1985) **25**: 325–8.
10. SAYLE BA, FAWCETT HD, WILKEY DJ et al, Indium-111 chloride imaging in chronic osteomyelitis, *J Nucl Med* (1985) **26**: 225–9.
11. AZOUZ EM, Computed tomography in bone and joint infections, *Journal de l'Association Canadienne des Radiologistes* (1981) **322**:102–6.
12. WING VW, JEFFREY RB, FEDERLE MMP et al, Chronic osteomyelitis examined by CT, *Radiology* (1985) **154**:171–4.
13. RAM PC, MARTINEZ S, KOROBKIN M et al, CT detection of intraosseous gas: a new sign of osteomyelitis, *AJR* (1981) **137**:721–3.
14. TANG JSH, GOLD RH, BASSETT LW et al, Magnetic resonance evaluation of musculoskeletal infection of the extremities, *Radiology* (1988) **166**:205–9.
15. MODIC MT, PFLANZE W, FEIGLIN DHI et al, Magnetic resonance imaging of musculoskeletal infections, *Radiol Clin North Am* (1986) **24**(2):247–58.
16. BERQUIST TH, BROWN ML, FITZGERALD RH et al, Magnetic resonance imaging: application in musculoskeletal infection, *Magnetic Resonance Imaging* (1985) **3**:219–30.
17. REIS ND, LANIR A, BENMAIR J et al, Magnetic resonance imaging in orthopaedic surgery, *J Bone Joint Surg (Br)* (1985) **67**(4):659–64.
18. ATLAN H, SIGAL R, HADAR H et al, Nuclear magnetic resonance proton imaging of bone pathology, *J Nucl Med* (1986) **27**: 207–15.
19. BARRY D, SCOLES PV, NELSON AD, Osteomyelitis in children: detection by magnetic resonance, *Radiology* (1984) **150**:57–60.
20. MODIC MT, PAVLICEK WP, WEINSTEIN MA et al, Magnetic resonance imaging of intervertebral disk disease, *Radiology* (1984) **152**:103–11.
21. MODIC MT, FEIGLIN DH, PIRAINO DW et al, Vertebral osteomyelitis: assessment using MR, *Radiology* (1985) **157**:157–66.
22. PAUSHTER DM, MODIC MT, MASARYK TJ, Magnetic resonance imaging of the spine: applications and limitations, *Radiol Clin North Am* (1985) **23**(3):551–62.
23. ROOS AD, VAN PERSIJN VAN MEERTEN EL, BLOEM JL et al, MRI of tuberculous spondylitis, *AJR* (1986) **147**:79–82.
24. SMITH FW, RUNGE V, PERMEZEL M et al, Nuclear magnetic resonance (NMR) imaging in the diagnosis of spinal osteomyelitis, *Magnetic Resonance Imaging* (1984) **2**:53–6.
25. RICHARDSON ML, Magnetic resonance imaging of the musculoskeletal system, *Radiol Clin North Am* (1986) **24**(2):137–42.
26. BELTRAN J, NOTO AM, HERMAN LJ et al, Joint effusions: MR imaging, *Radiology* (1986) **158**:133–7.
27. UNGER EC, GLAZER HS, LEE JKT et al, MRI of extracranial hematomas: preliminary observations, *AJR* (1986) **146**: 403–7.
28. ZIMMER WD, BERQUIST TH, McLEOD RA et al, Bone tumors: magnetic resonance imaging versus computed tomography, *Radiology* (1985) **155**:709–18.
29. VANEL D, DI PAOLA R, CONTESSO G, Magnetic resonance imaging in musculoskeletal primary malignant tumors, In: *Magnetic Resonance Annual* (Raven Press: New York, 1987) 237–61.

384 MRI atlas of the musculoskeletal system

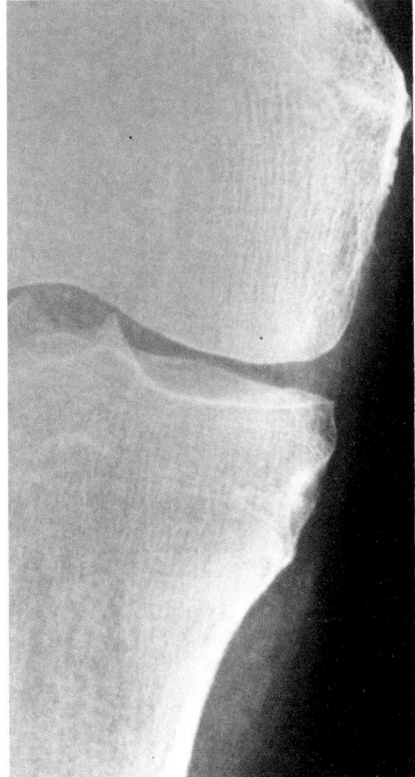

a

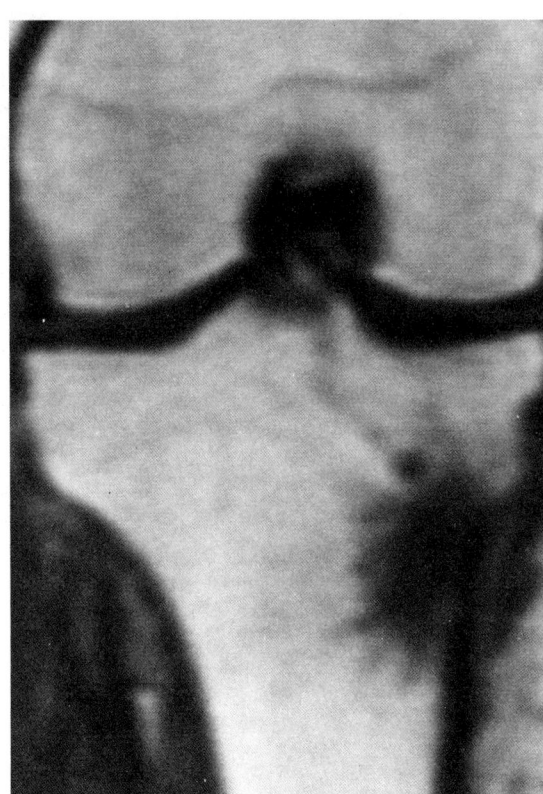

b

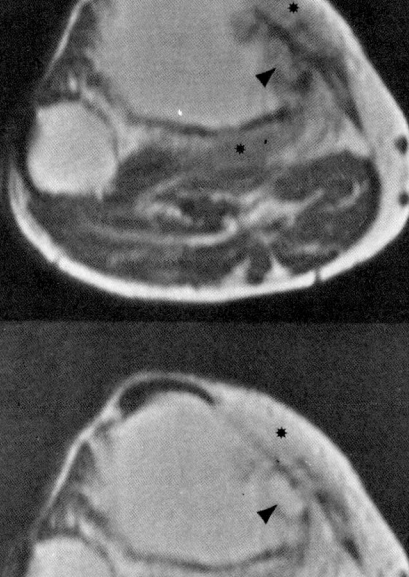

c

Figure 12.1

A 76-year-old man with acute pyogenic osteomyelitis of the right proximal tibia.
(**a**) Anteroposterior radiograph of area of interest in the right tibia reveals subtle irregularity of the cortex below the medial tibial condyles, which suggests a periosteal reaction and soft-tissue swelling. (**b**) Coronal MR image (SE 500/28) of the right knee reveals a lesion in the tibia manifesting homogeneous decrease in signal intensity and an irregular, ill-defined margin. Adjacent cortex and soft tissue have abnormal, slightly increased signal intensity.
(**c**) *Above*, axial MRI (SE 1000/28) at the level of the right tibial tubercle. Signal intensity of the marrow lesion (arrowhead) is abnormally decreased while signal intensity of the adjacent cortex and soft tissue (★) is slightly increased. *Below*, axial MR image (SE 2000/60) at the same level. Signal intensity of the lesion is homogeneous, and abnormally increased (arrowhead), as is the signal in the adjacent soft tissue (★).

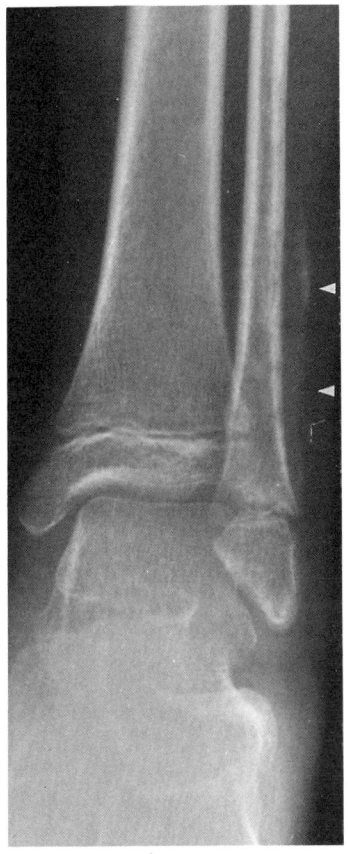
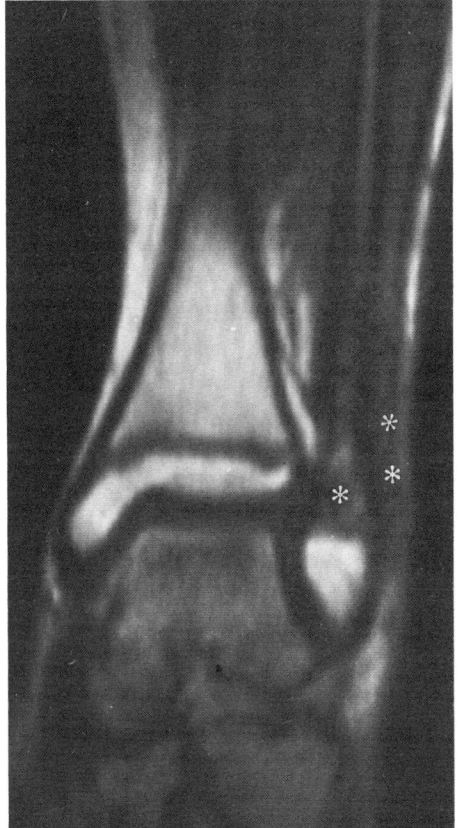
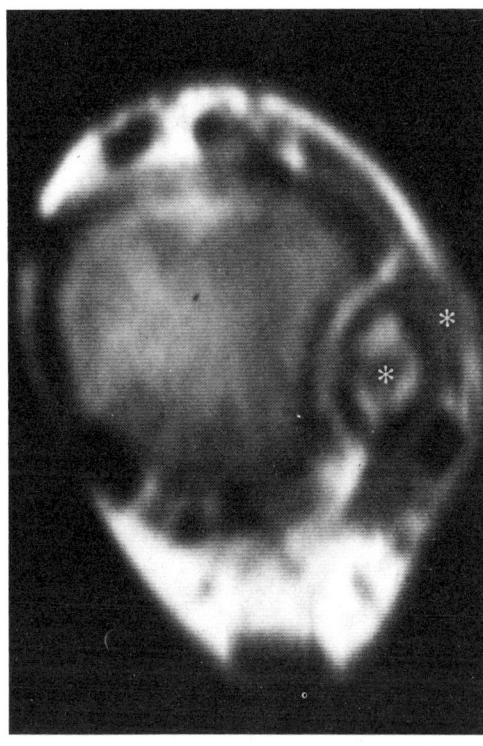
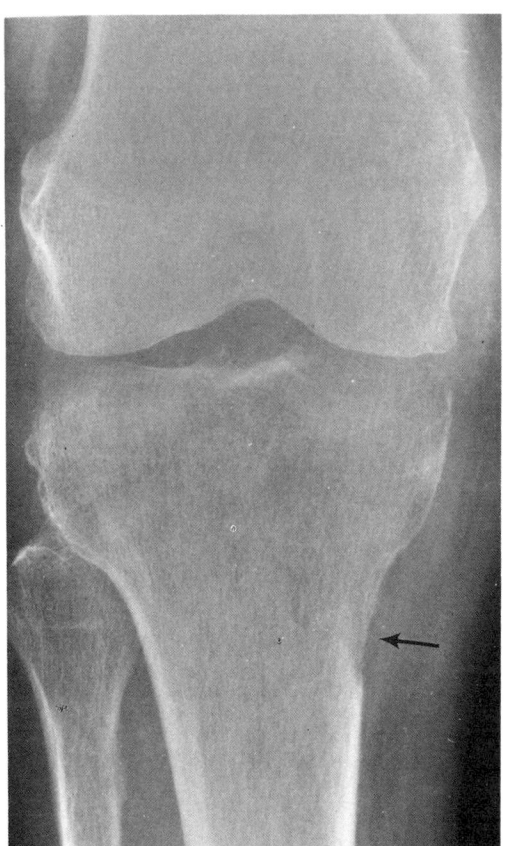

Figure 12.2

A 10-year-old girl with acute pyogenic osteomyelitis of the right distal fibula. (**a**) Anteroposterior radiograph of the right ankle reveals an ill-defined lytic lesion in the right distal fibula, associated with well-defined periosteal new bone formation (arrowheads). (**b**) Coronal MR image (SE 500/28) of the right ankle, and (**c**) axial MR image (SE 500/28) at the level of the metaphysis of the right distal fibula, disclose ill-defined decreased signal intensity in metaphysis of the right fibula (★), and abnormal, slightly increased signal intensity in the adjacent cortex and soft tissue (★). Normal bright subcutaneous fat is interrupted.

Figure 12.3

A 70-year-old woman with acute pyogenic osteomyelitis of right proximal tibia. (**a**) Anteroposterior radiograph of the right knee reveals ill-defined resorption of the proximal tibia (arrow) and adjacent soft-tissue swelling. (**b,c**) Contiguous coronal MR images (SE 500/28) disclose ill-defined, decreased signal intensity in the area (arrow) corresponding to the radiographic lesion shown in (**a**). Signal intensity of the adjacent cortex and soft tissue is abnormal and slightly increased.

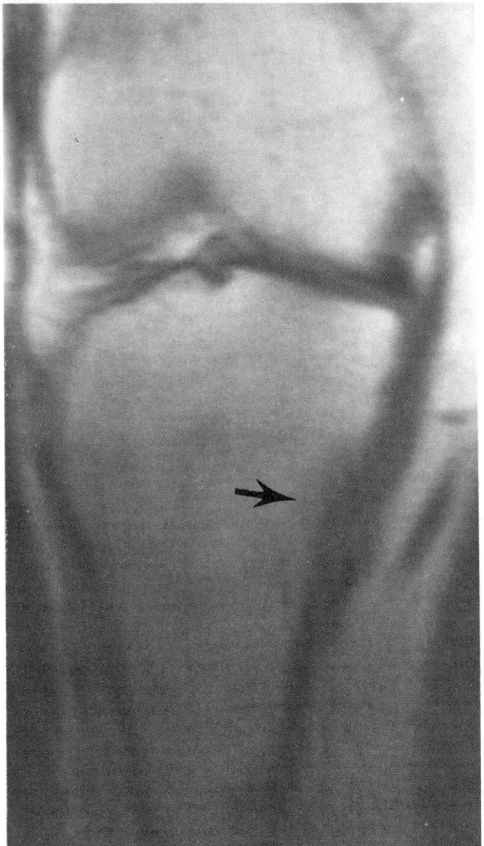

b

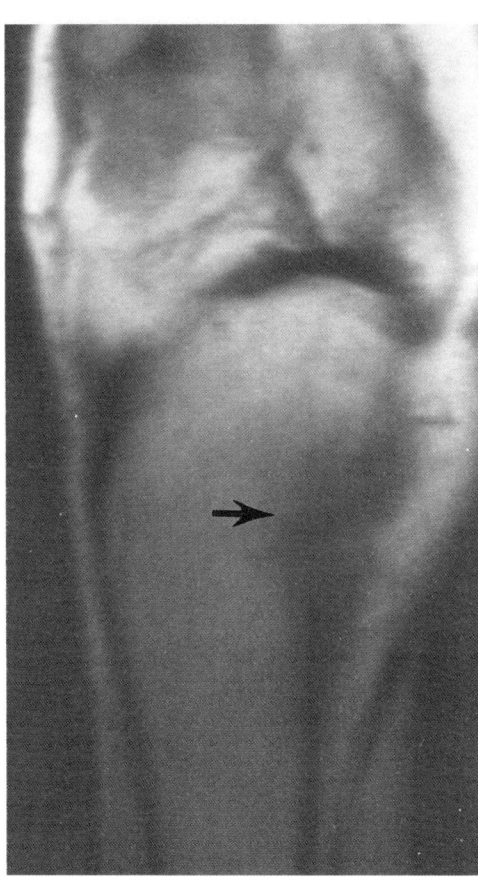

c

Figure 12.3 continued

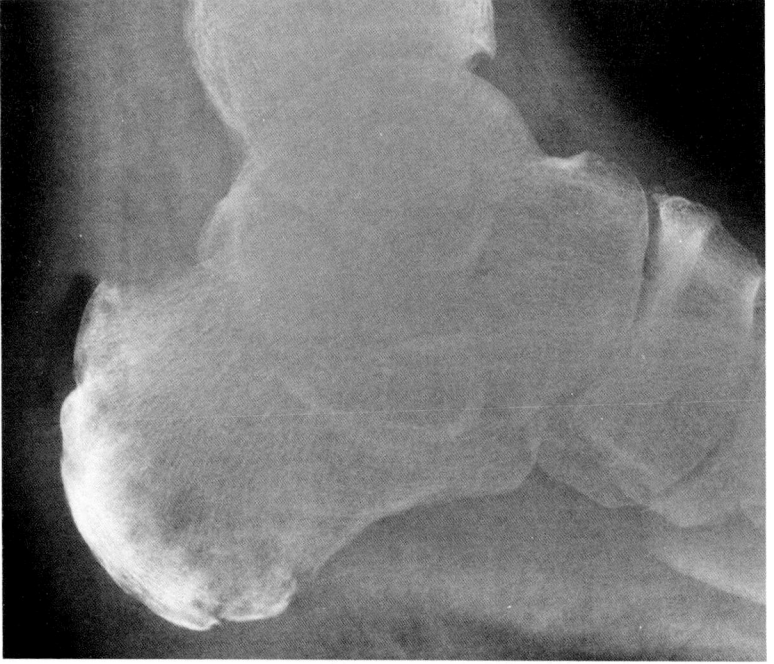

a

Figure 12.4

A 64-year-old diabetic man with acute pyogenic osteomyelitis of the left calcaneus and severe cellulitis of the left ankle. (a) Lateral radiograph of the left ankle shows ill-defined lytic lesions in the calcaneus. A deep soft-tissue ulcer is present posterior to the calcaneus. (b) Axial MR images of the bilateral cancanei (SE 1500/68). Signal intensity of posterior portion of the left calcaneus (★) is abnormally increased. The adjacent soft-tissue ulcer is coated with exudate that manifests very bright signal intensity (arrow).

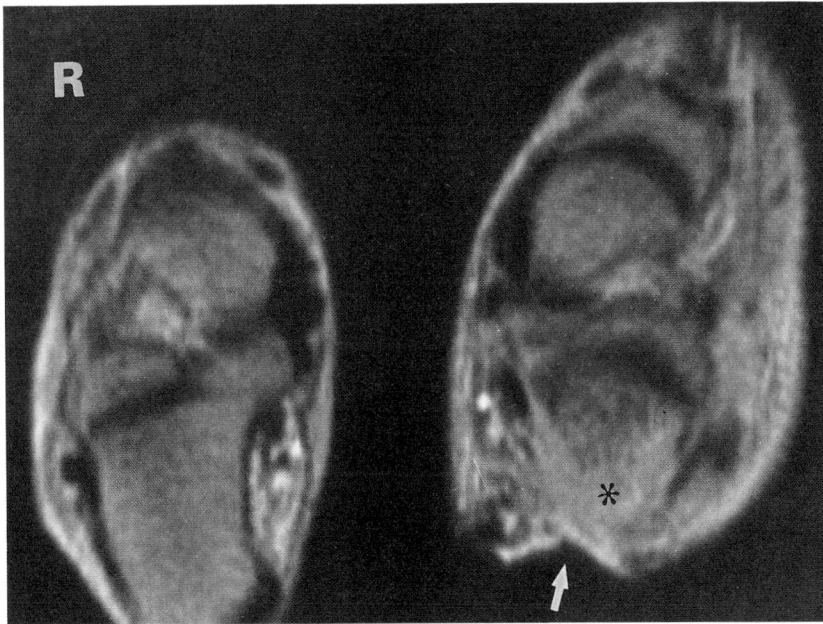

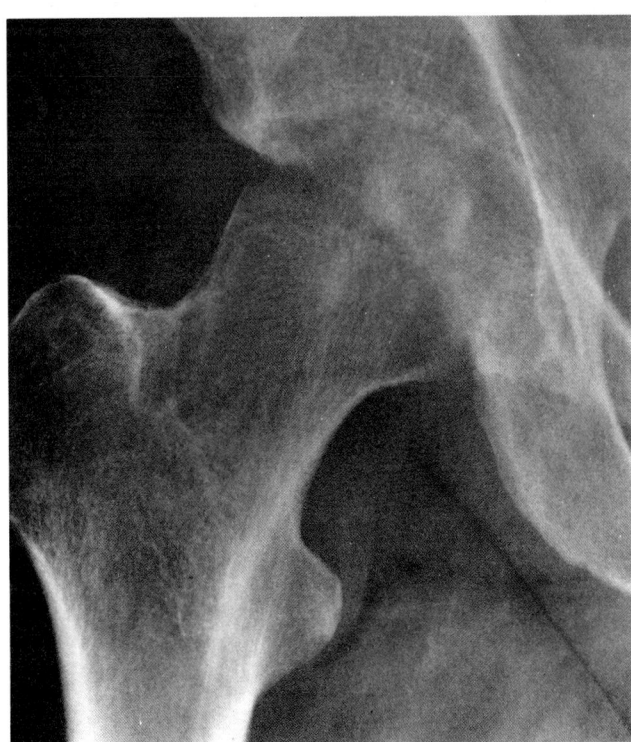
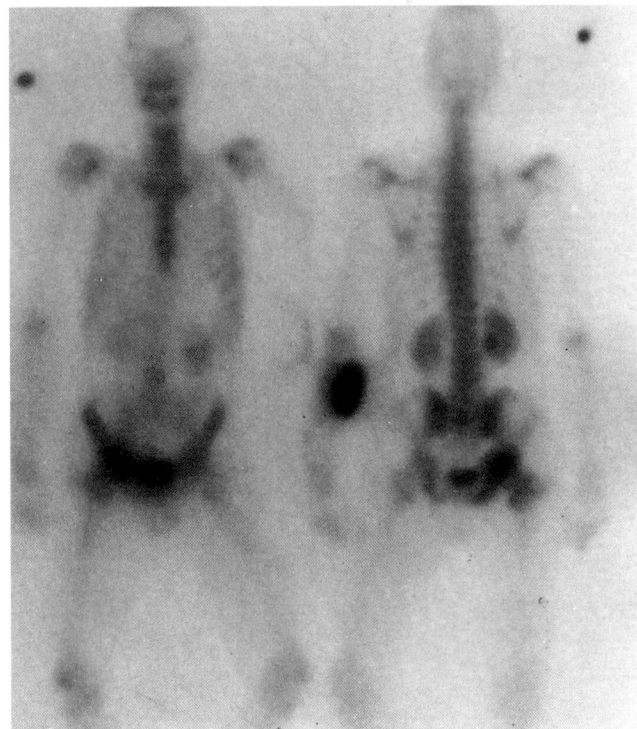

Figure 12.5

A 45-year-old man with a septic hip, confirmed by joint aspiration, and aplastic anemia. (**a**) Anteroposterior radiograph of the right hip shows resorption of the femoral head and narrowed joint space. (**b**) ^{99m}Tc bone scan reveals increased uptake of radionuclide by the right hip. (**c**) Coronal MR image (SE 499/28). Ill-defined, abnormally low-signal intensity is present in the medial aspect of the femoral head (arrow), extending below the fovea capitis, and the acetabulum. The distended joint capsule (★), owing to exudate, manifests intermediate signal intensity. (**d**) MR image in the same coronal plane as (**c**) (SE 1999/56) discloses high-signal intensity of fluid (★) distending the capsule and medial joint space. High-signal intensity of the femoral head (arrow) is due to osteomyelitis, and corresponds to the area shown in (**a**). Signal intensity of the acetabulum remains relatively low compared to the femoral head, suggesting a change in the marrow consistent with the patient's known myelofibrosis.

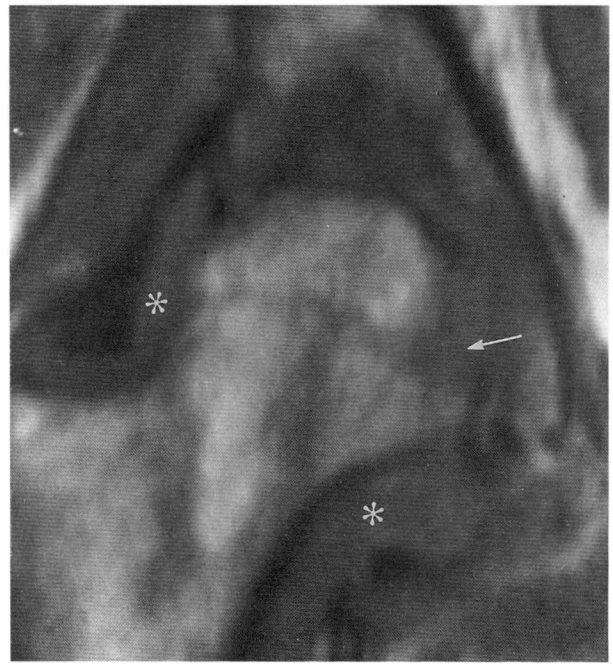

c

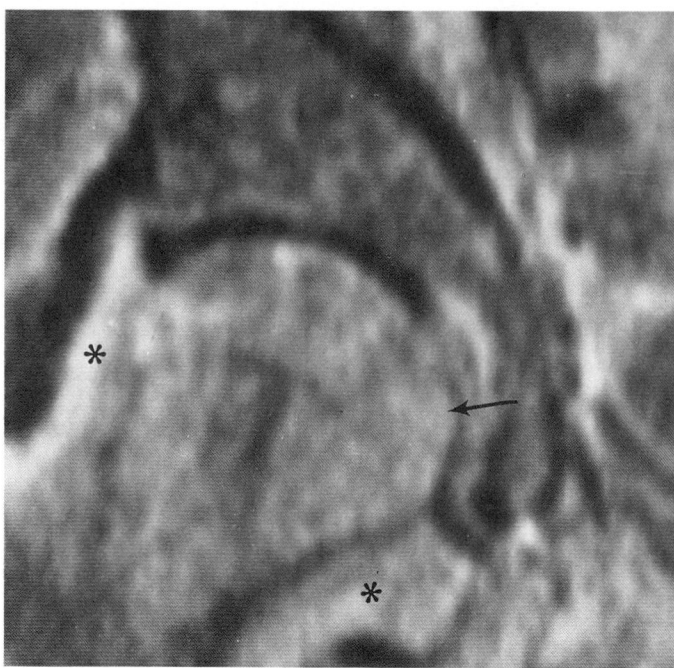

d

Figure 12.5 continued

Figure 12.6

A 53-year-old man with pyogenic osteomyelitis of the humerus and post-surgical septic arthritis of the shoulder. (**a**) Anteroposterior radiograph of the left shoulder shows destruction of the humeral head (arrowhead). (**b**) ^{99m}Tc bone scan reveals increased uptake of radionuclide in the right shoulder. (**c,e**) Axial and contiguous coronal MR images of the left shoulder (SE 500/28). Signal intensity of osteomyelitis (★) is abnormally decreased in the humeral head. Large focus of abnormal intermediate signal intensity involves regional soft tissue, the glenohumeral joint and axillary recess (arrowheads). Normally bright subcutaneous fat is interrupted (arrow). (**f,g**) MR images in the same planes as (**c–e**) (SE 1999/58). Homogeneous increased signal intensity is present in foci of osteomyelitis, corresponding to areas shown in (**d**) and (**e**). High-intensity effusion (arrowheads) is present in the glenohumeral joint, including the axillary recess. Diffuse, ill-defined increased signal intensity is also present in the acromioclavicular joint and adjacent soft tissue. Bright, well defined signal of the draining sinus tract characterizes its pus content (arrow).

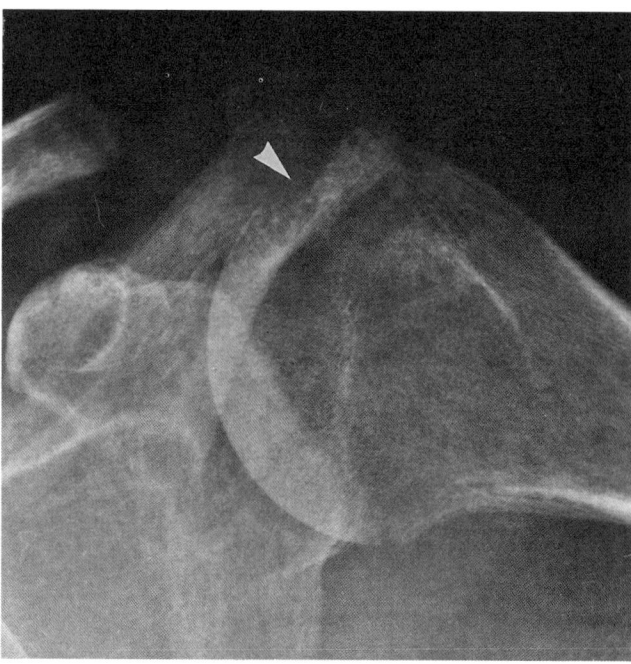

a

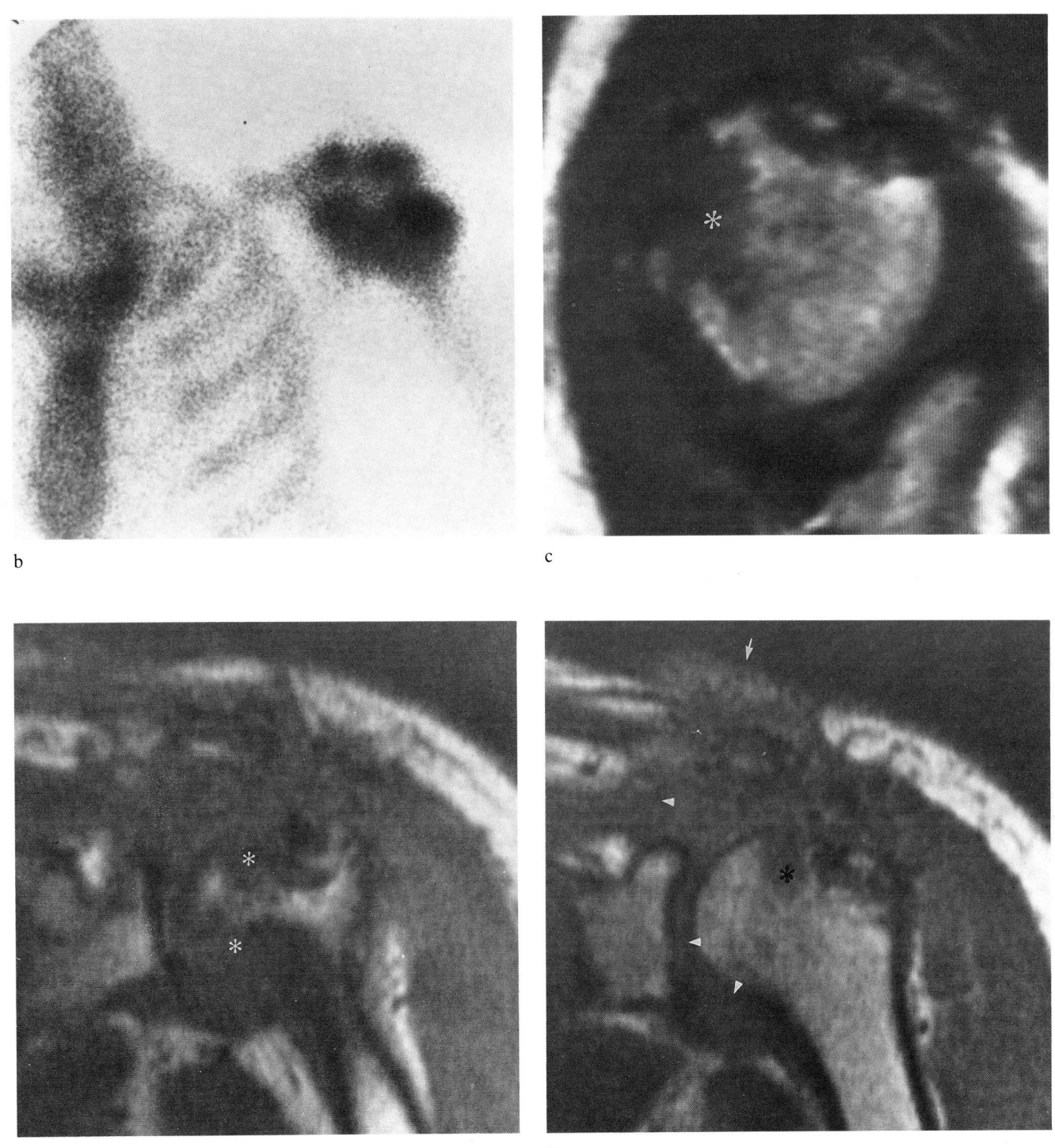

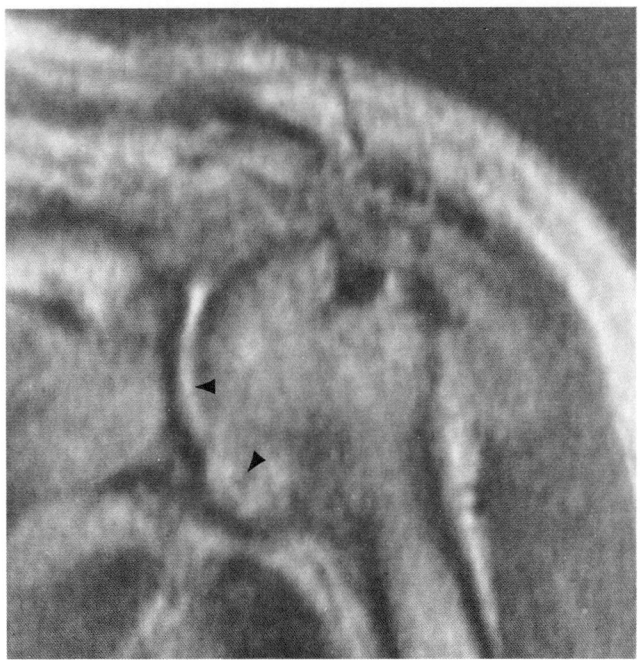
f

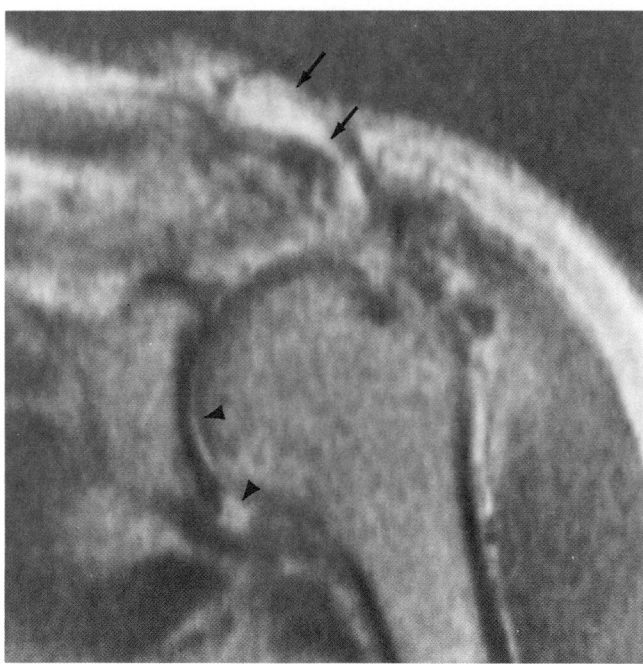
g

Figure 12.6 continued

Figure 12.7

An 84-year-old woman with cellulitis and clinically suspected osteomyelitis of the left foot (**a**) Anteroposterior radiograph of the foot shows striking osteoporosis and possible erosion of the 2nd metatarsal head. (**b**) ^{99m}Tc bone scan reveals increased uptake of radionuclide at the 1st and 2nd metatarsophalangeal joints. (**c**) Coronal MR image (SE 500/28) of the foot and (**d**) coronal MR image (SE 1500/56) at the same plane. Signal intensity is normal in the bones of the foot. (**e**) T_1-weighted axial MR image (SE 500/28) at the level of the 2nd metatarsal head, and (**f**) T_2-weighted axial MR image (SE 1500/56) at the same level. Signal intensity of the bones remains normal. Regional soft tissues manifest increased signal intensity in (**f**), suggesting cellulitis. Periostitis is characterized by increased signal intensity at the boundary of periosteum and cortex (★) on both (**e**) and (**f**). Since the bone marrow still appears normal, osteomyelitis was excluded.

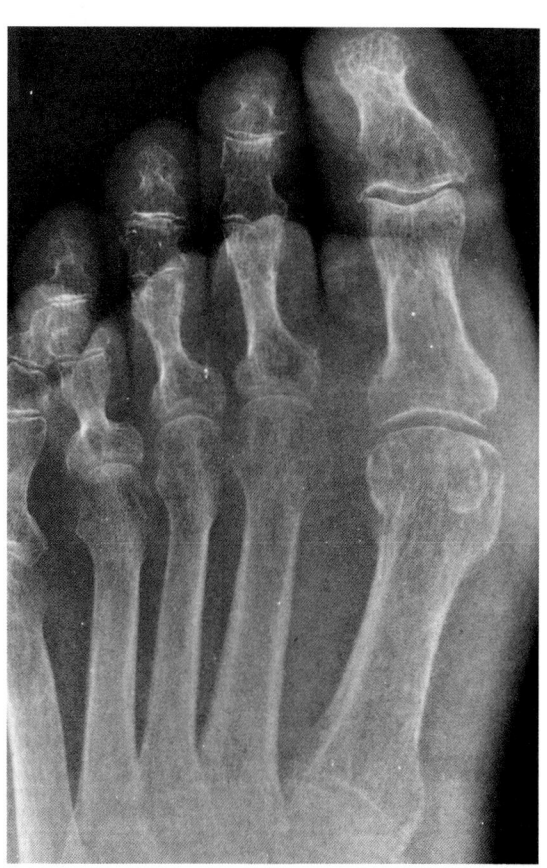
a

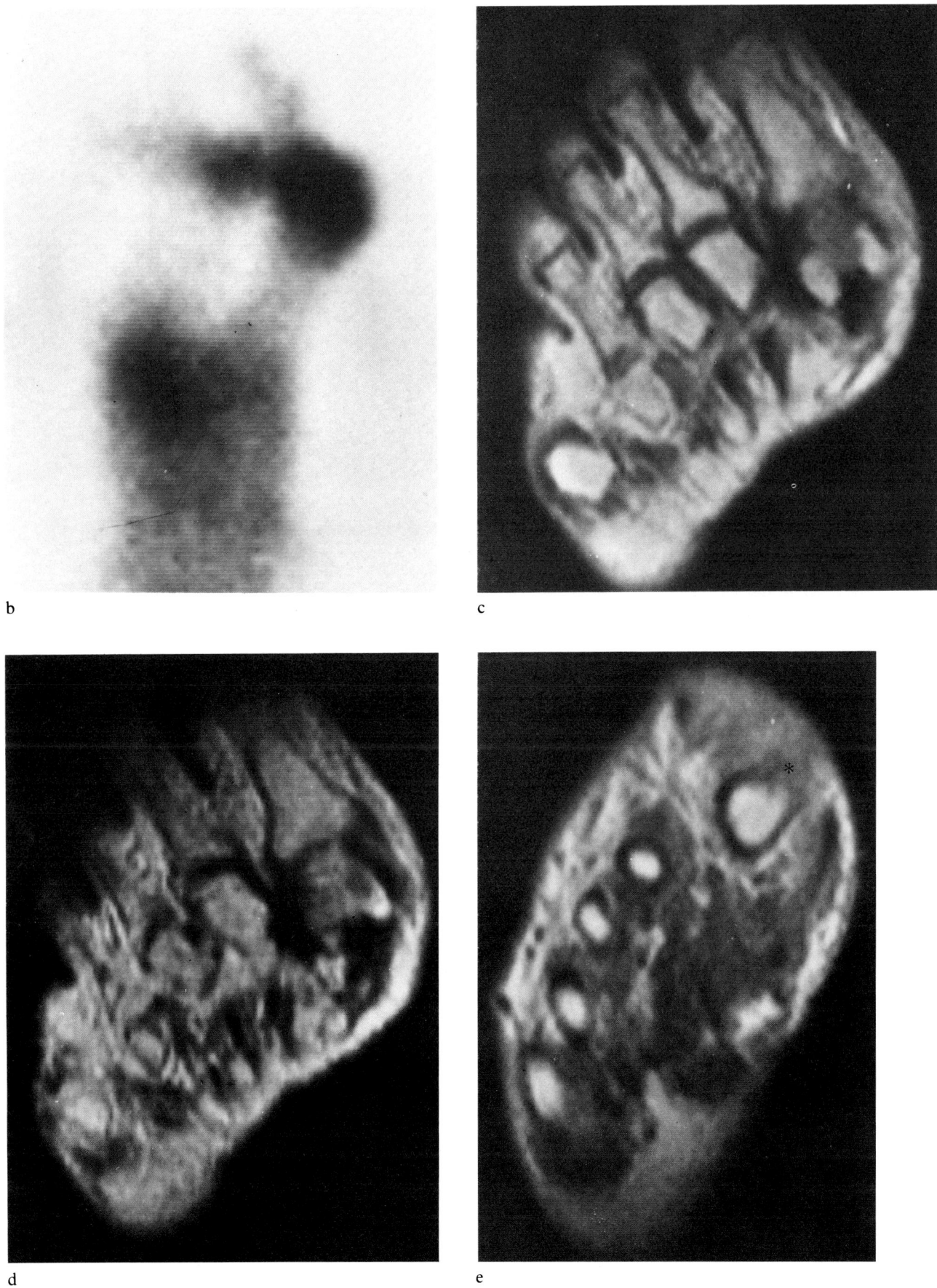

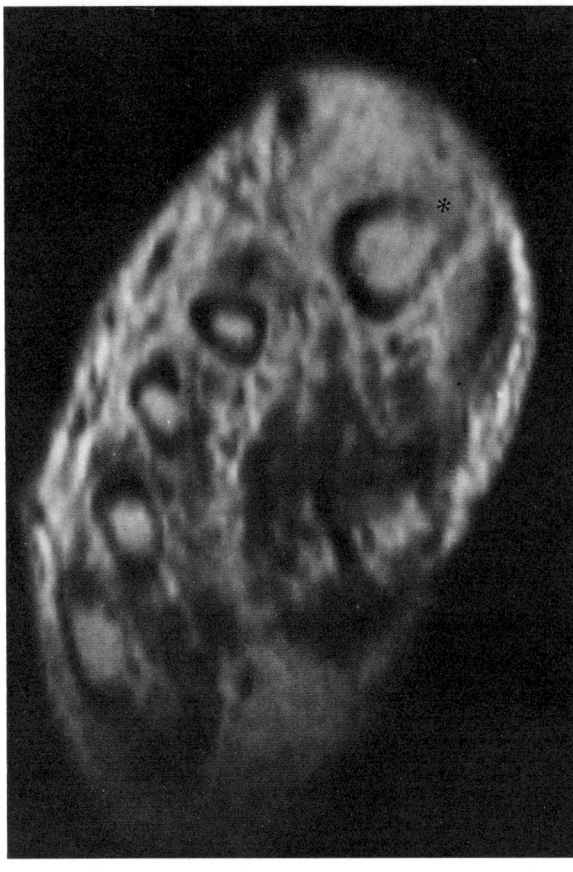

f

Figure 12.7 continued

Figure 12.8

A 15-year-old boy with subacute pyogenic osteomyelitis and Brodie's abscess of the right humerus. (**a**) Plain radiograph shows a well circumscribed focus of destruction in the metaphysis of the proximal humerus. (**b**) Coronal MR image (SE 500/28) of the shoulder reveals large focus of decreased signal intensity of the metaphysis, with ill-defined distal margin. The epiphysis is normal. (**c–e**) Contiguous axial CT images of the proximal humerus. A metaphyseal abscess cavity (★) and small extraosseous sequestrum (arrow) are depicted. The adjacent soft-tissue lesion is obscured by bony streak artifact. (**f–h**) Contiguous axial MR images at same levels (**f**, SE 500/28, **g,h**, SE 1500/56). The intra- and extraosseous lesions (★) have relatively high intensity signal on T_1-weighted image (**f**), the intraosseous lesion being outlined by a low-intensity rim. T_2-weighted images (**g,h**) disclose a very bright, homogeneous signal typical of Brodie's abscess (★), with soft-tissue extension (★). Small sequestrum cannot be depicted in MR images.

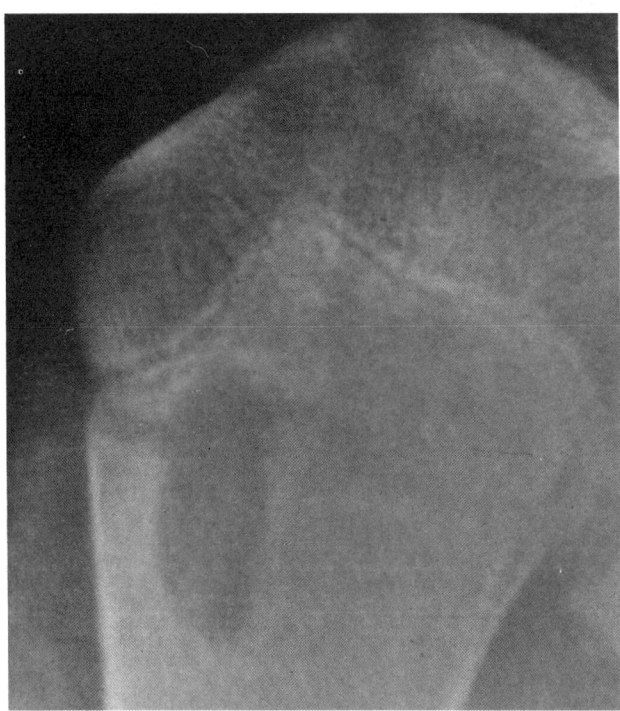

a

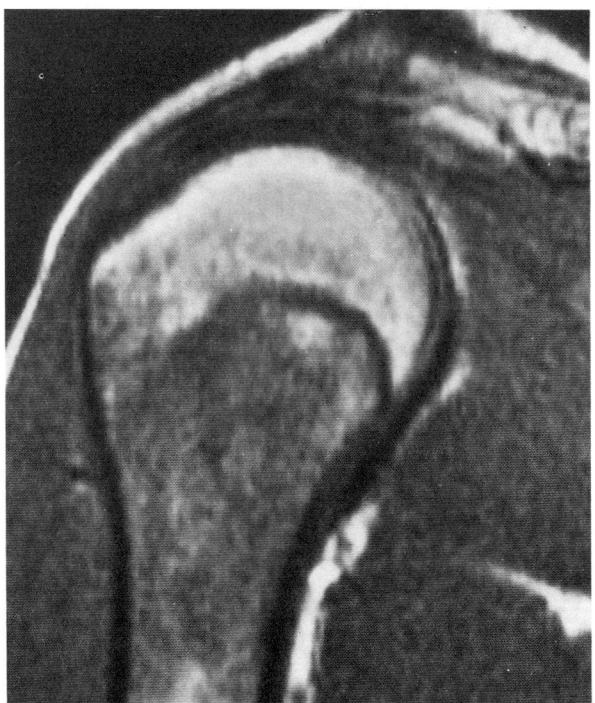

b

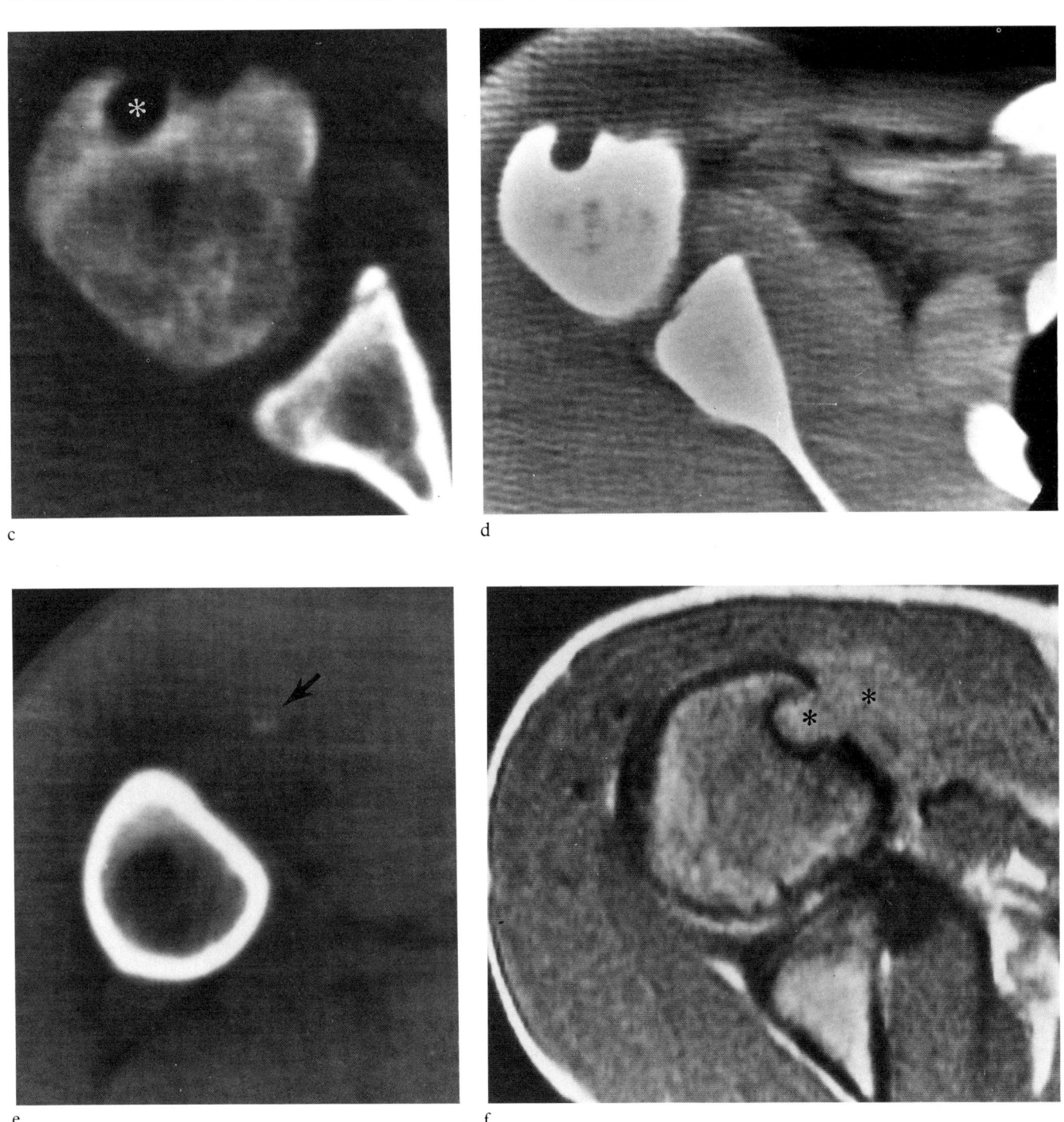

394 MRI atlas of the musculoskeletal system

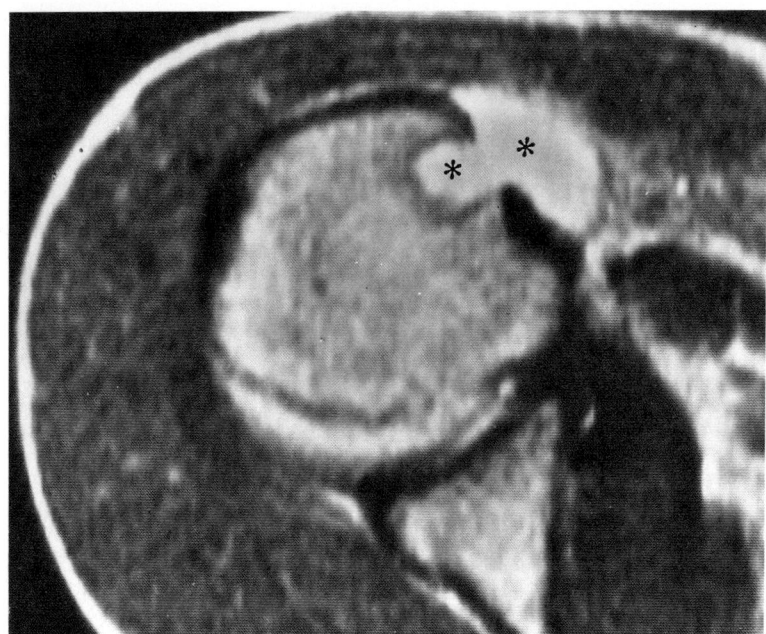

g

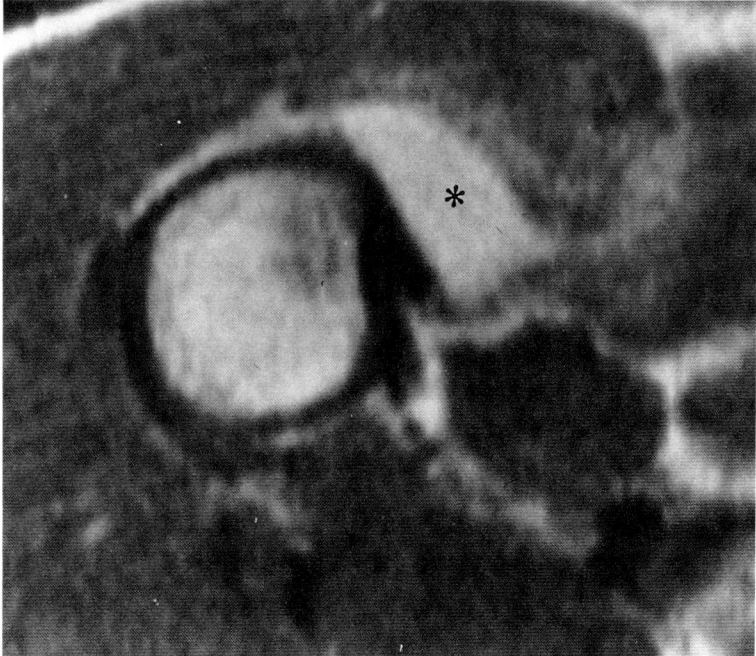

h

Figure 12.8 continued

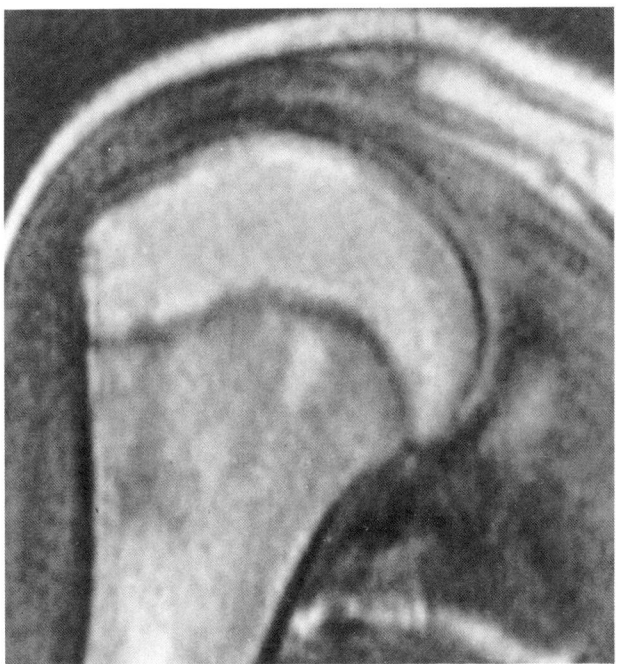

a

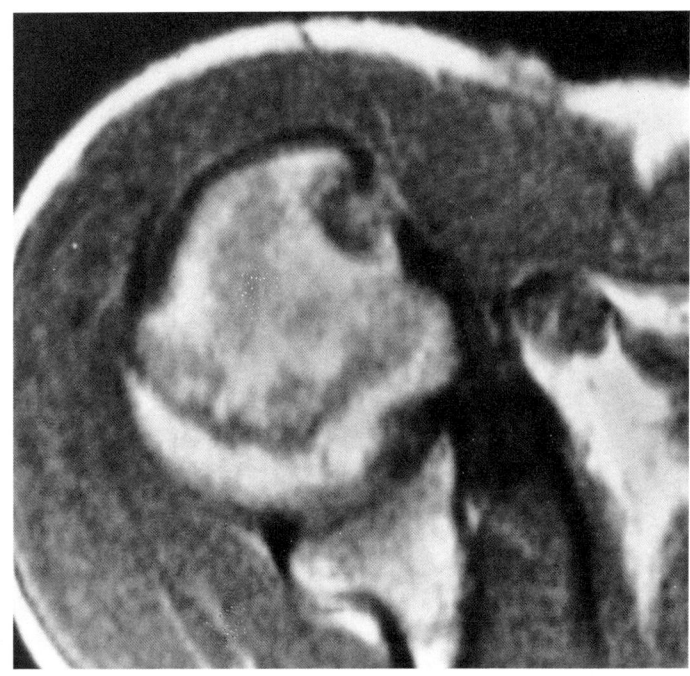

b

Figure 12.9

The same case as shown in Figure 8, 3 months after excision and drainage of the abscess. (**a**) Coronal MR image (SE 500/28) of the right shoulder. Almost normal signal intensity of the marrow is now present. (**b**) Axial MR image (SE 500/26). Signal intensity of the intraosseous lesion has decreased. (**c**) Axial MR image at the same level (SE 1500/56). Increased signal intensity of the lesion is far less prominent than before surgery, consistent with reparative granulation tissue. No abnormal signal intensity is present in soft tissue.

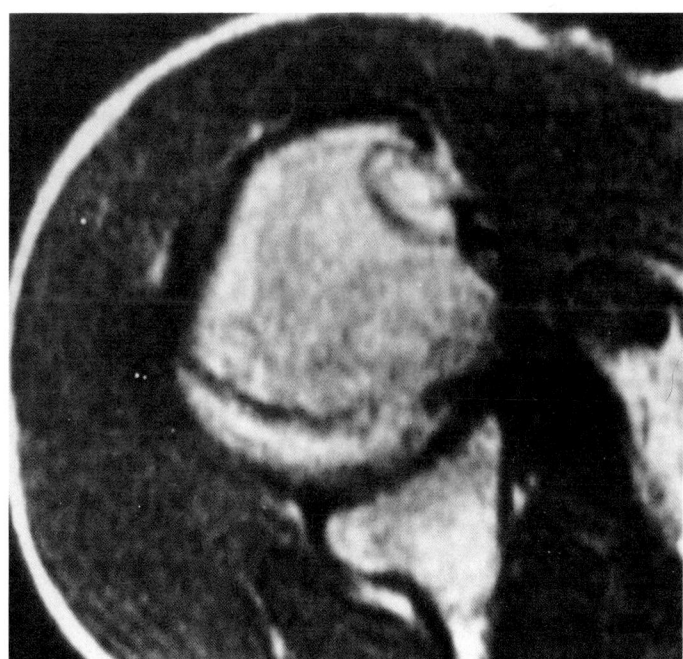

c

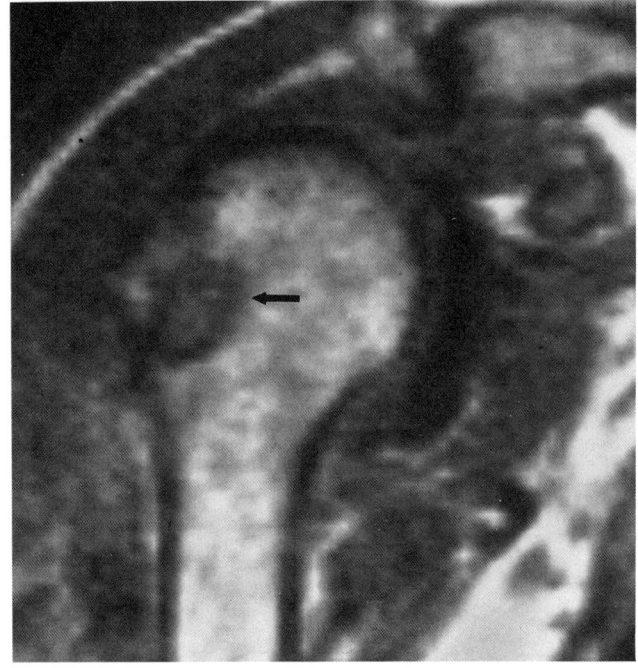

a

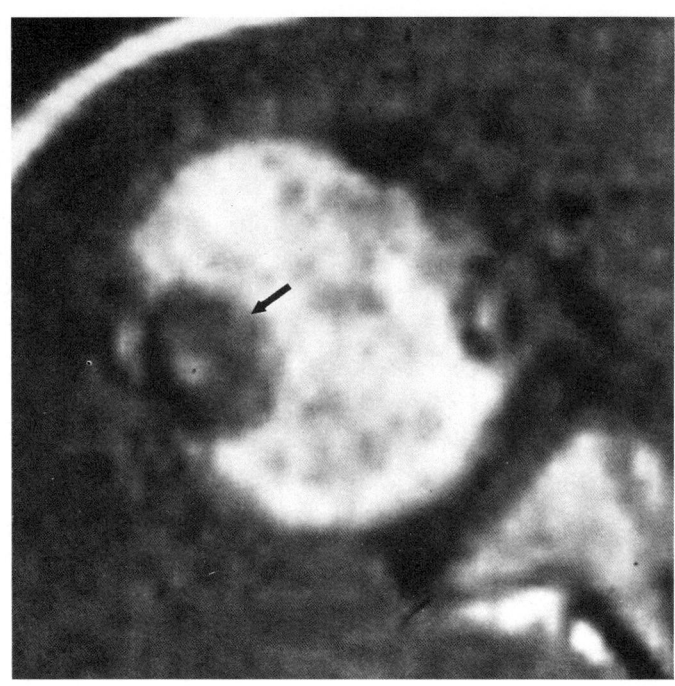

b

c

Figure 12.10

A 24-year-old man with subacute osteomyelitis of the humerus and disseminated coccidioidomycosis. (**a,b**) Coronal and axial MR images (SE 816/26) reveal a circumscribed signal of low intensity (arrow) in the proximal humerus. (**c**) Axial MR image at the same level (SE 2200/100) discloses low-intensity sequestrum of the cortical bone (★) surrounded by a zone of bright signal intensity representing abscess (arrow), well-demarcated by a low-intensity rim of reactive bone. (*Courtesy of Lee Chiu*)

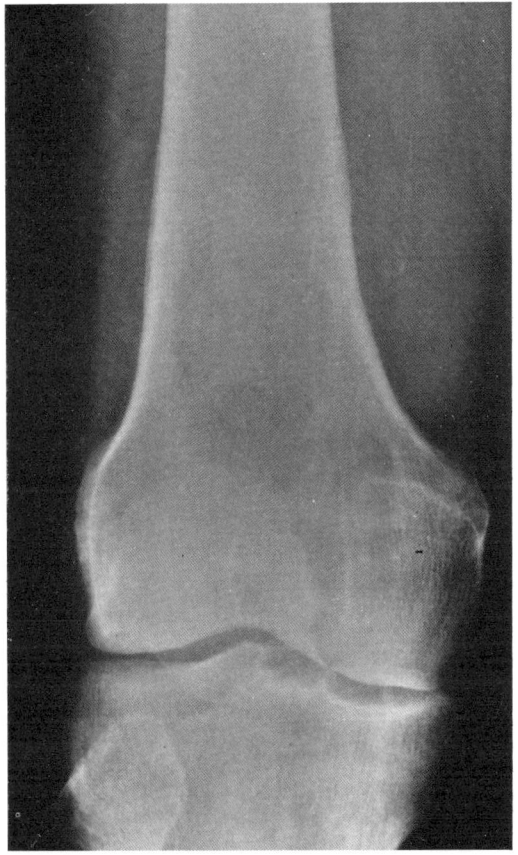
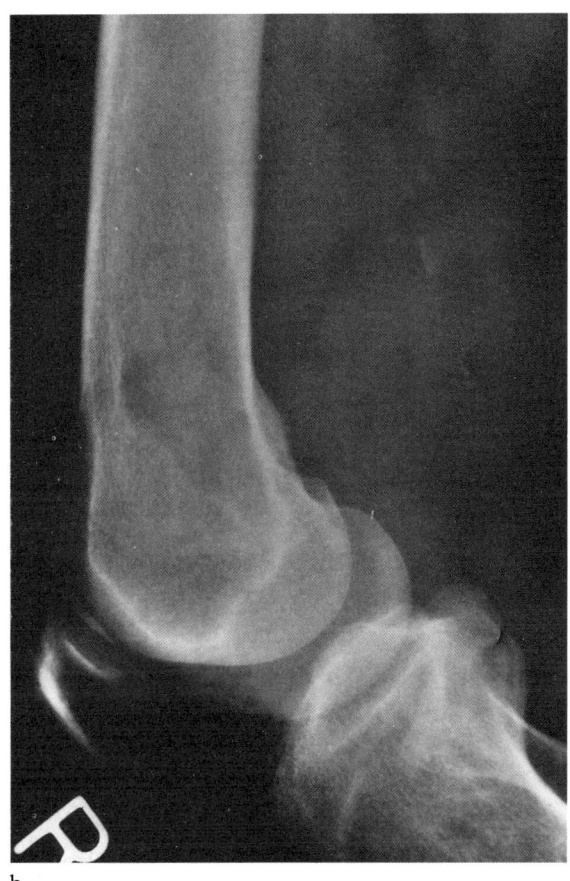
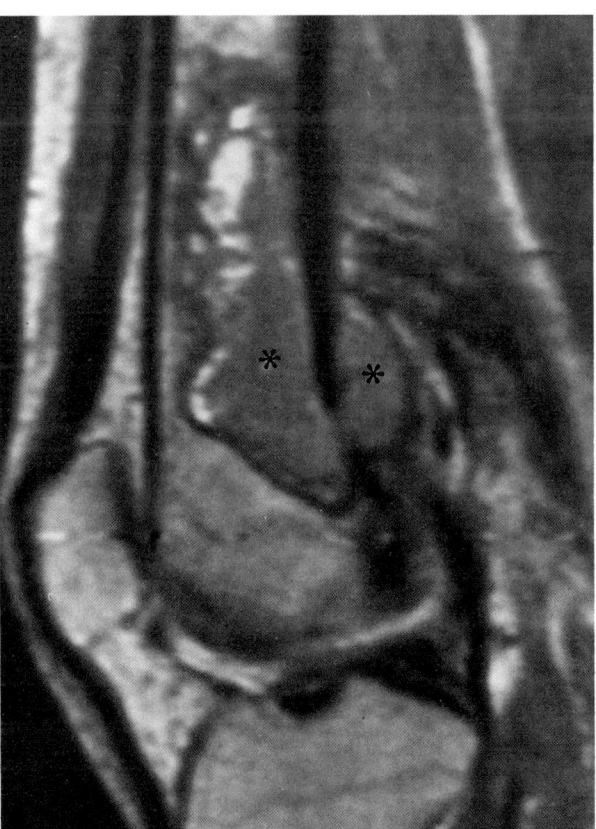

Figure 12.11

A 52-year-old man with chronic renal disease, who developed pyogenoc osteomyelitis complicating a medullary infarct of right femur.
(**a,b**) Anteroposterior and lateral radiographs of the femur reveal an ill-defined, scalloped, lytic lesion in the distal femur. Soft-tissue swelling is present along the posterior aspect of the femur. (**c**) First of double-echo sagittal MR image (SE 2000/25) discloses a well defined focus of necrotic bone composed of a large, central component of low signal intensity (★) surrounded by peripheral high-intensity signal. A well demarcated, low-intensity, posterior soft-tissue mass (★) communicates with the intraosseous lesion.
(**d,e**) Second echo sagittal and axial MR images (SE 2000/85). Signal intensity of the intra-and extraosseous abscesses (★) is characterized by the striking increase and homogeneity, well circumscribed by a dark rim. Effusion (arrowheads) is depicted in the knee joint.
(**f**) Subsequent CT scan with intracavitary injection of contrast medium shows communication between the extra- and intraosseous abscesses. (*Courtesy of Jerrold Mink*)

398 MRI atlas of the musculoskeletal system

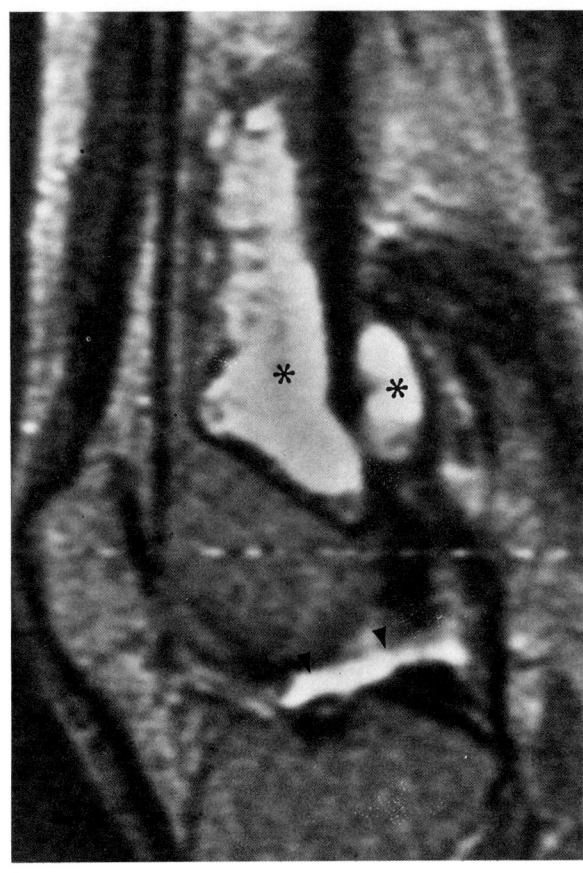

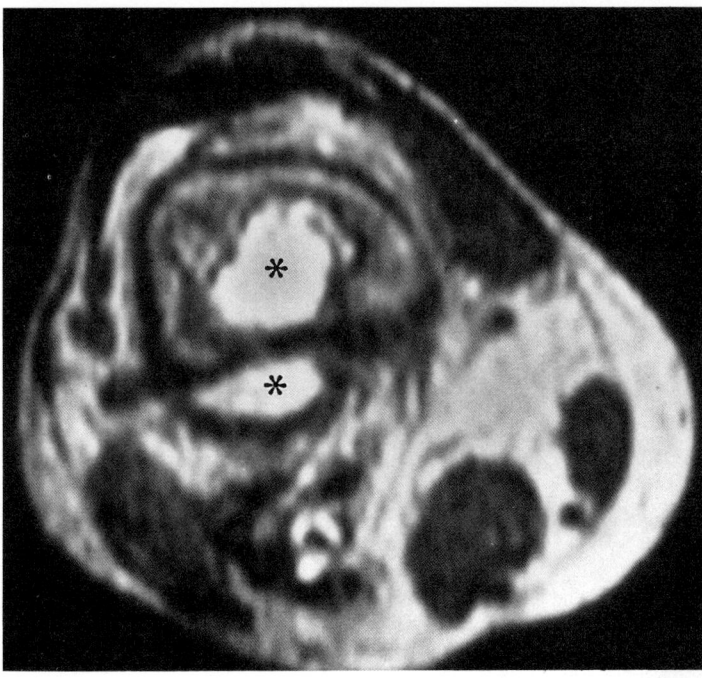

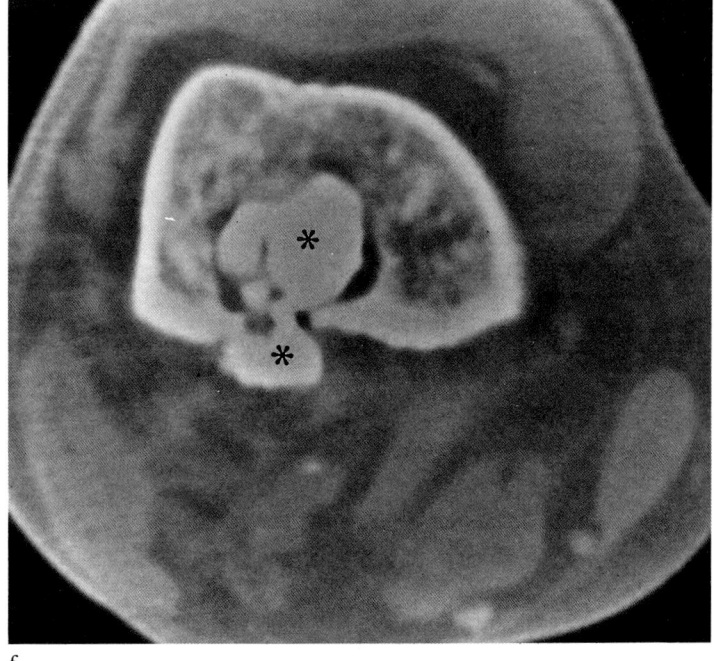

Figure 12.11 *continued*

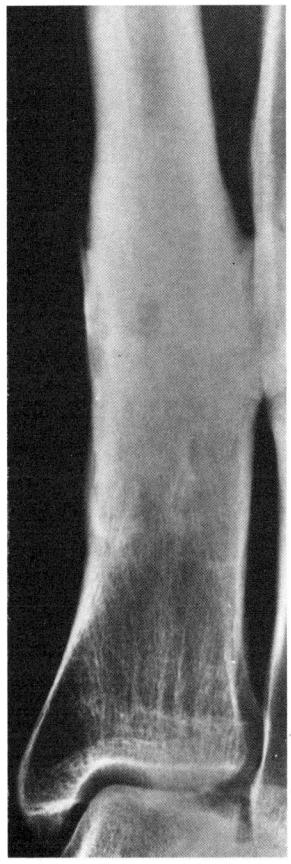
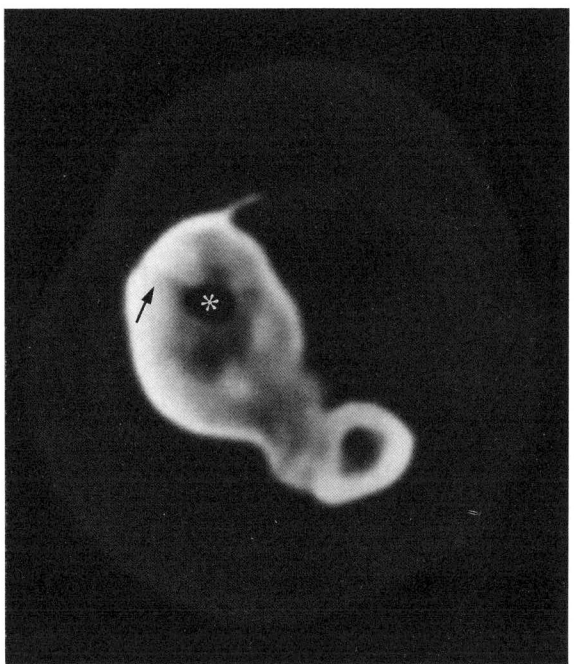
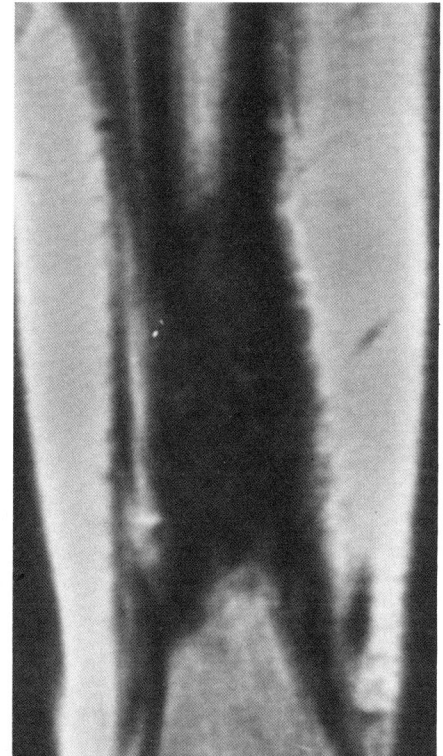

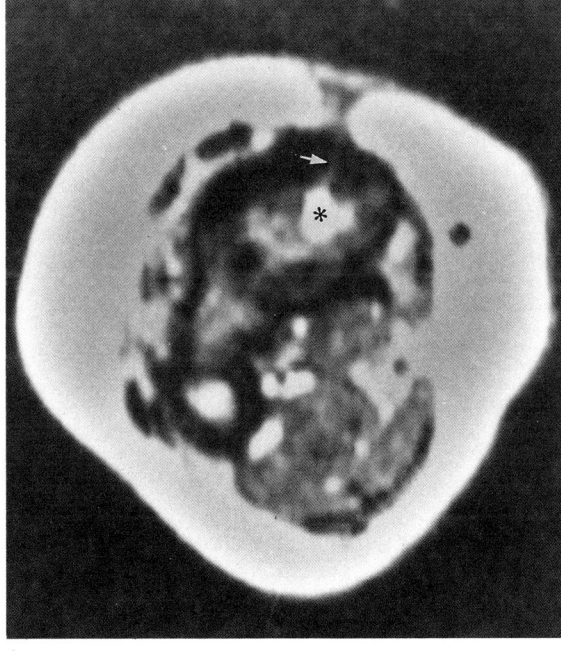

Figure 12.12

A 38-year-old woman with reactivation of chronic osteomyelitis and healed comminuted fracture of the tibia. (a) Plain radiograph reveals a deformed, widened, sclerotic region in the diaphysis of the tibia, the surrounding central radiolucent focus suggestive of an abscess cavity. (b) ^{111}Indium-labeled leukocyte scan reveals focus of increased activity in region of radiolucent focus seen in (a). (c) CT scan at the level of the abscess cavity (★) and cloaca (arrow). (d) Axial MR image at same level (SE 1500/56). Bright pus (★) is surrounded by intermediate-intensity signal of the repaired medullary bone, and low-intensity signal of thickened cortex. Linear focus of intermediate signal intensity extending from the intraosseous abscess to the skin represents the cloaca (arrowhead) and sinus tract. (e) Coronal MR image (SE 500/26). The extensive, well defined region of inhomogeneously decreased signal intensity incorporates the medullary cavity and thickened cortex.

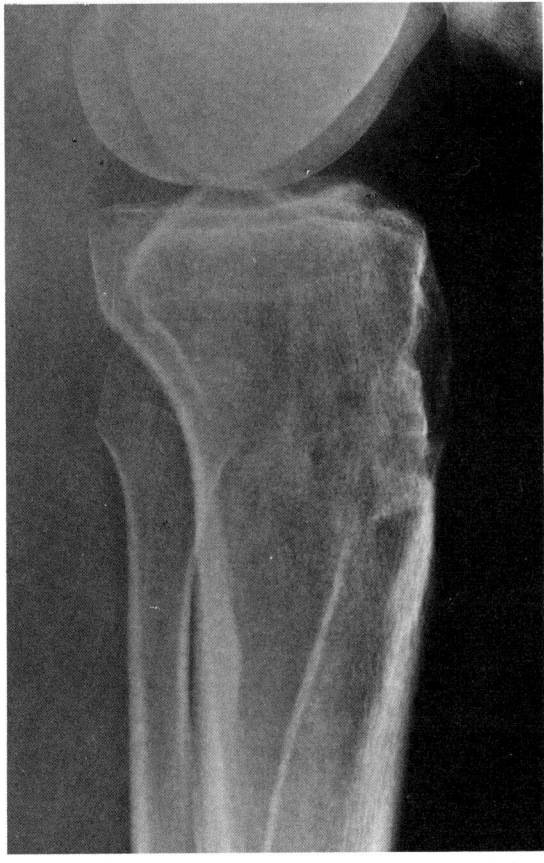
a

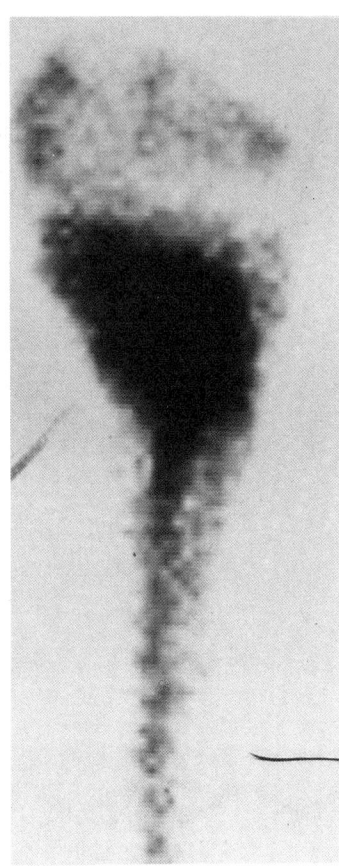
b

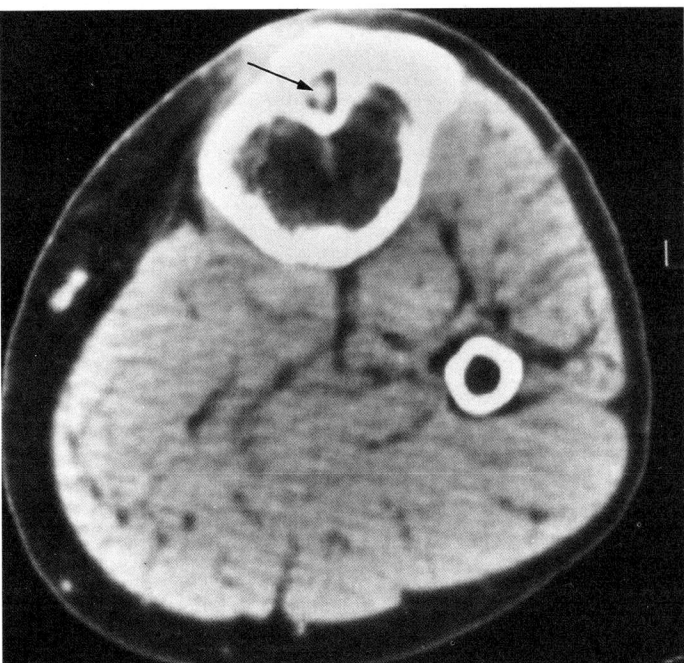
c

Figure 12.13

A 30-year-old man with chronic osteomyelitis and sequestrum of the proximal tibia resulting from an infected distal tibial fracture 6 years earlier.
(a) Lateral radiograph of the proximal end of the tibia discloses a deformity in the region underlying the tibial tubercle, and sclerotic reactive bone along the tract of the former intramedullary rod.
(b) ^{99m}Tc-diphosphonate bone scan discloses striking increased activity in the proximal tibia. (c) CT scan of the proximal tibia. The rod tract is bounded by a sclerotic rim and contains a central sequestrum (arrow). The cortex is thickened and irregular in contour, and the density of the pretibial soft tissue is increased.
(d) Axial MR image at the same level (SE 550/28). Irregular area of abnormally low-signal intensity is present in the tibia and pretibial soft tissue.
(e) Axial MR image (SE 1500/56) depicts bright, high-intensity abscess (arrow) with central dark sequestrum and rim of low intensity sclerotic bone. Soft-tissue signal remains unchanged, indicating absence of active infection and presence of scar tissue.

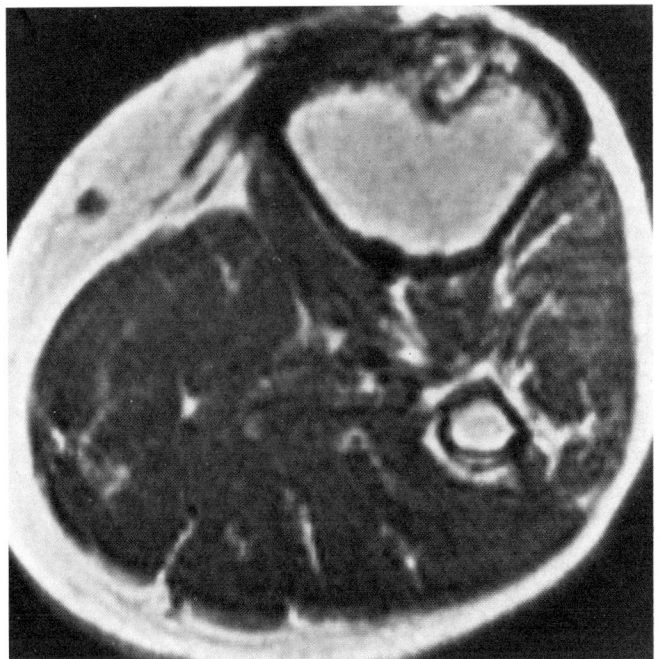
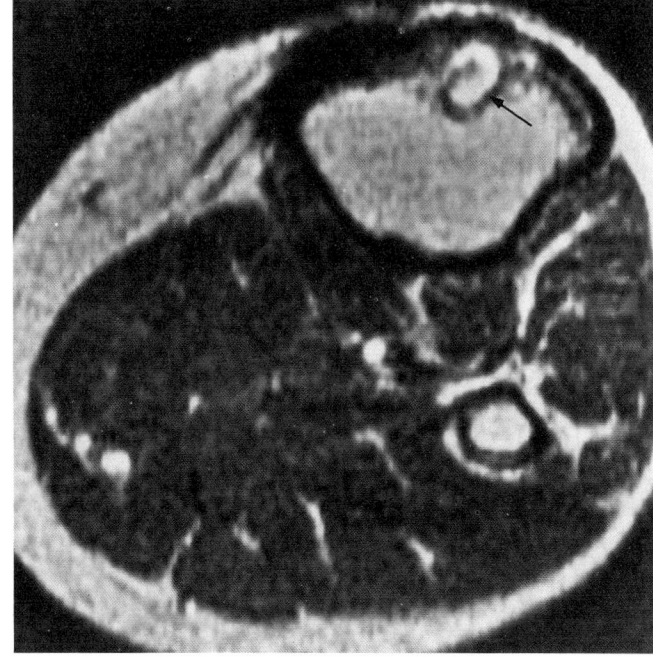

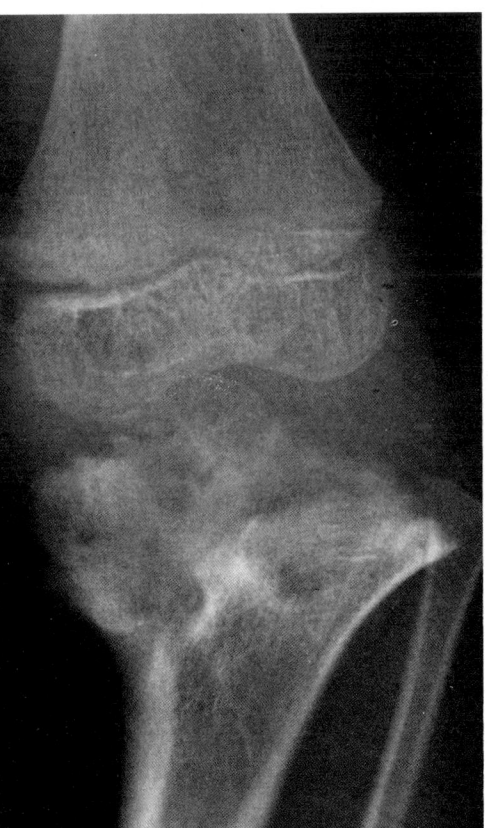
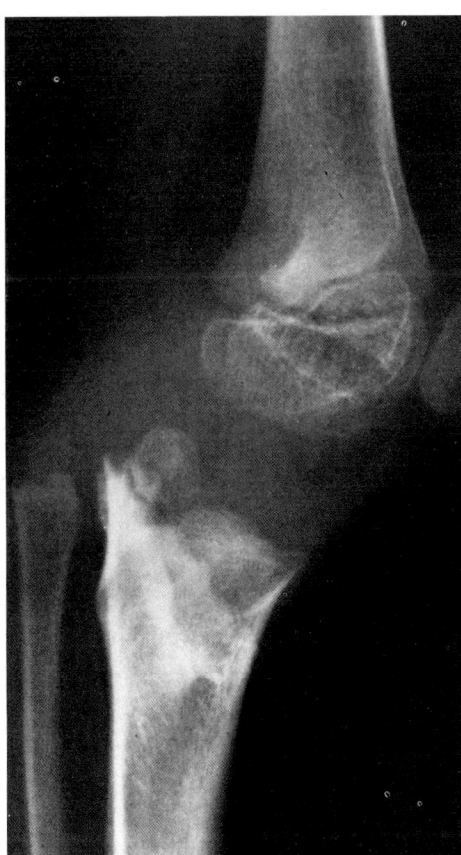

Figure 12.14
A 6-year-old boy with chronic osteomyelitis of the left proximal tibia and associated pyarthrosis of the knee. (**a,b**) Anteroposterior and lateral radiographs of the left knee show genu varus deformity associated with irregular, sclerotic new bone formation and lytic changes in the tibial metaphysis and epiphysis. (**c,d**) Coronal and (**e**) sagittal MR images (SE 500/28) disclose inhomogeneous, irregular signal intensity of the tibial epiphysis and metaphysis. Obvious destruction of the articular cartilage is characterized by loss of normal intermediate signal intensity. The foci of active infection cannot be depicted without T_2-weighted images.

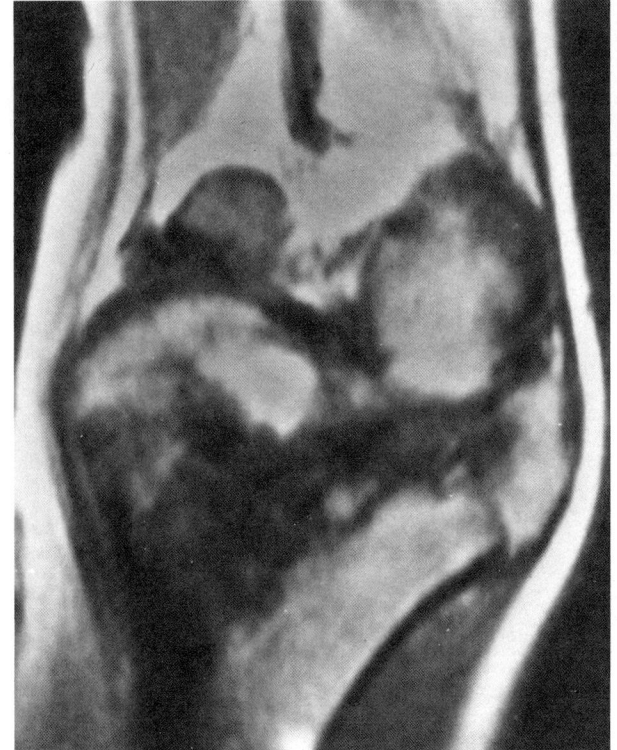

c

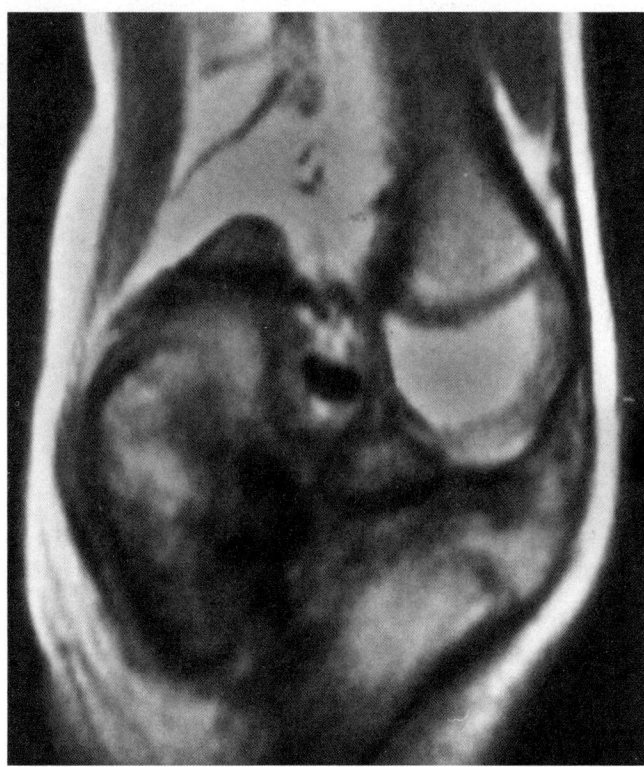

d

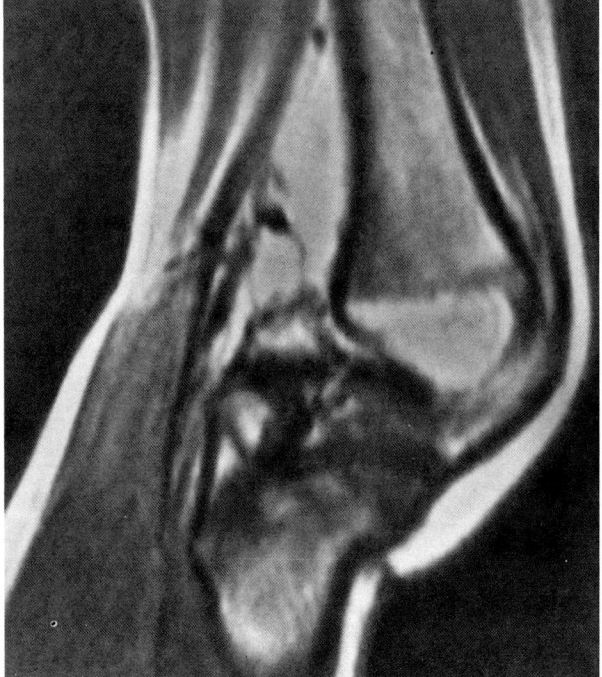

e

Figure 12.14 *continued*

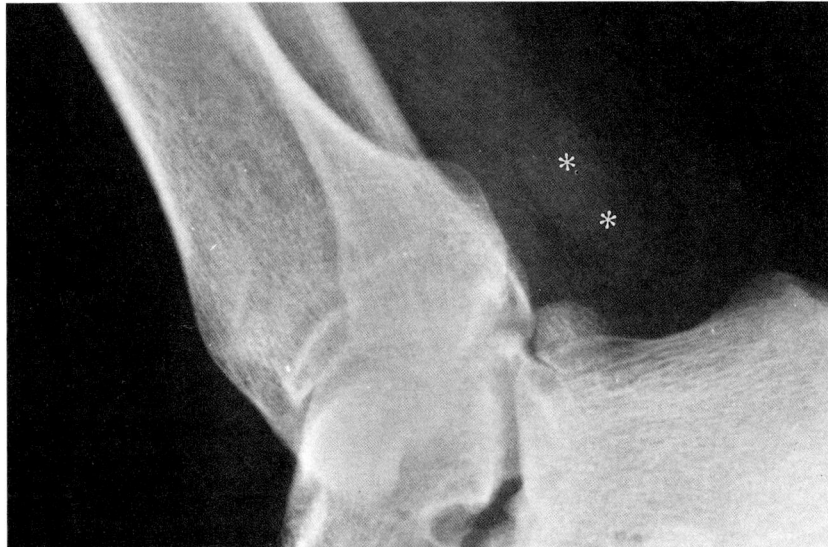

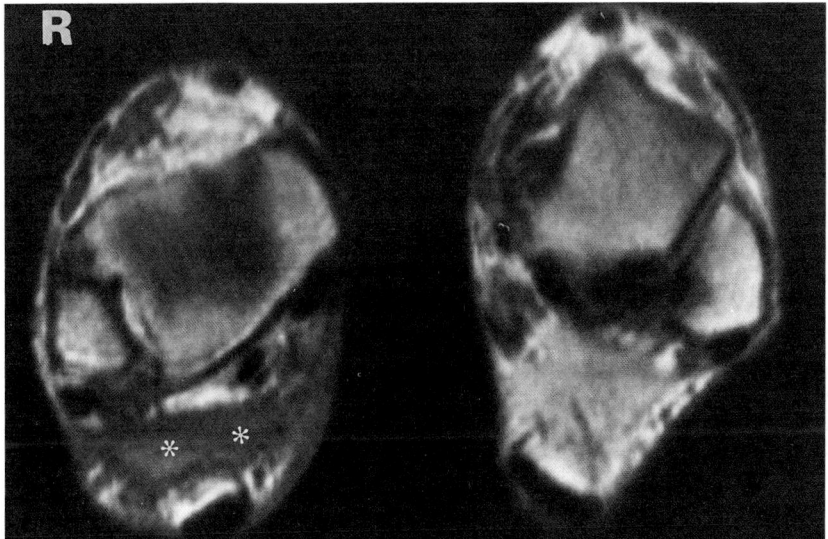

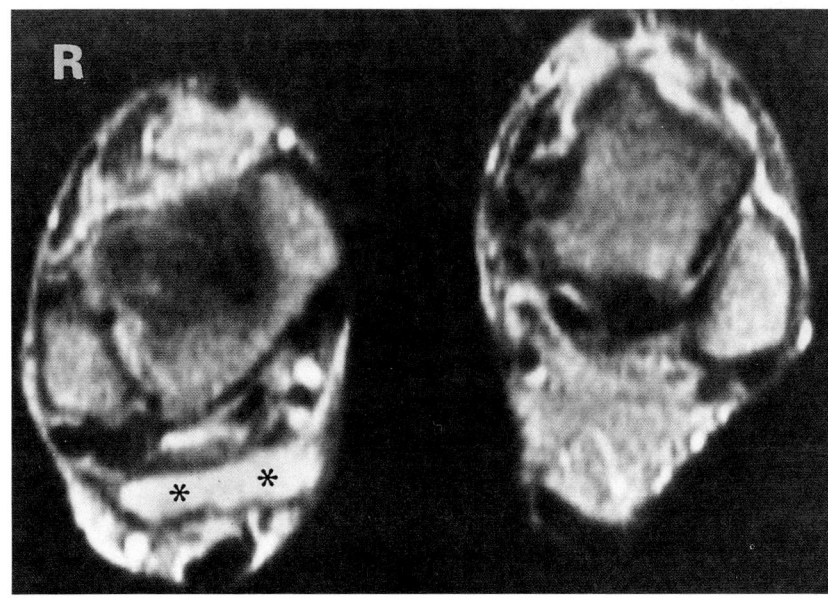

Figure 12.15

A 50-year-old man with a soft-tissue abscess of the right ankle. (**a**) Lateral radiograph reveals increased soft-tissue density (★) anterior to the Achilles tendon. (**b**) Axial MR image (SE 500/28). A fairly well defined abscess of intermediate signal intensity (★) lies anterior to the Achilles tendon. Diffuse areas of intermediate signal intensity are present in the adjacent skin (compare with the contralateral normal ankle). (**c**) Axial MR image (SE 2000/84). A soft-tissue abscess (★) is very bright, well-demarcated, and covered by bright, edematous skin. The underlying bone is normal.

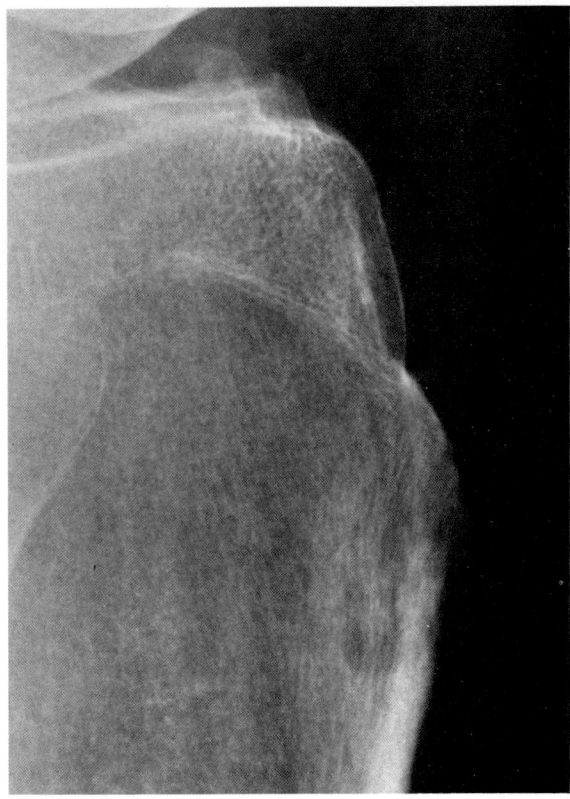
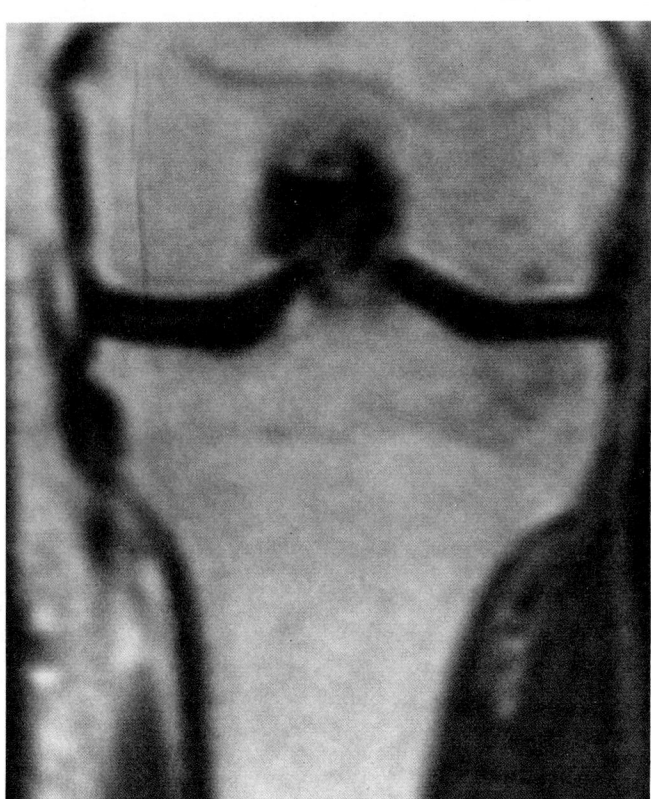
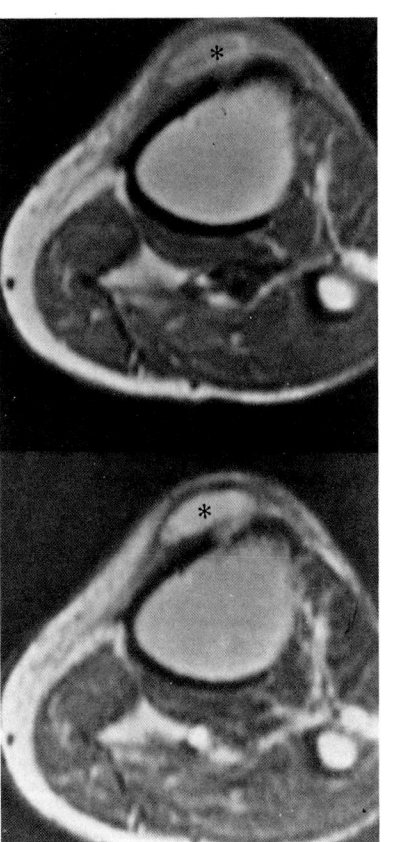

Figure 12.16

A 76-year-old man with a subcutaneous abscess and subjacent patellar tendinitis. (**a**) Lateral radiograph reveals ill-defined osteolytic foci in the region of the tibial tubercle. (**b**) Apparently normal coronal MR image (SE 500/28). (**c**) Axial MR images. *Above*, (SE 1000/20) reveals a well defined region of relatively high-signal intensity (★) representing an abscess anterior to the abnormally increased signal of the inflamed patellar tendon. The bone remains normal. *Below* (SE 2000/60) discloses a subcutaneous abscess with bright homogeneous signal (★), while the patellar tendinitis manifests irregular, high-signal intensity.

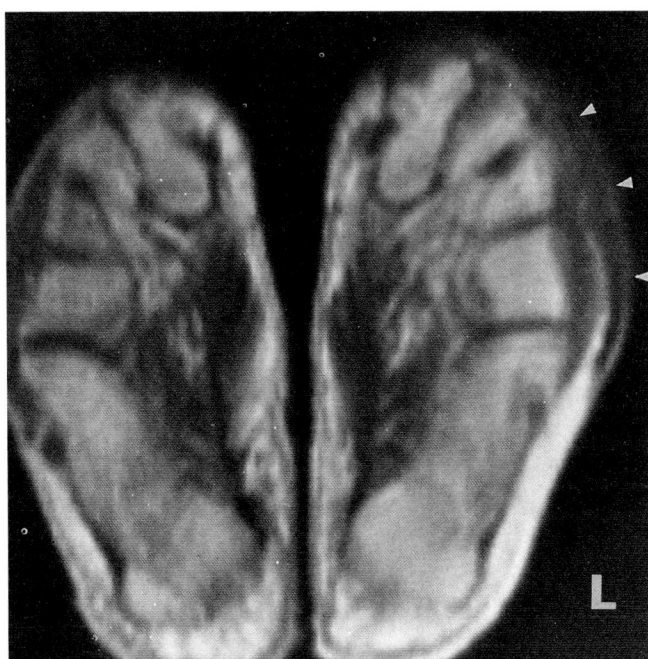

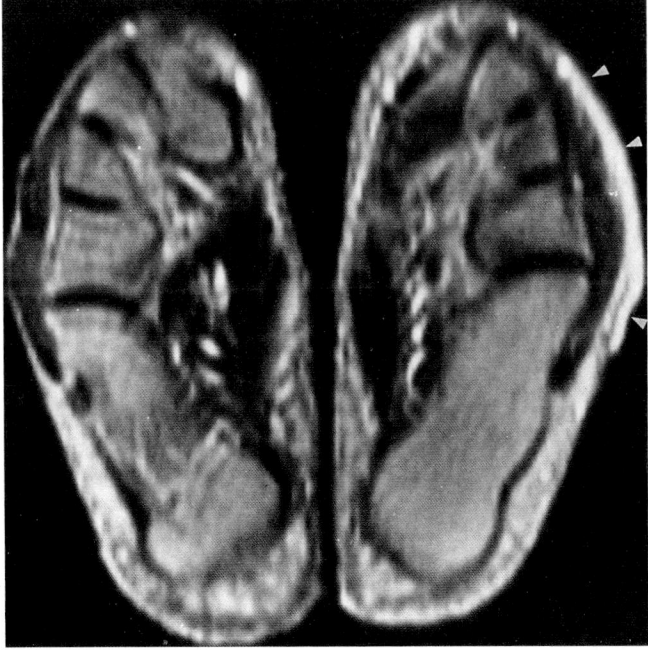

Figure 12.17

A 34-year-old woman with cellulitis of the left foot. (**a**) Axial MR image (SE 500/28) of both feet. Signal intensity at the lateral aspect of the left foot is abnormal, being slightly increased, with mild interruption of the subcutaneous fat (arrowheads) when compared with the contralateral normal side. (**b**) Axial MR image (SE 500/28) of soft-tissue areas (arrowheads) corresponding to those in (**a**) indicates an inflammatory process consistent with clinical diagnosis of cellulitis. The underlying bone is normal.

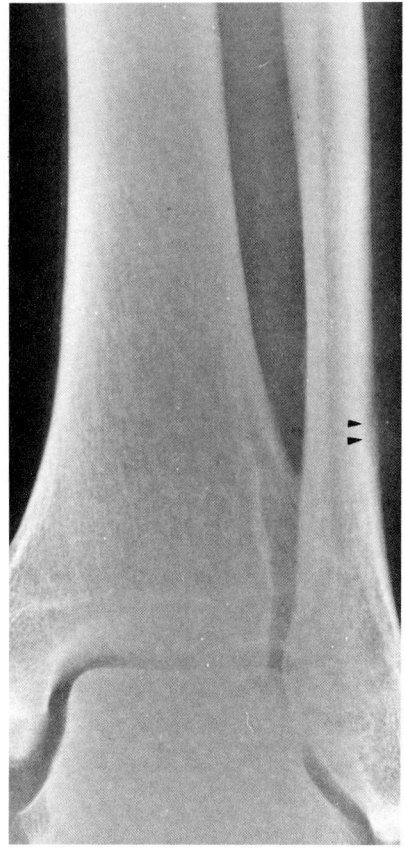
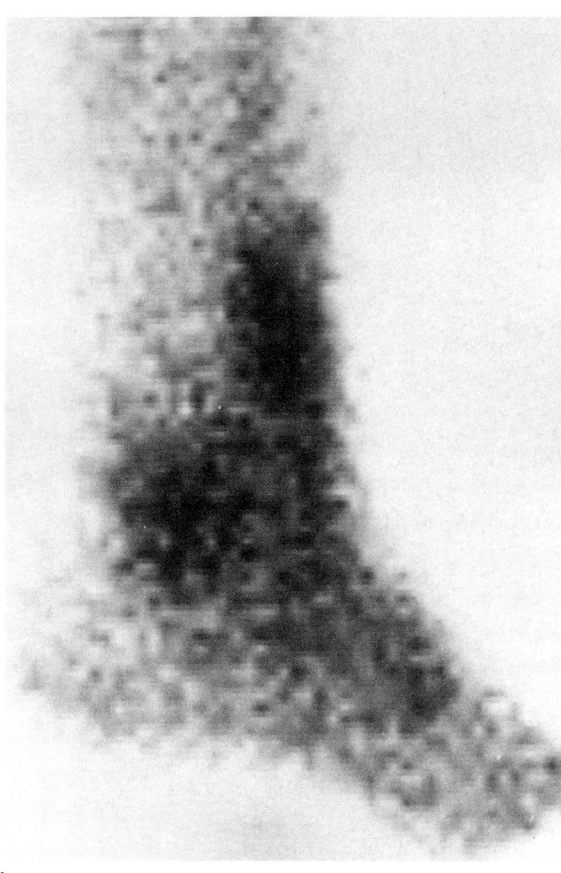
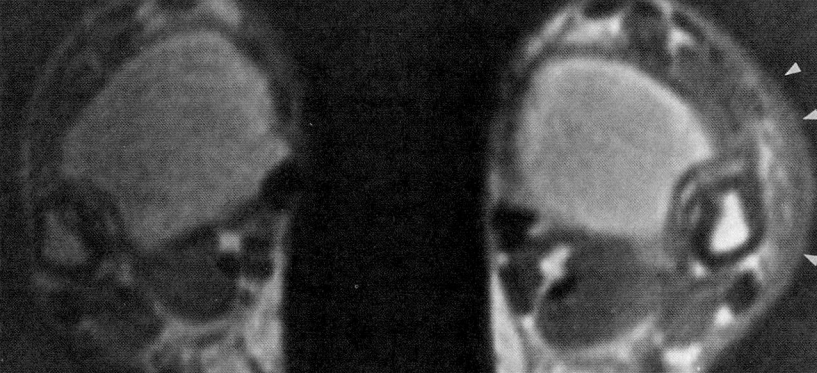

Figure 12.18

A 33-year-old man with cellulitis of the lower part of the left leg and periostitis of the left distal fibula. (**a**) Plain radiograph reveals periosteal new bone along the distal fibula (arrowheads) and adjacent soft-tissue swelling (**b**) ^{99m}Tc bone scan shows increased uptake at the left ankle. (**c,d**) Axial MR images at the level of the ankle. (**c**, SE 500/28 and **d**, SE 1500/56). Abnormal, slightly increased signal intensity of the soft tissue and interruption of normal, bright subcutaneous fat (arrowheads) are depicted adjacent to the lateral malleolus. The bone marrow is normal. (**e,f**) Axial MR images proximal to the ankle (**e**, SE 500/28 and **f**, SE 1500/56). Cellulitis and periostitis cause the same pattern of signal intensity as in (**c**) and (**d**) lateral to the distal fibula (arrowheads). The sinus tract, coated with exudate, is clearly manifested as a very bright signal on the T$_2$-weighted image (arrow).

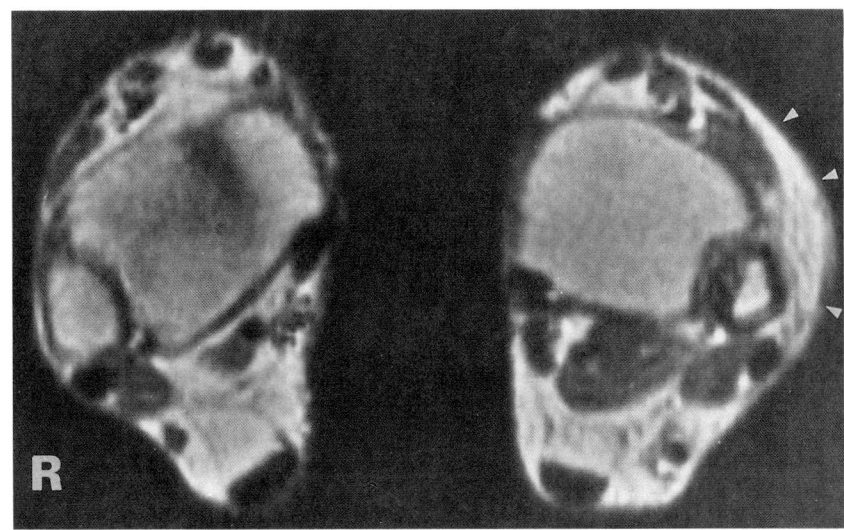

d

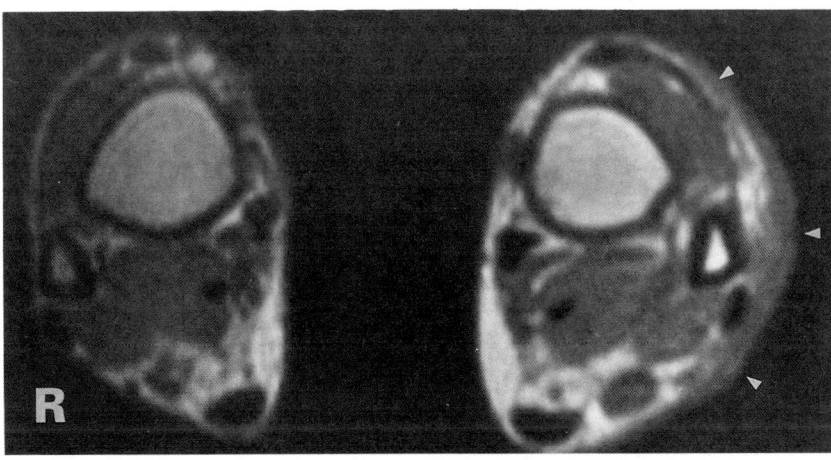

e

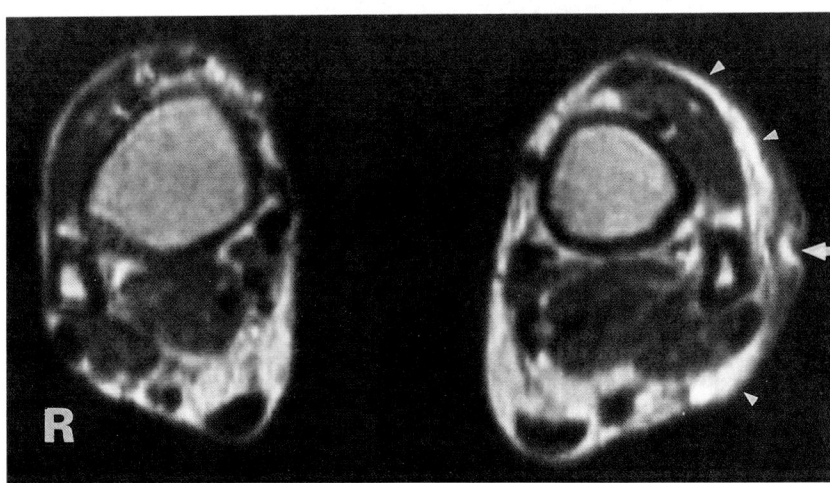

f

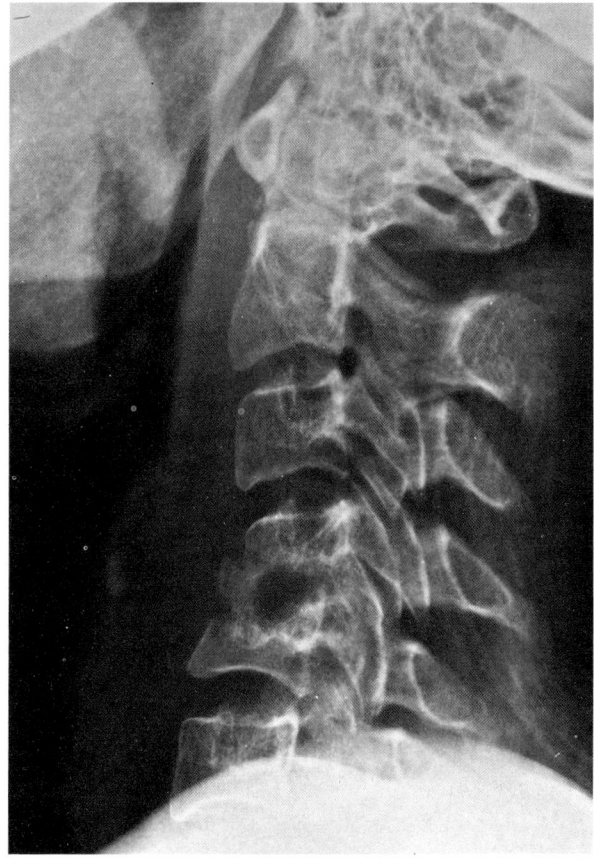
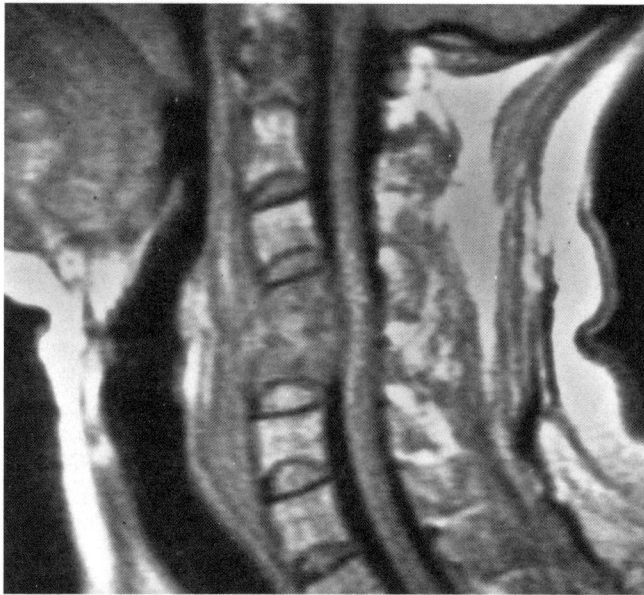
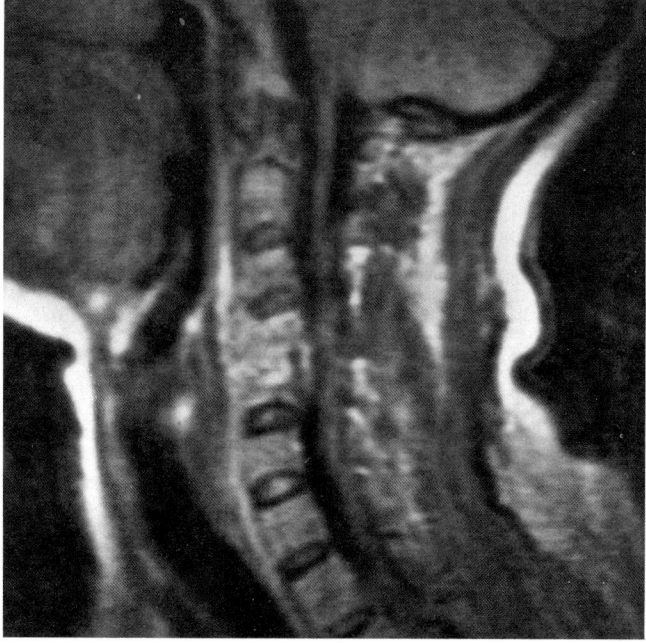

Figure 12.19

A 26-year-old woman with pyogenic osteomyelitis of C5–C6. (**a**) Lateral radiograph reveals destruction of bodies of C5 and C6 as well as the intervening disk. (**b**) Sagittal MR image (SE 500/28) discloses loss of C5–C6 intervertebral disk and decreased signal of the adjacent vertebral bodies extending posterior to involve the thecal sac. (**c**) Sagittal MR image (SE 1500/56) reveals increased signal intensity of the involved vertebrae. The destroyed intervertebral disk cannot be identified.

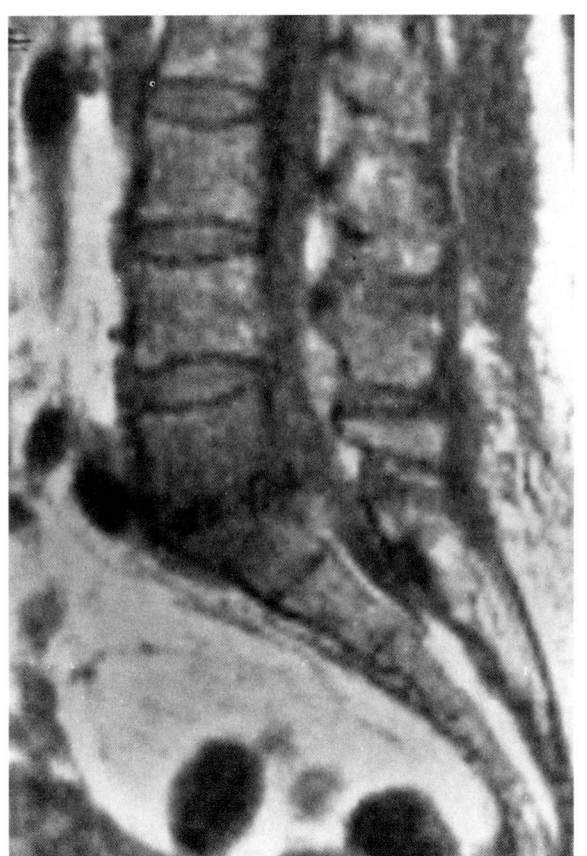
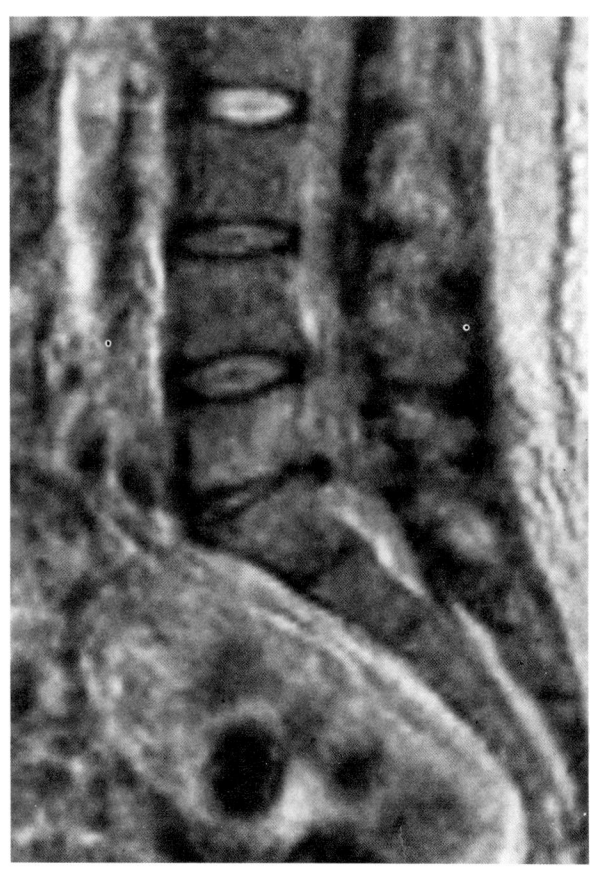

Figure 12.20

A 68-year-old woman with pyogenic osteomyelitis at the L5–S1 region. (**a**) T_1-weighted sagittal MR image (SE 700/30) reveals diminished signal of the L5-S1 intervertebral disk and adjacent portions of vertebral bodies. The adjacent end plates are indistinct. (**b**) T_2-weighted sagittal MR image (SE 1800/80) discloses increased signal of the L5-S1 disk and adjacent portions of the vertebral bodies, implying active infection. In addition, the normal intranuclear cleft, seen in the more superior disks, has disappeared. (*Courtesy of Lee Chiu*)

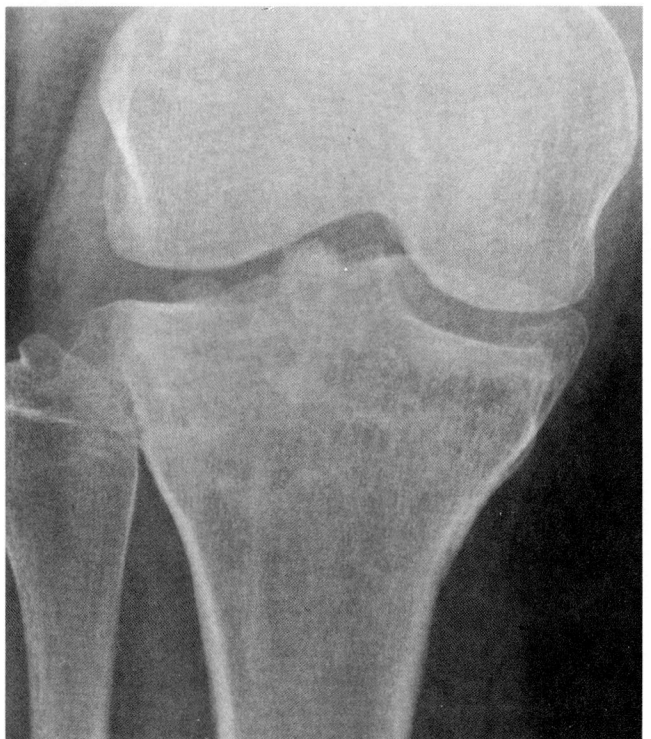

a

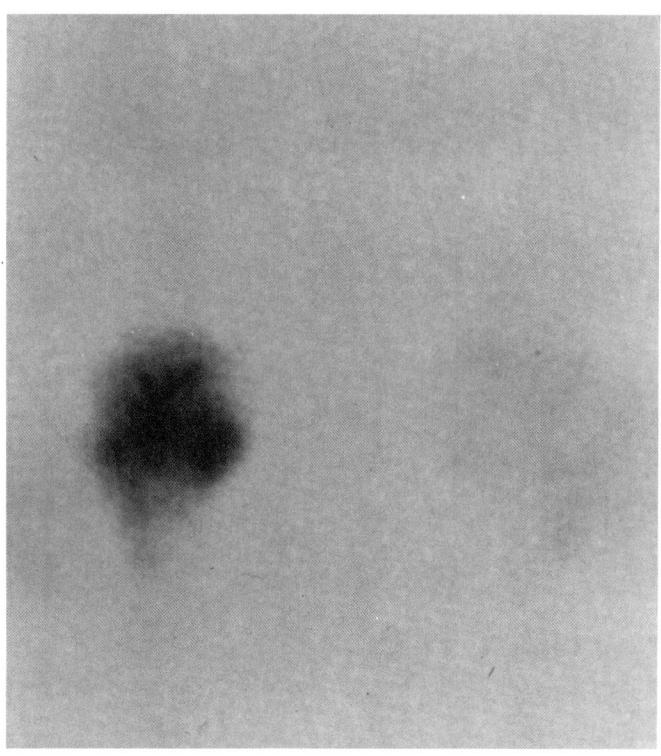

c

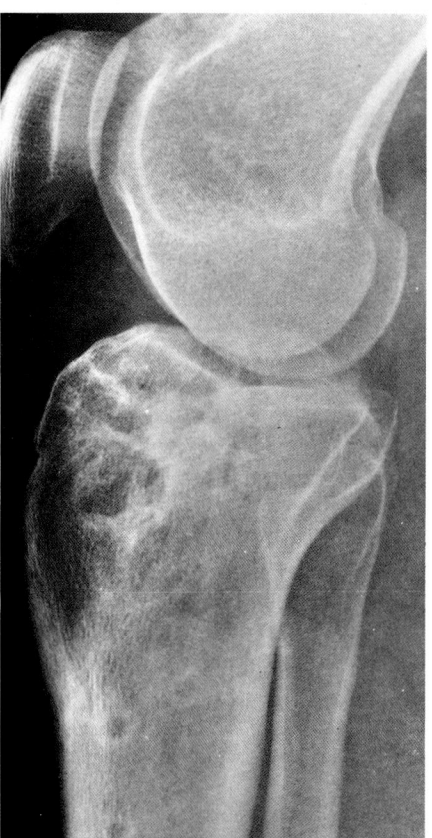

b

Figure 12.21

A 22-year-old woman with a healed fracture of the tibial plateau and adjacent localized symptoms and signs suggesting the possibility of osteomyelitis. (**a,b**) Anteroposterior and lateral radiographs reveal the healed fracture of the tibial plateau. Pin tracts are seen in the lateral view. (**c**) ^{99m}Tc-diphosphonate bone scan discloses intense activity in the region of the right knee. (**d**) Coronal MR image (SE 500/26) discloses abnormal, diffuse intermediate signal intensity of the proximal tibia with disappearance of the lateral compartment joint space. Only the articular cartilage of the medial femoral condyle (arrowheads) remains. Pin tracts appear as circular areas of low signal intensity with occasional bright, streaky artifacts owing to retained metallic microfragments. (**e**) Coronal MR image (SE 1500/56) discloses only slight increase in signal intensity compared to (**d**), still lower than that of normal marrow. Pin tracts appear unchanged. Severe narrowing with loss of the articular cartilage of the lateral compartment is clearly demonstrated. As seen in (**d**), only articular cartilage of the medial femoral condyle (arrowheads) remains. No focus of bright signal intensity is present to suggest inflammatory tissue or pus. The absence of inflammatory disease was confirmed at surgery.

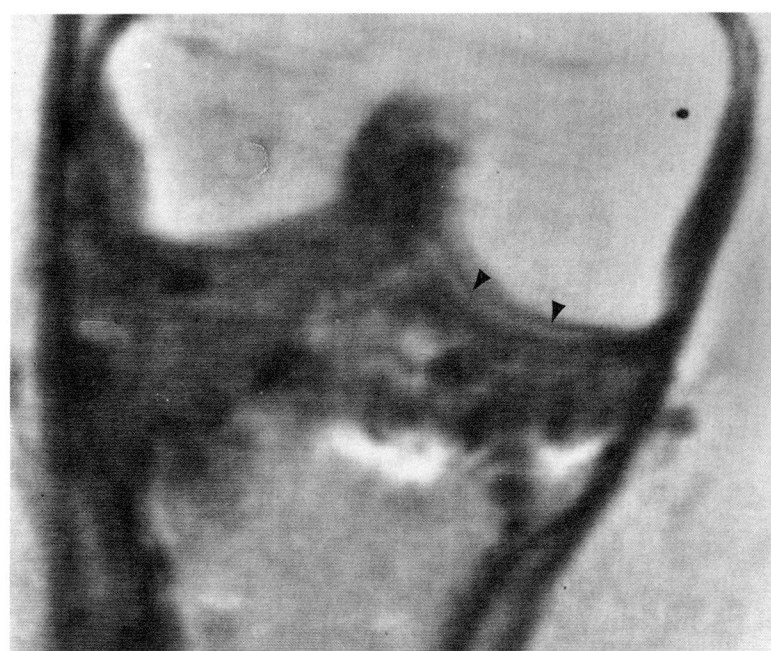
d

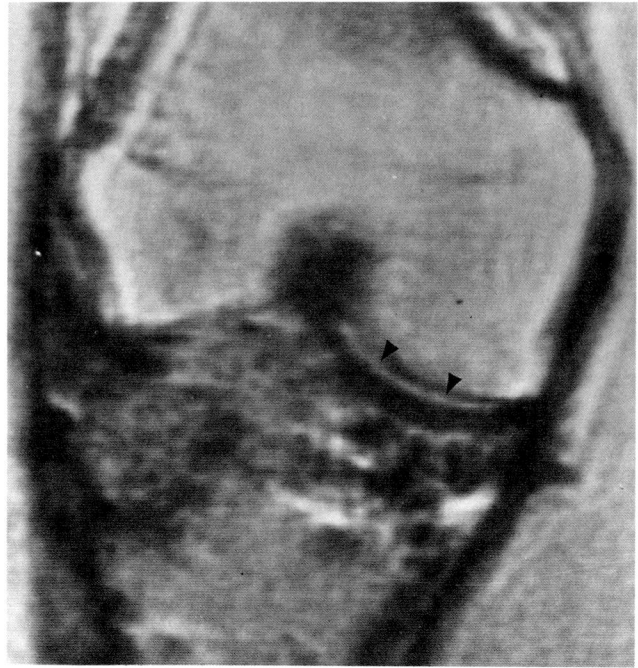
e

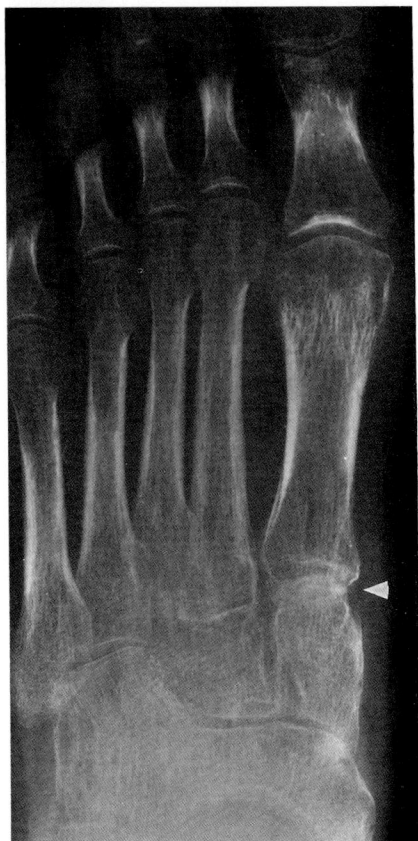
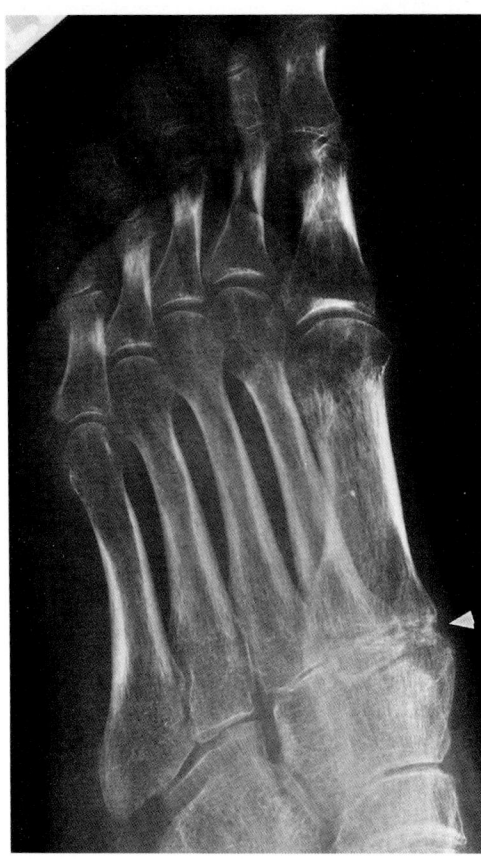
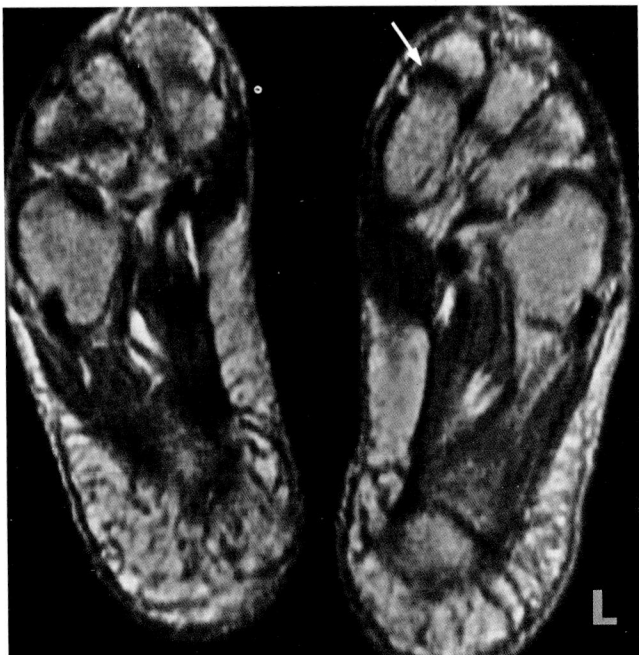

Figure 12.22
A 34-year-old man with pain in the left foot, aplastic anemia and healed pyarthrosis of the left 1st tarsometatarsal joint. (**a,b**) Anteroposterior and oblique radiographs of the left foot reveal narrowed 1st tarsometatarsal joint (arrowhead), minimal sclerosis of the subchondral bone and no adjacent soft-tissue swelling. (**c**) T_2-weighted axial MR image (SE 1500/56) reveals a broadened, more prominent and more distinct focus of low-signal intensity at the left 1st tarsometatarsal joint, consistent with loss of articular cartilage and thickened subchondral cortex (arrowhead). No significant abnormal increased signal intensity in soft tissue or bone is present to suggest active infection.

Index

abscess: Brodie's, 381, 382, 392–5
  subcutaneous, 404
Achilles tendon, ruptured, 277
acromegaly, 143
acromioclavicular joint: capsule hypertrophy, 108
  degenerative disease, 110
  fluid, 109
adenocarcinoma, 52, 347, 373
adhesions, temperomandibular joint, 74
adhesive tendonitis, 142, 171
alcoholism, 176
aliasing artifacts, 15, 24, 27, 41
anemia, 175, 177, 280, 284, 371, 387, 412
aneurysm, 320, 376–7
ankle, 3, 266–78
  normal anatomy, 267–76
ankylosing spondylitis, 61
aplastic anemia, 280–1, 284, 371, 387, 412
aplastic marrow, 280–1
arachnoiditis, 30, 51
arteriovenous malformations, 320, 378
arthritis, 143
  osteoarthritis, 177, 201, 202–3, 211, 218
  psoriatic, 61, 94
  rheumatoid, 31, 61
  septic, 205, 379, 381
arthrography, 95
arthropathy, hemophiliac, 290–1
arthroplasty, hip, 213–14
arthroscopy, 219
arthrotomography, 219
articular disk, temperomandibular joint, 60–1, 64–8, 75–87
artifacts, 1, 14–15, 22–3, 27–8, 41, 177, 212–14
aseptic necrosis, 175

astrocytomas, 30
avascular necrosis: hip, 175
  humeral head, 99, 125–7
  knee, 219, 263

Baker's cyst, 297
benign hypertrophy, 293–4
bone: metastases, 318, 342–51
  primary neoplasms, 281–2
  primary tumors, 2, 317–18, 321–41
bone islands, 318, 320, 339, 373
bone marrow, 1–2
  aplastic, 280–1
  in children, 280–1
  hematopoietic, 175, 185
  hips, 175
  hyperplasia, 175, 280
  infiltrative disorders, 319, 352–4
  osteomyelitis, 380–1
  primary tumors, 317–18
  vertebrae, 26, 30
breast carcinoma, 209, 345, 348, 350, 368–9, 375
Brodie's abscess, 381, 382, 392–5
bruxism, 59
'buffalo hump', 319–20, 367
bursa, infrapatellar, 255–6
bursitis, 98, 110–11
bursography, 95

calcium pyrophosphate dihydrate deposition disease, 61
carbon, 11
carcinoma: breast, 209, 345, 348, 350, 368–9, 375
  colon, 7, 346
  laryngeal, 349
  lung, 20, 341, 344, 347, 351
  prostate, 342–3, 345
cardiac-gated spine work, 28
carpal tunnel, 140, 146, 149–55, 161

carpal tunnel syndrome (CTS), 3, 139, 142–4, 164–8, 170, 172–4
cartilage disorders, knee, 218–19
cellulitis, 405–7
cerebrospinal fluid, 26
cervical spine, 34–7
cervical spondylosis, 143
chemical shift artifacts, 15, 27, 41
chemical shift selective sequence (CHESS), 218–19, 279
Chiari malformation, spine, 28, 42
children, 279–316
  bone marrow, 280–1
  primary bone neoplasms, 281–2
  primary soft-tissue neoplasms, 282–3
  trauma, 283
chloral hydrate, 279–80
chondromalacia patellae, 218, 219
chondrosarcoma, 318, 330–1
coccidioidomycosis, 396
collapsed vertebrae, 20
collateral ligaments, 216
  tears, 217, 243–6, 249–50, 253–4
Colles' fracture, 143
colon carcinoma, 7, 346
compression fractures, 49, 52, 315
congenital abnormalities, spine, 282–3
corticosteroids, 176, 186, 189–91, 195, 200, 264
cruciate ligaments, 215
  tears, 217, 237, 240–8, 252–4
crush fractures, 49, 52, 315
CT scanning, 1, 25, 29, 95, 379–80
Cushing's disease, 319–20
cystic myelomalacia, 28
cysts: ganglion, 142, 143, 319, 363–5
  meniscal, 3, 319, 366
  popliteal, 218, 258, 297

cysts (cont.)
  synovial, 3, 218, 258

degenerative disease:
  acromioclavicular joint, 110
  pediatrics, 283
  spine, 28–9, 46–8
dermal grafts, temperomandibular joint, 93
desmoid tumour, 360
developmental abnormalities, spine, 28, 42–5
diabetes, 143
diastematomyelia, 28, 305
disk disease, 28–9, 31, 46–8, 315
dislocation, shoulder, 122
double-contrast arthrography, 95
Down's syndrome, 314

elbow, 3, 129–38
  normal anatomy, 129–30, 131–6
enchondroma, 318, 335–7
energy windows, 11, 17
enostosis, 318, 339, 373–5
ependymomas, 30
Ewing's sarcoma, 55, 281–2, 310–12, 318, 327–9
exostoses, 338

fast low-angle shot sequence (FLASH), 218–19
femur: giant cell tumor, 24
  ischemic necrosis, 3, 175–6, 177, 185–200
  medullary infarcts, 219
  normal anatomy, 175, 179–80, 182
  osteogenic sarcoma, 19
  supracondyler fracture, 22–3
fibroma, 318, 335
fibrosarcoma, 318
fibrosis, 171

413

fibrous histiocytoma, 332
fibula, stress fractures, 251
field strengths, 1
fixation devices, 14–15, 22–3
foot, 266–78
   normal anatomy, 267–76
Fourier transformation technique, 60
fractures: compression, 49, 52, 315
   stress, 251
   tibial plateau, 252–4
   vertebrae, 29–30

ganglion cysts, 142, 143, 319, 363–5
Gaucher disease, 138, 176
giant cell arteritis, 193
giant-cell reaction, Proplast implants, 61, 88–92
giant cell tumor, 24, 163, 318, 333–4
gout, 61, 143
gradient field artifacts, 14, 15
Graves' disease, 143
growth plates, 279

hairy cell leukemia, 353
hand, 3, 139–74
hemangioblastoma, 30, 57
hemangioma, 30, 53–4, 143, 282, 298, 320
hematoma, 14, 143, 282, 371–2
hematopoietic marrow, 175, 185
hemophilia, 280, 290–2
hemosiderin deposits, 319, 320, 371–2
herniation, disks, 28–9, 46–8
Hill–Sachs lesions, shoulder, 97, 99, 124
hip, 3, 175–214
   ischemic necrosis, 175–6, 177, 185–200
   normal anatomy, 175, 179–83
histiocytoma, 332
humerus, necrosis, 96, 99, 125–7
hydrocephalus, 28
hydrogen atoms, 11–12
hydromyelia, 28
hyperplastic marrow, 175, 280
hypertrophic facet disease, 29
hypertrophic pulmonary osteoarthropathy, 341

image contrast, 13–14, 26
impingement syndrome, shoulder, 3, 95, 97–8, 107–18
incisional neuroma, 142, 143, 169
infarcts, bone, 285–7
infections, 177, 379–412
inflammatory disease, spine, 31, 58
infraspinatus tendonitis, 118
instability, shoulder, 95, 98–9, 119–24
intra-articular osteochondral fragments, 219–20, 259–60
intra-articular tumors, knee, 3
intramedullary neoplasia, 30
ischemic necrosis, 143
   femoral head, 3, 175–6, 177, 185–200
   humeral head, 96
   scaphoid fractures, 142, 169
   talus, 3

joint effusions, knee, 218, 237–8, 243–6, 252–7

Kienbock's disease, 142
knee, 3, 215–65
   avascular necrosis, 219, 263
   cartilage disorders, 218–19
   collateral ligament tears, 217, 243–6, 249–50, 253–4
   cruciate ligament tears, 217, 237, 240–8, 252–4
   intra-articular osteochondral fragments, 219–20, 259–60
   joint effusions, 218, 237–8, 243–6, 252–7
   medullary infarcts, 219, 264–5
   meniscal injuries, 216–17, 229–39, 241–4, 247
   normal anatomy, 215–16, 221–8
   osteochondritis dissecans, 218, 219, 261–2, 264
   patellar tendon injuries, 217–18, 245–6, 250
   pediatrics, 284–98
   synovial cysts, 218, 258

labral tears, 119–23
laryngeal carcinoma, 349
Legg-Perthes disease, 198, 295
leiomyosarcoma, 355–7
leukemia, 281, 284–5, 309, 319, 353
ligaments, 2
   ankle and foot, 278
   knee, 215, 216, 217, 237, 240–50, 252–4
   wrist, 139
ligamentum flavum hypertrophy, 29
lipoblastoma, 282, 312
lipoma, 28, 143, 282, 303, 317, 319–20, 366–7
lipomeningocele, 306–7
lipomeningomyelocele, 44–5
liposarcoma, 357–8
longitudinal relaxation, 12–13
lumbar spine, 38–40
lunate, Kienbock's disease, 142
lung carcinoma, 20, 341, 344, 347, 351
lupus erythematosis, 186–7, 189, 263–5
lymphangiomas, 282
lymphoblastic leukemia, 309
lymphocytic leukemia, 353
lymphoma, 209, 308, 319, 354

magnetic resonance imaging (MRI), 1–3
   physical principles, 11–18
malignant fibrous histiocytoma, 332
median nerve, 140, 145
   carpal tunnel syndrome, 142–3
medullary infarcts, 219, 264–5
melanoma, 350
meningioma, 30, 56
meningocele, 28, 302
meniscus, 215
   injuries, 3, 14, 216–17, 229–39, 241–4, 247
   meniscal cyst, 3, 319, 366
metabolic bone disease, 313
metacarpals, 140–1, 156–8
metal artifacts, 177, 212–14
metastases: bone, 318, 342–51
   hip, 177, 209
   spine, 20, 26, 30–1, 52
   wrist, 142

methylmethacrylate cement, 15, 24
motion artifacts, 14, 15
multiple myeloma, 352
multiple sclerosis (MS), 31
muscles, 2
myelocystocele, 302
myelofibrosis, 175
myeloma, 175, 319
myelomeningocele, 28
myocitis, 282
myxedema, 143

necrosis: femoral head, 175–6, 177, 185–200
   humeral head, 99, 125–7
   knee, 219, 263
   scaphoid fractures, 142, 169
   wrist, 143
neoplasms: primary bone, 281–2
   primary soft-tissue, 282–3
   spine, 30–1, 52–8
neuritis, 143
neuroblastoma, 288–9
neuroepithelial tumors, 288
neurofibroma, 30, 56, 297, 319, 361–2
neuroma, incisional, 142, 169
neutrons, 11
nuclear magnetic resonance (NMR), 11
nucleus, 11

osseous trauma, 99–100, 128
osteoarthritis, 177, 201, 202–3, 211, 218
osteochondritis dissecans, 3, 218, 219, 261–2, 264
osteochondroma, 211, 318, 338
osteogenic sarcoma, 19
osteoid osteoma, 318, 340
osteomyelitis, 2, 3, 31, 58, 379, 380–2, 384–6, 388–402, 408–11
osteonecrosis: ankle and foot, 276
   hip, 175, 193
   knee, 219, 263, 295
osteopenia, 5
osteophytes: hip, 177
   spine, 29
   temperomandibular joint, 73
osteoporosis, 7, 30, 49, 177, 206–8
osteosarcoma, 281, 296, 318, 321–6

Paget's disease, 177, 208, 320, 368–78
pancreatitis, 176, 265
parosteal osteosarcoma, 326
patellar tendon, 216
   injuries, 217–18, 245–6, 250, 404
pediatric spine, 282
pediatrics, 279–316
   bone marrow, 280–1
   primary bone neoplasms, 281–2
   primary soft-tissue neoplasms, 282–3
   trauma, 283
Pellegrini–Stieda disease, 250
periosteal osteosarcoma, 281
Phalen's sign, carpal tunnel syndrome, 142
Phemister procedure, 200
phosphorus, 11
physics, magnetic resonance imaging, 11–18
pigmented villonodular synovitis, 142, 319, 363

popliteal aneurysm, 376
popliteal cysts, 218, 258, 297
popliteus tendon, 215
primary soft-tissue neoplasms, 282–3
primary tumors of bone, 281–2, 317–18, 321–41
Proplast implants, temperomandibular joint, 87–92
prostate carcinoma, 342–3, 345
prostheses, 14, 177
protons, 11–12
pseudoganglions, carpal tunnel syndrome, 143
pseudomeningoceles, 30
psoriatic arthritis, 61, 94
pulse sequencing, 12–13
pyoarthrosis, 379, 401, 412

quantum theory, 12

radiofrequency magnetic field artifacts, 14, 15
radiofrequency (RF) pulse, 12–13, 15
radioulnar joint, 139–40
renal osteodystrophy, 313
resonant frequency, 12
retrodiscal hyperplasia, temperomandibular joint, 83–7
retrodiscitis, temperomandibular joint, 80–2
Reynold's sign, sickle-cell anemia, 313
rheumatoid arthritis, 31, 61
rotator cuff disease, 3, 95, 97–8, 107–18

sarcoma, 19, 359
scaphoid fractures, 142
scars, 282
sedation, children, 279–80
septic arthritis, 205, 379, 381
shoulder, 3, 95–128
   avascular necrosis, 99, 125–7
   impingement syndrome, 95, 97–8, 107–18
   instability, 95, 98–9, 119–24
   normal anatomy, 96–7, 101–6
   osseous trauma, 99–100, 128
   rotator cuff disease, 95, 97–8, 107–18
sickle-cell anemia, 176, 177, 204, 280, 287, 313
sodium, 11
soft-tissue tumors, 142, 319–20, 355–68
spin lattice relaxation, 12–13
spin-spin relaxation, 13
spine, 2, 25–58
   artifacts, 27–8
   degenerative changes, 28–9, 46–8
   developmental abnormalities, 28, 42–5, 282–3
   inflammatory disease, 31, 58
   metastases, 20, 26, 30–1, 52
   neoplasia, 30–1, 52–8
   normal anatomy, 25–7, 33–40
   osteomyelitis, 381
   pediatrics, 282, 299–316
   postoperative changes, 30
   trauma, 29–30, 49–51, 283
   vascular malformations, 31
   see also vertebrae

spondylitis, 381
spondylolisthesis, 29
spondylosis, 2, 29, 143, 382
static magnetic field artifacts, 14
stenosis, spinal canal, 29
stress fractures, fibula, 251
stress trabeculae, 175, 183–4
subacromial bursitis, 98, 110–11
subacromial spurs, 107–8
subscapularis tendon rupture, 124
supraspinatus tendonitis, 98, 112–18
surface coils, 25, 59–60, 96, 129, 317
synovial cysts, 3, 218, 258
synovial fluid, joint effusions, 218, 237–8, 243–6, 252–7
syringohydromyelia, 28, 30, 42–3, 49
systemic lupus erythematosus, 186–7, 189, 263–5

talus, ischemic necrosis, 3
TE (echo time), 13, 14, 18, 19
telangiectatic osteosarcoma, 296
temperomandibular joint, 2, 59–94
tendinitis, patellar, 217–18, 245–6, 250, 404
tendonitis, 95, 98, 112–18
  adhesive, 142, 171
  wrist, 139
tendons, 2
  rupture, 8–10, 277
tenosynovitis, 143, 165
  carpal tunnel syndrome, 143, 144
tethered cord, 28, 283, 302–5
thrombocytopenia, 371
tibia: medullary infarcts, 219
  periosteal osteosarcoma, 281
  tibial plateau fractures, 252–4
Tinel's sign, carpal tunnel syndrome, 142

T_1-relaxation, 12–13, 18
T_2-relaxation, 12, 13, 18
TR (repetition time), 13, 14, 18, 19
trabeculae, stress, 175, 183–4
transverse relaxation, 13
trauma: pediatrics, 283
  spine, 29–30, 49–51
tuberculous spondylitis, 381
tumor-like conditions, 320, 368–78
tumors, 2, 317–78
  marrow infiltrative disorders, 319, 352–4
  metastases in bone, 318, 342–51
  primary soft-tissue neoplasms, 282–3
  primary tumors of bone, 281–2, 317–18, 321–41
  soft tissue, 319–20, 355–68
  spine, 30–1, 52–8

ulna, giant cell tumor, 163

ulnar nerve palsy, 137
ultrasonography, 95

vascular malformations, spine, 31
vertebrae: bone marrow, 26, 30
  collapsed, 20
  compression fractures, 49, 52
  fractures, 29–30, 315–16
  osteomyelitis, 381
  pediatrics, 282, 299–316
  see also spine

wrap-around artifact, 15
wrist, 3, 139–74
  carpal tunnel syndrome, 139, 142–4, 164–8, 170, 172–4
  normal anatomy, 139–41, 145–62